Managing Global Warming

An Interface of Technology and Human Issues

Edited by

Trevor M Letcher

ELSEVIER

ACADEMIC PRESS

An imprint of Elsevier

Academic Press is an imprint of Elsevier
125 London Wall, London EC2Y 5AS, United Kingdom
525 B Street, Suite 1650, San Diego, CA 92101, United States
50 Hampshire Street, 5th Floor, Cambridge, MA 02139, United States
The Boulevard, Langford Lane, Kidlington, Oxford OX5 1GB, United Kingdom

Library of Congress Cataloging-in-Publication Data
A catalog record for this book is available from the Library of Congress

British Library Cataloguing-in-Publication Data
A catalogue record for this book is available from the British Library

ISBN 978-0-12-814104-5

For information on all Academic Press publications visit our website at https:/www.elsevier.com/books-and-journals

Publisher: Joe Hayton
Acquisition Editor: Lisa Reading
Editorial Project Manager: Joshua Bayliss
Production Project Manager: Sruthi Satheesh
Cover Designer: Mark Rogers

Typeset by SPi Global, India

Contents

List of contributors

Tony Amis GI Energy, Chicago, IL, United States

Masoud Babaei School of Chemical Engineering and Analytical Science, The University of Manchester, Manchester, United Kingdom

Subhamoy Bhattacharya University of Surrey, Guildford, United Kingdom

Miguel Brandão International Energy Agency (IEA) Bioenergy Task 38; KTH—Royal Institute of Technology, Stockholm, Sweden

Antoine Buchard Department of Chemistry and Centre for Sustainable Chemical Technologies, University of Bath, Bath, United Kingdom

Sylvain Caillol Institute Charles Gerhardt, University of Montpellier, Montpellier, France

Robert Chris Visiting Fellow, Department of Geography, Open University, Milton Keynes, United Kingdom

Declan Clausen Deputy Lord Mayor, City of Newcastle, Newcastle, NSW, Australia

Annette L. Cowie NSW Department of Primary Industries/School of Environmental and Rural Science, University of New England, Armidale, NSW, Australia; International Energy Agency (IEA) Bioenergy Task 38

Liang Cui University of Surrey, Guildford, United Kingdom

Renaud de Richter Institute Charles Gerhardt, University of Montpellier, Montpellier, France

Hasan Emre Demirci University of Surrey, Guildford, United Kingdom

R. Duffey Idaho Falls, Idaho, ID, United States

Mojgan Hadi Mosleh School of Mechanical, Aerospace and Civil Engineering, The University of Manchester, Manchester, United Kingdom

Anthony Harding School of Economics, Georgia Institute of Technology, Atlanta, GA, United States

Clare Heyward Institute for Advanced Sustainability Studies, Potsdam, Germany

Damon Honnery Department of Mechanical and Aerospace Engineering, Monash University-Clayton Campus, Clayton, VIC, Australia

Peter H. Howard Economics Director, Institute for Policy Integrity, New York University School of Law, New York, NY, United States

Masayuki Hyodo Yamaguchi University, Japan

Phil Ireland Online Progressive Engagement Network, Newcastle, NSW, Australia

Richard Kembleton EUROfusion PMU, Culham Centre for Fusion Research, UKAEA, Abingdon, United Kingdom

Ånund Killingtveit Dept of Civil and Environmental Engineering, Norwegian University of Science and Technology (NTNU), Trondheim, Norway

Prashant Kumar University of Surrey, Guildford, United Kingdom

Nicolas H. Lampkin The Organic Research Centre, Hamstead Marshall, Newbury, United Kingdom

Trevor M Letcher University of KwaZulu-Natal, Durban, South Africa; Laurel House, FosseWay, Stratton on the Fosse, Somerset, United Kingdom

Johannes Lindorfer Energy Institute at the Johannes Kepler University Linz, Linz, Austria

Tingzhen Ming School of Civil Engineering and Architecture, Wuhan University of Technology, Wuhan, China

Azizul Moqsud Yamaguchi University, Japan

Juan B. Moreno-Cruz School of Economics, Georgia Institute of Technology, Atlanta, GA, United States

Patrick Moriarty Department of Design, Monash University, Caulfield East, VIC, Australia

Georgios Nikitas University of Surrey, Guildford, United Kingdom

Nikolaos Nikitas University of Leeds, Leeds, United Kingdom

Lee Phillips Alogus Research Corporation, McLean, VA, United States

I. Pioro Faculty of Energy Systems and Nuclear Science, University of Ontario Institute of Technology, Oshawa, ON, Canada

Gerda Reiter Energy Institute at the Johannes Kepler University Linz, Linz, Austria

Mirjam Röder Supergen Bioenergy Hub, Tyndall Centre for Climate Change Research, School for Mechanical, Aerospace and Civil Engineering, The University of Manchester, Manchester, United Kingdom

Abubakar Kawuwa Sani Department of Civil and Environmental Engineering, Faculty of Engineering and Physical Sciences, University of Surrey, Guildford, United Kingdom

Janet L. Scott Department of Chemistry and Centre for Sustainable Chemical Technologies, University of Bath, Bath, United Kingdom

Majid Sedighi School of Mechanical, Aerospace and Civil Engineering, The University of Manchester, Manchester, United Kingdom

Andrea C. Simonelli Department of Political Science, Virginia Commonwealth University, Richmond, VA, United States

Rao Martand Singh Department of Civil and Environmental Engineering, Faculty of Engineering and Physical Sciences, University of Surrey, Guildford, United Kingdom

Laurence G. Smith The Organic Research Centre, Hamstead Marshall, Newbury, United Kingdom

Sampo Soimakallio International Energy Agency (IEA) Bioenergy Task 38; Finnish Environment Institute (SYKE), Helsinki, Finland

Horst Steinmüller Energy Institute at the Johannes Kepler University Linz, Linz, Austria

Peter Styring UK Centre for Carbon Dioxide Utilisation, Department of Chemistry, Dainton Building, Brook Hill; Department of Chemical & Biological Engineering, Sir Robert Hadfield Building, The University of Sheffield, Sheffield, United Kingdom

Robert Tichler Energy Institute at the Johannes Kepler University Linz, Linz, Austria

Matthew Turner IHS Global Limited, Enterprise House, Tetbury, United Kingdom

Alice Venn University of Bristol Law School, Bristol, United Kingdom

Nathan Vimalan VJ Tech Ltd, Reading, United Kingdom

Ivo Wallimann-Helmer Advanced Studies in Applied Ethics, Center for Ethics, University of Zurich, Zurich, Switzerland

Andrew Welfle Supergen Bioenergy Hub, Tyndall Centre for Climate Change Research, School for Mechanical, Aerospace and Civil Engineering, The University of Manchester, Manchester, United Kingdom

Section A

Introduction

Why do we have global warming? 1

Trevor M Letcher[*,†,1]
*Laurel House, FosseWay, Stratton on the Fosse, Somerset, United Kingdom,
†University of KwaZulu-Natal, Durban, South Africa
[1]Corresponding author: trevor@letcher.eclipse.co.uk

Chapter Outline

1.1 The greenhouse effect

The concept of the greenhouse effect goes back to the 1820s, when Joseph Fourier suggested that some component of the earth's atmosphere was responsible for the temperature at the surface of the earth. He was researching the origins of ancient glaciers and the ice sheets that once covered much of Europe [1].

Decades later, Tyndall followed up the Fourier's suggestion, and used an apparatus designed by Macedonio Melloni to show that CO_2 was able to absorb a much greater amount of heat than other gases. This fitted in with Fourier's concept and pointed to CO_2 as the component in the atmosphere that Fourier was looking for. The Melloni apparatus was called a thermomultiplier and was reported in 1831 [2,3]. Tyndall's results were published in references [4,5]. As a result, Tyndall can be named as the discoverer of the CO_2 greenhouse gas effect.

Linking CO_2 in the atmosphere to the burning of fossil fuels was to be the last link in the chain in understanding the reasons for the ice ages and also our own climate change. In the 1890s, Svante Arrhenius, an electrochemist, calculated that by reducing the amount of CO_2 in the atmosphere by half, the temperature of Europe would be lowered by about 4–5°C. This would bring it in line with ice age temperatures. This idea would only answer the question of why the ice age formed and then retreated, **if** there were large changes in atmospheric composition and in particular, changes in CO_2 concentration. At much the same time, also in Sweden, a geologist, Arvid

Managing Global Warming. https://doi.org/10.1016/B978-0-12-814104-5.00001-6

Högbom, had estimated that CO_2 from volcanic eruptions, together with the ocean uptake of CO_2, could explain how the CO_2 concentrations in the atmosphere could change and hence provide some explanation for the ice ages. Along the way, Högbom stumbled on a strange and new idea that the CO_2 emitted from industrial coal burning factories might influence the atmospheric CO_2 concentration. He did indeed find that human activities were contributing CO_2 to the atmosphere at a rate comparable to the natural geochemical processes. The increase was small compared to what was already in the atmosphere, but if continued, it would influence the climate. Arrhenius took up this concept, and his calculations are published in reference [6]. Arrhenius concluded that the emissions from human industry might someday bring on global warming. As a result, Arrhenius's name is forever linked to the greenhouse theory of global warming. However, thanks must also go to those who paved the way—Fourier, Melloni, Tyndall, Högbom, and probably many others.

Arrhenius's calculations were at first dismissed as unimportant or at worst faulty. A similar fate was met by G.S. Callendar who, in 1938, made the point that CO_2 levels were indeed climbing [7]. It was only in the 1960s, after C D Keeling measured the CO_2 concentration in the atmosphere and showed that it was rising rapidly, that scientists woke up to the fact that global warming was real and that anthropogenic activity was to blame.

Water vapor is an even more effective greenhouse gas than CO_2. Furthermore, its concentration in the atmosphere is very much higher than that of CO_2 (of the order of a hundred times higher) and probably contributes about 60% of the global warming effect. The amount of water vapor in the atmosphere is controlled by the temperature. An increase in the CO_2 concentration in the atmosphere will increase the global temperature only slightly, but that change is enough to increase the amount of water vapor in the air, through evaporation from the oceans. It is this feedback mechanism that has the greatest influence on global temperature. In a sense, paradoxically, the concentration of CO_2 acts as a regulator for the amount of water vapor in the atmosphere and is thus the determining factor in the equilibrium temperature of the earth. Without CO_2 in the atmosphere, the temperature of the earth would be very much cooler than it is today.

The amount of solar energy shining on the earth (with wavelengths ranging from 0.3 is 5 μm) is vast. It heats our atmosphere and everything on the Earth and provides the energy for our climate and ecosystem. At night, much of this heat energy is radiated back into space but at different wavelengths, which are in the infrared range from 4 to 50 μm [8]. This energy heats the greenhouse gas molecules (such as H_2O, CO_2, CH_4, etc.) in the atmosphere. The explanation is as follows: using CO_2 and H_2O as examples, this heating process takes place because the radiated IR frequency is in sync (resonates) with the natural frequency of the carbon-oxygen bond of CO_2 (4.26 μm being the asymmetric stretching vibration mode and 14.99 μm being the bending vibration mode) and the oxygen-hydrogen bond of H_2O. The increased vibration of the bonds effectively heats the CO_2 and H_2O molecules. These heated molecules then pass the heat to the other molecules in the atmosphere (N_2, O_2), and this keeps the earth at an equitable temperature. The vibrating frequencies of the O—O bond in oxygen and the N—N bond in nitrogen molecules are very different from the radiation frequencies and so are relatively unaffected.

1.2 The root cause of global warming

The scientific evidence that global warming is largely due to the rising CO_2 levels in the atmosphere is overwhelming and, furthermore, that the rising CO_2 concentration is due to human activities. Every scientific society and every research organization working in the field of climate change accepts this view. The atmospheric CO_2 concentration has increased from 280 ppm (280 ppm or 280 molecules per million molecules) [9] before the industrial revolution, to 413 ppm (observed at Mauna Loa Observatory on April 26, 2017) [10], and it is this increase of almost 50% that has triggered the present increase in global temperature.

The most compelling evidence that the increase in CO_2 is the most likely cause of global warming can be seen in the related graphs of CO_2 concentration in the atmosphere and the global average temperature as functions of time over the past many decades (see Figs. 1.1 and 1.2). The CO_2 increase is mirrored by an increase in the relative increase in average global temperatures over the past 60 years.

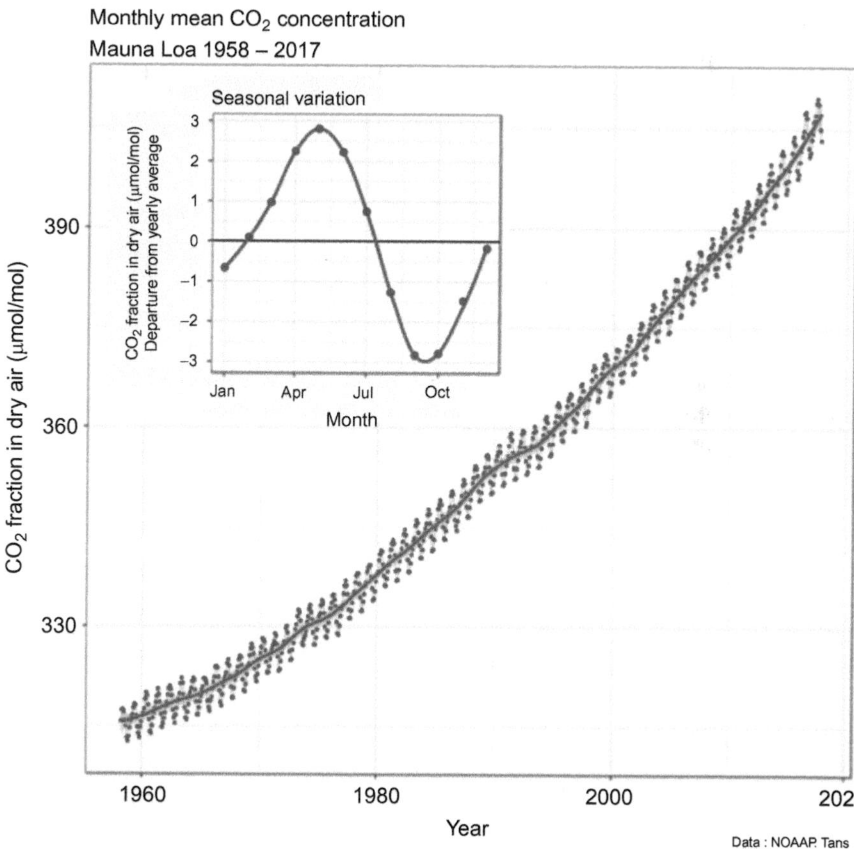

Fig. 1.1 The increase of CO_2 concentration over the past 60 years.
Data from Dr. Pieter Tans, NOAA/ESRL and Dr. Ralph Keeling, Scripps Institution of Oceanography. https://en.wikipedia.org/wiki/Keeling_Curve.

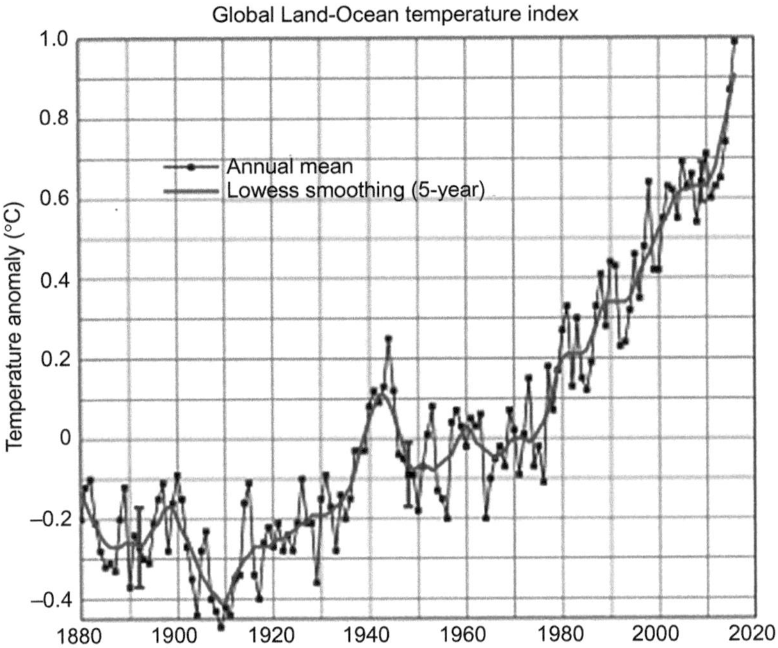

Fig. 1.2 The relative increase in the world's average surface air temperature from 1880 to 2009. Original data produced by ASA's Goddard Institute for Space Studies. http://data.giss.nasa.gov/gistemp/graphs/.

A question that needs answering is this: we know that the CO_2 level in the atmosphere is rising steadily: but is the CO_2 increase due to human activity? The evidence that it is indeed due to human activity is based on the relative ratios of carbon isotopes. The relative amount of ^{13}C in the atmosphere has been declining and that is because the ratio of ^{13}C in fossil fuel-derived CO_2 is much lower than the CO_2 produced from present-day decaying plants [11].

From the properties of each of the greenhouse gases (such as the wavelengths of energy), scientists can calculate how much of each gas contributes to global warming. The results show that CO_2 is responsible for about 20% of the earth's greenhouse effect, water vapor about 60% [12]. The rest is caused by minor greenhouse gases such as methane and chlorinated hydrocarbons. The relative concentration of the major greenhouse gases emitted by human activity is: CO_2, 63%; CH_4, 18% and N_2O, 6% [13].

It is perhaps of interest to note that it is not possible to obtain absolute proof that it is CO_2, which is largely responsible for global warming because we cannot do the definitive experiment of suddenly stopping the use of fossil fuels. And even if we could do this experiment, it would take decades to obtain a definite conclusion because of the long life CO_2 has in the atmosphere [13,14].

Most of the anthropogenic CO_2 entering the atmosphere comes from fossil fuels. The relative fraction of energy produced by fossil fuels has remained at over 86% over

Table 1.1 Worldwide primary energy consumption percentages from 2005 to 2015 [15]

	Energy (%)		
	2005	2010	2015
Oil	36	33.5	32.9
Coal	28.6	29.8	29.2
Natural gas	22.9	23.7	23.9
Nuclear	5.7	5.1	4.4
Hydropower	6.1	6.4	6.8

Table 1.2 The quantities of oil, coal, and natural gas mined pumped over the decade, 2005–15

	2005	2010	2016
Oil volume (10^6 barrels) per day [16]	81,908	83,251	92,150
Coal (10^6 t oil equivalent) per year [17]	3039	3633	3656
Natural Gas (10^9 m^3) per year [18]	2773	3192	3551

the past decade as is illustrated by worldwide primary energy consumption listed in Table 1.1. However, the quantity of fossil fuel extracted from the earth has increased significantly over the past 11 years, as seen in Table 1.2. This is reflected in the steadily increasing amount of CO_2 entering the atmosphere. However, between 2015 and 2016, world oil production increased by only 0.4%, world coal production fell by 6.2%, and natural gas increased by only 0.3%. This is the first sign that fossil fuel usage is slowing down.

The Global Carbon Project (GCP) [19] has reported that emissions in 2015 from burning fossil fuels and also from industry (especially cement production) account for 91% CO_2 caused by human activity with 9% from land use changes. In 2015, the GCP has reported that 9.9×10^9 t of carbon in the form of CO_2 from burning fossil fuels entered the atmosphere. Nevertheless, the GCP felt that there were signs that the emission of CO_2 from human activity was indeed showing signs of peaking.

1.3 Other causes of global warming and climate change including global cooling

In spite of the evidence presented, there has been much debate as to whether our present global warming and climate change could in fact be due to effects other than atmospheric gases [13]. These include the variation in the sun's energy; volcanic activity; changes in the earth's orbital characteristics, including the Malankovitch cycles; cosmic ray effects; and atmospheric aerosols. The first of these, relating to possible variations in solar radiation, has been investigated by many scientists, and all agree that this is not the main cause of our present situation [20]. Volcanic activity on earth has in the past resulted in short-lived climate changes, but experts working in the field state that this could not possibly be responsible for our present climate change [21]. The earth's wobble as it moves around the Sun is another possible culprit for inducing climate change. However, Lourens and Truter, scientists working in this area, have stated that from all the evidence, it is most likely that the climate change that we are currently experiencing is not due to variations of the earth's orbital movements [22]. Solar activity and cosmic ray bombardment from space is yet another possible cause of climate change, but experts S. Cohen and G. Stanhill feel that these effects cannot possibly be responsible for our present global warming [23].

Recent work by Macott et al. has prompted these researchers to write: "The earth's climate is complex and responds to multiple forcings, including CO_2 and solar insolation. Both of those have changed very slowly over the past 11,000 years. But in the last 100 years, the increase in CO_2 through increased emissions from human activities has been significant. It is the only variable that can best explain the rapid increase in global temperatures" [24].

Superimposed on the scenario of global warming is the effect of global dimming [20]. This effect was especially noticeable between the 1950s and 1980s when it was found that the sun's energy reaching places on the surface of the earth were less than in previous years—the reduction being of the order of a few percent. It was attributed to aerosol formation and particulates in the atmosphere resulting from the burning of coal and other hydrocarbons [25]. These particles were reflecting the sun's rays back into space, resulting in a dimming of the solar radiation. This effect does partially compensate for global warming. This is an important issue as the presence of particulate matter is responsible for major smog problems in some parts of the world, especially in China. Furthermore, particulates in the atmosphere are also related to illnesses suffered by asthmatics and other people with lung problems. These two issues have precipitated a drive to reduce particulates in the air by introducing precipitators at coal-burning power stations and replacing coal burning with renewable forms of energy. Reducing particulates in the atmosphere would reduce the global cooling effect and as a result could trigger a major increase in global warming unless there is a serious commitment to replacing fossil fuel burning processes with nonfossil fuel forms of energy.

In this chapter, the major greenhouse gas discussed has been CO_2, but there are other gases that are far more effective greenhouse gases. These include chlorofluorocarbons (CFCs) and nitrogen oxide (NOx) compounds. Fortunately, these are in very low concentrations in the air. These have been discussed by Tuckett [13]. Furthermore, a recent study of the effects of CFCs on global warming by Quing-Bin Lu of the University of Waterloo, has confirmed that these compounds are serious contenders for the most potent of greenhouse gases. It is not only their potency that is an issue but also their long lifetime in the atmosphere (see Table 1.3). In spite of their present low concentrations, they should be banned from production as their build-up in the atmosphere could be very damaging to future generations [26].

Table 1.3 Properties of some GHGs related to global warming

Property	GHG			
	CO_2	CH_4	N_2O	CFCs
Concentration (ppm)	413	1.7	0.32	0.0009
Lifetime (years)	50–200	12	120	45–1700
Contribution to overall Greenhouse effect (%)	63	18	6	13
Global warming potential	1	25	120	6130–14,400

Data from Table 2 of reference Tuckett RP. The role of atmospheric gases in global warming. In: Letcher TM, editor. Climate change, observed impacts on planet earth. 2nd ed. Oxford: Elsevier; 2016; p. 375–98.

The table also highlights the atmospheric concentrations, estimated life time in the atmosphere, greenhouse effect percentage, and the potential for global warming of greenhouse gases, taking CO_2 as unity.

Overall, human sources of CO_2 are much smaller than the natural emissions from animals, plants, decaying animals, vegetation, and volcanoes [27]. However, human activity has upset the balance in a cycle that has existed for thousands of years. The amount of carbon on the earth and in the atmosphere is fixed, but it is in a dynamic and equilibrium cycle, moving between living and nonliving things, and changing into different carbon compounds such as carbon dioxide in the air and in the oceans, solid carbonate rock, and the living cells of plants and animals. In the first step of the cycle, plants take up CO_2 from the air through the process of photosynthesis and release oxygen. The CO_2 is then converted into living cells. In the next step, the animals eat the plants, and the carbon in the plant cells are used to build animal tissue and cells. Animals also breathe in oxygen and exhale carbon dioxide, which in turn enter the atmosphere and the cycle continues. Dead plants and dead animals decompose and carbon is either released as CH_4 and CO_2 or stored in the soil. Superimposed on this cycle is the exchange of CO_2 between the atmosphere and the oceans. These processes were in near-perfect equilibrium before the industrial revolution [27,28].

1.4 Indicators of climate change

The evidence that global warming is altering our climate is very well documented, and almost no day goes by without more evidence for climate change. The indicators include more extreme weather events in the future; melting of Arctic sea ice; Antarctic Sea changes; land ice behavior (including glaciers and ice sheets); weather pattern changes; bird ecology changes (including migration); mammal and insect ecology changes and biodiversity loss; sea life and coral reef changes; marine diversity and intertidal indicators; plant ecology and plant pathogen changes; rising sea levels; and ocean acidification [29].

1.5 Why we must act now

There is a growing threat of environmental collapse in the future, as a result of changes in our present climate. We are beginning to see this with extreme weather events such as flooding, droughts and water crises, high winds, runaway fires, washaways and mud flows from land denuded of its natural rain soaking properties, high seas in coastal areas, together with rising sea levels, to mention a few. One consequence of climate change is the migration of insects and animals to more hospitable climates. A more frightening involuntary mass migration has already begun: of humans from lands unable to support the growing of crops and from areas where rising sea levels are beginning to threaten livelihood. It is not only natural disasters that are a cause for concern but also man-made disasters, which result indirectly from global warming that are a cause for concern. These include the reduced ability of land to soak up rain water as a result of land clearances and urban development resulting in flooding; chemical pollution in the form of pesticides, endocrine disruptors and hormonally active agents used on farms to increase yields; nuclear disasters through extreme weather; land-use decisions for agriculture; oil fires, coal mine fires, and even tire fires, which add their own contribution to rising CO_2 levels [30].

1.6 What must be done to reduce global warming?

Most world governments have accepted the assessment of the United Nations Framework Convention on Climate Change (UNFCCC) that a 2°C rise in mean global temperature above the preindustrial level must be the maximum limit. In order to meet this objective, studies generally indicate the need for global greenhouse gas emissions to peak before 2020 with a substantial reduction in emissions thereafter.

We need to reduce the amount of CO_2 entering the atmosphere and if possible we should find ways of removing some of the CO_2 presently in the atmosphere. Present-day CO_2 levels in the atmosphere exceed the natural equilibrium of dissolved CO_2 in the oceans and with the CO_2 uptake by biota on land. Unfortunately, this rising nonequilibrium amount of CO_2 remains in the air for a very long time (see Table 1.3). The reason is that CO_2, unlike other greenhouse gases such as CH_4, is very unreactive. It does not naturally react with most chemicals and, in thermodynamic terms, it has a very high Gibbs Energy of Formation. In order to bring about a reaction of CO_2 with another chemical, a significant amount of energy must be given to the system (e.g., heat energy). This is also the reason why it is so difficult to get rid of waste CO_2 from chemical reactions (e.g., cement manufacture, or even from burning fossil fuels) and why it is rarely used as a chemical feedstock in industry.

There are massive reserves of coal oil and gas in the earth. These convenient sources of energy are not only easy to use for heating and for producing energy, but also exist in a stored form, which allows them to be used at any time in the future. The reserves are summarized in Table 1.4 together with the expected lifetime of usage. Our future mindset must however not be seduced by the convenient properties of fossil

Table 1.4 World reserves of fossil fuels (OPEC 2016 stats, BP stats 2016) [31]

	Amount	Lifetime (year)	Present consumption
Natural gas	200×10^{12} m^3	3×10^3	63×10^9 m^3 per year [18]
Oil	1.6×10^{12} barrels	17×10^3	96×10^6 barrels per year [32]
Coal	0.9×10^{12}	110	7876×10^6 t per year [33]

fuel, but for the sake of the planet, the reserves must stay forever below ground, and nonfossil fuel sources of energy should be embraced.

This book is largely devoted to suggestions for solving the problem of reducing the amount of CO_2 we pump into the air by: using renewable energies and other nonfossil fuel forms of energy; sequestering the CO_2 produced in industrial process; sequestering the CO_2 in the air and possibly using geoengineering.

1.7 Are we making progress in reducing global warming?

This past year was again one of the hottest on record. In November 2017, the World Meteorological Organization (WMO) issued the following statement: "from January to September this year the average global temperatures were approximately 1.1°C above the pre-industrial era, putting us uncomfortably close to the Paris Agreement goal of 1.5°C" [34].

In 2016, the World Meteorological Organization confirmed that 2011–15 were the hottest 5-year period on record; and in 2017 the National Aeronautics and Space Administration (NASA) the National Oceanic and Atmospheric Administration (NOAA) all stated that 2016 was the hottest year ever recorded, making 16 of the 17 years since 2000 the warmest ever. More recent reports (in 2018) make the claim the 17 of the past 18 years have been the warmest ever [35].

It shows that we are not doing enough to reduce the amount of CO_2 in the atmosphere. The only way to reduce global warming is to reduce the amount of CO_2 we are pumping into the air, and if possible, begin removing CO_2 from the atmosphere.

At present, less than 20% of all energy sources are either renewable (wind, solar, hydropower, biomass tide, and geothermal) or nuclear. Replacing fossil fuel to reduce significantly our CO_2 emissions is going to be a mammoth task. The solution to this problem is the *raison d' être* for this book.

It has been estimated that in 2015 human activities contributed to roughly 40×10^9 t of CO_2 through burning coal and other fossil fuels, cement production, deforestation, and other landscape changes,. It has also been estimated that since the Industrial Revolution, over 2000×10^9 t of CO_2 have been added to the atmosphere. Human activities emit 60 or more times the amount of carbon dioxide released by volcanoes each year [36].

The population of the world is increasing and so is the need for more energy with a greater demand for more electricity. The world population (it is now 7.6×10^9 according to the latest 2018 United Nation estimate) is expected to reach 9×10^9 in 2050. It is increasing at a rate of 1.09% per year at the moment (2018) down from 1.14% per year in 2016 and down from the peak in 1963 of 2.2% per year. The expected rate of growth in energy demand over the next decade is greater than the growth rate of the population; this is largely due to the increase demand for electricity in developing countries. Electricity generation is expected to increase from 25×10^{12} kWh in 2017 to 31.2×10^{12} kWh in 2030—an increase of almost 2% per year [37].

At the moment, coal is still the largest producer of electricity worldwide and is not expected to be overtaken by renewables until 2040. The relative breakdown of electricity producers and future predictions is given in Table 1.5. It illustrates the energy dilemma of our time—the positive and encouraging increase in the deployment of renewable forms of energy is masked by the increasing overall energy needs of the world and that increase is still being met by further increases in fossil fuel usage. The present and future world electricity generation is dominated by the burning of fossil fuels (over 60%), and the prediction for 2040 is not much better (58%). It is no doubt driven by a number of forces including the relative economics of fossil fuels versus renewable energy; the massive inertia linked to status quo situations; and the fear of things new as opposed to well-tried technologies.

Electricity production is not the only producer of CO_2 in our atmosphere. The various sectors responsible for CO_2 generated as a result of human activity are given in Table 1.6.

Table 1.5 Breakdown of electricity production worldwide and a prediction over the next two decades [38]

	2012	2020	2030	2040
	Electricity production energy (10^{12} kWh)			
Oil	1.06	0.86	0.62	0.56
Nuclear	2.34	3.05	3.95	4.50
Renewables	4.73	6.87	8.68	10.63
Natural gas	4.83	5.20	7.47	10.10
Coal	8.60	9.73	10.12	10.62

Table 1.6 CO_2 emissions from fossil fuel in 2012 [39]

	Carbon dioxide emissions (%)
Electricity	41
Transport	22
Industrial (including cement manufacture)	20
Residential	6
Other	11

If there is the necessary political will to do so, we can replace the fossil-fuel-derived electricity with renewable forms of energy or nuclear energy or hydropower. However, we do have a problem with replacing transport fuel. We could 1 day have electric cars replace petrol vehicles and possibly even diesel vehicles, but replacing fossil fuel for air travel and sea travel is difficult if not impossible. Furthermore, some industrial processes such as cement manufacture, involving the heating of $CaCO_3$ resulting in the waste product, CO_2, are also problematic. Attempts at replacing petrol in transport with renewable fuel derived from biomass (sugar cane as done in Brazil or corn as done in the United States for petrol, and palm oil in Malaysia for biodiesel) has had some success, but the overall contribution has been relatively small [40]. In 2016, the biofuels contributed 4% to the world's transport fuels. The United States, Brazil, and Malaysia are the world leaders in biofuels.

All of this does indicates that the world is not on top of solving the global warming problem, in spite of the steady increase in the deployment of renewable forms of energy. The changeover from fossil fuel to renewables is just too slow. It is predicted that renewables will increase their share of electricity production from 21.9% in 2012 to 29.2% in 2040 (<0.3% per year) (see Table 1.5). We will have to work very much harder to replace fossil fuel as the main driving force of our energy industry.

One slight glimmer on the horizon is the fact that natural gas, methane (including shale gas) is better for the planet than burning coal, and in many countries coal is being replaced by natural gas. The reason why natural gas is better than coal is that the amount of CO_2 produced from burning CH_4 per unit of energy (50 g/MJ) is less than it is for coal (92 g/MJ); moreover, coal burning produces particulates. Of course, the burning of CH_4 still produces CO_2:

$$CH_4 + 2O_2 = CO_2 + 2H_2O$$

1.8 Conclusions

We believe that we do understand the underpinnings of global warming and that greenhouse gases and CO_2, in particular, are the root causes. We know that renewable forms of energy and possibly nuclear MUST replace fossil fuel where possible and that this must be done soon. There are limits to what can be replaced as we unfortunately depend on fossil fuels for transport both now and in the foreseeable future. Other areas such as electricity production using fossil fuel can and should be phased out.

Much depends on governments around the world having the will and energy to drive a nonfossil fuel policy. It should be a basic moral decision for the sake of future generations. Governments should not be driven by short-range decisions that benefit the few (including parliamentarians and also shareholders in fossil-fueled industries). They should be bold and brave enough to create a legacy for our children and our children's children.

This book is about managing global warming. It looks at a range of technical issues from the *status quo* of energy production today, to the many ways we can reduce CO_2 production with nuclear fission (and 1 day fusion) and the many renewable options available to us today. There are chapters on new carbon-based chemical industries that do not use oil-based feedstocks, greener farming to reduce greenhouse gas emissions, and geoengineering. Furthermore, the book also looks at a range of environmental and human issues that include the ethics of geoengineering, the economics of geoengineering and of global warming implications; human migration; justice in managing climate change; and local actions involving all human beings.

References

[1] Fourier J. Remarques Générales sur les Températures Du Globe Terrestre et des Espaces Planétaires. Annales de Chemie et de Physique 1824;27:136–67.
[2] Nobili L, Melloni M. Le Thermo-multiplicateur. Ann Chim (Phys) 1831;48:198–9.
[3] Sella A. Melloni's thermomultiplier. Chem World 2018;15:70.
[4] Tyndall J. On the Absorption and Radiation of Heat by Gases and Vapours…. Philos Mag 1861;22(Series 4):169–94. 273–85.
[5] Tyndall J. On radiation through the Earth's atmosphere. Philos Mag 1863;25(Series 4):200–6.
[6] Arrhenius S. On the influence of carbonic acid in the air upon the temperature of the ground. Philos Mag J Sci 1896;41:237–76.
[7] https://www.rmets.org/sites/default/files/qjcallender38.pdf.
[8] earthguide.ucsd.edu/virtualmuseum/climatechange1/02_3.shtml.
[9] https://link.springer.com/article/10.1007/BF02423528.
[10] https://www.co2.earth/daily-co2.
[11] http://www.realclimate.org/index.php/archives/2004/12/how-do-we-know-that-recent-cosub2sub-increases-are-due-to-human-activities-updated/.
[12] https://www.acs.org/content/acs/en/climatescience/.../its-water-vapor-not-the-co2.html.
[13] Tuckett RP. The role of atmospheric gases in global warming. In: Letcher TM, editor. Climate change, observed impacts on planet earth. 2nd ed. Oxford: Elsevier; 2016. p. 375–98.
[14] https://www.epa.gov/ghgemissions/global-greenhouse-gas-emissions-data.
[15] https://www.worldenergy.org/wp-content/uploads/2016/10/World-Energy-Resources-Full-report-2016.10.03.pdf.
[16] https://www.bp.com/en/global/corporate/energy-economics/statistical-review-of-world-energy/oil/oil-production.html.
[17] https://www.bp.com/en/global/corporate/energy-economics/statistical-review-of-world-energy/coal/coal-production.html.
[18] https://www.bp.com/en/global/corporate/energy-economics/statistical-review-of-world-energy/natural-gas/natural-gas-production.html.
[19] http://www.globalcarbonproject.org.
[20] Dorman LI. The role of space weather and cosmic ray effects in climate change. In: Letcher TM, editor. Climate Change, observed impacts on planet earth. 2nd ed. Oxford: Elsevier; 2016. p. 513–46.
[21] Stenchikov G. The role of volcanic activity in climate and global change. In: Letcher TM, editor. Climate change, observed impacts on planet earth. 2nd ed. Oxford: Elsevier; 2016. p. 419–48.

[22] Lourens LJ, Tuenter E. The role of variations of the earth's orbital characteristics. In: Letcher TM, editor. Climate change, observed impacts on planet earth. 2nd ed. Oxford: Elsevier; 2016. p. 399–418.
[23] Cohen S. The role of widespread surface solar radiation changes. In: Letcher TM, editor. Climate change: observed impacts on planet earth. 2nd ed. Oxford: Elsevier; 2016. p. 491–512.
[24] Shakun JD, Clark PU, He F, Marcott SA, Mix AC, Liu Z, Otto-Bliesner B, Schmittner A, Bard E. Global warming preceded by increasing carbon dioxide concentrations during the last deglaciation. Nature 2012;484:49–54.
[25] Haywood J. Atmospheric aerosols and their role in climate change. In: Letcher TM, editor. Climate change: observed impacts on planet earth. 2nd ed. Oxford: Elsevier; 2016. p. 449–64.
[26] Lu Q-B. Cosmic-ray-driven reaction and greenhouse effect of halogenated molecules. Int J Mod Phys B 2013;27:1350073. 38 p. https://doi.org/10.1142/S0217979213500732.
[27] https://earthobservatory.nasa.gov/Features/CarbonCycle/page5.php.
[28] Denman KL, Brasseur G, Chidthaisong A, Ciais P, Cox PM, Dickinson RE, Hauglustaine D, Heinze C, Holland E, Jacob D, Lohmann U, Ramachandran S, da Silva Dias PL, Wofsy SC, Zhang X. Couplings between changes in the climate system and biogeochemistry. In: Climate change 2007: the physical science basis. contribution of working group I to the fourth assessment report of the intergovernmental panel on climate change. Cambridge/New York, NY: Cambridge University Press; 2007.
[29] Letcher TM, editor. Climate change: observed impacts on planet earth. 2nd ed. Oxford: Elsevier; 2016. p. 21–340 [chapters 2–21].
[30] Vallero DA, Letcher TM, editors. Unraveling environmental disasters. Oxford: Elsevier; 2013 [chapters 8, 12–14].
[31] https://www.google.co.uk/search?q=global+reserves+of+oil+gas+and+coal+are+still+abundant&tbm=isch&tbo=u&source=univ&sa=X&ved=0ahUKEwi7-P6mh9vYAhUIuBQKHd0MCcQQsAQIRg&biw=1205&bih=563.
[32] https://www.iea.org/about/faqs/oil/.
[33] https://www.iea.org/about/faqs/coal/.
[34] https://sciencealert.com/world-meteorological-organisation-2017-climate-and-extreme-weather-report.
[35] https://www.nasa.gov/.../nasa-noaa-data-show-2016-warmest-year-on-record-globally.
[36] https://www.climate.gov/news-features/climate-qa/which-emits-more-carbon-dioxide-volcanoes-or-human-activities.
[37] https://www.statista.com/statistics/238610/projected-world-electricity-generation-by-energy-source/.
[38] https://www.statista.com > Energy & Environmental Services > Electricity.
[39] https://whatsyourimpact.org/greenhouse-gases/carbon-dioxide-emissions.
[40] https://www.iea.org/etp/tracking2017/transportbiofuels/.

The Paris Agreement—Implications for greenhouse gas removal and zero emissions energy production

2

Robert Chris[1]
Visiting Fellow, Department of Geography, Open University, Milton Keynes, United Kingdom
[1]Corresponding author: robert.chris@open.ac.uk

Chapter Outline

2.1 Introduction

Achieving the Paris Agreement temperature target requires the virtual elimination of greenhouse gas emissions by mid-century. If emissions reduce more slowly, achieving the target will require the removal from the atmosphere of greenhouse gases already

Managing Global Warming. https://doi.org/10.1016/B978-0-12-814104-5.00002-8

emitted. The amounts of these gases and their dilution in the atmosphere create a sociopolitical and engineering challenge of unprecedented proportions. Climatically relevant scale is orders of magnitude greater than industrial scale, and the interdependence of the options available to policymakers means that portfolios of individually plausible policies may not be plausible in aggregate.

This chapter does not promote any particular net emissions technology (NET) to deliver greenhouse gas removal (GGR). Instead, it treats them as a generic set of technologies that have in common the capture and permanent sequestration of atmospheric CO_2. The question addressed here is how much GGR will be required, not which NETs will deliver it [1].

Solar radiation management (SRM) can assist in meeting the temperature targets by reducing net insolation, thereby allowing a more gradual path to zero emissions, but it cannot contribute to reducing greenhouse gases (GHGs), a necessary element of any long-term sustainable climate policy. SRM is therefore outside the scope of this chapter.

The chapter uses outputs from a calculator designed to show how the interdependent policy variables affecting energy consumption and emissions combine to produce future consumption and emissions scenarios extrapolated to 2100. The core assumptions and relationships that define the algorithms behind the calculations are described in Section 2.2.

A critical challenge for policymakers is to decide which of the infinite range of possible futures are policy relevant. The chapter deploys the complexity theory concept of plausibility to define the feasible range of policy options and assess the extent to which they rely upon zero emissions energy (ZEE) and GGR. The concept of plausibility is explained in Section 2.3. In Section 2.4, the interconnectedness of the input parameters and their relative sensitivities are discussed without considering their feasibility. In Section 2.5 plausibility is used to discard infeasible policy options, and Section 2.6 looks at the policy implications of a range of feasible policy portfolios, ranked according to their dependence on GGR. In Section 2.7, the plausibility of meeting the demand for ZEE is assessed. The chapter closes with conclusions in Section 2.8.

Notation used in this chapter follows the conventions set out in Box 2.1.

2.2 Methodology

This analysis is based on global aggregates without reference to geographic or national distributions of energy consumption or production. It relies on three core axioms. First, total final energy consumption (TFC) can be met either from fossil fuels (FF) or from ZEE (biomass, nuclear, solar, wind, geothermal, tidal, and hydro). Second, in a heterogeneous market-based global economy, global energy consumption is market driven and cannot be determined by government fiat. Third, public policy can influence the rate and direction of change in energy consumption, its primary sources, and energy efficiency.

The analysis also makes the assumption that the climatic effect of a unit of GHG removed is the same as a unit not emitted. However, the reduction in atmospheric aerosols resulting from reduced FF combustion may generate countervailing net

Box 2.1

Notation	SI equivalent	Description
Gt	Gt (10^9 t or 10^{15} g)	Gigatonne or 1000 million tonnes
Gt(C)/yr	Gt(C) a^{-1}	Gigatonne of carbon per year
Gt(CO_2)/yr	Gt(CO_2) a^{-1}	Gigatonne of carbon dioxide per year
ppmv		Parts per million by volume
Quadrillion	10^{15}	1 million billion
1 M km^3	10^{15} m^3	1 million cubic kilometers
1 km^3/sec	10^9 m^3 s^{-1}	1 cubic kilometer per second
1 M km^2	10^{12} m^2	1 million square kilometers
~		About *or* approximately
MJ	10^6 J	Megajoule (1 MJ = 0.2778 kWh)
EJ	10^{18} J	Exajoule (1 EJ = 277.78 MWh)
EJ/yr	10^{18} J a^{-1}	Exajoules per year
Btu		British thermal unit (1 Btu = 0.293 Wh)

radiative forcing that would correspondingly increase the requirement for GGR and/or ZEE [2]. It is unclear at present how the additional radiative forcing from ocean outgassing [3] might be different for GHGs avoided and those removed.

The relationship between global temperature change and total cumulative anthropogenic CO_2 emissions (TCRE) has been shown to be near linear for cumulative emissions at least to 2000 Gt(C). The current best estimate is that limiting global warming relative to the period 1861–80 to less than 2°C with greater than 66% probability will require cumulative CO_2 emissions since 1870 to not exceed 1000 Gt(C) ([4]: Section 12.5.4.3). This analysis uses these values as the benchmark for the simulation of the emissions scenarios for different temperature targets.

The TCRE methodology is concerned only with CO_2 emissions. Non-CO_2 GHGs represent about one-third of the anthropogenic radiative forcing, and of this half is produced by methane, a short-lived GHG [5]. While these GHGs are climatically significant, even their complete removal from the atmosphere would not obviate a need for the removal of large amounts of atmospheric CO_2, which could only be delivered by radical changes to the global FF energy economy. Accordingly, because the policy implications of these changes are the primary focus of this chapter, non-CO_2 GHGs are ignored in this analysis. The IPCC estimates that accounting for non-CO_2 GHGs would reduce the carbon budget from 1000 to 790 Gt(C) ([4]: Section 12.5.4.3). This would make even greater both the scale and the urgency of the need for GGR identified in this chapter.

Long-term historical data for global energy consumption and emissions are extrapolated to 2100 using policy relevant variables. These variables, using averaged global values, include the timing and extent of the transition away from FF, reductions in TFC, and incremental improvements in energy efficiency. The dependent variables are the ZEE required to meet future energy demand, and where necessary, the GGR needed to remain within the available carbon budget. The extrapolations of historical data are used as benchmarks from which to assess the scale of change implied by alternative policy scenarios; they are not intended to be forecasts or predictions. A full explanation of the eleven policy variables is given in Section 2.5.

The calculator assumes that all transitions follow standard S-curve trajectories. The S-curve coefficients are calculated dynamically to generate smooth paths over the relevant periods.

The charts in Fig. 2.1 illustrate the calculator outputs. Each scenario is represented by two charts for energy consumption and emissions, respectively, each covering the period 2014–2100 (actual data for 2014). The energy chart shows the annual global TFC divided between FF energy and ZEE. The ZEE Index is the ratio of the slope of the imputed linear growth of ZEE from all sources from 2015 to 2100 to the slope of its growth during the base period 2004–14. The ZEE Index is a measure of the rate at which the development of ZEE must be accelerated to meet energy demand during the balance of the century. For example, a ZEE Index of 10 means that in every year from 2015 onwards, ZEE will need to grow by 10 times the average annual amount it grew in the base period. The actual growth of ZEE is unlikely to be linear, but for the purposes of assessing the policy implications of the incremental demand for ZEE, the ratio of the imputed linear growth rates is an adequate indicator. Energy charts overwritten with a Θ represent implausible policy portfolios (see discussion in Section 2.5).

The emissions chart shows annual (primary axis) and cumulative (secondary axis) emissions, including net emissions after GGR. The long-dashed line is the carbon budget for the chosen temperature target; the short-dashed line shows the trajectory of cumulative emissions converging to the carbon budget by 2100. Total emissions are shown by the solid white line with the slanted brick area being net emissions after GGR (horizontal bricks).

The UNFCCC use a fuzzy temperature increase target of "well below 2°C above preindustrial levels and pursuing efforts to limit it to 1.5°C." In this chapter 1.5°C and 2°C are used as outer boundaries, and 1.8°C is the value within these boundaries used to define "well below 2°C." Fig. 2.1 illustrates the IPCC's RCP2.6 scenario with some help from GGR. Here the temperature increase is 1.8°C, emissions fall to 40% of their 2010 value by 2050 and then reduce to zero by 2100, and GGR is deployed from 2050 [1]. In this scenario, GGR must grow to 9.7Gt(C)/yr. Cumulative emissions overshoot the carbon budget in 2043, reaching an excess of 11% in 2071 before being brought back on target by 2100. The ZEE Index is 15.

This analysis focuses throughout on energy consumption rather than primary production (TPES) or energy generation. Climate change has meant that sources of energy are under scrutiny creating an emerging consensus largely to abandon conventional FF energy in favor of various forms of renewable energy and nuclear. This requires that future energy policy not be predetermined by the historical reliance on FF. Accordingly, the future energy source profile that emerges from this analysis is based on end-user consumption and not on prior assumptions about primary energy sources. This treatment also accounts for transformation and other losses intrinsic to the current global energy economy in which most production of electricity wastes about two-thirds of the primary energy consumed in its manufacture [6].

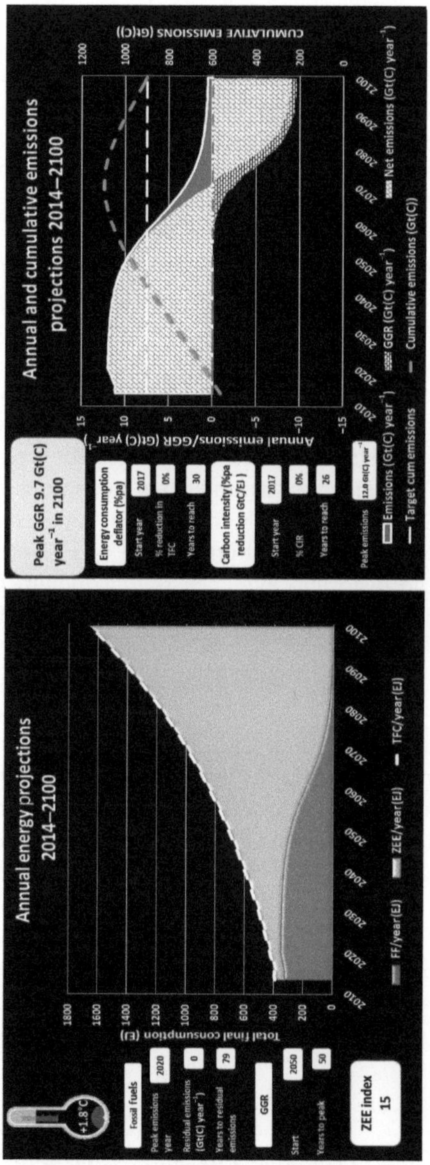

Fig. 2.1 Illustrative calculator outputs—scenario corresponding to IPCC RCP2.6.

The distinction in this calculator between FF energy and ZEE has required the energy losses that account for the bulk of the difference between global TPES and TFC to be distributed between the different primary energy sources from which they arise. This has been done in proportion to their consumption in the manufacture of electricity, thereby replacing electricity in the TFC data with its underlying primary energy sources.

Power generation, whether measured in terms of financial investment or megawatts, is an emergent property of future energy policy not a forward indicator of it. Considerable differences exist between the useful energy that is generated by different ZEE technologies. These differences arise largely from their intermittent nature and their relative efficiencies. For example, solar energy is a function of the available sunlight; wind energy depends on the wind; but nuclear energy is produced more or less continuously, albeit with thermal efficiency of only about one-third. This means that to supply any given amount of useful end-user energy requires significantly different amounts of generating capacity according to its primary source.

Technological advances that, for example, have dramatically improved the efficiency of solar photovoltaic cells have resulted in lower capital cost per unit of power, and this will have been translated by the rebound effect into increased end-user demand for energy [7]. This calculator assumes that in coming decades this increase in efficiency-driven ZEE demand will follow the long-established pattern of efficiency-driven increases in the production and use of FF energy. As the graph in Fig. 2.4 (*infra*) shows, the historical evidence is that incremental ZEE may have slowed the growth in FF production but has not yet begun to reduce it.

2.3 Plausibility

Policy robustness is central to building resilience, and building resilience is central to minimizing the likelihood that systemic catastrophe might prevent future generations from flourishing according to their own lights [8]. Robustness rests on the concept of plausibility, and here means specifically that:

> *... rather than seeking strategies that are optimal for some set of expectations about the long-term future, [policy robustness] seeks near-term strategies [...] that perform reasonably well compared to the alternatives across a wide range of plausible scenarios evaluated using the many value systems held by different parties to the decision.*
>
> *([9]: p. xiv).*

Plausible has its Oxford English Dictionary meaning—"an argument, an idea, a statement, etc.: seeming reasonable, probable, or truthful; convincing, believable." This subjective notion accommodates different and emerging perceptions of the world about us. What may seem plausible to one person may not to another, or even to the same person later when the context may have changed or she has become better informed. But plausibility cannot be capriciously determined; it must be based on the exercise of informed judgment rooted in evidence from the past and present to accommodate the uncertainties and surprises of the future. Plausibility is continuously and heuristically reconstructed.

In Section 2.5, the range of values of each independent variable is considered through a plausibility lens. But first, without being constrained by reasonableness, I examine the mathematical sensitivity of the two dependent variables, amounts of GGR and ZEE, to each input parameter.

2.4 The numbers

Table 2.1 describes five scenarios, a small but indicative subset of the entire range of putative futures. Where appropriate, results are calculated for three global temperature increase targets, 1.5°C, 1.8°C, and 2°C. All scenarios provide that in addition to the assumed FF emissions (including those from cement production), land use emissions amounting to some 1.5 Gt(C)/yr will continue indefinitely into the future [10]. All these scenarios can be adapted by assuming varying degrees of increased carbon intensity and reduced TFC.

No consideration is given to error ranges. The calculator produces unique mathematically correct outcomes for each set of input parameters. Alternatives and sensitivities can be explored by varying these input values. The purpose is not to produce a supposedly accurate set of predictions, but to identify the scale of the policy challenge implied by each scenario.

2.4.1 Business as usual

Under *BAU*, by 2100 cumulative emissions reach almost three times the 2°C carbon budget and more than four times the 1.5°C budget. Global average temperature rise is more than 5°C. The 1.5°C, 1.8°C, and 2°C carbon budgets are exceeded respectively in 2029, 2039, and 2045 (Scenario 1—there are 22 scenario illustrations referred to in the following discussion; they are shown in Box 2.2). In this scenario there is no attempt to stay within the carbon budget.

Table 2.1 Indicative scenario groups

BAU	Business as usual (BAU) assumes growth in TFC and ZEE, both following their unabated long-term trends, with no GGR
ZEE only	TFC follows long-term trend; no GGR and accelerated transition to ZEE
ZEE/REC	Reduced energy demand (REC) with accelerated transition to ZEE; no GGR
ZEE/REC/CIR	Mixed portfolios of emission reduction policies; no GGR
GGR+	Combined response including GGR

Box 2.2 Scenario charts

Scenario 1: BAU: follow historical trends, no attempt to meet UNFCCC temperature target.

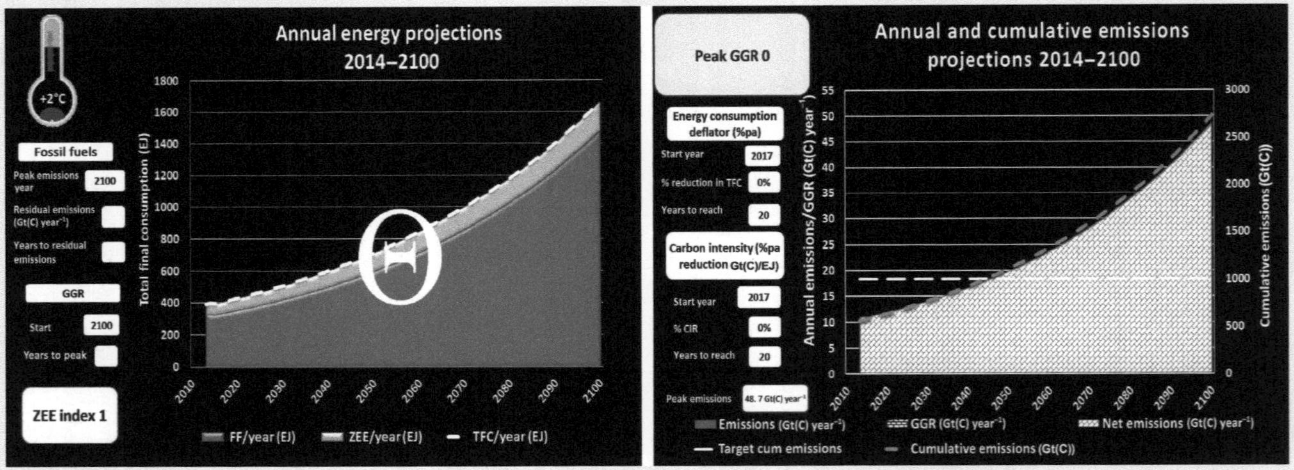

Box 2.2 —cont'd

Scenario 2: 1.8°C: ZEE only with emissions reducing to zero in 30 years from peak in 2022.

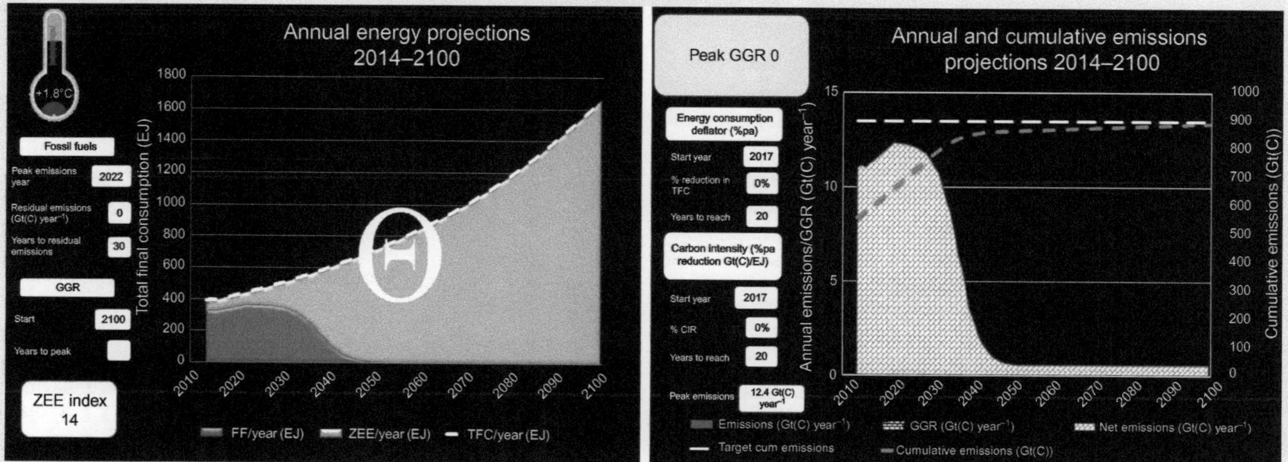

Continued

Box 2.2 —cont'd

Scenario 3: 2°C: ZEE only with emissions peaking in 2025 and stabilizing at 2 Gt(C)/yr by 2044.

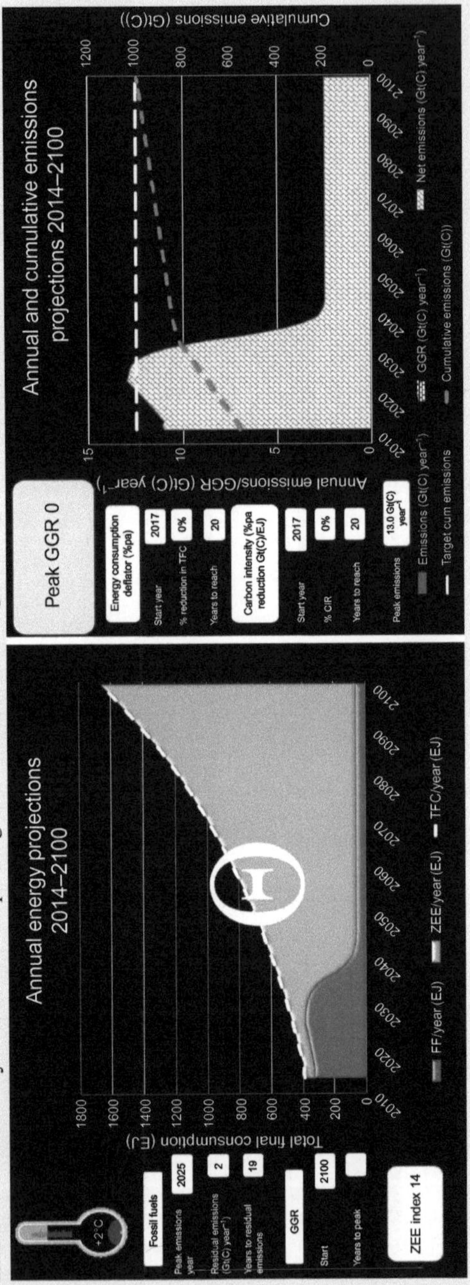

Box 2.2 —cont'd

Scenario 4: 1.8°C: ZEE + 50% REC with emissions peaking in 2020 and stabilizing at 1 Gt(C)/yr by 2045.

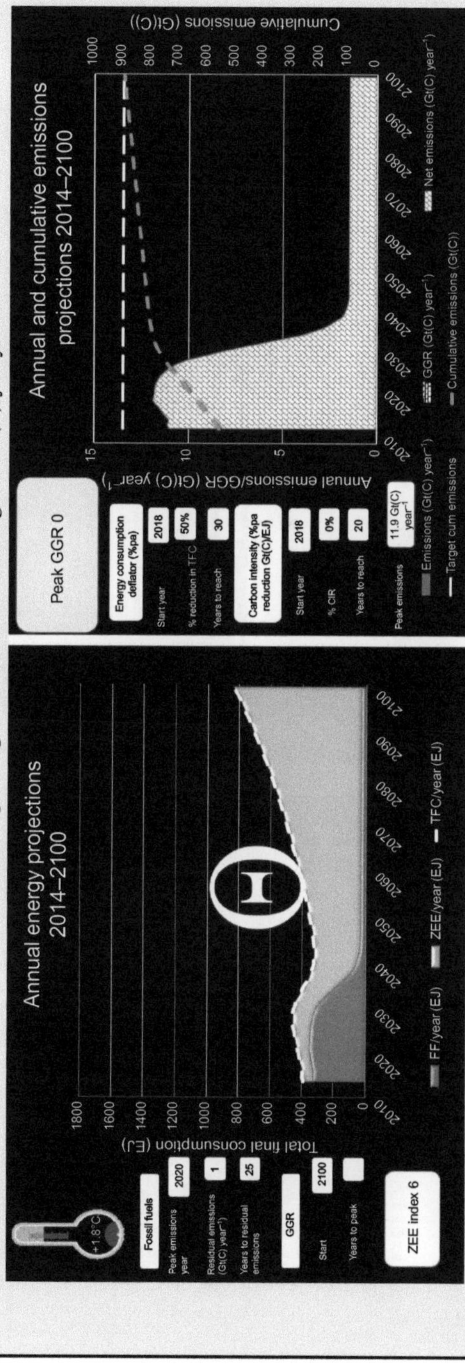

Continued

Box 2.2 —cont'd

Scenario 5: 1.8°C: ZEE + 50% REC with emissions peaking in 2020 and stabilizing at 2 Gt(C)/yr by 2032.

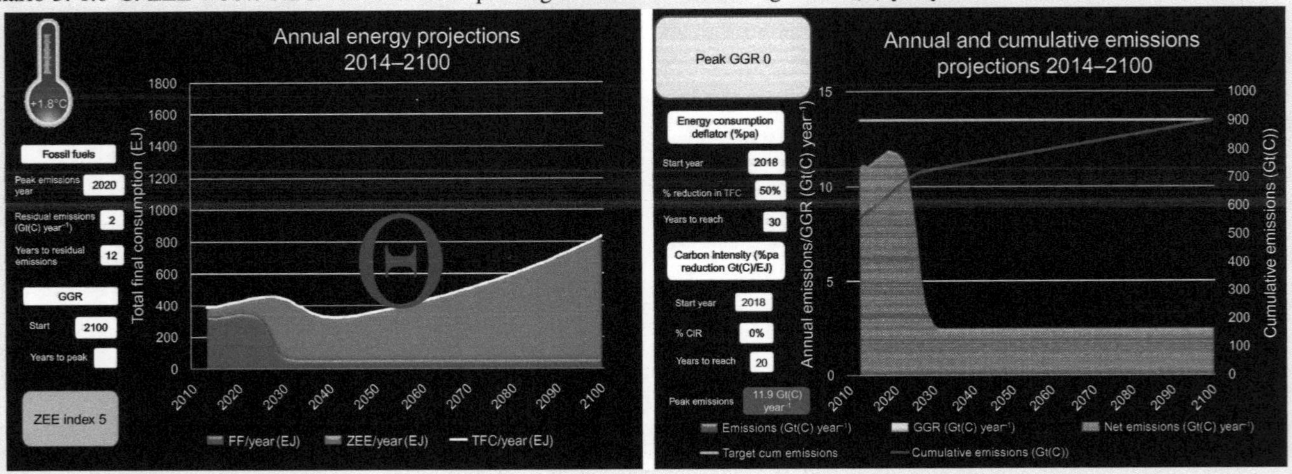

Box 2.2 —cont'd

Scenario 6: BAU + 98% CIR.

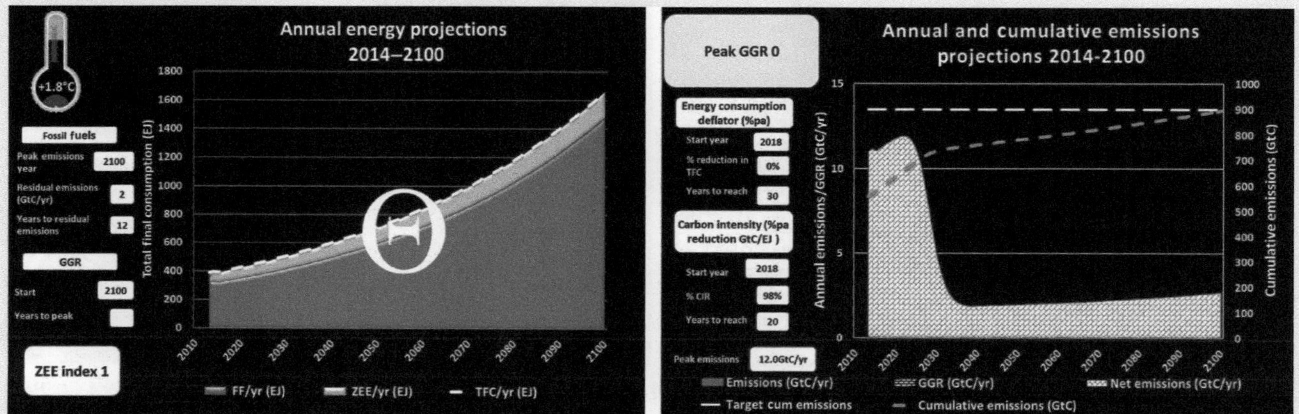

Continued

Box 2.2 —cont'd

Scenario 7: 1.8°C: ZEE + 40% REC + 70% CIR—emissions peak in 2025 and stabilize at 2 Gt(C)/yr by 2050.

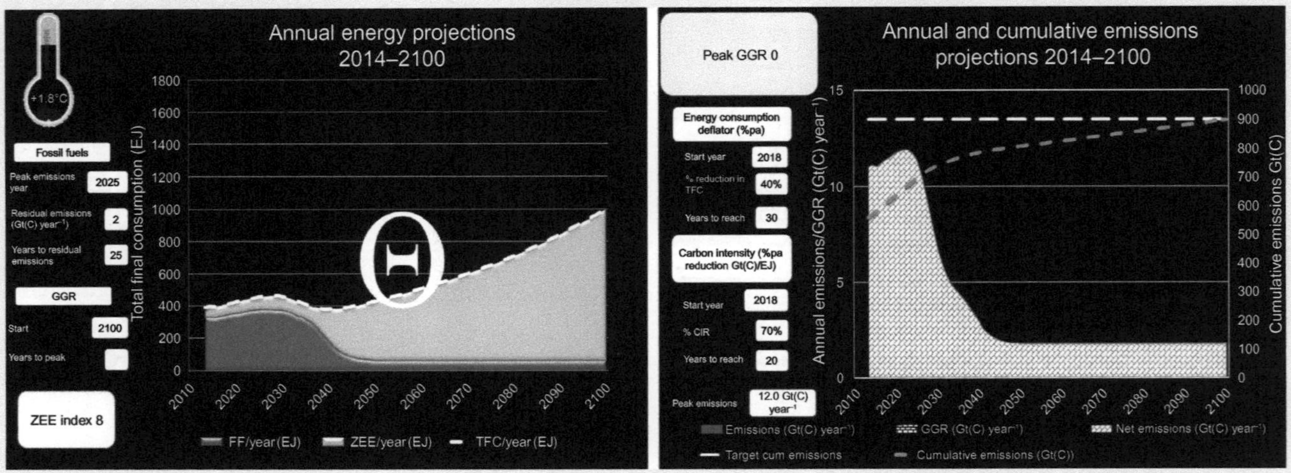

Box 2.2 —cont'd

Scenario 8: BAU + GGR from 2050.

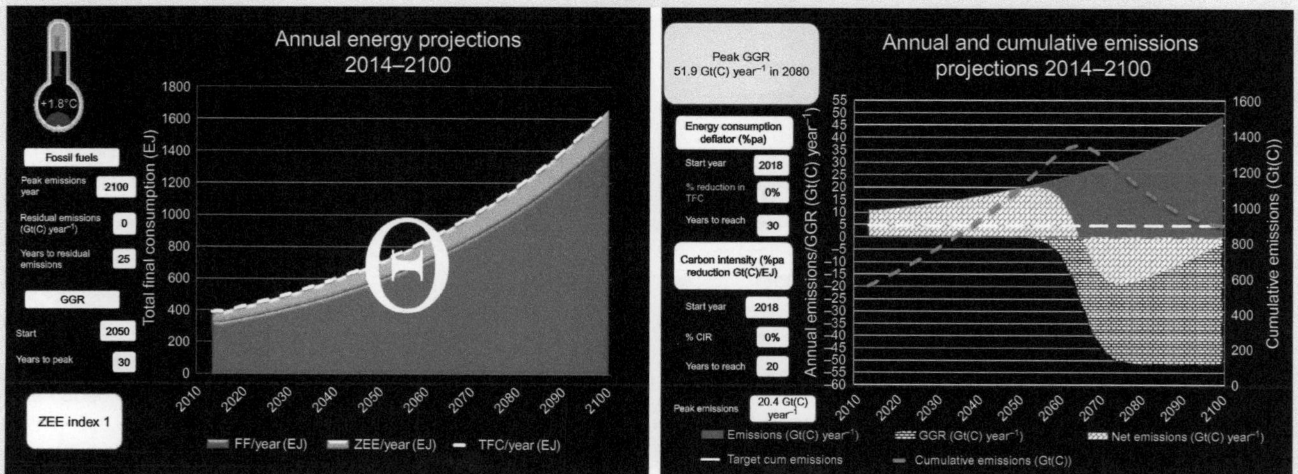

Continued

Box 2.2 —cont'd

Scenario 9: 2°C: Worst plausible outcome—emissions peak in 2040 and stabilize at 10 Gt(C)/yr by 2056, GGR starts in 2030 and reaches 14.7Gt (C)/yr by 2089.

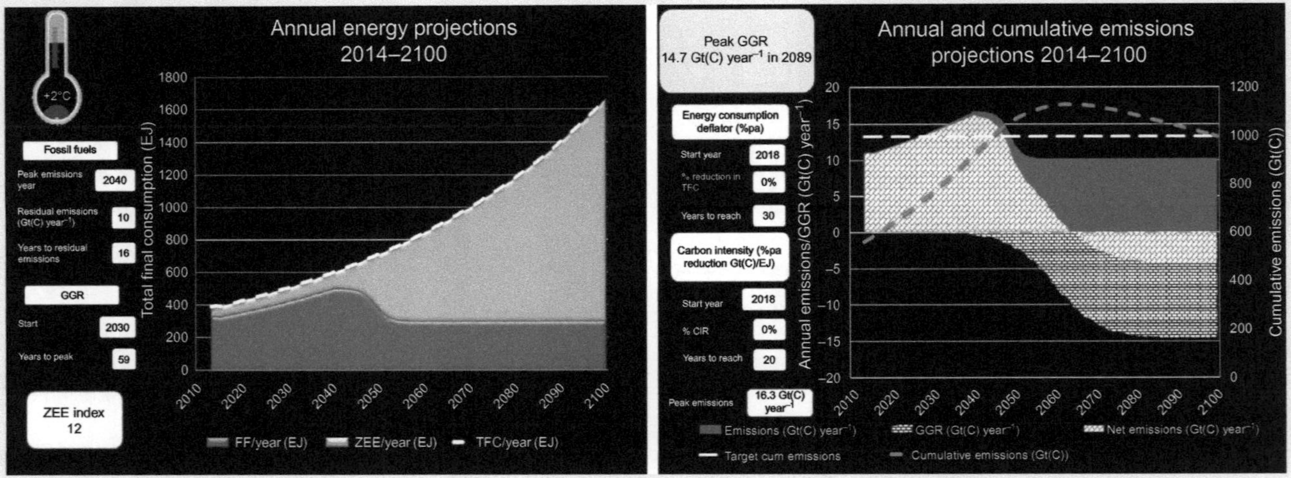

Box 2.2 —cont'd

Scenario 10: 1.8°C: Best plausible outcome—20% REC + 25% CIR with emissions peaking in 2017 and stabilizing at zero by 2057 and no GGR.

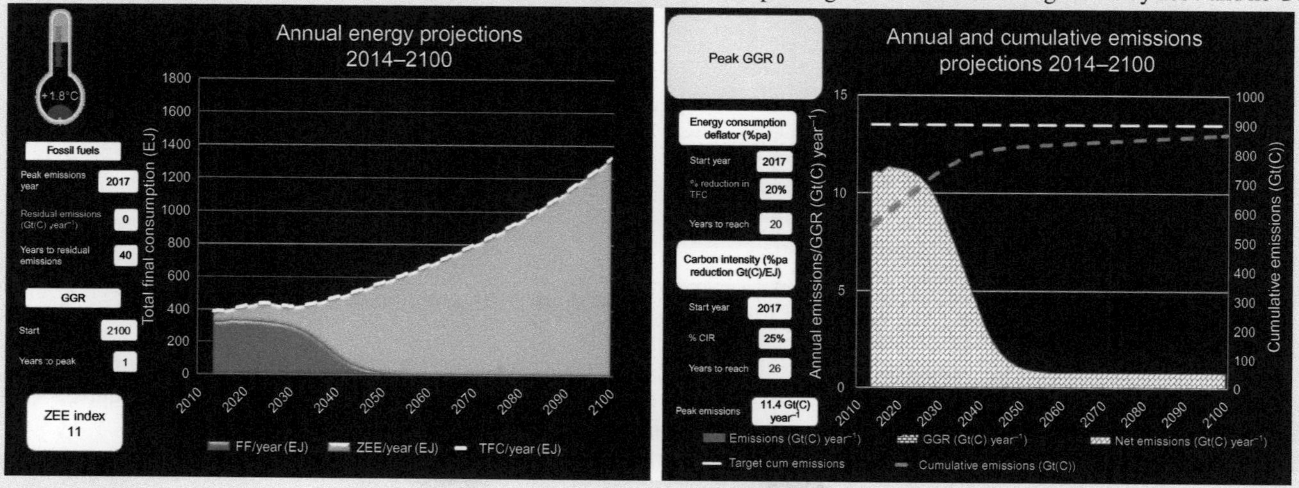

Continued

Box 2.2 —cont'd

Scenario 11: 1.5°C: Best plausible outcome—no REC or CIR with emissions peaking in 2017 and stabilizing at zero by 2057 with GGR starting in 2025 reaching 1.7 Gt(C)/yr by 2032.

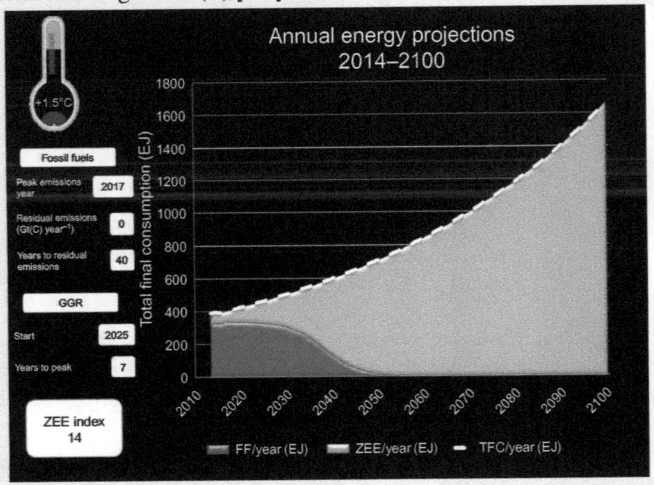

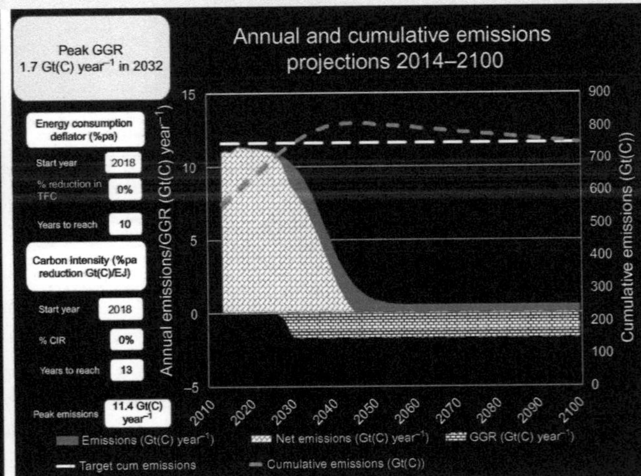

Box 2.2 —cont'd

Scenario 12: 1.8°C: ZEE only with no REC, CIR or GGR—emissions peak in 2018 and stabilize at zero by 2058.

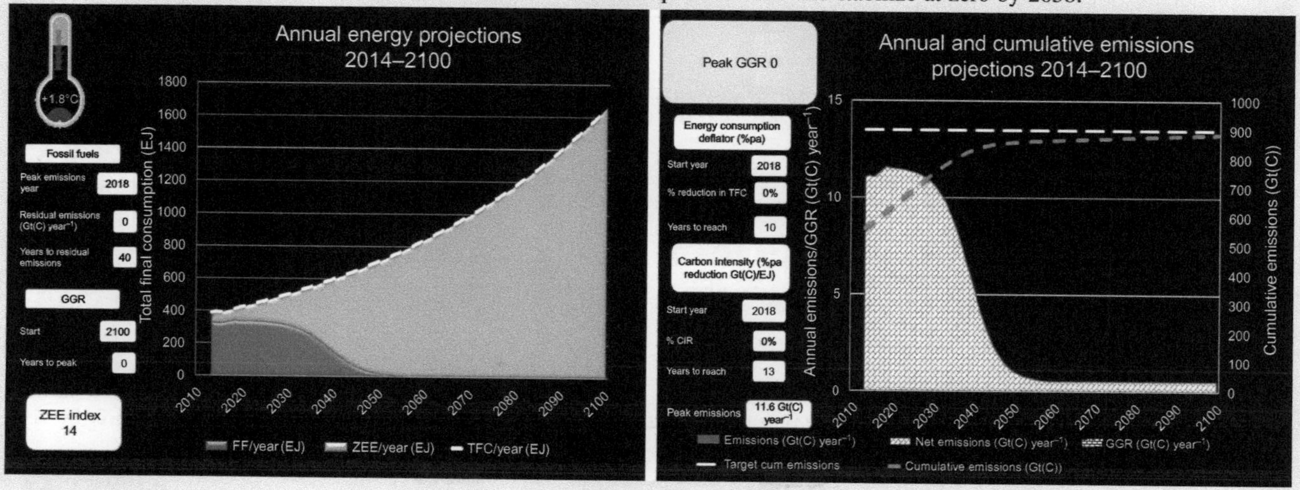

Continued

Box 2.2 —cont'd

Scenario 13: 2°C: ZEE + no GGR—emissions peak in 2018 and stabilize at 2 Gt(C)/yr by 2058.

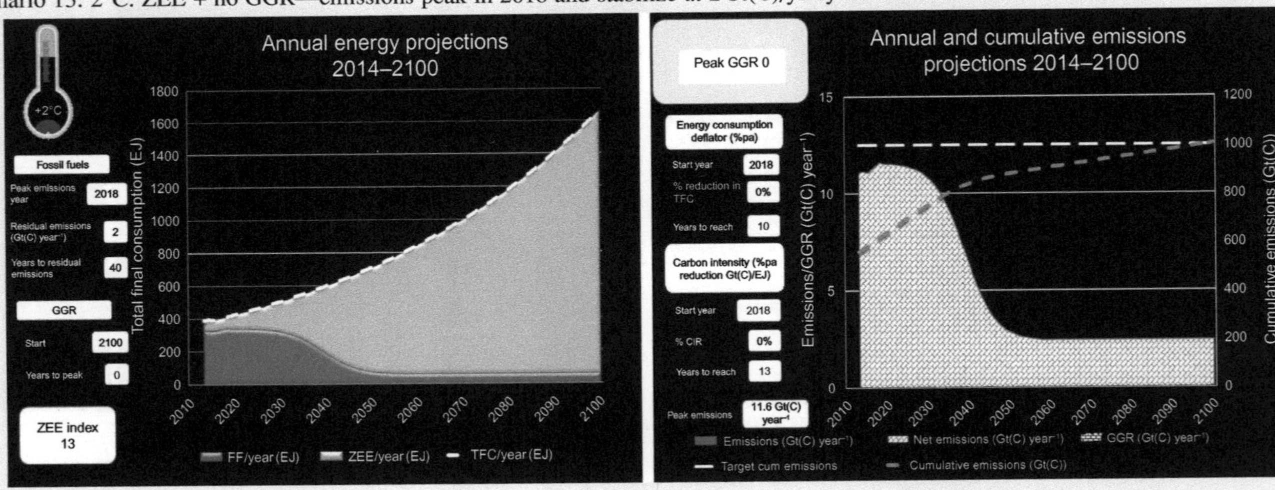

Box 2.2 —cont'd

Scenario 14: 1.8°C: ZEE + 5% CIR + 1.5 Gt(C)/yr GGR starting in 2025—emissions peak in 2020 and stabilize at 2 Gt(C)/yr by 2054.

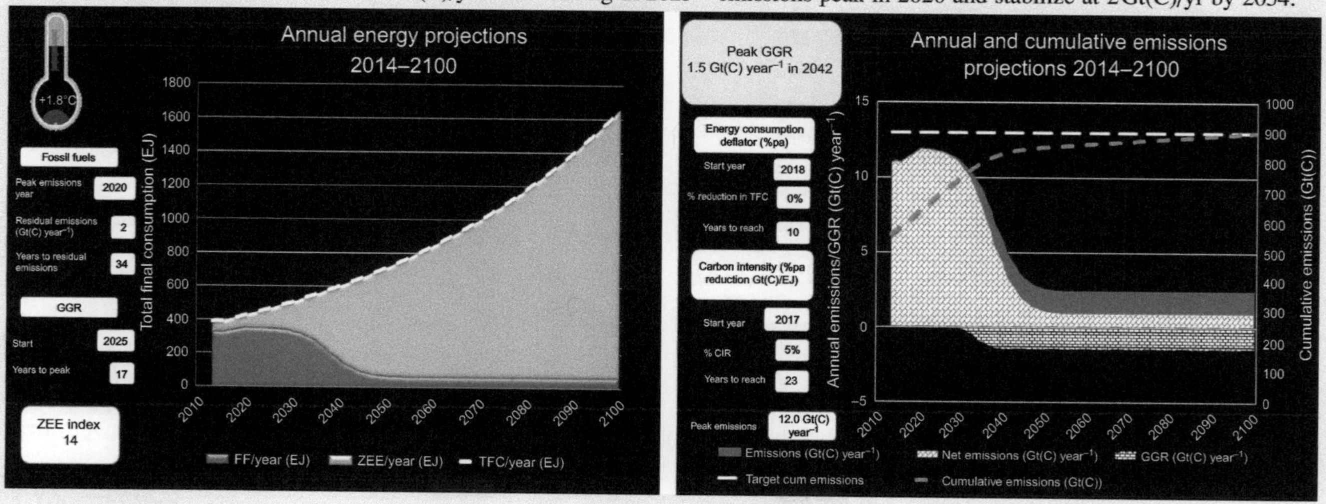

Continued

Box 2.2 —cont'd

Scenario 15: 1.5°C: ZEE + 25% CIR + 1.5 Gt(C)/yr GGR starting in 2025—emissions peak in 2017 and reduce to zero by 2057.

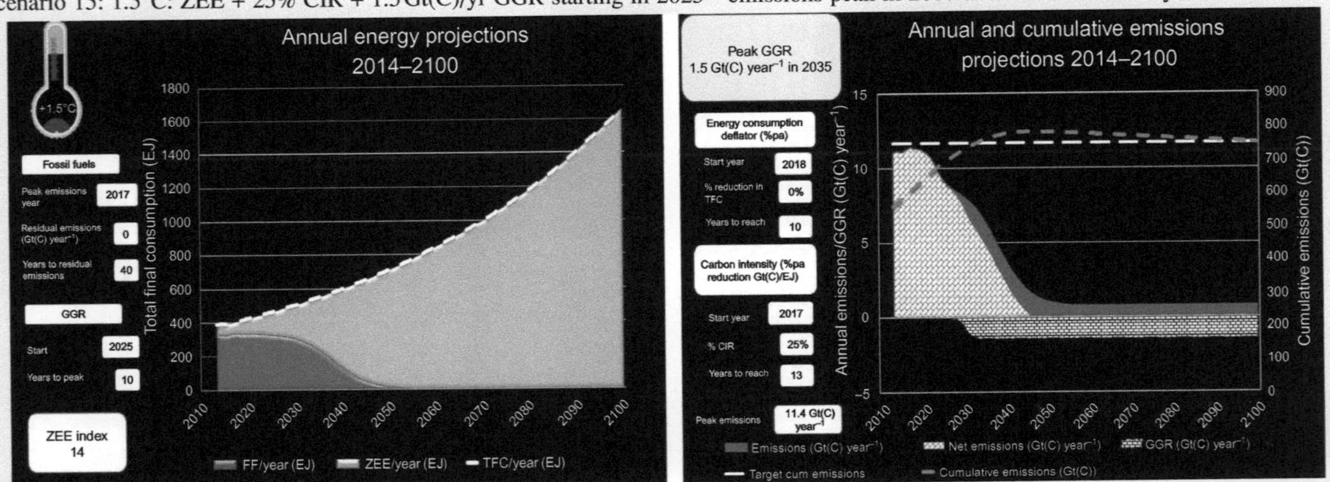

Box 2.2 —cont'd

Scenario 16: 2°C: ZEE + 20% REC + 1.1 Gt(C)/yr starting 2025—emissions peak in 2025 reduce to 2 Gt(C)/yr by 2065.

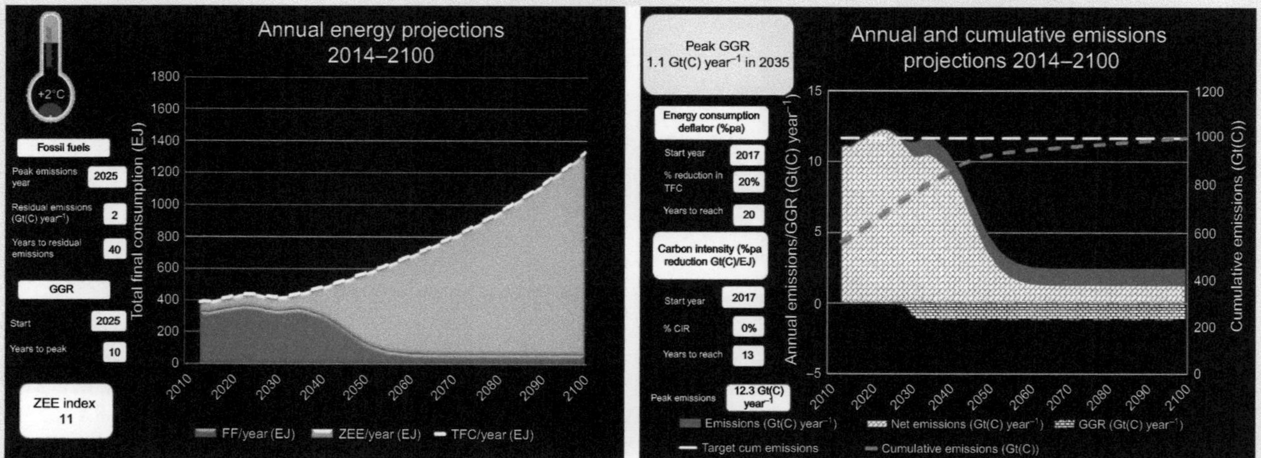

Continued

Box 2.2 —cont'd

Scenario 17: 1.8°C: ZEE + 5 Gt(C)/yr GGR from 2025 peaking in 2051—emissions peaking in 2025 reducing to 4 Gt(C)/yr by 2065.

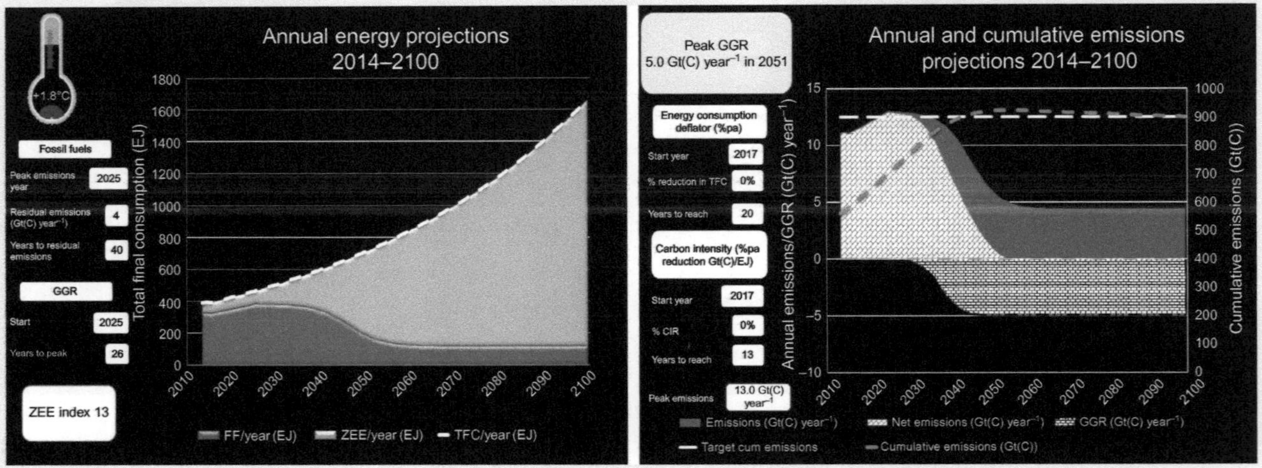

Box 2.2 —cont'd

Scenario 18: 1.5°C: ZEE + 20% REC +15% CIR + 5 Gt(C)/yr GGR by 2050—emissions peak 2025 and stabilize at 3 Gt(C)/yr by 2056.

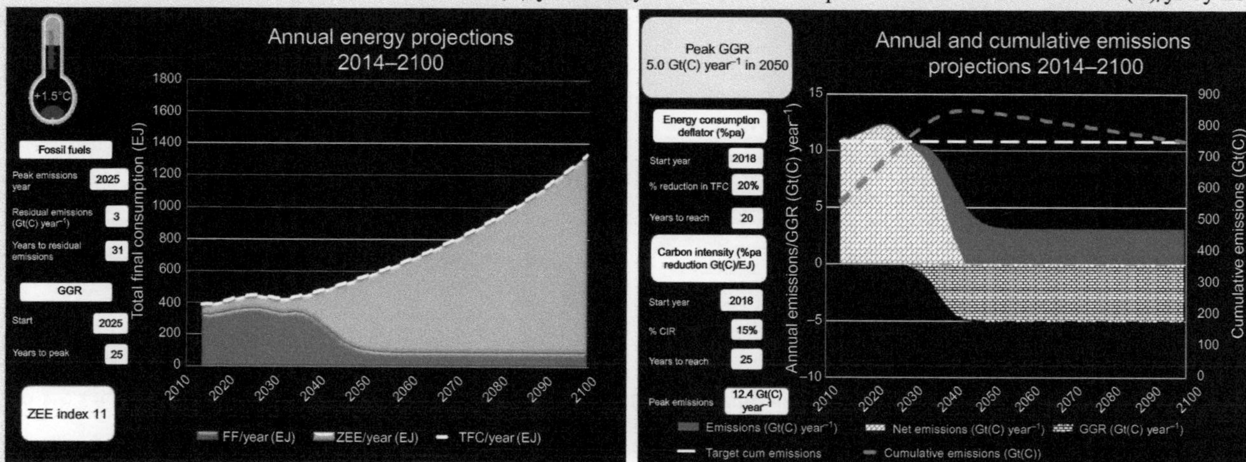

Continued

Box 2.2 —cont'd

Scenario 19: 2°C: ZEE + 5 Gt(C)/yr GGR starting 2040 and peaking in 2066—emissions peak in 2025 stabilize at 5 Gt(C)/yr by 2058.

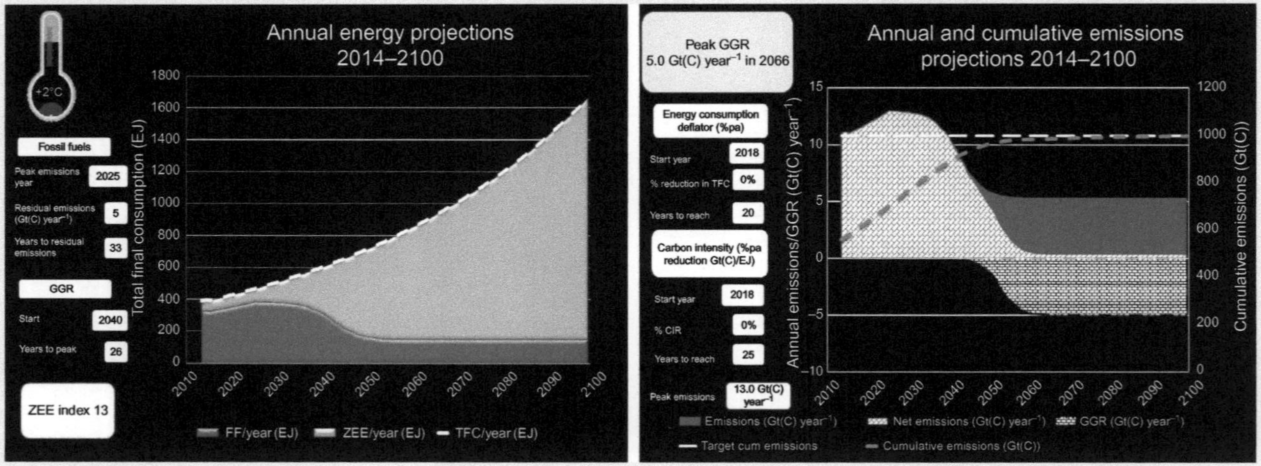

Box 2.2 —cont'd

Scenario 20: 1.5°C: ZEE + 15% CIR + 10 Gt(C)/yr GGR starting 2025 and peaking in 2066—emissions peak 2028 reducing to 7 Gt(C)/yr by 2062.

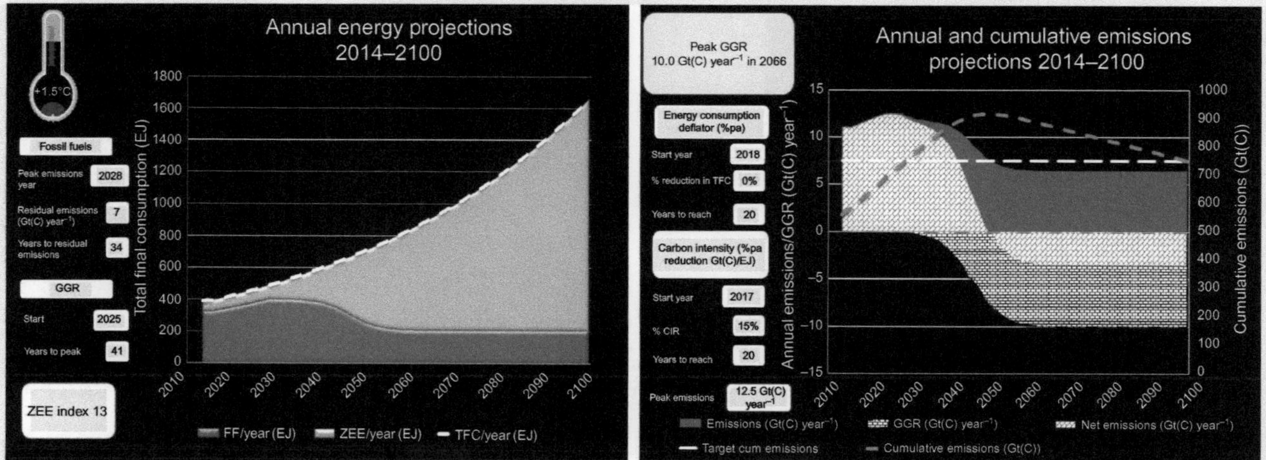

Continued

Box 2.2 —cont'd

Scenario 21: 1.8°C: ZEE + 10 Gt(C)/yr GGR starting 2025 and peaking in 2070—emissions peak 2030 and stabilize at 8 Gt(C)/yr by 2056.

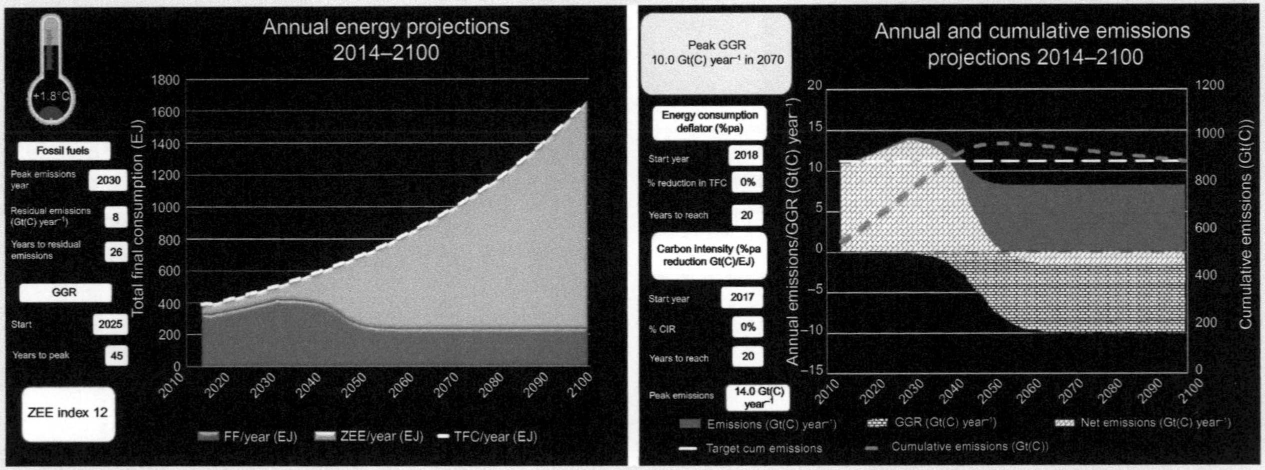

Box 2.2 —cont'd

Scenario 22: 1.8°C: ZEE + 20% REC + 25% CIR + 10 Gt(C)/yr GGR starting 2040 and peaking in 2090—emissions peak 2030 and stabilize at 8 Gt(C)/yr by 2060.

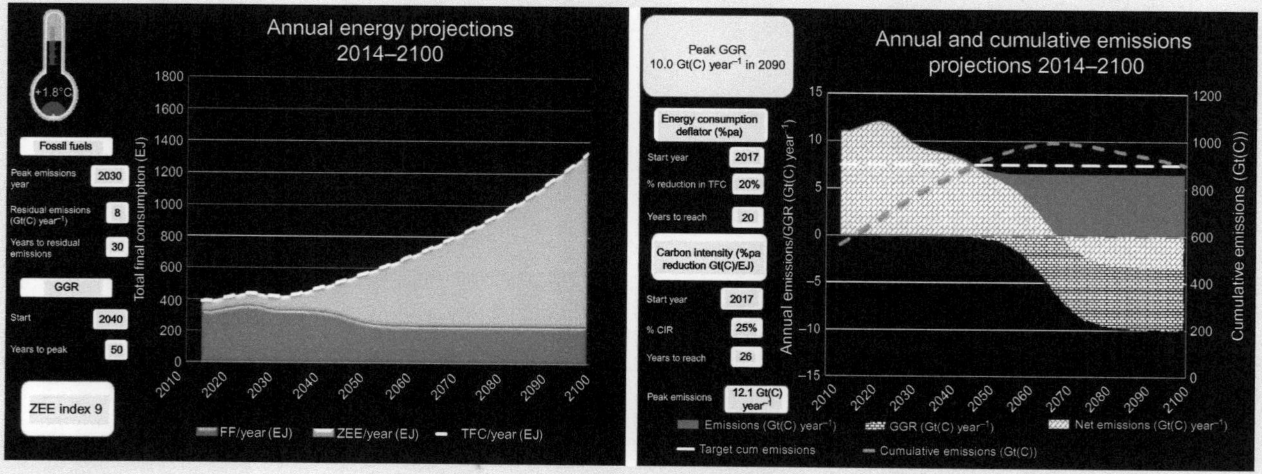

2.4.2 Zero emissions energy only

The *ZEE-only* scenario represents a policy of maintaining TFC growth at historical levels and relying on reducing emissions by transitioning to ZEE to stay within the carbon budget. The amount of ZEE required varies according to assumptions about the level of residual FF emissions. If these emissions are brought to zero within 30 years, the reduction must begin in 2022 (Scenario 2) or 2028 to meet the 1.8°C and 2°C targets, respectively. The 1.5°C target requires emissions to peak in 2017 and reduce to zero within 16 years.

If FF emissions are assumed to stabilize at 2 Gt(C)/yr, about 20% of 2015 levels, and peak in 2025, meeting the 2°C target would require them to reduce to this level by 2044 (Scenario 3). 1.5°C is unachievable and 1.8°C requires emissions to be reduced to 2 Gt(C)/yr within 1 year. With residual emissions of 4 Gt(C)/yr, the 2°C target requires that emissions peak in 2019 and reduce to 4 Gt(C)/yr within 1 year.

In all cases, the continuing land use emissions mean that at 2100 the cumulative emissions remain on a rising trend breaching the carbon budget thereafter. Only the introduction of GGR can ensure that cumulative emissions plateau below the relevant carbon budget.

It is also noteworthy that in all the variations of the *ZEE-only* scenario, the ZEE Index exceeds 12. The decline in emissions is so rapid that the bulk of TFC must come from ZEE.

2.4.3 Zero emissions energy/Reducing energy consumption

ZEE/REC introduces the policy of reducing energy consumption (REC). The impact of REC depends upon when it begins and how deep it is. In scenarios with an early, rapid, and deep transition away from FF, REC has almost no impact on cumulative emissions, but it does significantly reduce the amount of ZEE required later in the century. For example, for the 1.8°C scenario with emissions peaking in 2020 and reducing to 1 Gt(C)/yr within 25 years, a reduction in energy consumption starting in 2018 and reaching 50% within 30 years has no significant impact on cumulative emissions but reduces the ZEE Index from 14 to 6 (Scenario 4). This arises because the calculator assumes that if energy consumption falls, established energy suppliers with low marginal costs will respond to the lower demand by reducing prices, in line with conventional supply and demand market economics. Those investing in new energy sources do not have this price flexibility because their marginal costs will be less elastic. For this reason, short-term reductions in energy consumption are assumed to reduce the consumption of ZEE in preference to FF, and therefore have little impact on emissions.

There are two scenarios in which these assumptions would be invalid. First, if the marginal cost of ZEE falls consistently below that of FF; and second, if governments use policy levers to force the early retirement of FF energy production. The impact of these options on emissions can be examined in the calculator by varying the FF reduction rate parameter. In the last example, if emissions were reduced over 12 rather than 25 years, the 1.8°C target could still be met either with residual emissions reducing to

2 Gt(C)/yr rather than 1 Gt(C)/yr (Scenario 5), or peak emissions being deferred for 5 years to 2025. These variations make no appreciable difference to either cumulative emissions by 2100 or the demand for ZEE during the remainder of the century.

In the *BAU* scenario, the same energy demand reductions reduce cumulative emissions to 2100 by 40%. However, they still exceed the 1.5°C,1.8°C and 2°C carbon budgets by 117%, 81% and 63% respectively; there is no impact on the ZEE Index that remains at 1.

2.4.4 Zero emissions energy/Reducing energy consumption/ Carbon intensity reduction

Adding CIR (carbon intensity reduction) into the policy mix renders the targets less demanding. CIR refers specifically to reducing emissions from using FF by switching from high emitting to cleaner FF, for example from coal to natural gas. (Reducing emissions by replacing FF with ZEE is included in ZEE and not in CIR.) Moreover, CIR refers to the incremental reductions in carbon intensity beyond those subsumed in the long-term historical trend. The long-term changes in the shares of coal, oil, and gas are already reflected in the historical correlation between TFC and emissions (Fig. 2.4 *infra*), and therefore the continuation of these benefits is equally reflected in their extrapolations into the future. To avoid double counting, the CIR must be limited only to incremental improvements beyond those subsumed in the historical data.

Although carbon capture and storage (CCS) has the practical effect of reducing the carbon intensity of the fuels being combusted, because it also involves the capture and sequestration of carbon, in this chapter it has been included among NETs (see discussion in Section 2.5).

In the extreme case of *BAU*, CIR must begin in 2018 and reach 98% within 20 years to achieve the 1.8°C target (Scenario 6) (93% for 2°C, 1.5°C is unachievable). This is a world in which technology is focused almost entirely on reducing the emissions from the growing consumption of FF, rather than on replacing them with ZEE. Adding reductions in energy consumption has minimal additional impact. As will become apparent in Section 2.5, this is not a plausible future.

Bringing some CIR into the mix makes the targets less challenging in terms of other policy actions. For example, the 1.8°C target could be met by combining emissions peaking in 2025 and falling to 2 Gt(C)/yr in 25 years, TFC reductions commencing in 2018 and reaching 40% within 30 years, and CIR starting in 2018 and reaching 70% in 20 years (Scenario 7). If the emissions fall instead to 1 Gt(C)/yr, the requirement for CIR can be reduced to 55%. In these scenarios the ZEE Index is 8.

2.4.5 Greenhouse gas removal

In theory, any excess in cumulative emissions can be removed by GGR. For example, for *BAU* even the 1.8°C target can be met by 2100 with GGR commencing in 2050 and increasing over 30 years to 52 Gt(C)/yr (Scenario 8). However, while this scenario may be numerically coherent, GGR at that scale is implausible. Nevertheless, at lower levels, GGR can be a balancing factor that provides flexibility in generating plausible portfolios of policy options. Thus, if the combination of *ZEE/REC/CIR* falls short, it can be brought back on target by the addition of some GGR. This is explored in Section 2.5.

2.5 Applying plausibility

For a portfolio of interdependent policy options to be plausible, it is not sufficient that each of its components be plausible, their combination must also be plausible. For example, aggressive policies to retire FF energy are dependent on either there being an adequate amount of ZEE to take its place, or commensurate reductions in TFC. In this section, the policy levers are examined to assess their plausible ranges and determine the mix that would deliver the UNFCCC temperature targets.

There are eleven policy-driven variables that are mathematically but not necessarily politically independent. Their plausibility ranges are examined in four groups (Table 2.2). The suggestions given in the following paragraphs for the ranges of plausible values are indicative and not prescriptive.

As noted, if the timely scaling of GGR and ZEE to their required levels is implausible, then the combination of policy options from which they flow is also implausible, irrespective of the plausibility of the individual policy components (scenario energy charts marked with a Θ are implausible policy portfolios).

Table 2.2 List of calculator variables

Driver	Variable
FF emissions	• Year of peak emissions • Level at which emissions eventually stabilize (Gt (C)/yr) • Years taken from peak emissions to reach stabilization level
Reduction in energy consumption (REC)	• Target percentage reduction in TFC from its peak • Year REC starts • Years taken from start to reach REC target
FF carbon intensity (CIR)	• Target incremental percentage reduction in carbon intensity • Year CIR starts • Years taken from CIR start to reach target level
GGR	• Year GGR starts • Years taken from GGR start to reach calculated level

2.5.1 Fossil fuel emissions

The IEA [11] reports that GDP and emissions growth have become delinked since 2010 as the annual percentage increase in global emissions has consistently been lower than that for GDP. They argue that preliminary data for 2015 and 2016 suggest that emissions have stabilized. They note that this deviates from the long-term trend in which emissions and GDP have been closely correlated.

This trend combines four drivers: changes in carbon intensity (C/TFC); energy efficiency (TFC/GDP); consumption per capita (GDP/Population); and the growth in population. Algebraically this may be expressed as follows:

$$C = \frac{C}{TFC} \times \frac{TFC}{GDP} \times \frac{GDP}{Population} \times Population$$

where C represents the global carbon emissions, TFC is the total final energy consumption, and GDP and Population are their respective global aggregates.

Over the 56 years of this dataset (Fig. 2.2), population and consumption per capita have both increased almost threefold while carbon intensity (not shown but

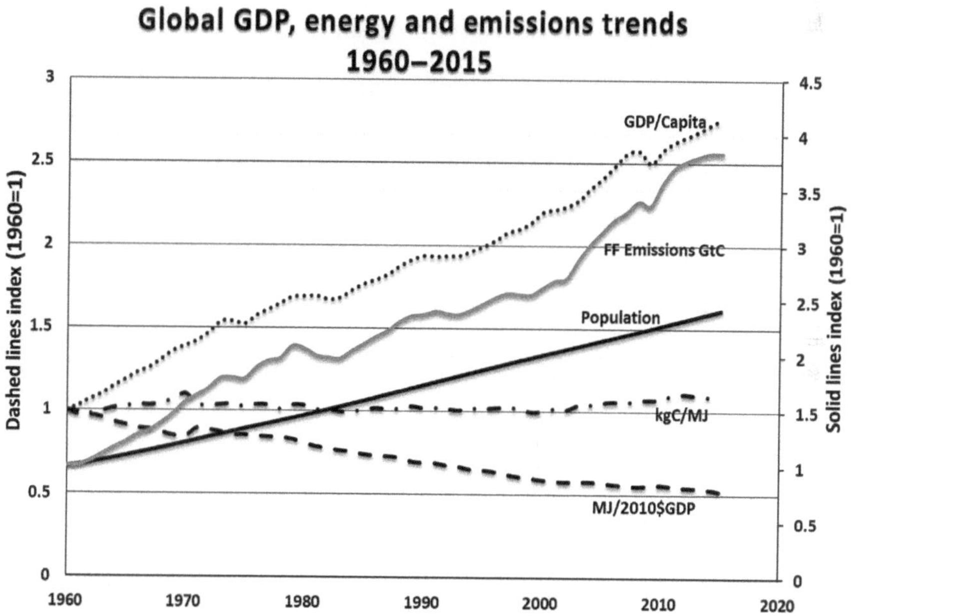

Fig. 2.2 Global GDP, energy, and emissions trends 1960–2015.
Based on data from the following sources: GDP and population data from World Bank; emissions data from CDIAC; energy consumption data from IEA and Pbl Netherlands Environmental Assessment Agency. GDP is stated in 2010 US$; emissions are stated in Gt(C) year^{-1}, energy efficiency is measured in units of megajoules per US dollar (2010) of GDP; and carbon intensity is measured in units of kilograms of carbon per megajoule and carbon efficiency is shown in units of grams of carbon per US dollar (2010) of GDP. Each data set is indexed to its 1990 value.

trend virtually coincides with energy efficiency) has marginally deteriorated; these three combine to increase emissions more than sevenfold. Only energy efficiency has shown a countervailing tendency, having almost doubled in the period. The drivers when combined account for the almost quadrupling of carbon emissions over the 56 years to 2015. (See also IEA data [12].)

If the trend established since 2010 is maintained, we may be on the threshold of peak emissions as the shift to ZEE gains momentum, but more data are required before such a conclusion can be reached with confidence. Preliminary data suggest a 2% increase in emissions in 2017 after no increase in emissions in 2016 [13]. Another reason for caution is that since the 1980s, the proportion of TFC delivered by FF remained virtually unchanged at about 84% (Fig. 2.3 and [11]). Indeed, even since 2005, incremental ZEE has been only 20% of incremental TFC, indicating that growth in ZEE continues to fall far short of satisfying the increasing demand for energy, let alone displacing legacy FF supplies.

These trends suggest that a range for the plausible value for the year in which emissions peak might be 2017–40. For peak emissions to occur later than 2040 would signify a major collapse of global climate policy. This cannot be dismissed as a possibility, but since at that point it would be impossible to stay within the UNFCCC temperature targets without an aggressive program of geoengineering, possibly also including SRM, it is not considered further here.

The level at which emissions eventually stabilize is more challenging to assess as it depends on a wide range of impossible-to-predict political and behavioral changes at

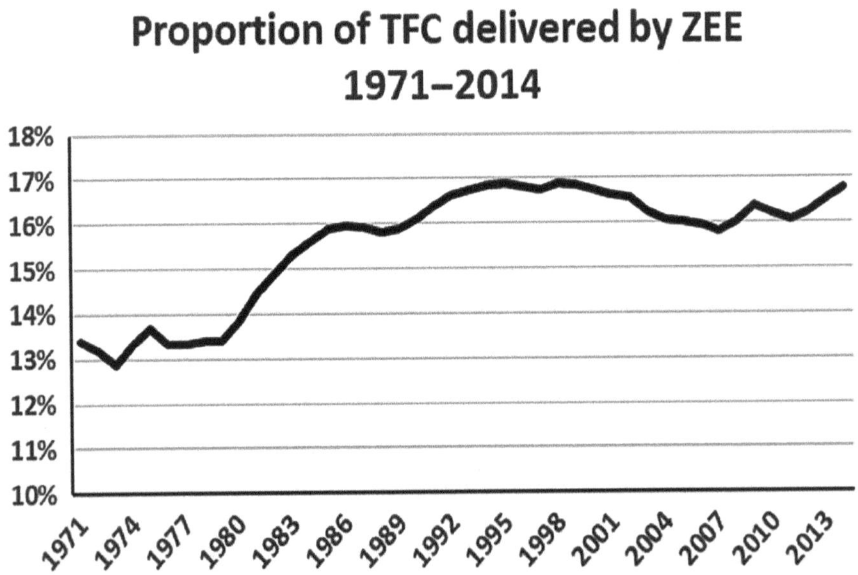

Fig. 2.3 Proportion of total final energy consumption delivered by zero emissions energy sources 1971–2014.
Source: Graph based on data from IEA.

global scale over many decades. Although reducing emissions to zero is climatically the optimum outcome, this may not be politically or socially feasible given the extent to which FF are embedded in modern life, and our commitment to the never-ending increase in consumption implicit in the prevailing imperative for continuing economic growth. I suggest a range between 0 and 10 Gt(C)/yr for the plausible value of the eventual stable level of emissions. This wide range reflects the uncertainty, and provides an opportunity to examine the policy implications of values at the extremes of the range. It does not preclude emissions peaking at higher levels before stabilizing within this range, although it ignores the potential negative impacts of hysteresis in overshoot scenarios.

The final variable in this group is the time it takes for emissions to stabilize at the chosen level. The drivers here are sociotechnological and political. Many historical precedents suggest that major structural changes to deeply embedded systems take more than a single generation. For example, Fig. 2.4A and B show the transition from wood through FF to modern renewables. These figures are based on US data but a similar, albeit shorter, story is apparent from global data [14]. As much wood and coal are burned today, as were ever burned. New fuels have struggled to keep pace with economic growth and never supplanted previous energy sources. Fig. 2.4A shows that coal took 60 years to reduce wood's market share from 100% to 10%, oil took 70 years to reduce coal's share from 70% to 15%, and the combined share of coal, oil, and natural gas has reduced only marginally in the last 50 years, a period in which energy consumption has more than doubled. There is clearly a great deal of inertia in the global energy economy. This is unsurprising given the high levels of investment in relatively long-lived energy assets and the costs of retrofitting legacy assets to cope with new fuels. It seems prudent to assume that it is not only plausible but also probable that new fuels will follow these repeated historical trends (see Section 2.7).

In setting a range of plausible values for the stabilization period, a reasonable supposition is that the lower the eventual level of emissions, the longer it will take to reach. Fig. 2.5 illustrates a plausibility range according to the degree to which emissions are to be reduced below their peak. This plausibility range is difficult to specify. Given the slow progress of the international community in reaching the Paris Agreement, it seems implausible that, at one extreme global emissions could be eliminated in less than 40 years, or at the other, that it could not be done within 120 years.

2.5.2 Reduction in energy consumption (REC)

Population and consumption per capita are the primary drivers of TFC, but it is also influenced by energy efficiency. As already noted (see Fig. 2.2 *supra*), the long-term trend in energy consumption reflects an annual reduction of about 1% p.a. in the energy required to produce each unit of GDP. This trend is reflected in the extrapolation of future energy demand and therefore the factor under consideration here is the potential for an incremental increase in efficiency. It is unclear to what extent policy interventions can further improve the underlying trend toward greater energy efficiency and simultaneously prevent it leading to greater consumption through the rebound effect [7].

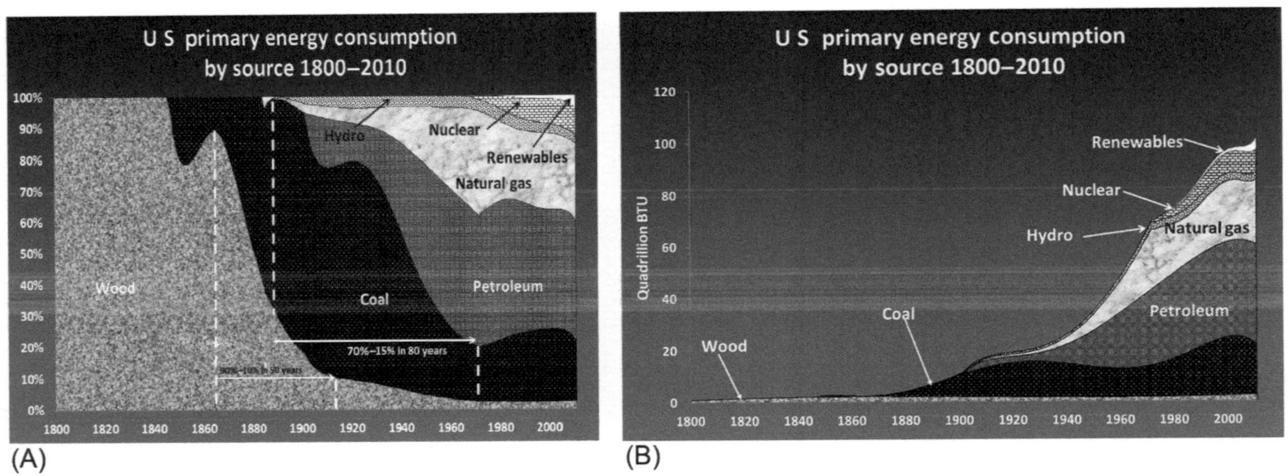

Fig. 2.4 US transition to new sources of energy 1800–2000. (A) Stacked percentages showing relative proportions. (B) Stacked areas showing absolute amounts in units of Quadrillion Btu. (Note: 1 quadrillion Btu = 1.044×10^{18} J.)
Based on data from US Energy Information Administration online at www.eia.gov/totalenergy/data/annual/perspectives.cfm accessed 23 March 2015.

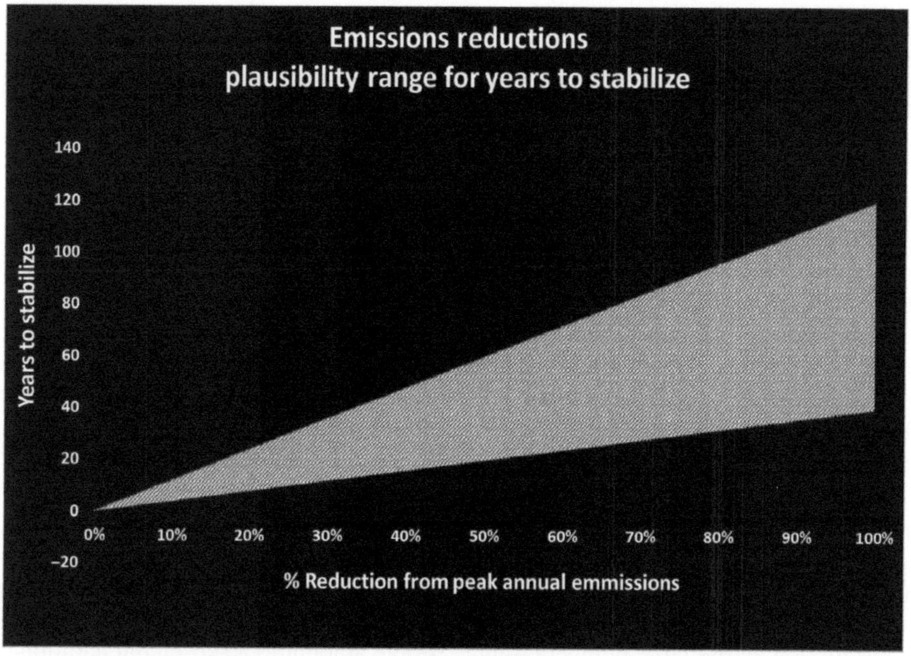

Fig. 2.5 Plausibility range for years to stabilize emissions.

Unrestrained market forces routinely translate improved efficiency into lower prices that encourage increased demand. This is evident in the long-term exponential growth of energy consumption despite consistent deep gains in energy efficiency. Policy intervention is needed to prevent increased energy efficiency translating into higher absolute energy demand. However, if policymakers intervene in this way, innovators could only profit from their energy-efficiency innovations if policymakers arranged for them to be remunerated in relation to the energy consumption saved. This would run counter to the neoliberal Zeitgeist and it is difficult to envisage such a novel change being implemented at global scale quickly enough to have a climatically significant impact on energy consumption. Without such a mechanism for remunerating the innovators, they would have little incentive to innovate. I suggest that significant efficiency gains, beyond those already accounted for in the long-term trend, are implausible, and the rebound effect might even make them negative. I shall therefore ignore the possibility of policy-driven efficiency gains beyond the continuation of the 1% p.a. subsumed in the data.

Only apocalyptic visions of the future contemplate any significant reduction in population. In such a future, implementing long-term policies to combat global warming is likely to be low on policymakers' to-do lists and may even be unnecessary. A population collapse to say, 1 billion, would probably reduce emissions disproportionately as the means of production implode through lack of labor. Supporters of Lovelock's Gaia Theory might find such an outcome unsurprising [15]. However, the mainstream view is that by 2100 population is forecast to be anywhere between 7 and 16 billion [16].

Although there may be ample scope for consumption per capita to be reduced in OECD and other developed countries, Kahneman et al. have shown [17] that we are reluctant willingly to forego benefits we already enjoy. Substantial resistance may therefore be assumed should policymakers attempt to impose significant cuts in energy consumption.

More than 70% of the global population are poor or on low income, defined as living on less than 10 dollars per day (2011 PPPUS$) [18]. For those 5 billion people, there are both moral and economic imperatives that will drive them to increase energy consumption per capita. Hitherto, energy for the world's poor has largely come from traditional biomass. Notwithstanding efforts to facilitate the transition to low and zero emission energy sources, extensive reliance on biomass seems likely to continue. Fig. 2.4 (*supra*) illustrates that even in the wealthy United States, the consumption of traditional biomass has hardly changed in absolute terms in 200 years, a period that has seen a thirtyfold increase in its population and several generations of new primary energy sources.

Given the current distribution of fertility rates [16], if population were to approach the higher estimates, the likelihood is that most of that growth would be in societies with low-energy consumption per capita. Nevertheless, even low levels of personal consumption across large numbers of new people amount to a substantial additional demand for energy. In short, a reversal of the long-term historical trend of growth of energy consumption per capita is improbable and if it does happen, is likely to be modest; indeed, the possibility cannot be ignored that the historical growth trend might even accelerate.

Combining the factors affecting energy consumption, it is hard to imagine that in any orderly transition, significant global reductions would be feasible. As fast as the wealthy reduce consumption (and that might not be very fast at all and may even continue to follow the historical growth trend), the world's poor would be increasing it. The idea that, save *in extremis*, governments across the world would act in concert to force deep reductions in energy consumption that negatively impact their citizens' lifestyles, seems unlikely. For these reasons, I set the plausibility range for reductions in TFC from its historical trend, between 0% and a generous 20%.

The start date for these efforts to reduce energy consumption is a matter of public policy. There seems little reason to prefer one start date against another and therefore any year is plausible.

2.5.3 Reduction in FF carbon intensity (CIR)

As already noted, reduction in FF carbon intensity (CIR) refers only to reducing emissions by burning FF with lower carbon emissions. Natural gas has the lowest emissions of all FF, about half those of coal when used to produce electricity [19,20]. Even if all FF were replaced with gas and used to produce electricity to power a wholly electrified global economy, the reduction in emissions could not exceed 50% and given that 30% of energy consumption is already from natural gas and ZEE, the theoretical upper bound for reduced emissions is about 35%. I propose a range of plausible values for CIR between 0% and 25%.

2.5.4 Greenhouse gas removal (GGR)

An important distinction must first be made between carbon capture and storage (CCS) of FF emissions and CCS of biomass emissions. CCS of FF emissions is not a NET because the GHGs already in the atmosphere remain there. CCS applied to biomass combustion is a NET, the most discussed of which are BECCS (bioenergy CCS) and biochar (the production of charcoal for incorporation into the soil). Despite CCS of FF emissions not being a NET, it does require the capture and sequestration of the emitted GHGs. These processes compete for resources with genuine NETs and therefore when considering the plausibility of these technologies, it is useful to combine them.

The key dependent variable from this calculator is the annual amount of GGR, measured in terms of gigatonnes of carbon to be sequestered to meet the chosen temperature target in the relevant scenario. For GGR, the two independent variables are the year in which it commences and the period in years taken to reach the calculated annual amount of sequestration. Given that GGR technologies do not yet exist in any scalable form, it will require concerted action by policymakers before their deployment can begin in earnest. I suggest this is implausible prior to 2025, and even this timing assumes considerable investment in research and development in the immediate future.

The second variable is the rapidity with which GGR is up-scaled after its commencement. This will also be policy rather than market dependent because the scale at which GHGs must be extracted for climate relevance exceeds any likely market capacity for them. Again, this may change if any currently nascent technologies emerge with unexpected potential. However, the time it would take the capacity of these markets to reach the necessary scale makes it relatively unlikely that they will be able to make a major contribution to meeting the climate targets (refer to Fig. 2.4).

The time it takes for GGR to be scaled is likely to be longer, the higher its eventual level. The minimum plausible period I propose is illustrated in Fig. 2.6.

All approaches to GGR entail two core processes: the capture of GHGs from the free atmosphere, and their sequestration in secure and permanent storage. While at laboratory, and even at industrial scale, there is already a considerable body of expertise with both, their scale is more than two to three orders of magnitude smaller than current global CO_2 emissions [21,22]. To grasp the scale, consider current annual emissions of ~35 Gt(CO_2)/yr (containing ~10 Gt(C)). This CO_2 is contained within 45×10^{15} m^3 (45 Mkm^3) of air (assuming 400 ppmv at 1 atm and 20°C). Processing air at 1 km^3/sec (assuming 100% efficiency) would remove 70% of current emissions and would, for example, produce 32 km^3 of liquefied CO_2 or 70 Gt$MgCO_3$ if mineralized. These are quantities of material that are one or two orders of magnitude greater

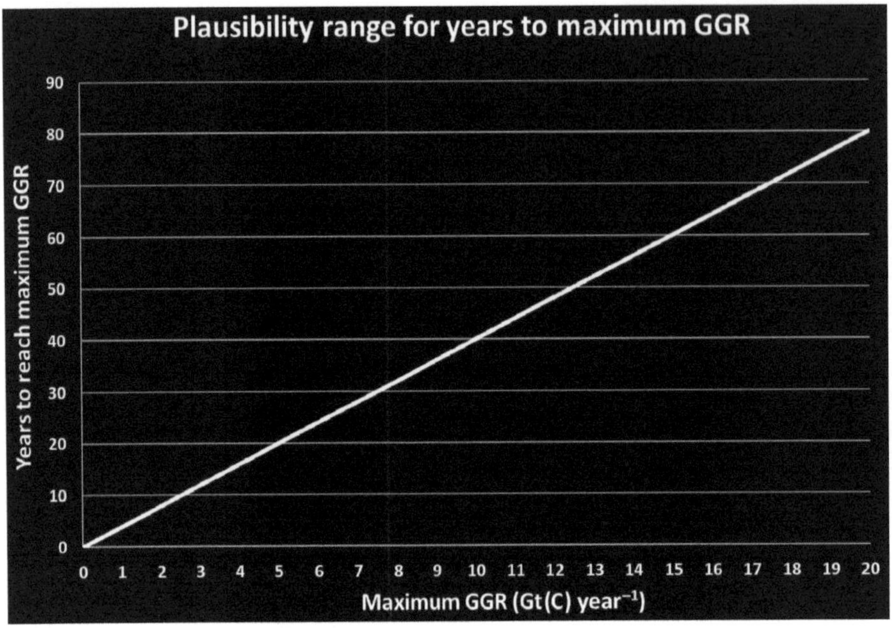

Fig. 2.6 Assumed linear relationship between eventual required annual amount of GGR (Gt(C)/yr) and the minimum time taken to reach it. (This graph determines the period over which GGR is scaled. The annual amounts of GGR within that period are dependent variables.)

than current industrial scale. The global economy would take some decades to reach these levels of activity.

NETs access large-scale mass airflow, broadly by one of two routes: the wind or powered fans. But each also requires other resources in different degrees, including land, water, minerals, fertilizer, labor, transport, environmental management, and energy. Novel technologies, such as solar towers, have the potential to industrialize the acquisition of air on a small footprint and with a minimal call on other resources but at the expense of high initial capital cost. By way of comparison, to capture 10 Gt(C) by photosynthesis would require at least 10×10^{12} m^2 (10 Mkm^2) of land, which represents ~60% of the available land (Fig. 2.7), whereas to capture the same amount using solar towers, assuming a 50% capture fraction, would require less than 0.2×10^{12} m^2 (0.2 Mkm^2).

The resource implications of the necessary scale cannot be underestimated. Biomass is typically ~50% carbon, so even assuming 100% efficiency, every gigatonne of sequestered carbon requires 2 Gt of dry biomass. If the process whereby the carbon is extracted from the biomass and sequestered is, say, 50% efficient, the amount of biomass feedstock doubles again to 4 Gt. Moreover, many processes involve multiple handling. Biochar requires biomass to be harvested, dried, chipped, transported to biochar plants, and then the biochar and the residues to be distributed

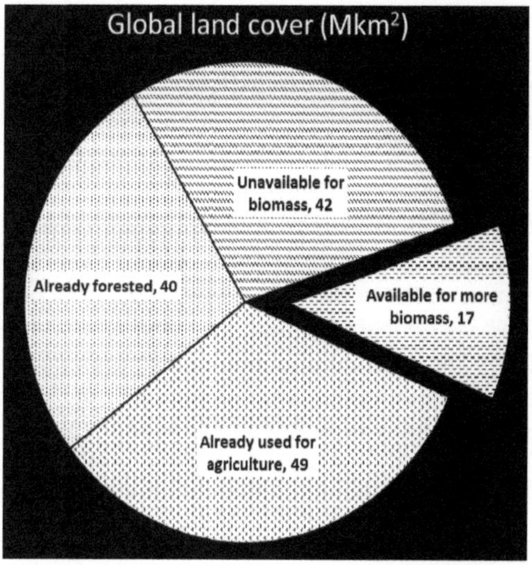

Fig. 2.7 Global land cover.
Based on data from Latham J, Cumani R, Rosati I, Bloise M. Global land cover share (GLC-SHARE) databasebeta-release version 1.0-2014. Rome: FAO; 2014.

and disposed of. Each step increases the handling requirement. For mineralization (e.g. $Mg_2SiO_4 + 4CO_2 + 4H_2O = 2MgCO_3 + SiO_2 + 2CO_2 + 4H_2O$), the multiples are even higher with each gigatonne of sequestered carbon requiring the handling of more than 15 Gt of minerals, and 6 Gt of water.

The second challenge concerns the permanent sequestration of the GHG. To give some sense of scale, 1 Gt(C) stored as liquefied CO_2 requires almost the same space as current global annual oil production ($\sim$5 km^3). If the chosen route is biochar, the same 1 Gt(C) implies about 3.5 Gt biomass, more than 4×10^{12} m^2 (4 Mkm^2) of land to grow it (ignoring the use of recycled waste), some 20 million standard container shipments, and if this level of production were to be reached in say, 50 years, the construction of six biochar plants a week until 2067 [8]. (These figures should be treated with some caution as they are dependent on a number of highly variable input parameters, including yields per hectare, pyrolysis technologies and efficiencies, and access to waste biomass. The figures presented here are median figures.)

It is premature to establish plausibility ranges for such an inchoate set of technologies. The amount of GGR in terms of captured and sequestered carbon is a dependent variable in the calculator. It is for policymakers to determine what is plausible since they control the determining factors, namely, research and investment, and the access to key resources such as land, labor, energy, and finance. The amount of GGR that is feasible is likely to increase as the science progresses. References to high values of GGR should not be taken to imply that they are, or might ever, become feasible. On the other hand, much that we do today would not have been thought feasible, or even conceivable, a century or two ago, and in many cases, even 50 years ago. If something similar to Moore's Law were to apply to this new technology, it seems plausible that in a few decades, volumes of GGR that today we might consider fanciful, could become a reality.

As a somewhat arbitrary lower limit, I have assumed that whatever the GGR target may be, it is implausible that it could be achieved in fewer years than the target in Gt(C)/yr multiplied by 4. Thus, to reach a GGR target of 10 Gt(C)/yr would take at least 40 years. The primary constraints are political, scientific, and engineering. Without the political will to provide the funding for research and development, the science will not progress fast enough. Even once the science points the way, the scale of the infrastructure required to deliver any form of GGR at climate relevant quantities will become constrained by the global capacity to build at the necessary pace.

The final parameter whose plausibility range must be considered is the requirement for ZEE. This is discussed in Section 2.7 where I conclude that rather than restricting the range of plausible future values for ZEE, the ZEE requirement can be seen as an emergent global policy objective.

The upper and lower plausibility bounds of the eleven independent variables are set out in Table 2.3.

Table 2.3 Summary of plausibility ranges for the eleven independent variables

Parameter	Best plausible value	Worst plausible value
Emissions (FF)		
Peak emissions year	2017	2040
Residual FF and cement emissions (RE) (Gt (C)/yr)	0	10
Years to reach residual emissions (PE = peak emissions in Gt(C) – dynamically determined)	40 × (PE – RE)/PE	120 × (PE – RE)/PE
Energy consumption (TFC)		
Energy consumption reductions start (ERSY)	2017	2100
Percentage reduction of energy consumption (PREC) (%)	20	0
Years to deliver incremental reduction of energy consumption	PREC × 50	2100 – ERSY
Carbon intensity (CIR)		
Carbon intensity reductions start (CRSY)	2017	2100
Percentage CIR (PCIR) (%)	25	0
Years to deliver incremental CIR	PCIR × 50	2100 – CRSY
GGR		
GGR start year (SSY)	2025	2050
Years to reach maximum GGR (MSY) (MSA = maximum GGR amount (Gt(C)/yr) – dynamically determined)	MSA/4	MSA/4

2.6 Policy implications

Using the worst plausible values for all the input parameters, there is no amount of GGR that will deliver even the 2°C target if it does not start before 2050. If the GGR starts in 2025 this target is achievable but only if GGR grows to an annual amount of 11.4 Gt(C)/yr within 46 years, or if it starts in 2030, growing to 14.7 Gt(C)/yr within 59 years (Scenario 9). These amounts of GGR might become plausible in the future if sufficient investment is made in the appropriate technologies. This corresponds to more than 50 Gt(CO_2)/yr, a mass of material about double the combined total of current global mining of all fuel and mineral resources [23]. In the absence of a credible plan to deliver GGR at this scale, for now at least, it must be considered implausible. This scenario also implies a ZEE Index of 12, illustrating the challenge of simultaneously scaling GGR and ZEE technologies.

Using the best plausible values for all the input variables, with no GGR, the cumulative emissions by 2100 fall 34 Gt(C) short of the 1.8°C target with a ZEE Index of 11 (Scenario 10). This shows that we have within our reach the possibility of keeping global warming to well below 2°C without recourse to GGR albeit with only a little headroom. However, in scenarios where the transition from FF to ZEE is rapid and deep, the contribution achievable from REC and CIR is minimal. Even excluding any benefit from REC and CIR, reducing emissions to zero by 2057 still keeps cumulative emissions within the 1.8°C carbon budget without any help from GGR, and the 1.5°C target requires less than 2 Gt(C)/yr of GGR by mid-century (Scenario 11). The ZEE Index remains at a challenging 14.

Demonstrably there exist multiple combinations of input values compatible with the Paris Agreement. In the following paragraphs, a few of these will be examined by reference to increasing levels of GGR.

2.6.1 Zero GGR

Zero GGR requires that the available carbon budget be rationed to cover the elimination of FF. For example, to meet the 1.8°C target, emissions would need to peak by 2018 and reduce to zero within 40 years (the minimum plausible period for a reduction to zero emissions) (Scenario 12). This requires that the almost flatlining of emissions for 2013–15 proves to be the peak, rather than, as recent data suggest [13], another soon-to-be corrected anomaly of which there have been many in the past 250 years. The ZEE Index is 14.

The 2°C target could be reached in the same time span but this higher temperature target would allow emissions to settle at 2 Gt(C)/yr rather than be totally eliminated. This relaxes very slightly the demand for ZEE with the Index reducing to 13 (Scenario 13).

There is no plausible scenario, even with reductions in energy demand and improvements in carbon intensity that would deliver either 1.5°C or 1.8°C with residual emissions at 2 Gt(C)/yr or more.

It is apparent that GGR can only be avoided if FF emissions have already peaked or will do so imminently, and are subject to rapid reduction to almost zero. When FF is retired so rapidly, reductions in energy consumption (REC) and transitions to low emissions FF (CIR) have little impact on cumulative emissions although they do reduce significantly the requirement for ZEE. The rapid reduction in emissions must therefore be accomplished in parallel with building ZEE resources to fill the vacated energy space. The implications for ZEE are considered in Section 2.7. This policy complex requires resolute and urgent action for it to be deliverable.

2.6.2 GGR of 1.5 Gt(C)/yr

With GGR starting in 2025 and building to 1.5 Gt(C)/yr within 17 years, the 1.8°C target can be met if FF emissions peak in 2020 and fall to 2 Gt(C)/yr by 2054 with a modest contribution of 5% from CIR by 2040. In this scenario, no assistance is needed from REC (Scenario 14).

However, even with maximum support from CIR (implying a concerted global shift to replace coal with natural gas), the 1.5°C target can only be met if FF emissions have already peaked and are reduced to zero in 40 years (Scenario 15). In this scenario, the GGR target must be reached in only 10 years, and even maximum REC makes no material difference to cumulative emissions although it does reduce the ZEE Index from 14 to 11.

The 2°C target is achieved with 1.5 Gt(C)/yr of GGR by 2042 and a more gradual decline in FF emissions to 4 Gt(C)/yr within 40 years together with modest assistance from CIR of 3% achieved over 25 years. Plausible TFC reductions (REC) are too small to have a material impact on these outcomes; however, REC of 20% achieved in 10 years from 2017 would reduce the ZEE Index from 13 to 10.

If, on the other hand, FF emissions do not peak until 2025, to deliver 2°C with no more than 1.5 Gt(C)/yr of GGR by 2035, emissions must stabilize at 2 rather than 4 Gt (C)/yr within 40 years. The ZEE Index remains at 14. If this scenario is enhanced by TFC reductions of 20% over 20 years, while the impact on emissions is negligible, the ZEE Index reduces to 11, considerably less demanding than 14 but still a challenging growth target. The requirement for GGR also reduces to 1.1 Gt(C)/yr (Scenario 16). Alternatively, the growth in GGR could be slowed to reach a peak of 1.5 Gt(C)/yr by 2065.

GGR of 1.5 Gt(C)/yr allows policymakers more flexibility in reducing FF emissions, but they must still have peaked within the next decade and reduce to close to zero within 50 years. REC has almost no impact on cumulative emissions because to keep the requirement for GGR at or below 1.5 Gt(C)/yr, the reduction in emissions must be so fast that accelerating it further produces only a modest additional advantage. While CIR allows residual emissions to settle at a higher level, this is a chimerical advantage since it merely shifts exceeding the carbon budget into the early part of the next century. The ZEE Index in all these scenarios is more than 10. However, the bulk of this ZEE is required from 2040 onwards, giving policymakers a couple of decades to lay the foundations for increased ZEE resources at the necessary scale (see Section 2.7).

2.6.3 GGR of 5 Gt(C)/yr

GGR starting in 2025 and growing to 5 Gt(C)/yr within 26 years provides even greater flexibility for policymakers. For example, the 1.8°C target could be delivered with emissions peaking in 2025 and falling to 4 Gt(C)/yr by 2065 while allowing TFC growth to follow its long-term trend (Scenario 17).

Even the 1.5°C target can be achieved although it requires emissions to stabilize at only 3 Gt(C)/yr by 2056 and requires help from REC (20% in 20 years) and CIR (15% in 25 years). However, in this scenario, the carbon budget is breached in 2030 and is gradually brought back on target through the remainder of the century (Scenario 18). There is no combination of plausible policies that would avoid breaching the 1.5°C carbon budget.

If the commencement of GGR is deferred until 2040, the 1.8°C target would require emissions to settle 5 years earlier at the lower level of 3 Gt(C)/yr. This overshoots the carbon budget in 2042. No combination of plausible policies can eliminate this overshoot. For the less demanding 2°C target, emissions could be reduced to 5 Gt(C)/yr by 2058 and if GGR grows to 5 Gt(C)/yr over 26 years starting in 2040, there would be no overshoot of the carbon budget (Scenario 19). The ZEE Index remains high at 13. Reducing TFC by 20% in 20 years from 2018 reduces the ZEE Index to 10 and reduces marginally the requirement for GGR to 4.7 Gt(C)/yr. This is a world in which GGR removes atmospheric CO_2 as fast as FF consumption puts it there.

2.6.4 GGR of 10 Gt(C)/yr

This set of scenarios describe a world in which FF energy production is much as it is today, and as fast as the GHG emissions are produced, they are captured and permanently sequestered. At this level of GGR, policymakers have great freedom to manage the transition from FF to ZEE. One scenario that delivers the 1.5°C target with 10 Gt(C)/yr of GGR reached within 41 years from 2025, is for emissions to peak by 2028 and be reduced to 7 Gt(C)/yr by 2062 together with a contribution of 15% in CIR delivered over 20 years to 2037 (Scenario 20). Without any REC, this still produces a ZEE Index of 13, although this can be reduced to 10 with the maximum plausible REC; this also reduces the requirement for GGR to 9.5 Gt(C)/yr. In this scenario, there is a significant overshoot of the carbon budget that peaks in 2050 and is gradually recovered during the remainder of the century.

For 1.8°C there is even more latitude with emissions peaking in 2018 and then returning to today's level of 10 Gt(C)/yr within 10 years and GGR commencing in 2025 and reaching 10 Gt(C)/yr by 2065; no assistance is required from CIR or REC. Alternatively, emissions could peak in 2030 and take 26 years to stabilize at 8 Gt(C)/yr, together with GGR commencing in 2025 and taking 45 years to reach 10 Gt(C)/yr (Scenario 21). Here there is a modest overshoot and the ZEE Index without REC is 12. With 20% REC, the ZEE Index reduces to 9 and the requirement for GGR reduces to 9.1 Gt(C)/yr.

Finally, with GGR of 10 Gt(C)/yr and maximum REC and CIR, emissions could peak in 2030 and reduce to 8 Gt(C)/yr by 2060 provided GGR commences in 2040 and peaks in 2090 (Scenario 22). However, even in this relatively fossil-fuel-friendly world, the consumption of ZEE will be considerable with a ZEE Index of 9, implying the need for a total of more than 38,000 EJ of ZEE during the remainder of the century.

GGR of 10 Gt(C)/yr is demonstrably capable of delivering the UNFCCC global warming targets but at this scale is testing the limits of plausibility. In all these scenarios, the GGR industry will rival global mining in size and resources and must be developed alongside a rapidly growing ZEE sector. However, even with this relatively undemanding target for the reduction of FF emissions, by 2100 the current proportions of FF and zero emissions energy will need to be reversed with FF delivering less than 20% of TFC and the annual consumption of ZEE being almost 17 times its 2014 figure.

2.7 Delivering ZEE

As already noted, never has a novel energy source resulted in less energy production from legacy sources (Fig. 2.4 *supra*). In the decade to 2014, a period during which the impetus toward green energy began to gather pace, the total amount of ZEE grew annually by about 35% of the 2004 figure, reaching 66 EJ by 2014. Fig. 2.8 shows the relatively small contribution so far made by new ZEE technologies such as solar and wind.

Fig. 2.9 shows the greatest likely annual and cumulative consumption of ZEE assuming that TFC will continue to increase at its historical rate and that the transition away from FF will be rapid. Even in this high growth scenario, it is apparent that some 80% of the ZEE necessary to meet demand during the balance of the century will not be required until after 2050. From a policymaking perspective, this growth profile would seem gradual enough that given the political will, sufficient ZEE could be developed in a timely fashion.

A ZEE Index of 14 implies an aggregate ZEE demand between 2015 and 2100 of more than 60,000 EJ. In political, finance, and engineering terms, this is challenging given that current annual ZEE isless than 70 EJ. However, this increase in ZEE will not occur proportionately across all ZEE sources. The scope for increasing biomass (64% of ZEE in 2014) is very much less than that for solar and wind. Nuclear has considerable growth potential but its political sensitivity and concerns about the disposal of radioactive waste may be severe limiting factors [24]. The ZEE Index understates the challenge in scaling ZEE if most of the growth to about 1500 EJ/yr by 2100 is to come from solar and wind, which in 2014 delivered less than 5 EJ (Fig. 2.8).

Novel technologies may have a role but the time from conception to global scale is such that they are unlikely to have a major impact until the latter part of the century, by which time those ZEE technologies currently at hand will have to have already secured the delivery of our climate change targets.

While the virtual elimination of dominant legacy energy sources may be unprecedented, these figures suggest that it is plausible for the UNFCCC target to be

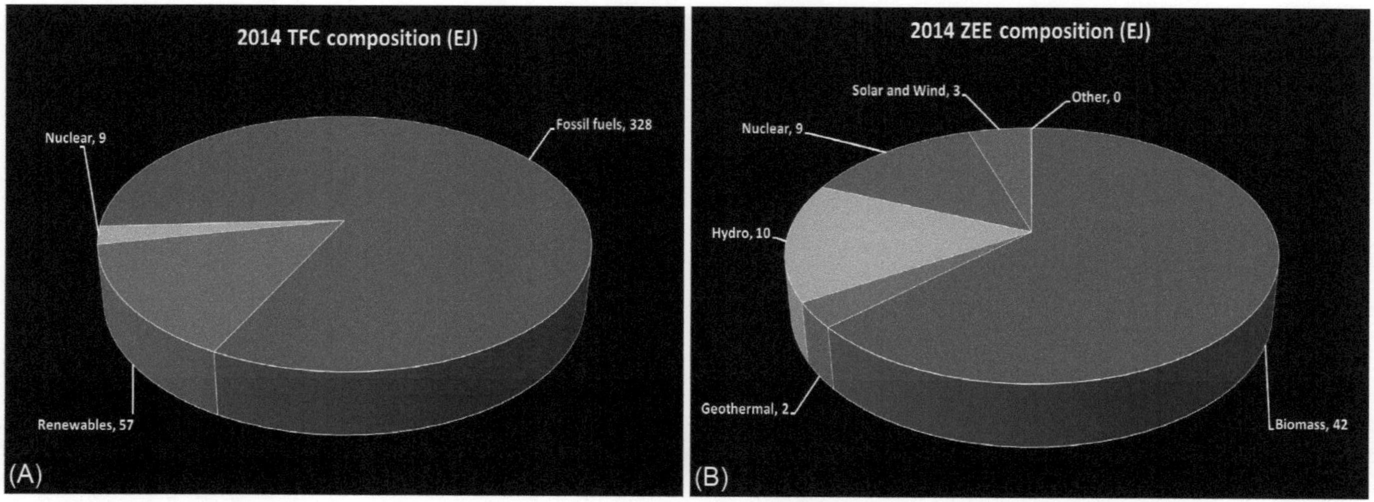

Fig. 2.8 (A) TFC composition; (B) ZEE composition.
Source: From 2014 data provided by IEA.

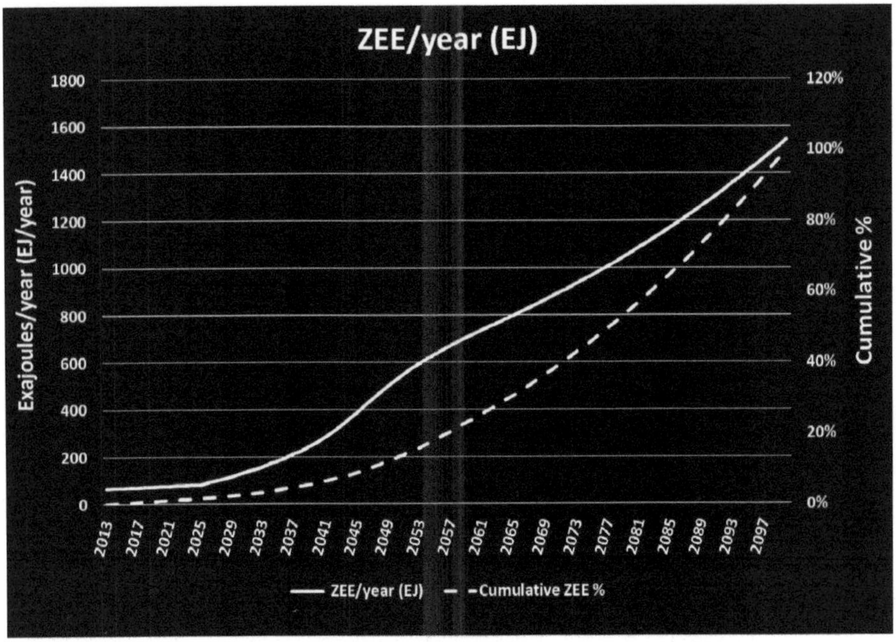

Fig. 2.9 Growth of ZEE to 2100 assuming rapid transition away from FF, and TFC growth in line with historical trend.

achieved provided the infrastructure is laid down in the next couple of decades to deliver ZEE at the necessary global scale in the second half of the century. This would be consistent with the two to three generations needed for the global penetration of major new technological infrastructure, but requires the alignment within the next decade of the political, market, and social forces necessary to make it happen.

2.8 Conclusion

Robust policies are those that deliver at least adequate outcomes in the broadest range of plausible futures. Robust policies rarely deliver optimal outcomes in what might from time to time be considered the most probable futures; this is the price to be paid for the system redundancy needed to protect against the undesirable futures currently thought less likely, or not even contemplated, yet eminently possible. Climate policy that is based on the more, rather than the less challenging plausible futures, will minimize a wider range of risks and will therefore be more robust.

Greater amounts of greenhouse gas removal allow for a more gradual transition from fossil fuel energy to zero emissions energy. Too little greenhouse gas removal forces the pace of the transition, possibly beyond the capacity of the global energy infrastructure to cope; too much greenhouse gas removal could not only have significant cost implications but might also promote complacency among policymakers, misleading them into regarding greenhouse gas removal as the solution to the problem of climate change rather than an adjunct in a complex policy regime aimed at managing it heuristically. Balancing emissions reductions and greenhouse gas removal will be a complex process involving a range of interdependent political, social, and economic factors. However, substantial urgent investment in empirical greenhouse gas removal research and development is a no-regrets policy given that even though the ultimate requirement for, and means of delivery of greenhouse gas removal are presently uncertain, the need for a significant amount of it is not. It is the responsibility of today's policymakers not to deny their counterparts in coming decades the ability to fine tune greenhouse gas removal policy by recklessly failing to undertake the initial empirical research on which all future greenhouse gas removal must depend.

The interdependence of policies highlights the link between the reduction of fossil fuel consumption and the increase of zero emissions energy consumption. In almost every plausible scenario, the ZEE Index is between 10 and 14 implying a quantum change in the dependence on zero emissions energy. In numeric terms, the 2014 zero emissions energy consumption of 66 EJ, of which less than 5 EJ came from solar and wind, must increase to something of the order of 1500 EJ/yr by the end of the century.

The resources required for both greenhouse gas removal and the zero emissions energy are each of the same order as those currently devoted to global mining of all mineral and fuel resources. To develop simultaneously both these new activities at scale will present an unprecedented challenge for proximate generations. Today's policymakers have a responsibility to lighten that burden to the extent they can. And today's consumers and taxpayers have an obligation not to frustrate the shift away from fossil fuels or the investment of public funds in zero emissions energy and greenhouse gas removal research and development, by resisting the market interventions necessary to make them happen.

References

[1] van Vuuren DP, Hof AF, van Sluisveld MA, Riahi K. Open discussion of negative emissions is urgently needed. Nat Energy 2017;2:902–4.

[2] Matthews HD, Zickfeld K. Climate response to zeroed emissions of greenhouse gases and aerosols. Nat Clim Chang 2012;2:338–41.

[3] Vichi M, Navarra A, Fogli PG. Adjustment of the natural ocean carbon cycle to negative emission rates. Clim Chang 2013;118:105–18. https://doi.org/10.1007/s10584-012-0677-0.

[4] IPCC. Climate change 2013—the physical science basis: Working Group I contribution to the fifth assessment report of the Intergovernmental Panel on Climate Change. New York: Cambridge University Press; 2014.

[5] US Department of Commerce N. NOAA/ESRL Global Monitoring Division—The NOAA Annual Greenhouse Gas Index (AGGI) n.d. https://www.esrl.noaa.gov/gmd/aggi/aggi.html [Accessed 12 December 2017].

[6] IEA. Key World Energy Statistics 2017. International Energy Agency; 2017.

[7] Saunders HD. Historical evidence for energy efficiency rebound in 30 US sectors and a toolkit for rebound analysts. Technol Forecast Soc Chang 2013;80:1317–30. https://doi.org/10.1016/j.techfore.2012.12.007.
[8] Chris R. Systems thinking for geoengineering policy: how to reduce the threat of dangerous climate change by embracing uncertainty and failure. London: Routledge; 2016.
[9] Lempert RJ, Popper SW, Bankes SC. Shaping the next one hundred years, Santa Monica, CA: RAND Corporation; 2003. http://www.rand.org/pubs/monograph_reports/MR1626.html. Accessed 25 March 2003.
[10] Houghton RA. Carbon flux to the atmosphere from land-use changes 1850–2005. Oak Ridge, TN: Carbon Dioxide Information Analysis Center, Oak Ridge National Laboratory, U.S. Department of Energy; 2008.
[11] IEA. IEA finds CO_2 emissions flat for third straight year even as global economy grew, https://www.iea.org/newsroom/news/2017/march/iea-finds-co2-emissions-flat-for-third-straight-year-even-as-global-economy-grew.html; 2017. Accessed 7 November 2017.
[12] IEA. CO_2 emissions from fuel combustion 2016, International Energy Agency; 2016. https://www.oecd-ilibrary.org/co2-emissions-from-fuel-combustion-2016_5jlwz7khrbzv.pdf?itemId=%2Fcontent%2Fpublication%2Fco2_fuel-2016-en&mimeType=pdf. Accessed 6 April 2017.
[13] Quéré CL, Andrew RM, Friedlingstein P, Sitch S, Pongratz J, Manning AC, et al. Global carbon budget 2017. Earth Syst Sci Data Discuss 2017;1–79. https://doi.org/10.5194/essd-2017-123.
[14] BP. Statistical Review of World Energy 2017, https://www.bp.com/en/global/corporate/energy-economics/statistical-review-of-world-energy.html; 2017. Accessed 7 November 2017.
[15] Lovelock J. The revenge of Gaia: why the earth is fighting back and how we can still save humanity. 1st ed. London: Allen Lane; 2006.
[16] UNDP. World population prospects 2017, https://esa.un.org/unpd/wpp/Graphs/Probabilistic/POP/TOT/. Accessed 7 November 2017.
[17] Kahneman D, Slovic P, Tversky A. Judgment under uncertainty: Heuristics and biases. Cambridge: Cambridge University Press; 1982.
[18] World population by income. Pew Research Center's Global Attitudes Project 2015. http://www.pewglobal.org/interactives/global-population-by-income/ [Accessed 7 November 2017].
[19] MacKay DJ. Sustainable energy–without the hot air. England: UIT Cambridge; 2009.
[20] MacKay DJ, Stone TJ. Potential greenhouse gas emissions associated with shale gas extraction and use. London: Department of Energy and Climate Change; 2013.
[21] Berhens M. Utilization of CO_2, http://www.fhi-berlin.mpg.de/acnew/department/pages/teaching/pages/teaching__wintersemester__2012_2013/malte_behrens__utilization_of_carbon_dioxide__130125.pdf; 2013. Accessed 10 November 2017.
[22] Mazzotti M, Baciocchi R, Desmond M, Socolow R. Direct air capture of CO_2 with chemicals: optimization of a two-loop hydroxide carbonate system using a countercurrent air-liquid contactor. Clim Chang 2013;118:119–35. https://doi.org/10.1007/s10584-012-0679-y.
[23] Reichl C, Schatz M, Zsak G. World mining data 2017, http://www.wmc.org.pl/sites/default/files/WMD2017.pdf; 2017. Accessed 10 April 2017.
[24] Blowers A. The legacy of nuclear power. 1st ed. London, New York: Routledge; 2016.

Current status of electricity generation in the world and future of nuclear power industry

3

I. Pioro[*,1], *R. Duffey*[†]
[*]Faculty of Energy Systems and Nuclear Science, University of Ontario Institute of Technology, Oshawa, ON, Canada, [†]Idaho Falls, Idaho, ID, United States
[1]Corresponding author: igor.pioro@uoit.ca

Chapter Outline

Nomenclature

P, p	pressure, Pa
Q	amount of heat (J)
T	temperature (°C)
W	work (J)

Greek letters

η	efficiency (%)

Subscripts

com	compressor
cr	critical
el	electrical
in	inlet

Managing Global Warming. https://doi.org/10.1016/B978-0-12-814104-5.00003-X

out	outlet
s, sat	saturated or saturation
turb	turbine

Abbreviations

ABWR	Advanced Boiling Water Reactor
AECL	Atomic Energy of Canada Limited
AGR	Advanced Gas-cooled Reactor
av.	average
AP	Advanced Plant
BN	Fast Sodium (reactor) (in Russian abbreviations)
BWR	Boiling Water Reactor
CA	California (state)
CANDU	CANada Deuterium Uranium (nuclear reactor)
DOE	Department of Energy (United States)
EEC	Electrical Energy Consumption
EGP	Power Heterogeneous loop Reactor (in Russian abbreviations)
eff.	efficiency
EPR	European Pressurized-water Reactor
GE	General Electric
HDI	Human Development Index
LCOE	Levelized Cost Of Energy/electricity
LGR	Light-water-cooled Graphite-moderated Reactor
LMFBR	Liquid-Metal Fast-Breeder Reactor
LNG	Liquefied Natural Gas
LWR	Light Water Reactor
MHI	Mitsubishi Heavy Industries
NASA	National Aeronautics and Space Administration
NOAA	National Oceanic and Atmospheric Administration
NPP	Nuclear Power Plant
NY	New York (state)
O&M	Operation and Maintenance
PHWR	Pressurized Heavy-Water Reactor
PV	Photo Voltaic
PWR	Pressurized Water Reactor
RBMK	Reactor of Large Capacity Channel type (in Russian abbreviations)
rpm	revolutions per minute
SFR	Sodium Fast Reactor
TX	Texas (state)
UAE	United Arab Emirates
UK	United Kingdom
USA	United States of America
USSR	Union of Soviet Socialist Republic
WA	Washington (state)

3.1 Statistics on electricity generation in the world

This chapter is a logical continuation of our previous publications on this topic [1–4]. It is well known that electricity generation and consumption is the key factor for advances in industry, agriculture, technology, and the standard of living (see Table 3.1, and Figs. 3.1 and 3.2). Also, strong power industry with diverse energy sources is very important for a country's independence. In general, electricity (see Fig. 3.3) can be generated mainly from: (1) nonrenewable energy sources such as coal (see Fig. 3.4), natural gas, oil, and nuclear (see Fig. 3.5); and (2) renewable energy sources such as hydro (see Figs. 3.6–3.9), biomass, wind (see Figs. 3.10–3.12), geothermal (see Figs. 3.13 and 3.14), solar (see Figs. 3.15–3.20), and marine power (see Figs. 3.21 and 3.22).

Figs. 3.4–3.10, 3.12, 3.13, 3.15–3.17, 3.19, 3.21, and 3.22 show selected power plants of the world. Figs. 3.11, 3.14, and 3.18 show maps of wind-speed distribution over the United States, geothermal resources of the United States, and annual average direct normal solar-resource-data distribution over the United States. Operation of various energy sources in an electrical grid is discussed in Section 3.2; and modern thermal and nuclear power plants are discussed in Sections 3.3 and 3.4, respectively.

Today, the main sources for global electrical-energy generation (see Fig. 3.3A) are:

(1) Thermal power—primarily using coal (39.9%) and secondarily using natural gas (22.6%);
(2) "Large" hydroelectric plants (17.2%); and
(3) Nuclear power from various Nuclear Power Plants (NPPs) (11.2%).

The last 9.2% of the electrical energy is generated using oil (4.2%), and the remainder (5%) generated from biomass, geothermal, and intermittent wind, and solar energy. In addition, renewable energy sources such as marine power (tidal and wave power) are also intermittent.

Main sources for electrical energy generation in selected countries are shown in Fig. 3.3. Seventeen top power plants of the world by installed capacity are listed in Table 3.2, and largest operating power plants of the world (based on installed capacity) by energy source—in Table 3.3.

A comparison of the data in Fig. 3.3 with those data (mainly, related to 2013 or even earlier) presented in our previous paper from 2015 [2] shows that world usage of coal and oil for electricity generation has dropped by ~1%, and of nuclear by 2.3%; usage of gas and hydro power has increased by about 1%, and renewables by a couple percent. However, these changes are not so significant within even a number of years.

For China trends are different (Fig. 3.3B). China has significantly decreased the usage of coal for electricity generation from 69% to 65% (actually, not long ago China used up to 80% of coal-based electricity!); usage of gas has decreased by 2%; but usage of hydro power increased from 6% to 20%, nuclear from 1% to 4%, wind from 0% to 4%, and solar from 0% to 1%, which is a very good trend, that is, decreasing usage of "dirty" coal for electricity generation.

However, on opposite, India has significantly increased usage of coal for electricity generation from 52% to 77%; and, at the same time, usage of hydro power has

Table 3.1 Population, electrical energy consumption (EEC), and human development index (HDI) in selected countries [1, 5–7]

No.	Country	Population in millions (Jan 2018)	EEC[a] TW h (2016)	EEC[a] Watts per capita	Rank (2015)	HDI[b] (2015)
1	Norway	5.35	149.5	3190	1	0.949
2	Australia	24.77	256.9	1184	2	0.939
3	Germany (see Fig. 3.3E)	82.29	648.4	899	4	0.926
4	USA (see Fig. 3.3D)	326.77	4350.8	1520	10	0.920
5	Canada (see Fig. 3.3J)	36.95	663.0	2048	10	0.920
6	UK (see Fig. 3.3F)	66.57	338.6	581	16	0.909
7	Japan	127.19	999.6	897	17	0.903
8	France (see Fig. 3.3L)	65.23	553.4	968	21	0.897
9	Italy (see Fig. 3.3H)	59.29	286.3	551	26	0.887
10	Russia (see Fig. 3.3G)	143.96	1087.1	862	49	0.804
11	Brazil (see Fig. 3.3I)	210.87	581.7	315	79	0.754
12	Ukraine (see Fig. 3.3K)	44.00	163.7	425	84	0.743
13	China (see Fig. 3.3B)	1415.05	6142.5	496	90	0.738
14	World (see Fig. 3.3A)	7606.68	24 816.4	372	102	0.717
15	South Africa	57.40	251.9	501	119	0.666
16	India (see Fig. 3.3C)	1400.05	1400.8	114	131	0.624
17	Pakistan	200.81	115.4	66	147	0.550
18	Afghanistan	36.37	0.8 (2011)	2.5	169	0.479
19	Chad	15.35	0.1 (2005)	0.7	186	0.396
20	Niger	22.31	0.2 (2005)	1.0	187	0.353
21	Central African Republic	4.74	0.1 (2005)	2.4	188	0.352

Data for all countries in the world are listed in [1,5–7].

[a] $$\mathrm{EEC}/\frac{\mathrm{W}}{\mathrm{Capita}} = \frac{\left(\mathrm{EEC},\ \frac{\mathrm{TW\,h}}{\mathrm{year}}\right) \times \frac{10^{12}}{365\ \mathrm{days} \times 24\ \mathrm{h}}}{(\mathrm{Population,\ Millions}) \times 10^{6}}.$$

[b] HDI (Human Development Index) by United Nations (UN); HDI is a comparative measure of life expectancy, literacy, education, and standards of living for countries worldwide. HDI is calculated by the following formula: $\mathrm{HDI} = \sqrt[3]{\mathrm{LEI} \times \mathrm{EI} \times \mathrm{II}}$, where LEI, Life Expectancy Index, EI, Education Index, and II, Income Index. It is used to distinguish whether the country is a developed, a developing, or an underdeveloped country, and also to measure the impact of economic policies on quality of life.

decreased from 24% to 11%; of gas—from 11% to 4%, and of wind power—from 7% to 4%. Usage of other sources for electricity generation almost has not changed.

The United States have decreased usage of coal from 50% to 30%; increased usage of gas from 19% to 34% and renewables without hydro—from 3% to 7%. However, nuclear and hydro power have been left on the same level, that is, 20% and 7%, respectively, which is also a good trend (Fig. 3.3D).

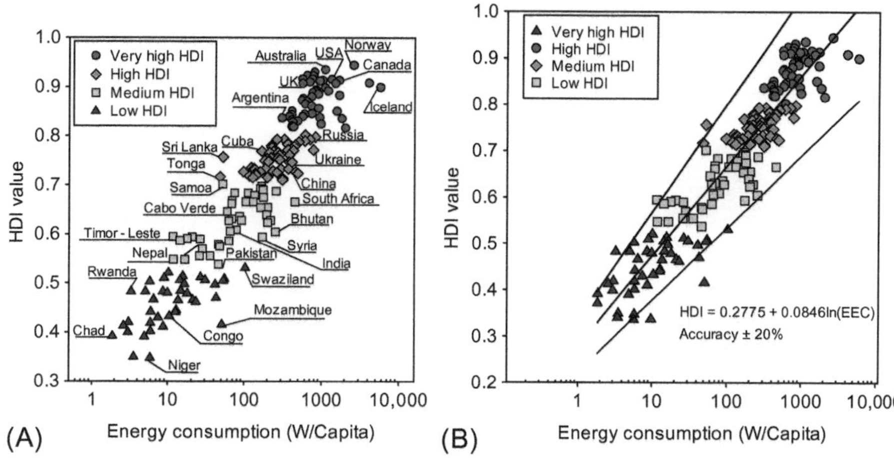

Fig. 3.1 Impact of electrical energy consumption (EEC) on human development index (HDI) for all countries of the world: (A) Graph with selected countries shown and (B) HDI correlation (in general, the HDI correlation might be an exponential rise to maximum (1), but based on the current data it is a straight line in regular—logarithmic coordinates).
Based on data from references Handbook of generation IV nuclear reactors. Duxford: Elsevier–Woodhead Publishing (WP); 2016. 940 pp. Handbook content is Available from: https://www.gen-4.org/gif/jcms/c_9373/publications; Central Intelligence Agency (CIA). The world factbook: electricity consumption, CIA.org; 2016. Available from: https://www.cia.gov/library/publications/the-world-factbook/rankorder/2233rank.html [Accessed 16 January 2016]; Central Intelligence Agency (CIA). The world factbook: population, CIA.org; 2016. Available from: https://www.cia.gov/library/publications/the-world-factbook/rankorder/2119rank.html [Accessed 16 January 2016]; United Nations (UN). Table 1: Human development index and its components, United Nations Development Programme; 2016. Available from: http://hdr.undp.org/en/composite/HDI [Accessed 16 January 2016].

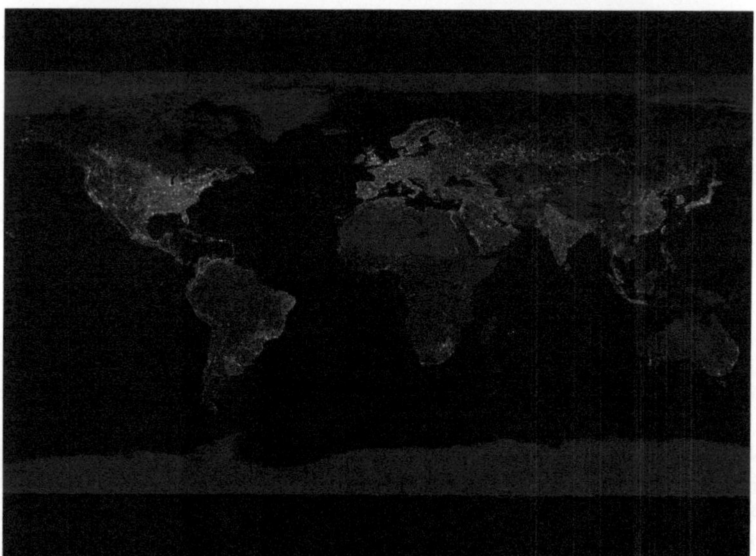

Fig. 3.2 This composite image, showing a global view of Earth at night, was compiled from over 400 satellite images.
Image credit: NASA/NOAA. https://www.nasa.gov/topics/earth/earthday/gall_earth_night.html. Last updated: August 4, 2017.

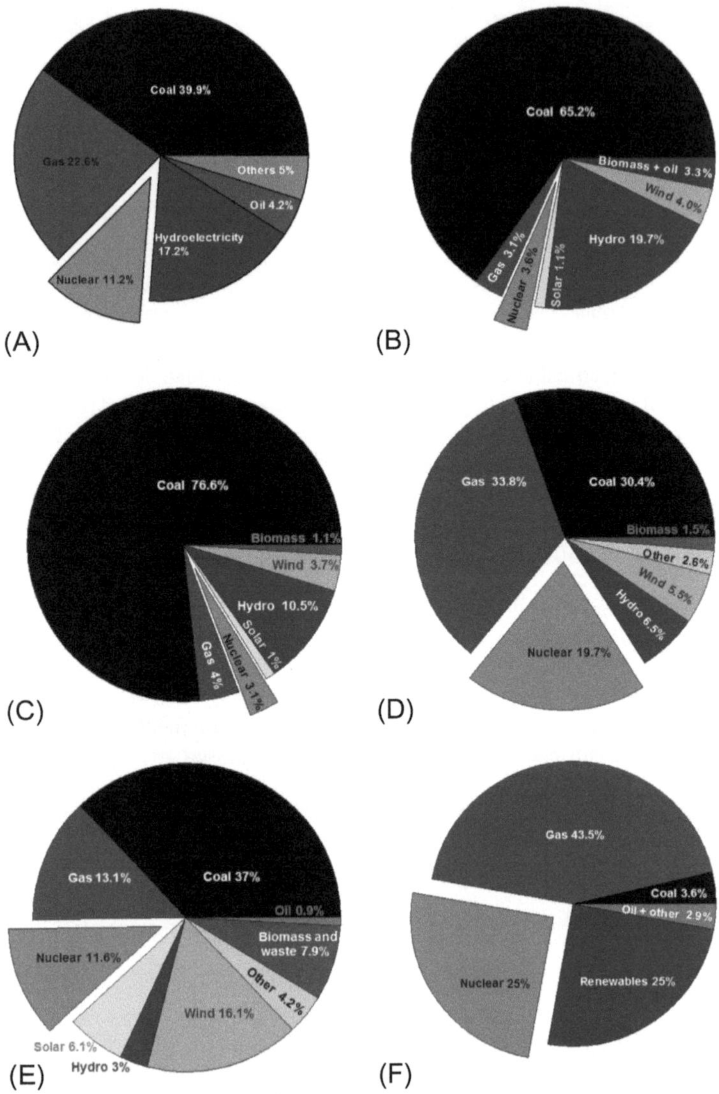

Fig. 3.3 Electricity generation in the world and selected countries by source: Data for sector diagrams and EEC—for 2016 (if not noted otherwise); population—Jan. 2018; and HDI—for 2015. (A) World: Population 7607 million; EEC 24,816 TW h per year or 372 W per Capita; HDI 0.717 or HDI Rank 102 (diagram based on [1]). (B) China: Population 1415 million; EEC 6143 TW h per year or 496 W per Capita; HDI 0.738 or HDI Rank 90. (C) India: Population 1400 million; EEC 1400 TW h per year or 114 W per Capita; HDI 0.624 or HDI Rank 131. (D) USA: Population 327 million; EEC 4351 TW h per year or 15120 W per Capita; HDI 0.920 or HDI Rank 10. (E) Germany: Population 82 million; EEC 648 TW h per year or 899 W per Capita; HDI 0.926 or HDI Rank 4. (F) UK: Population 67 million; EEC 339 TW h per year or 581 W per Capita; HDI 0.909 or HDI Rank 16.

Continued

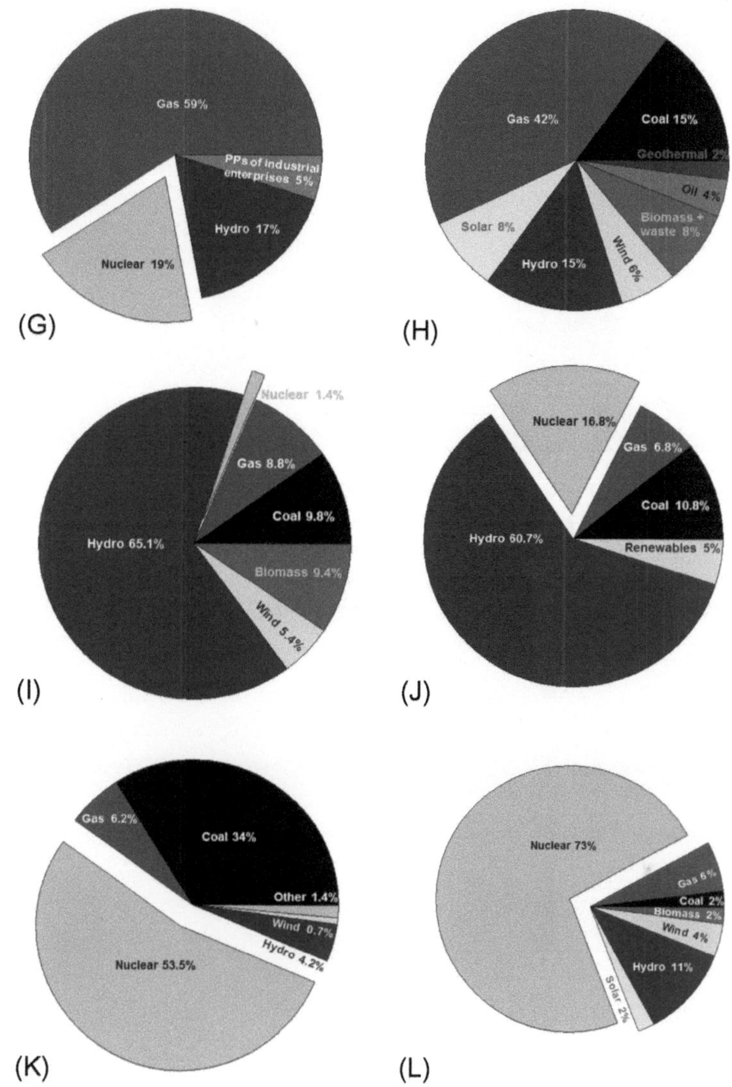

Fig. 3.3—cont'd (G) Russia: Population 144 million; EEC 1087 TW h per year or 862 W per Capita; HDI 0.804 or HDI Rank 49. (H) Italy: Population 59 million; EEC 286 TW h year or 551 W per Capita; HDI 0.887 or HDI Rank 26. (I) Brazil: Population 211 million; EEC 582 TW h per year or 315 W per Capita; HDI 0.754 or HDI Rank 79. (J) Canada (2015): Population 37 million; EEC 663 TW h per year or 2048 W per Capita; HDI 0.920 or HDI Rank 10. (K) Ukraine (2015): Population 44 million; EEC 164 TW h per year or 425 W per Capita; HDI 0.743 or HDI Rank 84. (L) France: Population 65 million; EEC 553 TW h per year or 968 W per Capita; HDI 0.897 or HDI Rank 21.

Fig. 3.4 Largest thermal power plant in the world—5780-MW_{el} Taichung coal-fired power plant (Taiwan). The plant equipped with 550 $MW_{el} \times 10$ coal-fired steam generators-turbines +70 $MW_{el} \times 4$ gas-fired steam generators-turbines, and is the world's largest emitter of carbon dioxide with over 40 million tonnes per year [1,8].
Photo taken from Wikimedia Commons, author/username 照片 阿爾特 斯.

Fig. 3.5 Aerial view of the largest NPP in the world—6384-MW_{el} Bruce NPP. (The Douglas Point NPP was Canada's first full-scale NPP and the second CANDU reactor. Its success was a major milestone for Canada to enter into the global nuclear power scene. Construction began on February 1, 1960 and the decommission date: May 4, 1984.)
Courtesy of Bruce NPP: www.brucepower.com.

(A)

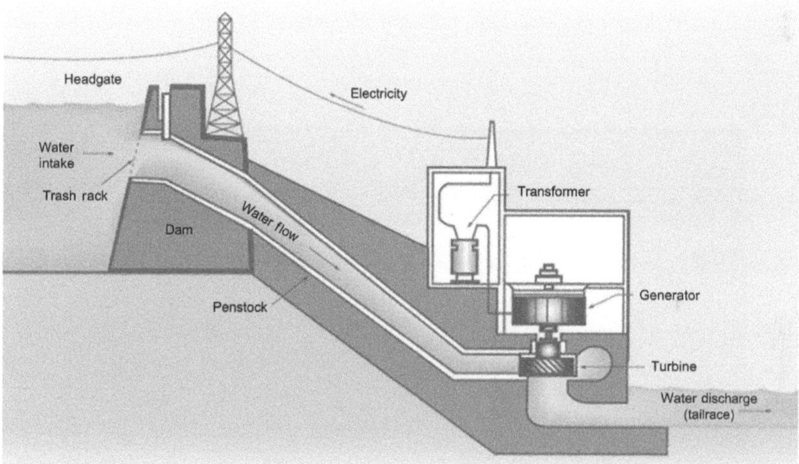

(B)

Fig. 3.6 (A) Largest hydro plant in the world by installed capacity—Three Gorges Dam (21,100 MW_{el} (planned power 22,500 MW_{el})—700 $MW_{el} \times 30 + 2 \times 50$ MW_{el} Francis turbines) (Yangtze River, China) [1]. The project costs \$26 billion. Height of gravity dam is 181 m, length: 2.335 km, top width: 40 m, base width: 115 m, flowrate: 116,000 m^3per s, artificial lake capacity: 39.3 km^3, surface area: 1045 km^2, length: 600 km, maximum width: 1.1 km, normal elevation: 175 m, and hydraulic head: 80.6–113 m. (B) Schematic of conventional hydroelectric power plant.
(A) Courtesy of Chinese National Committee on Large Dams; (B) Courtesy of Ontario Power Generation (OPG), Canada (www.opg.com).

Fig. 3.7 Photo of the tallest dam in the world—3015-MW_{el} Nurek hydroelectric power plant with 300-m dam (Tajikistan). The plant equipped with $335\,MW_{el} \times 9 = 3015\,MW_{el}$ Francis turbines [9].
Photo taken from Wikimedia Commons, author I. Rustamov.

Fig. 3.8 Photo of the largest run-of-the-river hydro plant in the world—2620-MW_{el} Chief Joseph hydroelectric power plant (i.e., an artificial lake behind the dam is not able to store large amounts of water) (Bridgeport, WA, United States). The plant is equipped with 27 Francis turbines ~100 MW_{el} each. Height of gravity dam is 72 m, length: 1.817 km, top width: 7 m, base width: 50 m, flowrate: 6030 m^3 per s, artificial lake capacity: 0.636 km^3, surface area: 34 km^2, and length: 82 km [10].
Photo taken from Wikimedia Commons, original photo by US Army Corps of Engineers.

(A) (B)

Fig. 3.9 Photos of 1932-MW_{el} pumped-storage Okutataragi hydroelectric power plant (Japan): (A) dam of Kurokawa reservoir (the upper reservoir) and (B) dam of Tataragi reservoir (the lower reservoir). The Kurokawa reservoir dam height is 98 m, length: 325 m, artificial lake capacity: 0.033 km^3, and surface area: 5.2 km^2. The Tattaragi reservoir dam height is 64.5 m, length: 278 m, artificial lake capacity: 0.019 km^3, and surface area: 13.4 km^2. Quite often, pumped-storage hydroelectric power plants work together with NPPs to keep them at full capacity during night hours.
Photos taken from Wikimedia Commons, authors/usernames Qurren and 663 highland; Source: Wikipedia, October 10, 2017.

Fig. 3.10 One of the largest wind farms in the world—781-MW_{el} onshore Roscoe wind turbine power plant (TX, United States). Plant equipped with 627 turbines: 406 MHI 1 MW_{el}; 55 Siemens 2.3 MW_{el}, and 166 GE 1.5 MW_{el}. The project cost more than 1 billion dollars, provides enough power for more than 250,000 average Texan homes, and covers area of nearly 400 km^2, which is several times the size of Manhattan, New York, NY, United States. In general, wind power is suitable for harvesting when average wind velocity is at least 6 m per s (21.6 km per h) (see wind-speed distribution over the United States in Fig. 3.11) [1].
Wikimedia Commons, photo by author/username: Fredlyfish4.

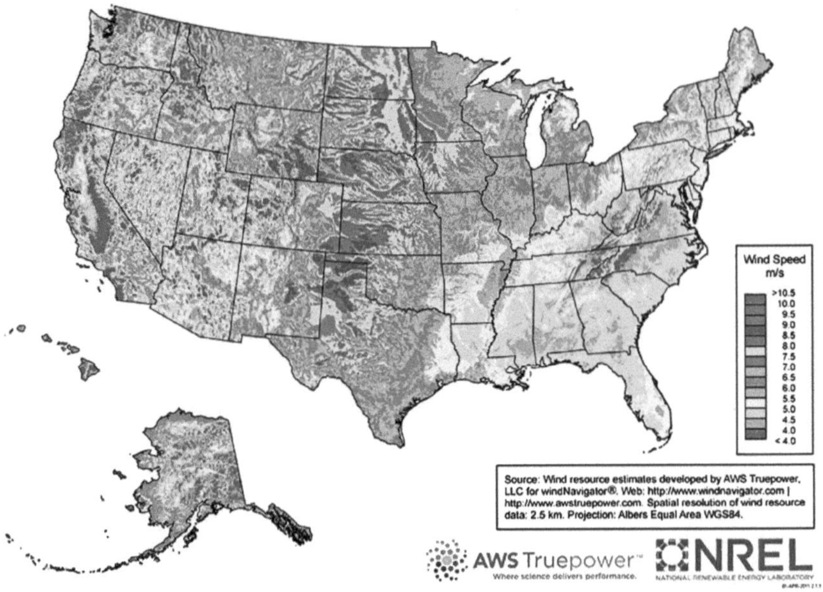

Fig. 3.11 Map of wind-speed distribution over the United States (shown here just for reference purposes and as an example). Figure shows that winds with average speed of 6 m per s and above (brown, red, and purple colors) have been uncovered only over the central part of the United States from North to South. However, average wind speed along the United States shores of Great Lakes, Atlantic and Pacific oceans, and Gulf of Mexico is usually higher than 6 m per s at the height of 90 m from the see level [1].
Courtesy of DOE, United States.

Fig. 3.12 Photo of offshore wind park, Horns Rev. II (Denmark). The facility, which is located in the North Sea, has the maximum output of about 210 MW_{el} (91 × 2.3 MW_{el} Siemens wind turbines), enough to meet the electricity needs of ~200,000 households [11]. In general, as of today offshore wind power plants are at least 2.5–3 times smaller by installed capacity than onshore wind power plants. However, offshore wind turbines can be significantly larger by power and by a rotor diameter compared to that of onshore turbines. For example, Alstom 6-$MW_{el\ net}$ wind turbine for the Haliade Offshore Platform has a rotor with the diameter of 150 m and tower 100-m high [12]. This Alstom turbine can operate within 3 m per s (11 km per h)—25 m per s (90 km per h). One of negative impacts from wind farms on environment is killing birds and bats with rotating blades and even with turbulence created by them.
Siemens press photo; copyright Siemens AG, Munich/Berlin, Germany.

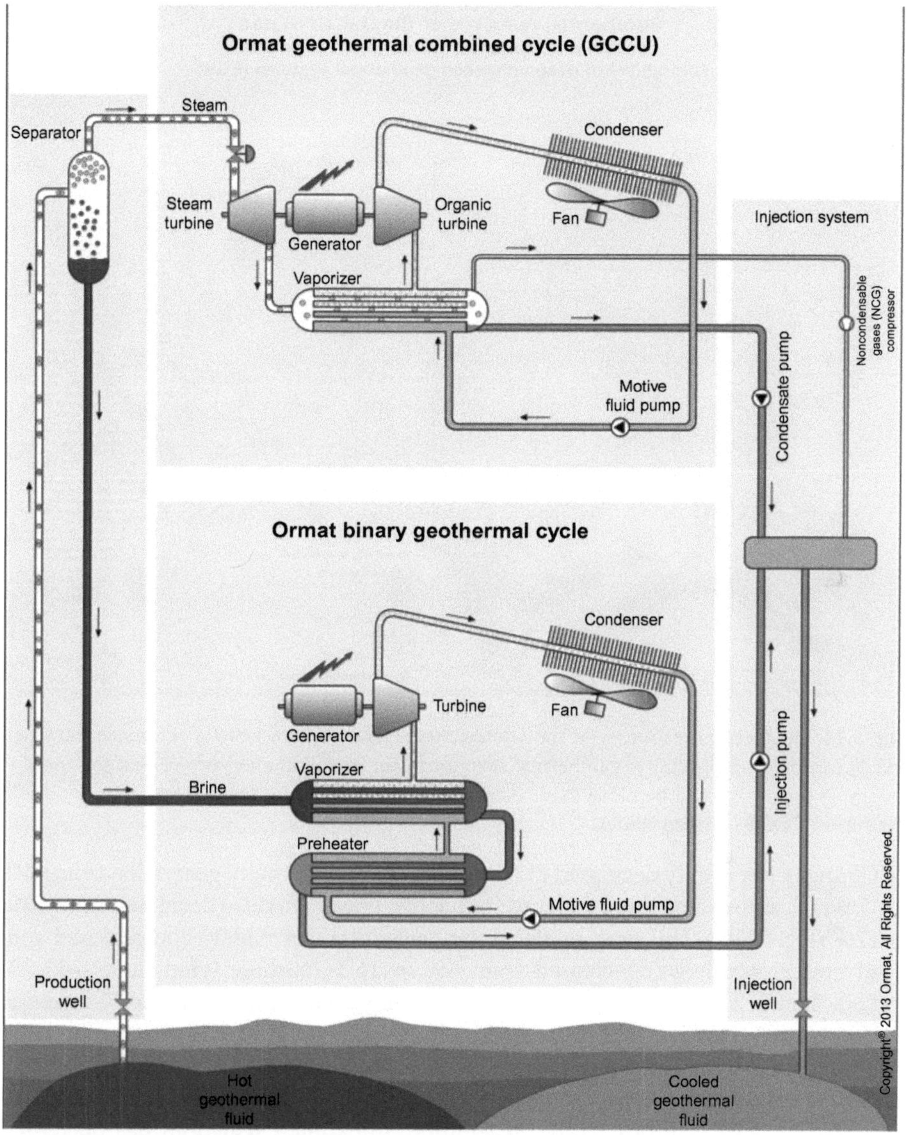

Fig. 3.13 One of possible schematics of integrated geothermal combined cycle (IGCC) power plant.
Courtesy of ORMAT, United States: http://www.ormat.com/geothermal-power.

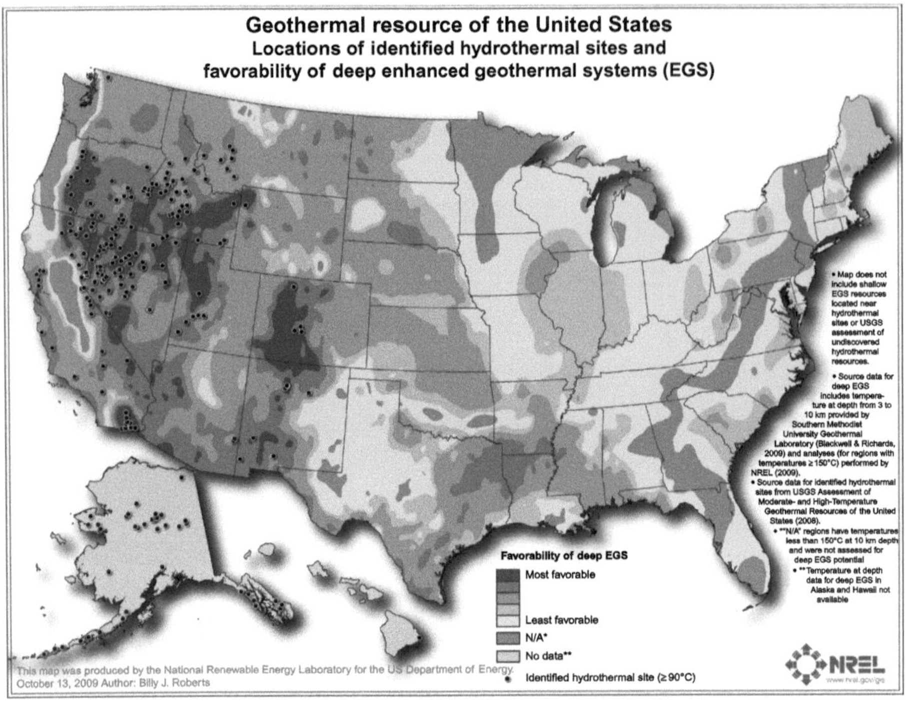

Fig. 3.14 Geothermal resources of the United States (shown here just for reference purposes and as example). Eventually, geothermal resources can be found everywhere, but the most favorable for use should be close to surface and to have higher temperatures.
Courtesy of DOE, United States.

Germany has visibly decreased the usage of coal for electricity generation from 44% to 37% and, at the same time, the usage of nuclear power was also decreased from 18% to 12% (Fig. 3.3E). This drop in electricity generation was mainly compensated with wind power, which was increased from 8% to 16% (onshore wind farms—13.3% and off-shore—2.8%), with biomass with 2% increase, and with solar up to 4% increase.

The United Kingdom has decreased their usage of coal for electricity generation even more significantly than Germany, that is, from 25% to 3% within 2013–2017 (Fig. 3.3F). At the same time, usage of oil for electricity generation was also decreased from 7% to ~3%. Unfortunately, in addition, the United Kingdom has decreased usage of nuclear—from 29% to 25%. This drop in the electricity generation from coal, oil, and nuclear was substituted mainly with gas (usage increased from 30% to 44%) and renewables (increase from 9% to 25%).

Russia has increased the usage of gas for electricity generation from 51% to 59%, nuclear from 16% to 19%, and hydro power from 16% to 17% (Fig. 3.3G). Due to these increases the usage of coal has substantially decreased from 16% to <5%. However, Russia is behind other developed countries in use of renewables such as wind and solar.

Italy, as a nonnuclear-power country, has decreased the usage of gas, coal, and oil for electricity generation by substituting these sources with renewables such as solar, wind, biomass, and geothermal (Fig. 3.3H).

Fig. 3.15 Ariel view of the largest in the world concentrated solar thermal power plant—the Ivanpah Solar Electric Generating System, Mojave Desert, CA, United States [1]. Installed capacities: Gross—392 MW_{el} and net—377 MW_{el}; capacity factor 31%; planned annual generation ~1040 GW h; site area 16 km^2 (4000 acres); deploys 173,500 sun-tracking heliostats, each has 2 mirrors (reflecting surface area is $7.02 \times 2 = 14.04\,m^2$; total reflecting area is 2.4 km^2). The intercepted average solar heat flux is about 310 W per m^2. However, after taking into consideration reflection, transmission, radiation and absorption losses, it is about 170 W per m^2 (efficiency is ~55%). The heliostat mirrors focusing sunlight on receivers located on solar power towers (~140-m height). The receivers generate steam to drive single-casing reheat turbines (~130 MW (174,000 hp)). Gross thermal efficiency of the plant is ~29%. The plant is equipped with air-cooled condensers. The project cost is 2.2 billion USD. The electricity generated by the complex is enough to serve more than 140,000 homes in California during peak hours of a day. The plant will reduce carbon dioxide (CO_2) emissions by more than 400,000 metric tonnes per year. Negative impact: (a) birds killed by burning and due to crashing into mirrors (~6000 birds were killed per year); and (b) cannot operate at night (no thermal storage system). Also, a couple of years ago, the plant has torched itself due to unexpected lowering high-intensity solar beams by computer at one tower.
Wikimedia Commons: photo by Craig Butz; Wikipedia, October 10, 2017.

Brazil is a very good example of using hydro power in a large scale (Fig. 3.3I). However, within last years, the usage of hydro power for electricity generation has decreased from 84% to 65%, but, at the same time, other renewables such as wind and biomass have started to be more visible sources for electricity generation. In spite of the latter, Brazil has also increased the usage of coal and gas.

Canada is another example after Brazil in terms of using hydro power on a large scale for electricity generation (about 61–63% for a number of years) (Fig. 3.3J). In addition, more electricity is coming from other renewables. Due to that Canada has managed to decrease usage of coal from 15% to 11%.

(A)

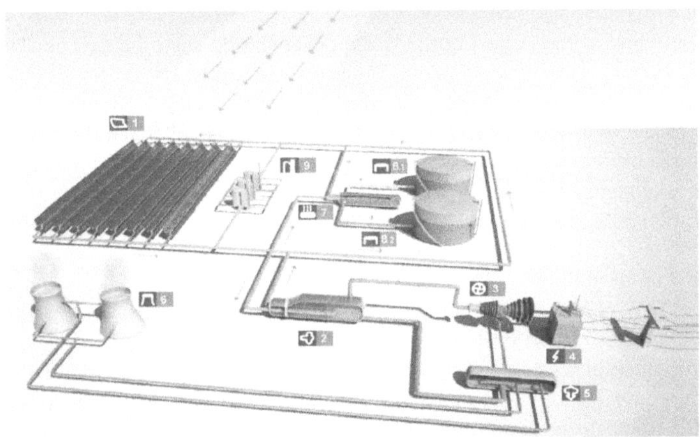

(B)

Fig. 3.16 (A) Aerial view showing portions of Solar Energy Generating Systems (SEGS) (CA, United States). SEGS is one of the largest solar energy power plants in the world. SEGS consist of 9 concentrated-solar-thermal plants with 354-MW_{el} installed capacity. The average gross solar output of SEGS is about 75 MW_{el} (capacity factor is ~21%). At night, turbines can be powered by combustion of natural gas. NextEra claims that the SEGS power 232,500 homes and decrease pollution by 3800 t per year (if the electricity had been provided by combustion of oil). The SEGS have 936,384 mirrors, which cover more than 6.5 km^2. If the parabolic mirrors would be lined up, they will extend over 370 km. In 2002, one of the 30-MW_{el} Kramer Junction sites required 90 million dollars to construct, and its O&M costs are about 3 million dollars per year, which are 4.6¢ per kWh. However, with considered lifetime of 20 years, the O&M and investments interest and depreciation triples the price to ~14¢ per kWh [13]. (B) General schematic of a typical concentrated solar power plant (CSPP): (1) SENERtrough collectors; (2) steam-generator system: (3) steam turbine with electrical generator; (4) electrical transformer; (5) condenser; (6) cooling towers; (7) heat exchanger; (8) thermal storage system; and (9) heat transfer fluid (HTF) boiler.
(A) Wikimedia Commons: photo by A. Radecki; (B) Based on Valle 1 and 2 CSPP built in the province of Cadiz, Spain by SENER/TORRESOL ENERGY (courtesy of SENER/TORRESOL ENERGY).

Fig. 3.17 Ariel view of the first of such kind Gemasolar—19.9-MW_{el} concentrated solar power plant with 140-m high tower and molten-salt heat storage system (Seville, Spain). The plant consists of 2650 heliostats (each 120 m^2 and total reflective area 304,750 m^2), covers 1.95 km^2 (195 ha), and produces 110 GWh annually, which equals to 30,000 t per year of carbon dioxide emission savings [14]. This energy is enough to supply 25,000 average Spanish houses. The storage system allows the power plant to produce electricity for 15 h without sunlight (at night or on cloudy days). Capacity factor is 75%. Solar receiver thermal power is 120 MW_{th}, and plant thermal efficiency is about 19%. Molten salt is heated in the solar receiver from 260°C to 565°C by concentrated sun light reflected from all heliostats, which follow the sun, and transfers heat in steam generator to water as working fluid in subcritical pressure Rankine steam power cycle [1].
Courtesy of SENER/TORRESOL ENERGY.

Ukraine is one of a few countries, which substantially relies on nuclear power (Fig. 3.3K). For many years NPPs contribute about 50–53% to the total electricity generation. However, due to lack of cheap gas, Ukraine has increased usage of coal up to 34%.

Over the same period, France has not significantly changed their usage of various sources for electricity generation (Fig. 3.3I). However, small decrease in usage of nuclear power from 78% to 73% was compensated with more reliance on renewables such as wind, biomass, and solar, and on gas power.

The world's largest coal-fired power plant is shown in Fig. 3.4, and the largest fully operating NPP—in Fig. 3.5. Some of the world's largest renewable power plants are shown in Figs. 3.6–3.9 (hydroenergy); Figs. 3.10 and 3.12 (wind energy); Figs. 3.13 and 3.14 (geothermal energy); Figs. 3.15–3.20 (solar energy); and Figs. 3.21 and 3.22

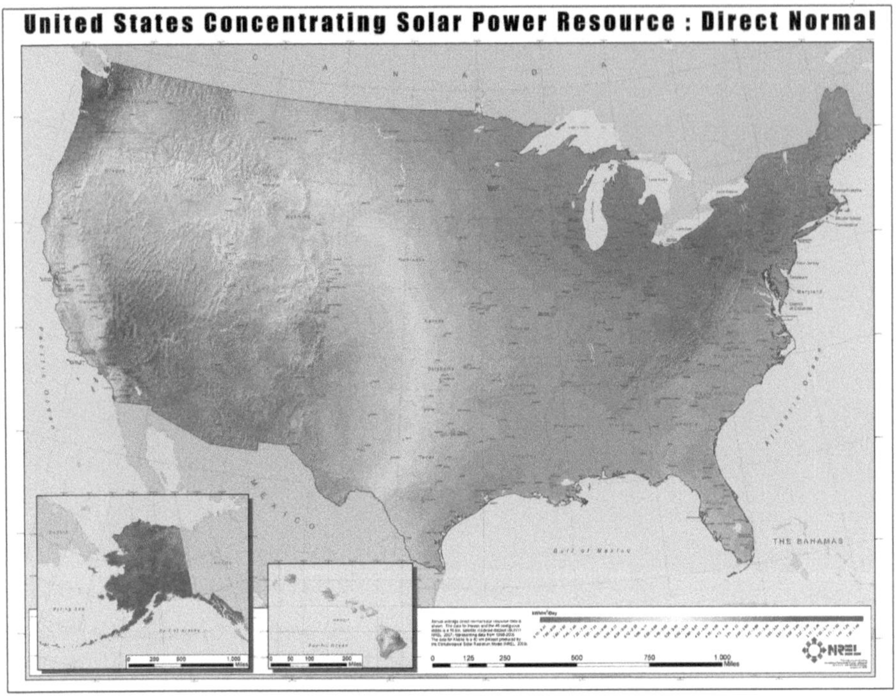

Fig. 3.18 Map of annual average direct normal solar-resource-data distribution over the United States (shown here just for reference purposes and as example). In general, the amount of solar radiation that reaches any one spot on the Earth's surface varies according to: (1) geographic location; (2) time of day; (3) season; (4) local landscape; and 5) local weather [1]. Courtesy of DOE, United States.

Fig. 3.19 Photo of typical flat-panel photovoltaic power plant (19 MW_{el}) located near Thüngen, Bavaria, Germany. Agricultural or recreational land is used for low-efficiency PV power plant. Wikimedia Commons: photo by OhWeh.

Fig. 3.20 "Proper" installation of PV panels: (A) Photovoltaic panels installed on roof of house. (B) Photovoltaic panels installed on roof of parking lot [1].
Photos by I. Pioro, Bavaria, Germany.

Fig. 3.21 Photo of 240-MW_{el} Rance tidal power plant (10 MW_{el} × 24 units) (France). The plant supplies on average 96 MW_{el} with capacity factor of 40%, providing annual output of ~600 GWh [15]. The barrage is 750 m long. The plant portion of the dam is 332.5 m long. The tidal basin measures 22.5 km^2.
Photo taken from Wikimedia Commons, author/username Dani 7C3.

(tidal and wave energy). Fig. 3.11 shows the average wind speeds across the United States, and Fig. 3.18—the average solar radiation across the United States.

Two very important parameters [1,2] of any power plant are:

(1) Overall (gross) or net efficiency: Gross efficiency of a unit during a given period of time is the ratio of the gross electrical energy generated by a unit to the energy consumed during the same period by the same unit. The difference between gross and net efficiencies is internal needs for electrical energy of a power plant, which might be not so small (5% or even more) and

Fig. 3.22 Photo of Pelamis Wave-Energy Converter on site at the European Marine Energy Test Centre (EMEC), Orkney, Scotland, 2007. This converter was rated at 750 kW_{el} and was the world's first offshore wave power machine to generate electricity into grid system [16]. The prototype is 120 m long and 3.5 m in diameter. It consisted of four tube sections linked by three, shorter, power-conversion modules.
Photo taken from Wikimedia Commons, author/username P123.

(2) Capacity factor of a plant: Net capacity factor of a power plant is the ratio of the actual output of a power plant over a period of time (usually, during a year) and its potential output if it had operated at a full nameplate capacity the entire time. To calculate the capacity factor, the total amount of energy a plant produced during a period of time should be divided by the amount of energy the plant would have produced at the full capacity. Capacity factors vary significantly depending on the type of a plant. Average capacity factors of various power plants are listed in Tables 3.2–3.4 [1,2].

3.2 Share and operation of various energy sources in an electrical grid

The fraction of total global energy used for electrical power generation is presently about 42% [21]. This fraction is expected to increase with increasing known gas reserves (especially, in United States, China, and Australia) and falling prices and resulting enhanced energy exports. Of that fraction some 85% is from gas, coal, and oil. Carbon resources are expected to continue to dominate despite the fuel switching from coal to natural gas and subsidies for noncarbon sources like wind and solar. The mix of sources deployed in any particular electrical generation system varies widely [22]. Electrical generation sources and the energy "mix" are dependent on a number of key factors:

(a) The availability locally or nationally of specific fuel resources, particularly, coal, natural gas, uranium, and/or oil. Therefore, net energy-importing countries and regions (e.g., Japan, United Kingdom, and mainland Europe) will likely see more diversification of supply alternatives and government policies for subsidies and net metering pricing.

Table 3.2 Seventeen top power plants of the world by installed capacity [17]

No	Plant	Country	Capacity (MW_{el})	Av. annual generation (TWh)	Capacity factor (%)	Plant type
1	Three Gorges Dam (Fig. 3.6A)	China	22,500	93.5	47	Hydro
2	Itaipu Dam	Brazil/ Paraguay	14,000	103.1	84	Hydro
3	Xiluodu	China	13,860	55.2	46	Hydro
4	Guri Dam	Venezuela	10,200	47	52	Hydro
5	Tucurui Dam	Brazil	8370	21.4	30	Hydro
6	Kashiwazaki-Kariwa (not in service)	Japan	7965	(60.3)	(86)	Nuclear
7	Grand Coulee Dam	USA	6809	20.2	34	Hydro
8	Xiangjiaba	China	6448	30.7	54	Hydro
9	Longtan Dam	China	6426	17.3	31	Hydro
10	Sayano-Shushenskaya	Russia	6400	27	48	Hydro
11	Bruce (Fig. 3.5)	Canada	6384	47.6	85	Nuclear
12	Kori	South Korea	6040	39.3	74	Hydro
13	Krasnoyarsk Dam	Russia	6000	23	44	Hydro
14	Hanul	South Korea	5881	48.2	94	Nuclear
15	Hanbit	South Korea	5875	47.6	92	Nuclear
16	Nuozhadu Dam	China	5850	23.9	47	Hydro
17	Zaporizhia	Ukraine	5700	48.2	97	Nuclear

(b) The generating price advantage for any given source determines the market share only in truly competitive markets. Therefore, for electrical generators in power markets controlled by central decisions or government policies (e.g., India, Korea, and Russia), other considerations such as national energy security and national investment priorities affect generating market share.

(c) The political environment may favor selecting noncarbon sources over concerns on potential Climate Change. Power-pricing tools may lead to preferential grid "feed in" or generating tariffs, allocating a "portfolio" or market share, or offering selective investment incentives, and altering the operating "merit order" so that low-cost generators may even be paid not to produce.

(d) The presence of regional or national Trade Treaties (e.g., the NAFTA in North America, or EU-Russia import policies) may allow energy preferential terms in or between certain partners, in addition to power grid interlinkages or gas pipelines and their routing.

(e) The advances in generating technology efficiency, safety improvements, and emissions reduction may be subsidized by governments. Indirect methods and tools include national

Table 3.3 Largest operating power plants of the world (based on installed capacity) by energy source [17]

Rank	Plant	Country	Capacity (MW_{el})	Plant type
1	Three Gorges Dam (Fig. 3.6A)	China	22,500	Hydro (dam)
2	Bruce NPP (Fig. 3.5)	Canada	6384	Nuclear
3	Taichung (Fig. 3.4)	Taiwan	5780	Coal
4	Shoaiba	Saudi Arabia	5600	Fuel oil
5	Surgut-2[a]	Russia	5597	Natural gas
6	Gansu	China	5160	Wind (onshore)
7	Bath County[b]	USA	3003	Hydro (pumped storage)
8	Chief Josef (Fig. 3.8)	USA	2620	Hydro (run-of-the-river)
9	Eesti	Estonia	1615	Oil shale
10	The Geysers	USA	1517	Geothermal
11	Shatura[a]	Russia	1500	Peat[a]
12	Kurnool Ultra Mega Solar	India	900	Solar (Photo-Voltaic (PV))
13	Ironbridge	UK	740	Biofuel[a]
14	London Array	UK	630	Wind (offshore)
15	IPP3[a]	Jordan	573	Internal combustion engines
16	Ivanpah (Fig. 3.15)	USA	392	Solar (concentrated thermal)
17	Sihwa Lake	South Korea	254	Tidal
18	Vasavi Basin Bridge	India	200	Diesel
19	Sotenäs	Sweden	3	Marine (wave)

[a] It should be noted that actually, some thermal power plants use multifuel options, for example, Surgut-2 (15% natural gas); Shatura (peat: 11.5%, natural gas: 78%, fuel oil: 6.8%, and coal: 3.7%) power plants.
[b] Pumped storage hydroelectricity (PSH), or pumped hydroelectric energy storage (PHES), is a type of hydroelectric power plant used by electric grids for load balancing. During off-peak hours (or during periods of lower electricity prices), usually at night, water is pumped from a lower elevation reservoir to a higher elevation one. During peak hours (or periods of high electricity prices), the plant is used as a regular hydroelectricity plant. It should be noted that such plants usually consume energy overall, but the plant increases revenue by selling more electricity during periods of peak demand, when electricity prices are highest (based on [18]).

energy R&D programs, nationalized or majority government-owned manufacturers and constructors (particularly in the nuclear arena), centralized control of labor market and engineering costs, or funding prototype or "demonstration" development and deployment. Hence, there may be "hidden' subsidies" that do not allow direct international comparisons.

(f) The use of regulated limits constraints on CO_2 emissions on the use of carbon fuels in energy systems, like coal and gas, manufacturing and transportation affect certain power generation markets. Such schemes (as in the EU or California, United States) include "cap and trade" markets in which permits or emissions rights are issued within some

Table 3.4 Average (typical) capacity factors of various power plants [19] (for US data, see [20])

No.	Power plant type	Location	Year	Capacity factor (%)
1	Nuclear (Fig. 3.5)	United States	2016	93
		Russia	2014	81
		World	2015	82
2	Geothermal (Fig. 3.13)	United States	2016	74
3	Bioenergy	United States	2016	47–70
		United Kingdom	2015	68
4	Combined cycle	United States	2016	56
		United Kingdom	2015	32
5	Coal fired (Fig. 3.4)	United States	2016	53
		United Kingdom	2015	39
6	Hydroelectric (Figs. 3.6–3.9)	United States	2016	38
		United Kingdom	2015	41
		World	2011–13	30–55
7	Wind (Figs. 3.10 and 3.12)	United States	2016	35
		United Kingdom	2015	34
		World	2011–13	20–40
8	Concentrated solar thermal (Figs. 3.15–3.17)	United States	2016	22
		Spain (molten salt with storage)	2014	63
9	PV solar (Fig. 3.20)	United States	2016	27
		United Kingdom	2015	12
10	Wave (Fig. 3.22)	United Kingdom	2015	3

quasipolitically determined emissions total or target, and currently are bought and sold at a price, say typically, \$10–\$20 per ton of carbon. Because these schemes are not globally uniformly adopted, the consequent embedded or hidden added cost of emissions causes power market distortions that may affect investment, generating mix, and industrial competitiveness.

(g) The perceived or actual return on investment and investor/customer risk exposure is sensitive to adverse future regulatory changes, the potential for project cost overruns, adverse power market conditions, and ensuring return on investment (ROI) via long-term contracts for difference or power purchase agreements. The presence or extent of explicit or implicit government price guarantees, construction subsidies or loans, or long-term commitments also determine strategic investments, and generally favor short-term and/or low-risk projects that therefore need to carry little "risk margin" or added interest rate "adders."

This intertwined confluence of market forces, political policies, and resource availability means that electricity energy price and market share is sometimes largely determined by national and local considerations.

Two examples of how various energy sources generate electricity in a grid can be illustrated based on the Province of Ontario (Canada) system. Fig. 3.23A shows installed capacity and electricity generation by energy source in Ontario (Canada) in 2012. Analysis of Fig. 3.23A shows that in Ontario the various installed capacities in 2012 were nuclear (34%), gas (26%), hydro (22%), coal (8%), and renewables (mainly wind) (8%). However, electricity (see Fig. 3.23B) was mainly generated by nuclear (56%), hydro (22%), natural gas (10%), renewables (mainly wind) (5%), and coal (2%).

Fig. 3.24A shows power generated by various energy sources in Ontario (Canada) on June 19, 2012 (a peak power on hot summer day, when major air conditioning was required) and corresponding to that Fig. 3.24B shows capacity factors of various energy sources. Analysis of Fig. 3.24 shows that electricity that day from midnight till 3 o'clock in the morning was mainly generated by nuclear, hydro, gas, wind, "other," and coal. After 3 o'clock in the morning, wind power fell, but electricity consumption started to rise. Therefore, "fast-response" gas-fired power plants and later hydro and coal-fired power plants plus "other" power plants started operation to compensate. After 6 o'clock in the evening, energy consumption dropped slightly in the province, and at the same time, wind power increased. As a result, gas-fired, hydro, and "other" power plants decreased energy generation. After 10 o'clock in the evening, energy consumption dropped even more. As a result, coal-fired power plants decreased their electricity generation followed by gas-fired and hydropower plants.

Currently, the Province of Ontario (Canada) has completely eliminated coal-fired power plants from their electrical grid. Some of them were closed, others—converted to natural gas. Fig. 3.25A shows installed capacity, and Fig. 3.25B—electricity generation by energy source in the Province of Ontario (Canada) in 2015. Analysis of Fig. 3.25A shows that in Ontario major installed capacities in 2015 were nuclear

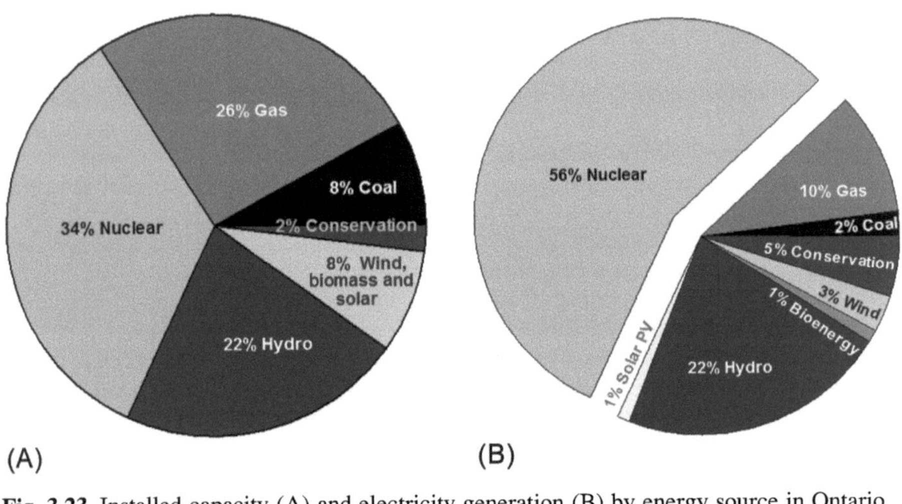

Fig. 3.23 Installed capacity (A) and electricity generation (B) by energy source in Ontario (Canada), 2012–13 [1].
Based on data from Ontario Power Authority: http://www.powerauthority.on.ca and Ontario's Long-Term Energy Plan.

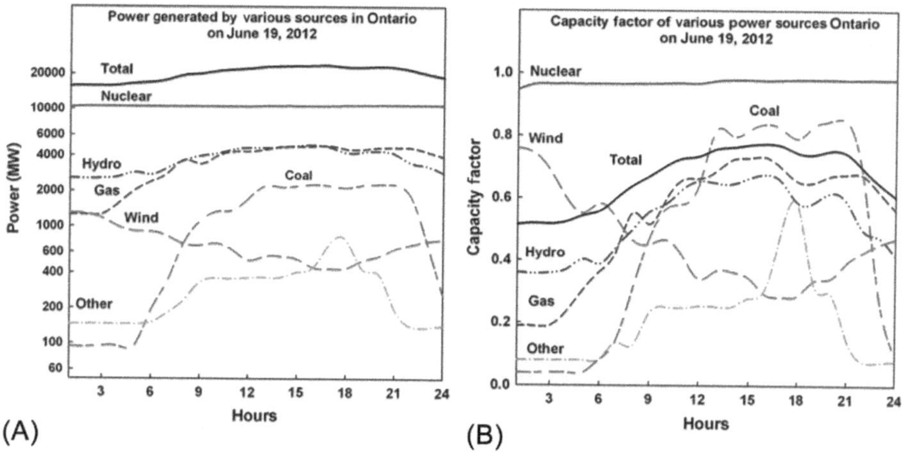

Fig. 3.24 Power generated (A) and capacity factors (B) of various energy sources in Ontario (Canada) on June 19, 2012 (shown here just for reference purposes) [1].
Based on data from http://ieso.ca/imoweb/marketdata/genEnergy.asp.

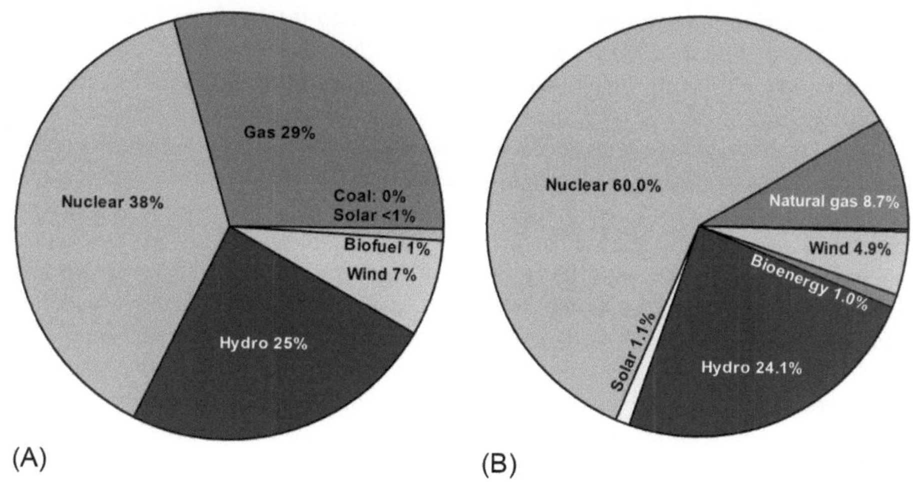

Fig. 3.25 Installed capacity (A) and electricity generation (B) by energy source in Ontario (Canada), 2014–15 [1].
Based on data from Ontario Energy Board: http://www.ontarioenergyboard.ca/ and Ontario Energy Report http://www.ontarioenergyreport.ca/.

(38%), gas (29%), hydro (25%), and renewables (mainly wind) (8%). However, electricity (see Fig. 3.25B) was mainly generated by nuclear (60%), hydro (24%), natural gas (8.7%), and renewables (mainly wind) (4.9%).

As a result, Ontario has committed to a massive $25B refurbishment and multiyear life extension of many of its existing NPPs, on the grounds that "There are currently no

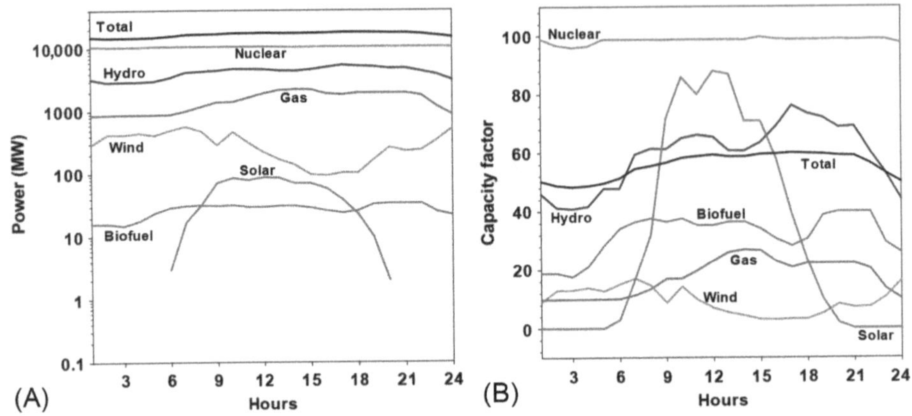

Fig. 3.26 Power generated (A) and capacity factors (B) of various energy sources in Ontario (Canada) on June 17 (Wednesday), 2015 (shown here just for reference purposes) [1]. Based on data from http://ieso.ca/imoweb/marketdata/genEnergy.asp.

alternative generation portfolios that could provide the same supply of low emissions baseload electricity generation at a comparable price to the Nuclear Refurbishment Plan." (Ontario Financial Accountability Office, "Nuclear Refurbishment Report" November 21, 2017: http://www.fao-on.org/en/Blog/Publications/FAO-NR-Report-Nov-2017.)

Fig. 3.26 shows power generated (A) and capacity factors (B) of various energy sources in Ontario (Canada) electrical grid on June 17, 2015. Analysis of data in Fig. 3.26 shows that electricity that day from midnight till 3 o'clock in the morning was mainly generated by nuclear, hydro, gas, wind, and biofuel. After 3 o'clock, biofuel power plants increased slightly electricity generation followed by hydro and gas-fired power plants. Also, at the same time, wind power plants started to generate electricity. However, after 7 o'clock, wind power started to fluctuate and, eventually, decreased significantly. After 6 o'clock in the morning, solar power plants started to generate electricity. During the day, hydro, gas-fired and biofuel power plants had variable electricity generation to compensate changes in consumption of electrical energy and variations in generating electricity from wind and solar power plants. After 9 o'clock in the evening, energy consumption started to drop in the province, and at the same time, wind power increased. Therefore, gas-fired, hydro, and biofuel power plants decreased energy generation accordingly. It showed be noted that NPPs operated at about 100% of installed capacity providing reliable basic power to the grid. This example shows clearly that any grid that includes NPPs and/or renewable energy sources must also include "fast-response" power plants such as gas-fired, coal-fired, and/or large hydropower plants to compensate changes in consumption of electrical energy per day and variations in electricity supply by wind and/or solar power plants.

Usually, NPPs operate continuously on the maximum load, because of a high capital cost and low operating costs. The relative cost of electrical energy generated by any system is not only dependent on building capital costs and operating expenses, but also dependent on the capacity factor. The higher the capacity factor—the better, as generating costs fall proportionally. However, some renewable energy sources with exception of large hydroelectric power plants can have significantly lower capacity factors compared to those of thermal and nuclear power plants. Consequently, in today's politico-socio-economic world, many governments subsidize selected low-capacity factor sources, like wind and solar, using preferential rates, enforced portfolios, artificial tariffs, market rules, and power purchase agreements to partly offset the competitive advantage of lower cost generation from natural gas, coal, and nuclear. It is against the market background, of low cost natural gas and of directly or indirectly subsidized alternates, that today's and tomorrow's NPPs must operate. Levelized/relative cost of electricity from various energy sources in United States, Germany, and Ontario (Canada) is listed in Tables 3.5–3.7.

Also, it should be noted here that countries having a large percentage of variable power sources, such as wind and solar, run the risk of an electrical grid collapse due to unpredicted power instabilities. Moreover, the following detrimental factors are

Table 3.5 Projected levelized cost of electricity (LCOE)[a] in United States by 2020 (as of 2015) [23]

		Projected LCOE (USD per MWh)		
No.	**Power-generating technology**	**Min**	**Ave**	**Max**
1	Geothermal	43.8	47.8	52.1
2	Natural gas (NG) advanced combined cycle	68.6	72.6	81.7
3	Wind onshore	65.6	73.6	81.6
4	Natural gas (NG) conventional combined cycle	70.4	75.2	85.5
5	Hydro	69.3	83.5	107.2
6	Conventional coal	87.1	95.1	119
7	Advanced nuclear	91.8	95.2	101
8	Natural gas (NG) advanced combined cycle with carbon capture and storage (CCS)	93.3	100.2	110.8
9	Biomass	90	100.5	117.4
10	Natural gas (NG) advanced combustion turbine	94.6	113.5	126.8
11	Integrated coal-gasification combined cycle (IGCC)	106.1	115.7	136.1
12	Solar photo-voltaic (PV)	97.8	125.3	193.3
13	Natural gas (NG) conventional combustion turbine	107.3	141.5	156.4
14	Integrated gasification combined cycle (IGCC) with carbon capture and storage (CCS)	132.9	144.4	160.4
15	Wind offshore	169.5	196.9	269.8
16	Solar thermal	174.4	239.7	382.5

[a] The LCOE is a measure of a power source, which attempts to compare different methods of electricity generation on a comparable basis. It is an economic assessment of the average total cost to build and operate a power-generating asset over its lifetime divided by the total energy output of the asset over that lifetime. The LCOE can also be regarded as the minimum cost at which electricity must be sold in order to break even over the lifetime of the project.

Table 3.6 Levelized cost of electricity (LCOE) in Germany in 2013 [23]

No.	Power-generating technology	LCOE in Euro per MWh		
		Min	Ave	Max
1	Coal-fired power plants (brown coal)	38	45.5	53
2	Coal-fired power plant (hard coal)	63	71.5	80
3	Onshore wind farms	45	76.0	107
4	Combined cycle gas turbine (CCGT) power plants (cogeneration)	75	86.5	98
5	Photo-voltaic (PV) systems	78	110.0	142
6	Offshore wind power	119	156.5	194
7	Biogas power plant	135	192.5	250

Table 3.7 Relative cost of electricity from various energy sources in Ontario in Canadian $ per MWh in 2013–14 [24]a

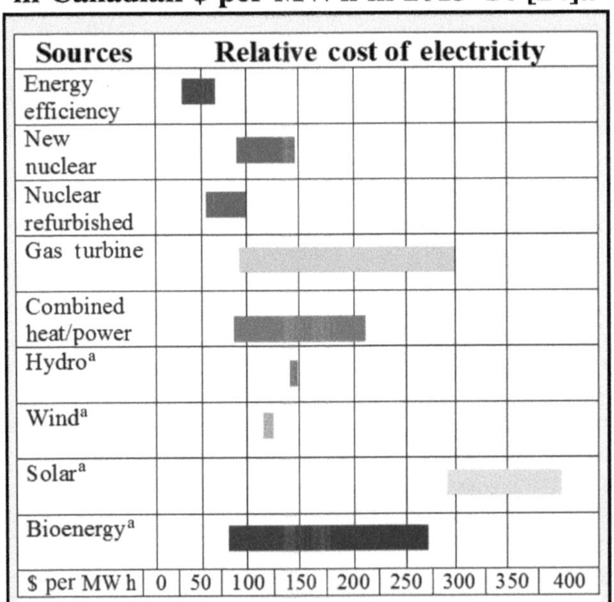

[a] Updated for January 2014 Feed-in-Tariff prices

[a] Updated for January 2014 Feed-in-Tariff prices.
Based on data from Ontario's Long-Term Energy Plan, Ministry of Energy, Toronto, Canada, 2013.

usually not considered during estimation of variable power sources costs: (1) costs of fast-response power plants with service crews on site 24/7 as a back-up power; and (2) faster amortization/wear of equipment of fast-response plants. Furthermore, considering the fast changes in climate, possible catastrophic events such as powerful hurricanes, melting ice caps in mountains, and changes in solar activity, countries should not rely on unreliable renewable sources such as hydro, wind, and solar, unless there is a significant backup energy source.

3.3 Modern thermal power plants

In general, the vast majority of thermal power plants [1,2,25] are based on the following thermodynamic cycles:

(1) Rankine steam-turbine cycle (the mostly widely used in various power plants; usually, for solid, gaseous and liquid fuels, but other energy sources can be also used, for example, geothermal, solar, etc.) (Fig. 3.27) with regeneration and steam superheat. Rankine cycle can be at subcritical pressures—older coal- and gas-fired power plants, also, geothermal and solar-thermal plants, and at supercritical pressures (for more details on supercritical pressures, see the following publications: [26–29])—at modern advanced coal- (gas-) fired power plants.

(2) Brayton gas-turbine cycle (the second one after the Rankine cycle in terms of application in power industry; only for clean gaseous fuels) (Fig. 3.28A and B, upper schemes);

(3) Combined cycle, i.e., combination of Brayton and Rankine cycles in one plant (only for gaseous fuels) (Fig. 3.28). Subcritical-pressure Rankine cycle is used at combined-cycle power plants.

(4) Diesel internal-combustion-engine cycle (for Diesel fuel used in Diesel generators). (The largest diesel generators in the world are 50-MW_{el} (~68,000 hp) Hyundai diesel engines installed at the 200-MW_{el} (50 $MW_{el} \times 4$ engines) Vasavi Basin Bridge Diesel-Generator Plant (Chennai (formerly Madras) India) (www.worldmaritimenews.com and www.hyundaicorp.com);

(5) Otto internal-combustion-engine cycle (usually, for natural or liquefied gas, but also, gasoline can be used for power generation; however, it is more expensive fuel compared to gaseous fuels), also used in internal-combustion-engine generators (Fig. 3.29).

It should be noted that internal combustion engines electrical generators are very important for emergency/backup power at nuclear and other power plants, and various installations.

The major driving force for all advances in thermal power plants is directed toward increasing thermal efficiency in order to reduce operating fuel costs and minimize specific emissions. Typical ranges of thermal efficiencies of modern thermal power plants are listed in Table 3.8 for reference purposes.

Despite all advances in thermal power plants design and operation worldwide, they are still considered as environmentally "unfriendly" due to significant carbon dioxide emissions (for example, the largest in the world 5780 MW_{el} Taichung coal-fired power plant (Taiwan) (Fig. 3.4) is the world's largest emitter of carbon dioxide with over 40 million tonnes per year [1,2]) and air pollution as a result of the combustion process. In addition, coal-fired power plants produce significant amounts of slag and ash, and other greenhouse gases such as SO_2, which contributes to acid rains.

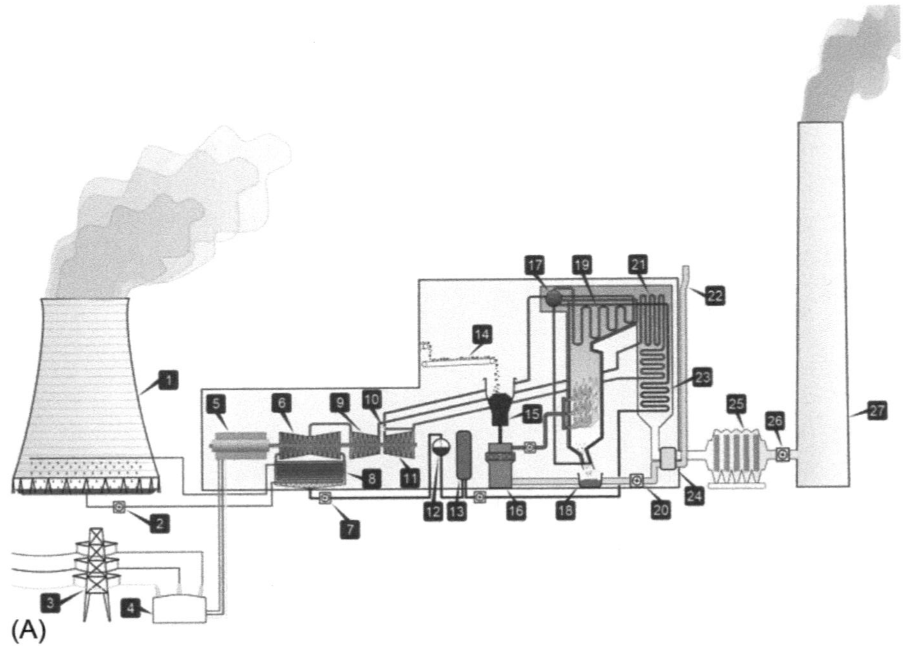

Fig. 3.27 (A) Typical scheme of coal-fired thermal power plant: (1) Cooling tower; (2) Cooling-water pump; (3) Transmission line (3-phase); (4) Step-up transformer (3-phase); (5) Electrical generator (3-phase); (6) Low-pressure steam turbine; (7) Condensate pump; (8) Surface condenser; (9) Intermediate-pressure steam turbine; (10) Steam control valve; (11) High-pressure steam turbine; (12) Deaerator; (13) Feedwater heater; (14) Coal conveyor; (15) Coal hopper; (16) Coal pulverizer; (17) Boiler steam drum; (18) Bottom-ash hopper; (19) Superheater; (20) Forced-draught (draft) fan; (21) Reheater; (22) Combustion-air intake; (23) Economizer; (24) Air preheater; (25) Precipitator; (26) Induced-draught fan; and (27) Flue-gas stack [1].

Continued

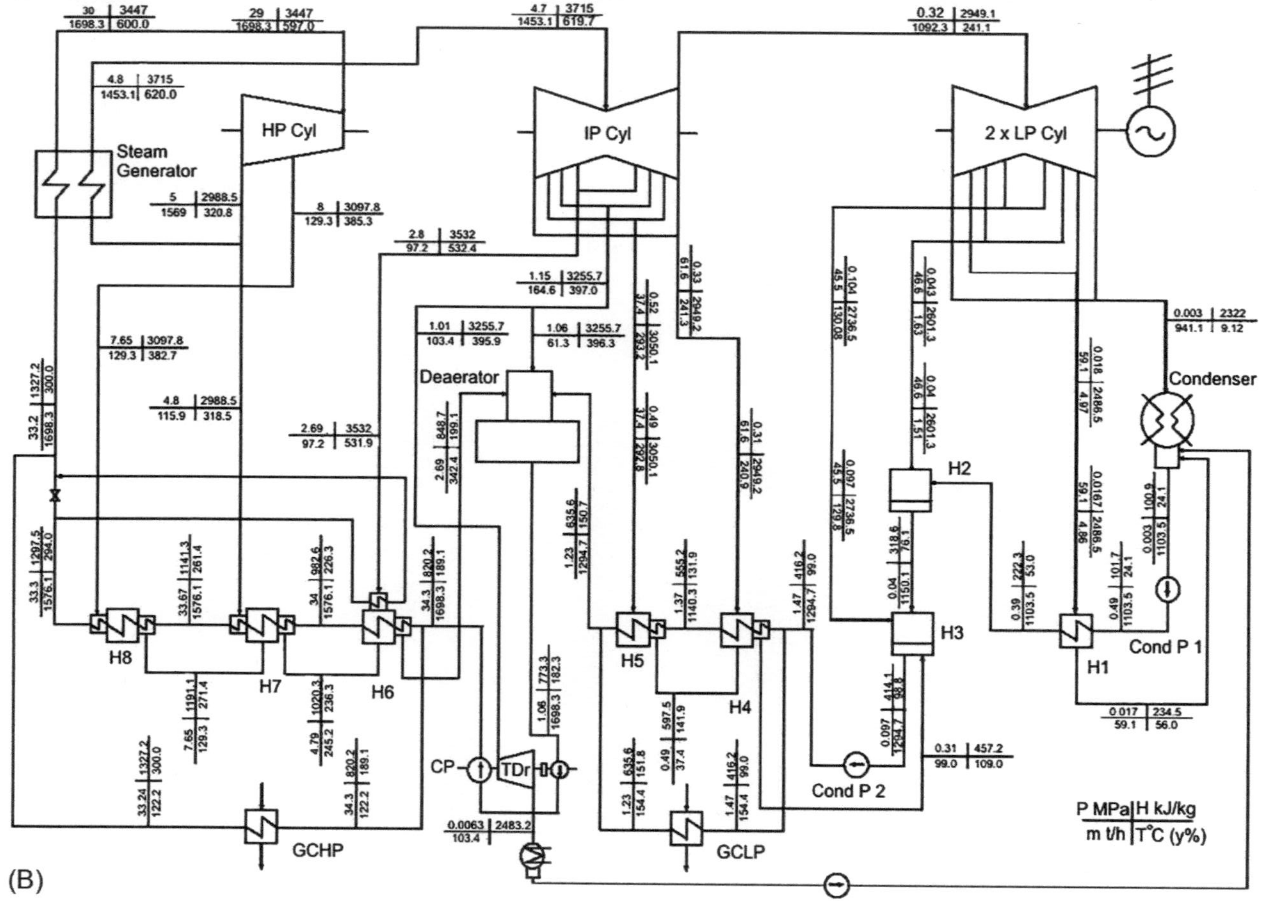

Fig. 3.27—cont'd (B) Single-reheat-regenerative cycle 600 MW_{el} Tom'-Usinsk thermal power plant (Russia) layout [1]: Cyl, cylinder; H, heat exchanger (feedwater heater); CP, circulation pump; TDr, turbine drive; Cond P, condensate pump; GCHP, gas cooler of high pressure; and GCLP, gas cooler of low pressure.

Continued

(C)

(D)

Fig. 3.27—cont'd (C) Photo of Tomato-Atsuma (Japan) coal-fired power plant Unit No. 4—supercritical-pressure steam turbine without cover: 1 high-pressure (HP), 1 intermediate pressure (IP) and 2 double-flow low-pressure (LP) cylinders: 700 MW_{el}, TC4F-43, 3000 rpm, steam parameters: 25.0 MPa pressure and 600°C/600°C (primary steam/reheat) temperature [1]. (D) Photo of low-pressure (LP) double-flow steam-turbine rotor with blades for supercritical-pressure coal-fired power plant [1].
(A) Author/user: BillC; https://commons.wikimedia.org/wiki/File:PowerStation2.svg, Accessed January 26, 2016; (B) Kruglikov et al., TsKTI, Russia, 2009; (C) Courtesy and copyright by Hitachi, Ltd.; (D) Siemens press photo; courtesy and copyright Siemens AG, Munich/Berlin, Germany.

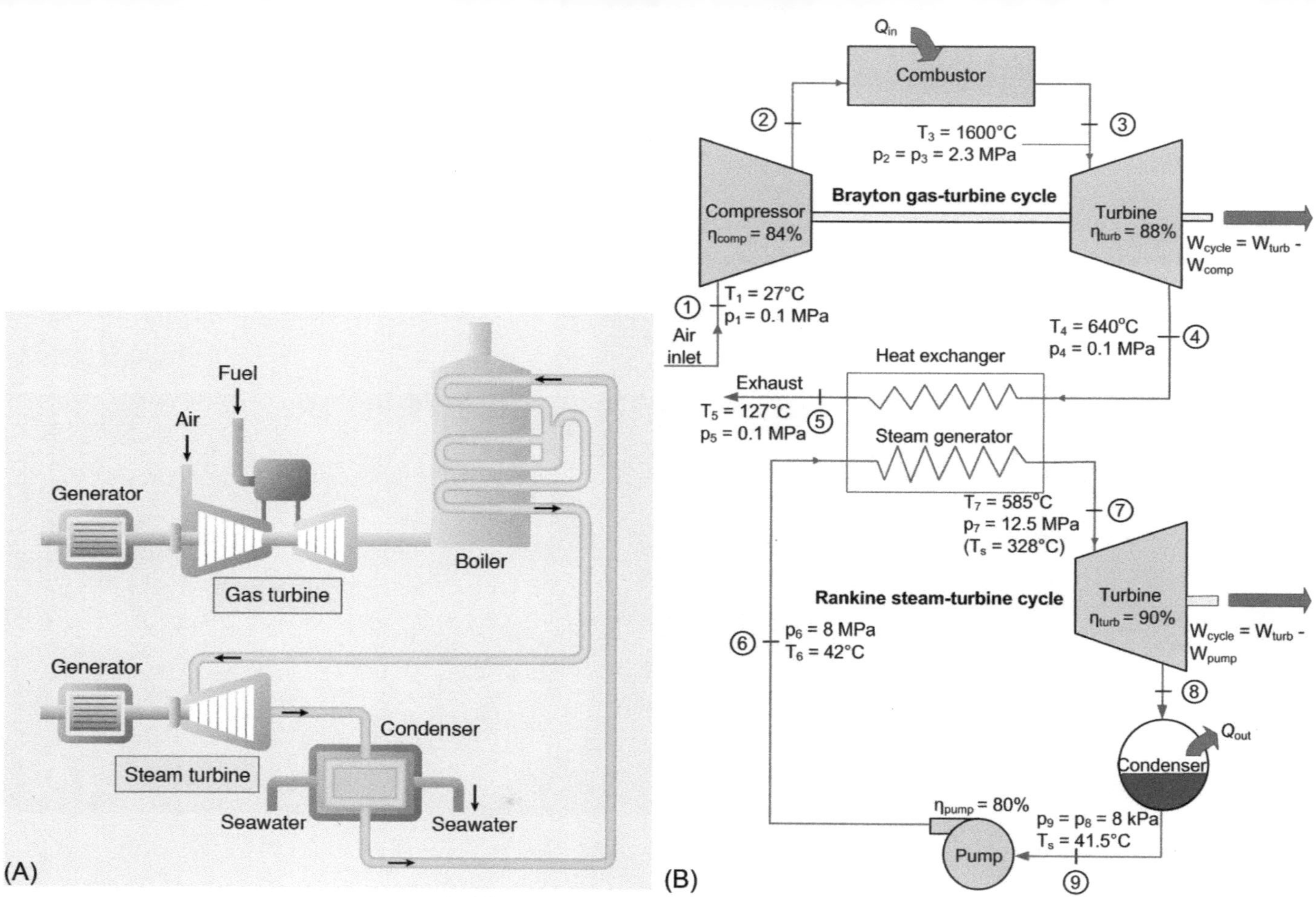

Fig. 3.28 (A) Simplified schematics of combined-cycle power plant [1]. Thermal efficiencies are the highest in power industry—up to 62% (Brayton cycle: 30% and Rankine cycle: 40%). Current level of inlet temperatures to gas turbine is about 1600–1650°C and to steam turbine—620°C. (B) Modern combined-cycle power plant schematic [1].

Continued

(C)

(D)

Fig. 3.28—cont'd (C) Photo of combined-cycle power plant gas-turbine rotor with compressor blades (at front) and turbine blades (at rear) [1]. (D) Photo of combined-cycle power plant steam turbine with open cover [1]. Single-cylinder reheat turbines are used.
(A) Courtesy and copyright of MHI; (B) Partially based on data from MHI and Siemens; (C) and (D) Courtesy and copyright of MHI.

Fig. 3.29 The largest in the world 18-cylinder Otto cycle natural gas engine (18V50SG) power-generating unit with output of up to 18.3 MW_{el}, 50 Hz and 500 rpm. The natural gas fueled, lean-burn, medium-speed engine is a reliable, high-efficiency, and low-pollution power source for flexible base load, intermediate peaking, and combined-cycle power plants. Efficiency—48.6%. Dimensions in m: Length: 18.8; width: 5.33; height: 6.34; bore: 500 mm; and weight: 360 t. Engine is 4-stroke spark ignition with prechamber.
Courtesy of WÄRTSILÄ, Finland.

Table 3.8 **Typical ranges of thermal efficiencies (gross) of modern thermal power plants [1,2]**

No	Thermal power plant	Gross Eff. (%)
1	Combined-cycle power plant (combination of Brayton gas-turbine cycle (fuel—natural gas or LNG; combustion-products parameters at the gas-turbine inlet: $T_{in} \approx 1650°C$) and Rankine steam-turbine cycle (steam parameters at the turbine inlet: $T_{in} \approx 620°C$ ($T_{cr} = 374°C$)) (Fig. 3.28)	Up to 62
2	Supercritical-pressure coal-fired power plant (Rankine cycle steam inlet turbine parameters: $P_{in} \approx 25$–38 MPa ($P_{cr} = 22.064$ MPa), $T_{in} \approx 540$–625°C ($T_{cr} = 374°C$) and $T_{reheat} \approx 540$–625°C) (Fig. 3.27)	Up to 55
3	Internal combustion engine generators (Diesel cycle and Otto cycle with natural gas as a fuel (Fig. 3.29))	Up to 50
4	Subcritical pressure coal-fired power plant (older plants) (Rankine cycle steam: $P_{in} \approx 17$ MPa, $T_{in} \approx 540°C$ ($T_{cr} = 374°C$) and $T_{reheat} \approx 540°C$) (Fig. 3.27A)	Up to 43
5	Concentrated solar thermal power plants with heliostats, solar receiver (heat exchanger) on a tower and molten-salt heat storage system (Figs. 3.15–3.17)	Up to 30

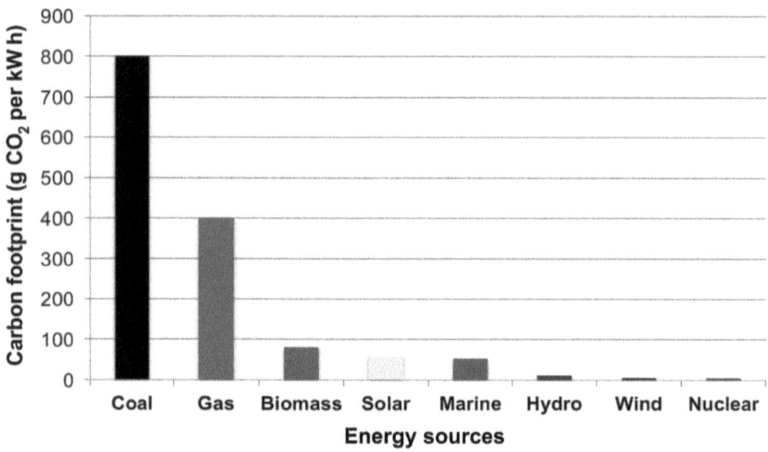

Fig. 3.30 Carbon footprint for various energy sources. If carbon capture and storage (CCS) is used, then the carbon footprint can be decreased for coal by about sixfold and for gas about sixfold. (For carbon footprint of NPPs—see Fig. 3.32.)
Courtesy of Dr. J. Roberts, University of Manchester, United Kingdom; based on data from http://researchbriefings.files.parliament.uk/documents/POST-PN-268/POST-PN-268.pdf.

Comparison of various electricity-generating power plants based on carbon footprint is shown in Fig. 3.30, and deaths per terawatt for various energy sources—in Fig. 3.31.

3.4 Modern nuclear power reactors and nuclear power plants

Chapter 4 is fully dedicated to nuclear fission power reactors and NPPs. Therefore, in this Section, only general statistics and discussion on nuclear power are provided.

Nuclear power is often considered to be a nonrenewable energy source as the fossil fuels, such as coal and gas. However, nuclear resources can be used for significantly longer than some fossil fuels, and in some cases almost indefinitely, if recycling of unused or spent uranium fuel, thoria fuel resources, and fast reactors are used. The major advantages of nuclear power [1,2,30] are:

(1) concentrated and reliable source of almost infinite energy and is independent of weather conditions;
(2) high capacity factors are achievable, often in excess of 90% with long operating cycles, making units suitable for continuous baseload operation (Table 3.4 and Figs. 3.24 and 3.26);
(3) essentially negligible operating emissions of carbon dioxide and relatively small amount of wastes compared to alternate fossil-fuel thermal power plants (Figs. 3.30 and 3.32, and Tables 3.9 and 3.10);
(4) relatively small amount of fuel required compared to that of fossil-fuel thermal power plants (Table 3.9); and

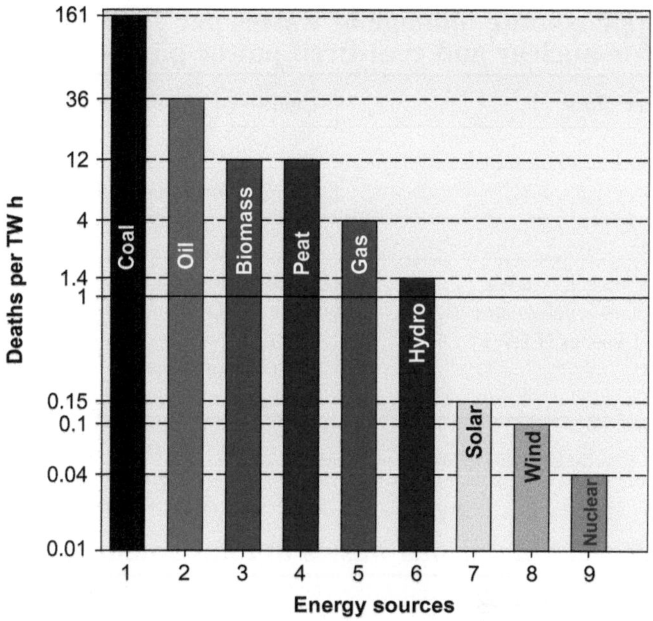

Fig. 3.31 Deaths per TWh for various energy sources.
Based on data from https://www.nextbigfuture.com/2011/03/deaths-per-twh-by-energy-source.html.

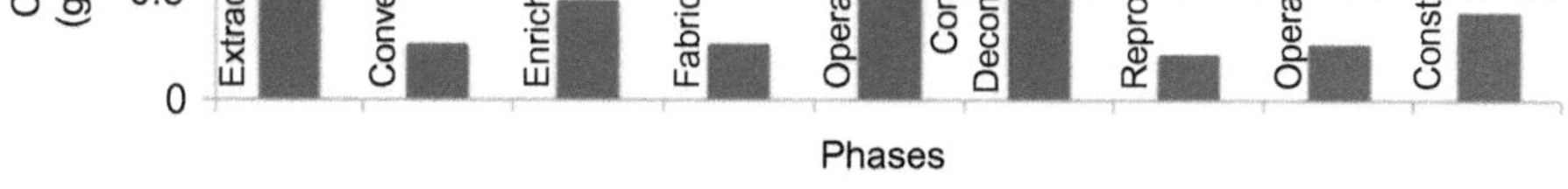

Fig. 3.32 Carbon footprint of nuclear power plant various phases.
Courtesy of Dr. J. Roberts, University of Manchester, United Kingdom; based on data from British Energy for Torness AGR NPP.

(5) NPPs can supply relatively cheap electricity for recharging of electrical vehicles during night hours as they usually operate on full load (capacity) day and night.

As a result, nuclear power source of energy is considered as the most viable one for electrical generation within next 50–100 years. Nuclear power must, however, operate and compete in energy markets based on the relative costs and strategic advantages of the available fuels and energy types. At the same time, it must be recognized that

Table 3.9 Approximate tonnage of wastes per 1000-MW_{el} power per year for nuclear and coal-fired power plants

Nuclear power plant	Coal-fired power plant
Fuel	
25t of UO_2	2.6 million t of coal (5 × 1400t trains a day)
Wastes	
35t high-level wastes (HLW) 310t intermediate-level wastes (ILW) 460t low-level wastes (LLW)	6,500,000t of CO_2 900t of SO_2 4500t of NO_x 320,000t of ash 400t of toxic heavy metals

Courtesy of Dr. J. Roberts, University of Manchester, United Kingdom.

Table 3.10 Percent of various wastes in total amount

No	Wastes	% of total amount
1	Mining and quarrying	27.30
2	Agriculture	20.13
3	Demolition and construction	18.51
4	Industrial	12.73
5	Dredged spoils	7.64
6	Household	6.94
7	Commercial	6.48
8	Sewage sludge	0.23
9	Radioactive	0.04

Courtesy of Dr. J. Roberts, University of Manchester, United Kingdom; partially based on data from: https://publications.parliament.uk/pa/cm200405/cmselect/cmenvfru/130/130we13.htm.

nuclear energy is highly and uniquely politicized, not only because of nonproliferation concerns, international uranium and thorium supply availability, but also because of strategic national energy policy considerations, and national concerns over energy independence, and positioning in global commercial markets. It was these national issues, with high dependency on Middle East oil, which initially drove the building of many NPPs in the past. This does not follow the orderly ideal of the evolution of reactor Generations I, II, III, and IV [1]. Therefore, the following key historical perspective and progression must be understood:

- During the initial phases of the Cold War, the United States, Russia, and United Kingdom built water- and gas-cooled nuclear power reactors/NPPs, plus under the "Atoms for Peace" initiative these light-water-reactor (LWR) designs were licensed to and built in France, South Korea, and Japan.

- In Europe, national variants of these LWR designs were then developed and deployed in Sweden, Germany, France, while Korea and Japan followed a similar independent development path.
- After being cut out of nuclear cooperation agreements, and without pressure-vessel and fuel-enrichment manufacturing capability, Canada and India developed their own Pressurized Heavy Water Reactor (PHWR) designs, which were also internationally deployed in Pakistan (1 unit in 1972), Argentina (1 unit in 1984), South Korea (4 units in 1983; 1997; 1998; and 1999), Romania (2 units in 1996 and 2007), and China (2 units in 2002 and 2003).
- Three major nuclear accidents and fear of radiation release severely and adversely affected nuclear development and deployment (for details, also see Fig. 3.33): (1) Following Three Mile Island core melt in the United States new plants essentially stopped in the United States; (2) After the Chernobyl reactor explosion in the USSR further plans for new reactors were halted and plant closures occurred in Europe, especially, in Germany and Italy; and Russia re-examined its development strategy; and (3) The latest Fukushima reactors explosions and meltdowns understandably paralyzed deployment in Japan, and led to enhanced political

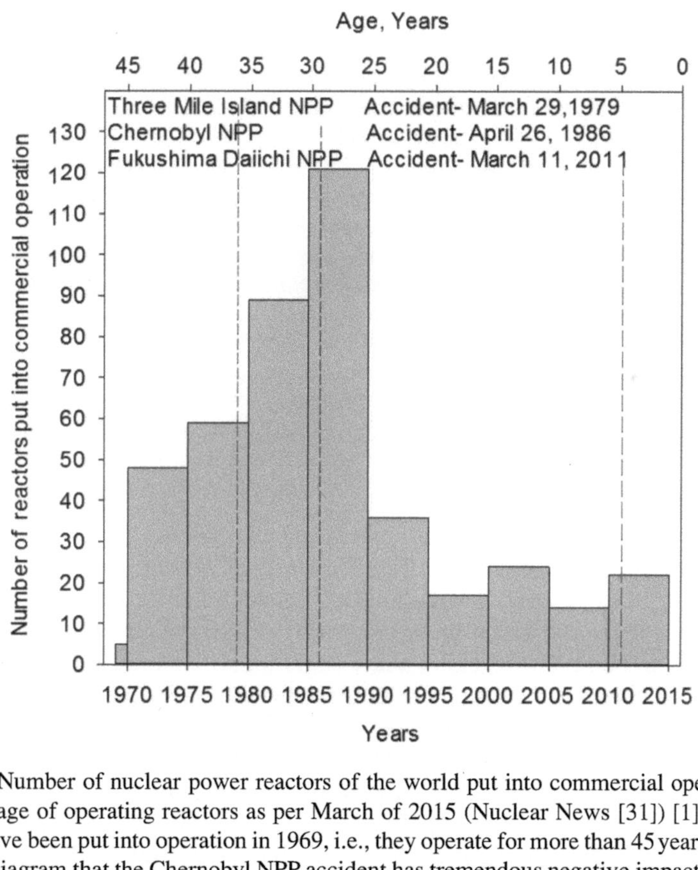

Fig. 3.33 Number of nuclear power reactors of the world put into commercial operation vs years and age of operating reactors as per March of 2015 (Nuclear News [31]) [1]. Several reactors have been put into operation in 1969, i.e., they operate for more than 45 years. It is clear from this diagram that the Chernobyl NPP accident has tremendous negative impact on nuclear power industry, which is lasting for decades. And currently, we have additional negative impact of the Fukushima Daiichi NPP accident.

opposition to nuclear power in Germany, Taiwan, South Korea, Switzerland, Spain, and other countries.

- Stagnation of reactor development and deployment enabled the loss of expertise, and the inability to complete the building of new nuclear facilities on time/schedule and within the agreed budget. Meanwhile, two key developments took place: Cheap natural gas arrived, with the United States quickly becoming a major energy exporter, and the rise of China as an economic engine, requiring major investment in power sources. This shift in the global energy scene coincided with postcommunist Russia consolidating its nuclear commercial complex; and with China emerging as a global economic force, not only importing all of the new designs to "digest" the technology, but aggressively moving forward indigenous development of nearly every reactor type.
- The future is unfortunately unpredictable. The emergence of potential climate change, partially driven by fossil-fuel (CO_2) emissions, is now of international concern and scale. However, individual markets do not allow so-called carbon credits or financial incentives for NPPs, despite their low CO_2 emissions (for details, see Figs. 3.30 and 3.32). Major future energy scenarios by international organizations see a role for nuclear, but usually focus on alternatives like conservation or "renewables." These projections and hopes avoid the inevitability of the global drive for economic prosperity for all those presently without it, meaning a manifold increase in energy use, and, hence, emissions of every kind. It has been and can be calculated that literally thousands of reactors would need to be built and deployed at a rate of a few per week over the next 20 years to address or help to stabilize the rising emissions. The world is not prepared, and neither are the nuclear and political establishments, to undertake this task unless massive investments of capital and political will occur.
- Limited additional deployment in Europe and the United States, but new builds are continuing in Russia, China, India, South Korea, and the Middle East. The inability of modern designs to compete or be built on schedule/within the agreed budget has led to major bankruptcies and restructuring of the existing major commercial vendors. Major build programs continue, where national investment and political support exist (China, Russia, and India), and they are turning toward export markets in Argentina, Bangladesh, Egypt, Middle East, Philippines, Turkey, and even toward gaining a foothold in the United Kingdom and elsewhere in Europe (e.g., Romania).
- China is aggressively pushing its own design variants, using financial investment as a lever in Argentina, Pakistan, United Kingdom, etc. to gain market entry. Russia, South Korea, and Japan are also focused on export markets, because their own domestic markets are saturated.

Current statistics of all world nuclear power reactors connected to electrical grids are listed in Tables 3.11–3.13, and Fig. 3.33. Analysis of the current statistical data on nuclear power reactors shows that, currently, 31 countries in the world have operating nuclear power reactors (within these countries: 17 plan to build new reactors and 14 don't plan to build new reactors) and 4 countries without nuclear power reactors (Bangladesh, Belarus, Turkey, and United Arab Emirates (UAE)) are working toward introducing nuclear energy on their soils [32].

The largest group of nuclear reactors by type is Pressurized Water Reactors (PWRs) (290 from 448 reactors or 65% of total), and quite significant number of PWRs are planned to be built (about 77) (for details, see Table 3.11). The second largest group of reactors is Boiling Water Reactors (BWRs)/Advanced BWRs (ABWRs) (78 reactors or 17% of total). The third group is Pressurized Heavy Water Reactors (PHWRs) (49 reactors or 11% of total). Considering the number of forthcoming

Table 3.11 Number of nuclear power reactors in operation and forthcoming as per March 2017 (Nuclear News [32]) and before the Japan earthquake and tsunami disaster (March 2011) (Nuclear News [33]) (additional data on reactors are shown in [1])

No	Reactor type (some details on reactors)	No. of units		Installed capacity, (GW_{el})		Forthcoming units	
		As of March 2017	Before March 2011	As of March 2017	Before March 2011	No. of units	GW_{el}
1	Pressurized water reactors (PWRs) (largest group of nuclear reactors in the world—65%)	290 ↑	268	273 ↑	248	77	84
2	Boiling water reactors (BWRs) or advanced BWRs (second largest group of reactors in the world—17%; ABWRs—the only ones Gen-III+ operating reactors)	78 ↓	92	76 ↓	84	6	8
3	Pressurized heavy water reactors (PHWRs) (third largest group of reactors in the world—11%; mainly CANDU reactor type)	49 ↓	50	25 ↓	25	9	6
4	Advanced gas-cooled reactors (AGRs) (3%) (UK, 14 reactors); (all these CO_2-cooled reactors will be shut down in the nearest future and will not be built again)	14 ↓	18	8 ↓	9	1[a]	0.2[a]
5	Light-water-cooled, graphite-moderated reactors (LGRs) (3%) (Russia, 11 RBMKs and 4 EGPs[b]; these pressure-channel boiling-water-cooled reactors will be shut down in the nearest future and will not be built again)	15	15	10	10	0	0
6	Liquid-metal fast-breeder reactors (LMFBRs) (Russia, SFRs—BN-600 and BN-800)	2 ↑	1	1.3 ↑	0.6	5	1.6
In total		448 ↑	444	393 ↑	378	82	88

Explanations to Table: Data in Table include 42 reactors in Japan, the vast majority of which are currently (i.e., October of 2017) not in service. Also, data in Table were updated with the latest information on nuclear reactors taken from other sources.

Arrows mean decrease or increase in a number of reactors.

[a] Forthcoming reactor is a helium-cooled reactor—High-temperature reactor pebble-bed modular (HTR-PM).

[b] EGP—Power heterogeneous loop reactor (in Russian abbreviations), channel-type, graphite-moderated, light-water coolant, boiling reactor with natural circulation.

Table 3.12 Number of nuclear power reactors by nation (11 nations with the largest number of reactors ranked by installed capacity) as per March of 2017 (Nuclear News [32]) and before the Japan earthquake and tsunami disaster (March of 2011) (Nuclear News [33])

No	Nation	No. of units (PWRs/BWRs)		Installed capacity, (GW_{el})		Changes in number of reactors from March 2011
		As of March 2017	Before March 2011	As of March 2017	Before March 2011	
1	USA	99 (65/34)	104	101	103	↓ Decreased by 5 reactors
2	France	58 (58/–)	58	63	63	No changes
3	Japan[a]	42 (19/23)	54	42	47	↓ Decreased by 12 reactors
4	China	37 (35/–/2[b])	13	32	10	↑ Increased by 24 reactors
5	Russia	35 (18/–/15[c]/2[d])	32	26	23	↑ Increased by 3 reactors
6	South Korea	25 (21/–/4[b])	20	23	18	↑ Increased by 5 reactors
7	Canada	19 (–/–/19[b])	22	13	15	↓ Decreased by 3 reactors
8	Ukraine	15 (15/–)	15	13	13	No changes
9	Germany	8 (6/2)	17	11	20	↓ Decreased by 9 reactors
10	Sweden	10 (7/3)	10	10	9	No changes
11	UK	15 (1/–/14[e])	19	9	10	↓ Decreased by 4 reactors
In total		362	364	342	331	↓ Decreased by 2 reactors (however, installed capacity increased by 10.7 GW_{el})

Data for all countries with nuclear power reactors are in Table 3.13.
Arrows mean decrease or increase in a number of reactors.
[a] Currently, i.e., in October of 2017, the vast majority of nuclear power reactors are not in operation. However, there are plans to put them into operation in the nearest future.
[b] PHWRs.
[c] No of LGRs.
[d] LMFBRs.
[e] AGRs.

Table 3.13 Nuclear power reactors by Nation (as per March of 2017) (Nuclear News, ANS, March 2017 [32])

No	Nation	# Units	Net MW_{el}	# Units	Net MW_{el}
		(in operation)		(forthcoming)	
1	Argentina	3 (PHWRs)	1627	1	717
2	Armenia	1 (PWR)	375	0	0
3	Bangladesh	2 (PWRs) (planned)	–	2	2400
4	Belarus	2 (PWRs) (planned)	–	2	2218
5	Belgium	7 (PWRs)	5913	0	0
6	Brazil	2 (PWRs)	1884	1	1245
7	Bulgaria	2 (PWRs)	1906	0	0
8	Canada	19 (PHWRs)	13,524	0	0
9	China	37 (35 PWRs; 2 PHWRs)	32,400	20	20,500
10	Czech Republic	6 (PWRs)	3930	0	0
11	Finland	4 (2 PWRs; 2 BWRs)	2758	2	2800
12	France	58 (PWRs)	63,130	1	1600
13	Germany	8 (6 PWRs; 2 BWRs)	10,799	0	0
14	Hungary	4 (PWRs)	1889	2	2400
15	India	21 (18 PHWRs; 2 BWRs; 1 PWR)	5308	8	5257
16	Iran	1 (PWR)	915	0	0
17	Japan	42 (19 PWRs; 18 BWRs; 5 ABWRs)	39,752	2	2650
18	Mexico	2 (BWRs)	1552	0	0
19	Netherlands	1 (PWR)	482	0	0
20	Pakistan	4 (3 PWRs; 1 PHWR)	1005	4	3343
21	Romania	2 (PHWRs)	1300	2	1440
22	Russia	35 (18 PWRs; 15 LGRs; 2 LMFBRs)	26,058	7	7832
23	Slovakia	4 (PWRs)	1816	2	880
24	Slovenia	1 (PWR)	688	0	0
25	South Africa	2 (PWRs)	1860	0	0
26	South Korea	25 (21 PWRs; 4 PHWRs)	23,044	5	6820
27	Spain	7 (6 PWRs; 1 BWR)	7121	0	0
28	Sweden	10 (3 PWRs; 7 BWRs)	9740	0	0
29	Switzerland	5 (3 PWRs; 2 BWRs)	3333	0	0
30	Taiwan	6 (2 PWRs; 4 BWRs)	5052	2	2600
31	Turkey	4 (PWRs) (planned)	–	4	4800
32	Ukraine	15 (PWRs)	13,107	3	2850
33	UAE	4 (PWRs) (planned)	–	4	5380
34	United Kingdom	15 (1 PWR; 14 AGRs)	8883	2	3200
35	USA	99 (65 PWRs; 34 BWRs)	102,198	6	7100
Total		448	39,349	82	88,032

Summary: 31 countries have operating nuclear power reactors, and 4 countries plan to build nuclear power reactors (in green color—countries, which plan to build nuclear power reactors). 14 countries don't plan to build nuclear power reactors (black color).

reactors, the number of BWRs/ABWRs and PHWRs will possibly decrease within next 20–25 years. Furthermore, within next 10–15 years or so, all advanced gas-cooled reactors (AGRs) and light-water-cooled graphite-moderated reactors (LGRs) will be shut down. However, instead of carbon-dioxide-cooled AGRs helium-cooled reactors will be built and put into operation.

Analysis of the data in Tables 3.12 and 3.13 shows that real nuclear "renaissance" is in China (22 reactors built and put into operation within the past 6 years!), in South Korea (addition of 5 reactors), and in Russia (addition of 3 reactors). Meanwhile, the most significant drop in a number of reactors is in Japan (12 reactors were shut down) (only several reactors out of 42 are currently in operation), in Germany (9 reactors), in United States (5 reactors), in United Kingdom (4 reactors), and in Canada (3 reactors). In addition, Germany and Canada have no plans to build new reactors.

Fig. 3.33 shows impact of the major NPPs accidents within last 50 years on new builds. Analysis of the data in this figure shows that we might face a very significant drop (up to 3 times) in a number of operating nuclear power reactors somewhere between 2030 and 2040; if we assume that current operating term of reactors is on average 45 years, and rate of building and putting into operation new reactors is ~21 reactors per 5 years (Fig. 3.34). Even with higher rates of new nuclear capacities addition, we will have a tangible decrease in a number of operating reactors. If this forecast(s) is correct, the nuclear power industry will face very difficult times ahead.

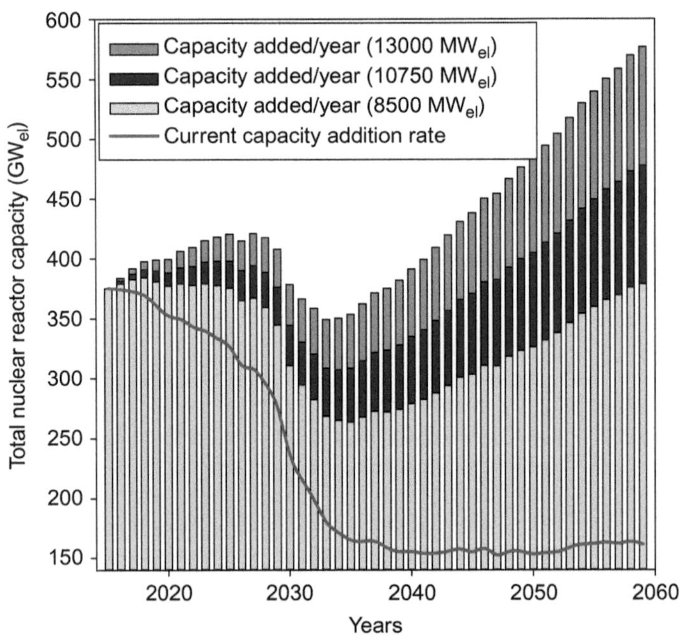

Fig. 3.34 Possible scenarios for future of nuclear power; based on 45 years in service of current reactors and adding new reactors with rate of ~21 reactor per 5 years (red line) [1].

Table 3.14 Typical ranges of thermal efficiencies (gross) of modern nuclear power plants [1]

No	Nuclear power plant	Gross efficiency, (%)
1	Carbon-dioxide-cooled reactor NPP (Generation III) (reactor coolant: $P=4$ MPa and $T=290$–650°C; steam: $P=17$ MPa ($T_{sat}=352$°C) and $T_{in}=560$°C)	Up to 42
2	Sodium-cooled fast reactor (BN-600/BN-800) NPP (steam: $P=14$ MPa ($T_{sat}=337$°C) and $T_{in}=505$°C)	Up to 40
3	Pressurized-water-reactor NPP (Generation III+, to be implemented within next 1–10 years) (reactor coolant: $P=15.5$ MPa and $T_{out}=327$°C; steam: $P=7.8$ MPa and $T_{in}=293$°C)	Up to 38
4	Pressurized-water-reactor NPP (Generation III, current fleet) (reactor coolant: $P=15.5$ MPa and $T_{out}=292$–329°C; steam: $P=6.9$ MPa and $T_{in}=285$°C)	Up to 36
5	Boiling-water-reactor NPP (Generation III, current fleet) ($P_{in}=7.2$ MPa and $T_{in}=288$°C)	Up to 34
6	Pressurized heavy water reactor NPP (Generation III, current fleet) (reactor coolant: $P=11$ MPa and $T=260$–310°C; steam: $P=4.6$ MPa and $T_{in}=259$°C)	Up to 32

It should be once more emphasized that, in general, current problems in the world nuclear power industry are significant delays in putting into operation new, mainly, Generation III+ reactors, e.g., EPRs by AREVA, APs-1000 by Toshiba-Westinghouse, etc.; indecision of governments in terms of support of nuclear-based electricity generation; and radioactive-waste management and safe storage.

In spite of all current advances in nuclear power, NPPs have the following deficiencies: (1) Generate radioactive wastes; (2) Have relatively low thermal efficiencies, especially, NPPs equipped with water-cooled reactors (up to 1.6 times lower than that for modern advanced thermal power plants (see Tables 3.14 and 3.8, respectively); (3) Risk of radiation release during severe accidents; and (4) Production of nuclear fuel is not an environment-friendly process. Therefore, all these deficiencies should be addressed in next generation—Generation IV reactors and NPPs [1].

3.5 Conclusions

1. It is well known that electrical power generation is the key factor for advances in industry, agriculture, technology, and standard of living. Also, strong power industry with diverse energy sources is very important for a country's independence.
2. Major sources for electrical energy generation in the world today are: (1) thermal—primary coal (39.9%) and secondary natural gas (22.6%); (2) "large" hydro (17.2%); and (3) nuclear

(11.2%). The remaining 9.2% of the electrical energy is generated using oil (4.2%) and renewable sources (biomass, wind, geothermal, and solar energy) (5%) in selected countries.

3. Other energy sources such as renewable wind and solar power have a significant visible impact in some countries, especially, where there are government incentives with electricity prices guaranteed by legislation and power-purchase contracts. However, these apparently attractive renewable-energy sources (wind, solar, tidal, etc.) are not reliable as full-time energy sources for industrial power generation. To overcome this problem, a grid must also include "fast-response" power plants such as gas- (coal-) fired and/or large hydro-power plants.
4. In general, the major driving force for all advances in thermal and nuclear power plants is thermal efficiency and generating costs. Ranges of gross thermal efficiencies of modern power plants are as the following: (1) Combined-cycle thermal power plants—up to 62%; (2) Supercritical-pressure coal-fired thermal power plants—up to 55%; (3) Carbon-dioxide-cooled reactor NPPs—up to 42%; (4) Sodium-cooled Fast Reactor (SFR) NPP—up to 40%; (5) Subcritical-pressure coal-fired thermal power plants—up to 43%; and (6) Modern water-cooled-reactor NPPs—30%–36%.
5. In spite of advances in coal-fired thermal power plants design and operation worldwide, they are still considered environmental "unfriendly" as a result of carbon dioxide emissions and significant production of slag and ash wastes. Recently, legislated measures have been proposed to limit such emissions, going beyond voluntary and regional emission credits and allowable portfolios.
6. Combined-cycle thermal power plants with natural-gas fuel are considered as relatively clean fossil-fuel-fired plants compared to coal and oil power plants, and are dominating new capacity additions, because of their lower carbon dioxide production and lower costs using natural gas derived from "fracking" processes.
7. Nuclear power is, in general, a nonrenewable source unless fuel recycling, thoria fuel, and/or fast reactors are adopted, which means that nuclear resources can be used significantly longer than some fossil fuels. Currently, this source of energy is considered as the most viable one for baseload electrical generation for the next 50–100 years.
8. However, all current Generations II and III and oncoming Generations III+ NPPs, especially, those equipped with water-cooled reactors, are not very competitive with modern thermal power plants in terms of thermal efficiency (30%–36% for NPPs with water-cooled reactors and 55%–62% for supercritical-pressure coal-fired and combined-cycle power plants, respectively).
9. Enhancements are needed beyond the current building plans for NPPs; many of new designs are coming from China. These new designs must compete in the world markets, and if possible, without government subsidies or power price guarantees. New-generation NPPs must have thermal efficiencies close to those of modern thermal power plants, i.e., within a range of at least 40%–50%, and incorporate improved safety measures and designs.
10. The major advantages of nuclear power are well known, including cheap reliable baseload power, high capacity factor, low carbon dioxide emissions, and minor environmental impact. However, these factors are offset today by a competitive disadvantage with natural gas and the occurrence of three significant nuclear accidents (Fukushima, Chernobyl, and Three Mile Island)—the latter having caused significant social disruption together with high capital costs.

Acknowledgments

The authors would like to express their great appreciation to unidentified authors of Wikipedia website articles and authors of photos from the Wikimedia Commons website for their materials used in this Chapter in the text, tables, and figures' captions. Materials and illustrations provided by various companies and used in this Chapter are gratefully acknowledged.

References

[1] Handbook of generation IV nuclear reactors, Duxford: Elsevier–Woodhead Publishing (WP); 2016. 940 pp. Handbook content is Available from: https://www.gen-4.org/gif/jcms/c_9373/publications.

[2] Pioro I, Duffey R. Nuclear power as a basis for future electricity generation, ASME J Nuclear Eng Rad Sci 2015;1(1). 19 pp. Available from: http://nuclearengineering.asmedigitalcollection.asme.org/article.aspx?articleID=2085849.

[3] Pioro I, Kirillov P. Current status of electricity generation in the World, In: Méndez-Vilas A, editor. Materials and processes for energy: communicating current research and technological developments. Energy book series #1Spain: Formatex Research Center; 2013. p. 783–95. Available from: http://www.formatex.info/energymaterialsbook/book/783-795.pdf.

[4] Pioro I. Nuclear power as a basis for future electricity production in the World, In: Mesquita AZ, Rezende HC, editors. Current research in nuclear reactor technology in Brazil and Worldwide. Rijeka: INTECH; 2012. p. 211–50. [chapter 10]. Available from: http://www.intechopen.com/books/current-research-in-nuclear-reactor-technology-in-brazil-and-worldwide/nuclear-power-as-a-basis-for-future-electricity-production-in-the-world-generation-iii-and-iv-reacto.

[5] Central Intelligence Agency (CIA). The world factbook: electricity consumption, CIA.org. Available from: https://www.cia.gov/library/publications/the-world-factbook/rankorder/2233rank.html; 2016. Accessed 16 January 2016.

[6] Central Intelligence Agency (CIA). The world factbook: population, CIA.org. Available from: https://www.cia.gov/library/publications/the-world-factbook/rankorder/2119rank.html; 2016. Accessed 16 January 2016.

[7] United Nations (UN). Table 1: Human development index and its components, United Nations Development Programme, Available from: http://hdr.undp.org/en/composite/HDI; 2016. Accessed 16 January 2016.

[8] https://en.wikipedia.org/wiki/Taichung_Power_Plant, [Accessed 10 October 2017].

[9] https://en.wikipedia.org/wiki/Nurek_Dam, [Accessed 10 October 2017].

[10] https://en.wikipedia.org/wiki/Chief_Joseph_Dam, [Accessed 10 October 2017].

[11] https://en.wikipedia.org/wiki/Horns_Rev_2, [Accessed 10 October 2017].

[12] http://www.alstom.com/power/renewables/wind/turbines/, [Accessed 10 October 2017].

[13] https://en.wikipedia.org/wiki/Solar_Energy_Generating_Systems, [Accessed 10 October 2017].

[14] https://en.wikipedia.org/wiki/Gemasolar_Thermosolar_Plant, [Accessed 10 October 2017].

[15] https://en.wikipedia.org/wiki/Rance_Tidal_Power_Station, [Accessed 10 October 2017].

[16] https://en.wikipedia.org/wiki/Pelamis_Wave_Energy_Converter, [Accessed 10 October 2017].

[17] https://en.wikipedia.org/wiki/List_of_largest_power_stations, [Accessed 10 October 2017].
[18] https://en.wikipedia.org/wiki/Pumped-storage_hydroelectricity, [Accessed 10 October 2017].
[19] https://en.wikipedia.org/wiki/Capacity_factor, [Accessed 10 October 2017].
[20] https://www.eia.gov/electricity/monthly/epm_table_grapher.php?t=epmt_6_07_b, [Accessed 10 October 2017].
[21] BP Energy Outlook. Available from: https://www.bp.com/content/dam/bp/pdf/energy-economics/energy-outlook-2017/bp-energy-outlook-2017.pdf; 2017.
[22] EIA International Energy Outlook. Available from: https://www.eia.gov/outlooks/ieo/pdf/0484(2017).pdf; 2017.
[23] https://en.wikipedia.org/wiki/Cost_of_electricity_by_source, [Accessed 10 October 2017].
[24] http://www.energy.gov.on.ca/en/files/2014/10/LTEP_2013_English_WEB.pdf, [Accessed 10 October 2017].
[25] Pioro I, Kirillov P. Current status of electricity generation at thermal power plants, In: Méndez-Vilas A, editor. Materials and processes for energy: communicating current research and technological developments. Energy book series #1Spain: Formatex Research Center; 2013. p. 796–805. Available from: http://www.formatex.info/energymaterialsbook/book/796-805.pdf.
[26] Pioro I, Mahdi M, Popov R. Application of supercritical fluids in power engineering. In: Chen L, Iwamoto Y, editors. Advanced applications of supercritical fluids in energy systems. Hershey, PA: IGI Global; 2017. p. 404–57 [chapter 13], 626 pp.
[27] Pioro I, Mokry S. Thermophysical properties at critical and supercritical conditions, In: Belmiloudi A, editor. Heat transfer. Theoretical analysis, experimental investigations and industrial systems. Rijeka: INTECH; 2011. p. 573–92. Available from: http://www.intechopen.com/books/heat-transfer-theoretical-analysis-experimental-investigations-and-industrial-systems/thermophysical-properties-at-critical-and-supercritical-pressures.
[28] Pioro I, Mokry S. Heat transfer to fluids at supercritical pressures, In: Belmiloudi A, editor. Heat transfer. Theoretical analysis, experimental investigations and industrial systems. Rijeka: INTECH; 2011. p. 481–504. Available from: http://www.intechopen.com/books/heat-transfer-theoretical-analysis-experimental-investigations-and-industrial-systems/heat-transfer-to-supercritical-fluids.
[29] Pioro IL, Duffey RB. Heat transfer and hydraulic resistance at supercritical pressures in power engineering applications. New York, NY: ASME Press; 2007. 334 pp.
[30] Pioro, I. and Kirillov, P., 2013. Current status of electricity generation at nuclear power plants, Materials and processes for energy: communicating current research and technological developments, Energy book series #1, Editor: A. Méndez-Vilas, Publisher: Formatex Research Center, Spain, pp. 806–817. Available from: http://www.formatex.info/energymaterialsbook/book/806-817.pdf.
[31] Nuclear News. Publication of American Nuclear Society (ANS); March 2015. p. 39–72.s.
[32] Nuclear News. Publication of American Nuclear Society (ANS); March 2017. p. 35–67.
[33] Nuclear News. Publication of American Nuclear Society (ANS); March 2011. p. 45–78.

Section B

Reducing CO_2: Fossil Fuels, Nuclear Energy

Current and future nuclear power reactors and plants

4

I. Pioro[*, 1], *R. Duffey*[†]
[*]Faculty of Energy Systems and Nuclear Science, University of Ontario Institute of Technology, Oshawa, ON, Canada, [†]Idaho Falls, Idaho, ID, United States
[1]Corresponding author: igor.pioro@uoit.ca

Chapter Outline

Managing Global Warming. https://doi.org/10.1016/B978-0-12-814104-5.00004-1

Nomenclature

c_p	specific heat at constant pressure (J (kg K)$^{-1}$)
D_{hy}	hydraulic-equivalent diameter (m)
G	mass flux (kg (m^2 s)$^{-1}$)
h_{fg}	latent heat of evaporation (J kg^{-1})
k	thermal conductivity (W (m K)$^{-1}$)
P, p	pressure (Pa)
q	heat flux (W m^{-2})
s	specific entropy (J (kg K)$^{-1}$)
T, t	temperature (°C)
v	specific volume (m^3 kg^{-1})

Greek symbols

Δ	difference
μ	dynamic viscosity (Pa s)
µ	prefix $\times 10^{-6}$ (e.g., µPa s)
ρ	density (kg m^{-3})

Nondimensional numbers

Pr	Prandtl number $\left(\frac{\mu \cdot c_p}{k}\right)$

Subscripts

c	condenser
cr	critical
el	electrical
f	fluid
fg	fluid-gas
fw	feed water
in	inlet
out	outlet
p	at constant pressure
sat	saturation
th	thermal
v	vapor

Abbreviations/acronyms

ABWR	Advanced Boiling Water Reactor
AECL	Atomic Energy of Canada Limited

AGR	Advanced Gas-cooled Reactor
AP	Advanced Plant
APWR	Advanced Pressurized Water Reactor
AR	Advanced Reactor
ave.	average
BN	fast sodium (nuclear reactor) (in Russian abbreviations)
BREST	fast reactor with lead coolant (in Russian abbreviations)
BWR	Boiling Water Reactor
CANDU	CANada Deuterium Uranium (nuclear reactor)
CEFR	Chinese Experimental Fast Reactor
CHF	Critical Heat Flux
DOE	Department of Energy
EC6	Enhanced CANDU 6 (nuclear reactor)
EGP	power heterogeneous loop reactor (in Russian abbreviations)
EPR	European Pressurized-water Reactor (France)
ESBWR	Economic Simplified BWR
Euratom	European Atomic Energy Community (EAEC)
FBR	Fast-Breeder Reactor
FOAK	First-Of-A-Kind
FW	Feed Water
GE	General Electric
gen	generation
GFR	Gas-cooled Fast Reactor
GIF	Generation IV International Forum
GNEP	Global Nuclear Energy Partnership
GT-MHR	Gas-Turbine Modular Helium Reactor
HP or H.P.	High Pressure
HPH	High-Pressure Heat (exchanger)
HPT	High-Pressure Turbine
HTC	Heat-Transfer Coefficient
HTR	High-Temperature Reactor
HTR-PM	High-Temperature Reactor-Pebble-bed Modules
HTSE	High-Temperature Steam Electrolysis
HTTR	High-Temperature Test Reactor
IAEA	International Atomic Energy Agency
ID	Inside Diameter
IHX	Intermediate Heat Exchanger
IPT	Intermediate Pressure Turbine
LBE	Lead-Bismuth Eutectic
LCOE	Levelized Cost Of Energy/Electricity
LEU	Low-Enriched Uranium
LGR	Light-water Graphite-moderated Reactor
LFR	Lead-cooled Fast Reactor
LMFBR	Liquid-Metal Fast-Breeder Reactor
LP or L.P.	Low Pressure
LPH	Low-Pressure Heater
LPT	Low-Pressure Turbine

LWR	Light-Water Reactor
max	maximum
MHI	Mitsubishi Heavy Industries (Japan)
MOX	Mixed OXide (nuclear fuel)
MSK-64	Medvedev-Sponheuer-Karnik (macroseismic intensity scale)
MSFR	Molten-Salt Fast Reactor
MSR	Molten-Salt Reactor or Moisture Separator and Reheater
NEA	Nuclear Energy Agency
NGNP	Next Generation Nuclear Plant
NIST	National Institute of Standards and Technology (United States)
NPP	Nuclear Power Plant
NRC	National Regulatory Commission (United States)
OBE	Operating Basis Earthquake (ground motion)
OD	Outside Diameter
PBMR	Pebble-Bed Modular Reactor
PCh	Pressure Channel (nuclear reactor)
PFBR	Prototype Fast Breeder Reactor (India)
PHWR	Pressurized Heavy-Water Reactor
PT	Pressure Tube (nuclear reactor)
PV	Pressure Vessel
PWR	Pressurized Water Reactor
RBMK	reactor of large capacity channel type (in Russian abbreviations)
REFPROP	Reference Properties (NIST software)
R&D	Research & Development
RPV	Reactor Pressure Vessel
rpm	revolutions per minute
SC	SuperCritical
SCW	SuperCritical Water
SCWR	SuperCritical Water-cooled Reactor
SFR	Sodium-cooled Fast Reactor
SKD	Supercritical pressure (in Russian abbreviations)
SMR	Small Modular Reactor
SSE	Safe Shutdown Earthquake (ground motion)
SVBR	lead bismuth fast reactor (in Russian abbreviations)
TRISO	tri-isotropic
TWR	Traveling Wave Reactor
VHTR	very high-temperature reactor
VVER	water water power reactor (in Russian abbreviations)
X or x	extraction (steam)

4.1 Introduction

The first successful use of nuclear power for electrical generation was achieved in several countries in the 1950s, and currently, Generation II and III nuclear power reactors are operating around the world (see Chapter 3). In general, definitions of nuclear reactor generations are as the following: (1) Generation I (1950–65)—early prototypes of nuclear reactors; (2) Generation II (1965–95)—commercial power reactors; (3) Generation III (1995–2010)—modern reactors and Generation III+ (2010–25)—reactors with improved parameters (evolutionary design improvements); and (4) Generation IV (2025)—future reactors with new parameters and enhanced safety features.

Furthermore, nuclear reactors can be classified, based on various parameters [1]:

1. Neutron spectrum: (a) thermal (the vast majority of current nuclear power reactors), (b) fast (currently, only two nuclear power fast-neutron-spectrum reactors are in operation in Russia: BN-600 and BN-800), and (c) interim or mixed spectrum.
2. Reactor-core design:
   - **I.** Neutron-core design: (a) homogeneous, i.e., the fuel and reactor coolant are mixed together (one of the Generation IV nuclear reactor concepts) and (b) heterogeneous, i.e., the fuel and reactor coolant are separated through a sheath or cladding (currently, all nuclear power reactors); and
   - **II.** General core design: (a) Pressure-vessel (PV) (the vast majority of current nuclear power reactors) and (b) Pressure-channel (PCh) or Pressure-tube (PT) reactors;
3. Coolant:
   - **I.** Light-water-cooled reactors;
   - **II.** Heavy-water-cooled reactors;
   - **III.** Gas-cooled reactors;
   - **IV.** Liquid-metal-cooled reactors;
   - **V.** Molten-salt-cooled reactors; and
   - **VI.** Organic-fluids-cooled reactors (existed only as experimental reactors some time ago).
4. Type of a moderator: (a) liquid moderator (H_2O (light water) and D_2O (heavy water) and (b) solid moderator (graphite, zirconium hydride (ZrH_2), beryllium (Be) and beryllium oxide (BeO)).
5. Application: (a) power reactors; (b) research reactors; (c) transport or mobile reactors (submarines and ships (icebreakers, aircraft carriers, etc.); (d) reactors for space applications; (e) industrial reactors for isotope production, etc.; and (f) multipurpose reactors.
6. Number of flow circuits: (a) single-flow circuit (once-through or direct-cycle); (b) double-flow circuit; and (c) triple-flow circuit.
7. Fuel enrichment: (a) Natural-uranium fuel (NU) (99.3$\%_{wt}$ of nonfissile isotope uranium-238 (U^{238}) and 0.7% of fissile isotope uranium-235 (U^{235})); (b) slightly enriched uranium (SEU) (0.8–2$\%_{wt}$ of U^{235}); (c) low-enriched uranium (LEU) (2–20% of U^{235}) (the vast majority of current nuclear power reactors); and (d) highly enriched uranium (HEU) (>20$\%_{wt}$ of U^{235}).
8. Fuel used: (a) Conventional nuclear fuels (low thermal conductivity): Uranium dioxide (UO_2, used in the vast majority of nuclear power reactors), mixed oxides (MOX) ($(U_{0.8}Pu_{0.2})O_2$, where 0.8 and 0.2 are the molar parts of UO_2 and PuO_2, used in some reactors); and thoria (ThO_2) (considered for a possible use instead of UO_2 in some countries, usually, with large resources of this type of fuel, for example, in India); and (b) alternative nuclear fuels (high thermal conductivity): Uranium dioxide plus silicon carbide (UO_2-SiC), uranium dioxide composed of graphite fiber (UO_2-C), uranium dioxide plus beryllium oxide (UO_2-BeO), uranium dicarbide (UC_2), uranium monocarbide (UC), and uranium mononitride (UN); the last three fuels are mainly intended for use in high-temperature Generation IV reactors.

4.2 Current nuclear power reactors and NPPs

This section is based on Appendix A1 from the *Handbook of Generation IV Nuclear Reactors* [1]; papers by Pioro and Duffey [2] and Dragunov et al. [3]; and chapters by Pioro and Kirillov [4–6].

This section is dedicated to modern nuclear power reactors and corresponding to that Nuclear Power Plants (NPPs), and includes typical operating conditions of selected water-cooled reactors (Fig. 4.1), NPPs layouts (Figs. 4.2–4.6, 4.8–4.10, 4.12–4.14, 4.16, 4.18–4.20, 4.22, 4.23, 4.25, and 4.26), *T-s* diagrams (Figs. 4.4, 4.11, 4.17, 4.21, 4.24, and 4.27), fuel assemblies (Fig. 4.7), fuel channel (Fig. 4.15), fuel element (Fig. 4.19B); and basic parameters of nuclear power reactors and NPPs (Tables 4.1–4.9). Nuclear power reactors and corresponding NPPs are listed in the following sequence: (1) Pressurized water reactors (PWRs)/ advanced PWRs (APWRs) (Figs. 4.2–4.7 and Tables 4.1–4.6); (2) boiling water reactors (BWRs)/ABWRs (Figs. 4.8–4.11 and Tables 4.7 and 4.8); (3) pressurized heavy-water reactors (PHWRs) (Figs. 4.12–4.18); (4) Advanced Gas-cooled Reactors (AGRs) (Figs. 4.19–4.21); (5) light-water graphite-moderated reactors (LGRs) (RBMKs (Figs. 4.22–4.24 and Table 4.1) and EGPs), and (6) liquid-metal fast-breeder reactors (LMFBRs) (Russian sodium fast reactors (SFRs)—BN-600

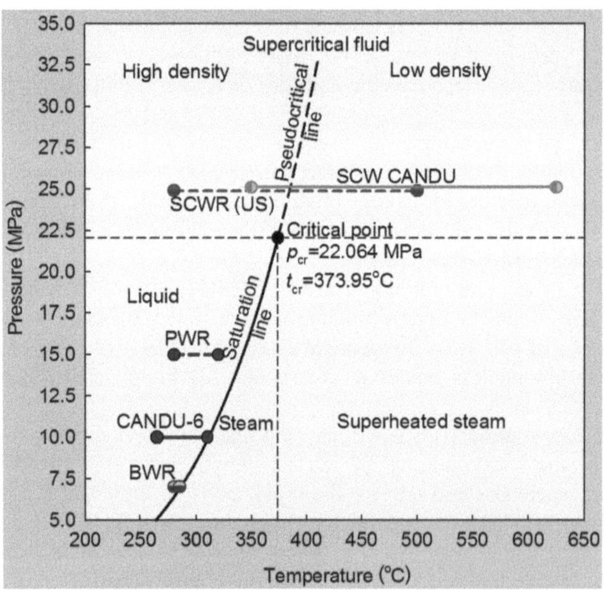

Fig. 4.1 Typical operating conditions (pressure drop is not accounted for) in *P-T* coordinates for BWRs, CANDU reactors, PWRs, and supercritical water-cooled reactors (SCWRs) (Generation IV concept) [1].

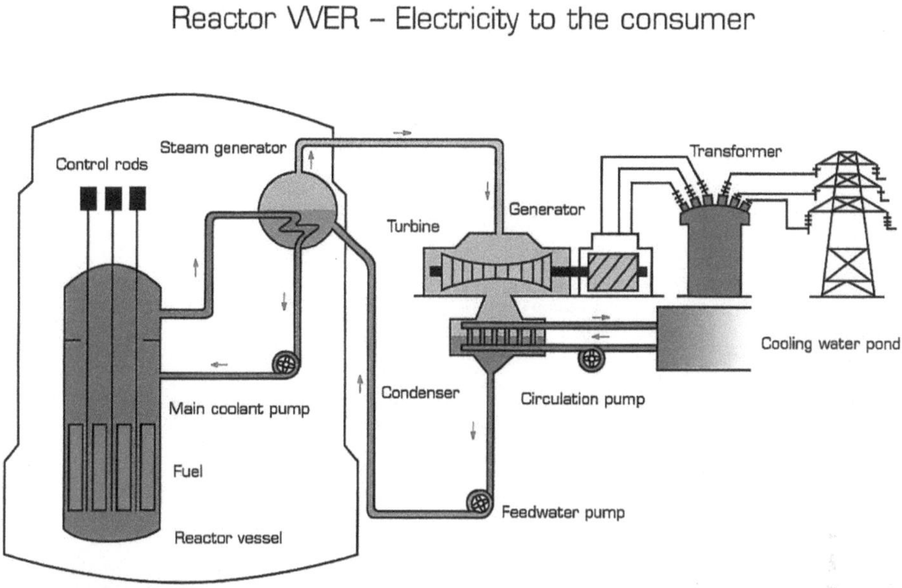

Fig. 4.2 Simplified scheme of typical pressurized water reactor (PWR) (Russian VVER-1000) NPP (ROSENERGOATOM, 2004) [1]: General basic features—(1) thermal neutron spectrum; (2) uranium dioxide (UO_2) fuel; (3) fuel enrichment about 4%; (4) indirect cycle with steam generator(s) (also, a pressurizer is required (not shown)), i.e., double-flow circuit (double loop); (5) reactor pressure vessel (RPV) with vertical fuel rods (elements) assembled in bundle strings cooled with upward flow of light water; (6) reactor coolant and moderator are the same fluid; (7) reactor-coolant outlet parameters: Pressure 15–16 MPa (T_{sat} = 342–347°C) and temperatures inlet/outlet 290–325°C; and (8) power cycle—subcritical-pressure regenerative Rankine steam-turbine cycle with single steam reheat (working fluid – light water, turbine steam-inlet parameters: Saturation pressure of 6–7 MPa and saturation temperature of 276–286°C).
Courtesy of ROSENERGOATOM. Russian nuclear power plants. 50 Years of nuclear power, Moscow, Russia; 2004. 120 pp.

(Figs. 4.25–4.27 and Tables 4.1 and 4.9)) and BN-800 (Table 4.9), i.e., the sequence is based on the decreasing number of particular types of reactors currently operated in the world (see Table 3.11 in Chapter 3). All of the power-conversion cycles for current NPPs are based solely on the subcritical-pressure regenerative and single-steam-reheat Rankine cycle (for thermal efficiencies of current NPPs, see Table 3.14 in Chapter 3).

Table 4.1 lists the parameters of Russian power reactors and NPPs, because Russia has a wide range of operating nuclear power reactors.

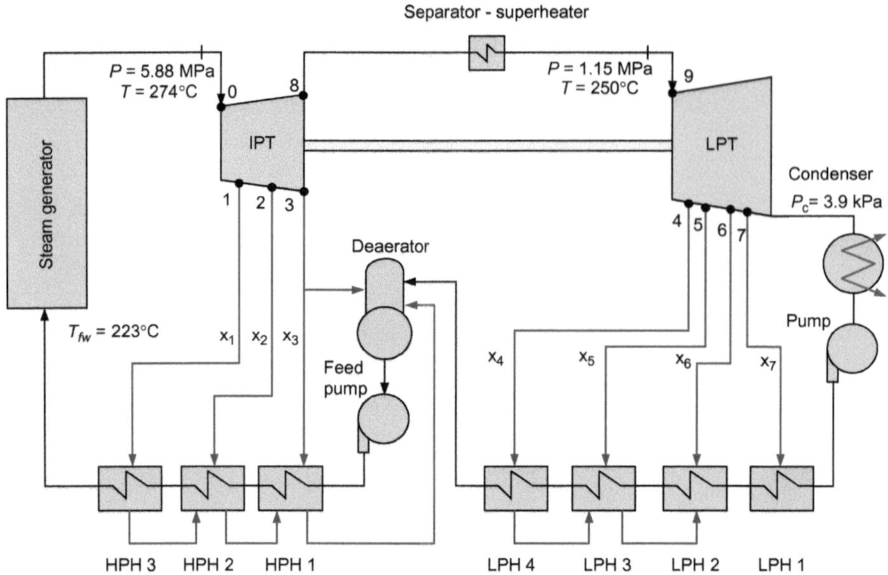

Fig. 4.3 Simplified thermodynamic layout of 1000-MW_{el} VVER-1000 PWR NPP [1]. Based on data from Grigor'ev VA, Zorin VM, editors. Thermal and nuclear power plants. Handbook, (in Russian). 2nd ed. Moscow: Energoatomizdat Publishing House; 1988. 625 pp; Margulova TC. Nuclear power plants, (in Russian). Moscow: Izdat Publishing House; 1955. 289 pp.

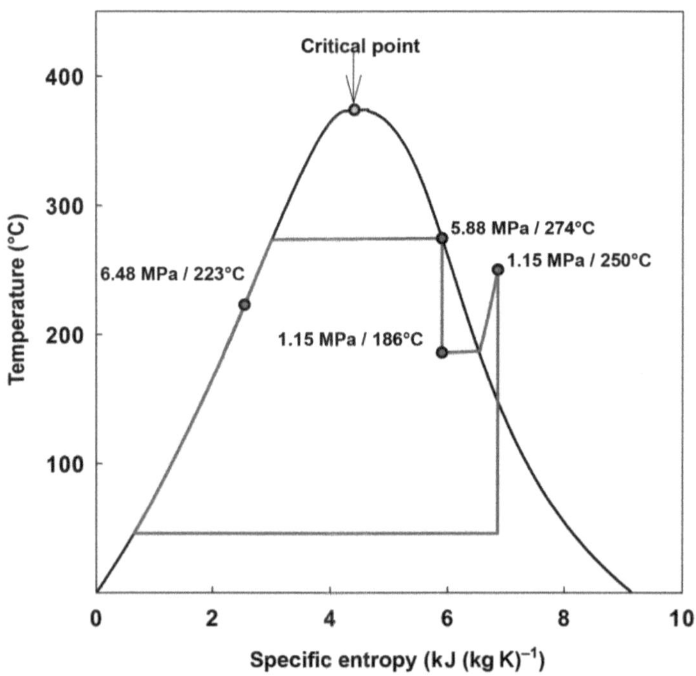

Fig. 4.4 Temperature-specific entropy diagram for VVER-1000 Rankine steam-turbine cycle [1].

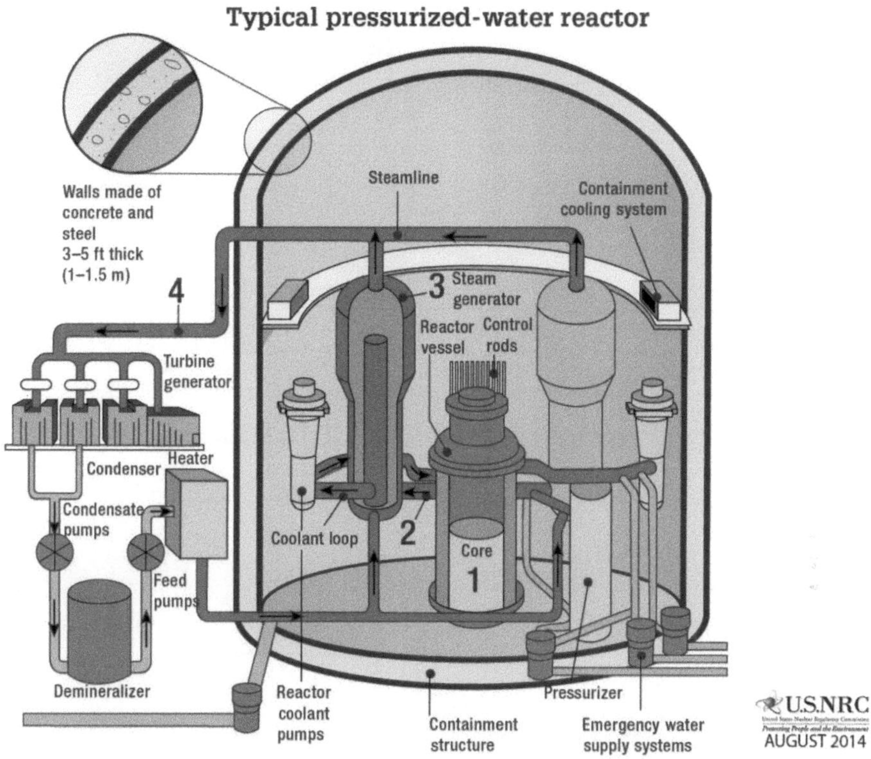

Fig. 4.5 Simplified layout of typical US pressurized water reactor (PWR) NPP [1]. Courtesy of US NRC.

4.2.1 Pressurized water reactors

Accounting on the available information in the open literature and figures provided by major nuclear vendors (ROSATOM/ROSENERGOATOM, MHI, and AREVA) and US NRC, it was decided to show the following reactors and NPPs:

(1) 1000-MW_{el} VVER-1000-reactor NPP (Russian design) (see Figs. 4.2–4.4). Basic parameters of VVER-1000 are listed in Table 4.1. Reference parameters of Generation III+ VVER are listed in Table 4.2 and additional parameters—in Table 4.3;
(2) Typical US PWR NPP (see Fig. 4.5). Basic parameters of US PWR NPP and AP-1000 NPP (Generation III+) are listed in Tables 4.4 and 4.5, respectively;
(3) MHI Advanced PWR NPP layout and typical PWR fuel assembly (see Figs. 4.6 and 4.7, respectively); and
(4) Main design and operating parameters for EPR (AREVA) (see Table 4.6).

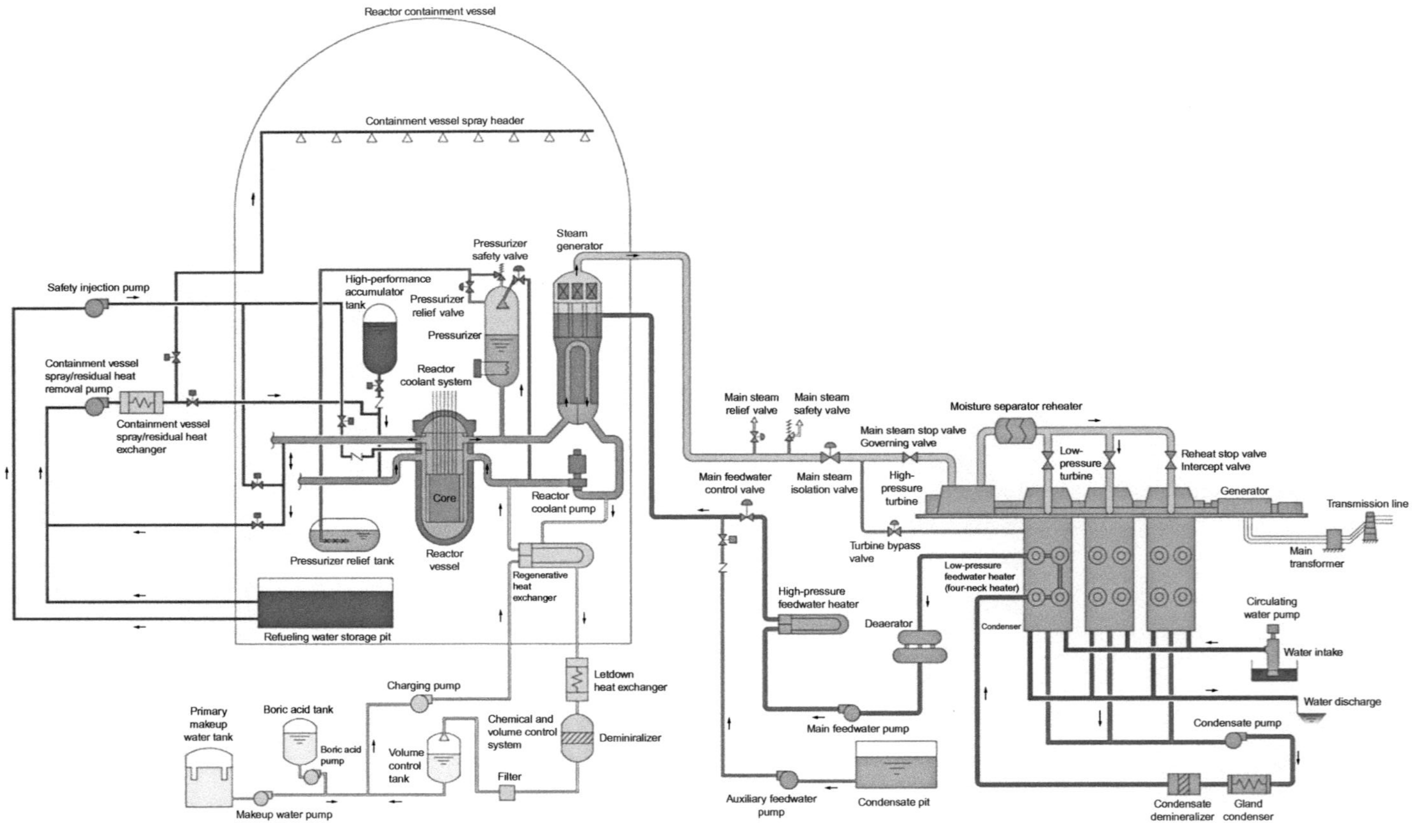

Fig. 4.6 Mitsubishi Heavy Industries (MHI) 1500-MW$_{el}$ advanced PWR NPP (Generation III+) simplified layout (two-stage reheat cycle) [1]. Courtesy and copyright of MHI.

(A) (B)

Fig. 4.7 Typical PWR fuel assembly [1]: A—Schematic of fuel assembly and B—photo. Courtesy and copyright by MHI.

It should be noted that in all NPPs with PWRs, ABWRs, BWRs, PHWRs, and LGRs subcritical-pressure regenerative and single-steam-reheat Rankine cycle with primary saturated steam is used. For a reheat of the secondary steam at lower pressure, the primary saturated steam is used. Therefore, the reheat temperature is lower than the primary steam temperature by 25–30°C. In general, the primary steam and secondary steam parameters at NPPs are significantly lower than those at thermal power plants. Due to this, thermal efficiencies of these NPPs equipped with water-cooled reactors (Table 3.14, Chapter 3) are lower than those of NPPs equipped with AGRs and LMFBRs (SFRs) (Table 3.14, Chapter 3), and significantly lower than those of modern advanced combine-cycle and supercritical-pressure thermal power plants (Table 3.8, Chapter 3).

Table 4.6 lists main design and operating parameters for EPR—French PWR of Generation III+ (AREVA company) [12].

Typical boiling-water reactor

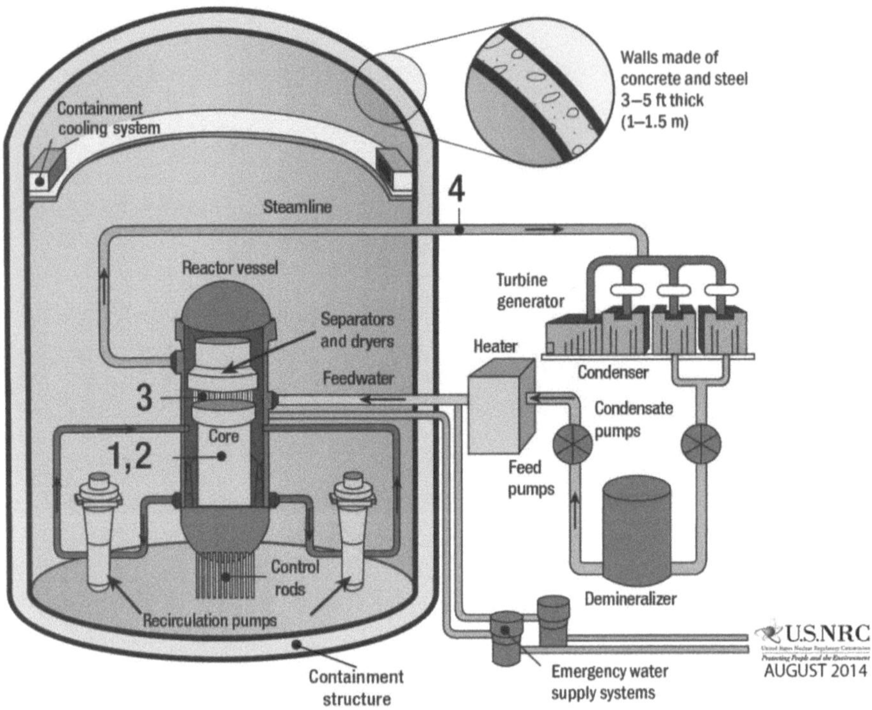

Fig. 4.8 Simplified layout of typical Boiling Water Reactor (BWR) NPP [1]: General basic features—(1) thermal neutron spectrum; (2) uranium dioxide (UO_2) fuel; (3) fuel enrichment about 3%; (4) direct cycle with steam separator (steam generator and pressurizer are eliminated), i.e., single-flow circuit (single loop); (5) RPV with vertical fuel rods (elements) assembled in bundle strings cooled with upward flow of light water (water and water-steam mixture); 6) reactor coolant, moderator, and power-cycle working fluid are the same fluid; (7) reactor coolant outlet parameters: Pressure about 7 MPa and saturation temperature at this pressure is about 286°C; and (8) power cycle—subcritical-pressure regenerative Rankine steam-turbine cycle with single steam reheat.
Courtesy of US NRC.

4.2.2 Boiling water reactors

From the information available in the open literature and figures provided by major nuclear vendors (Hitachi and Toshiba) and US NRC, the following reactors and NPPs are typical representatives of Generations III BWR and III+ ABWR NPPs:

(1) Typical US BWR NPP layout (see Fig. 4.8) and its parameters (see Table 4.7);
(2) Layout of ABWR Hitachi, Ltd. (see Fig. 4.9) and comparison of ABWR and BWR basic parameters (see Table 4.8); and
(3) ABWR NPP (see Figs. 4.10 and 4.11) (based on data from Toshiba company).

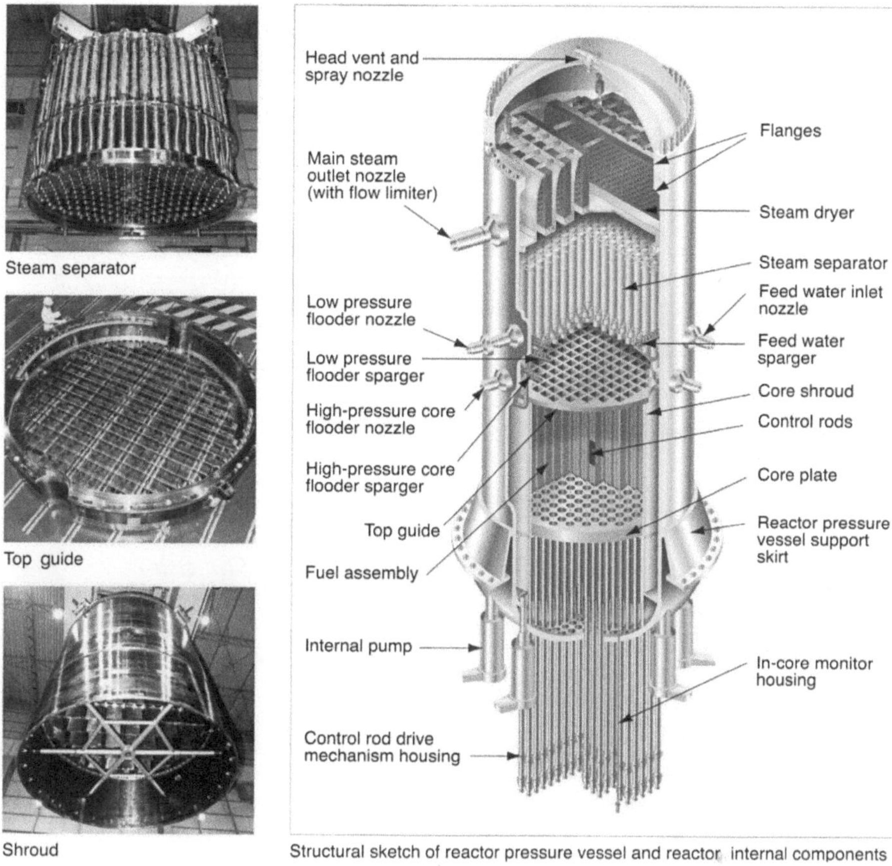

Fig. 4.9 Simplified layout of typical advanced boiling water reactor (ABWR) NPP [1]. Courtesy and copyright by Hitachi, Ltd.

Toshiba's ABWR has been in operation since 1996. In 1996, Unit 6 at the Kashiwazaki-Kariwa NPP was put into operation as the world's first ABWR plant, i.e., the first Generation III+ reactor and NPP. In 1997, Unit 7 began operation at the same plant. In 2005, another ABWR (Unit 5) was commissioned at the Hamaoka NPP. (As the result of the Fukushima disaster in Japan in March of 2011, all 42 Japanese reactors were shut down. Recently, in April of 2018, seven reactors have been restarted.)

An important feature in the ABWR is that the condenser plays a role of a deaerator. The system consists of the main deaerating condenser connected with air ejector and off-gas treatment system.

The corresponding layout and *T*-*s* diagram of a typical ABWR NPP are shown in Figs. 4.10 and 4.11, respectively.

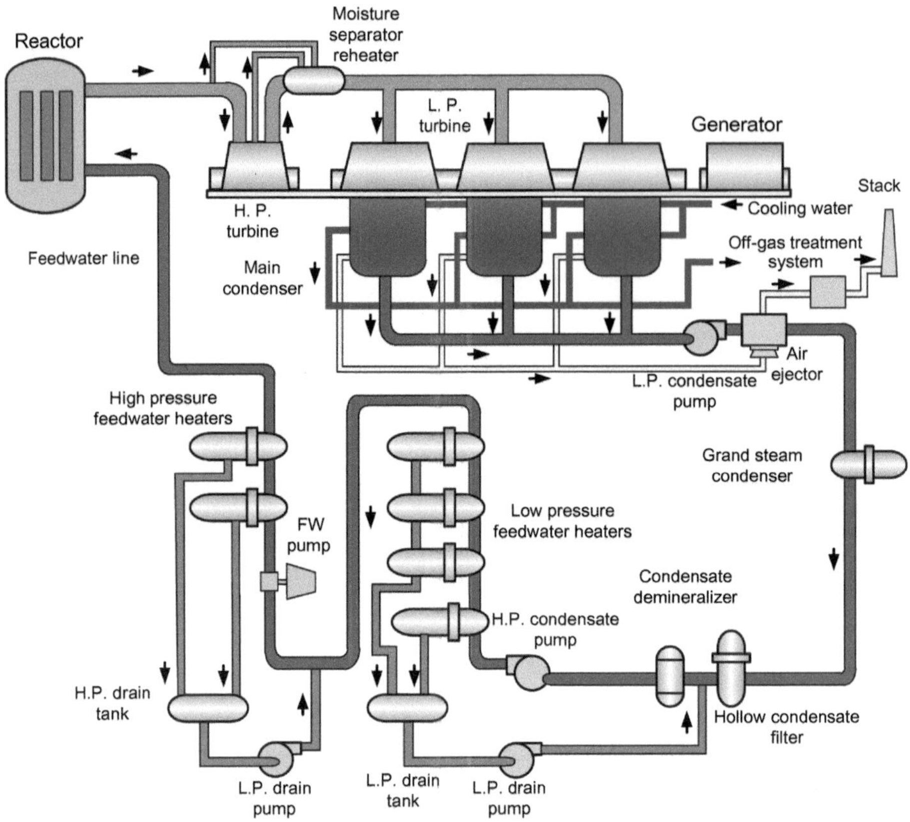

Fig. 4.10 Simplified layout of ABWR NPP [1].
Based on data from Toshiba. Leading innovation. Advanced Boiling Water Reactor (ABWR), Booklet, Toshiba Corporation, Japan; 2011.

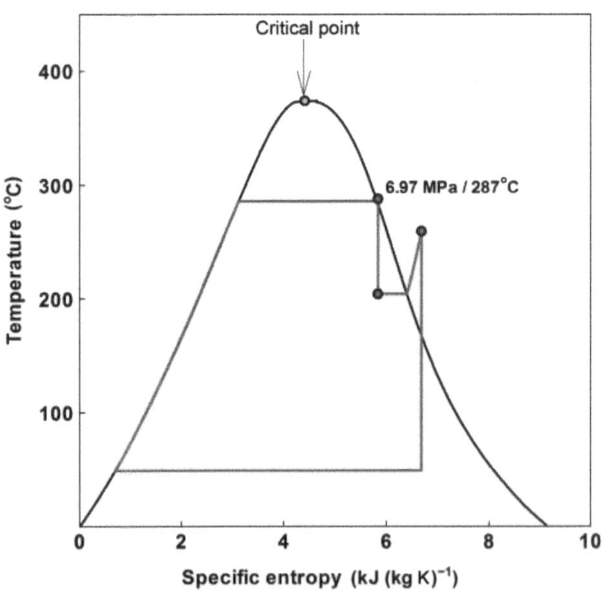

Fig. 4.11 *T-s* diagram for typical ABWR NPP Rankine steam-turbine cycle (reheat pressure was assumed to be ¼ of the main steam pressure) [1].

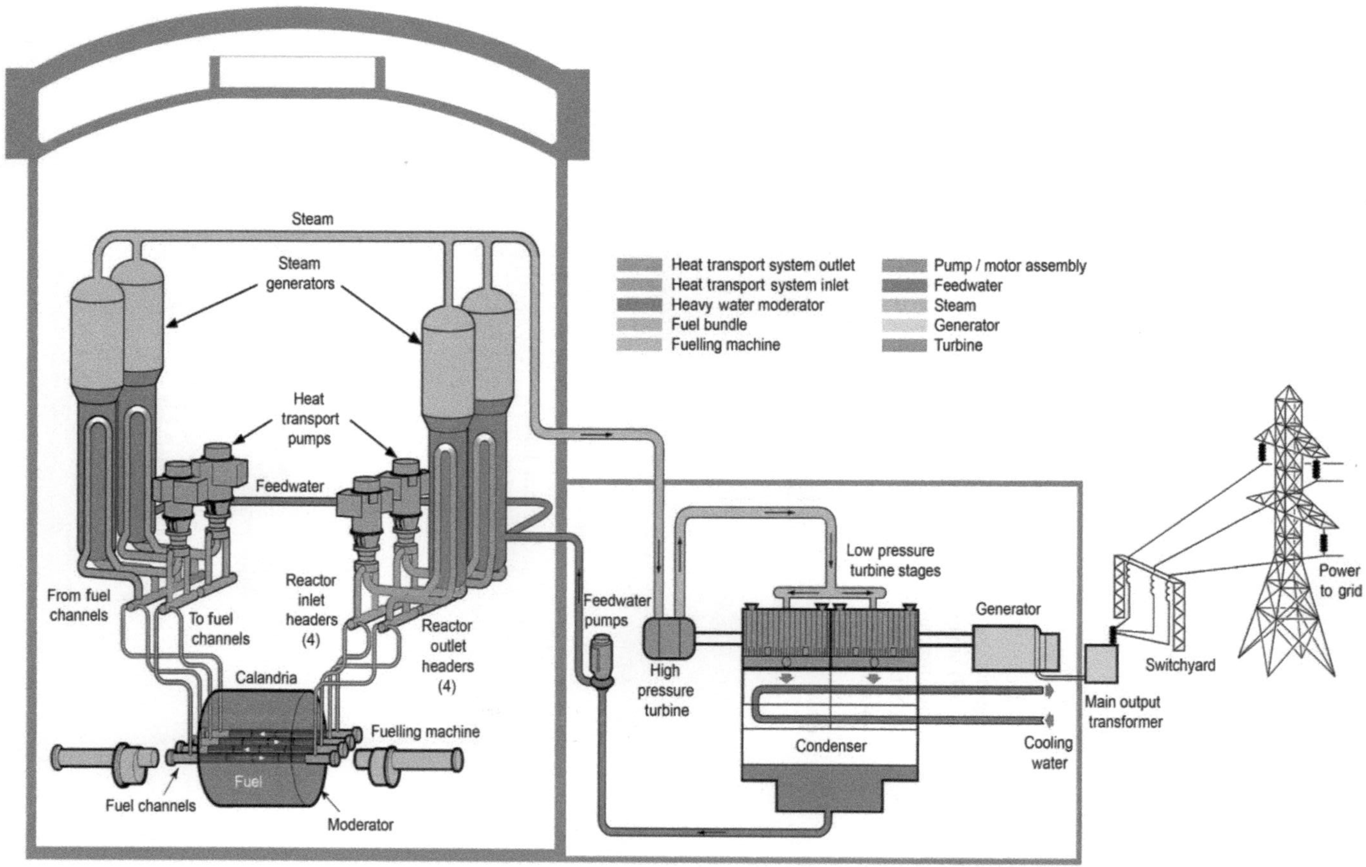

Fig. 4.12 Simplified CANDU NPP flow diagram [1]: Moisture separator and reheater (MSR) between HP and LP turbines are required, but not shown. Courtesy of SNC-Lavalin.

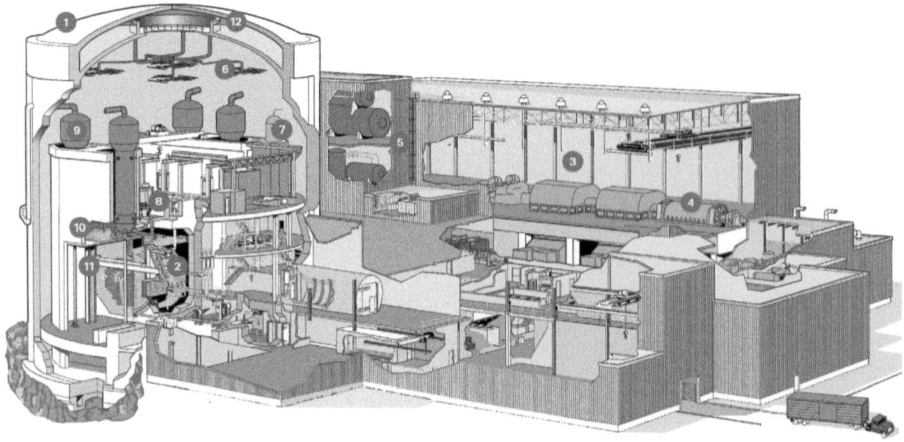

Fig. 4.13 Layout of EC6 NPP [1]: (1) Reactor building; (2) calandria vessel; (3) turbine building; (4) turbine generator; (5) service building; (6) spray system; (7) pressurizer; (8) heat-transport pumps; (9) steam generators; (10) heat-transport system; (11) fuelling machine for online refueling (one machine on one side of calandria vessel and another on another side); and (12) Reserve water tank.
Courtesy of SNC-Lavalin.

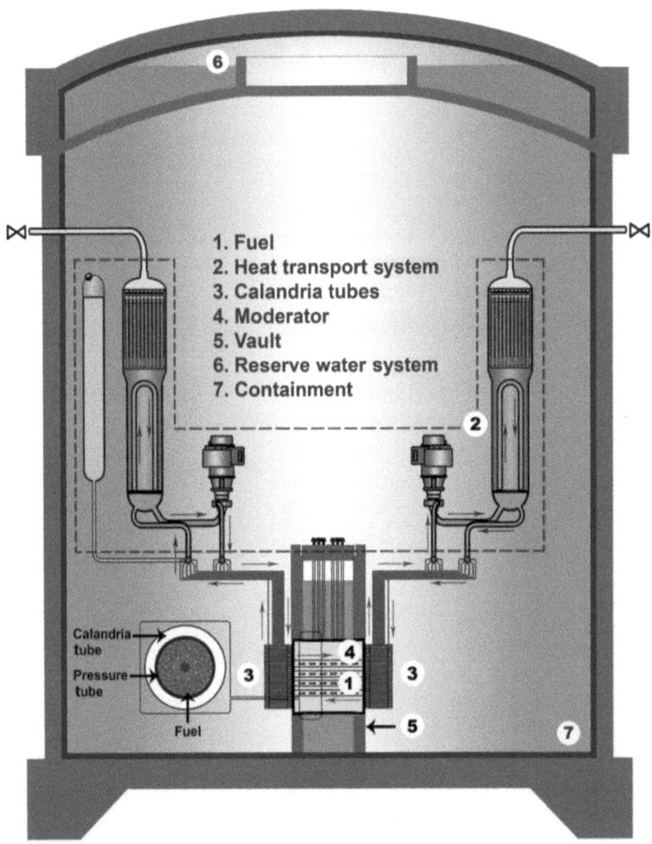

Fig. 4.14 Barriers for prevention of releases [1].
Courtesy of SNC-Lavalin.

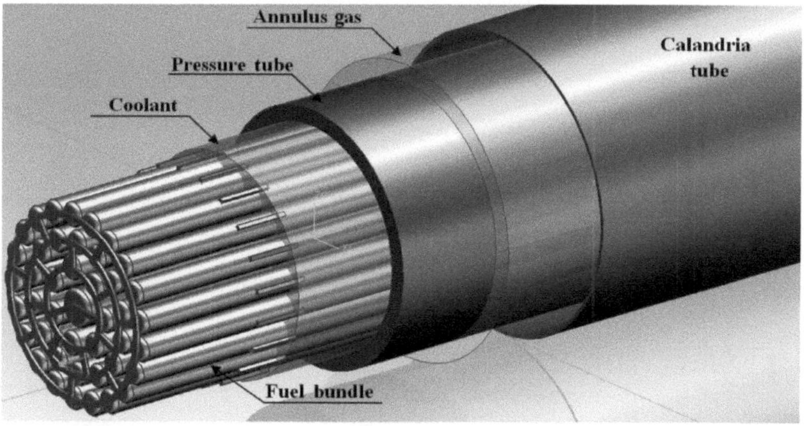

Fig. 4.15 3-D image of CANDU reactor fuel channel with fuel bundle [1].
Based on data from AECL.

4.2.3 Pressurized heavy-water reactors

According to the information in Nuclear News (March 2017) [13], the vast majority of PHWRs in the world are CANDU-type reactors designed by the Atomic Energy of Canada Limited (AECL). Currently, the CANDU reactor technology in Canada is being developed by the Candu Energy Inc. (http://www.candu.com/en/home/aboutcandu/default.aspx), a member of the SNC-Lavalin Group (Mississauga, Ontario, Canada). All information on CANDU reactors, including the figures (published with their permission) presented here, is based on brochures obtained from the Candu Energy Inc. For the latest data and developments in the CANDU reactor technologies, refer to the Candu Energy Inc. website (www.candu.com).

The brief description of PHWRs provided here is based on the Enhanced CANDU 6 (EC6) reactor (for details, see Figs. 4.12–4.15) and a brief description is provided later. The EC6 reactor is an evolutionary design, based on the operating CANDU 6 reactors. The CANDU 6 reactors were designed and constructed after the successful operation of the four Pickering NPP reactors (see Figs. 4.16 and 4.17).

The EC6 reactor is a 700-MW_{el}-class heavy-water-moderated and heavy-water-cooled pressure-tube (pressure-channel) reactor (see Figs. 4.12–4.15). The basic features and parameters of this reactor and corresponding NPP are as the following: (1) thermal neutron spectrum; (2) natural uranium dioxide (UO_2) fuel; (3) fuel enrichment about 0.71 wt.% (U^{235}), i.e., natural uranium; (4) indirect cycle with steam generator, i.e., double-flow circuit (double loop); (5) pressure-channel design, i.e., calandria vessel with horizontal fuel channels; (6) reactor coolant and moderator are separated, but both are heavy water; (7) reactor-coolant parameters: (a) inlet-header operating pressure 11.05 MPa and temperature 265°C; (b) outlet-header operating pressure 9.89 MPa and temperature 310°C (close to saturation temperature);

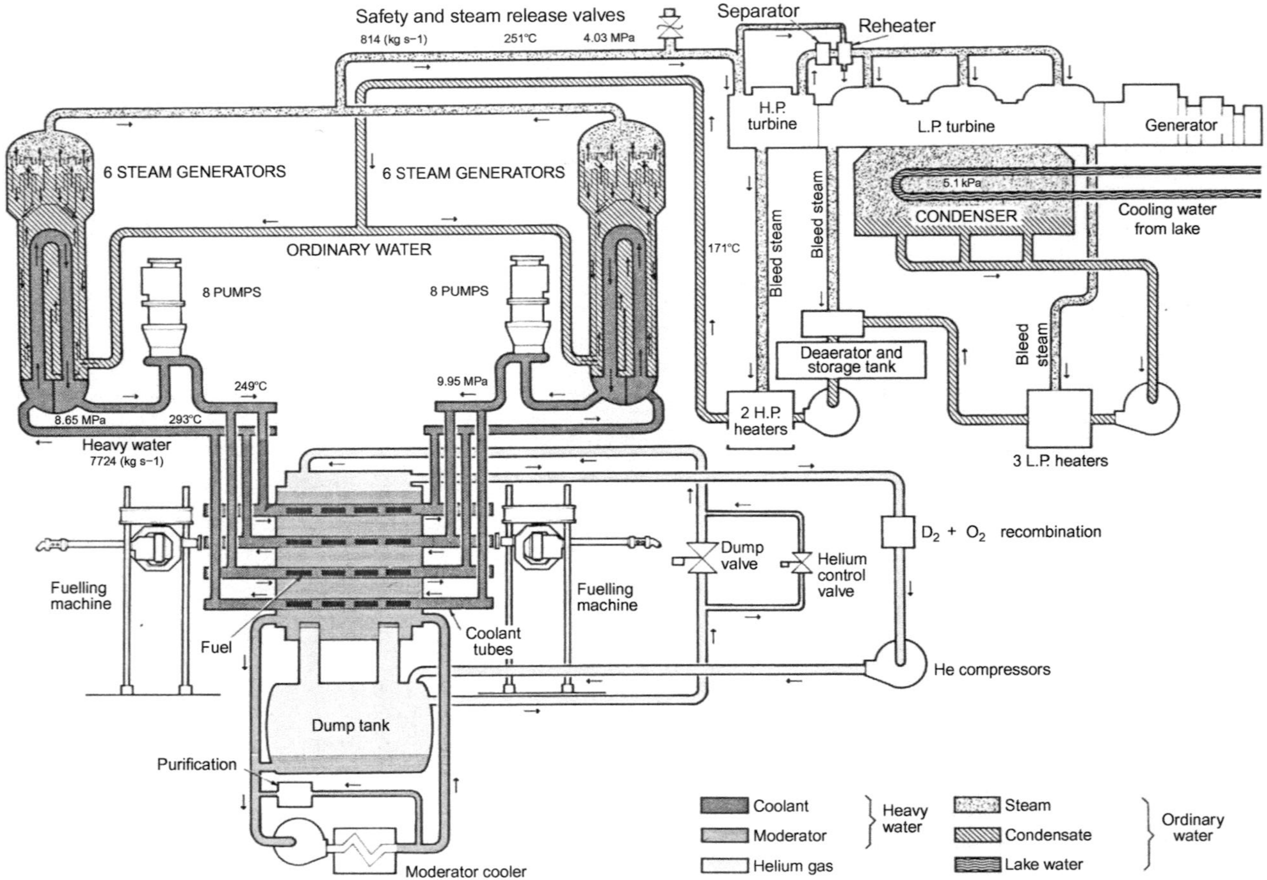

Fig. 4.16 Simplified flow diagram of 515-MW_{el} CANDU reactor NPP (Pickering NPP, Ontario, Canada) (AECL Report [9]) [1]. These 515-MW_{el} CANDU reactors are the smallest ones in Canada, and first two of them were put into operation in 1971. Also, a pressurizer is required, but not shown. (It should be noted that EC6 reactor has higher power-cycle working-fluid parameters.)

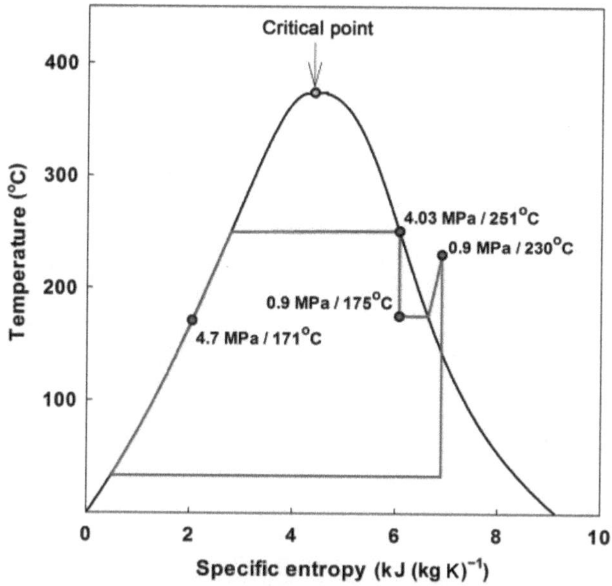

Fig. 4.17 *T-s* diagram of 515-MW$_{el}$ CANDU reactor Pickering NPP (Ontario, Canada) Rankine steam-turbine cycle [1]. (It should be noted that EC6 reactor has higher power-cycle working-fluid parameters.)

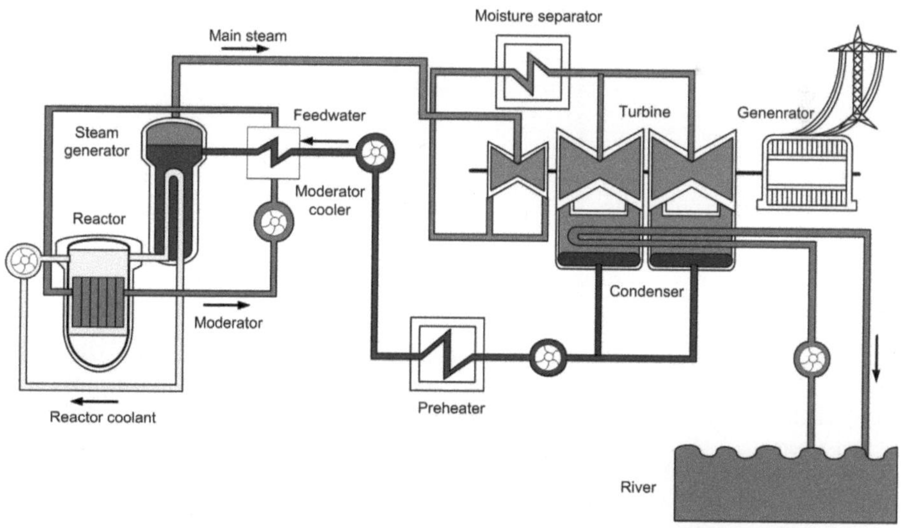

Fig. 4.18 Simplified PHWR NPP flow diagram (Siemens design) (Atucha, Argentina) [1]. Currently, both Unit 1 (335 MW$_{el}$) and Unit 2 (692 MW$_{el}$) are in operation.
Based on Nuclear Engineering International, September 1982, England.

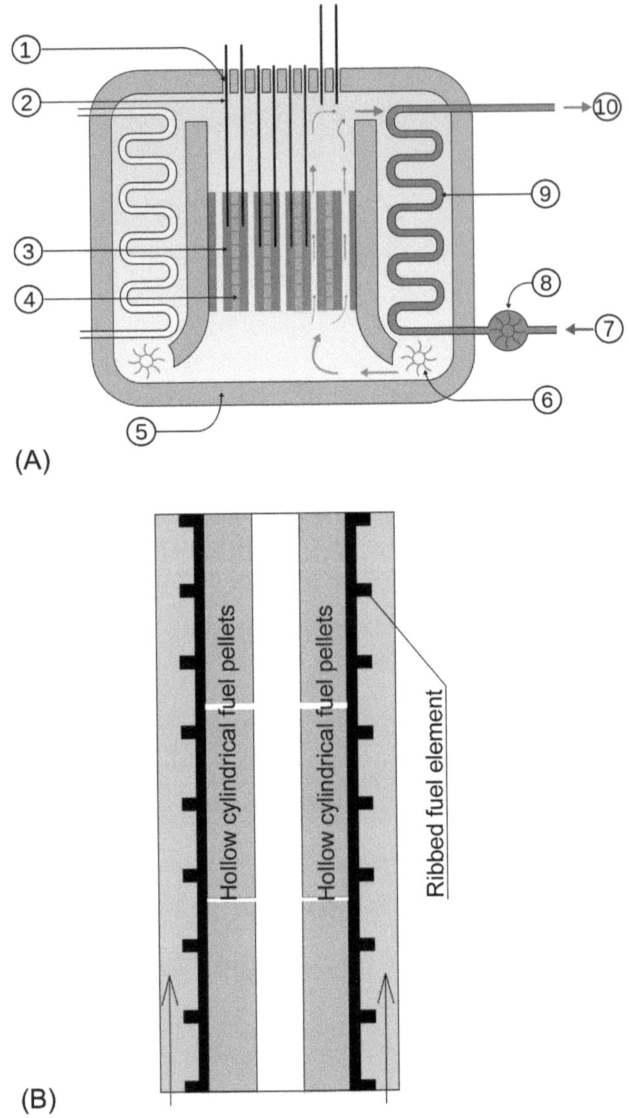

Fig. 4.19 (A) Simplified schematic diagram of Advanced Gas-cooled Reactors (AGRs) (Author: M. Woland, https://commons.wikimedia.org/wiki/File:AGR_reactor_schematic.svg; Accessed 28 January 2016) [1]: (1) Charge tubes; (2) control rods; (3) graphite moderator; (4) fuel assemblies; (5) concrete pressure vessel and radiation shielding; (6) gas circulator; (7) water; (8) water circulator; (9) heat exchanger; and (10) steam. Heat exchanger is contained within steel-reinforced concrete combined pressure vessel and radiation shield; and (B) AGR ribbed fuel element with hollow fuel pellet (Hewitt and Collier [10]) [1].

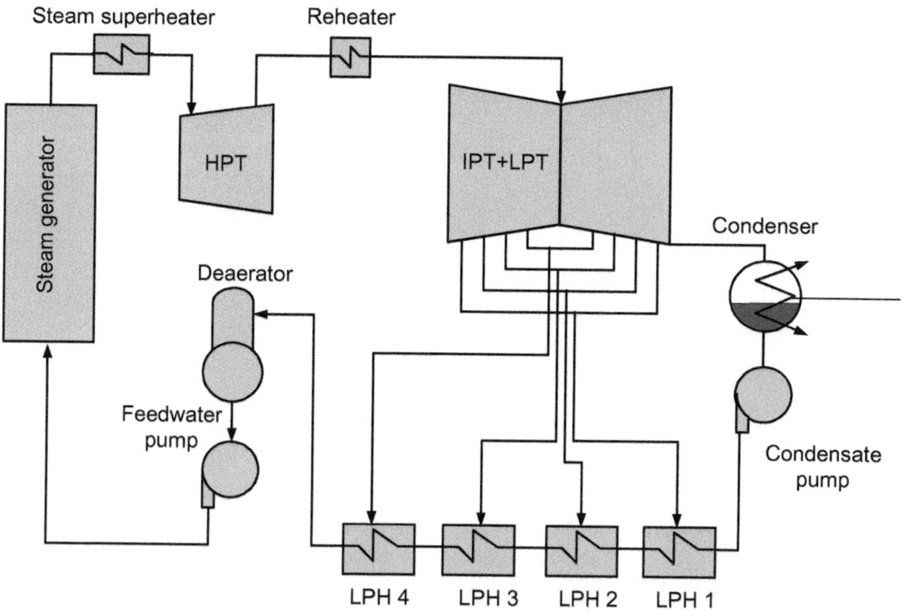

Fig. 4.20 Thermodynamic layout of AGR Torness NPP [1].
Based on data from Nonbel E. Description of the advanced gas cooled type of reactor (AGR), Report NKS/RAK2(96)TR-C2, November, Roskilde: RisØ National Laboratory; 1996. 88 pp.

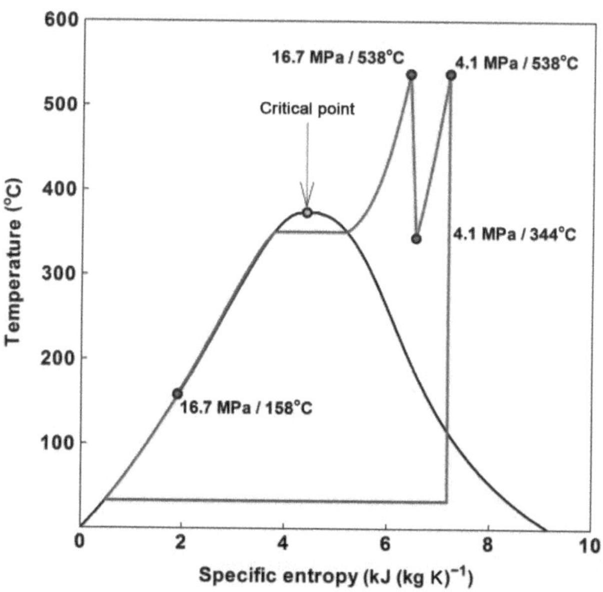

Fig. 4.21 *T-s* diagram for AGR Torness NPP Rankine steam-turbine cycle [1].

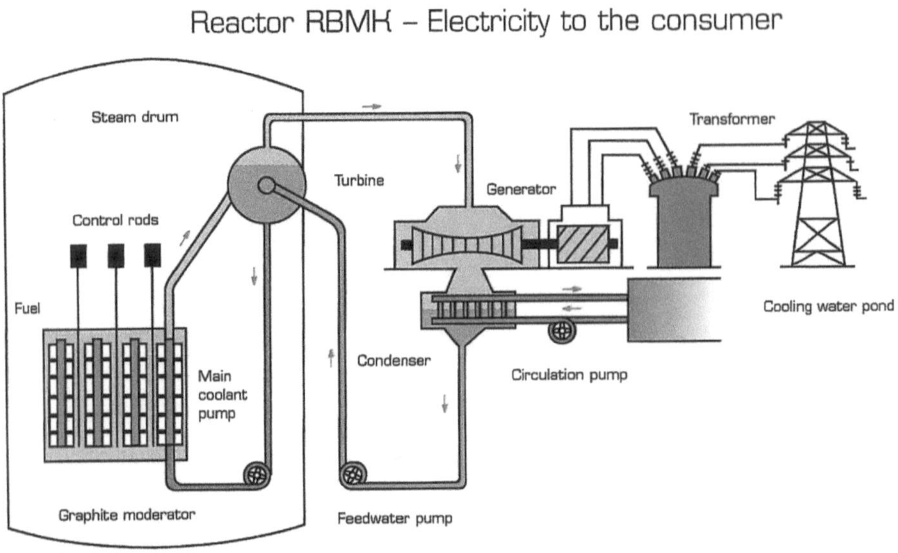

Fig. 4.22 Simplified schematic of 1000-MW_{el} Light-water-cooled Graphite-moderated Reactor (LGR) (Russian RBMK) NPP (ROSENERGOATOM, 2004) [1].
Courtesy of ROSENERGOATOM. Russian nuclear power plants. 50 Years of nuclear power, Moscow, Russia; 2004. 120 pp.

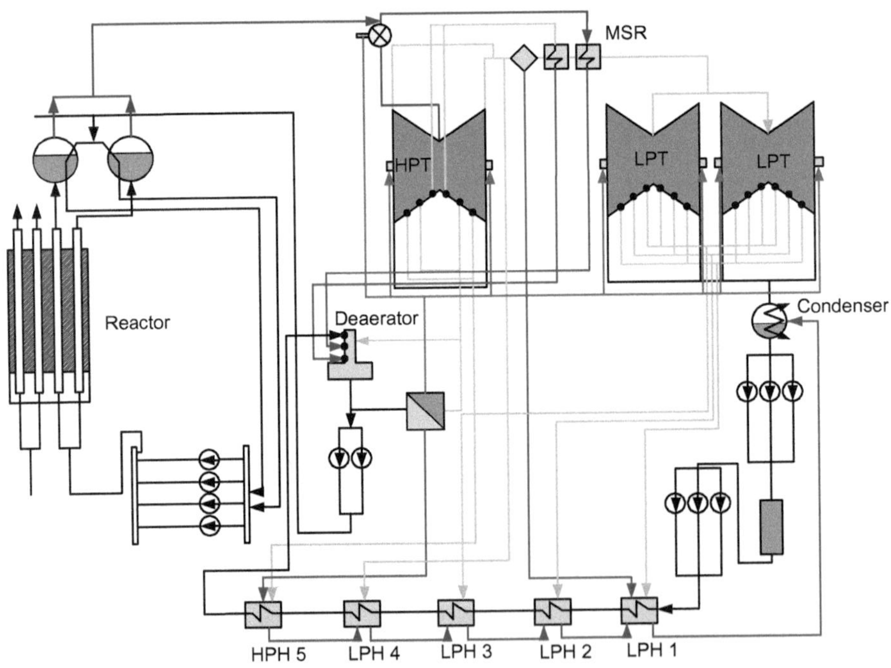

Fig. 4.23 Thermodynamic layout of RBMK-1000 NPP [1].
Adapted and simplified from Channel Nuclear Power Reactor (RBMK). Канальный Ядерный Энергетический Реактор РБМК (in Russian). In: Cherkashov YM, editor. Moscow: Publishing House GUP NIKIET; 2006. 631 pp.

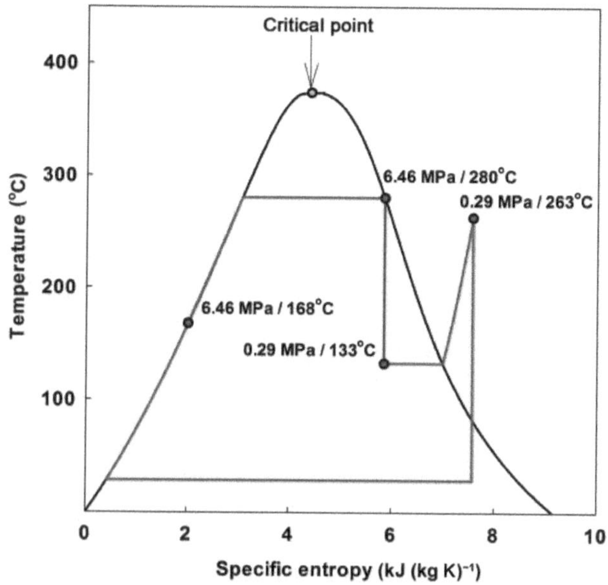

Fig. 4.24 *T-s* diagram for RBMK-1000 NPP Rankine steam-turbine cycle [1].

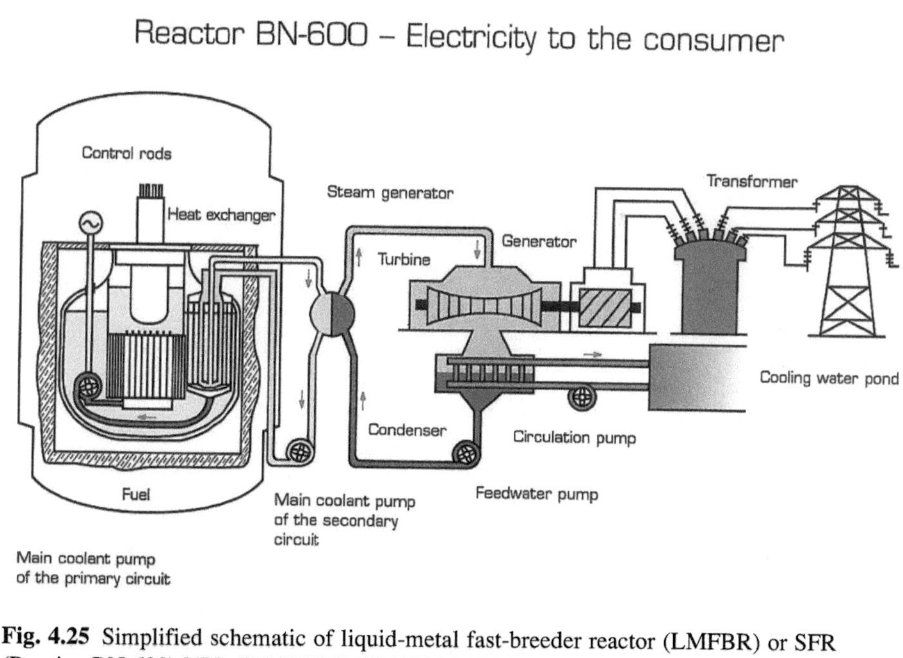

Fig. 4.25 Simplified schematic of liquid-metal fast-breeder reactor (LMFBR) or SFR (Russian BN-600) NPP (ROSENERGOATOM 2004 [11]) [1].
Courtesy of ROSENERGOATOM. Russian nuclear power plants. 50 Years of nuclear power, Moscow, Russia; 2004. 120 pp.

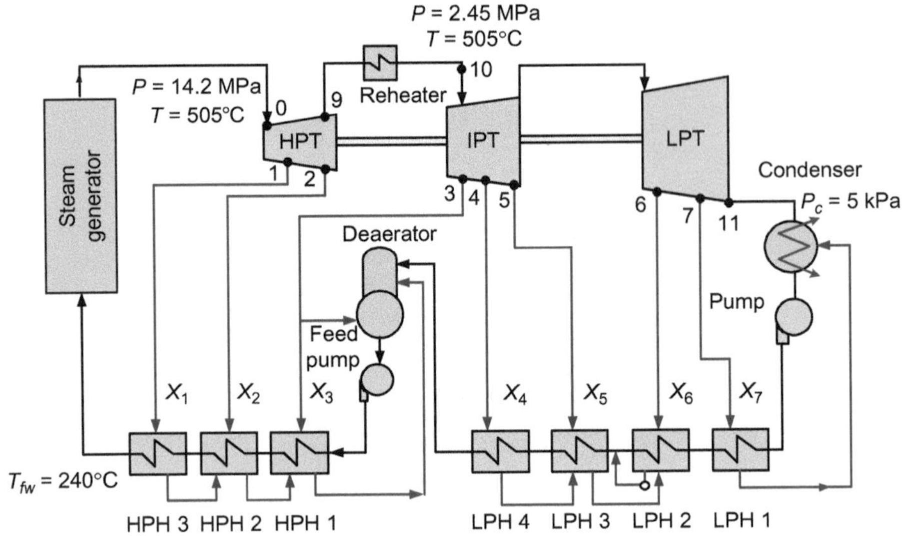

Fig. 4.26 Thermodynamic layout of 600-MW$_{el}$ BN-600 SFR NPP [1].
Based on data from Grigor'ev VA, Zorin VM, editors. Thermal and nuclear power plants. Handbook, (in Russian). 2nd ed. Moscow: Energoatomizdat Publishing House; 1988. 625 pp; Margulova TC. Nuclear power plants, (in Russian). Moscow: Izdat Publishing House; 1955. 289 pp.

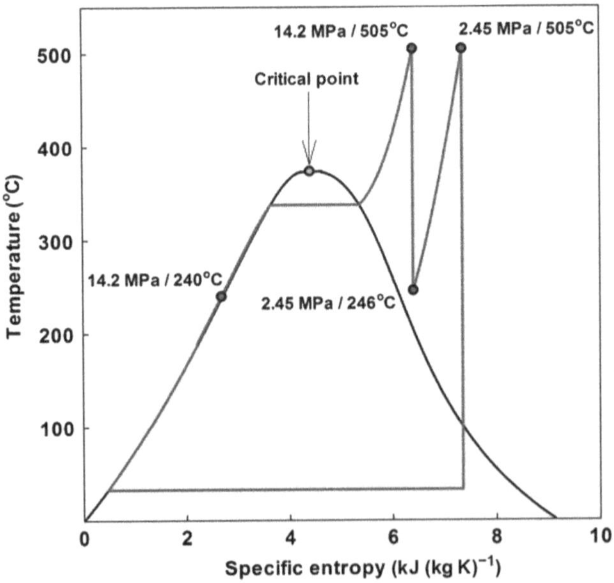

Fig. 4.27 *T-s* diagram for the 600-MW$_{el}$ BN-600 SFR NPP Rankine steam-turbine cycle [1].

Table 4.1 Major basic parameters of modern Russian power reactors and NPPs [1]

Parameter	VVER-1000 (Figs. 4.2–4.4)	RBMK-1000 (Figs. 4.22–4.24)	BN-600 (Figs. 4.25–4.27)
Thermal power (MW_{th})	3000	3200	1470
Electrical power (MW_{el})	1000	1000	600
Thermal efficiency (%)	33.3	31.3	40.8
Coolant pressure (MPa)	15.7	6.9	~0.1
Coolant flow ($t h^{-1}$)	64,800	32,000	21,800
Coolant temperature (°C)	290/322	284	377/550
Steam flow rate ($t h^{-1}$)	5880	5600	640
Steam pressure (MPa)	5.9	6.4	15.3
Steam temperature (°C)	276	280	505
Core: diameter/height (m/m)	4.5/10.9	11.8/7	2.05/0.75
Fuel enrichment (%)	4.3	2.0–2.4	17; 21; 26
No. of fuel assemblies	163	1580	371

Table 4.2 Reference parameters of Generation III+ VVER

Parameter	Value
Thermal power (MW_{th})	3200
Electrical power (MW_{el})	1160
NPP thermal efficiency (%)	36
Primary (reactor) coolant pressure (MPa)	16.2
Steam-generator pressure (MPa)	7.0
Coolant temperature at reactor inlet (°C)	298
Coolant temperature at reactor outlet (°C)	329
Steam-generator pressure/temperature (MPa/°C)	6.27/278
NPP service life (years)	50
Main equipment service life (years)	60
Replaced equipment service life (years, not less than)	30
Capacity factor (%)	up to 90
Load factor (%)	up to 92
Equipment availability factor	99
Length of fuel cycle (years)	4–5
Frequency of refuelling (months)	12–18
Fuel-assembly maximum burn-up (MW day $(kg(U))^{-1}$)	up to 60–70
Inter-repair period length (years)	4–8

Continued

Table 4.2 Continued

Parameter	Value
Annual average length of scheduled shutdowns (for refuelling, scheduled maintenance work) (days per year)	16–40
Refueling length (days per year)	≤16
Number of not scheduled reactor shutdowns per year	≤1
Frequency of severe core damage per year	$<10^{-6}$
Frequency of limiting emergency release per year	$<10^{-7}$
Efficient time of passive safety and emergency control system operation without operator's action and power supply (h)	≥24
OBE/SSE, magnitude of MSK-64 scale	6 and 7
Compliance with EUR requirements (yes/no)	Yes
RP main stationary equipment is designed for SSE of magnitude 8	

(Based on data from Ryzhov SB, Mokhov VA, Nikitenko MP, et al. Advanced designs of VVER reactor plant, In: Proceedings of the 8th International topical meeting on nuclear thermal-hydraulics, operation and safety (NUTHOS-8), Shanghai, China, October 10–14; 2010.

Table 4.3 Additional typical parameters of latest VVER-1000 series 300 and 400 (for basic parameters, see Table 4.2) [1]

Parameter	Value
Pressure vessel ID (m)	4.14
RPV wall thickness (m)	0.19
RPV height without cover (m)	10.9
Core equivalent diameter (m)	3.12
Core height (m)	3.56
Volumetric heat flux ($MW\,m^{-3}$)	110
Average volumetric flow rate in assembly ($m^3\,h^{-1}$)	515 ± 55
No. of fuel assemblies	163
No of rods per assembly	317
Fuel mass, ton of UO_2	80
Part of fuel reloaded during year	1/3
Fuel	UO_2
Fuel enrichment (%)	4

Table 4.4 Typical basic parameters of US PWR NPP

Parameter	Value
Thermal power (MW_{th})	3800
Electrical power (MW_{el})	1300
Thermal efficiency (%)	34
Specific power ($kW\ (kg(U))^{-1}$)	33
Power density ($kW\,L^{-1}$)	102
Average linear heat flux ($kW\,m^{-1}$)	17.5
Rod heat flux average/max ($MW\,m^{-2}$)	0.584/1.46
Core	
Length (m)	4.17
OD (m)	3.37
Reactor coolant system	
Pressure (MPa)	15.5
Inlet temperature (°C)	292
Outlet temperature (°C)	329
Mass flow rate (m) ($kg\,s^{-1}$)	531
Steam generators	
Total number	4
Outlet pressure (MPa)	6.9
Outlet temperature (°C)	284
Mass flow rate ($kg\,s^{-1}$)	528
Reactor pressure vessel (RPV)	
OD (m)	4.4
Height (m)	13.6
Wall thickness (m)	0.22
Fuel	
Fuel pellets	UO_2
Pellet OD (mm)	8.19
Rod OD (mm)	9.5
Zircaloy clad (wall) thickness (mm)	0.57
Rods per bundle (17×17)	264
Bundles in core	193
Fuel loading (ton)	115
Enrichment (%)	3.2
Reactivity control	
No. of control assemblies	68
Shape	Rod clusters
Absorber rods per assembly	24
Neutron absorber	Ag-In-Cd and/or B_4C
Soluble poison shim	Boric acid H_3BO_3

Based on data from Shultis JK, Faw RE. Fundamentals of nuclear science and engineering, 2nd ed. Boca Raton, FL: CRC Press; 2007. 591 pp.

Table 4.5 Basic parameters of AP-1000 US Generation III+ PWR NPP (Shultis and Faw [7,8])

Parameter	Value
Thermal power (MW_{th})	3415
Electrical power (MW_{el})	1110
Thermal efficiency (%)	33
Number of loops	2
Hot leg temperature (°C)	321
No. of fuel bundles	157
Type of fuel assembly	17 × 17
Active fuel height (m)	4.3
Linear heat flux ($kW\,m^{-1}$)	18.7
Control rod clusters assemblies	53 (16 Gy)
Reactor vessel ID (m)	3.99
Vessel flow rate ($m^3\,h^{-1}$)	68,100

Table 4.6 Main design and operating parameters for EPR (AREVA)

Parameter	Value
Thermal power (MW_{th})	4590
Electrical power (MW_{el})	1650
Gross thermal efficiency (%)	36–37
No. of loops	4
Operating/design pressure (MPa)	15.5/17.6
RPV inlet/outlet temperatures (°C)	295/330 (T_{sat} @ 15.5 MPa = 345°C)
Total flow/loop rate ($m^3\,h^{-1}$)	28,315
Main steam operating/design pressure (MPa)	7.72/10
Core	
Active height (m)	4.2
No. of fuel assemblies	241
No. of RCCAs (rod cluster control assemblies)	89
Fuel-assembly array (bundle)	17 × 17
No. of fuel rods	63,865
Average linear heat flux ($kW\,m^{-1}$)	16.7

Based on data from AREVA (France) http://www.areva.com/EN/operations-5444/epr-reactor-fact-sheet.html; [Accessed 15 October 2017].

Table 4.7 Typical parameters of US BWR (Shultis and Faw [7])

Parameter	Value
Thermal output (MW_{th})	3830
Electrical output (MW_{el})	1330
Thermal efficiency (%)	34
Specific power (kW $(kg(U))^{-1}$)	26
Power density ($kW L^{-1}$)	56
Average linear heat flux ($kW m^{-1}$)	20.7
Rod heat flux ave./max ($MW m^{-2}$)	0.51/1.12
Core	
Length (m)	3.76
OD (m)	4.8
Reactor-coolant system	
Pressure (MPa)	7.17
Feed-water temperature (°C)	216
Outlet steam temperature (°C)	290
Outlet steam flow rate ($kg s^{-1}$)	2083
Core flow rate ($kg s^{-1}$)	14,167
Core void fraction ave./max	0.37/0.75
Reactor-pressure vessel	
ID (m)	6.4
Height (m)	22.1
Wall thickness (m)	0.15
Fuel	
Fuel pellets	UO_2
Pellet OD (mm)	10.6
Rod OD (mm)	12.5
Zircaloy clad (wall) thickness (mm)	0.86
Rods per bundle (8 × 8)	62
Bundles in core	760
Fuel loading (ton)	168
Enrichment (%)	1.9
Reactivity control	
No. of control assemblies	193
Shape	Cruciform
Overall length (m)	4.42
Length of poison section (m)	3.66
Neutron absorber	Boron carbide
Soluble poison shim	Gadolinium

Table 4.8 Key specifications of ABWR and BWR-5 [1]

Parameter	Item	ABWR	BWR-5
Power (MW)	Electrical	1350	1100
	Thermal	3926	3293
Thermal efficiency (gross) (%)		34	33.4
Reactor core	Fuel assemblies	872	764
	Control rods	205	185
Reactor equipment	Recirculation system	Internal pump method	External recirculation type
	Control rod drive	Hydraulic/electric motor drive methods	Hydraulic drive
Reactor containment vessel		Reinforced concrete with built-in liner	Free-standing vessel
Residual heat-removal system		3 systems	2 systems
Turbine Systems	Thermal cycle	Two-stage reheat	Nonreheat
	Turbine (blade length) (m)	1.32	1.09
	Moisture-separation method	Reheat type	Nonreheat type
	Heater drain	Drain-up type	Cascade type

Courtesy of Hitachi-GE Nuclear Energy.

Table 4.9 Key design parameters of Russian SFRs—BN reactors (www.proatom.ru)

No.	Parameter	BN-600[a]	BN-800[a]	BN-1200[b]
1	Thermal power (MW_{th})	1470	2100	2800
2	Electrical power (MW_{el})	600	880	1220
3	Basic components:			
	No of turbines × type	3 × K-200-130	1 × K-800-130	1 × K-1200-160
	No of generators × type	3 × TГB-200-M	1 × T3B-800-2	1 × T3B-1200-2
4	Reactor vessel			
	Diameter (m)	12.86	12.96	
	Height (m)	12.60	14.82	20.72
5	No of heat-transfer loops	3	3	4
6	Temperature of reactor coolant: sodium, primary loop—T_{in}/T_{out} (°C)	377/550	354/547	410/550
7	Temperature of intermediate coolant: sodium, secondary loop—T_{in}/T_{out} (°C)	328/518	309/505	355/527

Table 4.9 Continued

No.	Parameter	BN-600	BN-800	BN-1200
8	Temperature of power-cycle working fluid: water/steam—T_{in}/T_{out} (°C)	240/505	210/490	275/510
9	Pressure at steam-generator outlet (MPa)	13.7	14.0	17.0
10	Scheme of steam reheat with	Sodium	Steam	Steam
11	Basic unchangeable components service term (years)	30	40	60
12	Number of fuel assemblies	371	565	432
13	Fuel enrichment (%)	17; 21; 26	20; 22; 25 (for MOX)	–
14	Average fuel burnup (MW day $(kg)^{-1}$)	70	70–100	up to 138
15	Continues reactor operation between refueling intervals (eff. days)	120–170	155	330
16	NPP thermal efficiency (gross) (%)	42.5	41.9	43.6
17	NPP thermal efficiency (net) (%)	40.0	38.8	40.5
18	Plant capacity factor	0.77–0.8	0.85	0.9
19	Service life (years)	45	45	60

[a]BN-600 and BN-800 are currently in operation at the Beloyarsk NPP; BN-600 commercial start—1981 and BN-800—2015.
[b]BN-1200—Generation IV concept of future Russian SFR.

and (c) maximum single-channel mass-flow rate $28.5\,kg\,s^{-1}$; (8) online refueling; and (9) power cycle—subcritical-pressure regenerative Rankine steam-turbine cycle with steam reheat (working fluid light water; turbine: one High-Pressure (HP) and two double-flow Low-Pressure (LP) cylinders; net thermal output 2080 MW_{th}; gross/net electrical output (nominal) 740/690 MW_{el}; turbine steam-inlet parameters: saturation pressure 4.69 MPa and temperature 260°C; feedwater: 5.8 MPa (T_{sat} = 273.4°C) and 187°C; also, condenser backpressure ranges from 3.74 to 4.9 kPa depending on condenser cooling temperature. The *T-s* diagram for an older CANDU reactor at the Pickering NPP (i.e., lower steam parameters compared with those of the EC6 reactor) is shown in Fig. 4.17.

A simplified layout of the PHWR NPP (Siemens design, Atucha, Argentina) is shown in Fig. 4.18. This is the PHWR design that uses a reactor vessel as the reactor-coolant pressure boundary with fuel channels inside, while the CANDU-type of PHWR uses the pressure tube in a fuel channel as the pressure boundary. (For PHWRs operated in India, see reference [1].)

4.2.4 Advanced Gas-cooled Reactors

A simplified schematic of an AGR (Torness NPP), ribbed fuel element, thermodynamic layout, and *T-s* diagram are shown in Figs. 4.19–4.21, respectively.

4.2.5 Light-water-cooled graphite-moderated reactors: RBMK and EGP

A simplified schematic, thermodynamic layout, and *T-s* diagram of a 1000-MW_{el} RBMK NPP are shown in Figs. 4.22–4.24, respectively. The basic data on the RBMK are listed in Table 4.1.

4.2.6 Sodium-cooled fast reactor: BN-600 and BN-800

A simplified schematic, thermodynamic layout, and *T-s* diagram of a 600-MW_{el} BN-600 NPP are shown in Figs. 4.25–4.27, respectively. The basic data on the Russian SFRs—BN reactors are listed in Table 4.9.

4.3 Generation IV International Forum

4.3.1 Introduction

This section consists of materials and figures taken directly from the Generation IV International Forum (GIF) website: https://www.gen-4.org/gif/jcms/c_9260/public (accessed 17 January 2016); with the permission of the GIF [1]. In general, the GIF and its six Generation IV nuclear reactor concepts are not the only next-generation or advanced reactors (ARs) currently under development in the world. Advanced small modular reactors (SMRs) can be considered under the class of ARs or next-generation reactors [1]. Other nuclear reactor concepts of the next generation or ARs are being researched and developed by various nuclear-engineering companies worldwide, for example, the Traveling Wave Reactor (TWR) by TerraPower (http://terrapower.com/pages/technology) in the United States. However, for the purpose of this book, we will mainly focus on the GIF six Generation IV nuclear reactor concepts (see reference [1] for further information).

4.3.2 Origins of the GIF

GIF meetings began in January 2000, when the US Department of Energy's (DOE) Office of Nuclear Energy, Science and Technology convened a group of senior governmental representatives from the original nine countries to begin discussions on international collaboration in the development of Generation IV nuclear energy systems.

This group, subsequently named the GIF policy group, also decided to form a group of senior technical experts to explore areas of mutual interest and make recommendations regarding both research and development (R&D) areas and processes by which collaboration could be conducted and assessed. This senior technical experts group first met in April 2000.

The founding document of the GIF, a framework for international co-operation in R&D for the next generation of nuclear energy systems, are set out in the GIF Charter, first signed in July 2001 by Argentina, Brazil, Canada, France, Japan, Republic of Korea, South Africa, the United Kingdom, and the United States.

The Charter has since been signed by Switzerland (2002), Euratom (2003), the People's Republic of China, and the Russian Federation in November 2006, and most recently by Australia (14th member of the GIF) in June 2016.

In July 2011, the 13 members agreed to sign an extension of the Charter signaling the wish to continue to co-operate in the research and development of Generation IV.

4.3.3 Generation IV goals

Eight technology goals have been defined for Generation IV systems in four broad areas: (1) sustainability, (2) economics, (3) safety and reliability, and (4) proliferation resistance and physical protection. These ambitious goals are shared by a large number of countries as they aim at responding to the economic, environmental, and social requirements of the 21st century. They establish a framework and identify concrete targets for focusing GIF R&D efforts.

Goals for Generation IV nuclear energy systems

Sustainability 1	Generation IV nuclear energy systems will provide sustainable energy generation that meets clean-air objectives and provides long-term availability of systems and effective fuel utilization for worldwide energy production
Sustainability 2	Generation IV nuclear energy systems will minimize and manage their nuclear wastes and notably reduce the long-term stewardship burden, thereby improving protection for the public health and the environment
Economics 1	Generation IV nuclear energy systems will have a clear life-cycle cost advantage over other energy sources
Economics 2	Generation IV nuclear energy systems will have a level of financial risk comparable to other energy projects
Safety and reliability 1	Generation IV nuclear energy systems operations will excel in safety and reliability
Safety and reliability 2	Generation IV nuclear energy systems will have a very low likelihood and degree of reactor-core damage
Safety and reliability 3	Generation IV nuclear energy systems will eliminate the need for offsite emergency response
Proliferation resistance and physical protection	Generation IV nuclear energy systems will increase the assurance that they are very unattractive and the least desirable route for diversion or theft of weapons-usable materials, and provide increased physical protection against acts of terrorism

These goals guide the cooperative R&D efforts undertaken by GIF members. The challenges raised by GIF goals are intended to stimulate innovative R&D covering all technological aspects related to design and implementation of reactors, energy-conversion systems, and fuel-cycle facilities.

In light of the ambitious nature of the goals involved, international cooperation is considered essential for a timely progress in the development of Generation IV systems. This cooperation makes it possible to pursue multiple systems and technical options concurrently and to avoid any premature down selection due to the lack of adequate resources at the national level.

4.3.4 Selection of Generation IV systems

For more than a decade, GIF has led international collaborative efforts to develop next-generation nuclear energy systems that can help meet the world's future energy needs. Generation IV designs will use nuclear fuel more efficiently, reduce wastes production, be economically competitive, and meet stringent standards of safety and proliferation resistance.

With these goals in mind, some 100 experts evaluated 130 reactor concepts before GIF selected six reactor technologies for further R&D. These reactor technologies include: (1) Gas-cooled fast reactor (GFR), (2) lead-cooled fast reactor (LFR), (3) molten-salt reactor (MSR), (4) supercritical water-cooled reactor (SCWR), (5) sodium-cooled fast reactor (SFR), and (6) very-high-temperature reactor (VHTR).

The latest information on the status of GIF-system arrangements and memoranda of understanding is shown in Fig. 4.28, and system development timelines as defined in the original Roadmap in 2002 and in the 2013 update—in Fig. 4.29.

The goals adopted by GIF provided the basis for identifying and selecting six nuclear energy systems for further development. The selected systems rely on a variety of reactor, energy-conversion, and fuel-cycle technologies. Their designs feature thermal- and fast-neutron spectra, closed and open fuel cycles as well as a wide range of reactor sizes from very small to very large. Depending on their respective degrees

System	CA	CN	EU	FR	JP	KR	RU	CH	US	ZA
SFR		✓	✓	✓	✓	✓	✓		✓	
VHTR		✓	✓	✓	✓	✓		✓	✓	
SCWR	✓		✓		✓		✓			
GFR			✓	✓	✓			✓		
LFR			P		P		P			
MSR			P	P			P			

✓ = Signatory to the system arrangement; P = signatory to the memorandum of understanding; Argentina, Brazil, and the United Kingdom are inactive.

Fig. 4.28 Status of the GIF system arrangements and memoranda of understanding (as of January 1, 2014). Correction to the Figure: Also, China has signed the SCWR System Arrangement in May of 2014. On June 22 of 2016, Australia signed the Charter, thus becoming the 14th member.
Courtesy of Generation IV International Forum.

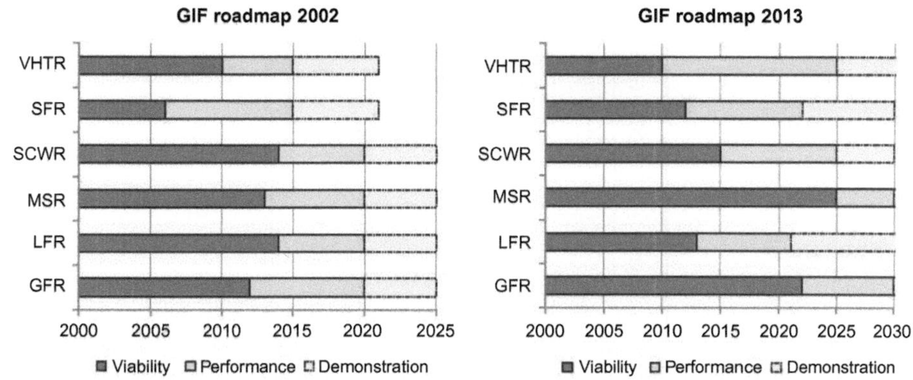

Fig. 4.29 System development timelines as defined in the original Roadmap in 2002 and in the 2013 update.
Courtesy of Generation IV International Forum.

of technical maturity, the Generation IV systems are expected to become available for commercial introduction in the period around 2030 or beyond. The path from current nuclear systems to Generation IV systems is described in a 2002 roadmap report entitled "A Technology Roadmap for Generation IV nuclear energy systems," which is currently being updated.

All Generation IV systems aim at performance improvement, new applications of nuclear energy, and/or more sustainable approaches to the management of nuclear materials. High-temperature systems offer the possibility of efficient process heat applications and eventually hydrogen production. Enhanced sustainability is achieved primarily through the adoption of a closed fuel cycle including the reprocessing and recycling of plutonium, uranium, and minor actinides in fast reactors and, also, through high thermal efficiency. This approach provides a significant reduction in waste generation and uranium-resource requirements. Table 4.10 summarizes the main characteristics of the six Generation IV systems.

On the GIF website (https://www.gen-4.org/gif/jcms/c_9260/public), the sequence of referencing six nuclear reactor concepts is based on alphabetical order, but in this chapter other sequences are used (see Figs. 4.28 and 4.29). Also, it was decided to list them according to the type of reactor coolant, i.e., first two reactors (VHTR and GFR)—helium cooled; next two concepts (SFR and LFR)—liquid-metal cooled; next one concept (MSR)—molten-salt cooled; and the last concept (SCWR)—supercritical-water cooled (see Table 4.10).

4.3.5 Six Generation IV nuclear energy systems

4.3.5.1 Very-high-temperature reactor

The very-high-temperature reactor (VHTR) (see Fig. 4.30) is a further step in the evolutionary development of high-temperature reactors (HTRs). The VHTR is a helium-gas-cooled, graphite-moderated, thermal-neutron-spectrum reactor with a core outlet

Table 4.10 Overview of Generation IV systems

No.	System	Neutron spectrum	Coolant	Outlet Temperature (°C)	Fuel cycle	Size (MW_{el})
1	VHTR	Thermal	Helium	900–1000	Open	250–300
2	GFR	Fast	Helium	850	Closed	1200
3	SFR	Fast	Sodium	500–550	Closed	50–150 300–1500 600–1500
4	LFR	Fast	Lead	480–570	Closed	20–180 300–1200 600–1000
5	MSR	Thermal/fast	Fluoride salts	700–800	Closed	1000
6	SCWR	Thermal/fast	Water	510–625	Open/closed	300–700 1000–1500

Courtesy of Generation IV International Forum.

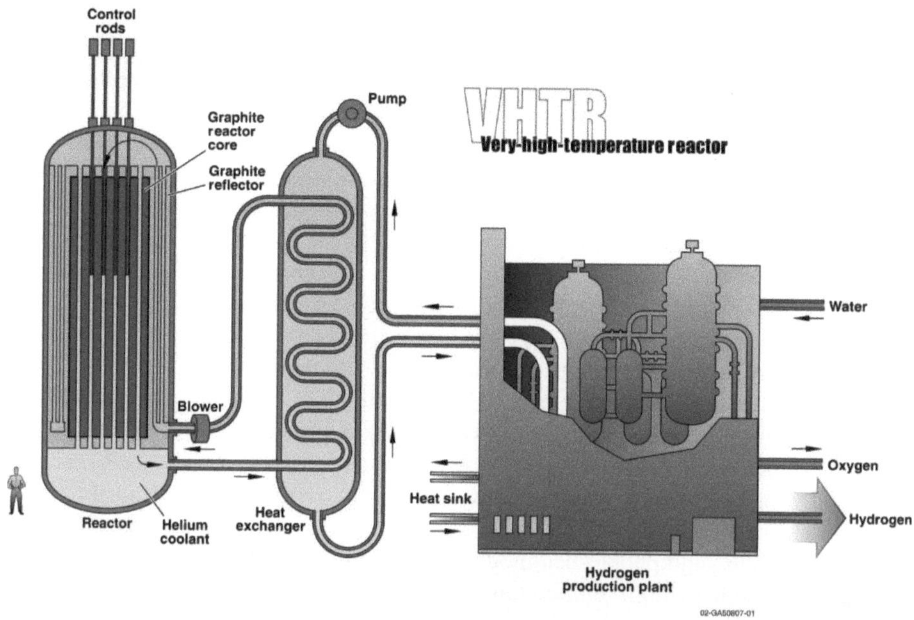

Fig. 4.30 Very-high-temperature reactor (VHTR): Helium-gas-cooled, graphite-moderated, thermal-neutron-spectrum reactor with core outlet temperature 900–1000°C (shown with hydrogen cogeneration).
Courtesy of Generation IV International Forum.

temperature >900°C, and a goal of 1000°C, sufficient to support high-temperature processes such as production of hydrogen through thermochemical processes. The thermal power of the reactor is set at a level that allows passive decay-heat removal, currently estimated to be about 600 MW_{th}. The VHTR is useful for cogeneration of electricity and hydrogen, as well as for other process-heat applications, related to the chemical, oil, and iron industries. It is able to produce hydrogen from water by using thermochemical, electrochemical, or hybrid processes with reduced emission of CO_2 gases. At first, a once-through low-enriched uranium (LEU) ($<20\%$ U^{235}) fuel cycle will be adopted, but a closed fuel cycle will be assessed, as well as potential symbiotic fuel cycles with other types of reactors (especially, light-water reactors (LWRs)) for waste-reduction purposes. The system is expected to be available for commercial deployment by 2020.

The technical basis for VHTR is the tri-isotropic (TRISO)-coated particle fuel. The VHTR has potential for inherent safety, high thermal efficiency, process-heat-application capability, low operation and maintenance costs, and modular construction.

In general, the reactor core of the VHTR can be a prismatic block core such as the Japanese high-temperature test reactor (HTTR), or a pebble-bed core such as the

Chinese HTR-10. For electricity generation, a helium-gas-turbine system can be directly set in the primary-coolant loop, which is called a direct cycle, or at the lower end of the outlet-temperature range, a steam generator can be used with a conventional Rankine cycle. For nuclear heat applications such as process heat for refineries, petrochemistry, metallurgy, and hydrogen production, the heat application process is generally coupled with the reactor through an intermediate heat exchanger (IHX), the so-called indirect cycle. The VHTR can produce hydrogen from only heat and water by using thermochemical processes (such as the Sulfur-Iodine (S-I) process or the hybrid-sulfur process), High-temperature steam electrolysis (HTSE), or from heat, water, and natural gas by applying the steam-reformer technology.

While the original approach for VHTR at the start of the Generation IV program focused on very high outlet temperatures and hydrogen production, current market assessments have indicated that electricity production and industrial processes based on high-temperature steam that require modest outlet temperatures (700–850°C) have the greatest potential for application in the next decade, and, also, reduce technical risk associated with higher outlet temperatures. As a result, over the past decade, the focus has moved from higher outlet temperature designs such as gas-turbine modular helium reactor (GT-MHR) and pebble-bed modular reactor (PBMR) to lower outlet temperature designs such as high-temperature reactor pebble-bed modules (HTR-PM) in China and the *Next-Generation Nuclear Plant* (NGNP) in the United States.

The VHTR has two typical reactor configurations, namely: (1) the pebble-bed type and (2) the prismatic-block type. Although the shape of a fuel element for two configurations is different, the technical basis for both configurations is same, such as the TRISO-coated particle fuel in the graphite matrix, full ceramic (graphite) core structure, helium coolant, and low power density, in order to achieve high outlet temperature and the retention of fission production inside the coated particle at normal operation condition and accident condition. The VHTR can support alternative fuel cycles such as U-Pu, Pu, MOX, and U-Th.

4.3.5.2 Gas-cooled fast reactor

The GFR (see Fig. 4.31) is a high-temperature helium-cooled fast-spectrum reactor with a closed fuel cycle. The core outlet temperature will be of the order of 850°C. It combines the advantages of fast-spectrum systems for long-term sustainability of uranium resources and waste minimization (through fuel multiple reprocessing and fission of long-lived actinides), with those of high-temperature systems (e.g., high thermal-cycle efficiency and industrial use of the generated heat for hydrogen production). It requires the development of robust refractory fuel elements and appropriate safety architecture. The use of dense fuel such as carbide or nitride provides good performance regarding plutonium breeding and minor-actinide burning. A technology demonstration reactor needed for qualifying key technologies could be in operation by 2020.

The GFR uses the same fuel-recycling processes as the SFR and the same reactor technology as the VHTR. Therefore, its development approach is to rely, in so far as

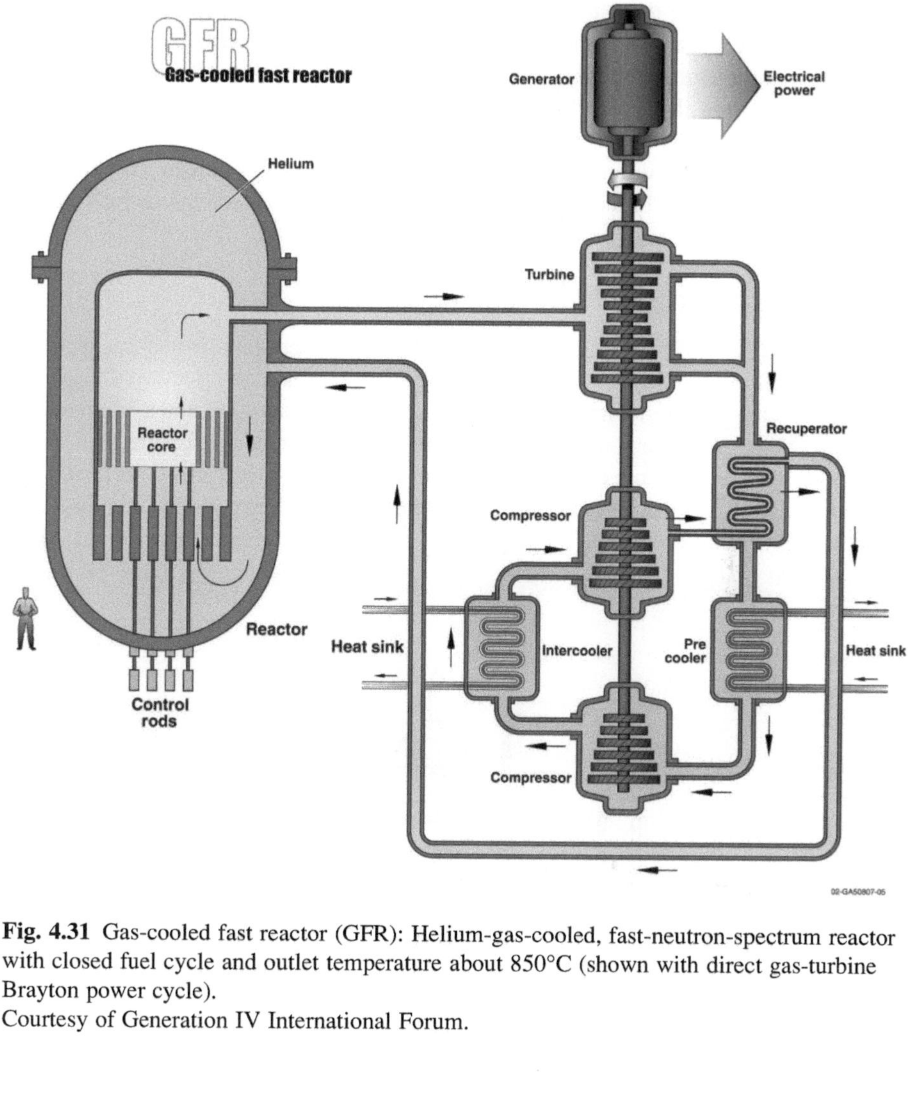

Fig. 4.31 Gas-cooled fast reactor (GFR): Helium-gas-cooled, fast-neutron-spectrum reactor with closed fuel cycle and outlet temperature about 850°C (shown with direct gas-turbine Brayton power cycle).
Courtesy of Generation IV International Forum.

feasible, on technologies developed for the VHTR for structures, materials, components, and power conversion system. Nevertheless, it calls for specific R&D beyond the current and foreseen work on the VHTR system, mainly on core design and safety approach.

The reference design for GFR is based around a 2400-MW_{th} reactor core contained within a steel pressure vessel. The core consists of an assembly of hexagonal fuel elements, each consisting of ceramic clad, mixed-carbide-fuelled pins contained within a ceramic hex tube. The favored material at the moment for the pin clad and hex tubes is silicon-carbide-fiber-reinforced silicon carbide.

As a new approach, the latest GFR concept will have the indirect cycle. A heat exchanger transfers the heat from the primary helium coolant to a secondary gas cycle containing a helium-nitrogen mixture, which in turn drives a closed-cycle gas turbine. The waste heat from the gas-turbine exhaust is used to raise steam in a steam generator, which is then used to drive a steam turbine. Such a combined cycle is common practice in natural gas-fired power plants, so it represents an established technology, with the only difference in the GFR case being the use of a closed-cycle gas turbine.

4.3.5.3 Sodium-cooled fast reactor

The SFR (see Fig. 4.32) uses liquid sodium as the reactor coolant. It features a closed fuel cycle for fuel breeding and/or actinide management (also, see Section 4.2.6). The two primary fuel-recycle-technology options are advanced aqueous and pyrometallurgical processing. A variety of fuel options are being considered for the SFR, with MOX preferred for advanced aqueous recycle and mixed metal alloy preferred for pyrometallurgical processing. Owing to the significant past experience accumulated with SFRs in several countries, the deployment of SFR systems is targeted for 2020.

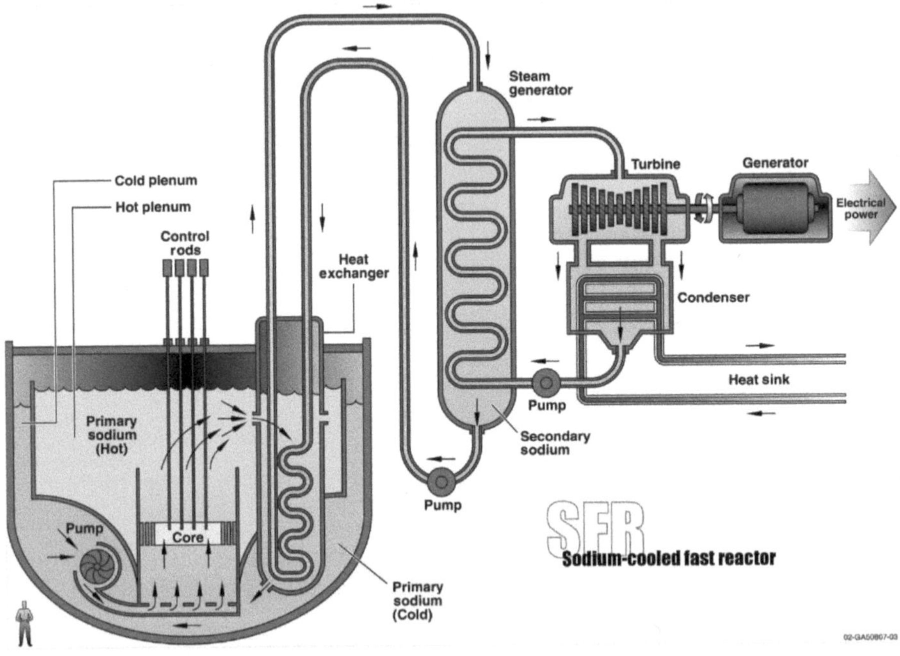

Fig. 4.32 Sodium-cooled fast reactor (SFR): Molten-sodium-cooled, fast-neutron-spectrum reactor with closed fuel cycle and outlet temperatures within 500–550°C (shown pool-type reactor with indirect steam-turbine Rankine power cycle).
Courtesy of Generation IV International Forum.

Using liquid sodium as the reactor coolant, a high power density with low coolant volume fraction and operation at low pressure can be achieved. While the oxygen-free environment prevents corrosion, sodium reacts chemically with air and water, and requires a sealed coolant system.

Plant size options under consideration range from small, 50 to 300 MW_{el}, modular reactors to larger plants up to 1500 MW_{el}. The outlet temperature is 500–550°C for the options, which allow the use of the materials developed and proven in prior fast-reactor programs.

The SFR closed fuel cycle enables regeneration of fissile fuel and facilitates management of minor actinides. However, this requires that recycle fuels be developed and qualified for use. Important safety features of the Generation IV system include a long thermal response time, a reasonable margin to coolant boiling, a primary system that operates near atmospheric pressure, and an intermediate sodium system between the radioactive sodium in the primary system and the power conversion system. Water/steam (Rankine cycle), supercritical carbon dioxide, or nitrogen (Brayton cycle) can be considered as working fluids for the power conversion system to achieve high performance in terms of thermal efficiency, safety, and reliability. With innovations to reduce capital cost, the SFR is aimed to be economically competitive in future electricity markets. In addition, the fast neutron spectrum greatly extends the uranium resources compared to thermal reactors. The SFR is considered to be the nearest-term deployable system for actinide management.

Much of the basic technology for the SFR has been established in former fast-reactor programs, and was confirmed by the *Phénix* (*French* for *Phoenix*) end-of-life tests in France, by operation of the Monju reactor in Japan, and the lifetime extension of BN-600 in Russia. New programs involving SFR technology include the Chinese experimental fast reactor (CEFR), which was connected to the grid in July 2011, India's prototype fast breeder reactor (PFBR), and the latest success in Russia with putting into operation the BN-800 reactor.

The SFR is an attractive energy source for nations that desire to make the best use of limited nuclear fuel resources and manage nuclear waste by closing the fuel cycle.

Fast reactors hold a unique role in the actinide management mission, because they operate with high energy neutrons that are more effective at fissioning actinides. The main characteristics of the SFR for actinide management mission are:

- Consumption of transuranics in a closed fuel cycle, thus reducing the radiotoxicity and heat load, which facilitates waste disposal and geologic isolation, and
- Enhanced utilization of uranium resources through efficient management of fissile materials and multirecycle.

High level of safety achieved through inherent and passive means also allows accommodation of transients and bounding events with significant safety margins.

The reactor unit can be arranged in a pool layout or a compact loop layout. Three options are considered:

(1) A large-size (600–1500 MW_{el}) loop-type reactor with mixed uranium-plutonium oxide fuel and potentially minor actinides, supported by a fuel cycle based upon advanced aqueous processing at a central location serving a number of reactors.

(2) An intermediate-to-large size (300–1500 MW_{el}) pool-type reactor with oxide or metal fuel and
(3) A small size (50–150 MW_{el}) modular-type reactor with uranium-plutonium-minor-actinide-zirconium metal-alloy fuel, supported by a fuel cycle based on pyrometallurgical processing in facilities integrated with the reactor.

4.3.5.4 Lead-cooled fast reactor

The LFR (see Fig. 4.33) is characterized by a fast-neutron spectrum, a closed fuel cycle with full actinide recycling, possibly in central or regional fuel-cycle facilities, and high-temperature operation at low pressure. The coolant may be either lead (preferred option) or lead-bismuth eutectic (LBE). The LFR may be operated as a breeder, a burner of actinides from spent fuel, using inert matrix fuel, or a burner/breeder using thorium matrices. Two reactor size options are considered: a small 50–150 MW_{el} transportable system with a very long core life, and a medium 300–600 MW_{el} system.

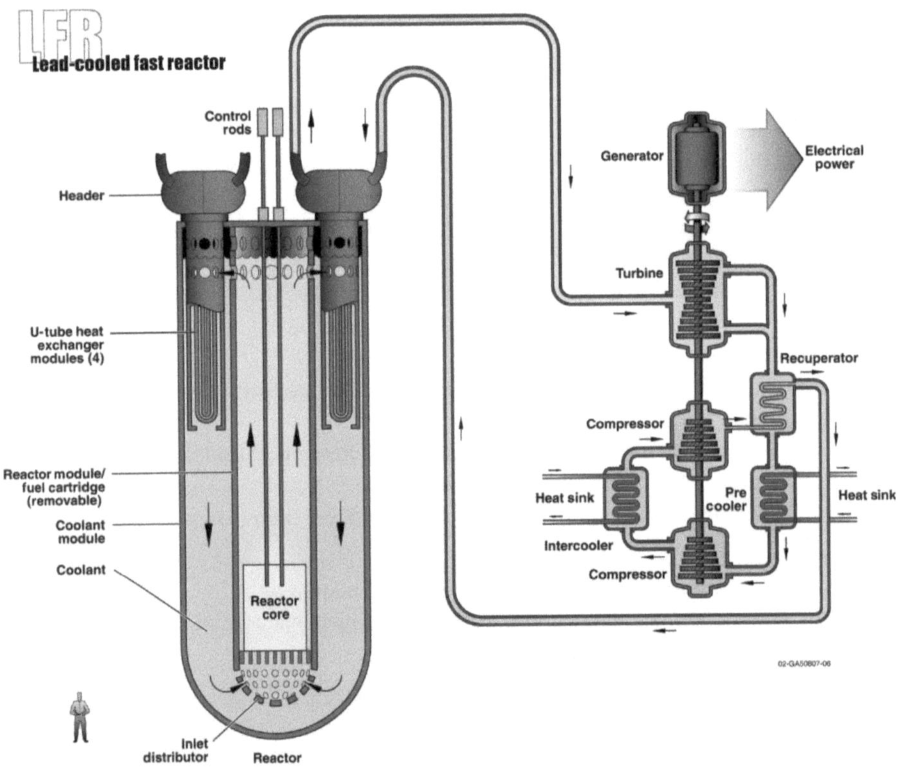

Fig. 4.33 Lead-cooled fast reactor (LFR): Molten-lead-cooled, fast-neutron-spectrum reactor with closed fuel cycle and outlet temperatures within 500–550°C (shown with indirect Brayton power cycle).
Courtesy of Generation IV International Forum.

In the long term, a large system of 1200 MW_{el} may be envisaged. The LFR system may be deployable by 2025.

Lead and LBE are relatively inert liquids with very good thermodynamic properties. The LFR would have multiple applications including production of electricity, hydrogen, and process heat. System concepts represented in plans of the GIF System Research Plan (SRP) are based on the European lead-cooled fast reactor (ELFR), Russia's BREST-OD-300 (fast reactor with lead coolant—Быстрый Реактор со Свинцовым Теплоносителем in Russian abbreviations), and the small secure transportable autonomous reactor (SSTAR) concept designed in the United States.

The LFR has excellent materials management capabilities, since it operates in the fast-neutron spectrum and uses a closed fuel cycle for efficient conversion of fertile uranium. It can also be used as a burner to consume actinides from spent LWR fuel and as a burner/breeder with thorium matrices. An important feature of the LFR is the enhanced safety that results from the choice of molten lead as a relatively inert and low-pressure coolant. In terms of sustainability, lead is abundant and, hence, available, even in case of deployment of a large number of reactors. More importantly, as with other fast systems, fuel sustainability is greatly enhanced by the conversion capabilities of the LFR fuel cycle, because they incorporate a liquid coolant with a very high margin to boiling and benign interaction with air or water. LFR concepts offer substantial potential in terms of safety, design simplification, proliferation resistance, and the resulting economic performance. An important factor is the potential for benign end state to severe accidents.

The LFR has development needs in the areas of fuels, materials performance, and corrosion control. During the next 5 years, progress is expected on materials, system design, and operating parameters. Significant test and demonstration activities are underway and planned during this time frame.

4.3.5.5 Molten-salt reactor

The MSR (see Fig. 4.34) embodies the very special feature of a liquid fuel. MSR concepts, which may be used as efficient burners of transuranic elements from spent LWR fuel, also have a breeding capability in any kind of neutron spectrum ranging from thermal (with a thorium fuel cycle) to fast (with a uranium-plutonium fuel cycle). Whether configured for burning or breeding, MSRs have considerable promise for the minimization of radiotoxic nuclear waste.

The MSR is distinguished by its core in which the fuel is dissolved in molten fluoride salt. The technology was first studied >50 years ago. Modern interest is on fast-reactor concepts as a long-term alternative to solid-fuelled fast-neutrons reactors. The onsite fuel-reprocessing unit using pyrochemistry allows breeding plutonium or uranium-233 from thorium. R&D progresses toward resolving feasibility issues and assessing safety and performance of the design concepts. Key feasibility issues focus on a dedicated safety approach and the development of salt redox potential measurement and control tools in order to limit corrosion rate of structural materials. Further work on the batchwise online salt processing is required. Much work is needed on molten-salt technology and related equipment.

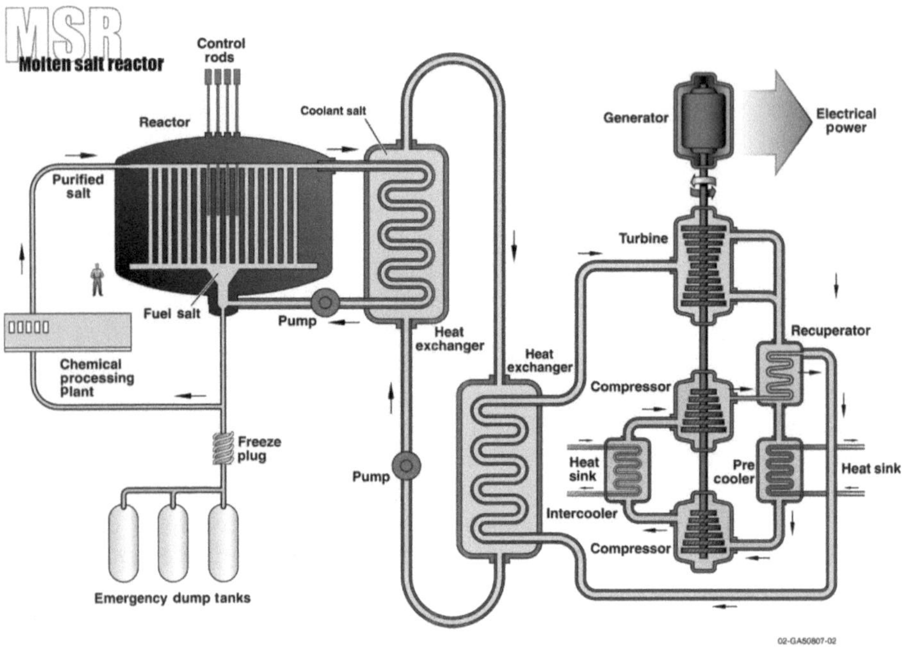

Fig. 4.34 Molten-salt reactor (MSR): Molten-slat-cooled reactor with outlet temperatures within 700–800°C (shown with indirect Brayton power cycle).
Courtesy of Generation IV International Forum.

MSR technology was partly developed, including two demonstration reactors, in the 1950s and 1960s in the United States (Oak Ridge National Laboratory). The demonstration MSRs were thermal-neutron-spectrum graphite-moderated concepts. Since 2005, R&D has focused on the development of fast-spectrum MSR concepts (MSFR) combining the generic assets of fast-neutron reactors (extended resource utilization, waste minimization) with those relating to molten-salt fluorides as fluid fuel and coolant (low pressure and high boiling temperature, optical transparency).

In contrast to most other molten-salt reactors previously studied, the MSFR does not include any solid moderator (usually graphite) in the core. This design choice is motivated by the study of parameters such as feedback coefficient, breeding ratio, graphite lifespan, and U^{233} initial inventory. MSFR exhibit large negative-temperature and void-reactivity coefficients, a unique safety characteristic not found in solid-fuel fast reactors.

Compared with solid-fuel reactors, MSFR systems have lower fissile inventories, no radiation-damage constraint on attainable fuel burn-up, no requirement to fabricate and handle solid fuel, and a homogeneous isotopic composition of fuel in the reactor. These and other characteristics give MSFRs potentially unique capabilities for actinide burning and extending fuel resources.

MSR developments in Russia on the molten-salt actinide recycler and transmuter (MOSART) aim to be used as efficient burners of transuranic (TRU) waste from spent

UOX and MOX LWR fuel without any uranium and thorium support and also with it. Other advanced-reactor concepts are being studied, which use the liquid-salt technology, as a primary coolant for Fluoride salt-cooled High-temperature Reactors (FHRs), and coated-particle fuels similar to high temperature gas-cooled reactors.

More generally, there has been a significant renewal of interest in the use of liquid salt as a coolant for nuclear and non-nuclear applications. These salts could facilitate heat transfer for nuclear-hydrogen-production concepts, concentrated-solar electricity generation, oil refineries, and shale-oil-processing facilities among other applications.

4.3.5.6 Supercritical water-cooled reactors

SCWRs (see Fig. 4.35) are a class of high-temperature, high-pressure water-cooled reactors operating with a direct (indirect) energy-conversion cycle and above the thermodynamic critical point of water (critical pressure: 22.064 MPa and critical temperature: 373.95°C). The higher thermodynamic efficiency and plant simplification opportunities afforded by a high-temperature, single-phase coolant translate into

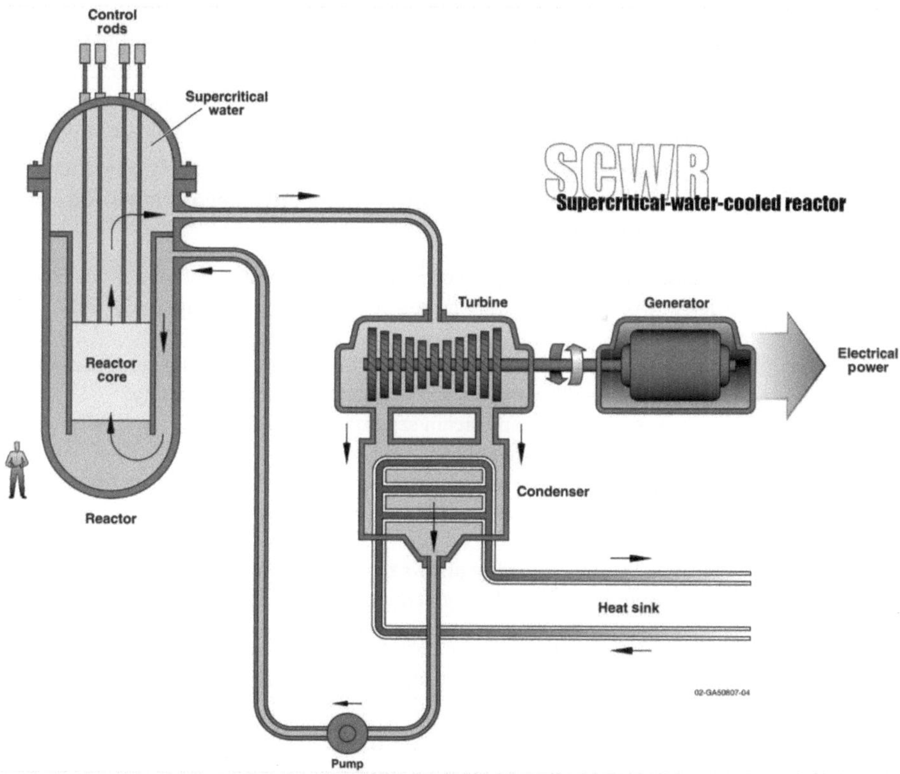

Fig. 4.35 Supercritical water-cooled reactor (SCWR): Supercritical water-cooled, thermal-neutron-spectrum reactor with outlet temperatures within 510–625°C (shown with direct steam-turbine Rankine power cycle).
Courtesy of Generation IV International Forum.

improved economics. A wide variety of options are currently considered: both thermal-neutron and fast-neutron spectra are envisaged; both pressure-vessel and pressure-tube configurations are considered, and, thus, use light water or heavy water can be used as a moderator. The operation of a 30–150 MW_{el} technology demonstration reactor is targeted for 2022.

Unlike current water-cooled reactors, the coolant will experience a significantly higher enthalpy rise in the core, which reduces the core mass flow for a given thermal power and increases the core outlet enthalpy to supercritical conditions. For both pressure-vessel and pressure-tube designs, a once-through steam cycle has been envisaged, omitting any coolant recirculation inside the reactor. As in a BWR, the supercritical "steam" will be supplied directly to the high-pressure steam turbine and the feed water from the steam cycle will be supplied back to the core. Thus, the SCWR concepts combine the design and operation experiences gained from hundreds of water-cooled reactors with those experiences from hundreds of fossil-fired power plants operated with supercritical water (SCW). In contrast to some of the other Generation IV nuclear systems, the SCWR can be developed incrementally step by step from current water-cooled reactors.

In general, SCWR designs have unique features that offer many advantages compared to state-of-the-art water-cooled reactors. However, there are several technological challenges associated with the development of the SCWR, and particularly the need to validate transient heat-transfer models (for describing the depressurization from supercritical to subcritical conditions), qualification of materials (namely, advanced steels for cladding), and demonstration of the passive safety systems.

SCWR designs have unique features that offer many advantages compared to the state-of-the-art water-cooled reactors:

- SCWRs offer increases in thermal efficiency relative to current-generation water-cooled reactors. The efficiency of an SCWR can approach 44% or more, compared to 30%–36% for current reactors.
- Reactor-coolant pumps are not required. The only pumps driving the coolant under normal operating conditions are feed-water pumps and condensate-extraction pumps.
- The steam generators used in PWRs and steam separators and dryers used in BWRs can be omitted since the coolant is supercritical in the core.
- Containment, designed with pressure-suppression pools and with emergency cooling and residual heat-removal systems, can be significantly smaller than those of current water-cooled reactors.
- The higher supercritical "steam" enthalpy allows to decrease the size of the turbine system and thus to lower the capital costs of the conventional island.

These general features offer the potential of lower capital costs for a given electrical power of the plant and of better fuel utilization, and thus a clear economic advantage compared with current LWRs.

Preconceptual core design studies for a core outlet temperature of >500°C have been performed in Japan, assuming either a thermal-neutron spectrum or a fast-neutron spectrum (Oka et al. [14]). Both options are based on a coolant heat-up in two steps with intermediate mixing underneath the core. Additional moderator for a thermal-neutron spectrum is provided by feed water inside water rods. The fast-spectrum option uses zirconium-hydride (ZrH_2) layers to minimize hardening of the neutron spectrum in case of core voiding. A preconceptual design of safety systems for both options has been studied with transient analyses.

A preconceptual plant design with 1700-MW-net electrical power based on a pressure-vessel-type reactor has been studied by Yamada et al. [15] and has been assessed with respect to efficiency, safety, and cost. The study confirms the target net efficiency of 44% and estimates a cost reduction potential of 30% compared with current PWRs. Safety features are expected to be similar to ABWRs.

A preconceptual design of a pressure-vessel-type reactor with a 500°C core outlet temperature and 1000-MW electrical power has been developed in Europe, as summarized by Schulenberg and Starflinger [16]. The core design is based on coolant heat-up in 3 steps. Additional moderator for the thermal-neutron spectrum is provided in water rods and in gaps between assembly boxes. The design of the nuclear island and of the balance of the plant confirms results obtained in Japan, namely, an efficiency improvement up to 43.5% and a cost reduction potential of 20%–30% compared with latest BWRs. Safety features as defined by the stringent European utility requirements are expected to be met.

Canada is developing a pressure-tube-type SCWR concept with a 625°C core outlet temperature at the pressure of 25 MPa. The concept is designed to generate 1200-MW electrical power (a 300-MW concept is also being considered). It has a modular fuel-channel configuration with separate coolant and moderator. A high-efficiency fuel channel is incorporated to house the fuel assembly. The heavy-water moderator directly contacts the pressure tube and is contained inside a low-pressure calandria vessel. In addition to providing moderation during normal operation, it is designed to remove decay heat from the high-efficiency fuel channel during long-term cooling using a passive moderator-cooling system. A mixture of thorium oxide and plutonium is introduced as the reference fuel, which aligns with the GIF position paper on thorium fuel. The safety-system design of the Canadian SCWR is similar to that of the Economic Simplified BWR (ESBWR). However, the introduction of the passive-moderator-cooling system coupled with the high-efficiency channel could reduce significantly the core-damage frequency during postulated severe accidents such as large-break loss-of-coolant or station black-out events.

Preconceptual designs of three options of pressure-vessel SCWRs with thermal-, mixed-, and fast-neutron spectrum have been developed in Russia.

Outside of the GIF framework, two conceptual SCWR designs with thermal- and mixed-neutron-spectrum cores have been established by some research institutes in China under framework of the Chinese national R&D projects from 2007 to 2012, covering some basic research projects on materials and thermohydraulics, the core/fuel design, the main system design (including the conventional part), safety systems design, reactor structure design, and fuel-assembly structure design. The related feasibility studies have also been completed, and show that the design concept has promising prospects in terms of the overall performance, integration of design, component structure feasibility and manufacturability.

Prediction of heat transfer in SCW can be based on data from fossil-fired power plants as discussed by Pioro and Duffey [17]. Computational tools for more complex geometries like fuel assemblies are available, but still need to be validated with bundle experiments. System codes for transient-safety analyses have been upgraded to include SCW, including depressurization transients to subcritical conditions. Flow stability in the core has been studied numerically. As in BWRs, flow stability can be ensured using suitable inlet orifices in fuel assemblies.

A number of candidate cladding materials have been tested in capsules, autoclaves, and recirculating loops up to 700°C at a pressure of 25 MPa. Stainless steels with >20% chromium (Cr) are expected to have the required corrosion resistance up to a peak cladding temperature of 650°C. More work is needed to develop alloys suitable for use at the design peak cladding temperatures of 850°C for the Canadian SCWR concept. Further work is also needed to better identify the coolant conditions that lead to stress-corrosion cracking. It has been shown that the creep resistance of existing alloys can be improved by adding small amounts of elements, such as zirconium (Zr). In the longer term, the steel experimental Oxide Dispersion Strengthened (ODS) alloys offer an even higher potential, whereas nickel-base alloys are being considered for use in ultra-supercritical fossil-fired plants are less favorable for use in SCWRs due to their high neutron absorption and associated swelling and embrittlement.

Key water chemistry issues have been identified by Guzonas et al. [18]: Predicting and controlling water radiolysis and corrosion product transport (including fission products) remain the major R&D areas. In this regard, the operating experience using nuclear steam reheat at the Beloyarsk NPP in Russia is extremely valuable [1,19].

4.3.6 Additional reactor classifications

In several countries, the term "SMR" has been introduced representing the initials of "Small Modular Reactor" or "Small-/Medium-sized Reactor," which addresses some of the Generation IV goals. Not presently deployed, these ideas or preliminary designs cover a size range from 20 to 600 MW_{el}), and feature claims for lower costs due to more factory-based construction techniques and the use of multiple "modules" to reduce investment and financial risk. SMRs include many design concepts that span from the entire range of the six Generation IV concepts and coolants (PBMRs, LMFBRS, etc.) plus others, which are essentially scaled-down LWR systems, e.g., as in the NuScale concept currently undergoing licensing review. The idea is to have a more flexible system that may also be suitable for remote regions, as being presently undertaken by Russia using "ice-breaker" reactor technology, or, of course, for military applications beyond nuclear-powered submarines. Various industry and investor groups have promoted these concepts, while seeking fiscal support, first-of-a-kind (FOAK) or prototype funding and market acceptance. Some of these SMR concepts were recently examined in detail by an extensive Australian Royal Commission study, which found them to be not economically competitive on a relative basis [20]. It is therefore premature to speculate here on the degree or extent to which these smaller units may be deployed. (More details on SMRs can be found in [1].)

4.3.7 Summary

In summary, Table 4.11 lists estimated ranges of thermal efficiencies (gross) of Generation IV NPP concepts for reference purposes (Pioro and Duffey [1,2]).

Table 4.11 Estimated ranges of thermal efficiencies (gross) of Generation IV NPP concepts (Generation IV concepts are listed according to thermal efficiency decrease)

No	Nuclear power plant	Gross eff. (%)
1	Very-high-temperature reactor (VHTR) NPP (reactor coolant—helium: $P=7\,MPa$ and $T_{in}/T_{out}=640/1000°C$; primary power cycle—direct Brayton gas-turbine cycle; possible back-up—indirect Brayton or combined cycles)	≥55
2	Gas-cooled fast reactor (GFR) or high-temperature reactor (HTR) NPP (reactor coolant—helium: $P=9\,MPa$ and $T_{in}/T_{out}=490/850°C$; primary power cycle—direct Brayton gas-turbine cycle; possible back-up—indirect Brayton or combined cycles)	≥50
3	Supercritical water-cooled reactor (SCWR) NPP (one of Canadian concepts; reactor coolant—light water: $P=25\,MPa$ and $T_{in}/T_{out}=350/625°C$ ($T_{cr}=374°C$); direct cycle; high-temperature steam superheat: $T_{out}=625°C$; possible back-up—indirect supercritical-pressure Rankine steam cycle with high-temperature steam superheat)	45–50
4	Molten-salt reactor (MSR) NPP (reactor coolant—sodium-fluoride salt with dissolved uranium fuel: $T_{in}/T_{out}=700/800°C$; primary power cycle—indirect supercritical-pressure carbon dioxide Brayton gas-turbine cycle; possible back-up—indirect Rankine steam cycle)	~50
5	Lead-cooled Fast Reactor (LFR) NPP (Russian design BREST-OD-300[a]: reactor coolant—liquid lead: $P\approx0.1\,MPa$ and $T_{in}/T_{out}=420/540°C$; primary power cycle—indirect subcritical-pressure Rankine steam cycle: $P_{in}\approx17\,MPa$ ($P_{cr}=22.064\,MPa$) and $T_{in}/T_{out}=340/505°C$ ($T_{cr}=374°C$); high-temperature steam superheat; (in one of the previous designs of BREST-300 NPP primary power cycle was indirect supercritical-pressure Rankine steam cycle: $P_{in}\approx24.5\,MPa$ ($P_{cr}=22.064\,MPa$) and $T_{in}/T_{out}=340/520°C$ ($T_{cr}=374°C$); also, note that power-conversion cycle in a different LFR design from other countries is based on a supercritical-pressure carbon dioxide Brayton gas-turbine cycle	~41–43
6	Sodium-cooled Fast Reactor (SFR) NPP (Russian design BN-600: reactor coolant—liquid sodium (primary circuit): $P\approx0.1\,MPa$ and $T_{in}/T_{out}=380/550°C$; liquid sodium (secondary circuit): $T_{in}/T_{out}=320/520°C$; primary power cycle—indirect Rankine steam cycle: $P_{in}\approx14.2\,MPa$ ($T_{sat}\approx337°C$) and $T_{in\ max}=505°C$ ($T_{cr}=374°C$); steam superheat: $P\approx2.45\,MPa$ and $T_{in}/T_{out}=246/505°C$; possible back-up in some other countries—indirect supercritical-pressure carbon-dioxide Brayton gas-turbine cycle)	~40

[a]BREST-OD-300 is fast reactor with "natural safety"-test demonstration.

4.4 Comparison of thermophysical properties of reactor coolants

This section is partially based on Appendix A2 from [1] and the paper by Dragunov et al. [21].

4.4.1 Introduction

4.4.1.1 Generation II, III, and III+ reactor coolants

The current fleet of nuclear power reactors (see also, Section 4.1 and [1]) uses the following reactor coolants:

(1) Light water (H_2O) at subcritical pressures and temperatures (critical pressure: 22.064 MPa and critical temperature: 373.95°C)—in PWRs (single-phase cooling, i.e., liquid cooling); BWRs (two-phase cooling, i.e., with flow boiling, outlet reactor steam quality is usually about 10%), and LGRs (two-phase cooling, i.e., with flow boiling, outlet fuel-channel steam quality is usually about 14% (maximum: 20%));
(2) Heavy water (D_2O) at subcritical pressures and temperatures (critical pressure: 21.671 MPa and critical temperature: 370.7°C)—in PHWRs (single-phase cooling; however, there is a possibility for flow boiling within some subchannels at the fuel-channel outlet, steam quality usually does not exceed 5%);
(3) Carbon dioxide (CO_2) at subcritical pressures, but at supercritical temperatures (critical pressure: 7.3773 MPa and critical temperature: 30.978°C)—in an AGRs; and
(4) Liquid sodium (Na) (melting temperature: 97.7°C and boiling temperature: 882.8°C)–in an SFR.

4.4.1.2 Generation IV reactor coolants

Generation IV nuclear reactor concepts (see also, Section 4.2 and [1]) will use the following reactor coolants:

(1) Light water (H_2O) at supercritical pressures and temperatures (critical pressure - 22.064 MPa and critical temperature: 373.95°C)—in SCWRs (single-phase cooling, because at supercritical pressures fluids are considered single-phase substances);
(2) Helium (He) at supercritical pressures and temperatures (critical pressure: 0.2276 MPa and critical temperature: −267.95°C)—in GFRs and VHTRs;
(3) Liquid sodium (Na) (melting temperature: 97.7°C and boiling temperature: 882.8°C)–in SFRs;
(4) Liquid lead (Pb) (melting temperature: 327.5°C and boiling temperature: 1750°C)—in LFRs;
(5) Liquid lead-bismuth eutectic (LBE) (44.5% Pb and 55.5% Bi) (LBE: Melting temperature: 123.5°C and boiling temperature: 1670°C)—in liquid metal-cooled reactors (LMRs), for example, in Russian SVBRs; and
(6) Molten fluoride salts (for example, FLiNaK (LiF-NaF-KF): Melting temperature: 454°C and boiling temperature: 1570°C)—in MSRs.

For a better understanding of the thermodynamic terms such as subcritical/supercritical pressures, supercritical fluid, superheated/saturated steam, etc., thermodynamic diagrams for light water, carbon dioxide, and helium are shown in Figs. 4.36–4.38.

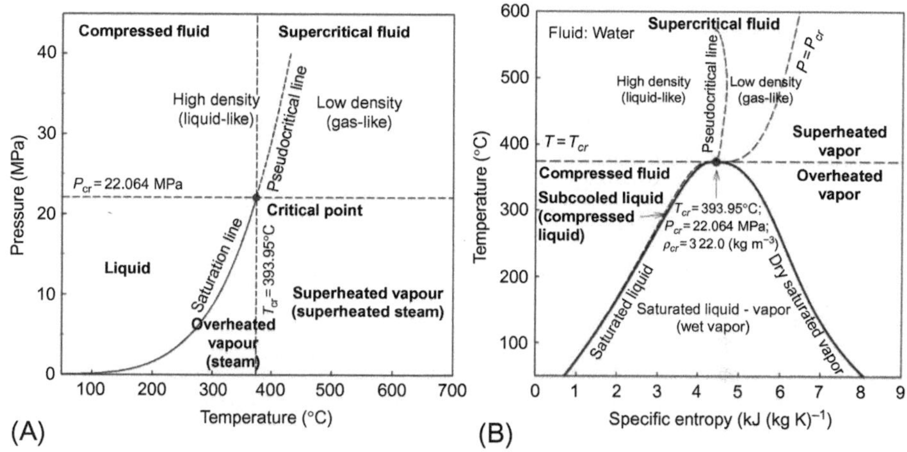

Fig. 4.36 Thermodynamic diagrams for light water [1]: (A) Pressure-temperature diagram; and (B) temperature-specific-entropy diagram.

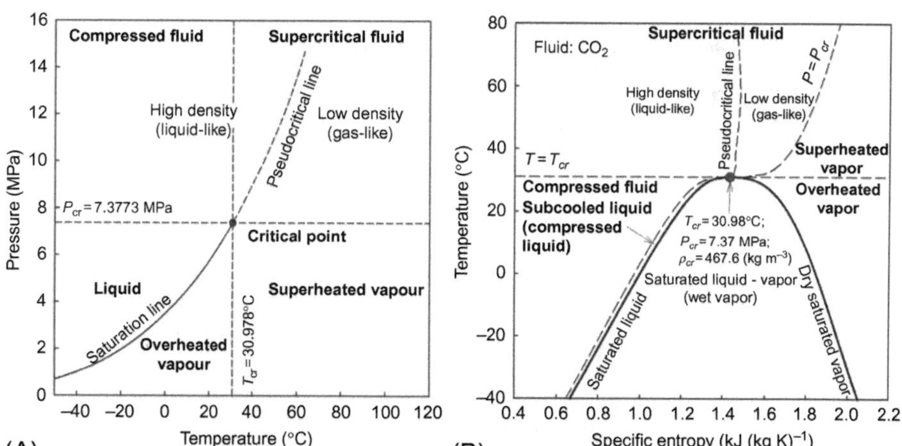

Fig. 4.37 Thermodynamic diagrams for carbon dioxide [1]: (A) Pressure-temperature diagram; and (B) temperature-specific-entropy diagram.

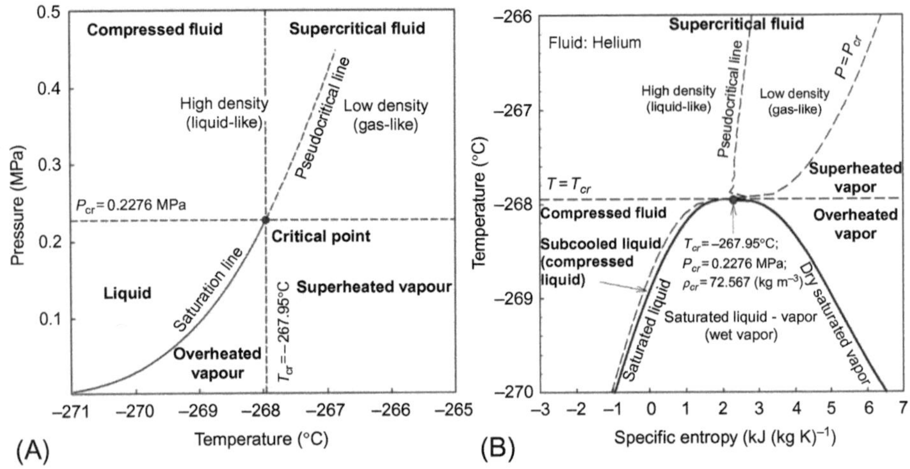

Fig. 4.38 Thermodynamic diagrams for helium [1]: (A) Pressure-temperature diagram; and (B) temperature-specific-entropy diagram.

A glossary of the terms used in Figs. 4.36–4.38 and elsewhere in the text is given now:

Compressed fluid is a fluid at a pressure above the critical pressure, but at a temperature below the critical temperature.

Critical point (also called a critical state) is a point at which the distinction between the liquid and gas (vapor) phases disappears, i.e., both phases have the same temperature, pressure, and specific volume or density. The critical point is characterized using the phase-state parameters T_{cr}, P_{cr}, and v_{cr} (or ρ_{cr}), which have unique values for each pure substance.

Pseudocritical line is a line, which consists of pseudocritical points.

Pseudocritical point (characterized with P_{pc} and T_{pc}) is a point at a pressure above the critical pressure, where the temperature ($T_{pc} > T_{cr}$) corresponds to the maximum value of the specific heat at this particular pressure.

Pseudocritical region is a narrow region around a pseudocritical point, where all thermophysical properties of a pure fluid exhibit rapid variations. For light water, it is about ±25°C from pseudocritical temperature.

Supercritical fluid is a fluid at pressures and temperatures that are higher than its critical pressure and critical temperature. However, quite often in various publications, a term supercritical fluid includes both terms—supercritical fluid and compressed fluid.

Overheated vapor is a dry vapor at a pressure and temperature below the critical pressure and temperature, respectively, but above the corresponding parameters of dry saturated vapor.

Supercritical "steam" is actually supercritical water, because at supercritical pressures the fluid is considered as a single-phase substance. However, this term is widely (and incorrectly) used in the literature in relation to supercritical "steam" generators and turbines.

Superheated vapor is a vapor at pressures below the critical pressure, but at temperatures above the critical temperature.

4.4.2 Reactor coolants by type

4.4.2.1 Fluid coolants

Subcritical-pressure light water is the most "popular" reactor coolant. It will be used in the subsequent comparisons as a reference case. In general, heavy water has many thermophysical properties and behaviors that are quite close to those of light water (for details, see Table 4.12). However, heavy water has a significantly lower neutron capture cross-section compared to that of light water, which allows for more thorough moderation. Therefore, only heat transfer characteristics of subcritical-pressure heavy water will be compared with those of other coolants.

One of the advantages of light/heavy water is high heat-transfer coefficients at forced convection and at flow boiling. However, there is a limit for efficient heat

Table 4.12 Comparison of selected thermophysical properties of heavy water and light water [1]

(a) At subcooled conditions: $P = 11$ MPa and $T = 260°C$ (approximately CANDU reactor fuel-channel inlet conditions)

	ρ	c_p	k	μ	Pr
Coolant	$kg\,m^{-3}$	$J\,(kg\,K)^{-1}$	$W\,(m\,K)^{-1}$	μPa s	–
D_2O	875.4	4657.1	0.536	114.7	0.997
H_2O	791.5	4887.5	0.618	103.7	0.820
Δ (%)	9.6	−5.0	−15.3	9.6	17.7

(b) At saturated conditions: $P = 10$ MPa and $T_{sat} = 310°C$ for heavy water (approximately CANDU reactor fuel-channel outlet conditions) and $T_{sat} = 311°C$ ($\Delta = 0.3\%$) for light water

	$\frac{\rho_f}{\rho_v}$	$\frac{c_{pf}}{c_{pv}}$	$\frac{k_f}{k_v}$	$\frac{\mu_f}{\mu_v}$	$\frac{Pr_f}{Pr_v}$	h_{fg}
Coolant	$kg\,m^{-3}$	$J\,(kg\,K)^{-1}$	$W\,(m\,K)^{-1}$	μPa s	–	$kJ\,kg^{-1}$
D_2O	$\frac{760.0}{62.5}$	$\frac{5823.6}{6730.7}$	$\frac{457.7}{82.0}$	$\frac{89.2}{20.2}$	$\frac{1.13}{1.66}$	1178.8
H_2O	$\frac{688.4}{55.5}$	$\frac{6123.7}{7140.8}$	$\frac{526.8}{76.6}$	$\frac{81.7}{20.2}$	$\frac{0.95}{1.88}$	1317.4
Δ (%)	$\frac{9.4}{11.2}$	$\frac{-5.2}{-6.1}$	$\frac{-15.1}{6.6}$	$\frac{8.4}{0}$	$\frac{16.3}{-13.5}$	−11.8

$\Delta = \frac{Property_{D_2O} - Property_{H_2O}}{Property_{D_2O}} \times 100\%$

Data based on National Institute of Standards and Technology. NIST reference fluid thermodynamic and transport properties–REFPROP. NIST standard reference database 23, Ver. 9.1. Boulder, CO: Department of Commerce; 2013; National Institute of Standards and Technology. NIST reference fluid thermodynamic and transport properties–REFPROP. NIST standard reference database 23, Ver. 9.1. Boulder, CO: Department of Commerce; 2010.

transfer, called critical heat flux (CHF), which usually cannot be exceeded during a nuclear reactor operation.

Supercritical water (SCW) is a coolant in an SCWR concept with an operating pressure of ~25 MPa, and reactor inlet and outlet temperatures of about 350°C and 625°C (max), respectively. Specifics of SCW thermophysical properties and heat transfer are discussed in Appendix A3 of reference [1] and in the following publications: IAEA-TECDOC-1746 [22]; Gupta et al. [23]; Pioro [24]; Pioro et al. [25]; Pioro and Mokry [26]; NIST REFPROP [27]; and Pioro and Duffey [17].

Main disadvantage of water as a reactor coolant is that to reach higher thermal efficiencies of NPP greater temperatures are needed, which require application of supercritical pressures.

4.4.2.2 Gas coolants

For comparison purposes in this Section, it was decided to consider subcritical-pressure carbon dioxide. Carbon dioxide at subcritical pressures is currently being used in the most efficient nuclear power reactors—AGRs. In general, carbon dioxide is not a strong absorber of thermal neutrons and does not become very radioactive. Other advantages of carbon dioxide are its chemical stability within the operating range of temperatures (292–650°C). In addition, carbon dioxide does not react with either the moderator or nuclear fuel.

Using helium as a reactor coolant at high outlet temperatures (850°C and 1000°C in GFR and VHTR, respectively) makes it possible to achieve very high thermal efficiencies of the plant that are close to those of modern advanced thermal power plants. The major advantages of helium are: (1) a relatively high thermal conductivity compared to that of other gases (the exception is hydrogen), which is close to that of liquids; and (2) its behavior as a noble or inert gas.

In general, the advantages of gaseous reactor coolants compared to water are a possibility to achieve high, or even very high, temperatures at the reactor outlet using significantly lower pressures; and there is no CHF phenomena at gas cooling, which limits heat transfer in fluid cooling. However, heat-transfer coefficients for gas forced-convection cooling are usually significantly lower than those for water cooling.

4.4.2.3 Liquid-metal coolants

Liquid sodium is currently used in the Russian BN-600 and BN-800 reactors—the only ones operating SFRs so far in the world—and is proposed to be used in Generation IV SFRs. Sodium is a well-known low-melting-point (97.7°C) alkali metal, which has the main advantages of high thermal conductivity and low neutron absorption cross-section. Also, the relatively high boiling point (882.8°C) of sodium allows a reactor to operate at pressures close to ~0.1 MPa. In addition, very high heat-transfer coefficients can be achieved with sodium cooling.

However, sodium is very reactive substance, which requires special precautions, when used as a reactor coolant. For improved reactor safety, a secondary sodium loop is utilized, which acts as a buffer between the radioactive sodium—reactor coolant in the primary loop and the water/steam in the third loop—a steam Rankine power cycle.

Lead is proposed for use in an LFR at pressures close to ~0.1 MPa. Lead has a higher melting point (327.5°C) and significantly higher boiling point (1750°C) compared to that of sodium, which significantly impacts on the reactor. Also, it is a more inert liquid metal than sodium. As a result, the LFR needs only two loops: (1) a primary loop with lead as a reactor coolant and (2) a secondary loop with water/steam as a Rankine power cycle.

LBE is an eutectic alloy of lead (44.5%) and bismuth (55.5%) and is being considered instead of lead as an option for the LFR. One of the main advantages of LBE is its melting point of 123.5°C, which is significantly lower than that of lead and quite close to that of sodium. Neither the lead nor the LBE react readily with water or air, in contrast to sodium, which allows for the elimination of the intermediate-coolant loop used in SFRs. Moreover, LBE is not a new technology—it has been proven by years of reliable experience as a coolant in nuclear-powered submarines operated by the Soviet Union since the 1970s.

A major advantage of liquid-metal reactor coolants is the low operating pressures inside a reactor (close to one atmosphere) with a possibility to achieve high temperatures. Also, all current liquid-metal reactors use a fast-neutron spectrum, which allows for more efficient fuel cycles.

Further information on liquid-metal reactor coolants can found in references [28–32].

4.4.2.4 Molten-salt coolants

Molten-salt fluorides, which are proposed as coolants in MSR, have promising thermophysical and thermohydraulic properties. Molten salts, similar to liquid metals, have a low vapor pressure even at high temperatures, which is quite attractive compared to water and gaseous coolants. Salts are less chemically reactive than sodium. In addition, salts can provide moderation due to their light-element composition such as F, Li, and Be in FLiBe.

Table 4.13 lists the main thermophysical properties of a number of coolants. The range of temperatures investigated covers the operating ranges of the corresponding reactors.

4.4.3 Thermophysical properties of proposed Generations II, III, III+, and IV reactor coolants

The basic thermophysical properties are shown within a wide range of temperatures (from 250°C to 1000°C) that cover the operating ranges of current and Generation IV reactors (see Table 4.13 and Fig. 4.39).

The properties of subcritical and supercritical water, carbon dioxide, and helium-4 were obtained from reference [27]. The properties of sodium were taken from reference [33], and the properties of other coolants were calculated using correlations presented in reference [34].

Table 4.13 **Basic reference parameters of selected Generations II, III, III+, and IV nuclear power reactors/concepts**

Reactor	Neutron spectrum	Core design		Reactor coolant	Moderator	Reactor cycle	No of circuits	P	T
								MPa	°C
PWR	Thermal	Heterogeneous	PV	Water		Indirect	2	~15.5	292–329
AGR	Thermal	Heterogeneous	PV[a]	CO_2	Graphite	Indirect	2	4	292–650
SFR	Fast	Heterogeneous	PV	Sodium	–	Indirect	3	~0.1	370–550
GFR	Fast	Heterogeneous	PV	Helium	–	Direct Indirect	1 2	9	490–850
VHTR	Thermal	Heterogeneous	PV	Helium	Graphite	Direct Indirect	1 2	7	490–1000
LFR	Fast	Heterogeneous	PV	Lead (LBE)	–	Indirect	2	~0.1	550–800 (420–540)
MSR	Epithermal	Homogeneous	PV	Sodium fluoride with dissolved uranium	Graphite	Indirect	3	~0.1	T_{out}=700–800
MSFR	Fast	Homogeneous	PV	Sodium fluoride with dissolved uranium	–	Indirect	3	~0.1	T_{out}=700–800
SCWR	Thermal	Heterogeneous	PV PCh (PT)	Water	Water Heavy water	Direct Indirect	1 2	~25	300–625
	Fast		PV PCh (PT)	Water	–	Direct Indirect	1 2		300–625

[a]Though coolant flows through individual channels inside graphite moderator, the actual pressure boundary is the vessel surrounding the moderator.

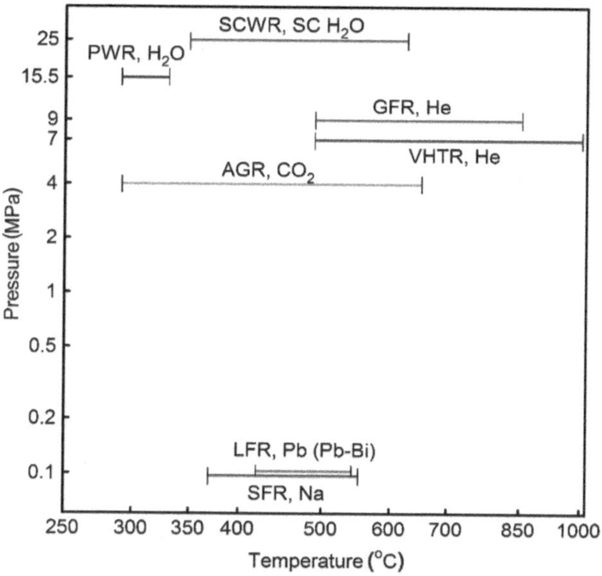

Fig. 4.39 Pressure-temperature diagram showing operating ranges of reactor coolants for PWR, AGR, SFR, and proposed Generation IV reactor concepts (pressure drop is not considered) [1].

Before comparing thermophysical properties of the reactor coolants, it is reasonable to have a general overview of the desired characteristics of a generic reactor coolant. Nuclear reactors have certain specific requirements for coolants, such as:

- high specific heat, thermal conductivity, and low viscosity;
- low corrosive and erosive effects on all the reactor materials;
- high boiling point and low melting point (not related to gaseous coolants);
- high thermal and radiation resistance;
- low neutron-absorption cross section;
- explosion-proof, noncombustible, nontoxic;
- widely available (not rare); and
- low neutron activation.

Fig. 4.40 shows densities profiles of reactor coolants vs. temperature. As expected, molten lead and lead-bismuth alloy have the highest densities following by those of molten salt and sodium. At ~250°C the densities of molten sodium, subcritical-pressure water, and SCW are quite close. However, with increasing temperature, the densities of water and SCW steadily decline. Within the pseudocritical range, the SCW density drops quite significantly due to the transition from "liquid-like" fluid to "gas-like" fluid (or from high-density fluid to low-density fluid). Gases, especially helium, have the lowest densities. The density of carbon dioxide is significantly higher than that of helium.

In general, the densities of the reactor coolants (with exception of SCW) decline almost linearly with increasing temperature (see Fig. 4.40). The densities of gases (helium and carbon dioxide) decrease about 1.6 times, but the density change for liquid metals and molten salt is not so significant. For SCW the density drops almost 8 times within the pseudocritical region.

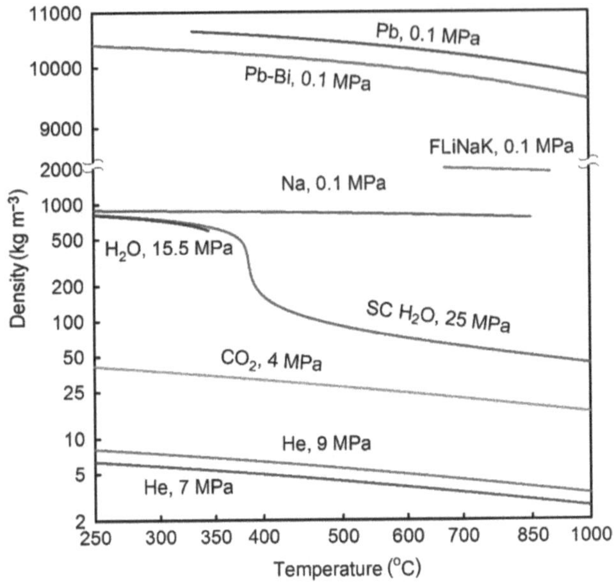

Fig. 4.40 Density of selected coolants vs temperature [1].

As one would expect, the thermal conductivity of liquid metals is significantly higher than that of gases (50–3000 times, see Fig. 4.41). The thermal conductivity of Na drops slightly, while that for Pb, LBE, He, and CO_2 increases linearly with the temperature. The thermal conductivity behavior of SCW is special. The thermal conductivity decreases almost linearly for temperature between 250°C and 350°C, then goes through a small peak in the pseudocritical point before decreasing smoothly from about 0.4 to 0.1 W $(\mathrm{m\,K})^{-1}$. As the temperature increases above 500°C, the thermal conductivity increases linearly to values higher than those of CO_2, but lower than those of He.

Most of the thermal properties of FLiNaK molten salt, calculated from references [35–38], have intermediate values between those of liquid metals and fluids. However, the viscosity of FLiNaK appears to be significantly higher than that of the rest of the coolants. This also causes the Prandtl number to be very high.

The temperature dependence of the dynamic viscosity of liquid metals is very different (Fig. 4.42). The viscosity of Na and Pb drops linearly over the whole range of temperature while the viscosity of Pb-Bi has a slower linear drop up to 600°C, and then the viscosity increases for temperatures between 600°C and 1000°C. Near 1000°C the viscosity returns to a value close to that measured at 250°C. The viscosities of gases increase linearly with temperature, and the viscosity of SCW at temperatures beyond the pseudocritical range behaves in a fashion similar to that of gases. In general, the shape of the viscosity-temperature curve for SCW is similar to that of its thermal conductivity. However, the viscosity does not exhibit a peak in the pseudocritical point.

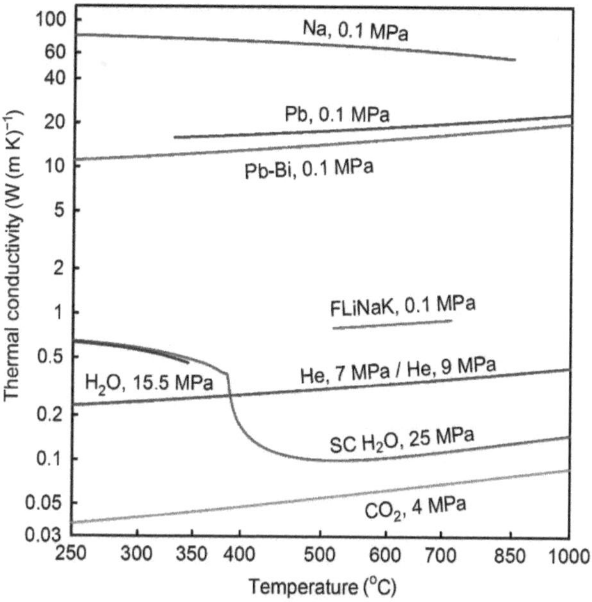

Fig. 4.41 Thermal Conductivity of selected coolants vs temperature [1].

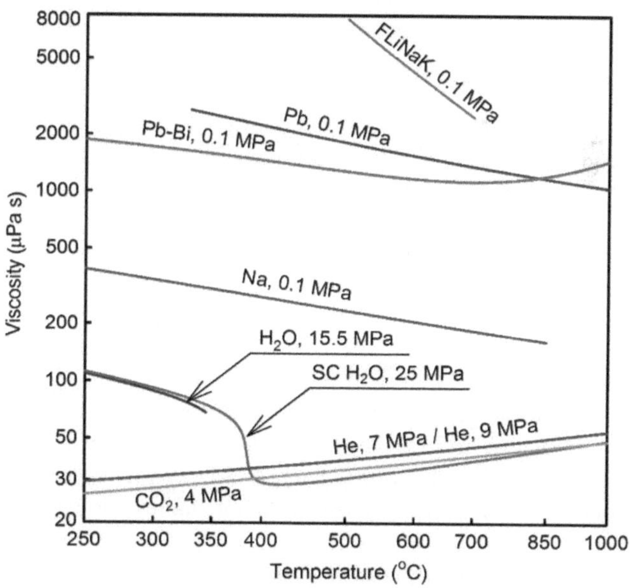

Fig. 4.42 Dynamic viscosity of selected coolants vs temperature [1].

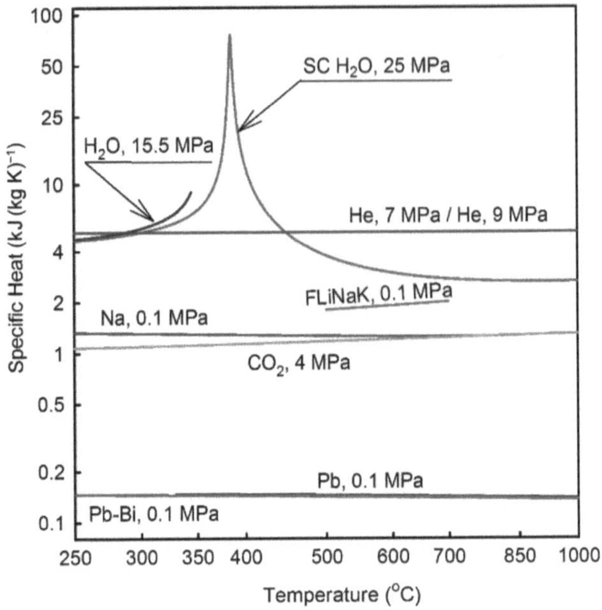

Fig. 4.43 Specific heat of selected coolants vs temperature [1].

The specific heat of He, Na, Pb, and Pb-Bi (Fig. 4.43) is nearly constant over the whole range of operational parameters. In the case of CO_2, the specific heat increases linearly and reaches the same value as Na at around 750°C. The specific heat of SCW goes through a peak (where its value increases almost 8 times) within the pseudocritical region. The specific heats of Pb and LBE are nearly identical and 10 times less than those of Na and CO_2, and almost 40 times less than that of He. At temperatures higher than 450°C, the specific heat of He is higher than that of SCW.

Fig. 4.44 shows the enthalpy increase vs. temperature for all reactor coolants. The enthalpy increase is straightforward and is related to the behavior of the specific heat. Therefore, the highest increase in enthalpy is in SCW, especially, within the pseudocritical range, where the specific heat has a peak. Eventually, SCW, water, and helium show quite close trends in enthalpy behavior. The enthalpy increases for sodium and carbon dioxide lie in the middle of the range. The lowest enthalpy increases are for the lead-bismuth alloy, and, also, for lead itself. The enthalpy rise for the molten salt is very sharp starting from relatively low values (below that for lead and lead-bismuth alloy) and almost reaching values for carbon dioxide and sodium at higher temperatures.

The dependence of the Prandtl number (*Pr*) (which is defined as a ratio of product of dynamic viscosity and specific heat to thermal conductivity) on temperature for different coolants is shown in Fig. 4.45. As follows from the definition, the profile

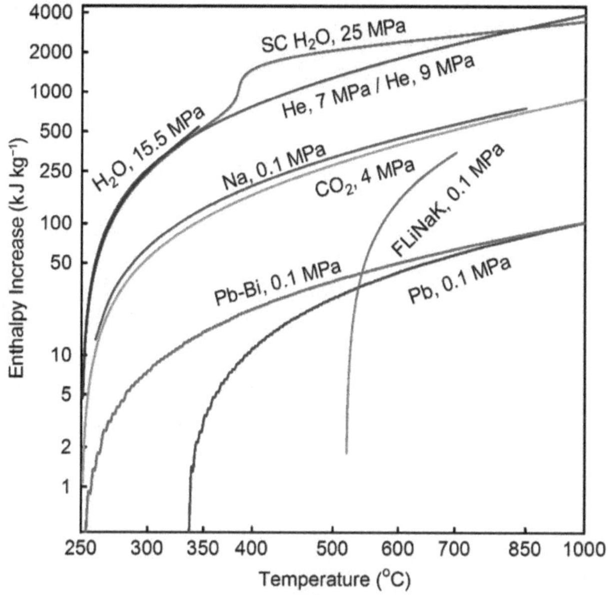

Fig. 4.44 Enthalpy of selected coolants vs temperature [1].

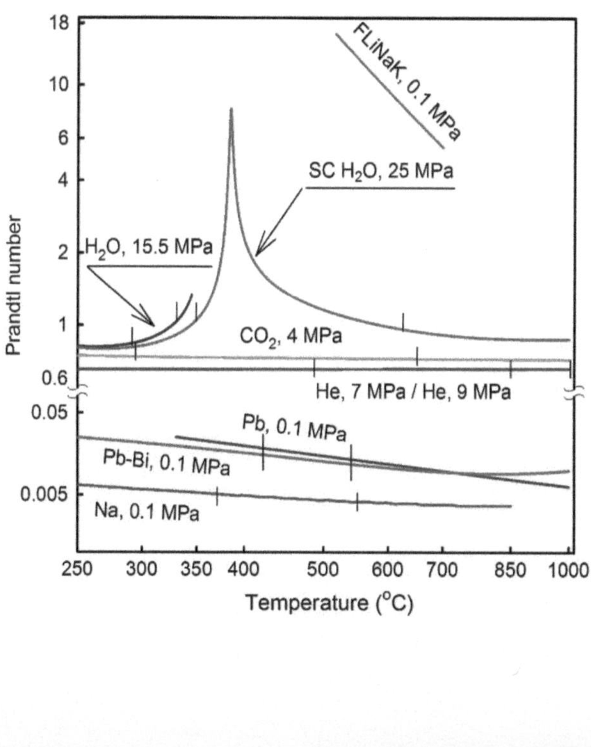

Fig. 4.45 Prandtl Number of selected coolants vs temperature [1].

of *Pr* is governed by the more significantly changing property of the coolant. We have established that the specific heat is nearly constant for all of the Generation IV reactors coolants except for SCW. Therefore, for most of the coolants, it is the ratio of the viscosity to the thermal conductivity that influences the shape of the *Pr*/temperature curve.

As we see from Fig. 4.45, the changes in the viscosity and thermal conductivity of the gases are such that they compensate each other, and the *Pr* of gases is virtually constant over most of the 750°C temperature span. However, for the liquid metals, the viscosity drops more significantly than the thermal conductivity increases. As a result, the *Pr* of liquid metals drops almost linearly with temperature. Due to an increase in viscosity of LBE at high temperatures, the corresponding value of *Pr* for Pb-Bi also increases. Since the specific heat of SCW goes through the most rapid changes compared with its other thermophysical properties, the *Pr* of SCW behaves similar to its specific heat. At high temperatures (>500°C), the Pr of SCW behaves similar to that of the gases.

The volumetric expansivity of liquid metals is much smaller than that of the remaining coolants and stays almost constant (see Fig. 4.46). The volumetric

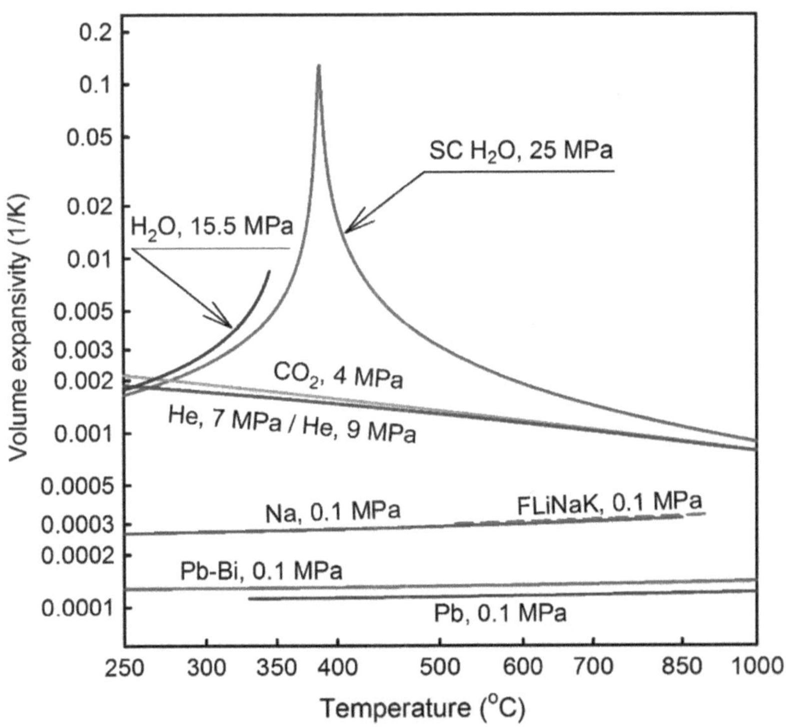

Fig. 4.46 Volume or volumetric expansivity of selected coolants vs temperature [1].

Table 4.14 Basic properties of helium, carbon dioxide, and water

No	Properties	Fluids		
		Helium	Carbon dioxide	Water
1	Chemical formula	He	CO_2	H_2O
2	Molar mass ($kg\,kmol^{-1}$)	4.0026	44.01	18.015
3	Triple point (°C)	−270.97	−56.558	0.01
4	Normal boiling-point temperature (°C)	−268.93	−78.464	99.974
5	Critical point temperature (°C)	−267.95	30.978	373.95
6	Critical point pressure (MPa)	0.2276	7.3773	22.064
7	Critical point density ($kg\,m^{-3}$)	72.567	467.6	322.0
8	Flammability; explosion hazard; and toxicity	–	–	–
9	Chemical reactivity	Inert gas	Moderate	Moderate-high
10	Corrosiveness	Inert gas	Yes	Very

Based on National Institute of Standards and Technology. NIST reference fluid thermodynamic and transport properties–REFPROP. NIST standard reference database 23, Ver. 9.1. Boulder, CO: Department of Commerce; 2013; National Institute of Standards and Technology. NIST reference fluid thermodynamic and transport properties–REFPROP. NIST standard reference database 23, Ver. 9.1. Boulder, CO: Department of Commerce; 2010.

expansivity of gases drops almost twice, in a linear fashion, within 250–1000°C range. Remarkably, the values of volumetric expansivity for SCW at temperatures below the pseudocritical point are close to those for gases. Near the pseudocritical point, the volumetric expansivity of SCW peaks. At higher temperatures, the volumetric expansivity of SCW gradually reaches values corresponding to those of gases.

To summarize, the thermophysical properties of liquid metals and gases experience only minor linear changes with increasing temperature. However, all the properties of water at pseudocritical conditions go through very rapid changes. The basic properties of helium, carbon dioxide, and water are summarized in Table 4.14. Basic properties of lead, molten salt (FLiNaK), and sodium are summarized in Table 4.15.

4.4.4 Heat-transfer coefficients in nuclear power rectors

Typical heat-transfer coefficient ranges for various reactor coolants are listed in Table 4.16. It shows that sodium has the highest heat-transfer coefficients among all the proposed coolants, making it a more competitive fluid for power conversion.

Fig. 4.47 shows the calculated heat-transfer coefficients at conditions corresponding to those of the operating reactors. The calculated values fall very close to those presented in Table 4.16. Among the coolants considered, sodium, in conditions close to SFR, has the highest heat-transfer coefficients of all the proposed coolants (70–80 $kW\,(m^2K)^{-1}$). Conditions achieved in a generic CANDU reactor

Table 4.15 Basic properties of lead, lead-bismuth, molten salt, and sodium

No	Properties	Fluids			
		Lead	Lead-bismuth	Fluoride salt[a]	Sodium
1	Chemical formula	Pb	44.5 Pb-55.5 Bi	FLiNaK	Na
2	Molar mass ($kg\,kmol^{-1}$)	207.2	~208	41.3	23
3	Density at 20°C ($kg\,m^{-3}$)	11,340	10,500	–	968
4	Melting-point temperature (°C)	327.5	123.5	454	97.8
5	Boiling-point temperature (°C)	1749	1670	1570	883
6	Heat of fusion ($kJ\,mol^{-1}$ ($kJ\,kg^{-1}$))	4.77 (23.0)	8.08 (38.8)	–	2.60 (113.0)
7	Heat of vaporization ($kJ\,mol^{-1}$ ($kJ\,kg^{-1}$))	179.5 (866.3)	178.1 (856.3)	–	97.42 (4236)
8	Flammability	Highly purified lead fine powder can ignite in air	Fine powder can ignite in air	–	Spontaneously ignites when heated above 115° C in air that has even modest moisture content; Generates flammable H_2 and caustic sodium hydroxide upon contact with water
9	Explosion hazard	Fine powder can ignite in air	Fine powder can ignite in air	–	Sodium powder is highly explosive in water and may spontaneously explode in presence of oxygen
10	Chemical reactivity	Reactive (oxidized in air)	Corrosive	–	Highly reactive
11	Toxicity	Poisonous	Poisonous	–	Can be poisonous
12	Corrosiveness	Yes Can embrittle metals	Yes	Yes	High

[a]Williams et al. [37].

Table 4.16 **Typical ranges of heat-transfer coefficients, heat fluxes, and sheath temperatures for reactors' coolants; and capacities per reactor-core volume**

Typical ranges of heat-transfer coefficients for reactors' coolants

Coolant	Heat-transfer coefficient (kW $(m^2K)^{-1}$
Na (forced convection) (~SFR conditions)	50–80
Boiling water (flow boiling) (~BWR conditions)	~40
CANDU reactor	~50
Water (single-phase forced convection)	~30
SCW (~SCWR conditions)	7–10
Pb (forced convection) (~LFR conditions)	25–35
Pb-Bi (forced convection) (~SVBR conditions)	20–30
He (rough surface)	10
CO_2 (high pressure) (~AGR conditions)	2–5

Typical ranges of heat fluxes for reactors' coolants

Coolant	Heat flux ($kW\,m^{-2}$)	T_{sheath}–$T_{coolant}$ (°C)
Na (forced convection) SFR	2000 (1800–2400)	25–30
Water (single-phase forced convection)	1500	50
Boiling water (flow boiling) BWR	1000	15
CANDU reactor	625	15
Boiling water in kettle (pool-boiling conditions)	150	15

Reactor type	Sheath temperature (°C)
AGR	750
SFR	700
PWR	390
BWR	300

Typical ranges of average capacity (kW) per reactor-core volume (L)

Reactor type	$kW\,L^{-1}$
AGR, HTGR, VHTR	6–10
RBMK	4–6
BWR	40–50
PWR (VVER)	100–150
SCWR (VVER-SKD[a])	100
SFR (BN)	400–550

[a]SKD, supercritical pressure in Russian abbreviations.

Based on Hewitt GF, Collier JG. Introduction to nuclear power, 2nd ed. New York, NY: Taylor & Francis; 2000. 304 pp. and data provided by P.L. Kirillov).

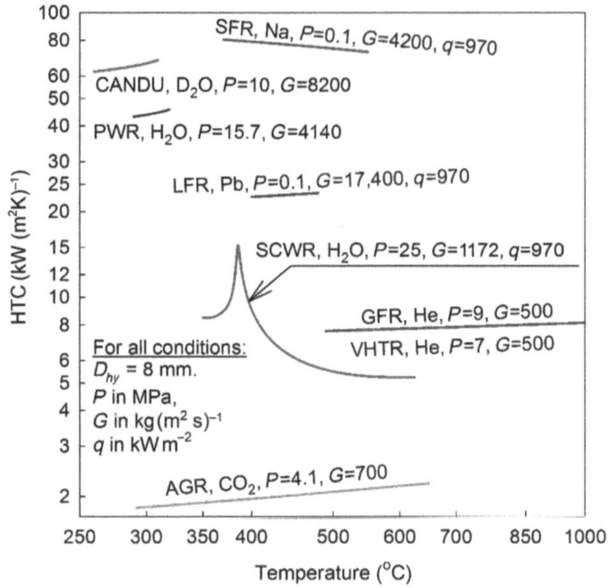

Fig. 4.47 Heat-transfer coefficients calculated for a flow of coolants in Generation IV, AGR, and PWR reactors in a bare tube at nominal operating pressures and at mass fluxes close to actual mass fluxes for the respective reactor [1].

(added for comparison purposes) allow heat-transfer coefficients above 60 kW $(m^2 K)^{-1}$. Calculations also showed that in a PWR, heat-transfer coefficients are about 45 kW $(m^2 K)^{-1}$. Lead, as expected, has heat-transfer coefficients around 25 kW $(m^2 K)^{-1}$, which is lower than that of another liquid-metal—sodium. Heat-transfer coefficients of SCW (5–15 kW $(m^2 K)^{-1}$) and CO_2 (1.8–2.5 kW $(m^2 K)^{-1}$) also lie within the typical ranges of values.

For calculations of subcritical H_2O, D_2O, CO_2, and He the value of heat flux was not taken into account, while for SCW, Pb, and Na the value of heat flux was assumed to be 970 kW m^{-2}. A hydraulic-equivalent diameter of 8 mm was used in the calculations for all the coolants.

Fig. 4.48 shows heat-transfer coefficients calculated for all coolants (including FLiNaK) for the generic conditions: $G = 1000$ kg $(m^2 s)^{-1}$, $q = 970$ kW $(m^2 K)^{-1}$, and $D_{hy} = 8$ mm.

It can be seen that at the chosen generic conditions, a sodium coolant has the highest heat-transfer coefficients, ranging from 58 to 96 kW $(m^2 K)^{-1}$, while CO_2 and FLiNaK have the lowest heat-transfer coefficients, ranging from 1 to 4 kW $(m^2 K)^{-1}$. The heat-transfer coefficient of SCW starts at ~5 kW $(m^2 K)^{-1}$, then goes through a peak within the pseudocritical region, where its value increases almost twice, and after that drops close to 4 kW $(m^2 K)^{-1}$ at temperatures above 450°C. Heat-transfer coefficients of the gases, water, heavy water, and lead increase slightly with temperature. Heat-transfer coefficients of the molten salt increase quite significantly with temperature. The heat-transfer coefficient of sodium drops linearly with temperature increase.

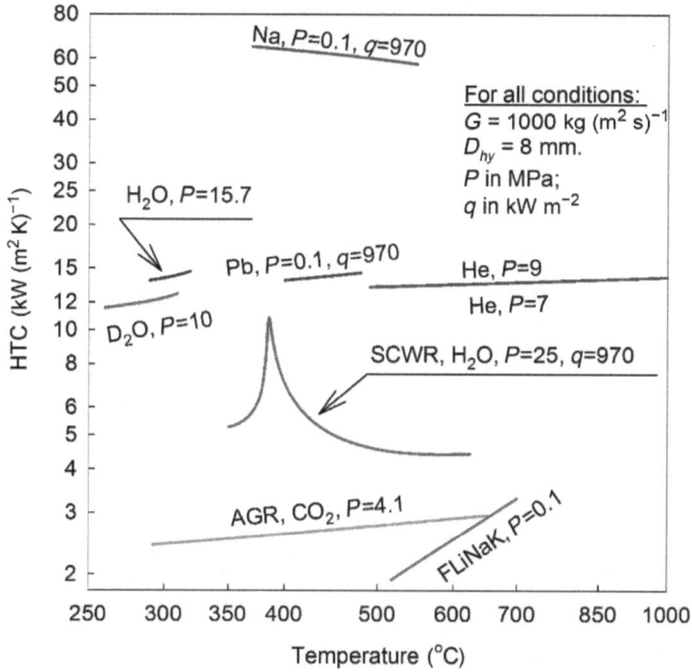

Fig. 4.48 Heat-transfer coefficients calculated for flow of coolants in Generation IV reactors, AGR, and PWR in a bare tube at generic operating conditions [1].

4.4.5 Conclusions

Based on the aforementioned, the following conclusions can be made.

- In general, liquid-metal coolants have high thermal stability, high boiling points, and very low saturated-vapor pressures, which distinguish them from other types of nuclear coolants.
- The specific heats of Pb and LBE are nearly identical and 10 times lower than those of Na and CO_2, and are almost 40 times lower than that of He. At temperatures higher than 450°C, the specific heat of He is even higher than that of SCW.
- As one would expect, the thermal conductivity of liquid metals is significantly higher than that of gases (50–3000 times). The highest thermal conductivity is for sodium with the value of 60–70 W $(\text{m K})^{-1}$).
- The volumetric expansivity of liquid metals is much lower than that of the other coolants examined, and stays almost constant.
- The thermophysical properties of liquid metals and gases show only small linear changes with temperature. However, all the properties of SCW go through very rapid changes within the pseudocritical range.
- The thermophysical properties of LBE, except for the thermal conductivity, are close to the average values of those of lead and bismuth.
- At high temperatures (>500°C), the Prandtl number of SCW behaves similar to gases.
- One of the least desirable properties of water is its high vapor pressure, which increases rapidly with temperature. Its relatively low critical temperature (~374°C) limits the

maximum temperature of the coolant and significantly limits the efficiency of the power conversion cycle.

- The specific heat of He is higher than that of CO_2 and liquid metals. The thermal conductivity of He is 10 times greater than that of CO_2. This characteristic facilitates heat transfer and reduces the size of heat exchangers. He is more inert than CO_2, does not absorb neutrons, and cannot become radioactive on its own.

4.5 Concise overview of conventional and alternative nuclear fuels

This Section is mainly based on Chapter 18 from [1] and Chapter by Peiman et al. [39].

4.5.1 Introduction

Nuclear fission is a reaction in which the nucleus of a heavy nuclide splits into smaller nuclides; a few new neutrons are created; gamma rays are emitted and a significant amount of energy is released. Nuclear fission has been used as a basis for production of heat in all the current nuclear reactors. Even though reactors can be categorized based on their cooling medium, pressure boundary, type of nuclear fuel, or neutron spectrum, they all have one common feature, which is the production of heat via a fission chain reaction in the nuclear fuel.

An important part of every reactor design involves the selection of a nuclear fuel and the design of fuel assemblies (bundles). As general requirements, a nuclear fuel should have a high melting point, acceptable thermal conductivity (higher is better), sufficient mechanical stability, good dimensional and irradiation stability as well as chemical compatibility with the cladding (sheath) and the coolant. Another important parameter that influences the design and selection of a nuclear fuel is the dominant neutron spectrum of a reactor. In this context, nuclear reactors can be categorized as fast-neutron spectrum, epithermal-neutron spectrum, and thermal-neutron spectrum. This classification is based on the energy group of neutrons that maintain the fission chain reaction. In a fast-neutron-spectrum reactor, the chain reaction is sustained mainly by fission of fast (e.g., high-energy) neutrons, while in an epithermal or thermal reactor, fission of epithermal (intermediate-energy) or thermal (low-energy) neutrons, respectively, maintain the chain reaction.

The neutron spectrum has an impact on the reactor design, selection of materials for the reactor core, the type of nuclear fuel, and the associated fuel cycle. Unlike fast-neutron-spectrum reactors, thermal-neutron-spectrum reactors utilize a moderator such as water, heavy water, or graphite in order to reduce the energy of high-energy neutrons. These two types of reactors use different coolants. Thermal-neutron-spectrum reactors use water or CO_2, which are composed of light elements, especially those having high scattering cross sections. Liquid-metal coolants, such as sodium or lead, are used in some fast-neutron-spectrum reactors. The neutron spectrum is dependent on the isotopic concentration of fissile and fertile nuclides in the fuel. As shown in Table 4.17, fast-neutron-spectrum reactors require a higher

Table 4.17 **Fuels, coolants/moderators, and neutron spectrums of various types of operating nuclear power reactors**

Reactor type	Main Countries	No.	GW_{el}	Fuel	U^{235} enrichment (wt.%)	Coolant/ moderator	Neutron spectrum
PWR	United States, France, Japan, China, Russia, South Korea	290	273	UO_2	2.1–3.1[a] 4.5–5.5[b]	Water/water	Thermal
BWR	United States, Japan, Sweden	78	76	UO_2	2.6–3.05[c]	Water/water	Thermal
CANDU or PHWR	Canada India	49	25	UO_2	0.71[d]	Heavy water/ heavy water	Thermal
AGR	United Kingdom	14	8	UO_2	2.3	CO_2/graphite	Thermal
LGR (RBMKs and EGPs)	Russia	11+4	10	Enriched UO_2	2.4[e]	Water/ graphite	Thermal
SFR: BN-600	Russia	2	1.3	UO_2 and PuO_2	17, 21, and 26	Liquid sodium/–	Fast

[a]Westinghouse design.
[b]VVER design.
[c]General Electric design initial enrichment is 1.7–2.0 wt.% U^{235}.
[d]EC6 design.
[e]RBMK design.

percentage of fissile nuclides compared to thermal-neutron-spectrum reactors. The majority of the current nuclear power reactors have been designed as thermal-neutron-spectrum reactors (PWRs, BWRs, PHWRs, AGRs, and LGRs), and only two reactors—BN-600 and BN-800 in Russia are fast reactors cooled with sodium (so-called SFRs).

In a nuclear fuel, the fission chain reaction is maintained by fission of fissile elements, which are capable of sustaining the fission reaction with neutrons of all energies. As such, fissile nuclides are used in the fuel of both thermal-neutron-spectrum and fast-neutron-spectrum reactors. The fissile nuclides of importance for nuclear reactors are U^{233}, U^{235}, Pu^{239}, and Pu^{241}. Among these fissile nuclides, only U^{235} is a naturally occurring nuclide, while others are produced by neutron capture of other nuclides during operation of a nuclear reactor. For instance, Pu^{239} is produced by neutron capture of U^{238}; Pu^{241}—by neutron capture of Pu^{240}; and U^{233}—by neutron capture of Th^{232}. The current nuclear reactors rely on U^{235} as the primary fissile element, often in an enriched proportion, but also use the Pu^{239} produced during irradiation by neutron capture in naturally occurring U^{238}. In fact, the fuel composition of most nuclear power reactors consists primarily of U^{238}, which undergoes fission only with very high-energy neutrons. In modern reactors, the degree of U^{235} enrichment is limited to $<20\%$ on the grounds of nonproliferation concerns, and the separation and recycling of fissile Pu^{239} from once-used fuel is subject to the international IAEA safeguards. As shown in Table 4.17, fast-neutron-spectrum reactors require a higher percentage of fissile elements than thermal-neutron-spectrum reactors as the probability of fission reaction of fissile elements such as U^{235} and Pu^{239} with fast neutrons is lower.

Even though the majority of nuclear power reactors are thermal-neutron-spectrum reactors, there has been a continuous scientific effort for design and operation of fast-neutron-spectrum reactors since the inception of the nuclear technology. There is a renewed interest in fast-neutron-spectrum reactors to be included as part of the overall nuclear fuel cycle, because of the advantages that these reactors offer. Currently, an international collaboration has focused on the development of six concepts of the Generation IV nuclear reactors. There are several fast-neutron-spectrum reactors among the selected designs (see Section 4.2).

Nuclear reactors can be designed on a basis of their fuel cycle, such that they breed more fissile nuclides than what they use. Breeder reactors can utilize uranium, thorium, and plutonium resources more efficiently. There are two types of breeder reactors: (1) fast-neutron-spectrum-breeder and (2) thermal-neutron-spectrum-breeder reactors, which are designed based on U^{238} (99.2% natural abundance) and Th^{232} (100% natural abundance), respectively. Fertile nuclides U^{238} and Th^{232} capture neutrons and transform, respectively, to fissile nuclides Pu^{239} and U^{233}. Through this process, which is known as breeding, the reactor produces more fissile nuclides than what it consumes. Fast-breeder reactors (FBRs) can also be used in order to transmute the long-lived minor actinides in the spent fuel to radionuclides with shorter half-lives. Thermal breeder reactors, on the other hand, produce less minor actinides in the spent fuel. Th^{232}-U^{233} breeding cycle can be utilized in both fast and thermal reactors. Even though both fast and thermal breeder reactors have been designed, FBRs are more efficient breeders. It is also notable that FBRs utilize the same fuel, UO_2-PuO_2, which has been used in some of the current nuclear power reactors.

The Th^{232}-U^{233} cycle is of interest due to the abundance of thorium in the earth's crust is between three to five times that of uranium. There are large thorium deposits in some countries such as India, Brazil, Australia, and United States. The U^{238}-Pu^{239} cycle is the most effective as far as fast neutrons are concerned. For Pu^{239}, the number of emitted neutrons in a fission reaction per absorbed neutrons is greater, when fission is induced by fast neutrons rather than thermal neutrons. The additional neutrons emitted can be utilized for transform more U^{238} nuclides to Pu^{239}. Hence, the fast-breeder reactors are based on U^{238}-Pu^{239} in which Pu^{239} and U^{238} in the core undergo fission. In a two-region reactor, the U^{238} nuclides in the core and in the blanket are transmuted to Pu^{239}. The blanket surrounds the reactor core and is the region containing the fertile nuclides.

Even though U^{235} is used as the primary fissile nuclide in the fuel of nuclear reactors, the fuel is designed in various geometrical configurations and chemical forms. In terms of geometrical configuration, nuclear fuels have been designed in the form of cylindrical pellets, annular pellets, pebbles, plates, and TRISO pellets. In the vast majority of current nuclear power reactors cylindrical pellets are used, but in AGRs annular pellets are used. In terms of the chemical structure, nuclear fuels can be classified into four categories: (1) metallic fuels, (2) ceramic fuels, (3) hydride fuels, and (4) composite fuels.

4.5.2 Metallic fuels

Uranium, plutonium, and thorium are the most common metallic fuels. These metallic fuels have high thermal conductivity, high fissile atom density, good neutron economy, and good fabrication. On the other hand, metallic fuels suffer from irradiation instability, poor mechanical properties and corrosion resistance, especially, at high temperatures, when exposed to air or water. The exposure of the metallic fuels to a neutron flux results in fuel swelling and has a negative effect on the thermal conductivity. In addition, metallic fuels suffer from low melting points, and, furthermore, the fuel undergoes a phase change. A phase transformation results in a volume change in the fuel. Consequently, the use of metallic uranium fuel is limited to temperatures below 660°C.

To improve these undesirable characteristics of the metallic uranium, uranium alloys such as uranium-aluminum, uranium-magnesium, and uranium-molybdenum have been developed. Metallic fuels have been used in some power reactors and research reactors with relatively low operating temperatures.

For use in high-temperature applications, a potential fuel must have a high melting point, high thermal conductivity, and good irradiation and mechanical stability. These requirements eliminate the use of metallic fuels mainly due to their low melting points and high irradiation creep and swelling rates. On the other hand, ceramic fuels have promising properties, which has made these fuels as the fuels of choice for the current nuclear power reactors and suitable candidates for high-temperature applications.

4.5.3 Ceramic fuels

Ceramic fuels have high melting points, good dimensional and radiation stability and are chemically compatible with most coolants and cladding (sheath) materials. In addition to melting point, the thermal conductivity of a fuel is a critical property that affects the operating temperature of the fuel (the highest temperature in a reactor is the fuel centerline temperature, for hollow pellets it will be the internal wall temperature). UO_2 has been used as the fuel of choice in PWRs, BWRs, and CANDU reactors. The thermal conductivity of UO_2 is between ~2 and 4 W $(m\,K)^{-1}$ within the operating temperature range of 1000°C and 2800°C. On the other hand, fuels such as UC_2, UC, and UN have significantly higher thermal conductivities compared to that of UO_2 and other oxide fuels as shown in Fig. 4.49. The higher thermal conductivities of these fuels result in lower fuel temperatures compared to those of UO_2 under the same operating conditions.

Considering chemical structures of ceramic fuels, these fuels can be categorized as oxide fuels, carbide fuels, and nitride fuels. Oxide fuels such as UO_2, MOX, and ThO_2 have low thermal conductivities compared to carbide and nitride fuels. Hence, from the heat-transfer point of view, oxide fuels can also be identified as low thermal-conductivity fuels. On the other hand, carbide (e.g., UC and UC_2) and nitride (e.g., UN) fuels are identified as high thermal-conductivity fuels. Table 4.18 lists basic properties of these fuels at 0.1 MPa and 25°C.

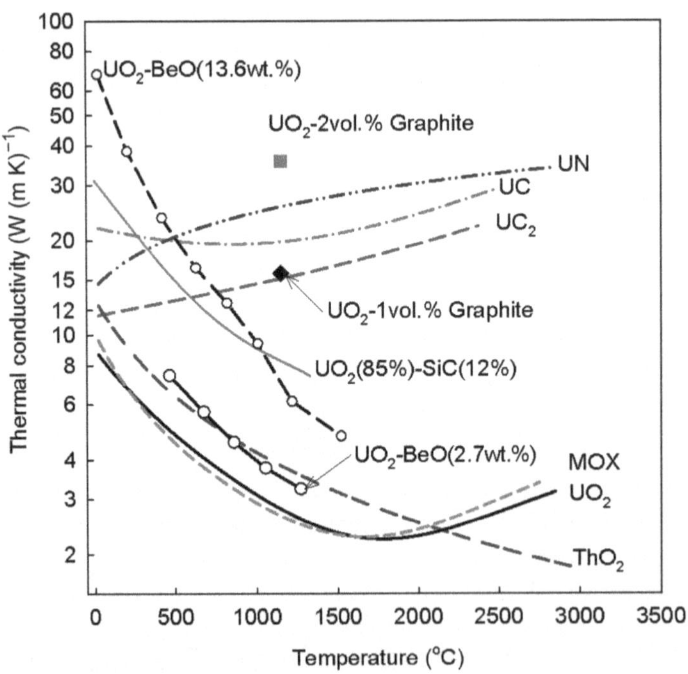

Fig. 4.49 Thermal conductivity of several nuclear fuels [1].

Table 4.18 Basic properties of selected fuels at 0.1 MPa and 25°C [1]

Property	Unit	UO_2	MOX	ThO_2	UC	UC_2	UN
Molecular mass	amu	270.3	271.2	264	250.04	262.05	252.03
Theoretical density	$kg\,m^{-3}$	10,960	11,074	10,000	13,630	11,680	14,420
Melting point	°C	2847 ± 30	2750 ± 50	3378 ± 17	2507	2375	2850 ± 30
					2520	2562	
					2532		
Heat capacity	$J\,(kg\,K)^{-1}$	235	240	235	203	233	190
Heat of vaporization	$kJ\,kg^{-1}$	1530	1498	–	2120	1975 ± 20	1144
							3325
Thermal conductivity	$W\,(m\,K)^{-1}$	8.7	7.8	9.7	21.2	11.57	14.6
Linear expansion coefficient ($\times 10^{-6}$)	K^{-1}	9.75	9.43	8.9	10.1	18.1	7.52
Electric resistivity ($\times 10^{-8}$)	$\Omega\,m$	7.32	–	–	250	120	146

4.5.3.1 *Oxide fuels*

Oxides of uranium, thorium, and plutonium have been used as nuclear fuels (i.e., UO_2, PuO_2, and MOX) have good corrosion resistance, high melting point, and excellent mechanical and irradiation stability. As such, UO_2, PuO_2, and MOX have been used as fuels in nuclear power reactors such as PWRs, BWRs, and CANDU reactors. In addition, oxide fuels are chemically compatible with the cladding (sheath) materials and water, which is used as the coolant in these reactors. On the other hand, the disadvantages of oxide fuels include low uranium-atom density, low thermal conductivity, and poor thermal-shock resistance.

Also, it should be noted that many physical properties of nuclear fuels have relatively large uncertainties and depend on many factors including porosity, manufacturing process, stoichiometric composition, and irradiation with neutron flux (see Fig. 4.50).

Currently, there is an interest in using thorium-based fuels in nuclear reactors. Thorium is widely distributed in nature and is approximately three times as abundant as uranium. However, ThO_2 does not have any fissile elements to fission with thermal neutrons. Consequently, ThO_2 must be used in combination with a "driver" fuel (e.g., UO_2, UC, or PuO_2), which has U^{235} as its initial fissile element. The presence of a "driver" fuel such as UO_2 in a nuclear reactor core results in the production of enough neutrons, which in turn start the thorium cycle.

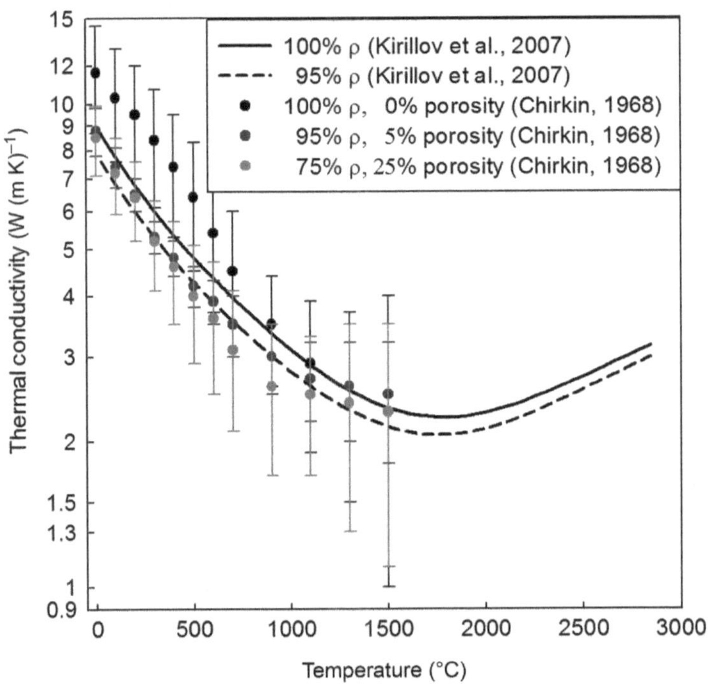

Fig. 4.50 Uncertainty in thermal conductivity of UO_2 [1].

4.5.3.2 Carbide fuels

Carbides of uranium and thorium have been considered as nuclear fuels. The use of carbides of plutonium has also been investigated, but as mixed carbides such as UC-PuC. Compared to MOX fuel, mixed carbide fuel has a higher thermal conductivity, higher heavy-metal density, and better neutron economy. Carbides of thorium, ThC and ThC_2, are the most stable compounds of thorium after thorium dioxide. ThC is stable up to temperatures close to its melting point. Carbides of uranium have desirable properties such as high thermal conductivities and high melting points. Uranium carbide (UC) and uranium dicarbide (UC_2) are two carbides of uranium, which can be used as nuclear fuels. Uranium sesquicarbide (U_2C_3) is another carbide of uranium. U_2C_3 cannot be manufactured through casting or compaction of a powder. But UC_2 may transform to U_2C_3 at high temperatures and under stress.

4.5.3.3 Nitride fuels

There are three compounds of uranium nitride system, namely, uranium mononitride (UN), uranium dinitride (UN_2), and uranium sesquinitride (U_2N_3). Among these compounds, UN has been considered as nuclear fuel for use in space nuclear reactors and sodium-cooled fast-breeder reactors, because of its superior properties such as high thermal conductivity, high melting point, and high uranium-atom density. The fuel residence time in the reactor core can be increased, when UN is used as a fuel.

4.5.4 Hydride fuels

A promising fuel for future use in LWRs is the hydride fuel, which has been used in TRIGA reactors. Uranium-thorium-zirconium fuels are also of interest due to a significant amount of thorium resources and utilization of the fuels in breeder reactors.

4.5.5 Composite fuels

Currently, there is a great interest in developing high thermal-conductivity fuel and/or improving the thermal conductivity of low thermal-conductivity fuels such as UO_2. High thermal conductivities result in lower fuel centerline temperatures and limit the release of gaseous fission products. As shown previously, UO_2 has a very low thermal conductivity, especially, at high temperatures compared to other fuels such as UC, UC_2, and UN. However, research has shown that the thermal conductivity of oxide fuels such as UO_2 can be increased by either adding a continuous solid phase or long, thin fibbers of a high thermal-conductivity material.

A high thermal conductivity material must have a low neutron absorption cross section depending on the reactor. In addition, it must have a high melting point and be chemically compatible with the fuel, the cladding, and the coolant. The need to meet these requirements narrows the potential materials to silicon carbide (SiC), beryllium oxide (BeO), and graphite (C).

4.5.6 Nuclear fuel cycles and global sustainability

Sustainable energy supply relies on essentially inexhaustible resources of fissionable fuel, notably uranium and thorium, and a benign or negligible impact on the Planet, notably reduced CO_2 emissions. Future nuclear power reactor development should focus on integrated and complementary reactor and fuel-cycle development. This is one of the major goals of the Generation IV/GIF Program [1], and requires nuclear fuel re-use. This is challenging, because of the implications for extensive recycling and reprocessing, while mandating and maintaining the aims of nuclear nonproliferation.

Nuclear energy is a global benefit, and its use has global consequences. Because of limited indigenous fuel deposits, some nations/countries (e.g., France, Japan, China, United States, India, Russia, and others) have proceeded to not only secure access to uranium fuel supplies on the open marketplace, but also embraced recycling as the major objective. In tandem with reactor deployment over the last 50 years, they have also pursued the development of so-called breeder reactors, in particular, variants of the sodium-cooled plutonium- or uranium-fuelled reactor (the so-called LMFBR). None of these reactors have been deployed commercially (exception is two LMFBRs: BN-600 and BN-800 (SFRs) in Russia), but are in full prototype testing in China, Russia, and India. This "closing-the-fuel-cycle" activity would remove reliance on imported or premanufactured enriched fuels, which are supplied by countries that already possess nuclear weapons and related fuel-processing capability (e.g., Russia, France, United States, and United Kingdom). The countries with sufficient and/or naturally abundant indigenous supplies (e.g., Canada, Australia, Africa, Kazakhstan, etc.) provide the "raw material," presently natural uranium.

There is no present shortage of uranium fuels, and the currently known or economic resources can fuel the existing and projected reactor numbers (<1000) with once-through no-recycling for 20–30 years, or more. Even if existing cheaper resources become depleted, or more "once-through-cycle" reactors are built, more exploration will undoubtedly occur as for any ore or mineral, and prices will simply depend on global demand and assured supply. Cheap natural gas will and must fill the global energy needs for a while as a bridge to a more sustainable future, displacing coal, where it makes economic and political sense, but there is no "free" solution. The constraining factors are primarily adequate nuclear expertise, national-political commitment, and the high NPP capital investment. Today, there is a large increase in the number of countries announcing plans for or expressing national interest in deploying nuclear power reactors, many for the first time (e.g., Chile, Egypt, Jordan, Philippines, Poland, Saudi Arabia, Turkey, UAE, etc.) and others proposing to add even more (e.g., Argentina, China, Finland, India, Korea, Romania, United Kingdom, and Russia).

Ultimately, in the future global energy context, there will continue to be inexorably growing demand for economic progress, increasing interest in alternative energy sources like renewables and hydrogen fuels, increasing desire for security and sustainability of energy supply, and rising concerns over the impact of increasing environmental emissions. These will all drive the need for even more energy, and eventually for more extensive deployment of nuclear energy worldwide as a "share" of the supply.

What that share will depend on actual sustainability and cost competitiveness. The international trends are for more effective uranium utilization, closed fuel cycles with reprocessing and recycle of spent fuel, and more effective and efficient management of spent fuel and reduction of eventual wastes, are becoming obvious. These trends require major exporters of nuclear reactors and uranium fuel with international commitments, to develop an effective international presence and new technical processes in order to keep technology relevant and competitive (e.g., as evidenced by the US GNEP efforts).

From any reactor or cycle, recycled or reused fuel is not cheap: but it is sustainable. Fig. 4.51 shows the schematic of possible fuel cycles. Of importance is fuel fabrication and the introduction of plutonium and thorium as fuels at the top level. This expands the possible fuel resources immensely, and suppliers should be considering proceeding with enrichment to add value to uranium and thorium resources in order to compete with other nations in global markets (e.g., in Australasia, Europe, and Asia).

The technical, economic, and nonproliferation issues and challenges associated with both enrichment and reprocessing, the issues associated with fuel supply guarantee together with spent fuel take back problems, and the issues associated with developing global technology partnerships must be pursued and resolved.

In the strategic economic context, it is necessary to develop a longer-term view that will synergistically benefit the economy and the global environment. Such a view will

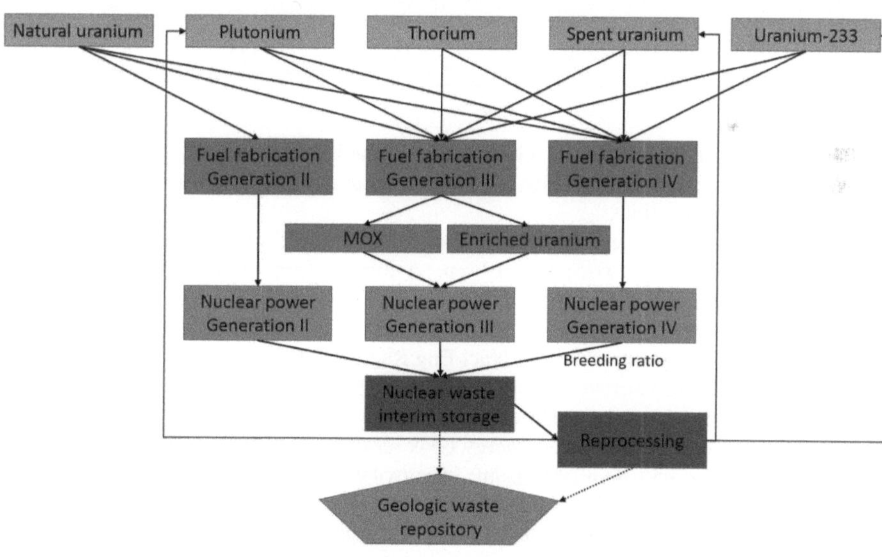

Fig. 4.51 Schematic of nuclear fuel cycles.
Based on figure from Edmonds JA, Wise MA, Dooley JJ, et al. Nuclear energy. A core element of global energy technology strategy to address climate change, report. United States: Battelle Memorial Institute; 2007 May. 22 pp.

provide a sound technological and scientific framework for the future, and must also address the collateral issues such as nuclear nonproliferation more effectively and realistically (e.g., India, Iran, Iraq, Israel, North Korea, and Pakistan positions).

Two fuel cycles are available now, which have both been extensively discussed over the past 50 years. One is the reuse of plutonium, having been literally produced in "breeder" reactors, but, which requires a commitment to a mixed fast-thermal reactor fleet. The other cycle is using thorium, which is more globally plentiful (perhaps three times more than uranium), so meets the future needs. With careful fuel design and recycling, a thermal reactor gives a near-breeding cycle, so is more sustainable with much lower (up to 10 times less) waste amounts and storage needs. This thorium cycle would enable more reactor deployment using today's reactor technologies. But there is no commitment today to full deployment, although France, India, Japan, and Russia are all studying the use of plutonium cycles; and China, India, and others are considering the deploying reactors using the integrated thorium cycle.

Acknowledgments

Authors would like to express their great appreciation to Mr. Sumio Fujii (Mitsubishi Heavy Industries (MHI)), Mr. Yuchiro Yoshimoto (Hitachi GE Nuclear Energy, Ltd.), Mr. Ala Alizadeh and Mr. Stephen Yu (Candu Energy Inc., SNC-Lavalin Group), US NRC, US DOE, ROSENERGOATOM, ROSATOM, Toshiba, and authors of the figures from the Wikimedia Commons website and Wikipedia for their materials used in this Chapter.
Also, authors would like to express his great appreciation to Dr. H. McFarlane, Dr. H. Paillere, Ms. G. Grosch, and Generation IV International Forum (GIF) Secretariat for their materials used in this Chapter.

References

[1] Pioro IL, editor. Handbook of Generation IV nuclear reactors. Duxford: Elsevier—Woodhead Publishing (WP); 2016. 940 pp. Available from: https://www.gen-4.org/gif/jcms/c_9373/publications.

[2] Pioro I, Duffey R. Nuclear Power as a Basis for Future Electricity Generation. ASME J Nucl Eng Rad Sci 2015;1(1). 19 pp. Available from: http://nuclearengineering.asmedigitalcollection.asme.org/article.aspx?articleID=2085849.

[3] Dragunov A, Saltanov E, Pioro I, Kirillov P, Duffey R. Power cycles of generation III and III$^+$ nuclear power plants. ASME J Nucl Eng Rad Sci 2015;1(2). 10 pp.

[4] Pioro I, Kirillov P. Current status of electricity generation in the world. In: Méndez-Vilas A, editor. Materials and processes for energy: communicating current research and technological developments. Energy book series #1. Spain: Formatex Research Center; 2013. p. 783–95. Available from: http://www.formatex.info/energymaterialsbook/book/783-795.pdf.

[5] Pioro I, Kirillov P. Current status of electricity generation at nuclear power plants, In: Méndez-Vilas A, editor. Materials and processes for energy: communicating current research and technological developments. Energy book series #1Spain: Formatex Research Center; 2013. p. 806–17. Available from: http://www.formatex.info/energymaterialsbook/book/806-817.pdf.

[6] Pioro I, Kirillov P. Generation IV nuclear reactors as a basis for future electricity production in the world, In: Méndez-Vilas A, editor. materials and processes for energy:

communicating current research and technological developments. Energy book series #1Spain: Formatex Research Center; 2013. p. 818–30. Available from: http://www.formatex.info/energymaterialsbook/book/818-830.pdf.
[7] Shultis JK, Faw RE. Fundamentals of nuclear science and engineering. 2nd ed. Boca Raton, FL: CRC Press; 2007. 591 pp.
[8] http://www.westinghousenuclear.com/New-Plants/AP1000-PWR; [Accessed 15 October 2017].
[9] AECL Report, Pickering generation station. Design description, Report prepared by Atomic Energy of Canada Limited; 1969. 229 pp.
[10] Hewitt GF, Collier JG. Introduction to nuclear power. 2nd ed. New York, NY: Taylor & Francis; 2000. 304 pp.
[11] ROSENERGOATOM, Russian nuclear power plants. 50 Years of nuclear power, Moscow, Russia; 2004. 120 pp.
[12] http://www.areva.com/EN/operations-5444/epr-reactor-fact-sheet.html; [Accessed 15 October 2017].
[13] Nuclear News. Publication of American Nuclear Society (ANS); 2017 March. p. 35–67.
[14] Oka Y, Koshizuka S, Ishiwatari Y, Yamaji A. Super light water reactors and super fast reactors. Germany: Springer; 2010. 416 pp.
[15] Yamada K, Sakurai S, Asanuma Y, Hamazaki R, Ishiwatari Y, Kitoh K. In: Overview of the Japanese SCWR concept developed under the GIF collaboration. Proc. ISSCWR-5, Vancouver, BC, Canada, March 13–16; 2011.
[16] Schulenberg T, Starflinger J, editors. High performance light water reactor. Design and analyses. Germany: KIT Scientific Publishing; 2012. 241 pp.
[17] Pioro IL, Duffey RB. Heat transfer and hydraulic resistance at supercritical pressures in power engineering applications. New York, NY: ASME Press; 2007. 334 pp.
[18] Guzonas D, Brosseau F, Tremaine P, Meesungnoen J, Jay-Gerin J-P. Water chemistry in a supercritical water-cooled pressure tube reactor. Nucl Tech 2012;179:205–19.
[19] Saltanov E, Pioro I. World experience in nuclear steam reheat. In: Tsvetkov PV, editor. Nuclear power: operation, safety and environment. Rijeka: INTECH; 2011. p. 3–28. Available from: http://www.intechopen.com/books/nuclear-power-operation-safety-and-environment/world-experience-in-nuclear-steam-reheat.
[20] Nuclear Fuel Cycle Royal Commission. Final report, quantitative analysis and initial business case; 2016 [chapter 6].
[21] Dragunov A, Saltanov E, Pioro I, Ikeda B, Miletic M, Zvorykina A. In: Investigation of Thermophysical and nuclear properties of prospective coolants for generation-IV nuclear reactors. Proceedings of the 21st International conference on nuclear engineering (ICONE-21), July 29–August 2, Chengdu, China, paper #16020; 2013. 11 pp.
[22] IAEA-TECDOC-1746. Heat transfer behaviour and thermohydraulics code testing for supercritical water cooled reactors (SCWRs). IAEA TECDOC series, September, Vienna, Austria, 496 pp. Available from: http://www-pub.iaea.org/books/IAEABooks/10731/Heat-Transfer-Behaviour-and-Thermohydraulics-Code-Testing-for-Supercritical-Water-Cooled-Reactors-SCWRs; 2014.
[23] Gupta S, Saltanov E, Mokry SJ, Pioro I, Trevani L, McGillivray D. Developing empirical heat-transfer correlations for supercritical CO_2 flowing in vertical bare tubes. Nucl Eng Des 2013;261:116–31.
[24] Pioro I. The potential use of supercritical water-cooling in nuclear reactors. In: Krivit SB, Lehr JH, Kingery TB, editors. Nuclear energy encyclopedia: science, technology, and applications. Hoboken, NJ: J. Wiley & Sons; 2011. p. 309–47.
[25] Pioro I, Mokry S, Draper S. Specifics of thermophysical properties and forced-convective heat transfer at critical and supercritical pressures. Rev Chem Eng 2011;27(3–4):191–214.

[26] Pioro I, Mokry S. Thermophysical properties at critical and supercritical conditions. In: Belmiloudi A, editor. Heat transfer. theoretical analysis, experimental investigations and industrial systems. Rijeka: INTECH; 2011. p. 573–92. Available from: http://www.intechopen.com/books/heat-transfer-theoretical-analysis-experimental-investigations-and-industrial-systems/thermophysical-properties-at-critical-and-supercritical-pressures.
[27] National Institute of Standards and Technology. NIST reference fluid thermodynamic and transport properties–REFPROP. NIST standard reference database 23, Ver. 9.1, Boulder, CO: Department of Commerce; 2010.
[28] Beznosov AV, Dragunov YG, Rachkov VI. Heavy-liquid metal coolants in nuclear engineering [in Russian]. Moscow: IzdAt Publishing House; 2007. 434 pp.
[29] Todreas NE, MacDonald PE, Heizlar P, Buongiorno J, Loewen E. Medium-power lead-alloy reactors: missions for this reactor technology. Nucl Tech 2004;147(3):305.
[30] Dement'ev BD. Nuclear power reactors [in Russian]. Moscow: Energoatomizdat Publishing House; 1990. 352 pp.
[31] IAEA. Status of liquid metal cooled fast reactors. Technical report series no. 246, Vienna: IAEA; 1985.
[32] Waltar AE, Reynolds AB. Fast breeder reactors. New York: Pergamon Press; 1981. 853 pp.
[33] Kirillov PL, Terentieva MI, Deniskina NB. Thermophysical properties of matcrials for nuclear engineering. 2nd ed. Moscow: IzdAT Publishing House; 2007. 192 pp.
[34] NEA No. 6195. Handbook on lead-bismuth eutectic alloy and lead properties, materials compatibility, thermal-hydraulics and technologies. France: Nuclear energy agency (NEA) and Organization for Economic co-operation and Development (OECD); 2007. 693 pp. Available from: https://www.oecd-nea.org/science/reports/2007/pdf/lbe-handbook-complete.pdf.
[35] Sohal M, Ebner M, Sabharwall P, Sharpe P. Engineering database of liquid salt thermophysical and thermochemical properties. INL/EXT-10-18297 report, 70 pp. Retrieved from http://www.inl.gov/technicalpublications/Documents/4502650.pdf; 2010 [Accessed 1 September 2014].
[36] Khokhlov V, Ignatiev V, Afonichkin V. Evaluating physical properties of molten salt reactor fluoride mixtures. J Fluor Chem 2009;130:30–7.
[37] Williams, D.F., Toth, L.M., and Clarno, K.T. ORNL-TM-12-2006. Assessment of Candidate Molten Salt Coolants for the Advanced High-Temperature Reactor (AHTR), ORNL/TM-2006/12 report, 86 pp. Retrieved from http://web.ornl.gov/~webworks/cppr/y2006/rpt/124584.pdf [Accessed 19 August 2014].
[38] Chrenkova M, Danek V, Vasiljev R, Silny A, Kremenetsky V, Polyakov E. Density and viscosity of the $(LiF\text{-}NaF\text{-}KF)_{eut}\text{–}KBF_4\text{–}B_2O_3$ melts. J Mol Liq 2003;102(1):213–26.
[39] Peiman W, Pioro I, Gabriel K. Thermal aspects of conventional and alternative fuels in super critical water-cooled reactor (SCWR) applications. In: Mesquita AZ, editor. Nuclear reactors. Rijeka: INTECH; 2012. p. 123–56. Available from: http://www.intechopen.com/books/nuclear-reactors/-thermal-aspects-of-conventional-and-alternative-fuels-in-supercritical-water-cooled-reactor-scwr-ap.
[40] Grigor'ev VA, Zorin VM, editors. Thermal and nuclear power plants. Handbook. 2nd ed [in Russian]. Moscow: Energoatomizdat Publishing House; 1988. 625 pp.
[41] Margulova TC. Nuclear power plants [in Russian]. Moscow: Izdat Publishing House; 1995. 289 pp.
[42] Ryzhov SB, Mokhov VA, Nikitenko MP, et al. In: Advanced designs of VVER reactor plant. Proceedings of the 8th International topical meeting on nuclear thermal-hydraulics, operation and safety (NUTHOS-8), Shanghai, China, October 10–14; 2010.

[43] Nonbel E. Description of the advanced gas cooled type of reactor (AGR). Report NKS/RAK2(96)TR-C2, November, Roskilde: RisØ National Laboratory; 1996. 88 pp.
[44] Channel Nuclear Power Reactor (RBMK). In: Cherkashov YM, editor. Канальный Ядерный Энергетический Реактор РБМК [in Russian]. Moscow: Publishing House GUP NIKIET; 2006. 631 pp.
[45] Edmonds JA, Wise MA, Dooley JJ, et al. Nuclear energy. A core element of global energy technology strategy to address climate change, report. United States: Battelle Memorial Institute; 2007. May, 22 pp.

Nuclear fusion: What of the future?

5

Richard Kembleton[1]
EUROfusion PMU, Culham Centre for Fusion Research, UKAEA, Abingdon, United Kingdom
[1]Corresponding author: Richard.kemp@ukaea.uk

Chapter Outline

5.1 The promise of fusion

Although nuclear fusion is frequently mentioned as an option for a future source of clean, almost inexhaustible energy, there is much misunderstanding of its nature and the problems of achieving it. This chapter aims to lay out the basis for fusion's claim to being part of a future energy system; the main routes and concepts toward delivering fusion power; problems remaining to be solved; and how fusion might contribute to a future energy market.

Nuclear fusion is the process of combining light elements to form heavier ones. This process releases energy all the way up to iron. However, the temperature and pressure required for fusion of heavier elements is beyond the reach of processes outside of a stellar collapse, so we must limit ourselves to isotopes of hydrogen

Managing Global Warming. https://doi.org/10.1016/B978-0-12-814104-5.00005-3

($^{1}_{1}H$, sometimes called *protium*; $^{2}_{1}H$, deuterium (D, stable); and $^{3}_{1}H$, tritium (T, unstable)); and other light elements such as helium (usually $^{3}_{2}He$) and boron. The reactivity (i.e., the probability that the collision of two fuel particles leads to a reaction) of some of these reactions is shown in Fig. 5.1.

Due to it being the reaction with the highest reactivity at achievable temperatures, most near-future fusion concepts are based on deuterium-tritium (D-T) fusion. The reaction is:

$$D + T = {}^{2}_{1}H + {}^{3}_{1}H \rightarrow {}^{4}_{2}He\,(3.5\,MeV) + n\,(14.1\,MeV)$$

This reaction results in a charged alpha particle ($^{4}_{2}He$) and a neutron. The alpha particle can be constrained by magnetic fields and interacts strongly with other charged particles (ions and electrons) in the fusion plasma to heat it up. The neutron is not well constrained and its energy is deposited in whatever material surrounds the plasma, usually causing considerable damage to the material at an atomic level. For energy production, it can be used to make more tritium, which has a short half-life of about 12 years and so does not occur naturally in any abundance:

$$^{6}_{3}Li + n \rightarrow {}^{4}_{2}He\,(2.05\,MeV) + {}^{3}_{1}T\,(2.75\,MeV)$$

$$^{7}_{3}Li + n \rightarrow {}^{4}_{2}He + {}^{3}_{1}T + n\,(-2.5\,MeV)$$

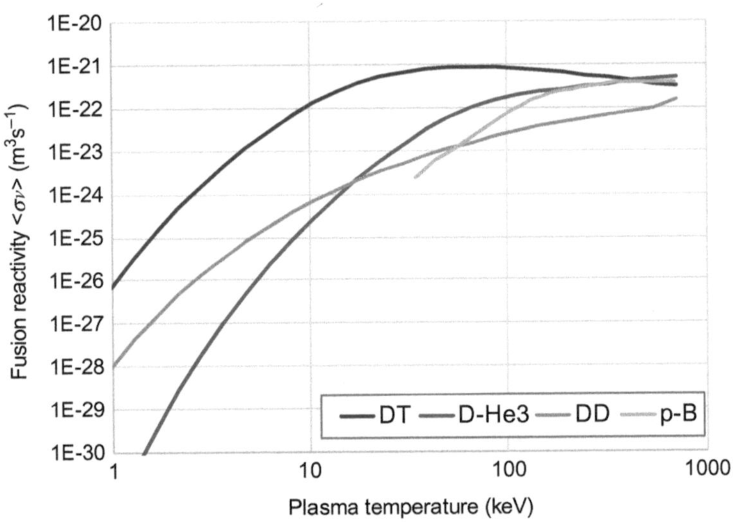

Fig. 5.1 Fusion reactivities for common reactions. The fusion power is then given by $P_{fus} = n_D n_T \langle \sigma v \rangle E_{fus}$. The deuterium-tritium (DT) reaction has a higher peak at a lower temperature than the alternatives, making it the most achievable option. The fully aneutronic hydrogen (proton)-Boron (p-B) reaction requires very high temperatures. See section 5.2 for more information.
Data from Bosch H-S, Hale GM. Improved formulas for fusion cross-sections and thermal reactivities, Nucl Fusion 1992; 32: 611–631.

The second reaction is endothermic and so most fusion concepts assume the tritium breeder is highly enriched with ${}^{6}_{3}Li$ to maximize energy output. In addition, a neutron multiplier such as lead or beryllium must be mixed with the lithium to ensure there are sufficient neutrons available to breed more tritium than is consumed by the reaction. D-${}^{3}_{2}He$ and p-B fusion reactions are aneutronic and do not require the breeding of tritium, so reactor concepts can be simpler (see Technology Challenges, Section 5.3) but require much higher temperatures to produce significant power. At these high temperatures, energy losses from the plasma through Bremsstrahlung radiation can cool the plasma more powerfully than the fusion products heat it, making sustained power production difficult.

The old joke is that nuclear fusion is 30 years in the future, and always has been. It is the case that research into fusion for energy has been a slower-than-expected process, but it has also been more difficult to realize than anticipated. Much of the work to date has revolved around controlling and heating plasmas to fusion-relevant regimes, discovering a wide range of instabilities and interesting behaviors on the way [1]. Recently, consideration has turned to the technological problems of actually building a fusion power plant, which has brought to light a new set of challenges, some of which are reviewed later. If constructed, a fusion power plant would be one of the most complex machines ever built, with environments requiring semiautonomous robotic maintenance. Engineering such a system to meet power-supply requirements of reliability, availability, maintainability, and inspectability is a considerable challenge.

The presentation of fusion in this chapter focuses mainly on D-T fusion in tokamaks (see Section 5.2), but most of the challenges and conclusions also apply to other concepts.

5.1.1 Fusion resources

As discussed earlier, the main fusion fuels are deuterium and lithium. Approximately one in 7000 hydrogen nuclei in the universe is deuterium, so 1 L of water contains 0.033 g of deuterium, which, when combined with tritium, releases the same energy as burning 280 L of oil. As there is no shortage of water, there is enough deuterium for billions of years of global energy supply. However, as discussed, tritium must be bred from lithium. Since the fusion reaction releases a lot of energy, it does not take much lithium to supply the needs of a single individual; a few grams of lithium, as found in a single laptop battery, can provide enough tritium to release 250,000 kWh, a typical European lifetime consumption of energy. Known land-based lithium resources provide enough lithium for around a thousand years of global energy demand if it is all used for fusion (see, for example, [2]). There is competition, though, from energy storage demands such as laptop batteries. However, although commercial methods do not yet exist, lithium is also extractable from seawater and in total the potential reserves from this source are orders of magnitude higher [3]. The lithium must, as previously stated, be enriched to ~90% ${}^{6}Li$. A number of methods to accomplish this exist at lab scale, but will require scaling up to commercial levels if fusion is to be successful as an energy technology.

Although these fusion fuels appear essentially unlimited, we must be careful not to be too dependent on other scarce resources of materials also required in the building and operation of nuclear fusion plants. Potential limitations arise from, for example, supplies of tantalum (an alloying element in low-activation steels), beryllium, tungsten (for plasma-facing surfaces), and helium (itself mostly obtained from natural gas wells and so currently dependent on the fossil-fuel industry). Estimates of the cost of fusion power plants also tend to put the use of common building materials such as concrete and steel at around a third of the total cost, and although this is nuclear-grade concrete and fusion-specific steel the costs of such basic raw materials can be very volatile, depending on international economic activity. It is worth remembering that fusion, like fission, is a capital-intensive energy source and these costs can far outweigh the costs of scarce materials, which are consumed in small quantities.

There is, however, a remaining fly in the ointment. To start up a fusion reactor will require at least several kilograms of tritium to sustain power (and neutron production) until the blanket is producing sufficient supply to maintain self-sufficiency (and also produce an excess to start other reactors), although this could theoretically be accomplished, albeit slowly, by using the neutrons produced from an initial D-D plasma [4]. Current global supplies of tritium are limited to that slowly and expensively produced in a small number of aging fission reactors [5]. This production could be stepped up if commercial demand was there, albeit on a multidecade lead time and considerable forethought would be required. As production fission reactors shut down, the availability of tritium is forecast to peak and decline following ITER (the next-step international fusion experiment currently under construction in France); there may only be enough tritium readily available for one or two power-plant-scale devices following ITER if they do not breed their own fuel. This bottleneck, and the low rate at which excess tritium might be produced in a fusion breeder blanket, may substantially limit the speed with which fusion power plants could be deployed even once the technology is successfully demonstrated.

5.1.2 Fusion safety

Nuclear fusion will of course require regulation as a nuclear technology. The main risk to the public arises from the potential for leaks of tritium, and this drives the safety design for conceptual plants. Tritium decays to ^{3}He through beta emission and as a hydrogenic species burns to form water, which is easily taken up by biological organisms. In addition, it easily permeates through many materials and systems exposed to it must be detritiated during decommissioning before they can be recycled. A fusion power plant may have >10 kg of tritium on-site in various systems—and much effort is devoted to the engineering of systems, which should minimize this.

The design goal for a fusion power plant is zero evacuation—that is, even under the worst conceivable series of system failures, the amount of released radioactive material should not be enough to require evacuation of any surrounding area [6]. This, of course, requires the engineering of multiple levels of safety systems. For example, it is possible that a plasma disruption could melt part of the interior of the reactor, rupturing coolant channels in the blanket and causing the release of tritiated steam at high pressure into the reactor. Safety engineers model such events and the resulting pressure rises in the components to calculate where to install rupture disks, expansion volumes, and condensation tanks to ensure the release would be safely contained with no risk of tritium leakage beyond a primary containment level—and in case this also failed, there is also secondary containment in line with conventional nuclear defense-in-depth principles [7].

Fusion plasmas, while containing a lot of energy, are intrinsically unstable and require active control. In any case of loss of control, the plasma may disrupt, possibly causing damage to the reactor, but it cannot sustain a runaway nuclear reaction. There is some residual nuclear decay heat in the materials following the collapse of the plasma (which also affects the maintenance of the plant), but this is dispersed across a large volume of material and even a failure of plasma control followed by a complete loss of coolant does not lead to melting of plant materials [8].

The neutrons produced by the D-T reaction cause activation of the materials surrounding the plasma to form low- and intermediate-level waste. It is possible to design materials (for example, so-called low-activation steels), which limit the quantity of long-lived radionuclides formed and are intended to permit recycling of most structural elements of a fusion reactor within hundreds of years, a far more tractable problem than the tens of thousands of years required for the storage of fissile waste [9]. However, this does require adequate detritiation and separation of materials extracted from the reactor and the full feasibility of these processes is not yet clear. The greatest production of long-lived nuclear material in common concepts is carbon-14, produced by neutron radiation of nitrogen, found in water (used as a coolant) and as a common alloying element in steel. Although ^{14}C occurs naturally in low levels (produced in the atmosphere by high-energy cosmic rays), it is readily taken up by biological organisms and can be incorporated into DNA, where its decay can be damaging (the half-life of ^{14}C is 5730 years). Therefore, the lifetime production of ^{14}C by a fusion power plant must be limited, leading to stringent manufacturing limits on structural alloys.

While a fusion reactor does not intrinsically use fissile material in its operation, the neutrons available from the reaction would allow the creation of nuclear isotopes if the right elements were incorporated into the blanket. This would not be a straightforward operation and would be immediately obvious to inspectors.

5.2 Fusion concepts

In order to generate significant fusion power, the reaction plasma needs high fuel densities and high reactivity:

$$P_{\text{fus}} = n_D n_T \langle \sigma v \rangle E_{\text{fus}}$$

where P_{fus} is the fusion power; n_D, n_T are the densities of deuterium and tritium fuel; $\langle \sigma v \rangle$ is the reactivity at the plasma temperature; and E_{fus} is the energy released by the fusion reaction. Fusion devices therefore aim at achieving high plasma densities and temperatures. The two main routes are *magnetic confinement* (for example, tokamaks and stellarators) and *inertial confinement*.

5.2.1 Magnetic confinement

A generic toroidal fusion reactor is shown in Fig. 5.2. The toroidal geometry is required due to the properties of magnetic fields (the so-called *hairy-ball theorem* [10]), although it makes the engineering more complex. Usually, in considering a power plant, steady-state operation is targeted to provide continuous power output and reduce cyclic loads on components, but this is not a fundamental requirement and some concepts are intrinsically pulsed.

The role of the magnetic field is to confine the plasma and hold it away from the reactor *first wall*, which provides protection to the structures behind it from direct radiation and particle loads. Behind the first wall is the *breeder blanket*, which removes the heat from the reactor for electricity generation and also creates new fuel. The blanket also provides shielding to the vacuum vessel, which seals the reactor against the outside world. Wrapped around the vacuum vessel are the *toroidal field (TF) coils*, the magnets that provide the main confinement field, and outside these may be *poloidal field (PF) coils* that provide plasma shaping and control capabilities. Finally, there is a set of plant infrastructure, which provides fuel extraction and injection, plasma heating, and thermal power extraction as well as maintenance for the reactor. Further considerations of the requirements on these systems are discussed in Section 5.3.

A purely toroidal magnetic field does not provide stable plasma confinement, so it is necessary to generate a poloidal component to make it "twist" around the plasma. In a tokamak, this poloidal field is generated by a current flowing in the plasma, and maintaining this current is one of the biggest challenges in creating a steady-state tokamak power plant. In a **stellarator**, the twist in the magnetic field is supplied by distorting the TF coils. Engineering such coil shapes is difficult and as the coils are moved further from the plasma to include blanket and shielding, the distortion must

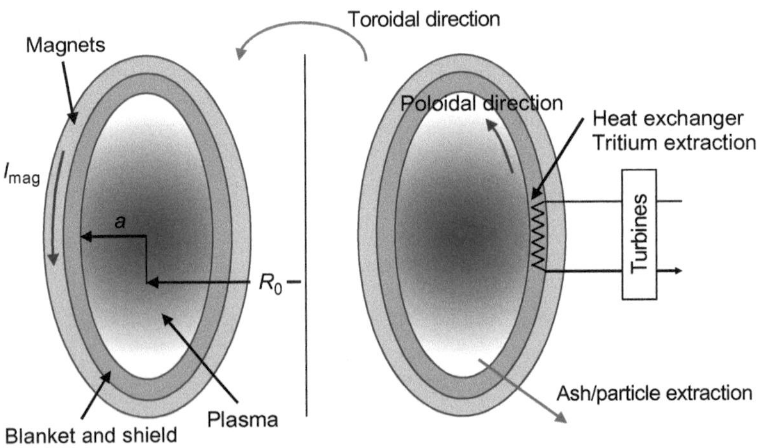

Fig. 5.2 A generic toroidal fusion reactor showing the plasma, blanket and shield, magnets, and power extraction.

increase. Gaining access to the interior of the vessel for maintenance and component replacement is complicated.

For a magnetized plasma, we need a figure of merit indicating how effective the confinement is. Usually the ratio of plasma pressure to magnetic field pressure, $\beta = \frac{\langle P \rangle}{B^2 / 2\mu_0}$, is used. Here $<P>$ is the volume-averaged plasma pressure, B is the magnetic field in the plasma, and μ_0 is the vacuum permeability. Although the transport processes in a magnetically confined plasma are complex, the simplest scaling of the confinement time is roughly $\sim a^2 B^2 T^{\frac{1}{2}} n^{-1}$, where aB is the number of diffusive steps to the edge of the plasma and $T^{\frac{1}{2}} n^{-1}$ is the collision rate. Due to the tokamak geometry, however, the magnetic field varies strongly within the plasma leading to additional classes of orbits much larger than the step size assumed here, and turbulent processes increase the transport further. Nevertheless, the easiest ways of achieving good energy confinement and hence high fusion power are to increase the machine size and magnetic field.

A **tokamak** is an axisymmetric torus in which the toroidal field (supplied by the magnets) is much larger than the poloidal field (supplied by a toroidal current flowing in the plasma). A tokamak geometry is defined by its major radius R_0, minor radius a (or aspect ratio $A = \frac{R_0}{a}$), the toroidal field in the plasma B, and the plasma current I_p. The plasma may also have elongation κ and triangularity δ, which change its cross section from a simple circle to a D-shape.

A useful additional figure of merit for a tokamak is the *normalized beta* $\beta_N = \beta \frac{aB}{I_p}$, which indicates how close the plasma is to destabilizing MHD (*magnetohydrodynamic*) activity for a particular geometry and operating regime.

Spherical tokamaks are more like a cored apple than a doughnut shape, with the consequence that the toroidal magnetic field varies much more across the plasma than in a conventional tokamak. This leads to a number of potential improvements in plasma performance. In particular, a higher pressure can be achieved for a given magnetic field, which, combined with the compactness of the device, could lead to more cost-effective approaches to fusion. The limited space on the inboard plasma side can make it hard to shield the center column and magnets from neutron radiation.

Stellarators attempt to avoid some of the issues of tokamaks by externally imposing the magnetic field twist through the use of distorted TF coils. This is an elegant physics solution but a complex engineering one (Fig. 5.3). The resulting plasma usually has an oval cross section, which twists as it moves around the device, meaning that the plasma chamber—whose walls must "hug" the plasma—also has a very variable cross section, making the robotic handling of components inside this volume extremely challenging. However, stellarators do not require a current to be driven in the plasma (although they do require auxiliary heating to achieve fusion burn) and the plasmas tend to be much more quiescent than tokamak plasmas, resulting in a number of promising simplifications.

A simple estimate of the size and requirements for performance of a magnetic-confinement-based fusion power plant can be made. If the plasma-facing components

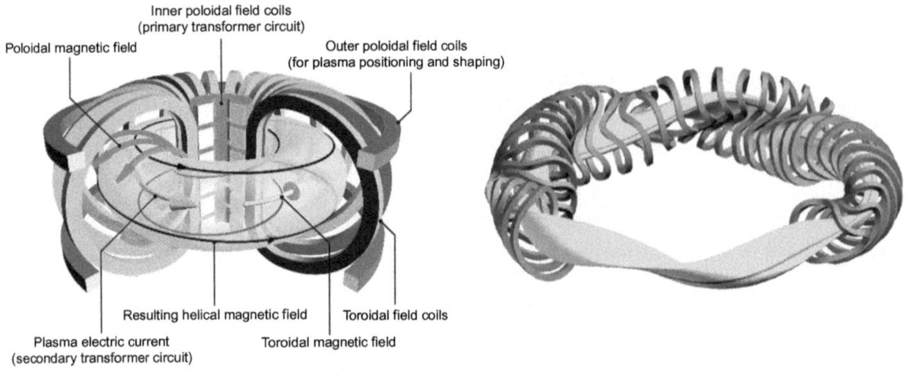

Fig. 5.3 The main magnets of a tokamak system (left) and a stellarator (right). Courtesy: EUROfusion, Max Planck Institute for Plasma Physics (IPP).

can handle a maximum neutron flux of $4\,\mathrm{MW\,m^{-2}}$, and a 1-GW (electrical) plant requires 3 GW (fusion) to generate that output (based on considerations of thermal efficiency and the power required to run plant systems [8], then that means the wall area of the reactor cannot be smaller than $600\,\mathrm{m}^2$. For a conventional tokamak of aspect ratio $A = 3$ and modest elongation ($\kappa = 1.6$), this means that the major radius $R_0 > 6\,\mathrm{m}$. Similar approaches applied to the plasma physics performance, divertor (plasma exhaust) power loading, etc. tend to lead to slightly larger machines than this, usually with R_0 about 8–9 m [11]. To achieve significant reductions in reactor size needs assumptions of advances in available material performance or plasma performance, which, while potentially possible in theory, are often not well experimentally demonstrated and would require a substantial increase in research to establish the experimental basis to provide the confidence needed to invest in a full-scale power plant designed on these principles.

5.2.2 *Inertial confinement*

Another possible method to achieve fusion is through the implosive compression of fuel. This can be accomplished by directing high-powered lasers onto a frozen D-T fuel pellet (so-called *direct drive*), or onto a casing around the pellet (a *hohlraum*) to generate a bath of X-rays to heat the pellet instead (*indirect drive*). The heating ablates the outer surface of the fuel pellet and explosively expands it, causing the interior to be rapidly compressed and heated and resulting in a burst of fusion. Any imperfection in the compression can cause fast-growing instabilities greatly reducing the peak core pressure and therefore fusion yield; the hohlraum technique intrinsically smooths the radiation field to give more reliable compression. The National Ignition Facility (**NIF**) uses 192 lasers to deposit around 20 kJ on the capsule in a few picoseconds to drive the pellet into fusion conditions [12]. As yet, this has not been as successful as hoped due to plasma instabilities limiting the density and temperature achievable in the fuel.

An inertial confinement fusion energy system would need to operate at several hertz to provide power-plant-scale electricity output. It would need to be made of similar structural materials to a magnetic confinement plant due to the same issues with neutron damage and activation. The plant would also require a breeder blanket to provide tritium fuel. It replaces plasma control problems with the problems of rapid replacement of a fuel capsule in the reaction chamber, and fast laser recharge and targeting.

5.3 Main technology challenges

Nuclear fusion presents a number of significant technological challenges, which must be solved before it can be made commercially available. A fusion reactor is a very complex device, and historical experimental systems have been chiefly physics experiments, designed for operational flexibility and, while experimental availability is a concern, the operating and maintenance regime required of a commercial power plant is very different to that of an experimental device.

5.3.1 Reactor materials

Fusion materials have a twofold problem. The first is resisting radiation damage from fusion neutrons to maintain their properties over the design lifetime—as well as exposure to high temperatures and high stresses. The second is the avoidance of elements, which form long-lived radionuclides under neutron radiation, giving rise to long-term radioactive waste—or even changing the nature and properties of an alloy through the production of transmutation elements. Additionally, the functional demands on materials vary widely: plasma-facing materials must be resistant against very high heat and particle loads, without sputtering; in-vessel structural materials must be dimensionally stable and remain ductile while installed; elements of the breeder blanket must multiply neutrons for tritium production—tritium itself is prone to permeating through materials and cannot be allowed to leach into the environment; diagnostic and heating systems may be directly exposed to radiation and sputtered dust, which will fog lenses and mirrors and affect the conductivity of electronics. Some of the materials used in reactor design and their issues are given in Table 5.1. It is difficult to conceive that a commercial reactor will be licensable by regulatory authorities without qualification of these materials in a nuclear environment, which implies that a high-energy neutron source is required for materials development.

5.3.2 Power exhaust

The power used to heat the plasma, either through self-heating from alpha particles or auxiliary systems for current drive or control, must either radiate away from the plasma or be conducted to the vessel walls somewhere. In a tokamak (and a stellarator), the structure designed for this is called the *divertor*. The divertor surface is directly exposed to the plasma and experiences the highest heat fluxes in the device,

Table 5.1 Materials used in fusion reactors, along with some of the problems faced [13–16]

Material	Used in...	Issues
Reduced-activation steels	In-vessel structural components	Radiation-induced swelling and embrittlement; temperature limits caused by microstructure; achieving alloying element limits imposed by waste considerations
Other structural steels	E.g., Superconducting magnets	E.g., High forces at cryogenic temperatures
Tungsten	Plasma-facing materials: first wall and divertor	Sputtering causing plasma contamination; radiation-induced embrittlement
Superconductors (Nb_3Sn, NbTi, etc.)	Magnets	Sensitive to radiation damage; potential for catastrophic quench if cryogenic temperatures are not maintained
Beryllium	Breeder blanket as neutron multiplier in some concepts	Toxic; limited global resources
Lithium	Breeder blanket as tritium breeder	Extraction from geological sources; efficient enrichment of lithium 6 to high levels
Lithium-lead eutectic fluid	Breeder blanket as neutron multiplier, breeder, and coolant in some concepts	Chemical compatibility, pumping
Optical materials	Lenses, windows, mirrors, and optical fibers in diagnostic systems	Neutron radiation leading to damage and fogging
Electrical insulators	Isolation of systems	Breakdown of dielectrical properties
Permeation barrier materials	Anything exposed to tritium	Prevention of tritium permeation through, and contamination of, materials
Material joints	Any manufactured subassembly	Welds and other materials joining techniques may result in microstructures more prone to neutron damage than parent material

causing erosion of the surface. It is a component, which must be replaced many times throughout the lifetime of the reactor (see Section 5.3.5). Divertor design is both a technology problem—designing materials and a structure to handle the loads and perform the exhaust functions required—and a physics problem as the plasma must also be designed and controlled in a way, which prevents very large loads falling directly on the plasma-facing materials.

5.3.3 Breeder blanket

The energy generated by the fusion plasma must be removed to generate electricity, and new tritium fuel must be created. This is the role of the *breeder blanket*, a structure that sits between the plasma and the rest of the machine to absorb neutrons and turns them into something useful. The blanket contains a neutron multiplier such as beryllium or lead, which allows a single fusion-produced neutron to create more than one new tritium atom, and a lithium-6-enriched breeder material, which produces that new tritium. The blanket also results in some energy multiplication due to the exothermic nature of the neutron multiplication and tritium-breeding reactions.

This is also a replaceable component. However, the lifetime is generally limited by neutron damage to the structural materials, not consumption of the neutron multiplier or breeder materials, which must be separated and recycled.

5.3.4 Superconducting magnets

The stability and confinement of a magnetized plasma is inherently limited by the magnetic field applied. Tokamaks use powerful magnets generating fields 10^5 times greater than Earth's natural magnetic field. This requires large magnetic coils, which carry high currents. In a standard conductor such as copper, these currents cause heating and, for a power reactor, this energy loss would be economically prohibitive. ITER, JT-60SA, and subsequent designs therefore use superconducting magnets, which can carry such currents without deleterious losses. However, designing these very large structures, which may be >20 m across and have to bear high (and cyclical) stresses at cryogenic temperatures, is a significant engineering challenge. Depending on the design and superconductor used, it may be necessary to sinter the entire coil, which must also be constructed on site as it will be too large to transport to a separate location. During operation, the coils must be maintained at near absolute-zero temperatures and shielded from neutron radiation, which would disrupt the superconductivity. High-temperature superconductors are starting to become available but cannot yet be produced in the quantity required for a large tokamak. These new superconductors may permit much higher fields, although as the forces on a magnetic coil scale with the magnetic field pressure ($\propto B^2$) this also requires stronger structural materials.

A large fraction of the estimated cost of a fusion power plant is the magnets and related systems (see Fig. 5.6).

5.3.5 Remote handling

Following a fusion burn, the interior of a fusion reactor will be a very hostile place awash with ionizing radiation. Robotic systems must be used to remove and replace components, which must then be detritiated and stored until safe to recycle. As these components must be moved around the plant into storage with the risk of tritium migration during operations, most of the space around and above the reactor and the storage are also required to be fully robotic in normal operations.

The placement of magnetic coils and the sizes of components mean that access to the reactor interior is difficult and the kinematics of component replacement complicated. However, the economics of energy production depend on high availability and this means that efficient maintenance systems must be designed. Basic designs for such systems vary considerably between power plant concepts, but most involve some sort of horizontal or vertical access port through which large sections of the blanket and divertor can be maneuvered (Fig. 5.4). Much of the maintenance time is spent cutting and welding connections to the in-vessel components, and the kinematics of removal can be strongly affected if, for example, the blanket remains full of coolant fluids after shutdown. The design and placement of the connections, and coolant flow though the blanket, is thus strongly impacted by remote handling considerations.

Fig. 5.4 also illustrates that each remote maintenance operation will require its own specialized casks and manipulators. For general, unforeseen events, a multipurpose boom will have to be deployed requiring extended shutdown.

Concepts optimized around remote handling and maintenance have also been developed, for example, the ARC reactor [18]. In this case, a number of advances in materials and other technology (for example, superconducting coils that can be split and rejoined, which is not possible with current commercial-scale materials) have been assumed to allow the design.

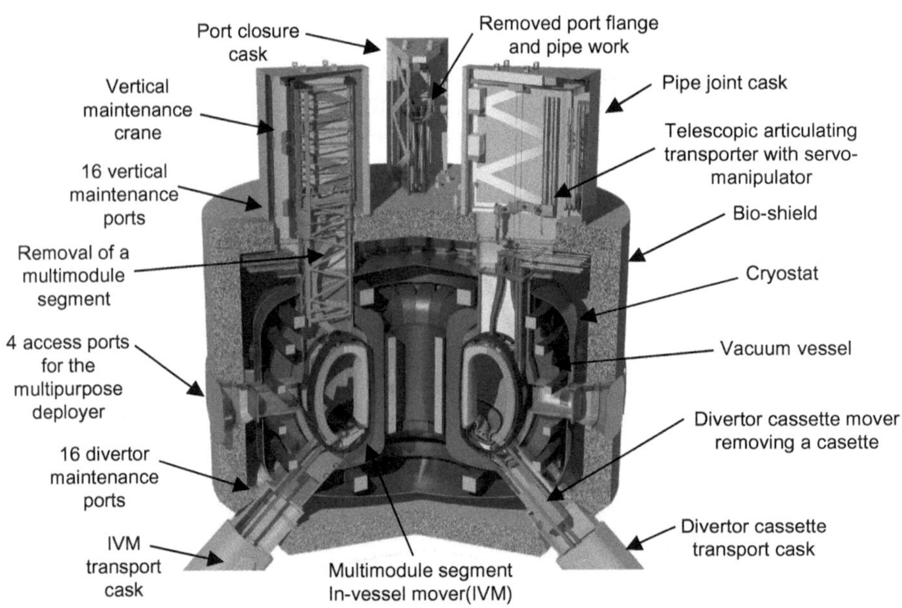

Fig. 5.4 A section of the DEMO torus and surrounding structures showing a possible remote handling scheme, and some of the associated equipment. This scheme is for a torus with 16 TF coils, hence the numbers of access ports, which need to run between them [17].
Courtesy: RACE, UKAEA.

5.3.6 Heating and current drive

Considerations from plasma power balances tell us that a magnetic confinement fusion reactor should operate at an average plasma temperature of around 15 keV and aim to achieve an energy confinement of $nT\tau > 8.3$ atm s (that is, a plasma pressure of several atmospheres for several seconds). A fusion plasma requires energy input to heat it to the point where significant self-heating from fusion alpha particles occurs and, for a tokamak, to provide a steady-state plasma current to maintain magnetic stability. A range of technologies exist, from radiofrequency schemes like electron- or ion-cyclotron resonances to the injection of high-energy neutralized particle beams. However, these methods are significantly inefficient, requiring at least double and often three times the coupled power in electrical power. They result in high recirculating power within the plant, meaning that a considerable proportion of the electricity generated is used to run the plasma control and heating systems. Some advanced tokamak concepts rely on plasma regimes with a high *bootstrap* current, a self-driven current in the plasma generated by pressure gradients within a toroidal geometry [19]. Conversely, modern stellarator designs are optimized to minimize this current to increase the plasma controllability.

5.3.7 Other plant issues

As well as the considerations outlined here, a fusion power plant requires other supporting systems. In particular, an isotope separation system/tritium plant to clean exhaust gases for reinjection and extract new tritium from the blanket coolant is needed, as is a cryogenic plant for maintaining the low temperatures of the magnets. Hot cells for storage and recycling of removed blanket and divertor components are required, as is space for the preparation of new components to be inserted. A river or ocean to provide water for the removal of waste heat from systems is also needed.

The plant power flow, to illustrate the demands imposed by plant systems, is shown in Fig. 5.5. The overall plant efficiency, given by the net electrical power produced divided by the fusion power, is strongly dependent on the effectiveness of the plant subsystems, particularly the heating and current-drive systems, and the primary coolant pumps used to drive the heat extraction from the in-vessel components. The energy conversion system efficiency is also affected by the temperatures achievable in those components, which is limited by the materials available for use in this high-neutron-irradiation environment.

Nuclear fusion has chiefly been a plasma physics experiment for much of the last 70 years. However, with renewed interest in alternative forms of energy, the development of ITER, and particularly the EUROfusion and CFETR programs (see Section 5.5), there is now a shift toward engineering the systems required for efficient power plant operation. However, as soon as significant neutron radiation—required to qualify the systems under plant conditions—is invoked specialist materials and remote handling is required; as the reactor is burning tritium to produce those neutrons, a breeder blanket is also required; to achieve significant radiation doses long pulses are required and so superconducting magnets are required: the best machine

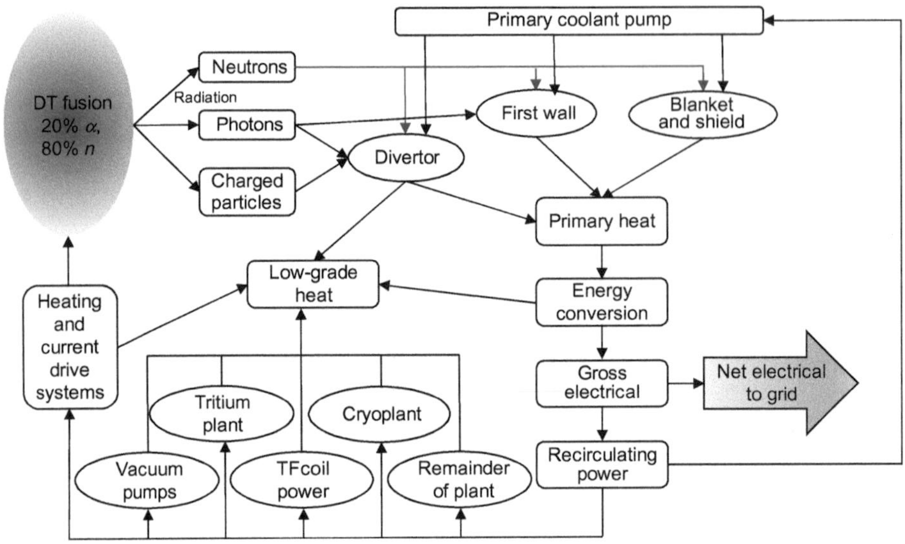

Fig. 5.5 Fusion power plant internal power flow, showing demands from different plant subsystems. The plant's efficiency (net electrical power divided by fusion power) hinges on reducing the recirculating power requirements of these systems, particularly the heating and current drive systems, and the primary coolant pumping. Low-grade heat is lost to the environment
Based on Kovari M, et al. PROCESS: a systems code for fusion power plants—part 2: engineering, Fusion Eng Des 2016; 104:9–20.

to test the technology is a nuclear tokamak, but to build a nuclear tokamak all the technology needs to be available. The only ways to square this circle are to attempt to build many options, which will likely fail in interesting and instructive ways, or to iteratively move toward an integrated design through careful modeling and theoretical development until sufficient confidence is gained from corroborative data (from fission, and from **IFMIF**, the International Fusion Materials Irradiation Facility [20] or other proposed facilities) to make the big step to building a fully nuclear fusion power plant.

As these technology issues and the range of potential fusion devices demonstrate, the road to a commercial fusion power plant is far from obvious. Nevertheless, considerable progress is being made in identifying how these problems link together and what the genuinely promising design options are.

5.4 Fusion's role in future energy markets

It is always hard to make predictions, and particularly so in any system that couples economics and technological innovation. But to explore the possible role of fusion in future energy markets different potential "storylines" can be created and the impacts

Table 5.2 EUROfusion future energy scenarios used to examine the possible roles of fusion power [21]

Storyline	Description
Harmony	Politicians from different world regions agree to work together to achieve very stringent global CO_2 emissions targets and energy market operators take a long-term view when deciding what technologies to invest in. This is a world of strong environmental responsibility
Paternalism	Differentiation between different world regions on their approaches to environmental responsibility, with operators taking a medium-term view of investments. Nevertheless, good political co-operation on global carbon emissions is assumed to meet tight targets
Fragmentation	A scenario of weak environmental responsibility, with short-term considerations dominating. Regional agreements on carbon emissions targets are achieved and a global energy trade is assumed to exist, but so does regional economic competition

of different policies toward emissions limits, energy prices, and international trade and co-operation can be explored (Table 5.2). In addition, models of particular energy systems can be created to examine the tradeoffs between intermittent sources such as renewables, energy storage systems, and baseload generation technologies such as coal or nuclear [22,23]. In these scenarios, fusion tends to displace fission generation depending on assumed costs and resources, and while energy storage combined with intermittent generation can play a very significant role in a mixed system, baseload sources are still required when accounting for seasonal and stochastic variations, albeit at much lower levels than currently used.

To fit fusion into these scenarios, we need an estimate of the costs of energy generated this way. Making estimates of the costs of a commercial release of a new technology is also difficult. Like fission, fusion systems tend to be capital-intensive (that is, expensive to build but relatively cheap to run) compared to fossil-fuel systems, which consume a great deal of fuel. Educated guesses can be made for the costs of different systems based on the amount of materials used and learning factors for scaling lab-based production to commercial production [8,24]. Such studies, usually based on tenth-of-kind projections (that is, the cost of the tenth power plant rather than the first, to account for some learning and development), lead us to believe fusion electricity can be competitive with fission costs. However, as, like fission, fusion internalizes a great many costs associated with safety, emissions, and waste management, it also appears potentially expensive compared to fossil fuels [25]. In the energy market scenarios, fusion only appears competitive where CO_2 emissions are tightly constrained and the resulting overall energy price is higher. This conclusion would change if projected fusion costs could be sharply reduced.

These cost estimates also allow us to see which systems contribute most to the cost of electricity (CoE) for fusion and therefore where improvements might be made

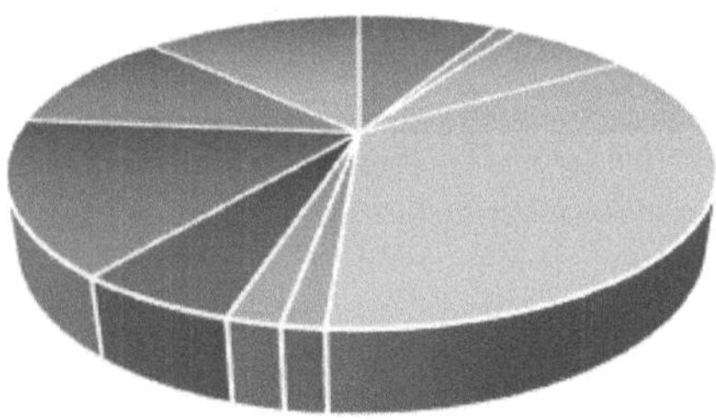

Fig. 5.6 Breakdown of projected costs for a fusion power plant. Approximately half the total costs are fusion-specific and dominated by the magnet systems. Around a quarter of the costs are construction and project management, with the balance conventional plant, land, and buildings. From the PROCESS reactor design code Kovari M, et al. PROCESS: a systems code for fusion power plants—part 1: physics, Fusion Eng Des, 2014; 89(12): 3054–3069.

(Fig. 5.6). In addition, we can use these models to try to assess the main global parameters upon which the cost of electricity depends [26]:

$$\text{CoE} \propto A^{-0.6}\eta_{th}^{-0.5}P_e^{-0.4}\beta_N^{-0.4}N^{-0.3}$$

where A is the plant availability; η_{th} is the plant thermal efficiency (electrical power over thermal power); P_e is plant output power; β_N is the plasma normalized beta; and N is the normalized density (the plasma density relative to the Greenwald density limit $n_G = \frac{I}{\pi a^2}$). The first of two of these are limited by available technology; the unit size is a design choice although strongly limited by technology and the capital funding available to build a power plant; and the available normalized beta and density are functions of the plasma physics scenario that can be achieved. What these studies reinforce is that technological developments are at least as important as physics for fusion power plant economics, and that therefore fusion for energy is as much a technology research program as a physics one.

5.5 Status of current research

The current largest tokamak and only magnetic device still operating, which has burnt D-T is **JET**, the Joint European Torus, located at Culham Centre for Fusion Energy, Oxfordshire, United Kingdom. It is the world's most powerful tokamak and is the focal point of the European fusion research program, collectively used by >40 European laboratories. Built in 1983, it holds the world record for fusion power (16 MW, in 1997 [27]). JET has been continuously upgraded to remain at the forefront of fusion research, most recently having a fully beryllium plasma-facing wall installed to mimic conditions in ITER. Other currently operational tokamaks include **ASDEX-U**, **EAST**, **WEST**, **DIII-D**, and **SST-1** [28–32]. **NSTX-U** [33] and **MAST-U** [34] are spherical tokamaks, with MAST-U focusing on divertor and exhaust physics.

ITER is the next-generation power-plant-scale (target: 500 MW fusion power) tokamak currently under construction at Caderache, France [35]. It is intended to be capable of long (100s of seconds) pulses and generate significant fusion power in a range of experimental scenarios to prove the controlled plasma physics required for commercial fusion power. ITER is an international collaboration between the EU, China, India, Japan, Korea, Russia, and the United States, covering nearly half the world's population. Much of the contribution is in-kind, with the partners supplying components rather than the ITER Organization carrying out conventional procurement. This is intended to allow each partner to build up industry interest and support for fusion as well as acquiring the engineering and development skills required for future work in the area. ITER is supported by a satellite tokamak, JT-60SA, being built in Naka, Japan. **JT-60SA** is a superconducting tokamak intended to investigate how to optimize the operation of fusion power plants following ITER through the study of advanced modes of plasma operation. ITER should be operational by 2025, with burning (thermonuclear D-T) plasmas by 2035 [36]. If successful, ITER will prove the plasma physics basis for fusion power plants. There is some scope, for example, the Test Blanket Module (TBM) program for testing of some fusion technology within ITER, which will provide essential information but is somewhat limited in scope due to lack of available space within the machine and the low lifetime neutron dose, which the materials will experience. A parallel materials program based around an International Fusion Materials Irradiation Facility (**IFMIF**) is also under development [20].

Following ITER, there will need to be a prototype power plant to demonstrate the integrated operation and development of technology suitable for commercialization. There are many concepts (generically called **DEMO**) in circulation, designed to fit into different local energy markets. One of the most ambitious—in terms of a fully integrated research program—is the European **EUROfusion** program [37] aimed at delivering a complete set of fusion power plant technology around the middle of the century. The progression of these systems is shown in Table 5.3.

Among the alternative magnetic fusion concepts, **W-7X** is a JET-scale stellarator, located at Greifswald, Germany [38]. Starting operation in 2015 and possessing superconducting coils, it is intended to demonstrate steady-state plasma operation and extend knowledge of stellarator performance toward the power plant domain. Additionally, the China Fusion Engineering Test Reactor (**CFETR**) program to develop a tokamak neutron source for materials and component testing is expanding [39].

There are also a number of alternative approaches being conducted by private companies, using concepts, which may permit smaller, higher power-density machines, although typically with a less comprehensive experimental bases and thus higher-risk (for example, [40–42]). In general, these companies are currently focused on the physics issues or a narrow technology option, such as high-field magnets using, for example, high-temperature superconductors, and do not have comprehensive

Table 5.3 **The path from the leading current machine, JET, to a fusion power plant based on the same concept**

Machine	Targets	
JET (current)	Fusion power ≈ plasma heating power ($Q = 1$) $R = 3\,m$, $B = 3.5\,T$, plasma volume $= 100\,m^3$ Resistive (copper) magnets, maintenance by deployable boom. No electricity or tritium production	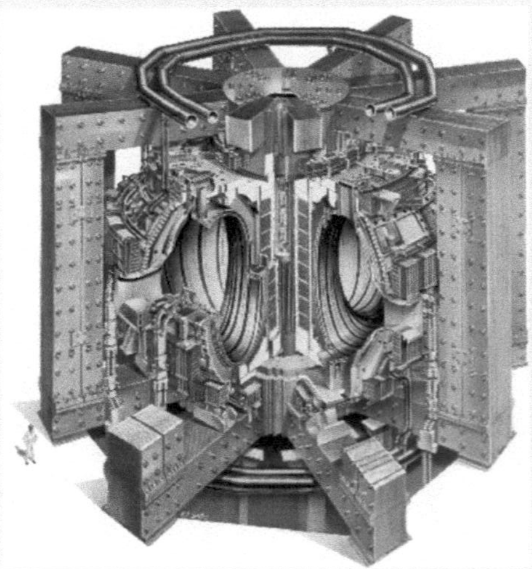

Table 5.3 Continued

Machine	Targets	
ITER (2025)	Fusion power ≈ electrical power required to run ITER ($Q=10$) $R=6\,\mathrm{m}$, $B=5.7\,\mathrm{T}$, plasma volume $=840\,\mathrm{m}^3$ Superconducting magnets, maintenance by deployable boom. Plasma pulses of many times confinement times. Test of tritium production technology (TBMs), no electricity production	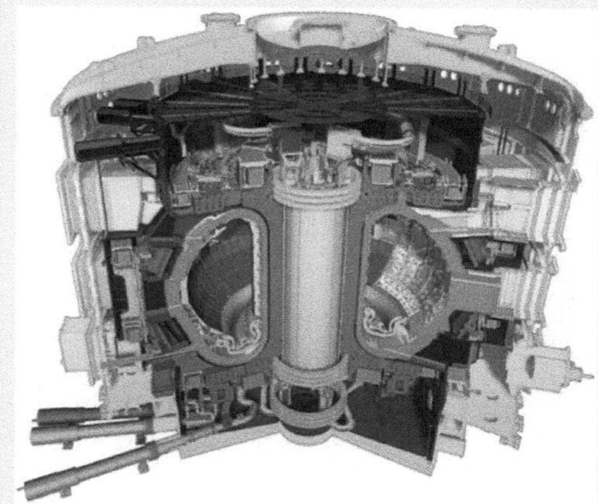
DEMO (2050s)	Electrical power generated > electrical power required to run DEMO ($Q>20$); low availability $R=\sim 9\,\mathrm{m}$, $B=\sim 5\,\mathrm{T}$, plasma volume $=\sim 2500\,\mathrm{m}^3$ Fully integrated power plant prototype technology capable of achieving high availability. Self-sufficient in tritium, low overall net electricity output. Very long pulses or steady-state operation	
Fusion Power Plant (2070s?)	Electrical power ≫ electrical power required to run FPP; competitive cost of electricity; high availability and reliability	

R is the tokamak major radius, *B* is the toroidal magnetic field, *Q* is the plasma power multiplication factor.
Courtesy: EUROfusion, ITER Organization, EUROfusion.

accompanying technology programs for a complete power plant design. Nevertheless, the exploration of wider conceptual space is welcome and may result in the discovery of new and useful regimes of operation, which can be adopted by the mainstream fusion program; alternatively, they will make use of the technologies and materials developed by national research programs in their own development.

Is enough being done to secure alternative, clean sources of energy, including fusion? Using figures available from the IEA on energy research, we can estimate that global public spending on R&D for fusion is currently around $\$2 \times 10^9\ a^{-1}$ ($2 bn per year), including ITER construction [43]. This is similar to the expenditure on renewables, and much less than that spent on fission. The global energy market is worth around $\$2 \times 10^{12}\ a^{-1}$ ($6 tn per year) [44] and total public-sector spending on energy R&D is around 0.15% of this, with fusion spend around 0.03%. Given the scale of the energy challenge faced in the immediate as well as the long-term future, it seems that not enough is being spent on energy research of any kind.

5.6 Summary

Fusion has a number of attractive properties, particularly in terms of resource availability, generation of waste and emissions, and low external costs, but has consistently proven to be technically challenging. Even the most well-developed current research programs do not lead to commercial fusion energy being widely available before the end of the century, meaning it is a long-term low-carbon solution not a near-term fix for current climate change issues. However, even with high use of renewables intermittency of generation leads to challenging requirements for energy storage, and it seems likely that there will always be a need for forms of baseload electricity generation in an optimized system; a need to which fusion is well suited. Enormous progress has been made in understanding the physics, technology, and materials problems in recent decades and, within the next 20 years, ITER and IFMIF should provide the final tokamak burning plasma physics and radiation-resistant materials basis allowing the next steps to be taken toward an electricity-producing reactor prototype around the middle of the century.

References

[1] Wesson J. Tokamaks. Oxford: Oxford University Press; 2011.

[2] Statista. Countries with the largest lithium reserves worldwide as of 2016. Available from: https://www.statista.com/statistics/268790/countries-with-the-largest-lithium-reserves-worldwide/ (Accessed 3 December 2017).

[3] Hoshino T. Innovative lithium recovery technique from seawater by using world-first dialysis with a lithium ionic superconductor. Desalination 2015;359:59–63.

[4] Zheng S, King DB, et al. Fusion reactor start-up without an external tritium source. Fusion Eng Des 2015;103:13–20.

[5] Kovari M, Coleman M, Cristescu I, Smith R. Tritium resources available for fusion reactors. Nucl Fusion 2017;58(2):026010.

[6] J. Raeder and et al, Review of the safety concept for fusion reactor concepts and transferability of the nuclear fission regulation to potential fusion power plants, GRS (gGmbH), GRS 389; 2016.
[7] IAEA. Fusion safety. Vienna: IAEA; 1981.
[8] Maisonnier D, et al. The European power plant conceptual study. Fusion Eng Des 2005;75–79:1173–9.
[9] Llewellyn-Smith C, Ward D. Fusion. Energ Policy 2008;36(12):4331–4.
[10] Wikipedia. Hairy ball theorem. Available from: https://en.wikipedia.org/wiki/Hairy_ball_theorem (Accessed 3 December 2017).
[11] Freidberg J. Plasma physics and fusion energy. Cambridge: Cambridge University Press; 2007.
[12] LLNL. National ignition facility and photon science. Available from: https://lasers.llnl.gov/ (Accessed 6 December 2017).
[13] Zinkle SJ, Busby JT. Structural materials for fission and fusion energy. Mater Today 2009;12(11):12–9.
[14] Zinkle SJ. Advanced materials for fusion technology. Fusion Eng Des 2005;74:31–40.
[15] Duffy DM. Fusion power: a challenge for materials science. Phil Trans R Soc A 2010;368:3315–28.
[16] Suri AK, Krishnamurthy N, Batra IS. Materials issues in fusion reactors. J Phys Conf Ser 2010;208:012001.
[17] Loving AB, et al. Pre-conceptual design assessment of DEMO remote maintenance. Fusion Eng Des 2014;89:2246–50.
[18] Sorbom BN, et al. ARC: a compact, high-field, fusion nuclear science facility and demonstration power plant with demountable magnets. Fusion Eng Des 2015;100:378–405.
[19] Kessel CE. Bootstrap current in a tokamak. Nucl Fusion 1994;34:1221.
[20] IFMIF. International fusion materials irradiation facility. Available from: http://www.ifmif.org/ (Accessed 3 December 2017).
[21] Cabal H, et al. Fusion power in a future low carbon global electricity system. Energ Strat Rev 2017;15:1–8.
[22] Romanelli F. Strategies for the integration of intermittent renewable energy sources in the electrical system. Eur Phys J Plus 2016;131:53.
[23] Wagner F. Feature of an electricity supply system based on variable input. Jülich: IPP; 2012.
[24] Kovari M, et al. PROCESS: a systems code for fusion power plants—part 1: physics. Fusion Eng Des 2014;89(12):3054–69.
[25] Cabal H, et al. In: Environmental externalities of a future fusion plant. 26th EPS conference on controlled fusion and plasma physics, Maastricht; 1999.
[26] Ward DJ, et al. The economic viability of fusion power. Fusion Eng Des 2005;75–79:1221–7.
[27] Rimini FG, JET Team. DT fusion power production in ELM free H modes in JET. Nucl Fusion 1997;39:1897.
[28] IPP. ASDEX upgrade. Available from: http://www.ipp.mpg.de/16195/asdex (Accessed 6 December 2017).
[29] IPP-CAS. EAST—experimental advanced superconducting Tokamak. Available from: http://english.ipp.cas.cn/rh/east/ (Accessed 6 December 2017).
[30] CEA. WEST—tungsten (W) environment in steady-state Tokamak. Available from: http://west.cea.fr/en/index.php (Accessed 6 December 2017).
[31] USDoE. DIII-D National fusion facility. Available from: https://science.energy.gov/fes/facilities/user-facilities/diii-d/ (Accessed 6 December 2017).

[32] IPR. Institute for plasma research. Available from: http://www.ipr.res.in/ (Accessed 6 December 2017).
[33] PPPL. National spherical torus experiment upgrade (NSTX-U). Available from: http://www.pppl.gov/nstx (Accessed 6 December 2017).
[34] UKAEA. MAST upgrade. Available from: http://www.ccfe.ac.uk/mast_upgrade_project.aspx (Accessed 6 December 2017).
[35] ITER. The way to new energy. Available from: https://www.iter.org/ (Accessed 3 December 2017).
[36] European Commission. Fusion as an energy source: the ITER project is back on track. Available from: https://ec.europa.eu/energy/en/news/fusion-energy-source-iter-project-back-track (Accessed 14 June 2017).
[37] EUROfusion. Home page. Available from: https://www.euro-fusion.org/ (Accessed 3 December 2017).
[38] IPP. Wendelstein 7-X. Available from: http://www.ipp.mpg.de/w7x (Accessed 3 December 2017).
[39] WanY Y. Overview of the present progress and activities on the CFETR. Nucl Fusion 2017;57:102009.
[40] Tokamak Energy. Tokamak energy. Available from: https://www.tokamakenergy.co.uk/ (Accessed 3 December 2017).
[41] TriAlpha Energy. TriAlpha Energy. Available from: https://tae.com/ (Accessed 3 December 2017).
[42] Helion Energy. Helion Energy. Available from: http://www.helionenergy.com/ (Accessed 3 December 2017).
[43] IEA. IEA World energy statistics and balances. Available from: http://www.oecd-ilibrary.org/energy/data/iea-world-energy-statistics-and-balances_enestats-data-en (Accessed 31 December 2016).
[44] BP. BP statistical review of world energy June 2017. 66th ed. BP; 2017.

Further reading

[45] Bosch H-S, Hale GM. Improved formulas for fusion cross-sections and thermal reactivities. Nucl Fusion 1992;32:611–31.
[46] Kovari M, et al. PROCESS: a systems code for fusion power plants—part 2: engineering. Fusion Eng Des 2016;104:9–20.

Global renewable energy resources and use in 2050

6

*Patrick Moriarty**,[1], *Damon Honnery*†
*Department of Design, Monash University, Caulfield East, VIC, Australia,
†Department of Mechanical and Aerospace Engineering, Monash University-Clayton Campus, Clayton, VIC, Australia
[1]Corresponding author: patrick.moriarty@monash.edu

Chapter Outline

6.1 Introduction

On the April 21, 2017, the United Kingdom, an early leader in the Industrial Revolution, had its first coal-free day for several centuries, according to the UK's National Grid [1]. In 1800, as the Industrial Revolution got underway in Western Europe, global primary energy use was around 20.35 EJ (EJ = exajoule = 10^{18} joule). Global fossil-fuel production, almost entirely coal, was only about 0.3 EJ or 10^6 t [2,3]. The rest was renewable energy, nearly all biomass energy. In year 2014, according

Managing Global Warming. https://doi.org/10.1016/B978-0-12-814104-5.00006-5

to International Energy Agency (IEA) statistics [4], total primary energy had risen to 574 EJ, with only 67 EJ from renewable energy (RE) sources, mostly biomass used as heating and cooking fuel in low-income countries.

Fig. 6.1 shows the growth of electricity produced from each of the five main RE sources considered in this chapter: biomass energy, hydroelectricity, wind, solar, and geothermal energy, together with global electricity production. What is surprising is that even after 1990, the year of the first Intergovernmental Panel on Climate Change (IPCC) report, the electricity share of RE continued to fall slowly until the mid-2000s [5].

The future of fossil fuels, still the dominant global energy source, is under threat for a variety of reasons. First is its impact on global climate change: most anthropic greenhouse gases (GHGs) annually released derive from fossil-fuel combustion [6]. The second concern relates to the future availability of fossil fuels at affordable prices. Already, much of the global annual production of oil and gas comes from non-conventional sources: deep water and polar oil, oil sands, heavy oils, and gas and oil from fracking. Most of these sources have higher costs—in terms of GHGs, environmental, and monetary—than their conventional equivalents for each joule of final energy.

The third reason is the pollution produced by both their production and especially from their combustion. Although technical solutions such as sulfur dioxide scrubbers and particulate traps are available for fossil-fuel power stations, and unleaded, low-sulfur fuels, and three-way catalytic converters for road vehicles, air pollution from oxides of nitrogen and from very fine particulate matter are still major health hazards, even in OECD cities. In many countries of the industrializing world, a combination of less stringent pollution standards, poor enforcement, rapid urbanization, and rapid

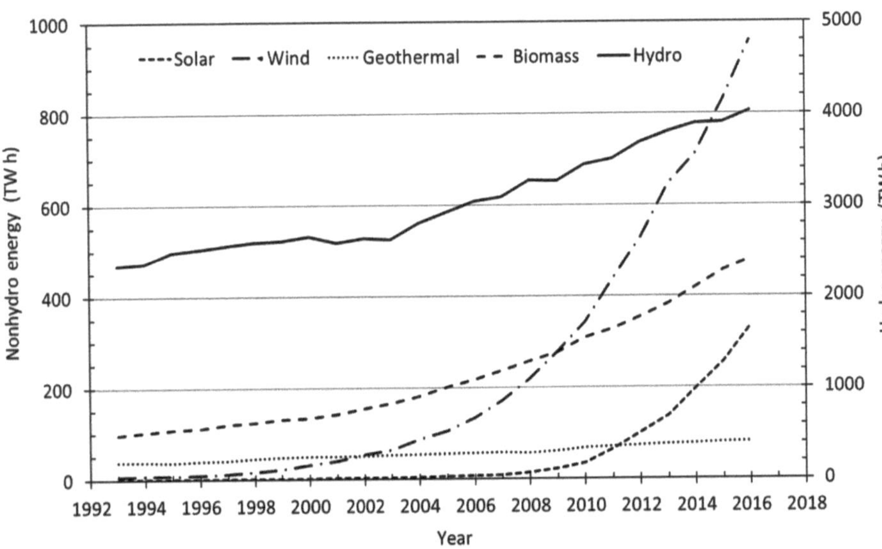

Fig. 6.1 Global RE electricity output vs year 1993–2016 [5].

growth in vehicle numbers makes air pollution, especially in cities, a major cause of sickness and mortality [7].

All energy sources—fossil, nuclear, or RE—must meet a basic criterion: the energy output must be larger (even several times larger) than the energy inputs, when both input and output energy are measured in a comparable way, usually as primary energy. Only for food energy, and novel energy sources still in the experimental stage, can inputs be larger than outputs. We live on a planet with finite resources, and for economic reasons, high-quality energy resources are usually developed first. Hence, as annual output of RE sources rise, or cumulative output of fossil fuels or uranium rises, progressively lower-quality sources must be tapped. For fossil fuels, this trend is already evident: a rising share of oil is from the deep ocean, the Arctic, and bitumen sands, and natural gas from fracking. The consequence is an early rise in the energy return on energy invested (EROEI) as the technology is developed and improved, then a peak value, followed by a steady fall in EROEI for each fossil fuel. Oil and natural have already passed their peak value, as have fossil fuels overall, but the EROEI for coal is still rising, as documented by Court and Fizaine [8].

Most RE sources can be expected to follow the same trend as for fossil fuels, with the possible exception of solar energy. This eventual fall in EREOI will occur for several reasons. First, the declining quality of the resource base (e.g., lower wind speeds, less suitable hydro sites, lower temperature geothermal fields) will tend to lower the energy output for a given energy conversion device compared with higher quality resources. Second, as we will argue, the technical potential for most RE sources is limited, even compared with present global primary energy use, with the exception of wind, solar, and wave energy. But the latter group are all *intermittent* sources of energy, necessitating energy storage and the inevitable energy losses this entails, if they are to replace fossil fuels. Further, since we need other forms of energy apart from electricity, conversion to other energy carriers (perhaps hydrogen or methanol) will also be needed. Third, even if the world soon acts decisively to mitigate climate change, further climate change will occur; which could adversely affect the EROEI of planned and even existing RE production. Most of the global estimates for the various RE sources have not considered EROEI values, and so will likely be overestimates of potential [9,10].

What is the future of global energy consumption? The oil company BP has projected global *commercial* energy use (i.e., excluding fuel wood) by fuel and region out to the year 2035 [11]. Overall primary energy use was expected to rise from 550 EJ in 2015 to 720 EJ in 2035, with nearly all the increase coming from non-OECD countries. All RE sources, including hydro, were forecast to roughly double from their 2015 level of 52.7 EJ to 119.6 EJ, in 2035, or one-sixth of the total. One difficulty with interpreting these numbers is that BP and the International Energy Agency (IEA) use conflicting methods for calculating the primary energy values of nonheat energy sources like wind and PV electricity. The IEA use a one-to-one conversion, while BP factor values by the inverse of modern thermal power plant efficiency [12]. The differences will become more pronounced if these nonthermal energy sources come to dominate total energy production. Use of each fossil fuel—coal, oil, and natural gas—was also projected to increase, with most of the growth coming from natural gas. Evidently, BP did not foresee any drop in CO_2 energy emissions by 2035. Bioliquids for transport would also remain marginal, rising from 3.1 to only 5.4 EJ over the period. Similarly, the US Energy Information Administration (EIA) projected global energy-related CO_2 emissions to continue to grow at 1.0% per year over the period 2010–40 [13]. Annual fossil-fuel use was likewise projected to continue to rise.

6.2 Biomass energy

6.2.1 Introduction

Bioenergy is by far today's leading RE source, but most consumption still takes the form of fuelwood, crop residues, and animal dung burnt at low efficiency in industrializing countries. Only about 4.3 EJ of the estimated total global consumption of 59 EJ is from modern forms of biomass, comprising liquid fuels (ethanol and diesel) for transport vehicles and fuel for electricity production, including bagasse burnt to provide power in cane sugar refineries [4,5,14].

Biomass is unique among RE sources in that, like fossil fuels, its energy is materially embodied. It can thus be stored and used at any time, unlike other RE sources such as wind or solar, which exist only as energy flows. Biomass is also different, at least in extent, from other RE sources in that its basic inputs of fresh water and fertile (and relatively flat) land, face increasing competition from other uses of biomass, namely, from agriculture (food, fodder for livestock, fibers such as cotton) and forestry (construction timber and products, fiber for paper, etc.).

6.2.2 Bioenergy in 2050

Bioenergy is the oldest form of nonfood energy, with its origins perhaps 500,000 years ago. Chances for further major technical breakthroughs in combustion appear small, with most research now concentrated on topics such as increasing crop productivity and converting cellulosic materials to liquid fuels for transport. At present, all such liquid fuels are derived from edible feedstocks—grains, sugar, and edible oils—which raises ethical questions in a world with an estimated 800 million people facing absolute food shortages [15]. The present effort put into cellulosic conversion may be misplaced: from an energy or GHG reduction viewpoint; it is better to use biomass directly to produce electricity, particularly if it displaces coal or oil as power plant fuels, rather than convert it to liquid fuels for transport [16].

Global land-based Net Primary Production (NPP) is the total mass of living plant matter produced annually, minus plant respiration. Terrestrial NPP, estimated at 2000 EJ [17], obviously fixes a theoretical upper limit on the human appropriation of NPP (HANPP). Estimates for global HANPP vary, depending upon which items are included or excluded, with values ranging from 10% to 50% [18]. Since nonhuman nature provides ecosystem services (apart from obvious ones like food and timber) on which we are dependent, any major rise in HANPP could merely be trading growth in one set of ecosystem services (e.g., food) with a corresponding loss in others (e.g., biodiversity, pest control). Axel Kleidon [19] estimated present HANPP as 40%, and based on a model of the vegetation-climate system, argued that for HANPP values beyond about 45%, any percentage rise would produce feedbacks that would reduce global NPP, and eventually, the absolute level of HANPP measured in energy or mass units.

The world population in 2015 stood at 7.35 billion; the UN [20] expect this to rise to 9.73 billion by 2050. If present trends for per capita income growth also continue, there will evidently be increased demand for all uses of biomass, raising the question of which ones should get priority. Clearly, from an ethical viewpoint, food should come first: we should produce enough food for the likely expanding human family. Nevertheless, this does not necessarily imply that present dietary trends are followed. If instead of expanding animal products as at present [18], the world moved toward a more vegetarian diet, agricultural production could be reduced, given that animal products need far more inputs (and produce more GHGs) per kilojoule of food energy than nonanimal products. Nevertheless, the FAO [19] anticipate that output of both grain and animal products will continue to rise, as will agricultural land area. Increasing agricultural yields can reduce the inputs of land needed for food production, but the FAO saw yields only growing slowly. Future productivity gains will be more difficult to make, particularly in the face of on-going climate, uncertainty about future phosphorus availability, biodiversity loss, and other changes. And although land productivity (output per hectare) has risen, the same is not true for energy productivity. Historically, food was produced with a much lower ratio of energy inputs to energy output—this ratio declined by more than an order of magnitude in the 20th century [20].

In 2014, although the nonfuel production of timber was 1836 Mt (1836×10^6 t), up from 1700 Mt in 1990 [19], timber is losing share as a construction material, as are natural fibers like cotton [18]. Yet in terms of meeting a given building function, such as the framing for a four-storey apartment block, timber has a much smaller carbon footprint than steel or reinforced concrete. This suggests that for climate mitigation, biomaterials should be expanding, rather than contracting as at present, their share of the market. Such a biomaterials expansion would lower the global potential for bioenergy, assuming that all biomass uses together face a global upper limit—that for HANPP.

Clearly, there is no simple answer to the question: What is the global technical potential for bioenergy? It all depends on future consumption of biomass for biomaterials and food. The input resources necessary to meet food needs in 2050 involve both ethical and technical questions. The huge range of estimates in the literature for 2050 bioenergy technical potential reflect this uncertainty: values vary from a low of 10–60 EJ to an upper value of over 1500 EJ [10]. Since this upper value is 75% of the entire global terrestrial NPP, it is clearly unrealistic. Given the superior GHG savings possible with using biomaterials as a substitute for more energy- and GHG-intensive materials, as well as rising food needs, values toward the lower end of the range are more likely.

However, the three human uses for biomass are not fully independent of each other. Human wastes can be used to produce methane from both sewage treatment works and from landfill gas. Some agricultural and forestry wastes can be used for bioenergy, although most will need to be retained in situ to maintain soil fertility and reduce wind or water erosion. Finally, biomaterials, particularly construction timber, after the end of its useful construction life—which may involve its reuse in some other construction project—can be combusted for energy.

6.3 Hydroelectricity

6.3.1 Introduction

In the early years of electricity, most was produced from hydro, but was soon overtaken by fossil fuels burnt in power stations. Today, despite rapid growth in recent decades of first wind, then solar electricity, hydro still dominates RE electricity production [5]. The hydro potential in OECD countries is now largely exploited; most of the remaining potential is in Asia, Africa, and South America [21]. Hydropower is a mature technology, and little in the way of technical advances can be expected. It also has EROEI values greater than other RE sources [22]. There are even indications that the energy ratio for new hydroconstruction is falling globally, in that the annual electricity output per megawatt of installed power over the period 1994–2011 was <40% of its value in the year 1993 [21].

6.3.2 Hydroelectricity in 2050

Hydroelectricity cannot be readily stored. But unless the hydroplant is a simple run-of-the-river installation, electricity can be generated on demand because of the gravitational energy of the water stored behind the dam wall.

It is likely that by 2050, most of the world's remaining technical potential for hydropower will be utilized. But some potential will remain unexploited—and so annual electric output will still be well below 30 EJ for several reasons. First, ongoing climate change will add to uncertainty about future catchment precipitation levels and the season distribution of annual river flows, although some areas (such as Arctic-draining rivers in northern Europe) will see conditions favorable for further hydro output. In some mountainous areas, such as the Himalayas, hydropotential could show a temporary increase, fuelled by continuing loss of glacier mass [6,23]. This uncertainty will impact on the economics of hydrodevelopment, given that dam structures can have an expected lifetime of 100 years or so.

Second, extreme rainfall events are expected to increase in frequency. The soil erosion potential varies nonlinearly with rainfall intensity (and so will catchment area landslides into the reservoir), so that sedimentation rates of hydroreservoirs will increase, shortening their service lives. In the Amazon basin, another factor could seriously affect the basin's vast hydropotential. Deforestation, which is still occurring in the Amazon, initially increases river flows, and thus hydropotential, because of less transpiration from trees. But beyond a certain level of basin forest loss, further deforestation reduces hydropotential. The modeled results of Stickler et al. [24] showed that hydropotential could be reduced to only a quarter of its full potential if 40% of forest cover is lost. This reduction occurs because in the Amazon the "rainfall systems are maintained, in part, by the forest itself through contribution of water vapor to the atmosphere through ET and through its associated influences on land–atmosphere energy exchange" [24]. Presumably any comparable loss of tree cover from climate change would have a similar effect.

6.4 Wind energy

6.4.1 Introduction

Wind energy is an ancient technology, with windmills several millennia old, and wind-powered sail boats much older. Modern wind turbines were first installed in numbers in the United States in the 1980s, but global output did not take off until the mid-1990s [5]. With strong growth in Germany, and then Spain, global wind output rose exponentially—albeit from a very small base—but since 2008, output (in terawatt hours of electricity) has only risen linearly (see Fig. 6.1). The steady growth in wind energy has been helped by new materials enabling much larger blade diameters, and the fact that virtually all commercial wind turbines have standardized on horizontal axis machines, although a vast number of alternative designs have been proposed [25]. Choice is still available for blade numbers (two or three), rated output (from fractional MW up to 6 MW or more), and tower height/blade diameter ratio.

The output from wind turbines has been found to deteriorate with age of installation: Staffell and Green [26] found a 1.6% drop in annual output for turbines installed in the United Kingdom. The result was that average load factors declined from 28.5% when installed to 21% after 19 years, which in turn would raise the cost of the electricity produced.

6.4.2 Wind energy in 2050

If the linear increases in output from 2008 to 2015 continue in coming decades, wind could be expected to supply 3985 TWh (14.3 EJ) globally in 2050, compared with the 2015 output of 841 TWh (3.0 EJ). Although this growth represents a more than fourfold increase, wind will still only be a marginal source in 2050, with estimates of total primary energy use then as high as 1000 EJ [12]. Even with strict land constraints on turbine placement, this value of 14.3 EJ is still well below the global technical potential of wind energy [2]. Slow growth, rather than resource limits, is what could prevent wind energy becoming a major energy source by 2050. In any case, wind turbines are now also being sited offshore, which increases the global wind resource base. Although more expensive to install and maintain, this cost is offset by higher wind speeds—and less public opposition.

Since the rebirth of wind turbines in the 1980s, blade diameters and rotor heights have steadily risen; the greater rotor heights (up to 200 m) enable both higher wind speeds and less wind speed variation across the blades. Even higher wind speeds could be obtained if the turbines could be placed at much greater heights, and hence greater electricity output per turbine of a given rating. Various proposals include turbines borne aloft by tethered balloons and tethered self-propelled turbines. Another proposed concept would be based on airborne kites [27].

As with ground-based turbines with an average rated output of, for example, 5 MW, many thousands would need to be deployed for significant electricity production. Great care would need to be taken to lessen the dangers to aircraft, birds, and even people on the ground from either falling cable or the turbines themselves. If wind energy continues to rise at only a linear rate, land and shallow sea locations will be ample, and air-based turbines are unlikely to be more than novelties, even in 2050.

6.5 Solar energy

6.5.1 Introduction

Solar energy is the only RE source with vast potential compared with any possible human energy consumption level, with annual global insolation estimated at 5600 ZJ. The insolation level at any location (in $W\,m^{-2}$) varies in a predictable way with latitude, season, and time of day, but also with cloud cover, which is harder to predict. The result is that solar energy technical potential per unit horizontal area is far greater for some countries than others, and that, like wind power, is only intermittently available. In 2015, only 253 TWh of solar electricity was produced, or slightly $<1\%$ of total electric output [5,28]. Most of this output was from photovoltaic (PV) cells with installations in a great many countries, and the rest from solar electricity energy conversion (STEC) plants in the United States and Spain [21]. Solar collectors are devices for heating water, and $652 \times 10^6 m^2$ of such devices have been installed worldwide by the end of 2016 [29]; the heat so generated (375 TWh in 2016) is included in RE statistics.

Not all solar energy needs to be converted by some energy device. *Passive* solar energy has been used for millennia, and its potential is vast. It can be used for both heating buildings and for natural lighting. Passive solar is most effective if buildings are designed with its use in mind, but existing buildings can be retrofitted. It is often very difficult to determine the difference between passive solar energy utilization and energy conservation, and passive solar energy is not presently included in RE production statistics. If the Earth's land and oceans did not annually absorb 39×10^6 EJ insolation from the sun, its temperature would be close to absolute zero, rather than its present average of 15–16°C. In one sense, solar energy already provides all our energy needs, except for a few hundred exajoules from fossil-fuel combustion, which makes any calculation of the contribution of passive solar energy difficult to interpret.

6.5.2 Solar energy in 2050

Recent growth in global solar electric production has been exponential, and given the possibility of technical breakthroughs, any predictions of future output are unlikely to be useful. It should be remembered that wind energy also experienced an exponential growth phase, before slowing to linear growth over the past decade.

Because of its intermittency, some researchers have put forward ambitious schemes to avoid this problem. One, first proposed in the mid-1970s, is solar satellite power (SSP). The proposal would involve a fleet of satellites, placed so as to receive 24-h insolation. Each would carry an array of PV cells on a lightweight frame. The electricity generated would be converted to microwave energy, beamed to Earth receiving stations, and then converted back to electricity. The costs of satellite placement would be high, and since energy conversion losses would occur at each stage from insolation through to electric power generation on Earth, EROEI values could be low.

Other proposals would see vast solar farms installed in each of the world's deserts, so that both seasonal and diurnal variations in insolation could be circumvented. Evidently this scheme would require a worldwide interconnected grid, the building of which would be the greater part of this hugely expensive Scheme [30]. A recent variant—the Desertrec scheme—would build large solar and wind farms in the deserts of north Africa and the Middle East to supply European as well as local energy needs [31]. It is doubtful, however, whether Europe would wish to become largely dependent for its electricity supply on distant countries. At present, very little electricity is exported across borders [4]. Furthermore, the project would still not solve the intermittency problem.

Solar electricity, particularly from PV cells, is different from other RE sources in several ways. Not only is its resource base orders of magnitude higher, but the monetary costs of PV cells of a given rated capacity have decreased exponentially with time, because of continuous improvement in the materials used. Further breakthroughs are promised [32]. But there are several factors that need consideration. The first is that the expected life of PV units may be much lower than anticipated—closer to 17 years rather than 30 years [33]. Second, the development of higher efficiency PV cells may rely on exotic materials, which have a low global resource base, and will prove increasingly costly to extract in monetary, energy, and environment terms. We may thus be substituting one environmental problem for another [34,35]. Third, PV cells are only part of the total system costs, and the other costs, particularly for supporting structures in solar farms, are less likely to see further cost reductions per installed watt.

6.6 Geothermal energy

6.6.1 Introduction

Geothermal energy derives both from the interior heat of the Earth, left over from its violent birth, and from radioactive decay of various isotopes in both Earth's interior and crust [2]. Total annual heat output at the planetary surface is estimated at some 1300 EJ, of which only about 280 EJ is from terrestrial surfaces. The first geothermal electric plant was built in Italy over a century ago, although the use of geothermal heat has a long history. The global technical potential for electricity production is thought to be small, with most estimates in the 1–3 EJ range, because field temperatures of above 150–200°C are needed [2].

For low-grade geothermal heat, however, published values for global potential are as high as 300,000 EJ [9]. Since, as discussed, only about 280 EJ is produced annually at the Earth's crust, clearly such large estimates rely on drawing down cumulative reservoirs of heat, and are not sustainable in the long run. Low-temperature geothermal heat can be used in two ways. First hot water from the field can be carried for a few km in heat-insulated pipes for use in homes, greenhouses, and swimming pools. One difficulty here is the potential mismatch between geothermal fields and population centers—particularly in the United States, where geothermal resources are concentrated in the Rocky Mountains.

Second, more recently, geothermal *heat pumps* are being increasingly used. By 2004, over a million were installed worldwide [36]. As with passive solar energy, it is difficult to classify geothermal heat pump energy. Useful energy for heating (or cooling) can be extracted from *any* large heat sink, which is above (or below) the ambient temperature. Even large lakes, which because of their thermal capacity are slow to heat or cool, can be used for cooling or heating. Much of what is considered geothermal energy is really solar energy stored in the ground [36].

6.6.2 Geothermal energy in 2050

Global growth in geothermal electricity output has been linear for the last 3–4 decades, and if this trend were to continue, geothermal installed capacity will rise from 13 GW in 2015 to roughly 20 GW by 2050 [5]. At the present average capacity factor output would only be about 115 TW h, a minor fraction of even the 2015 global electricity output of 24,100 TW h [5]. Such high rates of utilization evidently are not sustainable in the long run. Even for electricity production from high temperature sources, it is found economic to "mine" the heat source, resulting in depletion of the geothermal field, and the need for perhaps decades of recovery time before it can be used again.

The energy conversion devices for geothermal electric production, are like wind turbines, a mature technology. Just as some wind energy researchers are looking to go higher, to capture the stronger and more reliable winds, so geothermal researchers are considering utilizing the higher temperatures that progressively occur with depth below the surface—enhanced geothermal systems (EGS). In 2006 the Massachusetts Institute of Technology published a detailed study on EGS in the United States context, and gave a resource base for the United States in the millions of exajoules [37]. To assess its feasibility, two parameters are important to consider: the temperature and the depth of the geothermal heat source. The study showed that no heat sources above 200°C existed at depths above 4 km. Since the study also demonstrated that well monetary cost rises exponentially with depth, it is probable that input energy costs and final electricity costs would also rise nonlinearly with borehole depth. Disappointingly, no energy analysis of EGS was undertaken in this otherwise detailed report [38].

6.7 Other possible renewable energy sources

A number of other possible RE sources are under consideration, many of which can be grouped together as ocean energy sources. This list includes tidal energy, wave energy, ocean current energy, and ocean thermal energy conversion (OTEC).

The global scope for tides as an energy source is very small, since the total tidal energy resource base is only 75 EJ, of which about 73 EJ is dissipated at coastlines, but in most cases the tidal range is too small for effective utilization. The only large commercial plant for decades was on the Rance estuary in France, completed in 1966 and still supplying 0.5 TWh or 0.002 EJ of electricity annually. In 2011, a slightly larger plant (0.55 TWh) opened in South Korea, and several others are in the planning stage. None have a power output of more than about 250 MW [39]. The larger tidal energy projects dam a bay and function in a manner similar to low head hydro plants, but a less environmentally disruptive approach is to place turbines in tidal flows in river estuaries or straits. One such installation in a tidal flow strait near Belfast has a power output of 1.2 MW [40].

The main potential ocean energy source is wave energy, which occurs because some of the planet's wind energy is transferred to the surface waters by shear forces. Vast numbers of devices to capture wave energy have been invented, and several new designs have undergone ocean trials. López et al. [41], in their review article, listed the advantages and disadvantages of wave energy.

- Its power density, at 2–3 kW m^{-2}, is an order of magnitude greater than solar energy, and almost an order of magnitude greater than wind energy.
- Energy is available around 90% of the time, much greater than for wind or solar energy.
- Wave energy is well matched to demand, given that about 44% of the global population presently live within 150 km from a sea coast [40]. A related point is that wave energy is a potential RE source for nearly all countries with a sea coast, since it can be harvested on the open seas or at the coast.

Drawbacks of wave energy mainly result from the very variable height, frequency, and direction changes (for off-shore converters) of the waves, which all complicate design of the conversion devices. In many offshore locations, the converters must be designed to withstand the force of heavy waves, which increases both their cost and the difficulty of maintenance. At present (2018), despite several sea trials with prototype devices, no wave energy is being generated, despite the \$US 735×10^6 investment over the 2004–14 period [40]. As an example of recent designs, the Pelamis Wave Power device is segmented, and power is generated by the relative movement of the segments. It was the first grid-connected wave energy converter, with different versions deployed off both the Scottish and Portuguese coasts, but are now no longer in operation.

The temperature difference between the tropics and the polar regions drives the various ocean currents, such as the Gulf Stream. Some researchers have proposed tapping into the kinetic energy of these currents as they pass through constrictions like the Straits of Florida [42].

Another method of extracting energy from the ocean is through OTEC, which uses the temperature difference between the surface tropical oceans and the colder water 1 km down to run a heat engine to generate electricity. Since the deep cold waters are really part of the return flow of the major ocean currents, the global OTEC potential cannot be considered separately from that for energy extraction from surface (ocean) currents. Because the temperature difference is at best only about 20–25°C, electricity is generated at low efficiency, and very large volumes of water must be brought to the surface for each kWh produced. EROEI values will therefore be low.

Further, while small shore-based OTEC plants avoid this problem (and can also be used to coproduce fresh water), for globally significant production, OTEC plants will need to be ship-based and continuously move to maintain the necessary temperature difference. The electricity produced will have to be converted to some other energy form such as ammonia or hydrogen, and periodically shipped back to shore. The energy costs of water pumping and ship fuel, and of building the plant and the ship, and the energy losses in energy conversion (and reconversion if needed) will mean a very low EROEI value for OTEC electricity [43].

Each year ~35,000–40,000 km^3 of low salt content fresh water enters the world's oceans [44], which have an average salt content of about 3.5%. The resulting osmotic pressure at the interface is theoretically capable of generating 95 EJ of energy. [45]. At present this energy source is untapped, and there appear to be no plans for its development, possibly because of its technical difficulty, and its likely high environmental costs.

6.8 Discussion

In the year 2050, we have argued that RE will be operating under very different conditions than those that prevail in 2018. A high global carbon price is likely to be in place, with the price having progressively risen over time. In 2016, fossil-fuel use was still increasing globally. If further high output continues, it is likely that by 2050, the remaining fossil fuels will be much more costly to extract than at present, and will carry higher GHG and general environmental costs. Despite much research and even limited field trials, it is probable that in 2050 geoengineering will only be deployed locally. Despite its relatively low estimated monetary costs, it carries the risk of adverse environmental effects, particularly on regional precipitation, and hence lacks global political consensus. By 2050, carbon sequestration is likely to have been implemented on a small scale, but will probably be regarded as at best a minor technique for climate mitigation. All these factors will generally favor RE sources over fossil fuels, enabling RE output to steadily expand. On the other hand, over the next decade or so, it is possible that most carbon reductions will come from energy conservation and energy efficiency measures. The resulting spare capacity in fossil-fuel power stations could inhibit growth of RE output.

But as we have shown, ongoing climate, land use, and environmental changes will also adversely affect both the economics and technical potential for RE in 2050. The first factor is the expected growth of global population, which will, *ceteris paribus*, increase the demand for both agricultural and urban land. Already the growing urbanization of the world is leading to reductions in good farmland and increased need for water resources. It also, of course, will increase the demand for primary energy.

As production of RE continues its inevitable rise (see Fig. 6.1), the quality of the remaining resources will generally fall. Wind speeds for new wind farms will be lower than those being built today, new geothermal electricity plants will need to cope with lower field temperatures, and hydrosites with less favorable locations. Even for solar energy, with its near-inexhaustible resource base, high insolation areas will often be remote from load centers. Shortages of scarce minerals needed for PV cell and wind turbine production could also hinder RE output growth [46]. The result will be declines in the EROEI for RE, and hence rising costs per kWh produced.

Ongoing climate change will have mixed impacts on the fortunes of RE. Geothermal and tidal energy will not be directly affected, because they are independent of climate change. The most affected will be bioenergy, for two reasons. First, yield increases for key crops such as grain could stagnate or even reverse in some regions because of adverse growing conditions brought about by climate change [47], even with better crop selection to match the changing climate. Second, rising demand for food and biomaterials, the biomass-based products that compete with bioenergy, will likely lower the future global technical potential for bioenergy.

A countertrend will be advances in technology, particularly for solar energy. Several RE sources have been utilized for a century or more, and steady improvement rather than technical breakthroughs, are all that can be expected. Wind turbines have undergone great advances in materials, design, and output compared with those common in the 1920s in the rural areas of the United States, but no new breakthroughs are on the horizon, unless turbines can dispense with their earth-bound supporting tower and take to the skies (Section 6.4.2).

An additional consideration is the need for *energy storage*, as intermittent energy sources—mainly solar, but also wind and perhaps wave energy—together come to dominate the global energy supply. A variety of options are available for energy storage, ranging from the already firmly established pumped water storage, compressed air (or combustible gases) in underground caverns, batteries of various types including flow batteries, and even conversion of electricity to some other energy carrier, such as hydrogen [48,49]. Pumped water storage has high energy recovery, but the global possibilities for its expansion are limited. In contrast, underground storage of gas (either air or methane, for example) while lower in energy recovery, has perhaps the greatest potential.

There is at present no general agreement on the extent of energy storage necessary to support large-scale RE, with estimates examined for example ranging from 6% [50] to 35% [51] of annual supply. What is clear is that the storage method adopted will likely depend on the RE energy mix, grid connectivity, and the effect storage has on EROEI values [50–52]. For some energy sources, the storage problem is not so acute. For example, heat storage is possible for STEC because it is a thermal system, thus allowing some electricity production at times of low or zero insolation. In calculating EROEI, each storage method requires its own energy inputs for production and operation. Not only must inputs be factored into any assessment of RE supply, they must also be considered in relation to alternatives: increasing energy production (for example, increasing wind turbine numbers could return more energy than increasing storage), grid expansion to other networks, or connection to remote RE systems [53].

For some regions, though, particularly in the United States and western Europe, massive grid extensions will not be easy [10], so storage may be necessary with access to storage sites being a determining factor. For example, high-energy input cost systems such as batteries could be the only option for some isolated grid networks where there is little if any access to sites which have lower input energy, such as geological storage of compressed air [52]. Finally, as noted by [52], there is also the possibility that energy stored *outside* the energy sector as embedded energy could be more effective than storing energy in the energy sector; examples would include using excess energy to increase stocks of desalinated and purified water.

References

[1] Anon. New Sci 2017. 29 April: 5.
[2] Moriarty P, Honnery D. Rise and fall of the carbon civilisation. London: Springer; 2011.
[3] Smil V. Energy transitions: global and national perspectives. Westport, CT: Praeger; 2010.
[4] International Energy Agency (IEA). Key world energy statistics. Vol. 2016. Paris: IEA/OECD; 2016.
[5] BP. BP statistical review of world energy 2017. London: BP; 2017.
[6] Intergovernmental Panel on Climate Change (IPCC). Climate change 2014: synthesis report. Cambridge: Cambridge University Press; 2015.
[7] Lelieveld J, Evans JS, Fnais M, Giannadaki D, Pozzer A. Nature 2015;525:367–71.
[8] Court V, Fizaine F. Ecol Econ 2017;138:145–59.
[9] Moriarty P, Honnery D. Renew Sustain Energy Rev 2012;16:244–52.
[10] Moriarty P, Honnery D. Energ Policy 2016;93:3–7.
[11] BP. BP Energy Outlook 2017 edition. London: BP; 2017.
[12] Moriarty P, Honnery D. Energ Policy 2010;38:695–700.
[13] Energy Information Administration (EIA). International energy outlook 2016: US Dept of Energy, 2016. https://www.eia.gov/outlooks/ieo/pdf/0484(2016).pdf.
[14] Moriarty P, Honnery D. Sustainable energy resources: prospects and policy. In: Rasul MG, et al., editors. Clean energy for sustainable development. London: Academic Press/Elsevier; 2017 [chapter 1].
[15] Food and Agriculture Organisation (FAO). The state of food insecurity in the world, http://www.fao.org/3/a-i4646e.pdf; 2015.
[16] Campbell JE, Lobell DB, Field CB. Science 2009;324:1055–7.
[17] Schramski JR, Gattie DK, Brown JH. Proc Natl Acad Sci 2015; https://doi.org/10.1073/pnas.1508353112.
[18] Moriarty P, Honnery D. AIMS Energy J 2017;5(1):20–38.
[19] Kleidon A. Glob Planet Chang 2006;54:109–27.
[20] United Nations (UN) 2015. World population prospects: the 2015 revision. http://esa.un.org/unpd/wpp/Download/Standard/Population/ [Accessed 15 December 2015].
[21] World Energy Council (WEC). World energy resources: 2013 survey. London: WEC; 2013 (Also earlier surveys).
[22] Hall CAS, Lambert JG, Balogh SB. Energ Policy 2014;64:141–52.
[23] Pritchard HD. Nature 2017;545:169–74.

[24] Stickler CM, Coe MT, Costa MH, Nepstad DC, McGrath DG, Dias LCP. Proc Natl Acad Sci 2013;110:9601–6.
[25] Ragheb M. Wind energy converters concepts, http://mragheb.com/NPRE%20475%20Wind%20Power%20Systems/Wind%20Energy%20Converters%20Concepts.pdf; 2014.
[26] Staffell I, Green R. Renew Energy 2014;66:775–86.
[27] Richards G. Flying turbines. Engin Technol; 2010. 9–22 Oct: 40–41.
[28] Wikipedia. Solar power, https://en.wikipedia.org/wiki/Solar_power.
[29] International Energy Agency (IEA), http://www.iea-shc.org/solar-heat-worldwide; 2017.
[30] Seboldt W. Acta Astronaut 2004;55:389–99.
[31] Chatzivasileiadis S, Ernst D, Andersson G. Renew Energy 2013;57:372–83.
[32] Azvolinsky A. New Scientist 2017. 15 April: 28–31.
[33] Ferroni F, Hopkirk RJ. Energ Policy 2016;94:336–44.
[34] Van den Bergh J, Folke C, Polasky S, Scheffer M, Steffen W. Curr Opin Environ Sustain 2015;14:170–9.
[35] Moriarty P, Honnery D. Environ Sci Technol 2015;49:5255–6.
[36] Wikipedia. Geothermal heat pump, https://en.wikipedia.org/wiki/Geothermal_heat_pump.
[37] Massachusetts Institute of Technology (MIT). DOE Contract DE-AC07-051D14517, http://geothermal.nrel.gov; 2006.
[38] Moriarty P, Honnery D. Int J Hydrog Energy 2009;34(1):31–9.
[39] power-technology.com. Tidal giants—the world's five biggest tidal power plants. http://www.power-technology.com/features/featuretidal-giants—the-worlds-five-biggest-tidal-power-plants-4211218/ April 11 2014.
[40] Tollefson J. Nature 2014;508:302–4.
[41] López I, Andreu J, Ceballos S, Martínez de Alegría I, Kortabarria I. Renew Sust Energ Rev 2013;27:413–34.
[42] Wikipedia. Marine current power, https://en.wikipedia.org/wiki/Marine_current_power.
[43] Odum HT. Energy 2000;25:389–93.
[44] Wisser D, Fekete BM, Vörösmarty CJ, Schumann AH. Hydrol Earth Syst Sci 2010;14:1–24.
[45] Makarieva AM, Gorshkov VG, Li B-L. Ecol Complex 2008;5:281–8.
[46] Moriarty P, Honnery D. Energ Policy 2009;37:2469–74.
[47] Intergovernmental Panel on Climate Change (IPCC). Climate change 2014: Impacts, adaptation, and vulnerability. Cambridge: Cambridge University Press; 2014.
[48] Pickard WF. Proc IEEE 2014;102(7):1094–105.
[49] Lemmon JP. Nature 2015;525:447–9.
[50] Bussar C, Moos M, Alvarez R, Wolf P, Thien T, Chen H, et al. Energy Procedia 2014;46:40–7.
[51] Pleßmann G, Erdmann M, Hlusiak M, Breyer C. Energy Procedia 2014;46:22–31.
[52] Barnhart CJ, Dale D, Brandt AR, Bensen SM. Energy Environ Sci 2013;6:2804–10.
[53] Steinke F, Wolfrum P, Hoffmann C. Renew Energy 2013;50:826–32.

Further reading

[54] Food and Agriculture Organisation (FAO). Forestry products yearbook 2014. Rome: FAO; 2014.
[55] Mattick CS, Williams E, Allenby BR. IEEE Technol Soc Mag 2010. Fall:22–30.

Section C

Reducing Greenhouse Gases: Renewables and Zero Carbon/Carbon Neutral Forms of Energy and Electric Cars

Methane hydrate as a "new energy"

7

Liang Cui, Azizul Moqsud†, Masayuki Hyodo†, Subhamoy Bhattacharya*,1*
*University of Surrey, Guildford, United Kingdom, †Yamaguchi University, Japan
[1]Corresponding author: s.bhattacharya@surrey.ac.uk

Chapter Outline

7.1 Introduction

7.1.1 What is methane hydrate?

Methane hydrate (MH) is a solid compound in which a large amount of methane gas molecules (CH_4) are caged within a crystalline structure of water, as illustrated in Fig. 7.1, under low temperature and high pressure, forming a solid similar to ice [1]. It looks like ice, but starts burning when an open flame is brought close to it; methane hydrate is often called "fiery ice." As found in previous investigations [2], a great

Managing Global Warming. https://doi.org/10.1016/B978-0-12-814104-5.00007-7

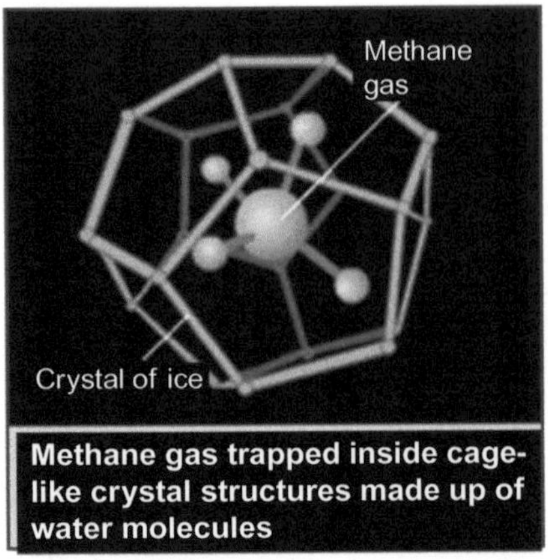

Fig. 7.1 Crystal structure of methane hydrate.

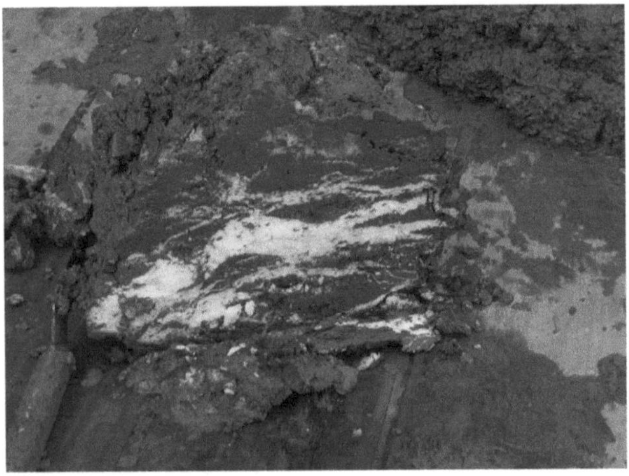

Fig. 7.2 Methane hydrate-bearing sediment [3].

amount of MH exists stably undersea. The natural methane hydrate may not be purely a white agglomeration like artificial methane hydrate. Since methane hydrate exists in between the sand particles of sandy sediments as shown in Fig. 7.2, the methane hydrate-bearing sediments do not appear white but rather look similar to soil.

7.1.2 Where is the reserve and how much?

MH exploitation has attracted great attentions since the total carbon content in MH is twice as large as that in petroleum or coal [4]. One cubic meter of MH dissociates to approximately 160–170 m^3 (at 0°C and 1 atm) of methane gas. MH is distributed almost all over the world (Figs. 7.3 and 7.4). The total worldwide estimated reserve is around $(1-5) \times 10^{15}$ m^3 [7]. It is considered as a type of promising and hugely

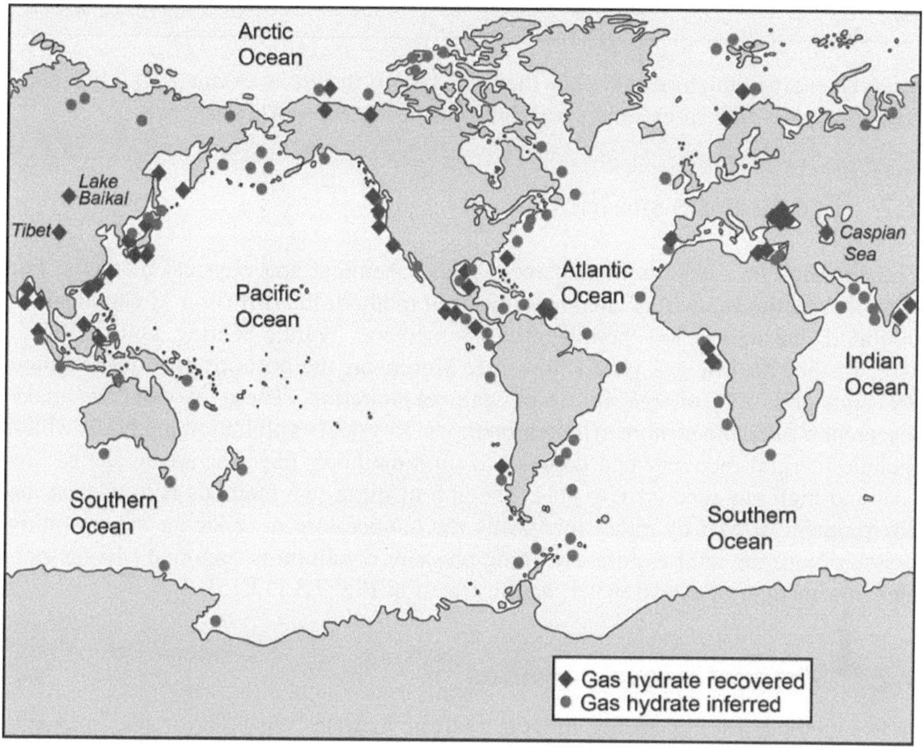

Fig. 7.3 Location of sampled and inferred gas hydrate occurrences worldwide [5].

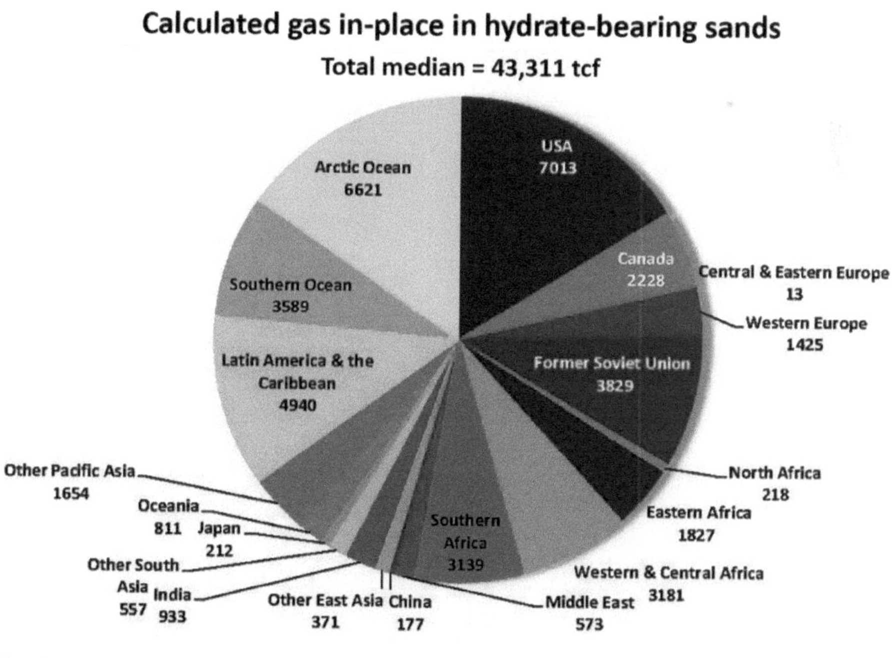

Fig. 7.4 Reserve of methane hydrate in different parts of the world [6].

reserved energy which can alleviate the energy crisis to certain extent [8,9]. Therefore, how to exploit MH safely and efficiently has become a worldwide focus.

7.2 Production methods

The exploitation methods mainly include the chemical and physical methods. For chemical methods such as chemical injection method, the diffusion of chemicals is inhibited due to the low permeability of methane hydrate-bearing sand (MHBS) [10], leading to low gas production rate. Moreover, the costs of chemical method are high and not suitable for large-scale exploitation. These two defects make researchers seek other more efficient methods. Physical exploitation methods, which include thermal recovery and depressurization methods, have the advantage of low cost and high gas productivity [11]. The aim of these two methods is to change the environment of MH by either increasing the temperature or reducing the confining pressure until the temperature-confining pressure condition is such that MH dissociates into methane gas and water, as illustrated in Fig. 7.5 [12,13].

7.2.1 Thermal recovery method

In this method, a well is drilled into the methane hydrate-bearing layer, and the methane hydrate is dissociated by heating, using a fluid (hot water or steam) heated at the surface in a boiler or similar device and circulated down through the well, as illustrated in Fig. 7.6. This causes methane hydrate to dissociate and generate methane gas.

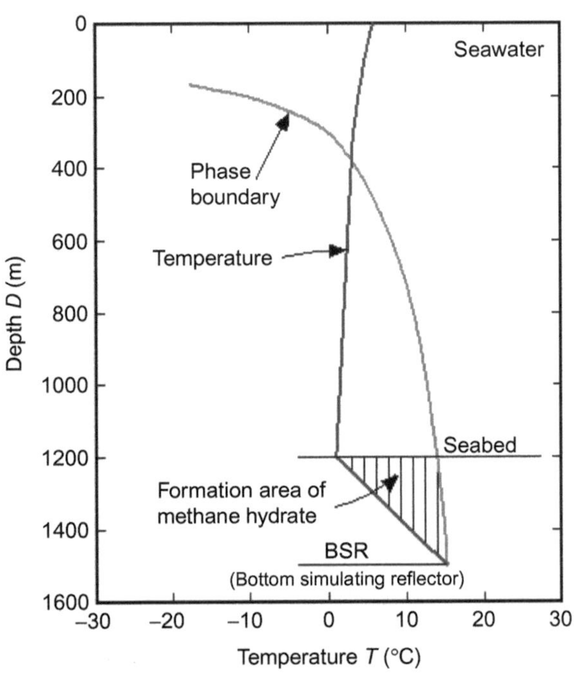

Fig. 7.5 The base condition to form MH.

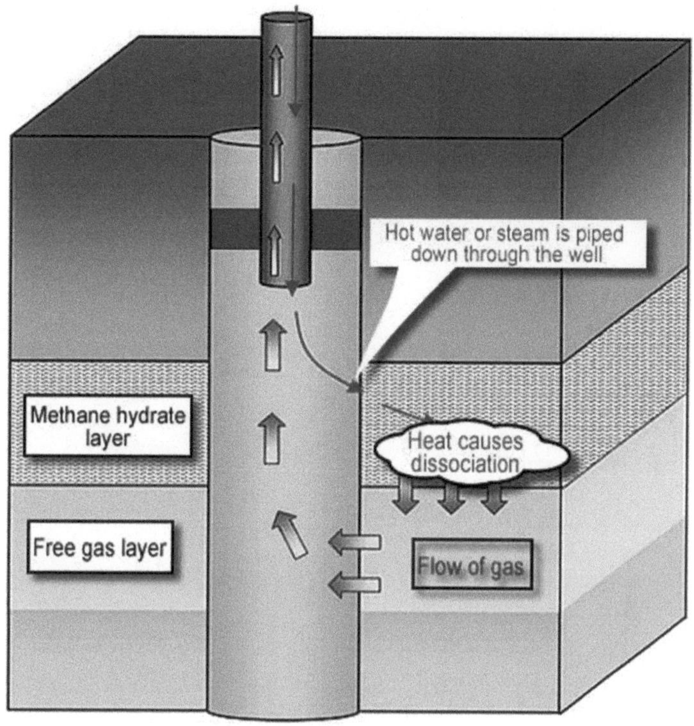

Fig. 7.6 Thermal recovery method for MH production.

7.2.2 Depressurization method

The depressurization method consists of lowering the pressure inside the well and encouraging the methane hydrate to dissociate into methane gas and water, as illustrated in Fig. 7.7. At the end of the production, the water is injected back and the water pressure recovered.

Promising progress has made in the MH gas production in both China and Japan [14]. Despite the obvious advantages of physical methods, there are potential problems associated with these exploitation processes: these include the stress changes caused by drilling, the settlement, landslide, and gas leakage. Due to these problems, some accidents have happened in these processes such as submarine landslides, platform foundation settlement, and failures of lifeline engineering projects [15,16]. Therefore, in order to safely extract the gas it is necessary to first understand the mechanical and dissociation properties of the methane hydrate deposits.

To study the dissociation process, previous in situ tests [17] have investigated the change of modulus by measuring the shear wave propagation velocities through the MHBS, while other mechanical properties, such as stress-strain relationships, could not be studied during the in situ MH dissociation. It is not practical to carry out laboratory tests on undisturbed samples either, because it is too difficult to maintain the

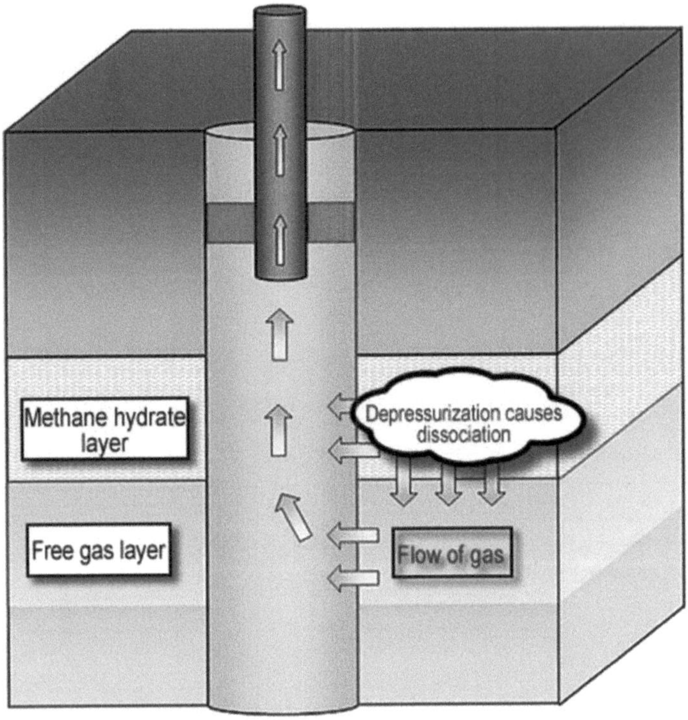

Fig. 7.7 Depressurization method for MH exploitation.

MH and moreover the samples are easily disturbed during transportation. As a result many previous laboratory tests were performed on artificial samples. In the next section the experimental studies of the changes in the mechanical properties of MHBS during MH dissociation using an improved triaxial apparatus carried out by Hyodo et al. [18] are described in detail.

7.3 Testing equipment and sample preparation

7.3.1 Triaxial testing apparatus

To reproduce the stress and temperature conditions in the deep seabed and to examine the mechanical behavior of MH-bearing sand specimens under the conditions at which MH is produced, a temperature-controlled high pressure triaxial testing apparatus was developed by Hyodo et al. [18]. With this apparatus the pore pressure and confining pressure could be controlled under various temperature and high pressure conditions. Fig. 7.8A shows the physical appearance of the device, while Fig. 7.8B shows a

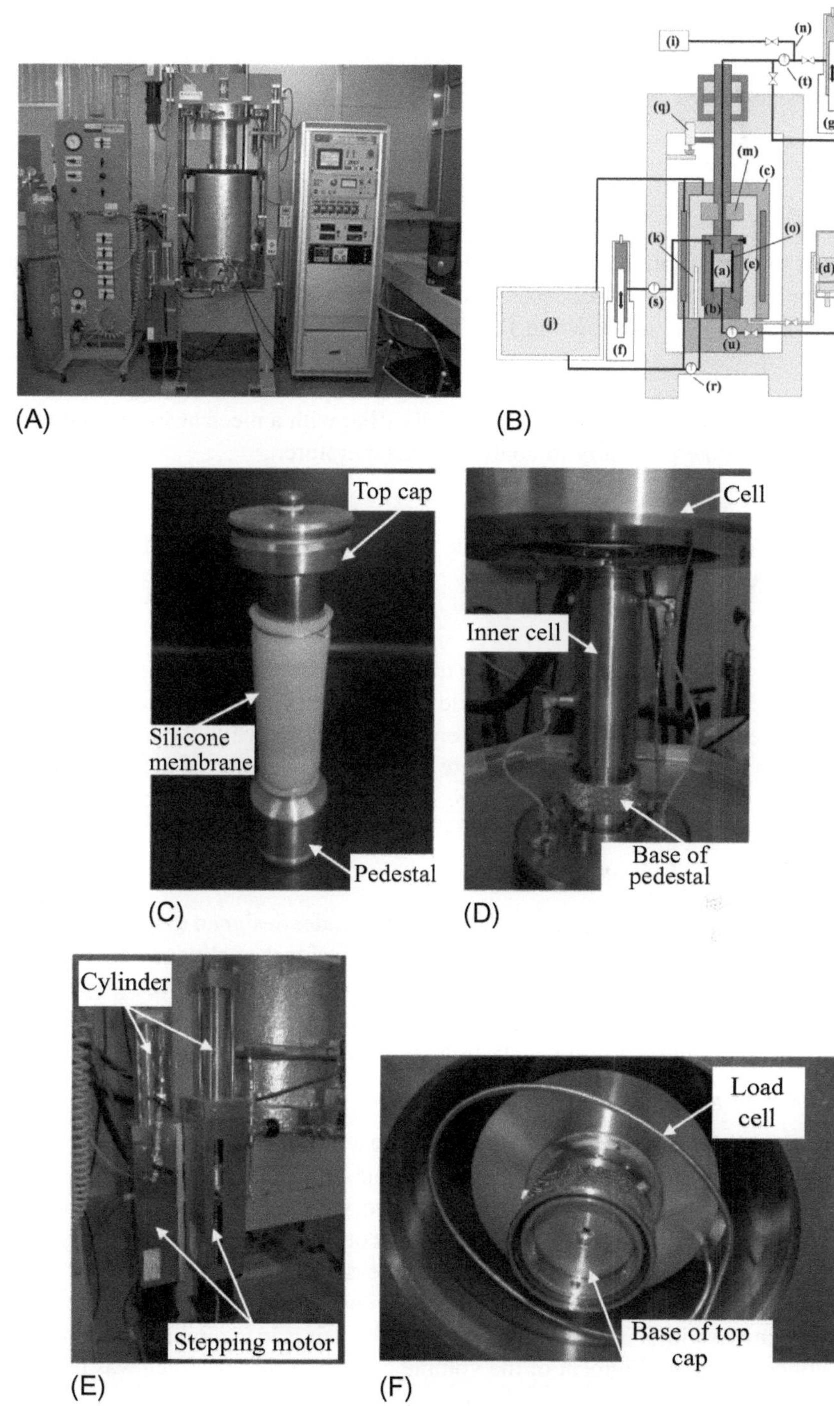

Fig. 7.8 View of triaxial test device for MH experiment [18].

schematic diagram of the testing system. A detailed description of each part shown in Fig. 7.8B is presented below:

(a) *Test specimen*

A cylindrical frozen specimen measuring 30 mm in diameter and 50 mm in height, or 50 mm in diameter and 100 mm in height was used.

(b) *Pedestal*

To prevent the dissociation of natural MH samples, test specimens had to be set up rapidly. For this purpose, the pedestal in this apparatus was designed to be a one-touch removable socket-type, as shown in Fig. 7.8C.

(c) *Cell*

The cell was designed to resist a pressure of 30 MPa, with a mechanism that allowed cell fluid to circulate internally to control the temperature.

(d) *Cell pressure generation device*

Cell pressures of up to 30 MPa controllable within the range of ± 0.1 MPa were supplied by oil pressure.

(e) *Inner cell*

Because of the possibility for the specimen to become unsaturated during a test, a double cell was adopted to measure the volume change. The change in volume could be obtained by measuring the water level difference in the cell, while the upper part was opened; however, since high pressures were used, the penetration of the piston in the sealed-up cylinder controlling the pressure made it possible to measure the volume change, as shown in Fig. 7.8D.

(f) *Syringe pump for inner cell*

To measure the volume change in the inner cell, a cylinder designed to resist pressures up to 30 MPa was installed and the change in volume inside the cylinder was measured by controlling the piston in the cylinder via a pulse-controlled method using a stepping motor. In addition, the change in volume of unsaturated specimens was measured by correcting the amount of piston penetration, as shown in Fig. 7.8E.

(g) *Upper syringe pump*

Up till now, high pressure triaxial compression test equipment used in geotechnical engineering has been used to reproduce the ground stress under high confining pressure. However, the pore water pressure associated with the high pressure condition needed to reproduce the large depths where MH could exist has not previously been investigated. To do this, a device similar to the syringe pump for the inner cell (Fig. 7.8F) was installed. At the extreme condition of 20 MPa, it was possible to control the pressure to within ± 0.05 MPa. Moreover, by using an incompressible solution in the cylinder, the measurement of the volume change of the specimen was possible by calculating the amount of penetration of the piston in the cylinder from the pulse. The pipe was made of stainless steel so that it could withstand high pressure. This was the main feature of this testing apparatus used to measure the high pore pressure in the depth of the seabed.

(h) *Lower syringe pump*

A mechanism similar to (g) was also installed.

(i) *Gas mass flow meter*

To calculate the MH saturation ratio, a mass flow meter for the gas was installed in the pipe. The gas that passed through the device was measured as a mass flow (in units of g min^{-1}) independent of temperature and pressure, and shown as flow volume (in units of mL min^{-1}) standardized at 20°C and 1 atm. In addition, the amount of gas could be calculated. The measurement range was 0–500 mL min^{-1}, and when MH was dissociated after the end of shearing, the amount of gas was measured while adjusting the valve in the pipe equipped with the mass flow meter connected to the specimen.

(j) *Device controlling the liquid temperature*

This device was used to adjust the temperature from −35°C to +50°C within the triaxial cell by circulating the cell fluid in the triaxial testing device from an external temperature controlled water tank. The temperature in the tank could be controlled within a range of ±0.1°C. To conduct tests at low temperatures, it was possible to use a fluid with a freezing point below −40°C or to use aurorabrine, which has excellent anticorrosion properties when in contact with various metals.

(k) *Thermometer*

A thermometer was installed near the side of the specimen in the triaxial room, so that the temperature in the triaxial compartment could be measured. The temperature in the cell was controlled based on the value indicated by the thermometer. When the temperature was changed while performing the tests, adequate time was provided until the temperature of the fluid and the specimen equalized.

(l) *Methane gas cylinder*

Methane for producing methane hydrate was stored in a gas cylinder which for safety reasons was installed outdoors.

(m) *Load cell*

To eliminate the influence of piston friction, a cylindrical-shaped loading cell that could not be affected by temperature and pressure was set up in the cell. The maximum permissible load was 200 kN, and it was possible to measure with accuracy of up to 1/1000 of the full-scale load, as shown in Fig. 7.8F.

(n) *Pipe*

Methane gas flowed inside the pipe after the test, and to prevent the accumulation of gas, the pipe was oriented vertically and as short a pipe as possible was used.

(o) *Membrane*

Because the specimens in the tests were subjected to low temperatures and high pressures, latex membranes conventionally used in triaxial tests were avoided; instead, silicon-type membranes were used because of their flexibility under low temperature and high pressure conditions. Note that because methane gas can permeate through silicon to some degree, butyl rubber was used in long-term tests, such as during MH dissociation.

The measuring apparatus was connected to a data collection system, with the vertical load, axial displacement, confining pressure, volume change of specimen, and pore water pressure automatically recorded by computer. To verify the accuracy of the measuring devices, the following shear tests were done using samples of silica sand which did not contain MH.

7.3.2 Specimen preparation

Based on visual observation of the undisturbed core samples obtained from the Nankai Trough around Japan, it is believed that MH in situ is buried within the pores between grains of the sand. Based on this, MH-bearing sand was artificially produced using the grain size distribution curve of the undisturbed core sample as shown in Fig. 7.9, and Toyoura sand was chosen as the host material. First, the amount of water for the target MH saturation was mixed with volume of sand which corresponded with the target density. The moist soil was placed in 15 layers in a mold measuring 30 mm in diameter and 60 mm high, with each layer compacted by a tamper 40 times. In order for the specimen to stand by itself, the mold containing the sand was placed in a freezer. The frozen specimen was then removed from the mold and placed on the pedestal,

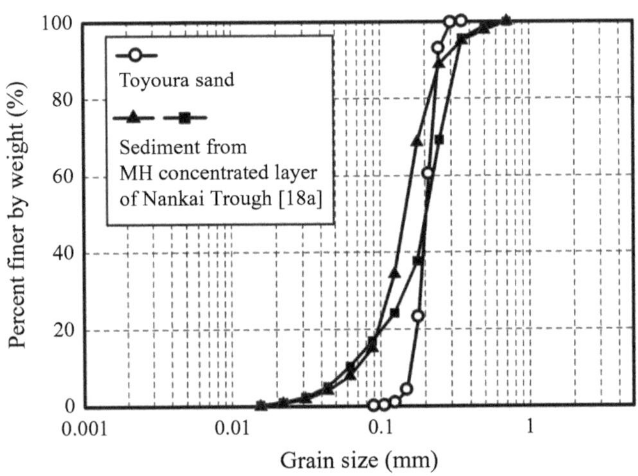

Fig. 7.9 Grain size distribution curves for Toyoura sand and sediments obtained from the Nankai Trough around Japan [18].

and the membrane was installed as shown in Fig. 7.8C. Because the specimens in the tests were subjected to low temperature and high pressure, rubber membranes conventionally used in triaxial tests were avoided; instead, silicon-type membranes were used because of their flexibility under low temperature/high pressure conditions and butyl rubber was used in long-term tests, such as MH dissociation because silicone is to some degree permeable to methane gas. The inner cell set up is shown in Fig. 7.8D.

7.3.3 Generation of MH and experimental procedure

After forming the specimen, it was subjected to a series of processes under specific temperatures and pressures, as depicted in Fig. 7.10. First of all, the frozen specimen (a) was thawed to room temperature inside the triaxial cell (b). Then, the back pressure was gradually increased to 4 MPa while methane was injected into the specimen (c) thus filling the pores of the specimen with methane. At this time, the gas pressure was increased over a period of time so that the specimen’s moisture content would not become nonuniform as a result of the pressurized injection. Next, the temperature in the triaxial cell was lowered to 1°C where the MH was stable, and the specimen environment was kept under constant temperature and pressure conditions for 24 h. By keeping the gas pressure constant in the connection between the specimen and the syringe pump and by observing the amount of gas flowing at various times, the transformation of water within the pores into hydrate was judged to be complete if there was no marked change in the amount of gas, as indicated in the Fig. 7.11. Note that the plots in the figure show some irregularities because in inducing the gas to flow into the

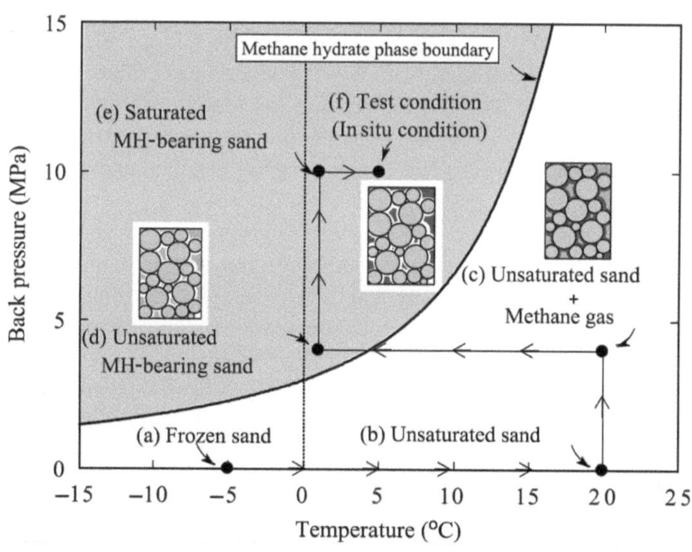

Fig. 7.10 State paths for pressure and temperature to produce MH-bearing sand [18].

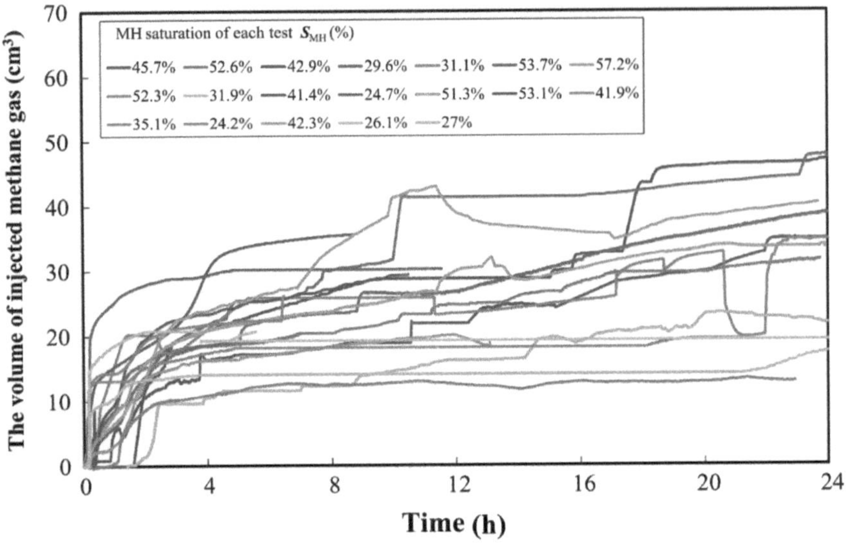

Fig. 7.11 Time histories of volume of methane gas injected into the MH-bearing sands [18].

specimen to promote the generation of MH, the volume of gas in the specimen was increased and decreased slightly while adjusting the upper and lower syringe pumps.

After the hydrate was generated, pure water under constant pressure was allowed to infiltrate the specimen. Here pure water is injected under stable MH conditions. The, residual methane gas then reacted with pure water to form MH, while, some MH dissolved into the pure water. Both MH formation and dissociation may happen during this process. The injected water was determined by the volume of the test specimen only. MH saturation was calculated from the dissociated gas volume of the specimen.

Then, the back pressure was applied (e) and the temperature was adjusted to the prescribed test condition (f). While keeping the pressure constant, consolidation was carried out until the specified effective stress was reached and shearing was conducted with a strain rate of 0.1% min^{-1}.

When the MH was to be generated after consolidation, the frozen specimen set up in the triaxial cell room was thawed immediately and consolidation was performed until the prescribed stress was reached. Then, methane gas was injected into the specimen, and when the hydrate was generated in the MH stable zone, the sample was sheared at the same temperature and back pressure as MH was to be generated before consolidation. These cases were used to examine the difference between the response of MH being generated in existing sedimentary soil layers and the case after MH was generated in the soil layers sediment.

After shearing, the temperature in the specimen was increased and MH dissociated; the amount of gas was measured using the gas mass flow meter shown in Fig. 7.8B. The amount of gas measured was then converted into MH saturation, assuming the density of MH was 0.913 mg m^{-3}.

7.3.4 Triaxial compression tests

7.3.4.1 Testing condition

The experimental conditions were considered to be equivalent to in situ conditions. In the experiments, the specimens were subjected to different levels of effective confining pressure (1, 3, 5 MPa), back pressure (5, 10, 15 MPa), and temperature (1°C, 5°C, 10°C). For all tests, a shearing rate of 0.1% min^{-1} was adopted. Tests were performed on two sets of specimens, porosity $n=40\%$ (relative density $D_r=90\%$) and $n=45\%$ ($D_r=40\%$), with various MH saturation ratios. The maximum deviator stress indicated in the table refers to the peak stress obtained from the deviator stress-axial strain relation; however, if a peak was not observed, the deviator stress corresponding to 15% axial strain was used.

7.3.4.2 Test results

The initial shear modulus increased with MH saturation due to the increased level of particle bonding (Fig. 7.12). This effect was still evident up to 15% axial strain. The Toyoura host sand demonstrated contractive behavior under high confining pressures.

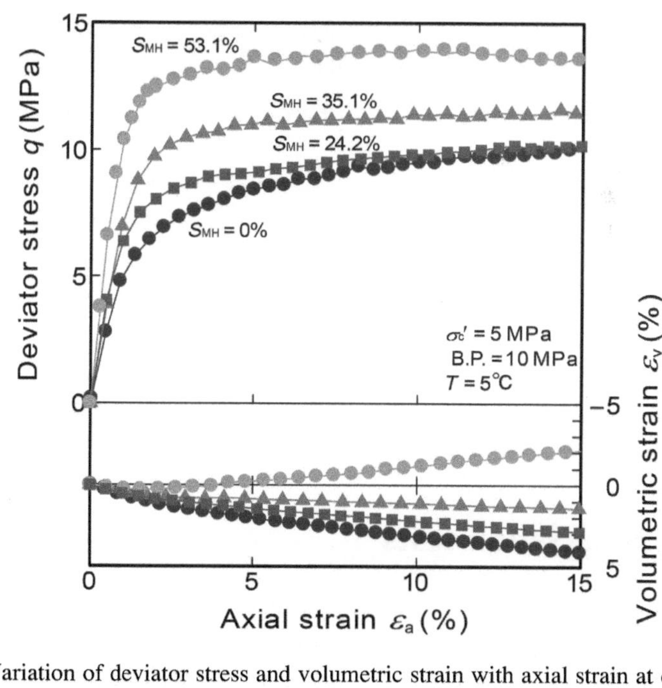

Fig. 7.12 Variation of deviator stress and volumetric strain with axial strain at different MH saturation [18].

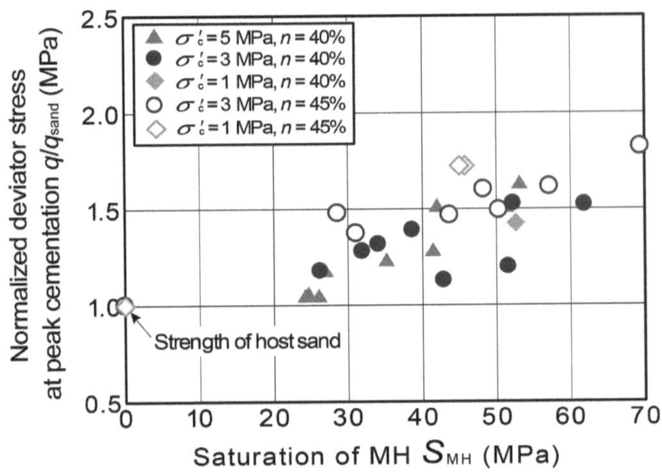

Fig. 7.13 Variation of normalized deviator stress at peak cementation with axial strain at different MH saturation [18].

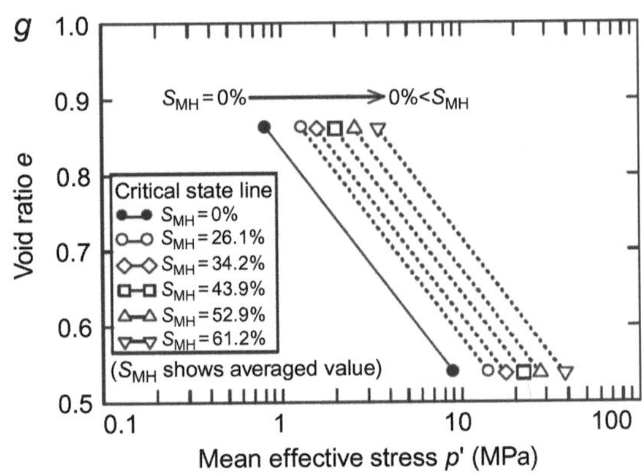

Fig. 7.14 Critical state lines for MH-bearing sand [18].

This became increasingly dilative as the MH saturation increased. The normalized strength of the MHBS sample increase evidently with the increasing MH saturation (Fig. 7.13), and the critical state line was shifted to the left with the increasing MH saturation (Fig. 7.14).

7.4 MH dissociation tests

After consolidation at an effective stress equivalent to an in situ condition, MH dissociation was performed by either increasing the temperature or decreasing the pressure. The purpose of this test was to study the effect of dissociation and consequent loss of particle bonding in the zone where failure could occur with unbonded sand.

7.4.1 Initial stress before MH dissociation started

Fig. 7.15 shows the stress paths followed for three different cases. In Case 1 there was no initial shear stress, and the temperature was increased to simulate thermal recovery; the mean principal effective stress was increased from a to a′ to simulate the depressurization method. In the case of depressurization, the back pressure was allowed to increase again, after dissociation, to ensure that the postproduction equilibrium conditions were being restored. In Case 2 an initial shear stress was applied before using thermal recovery or depressurization. Following this the temperature was allowed to increase at b; the back pressure was decreased from b to b′, and from b′ back to b after the dissociation of all the MH. Finally in Case 3 the stress path was taken into the metastable zone between the failure envelopes for pure sand and MH-bearing sand.

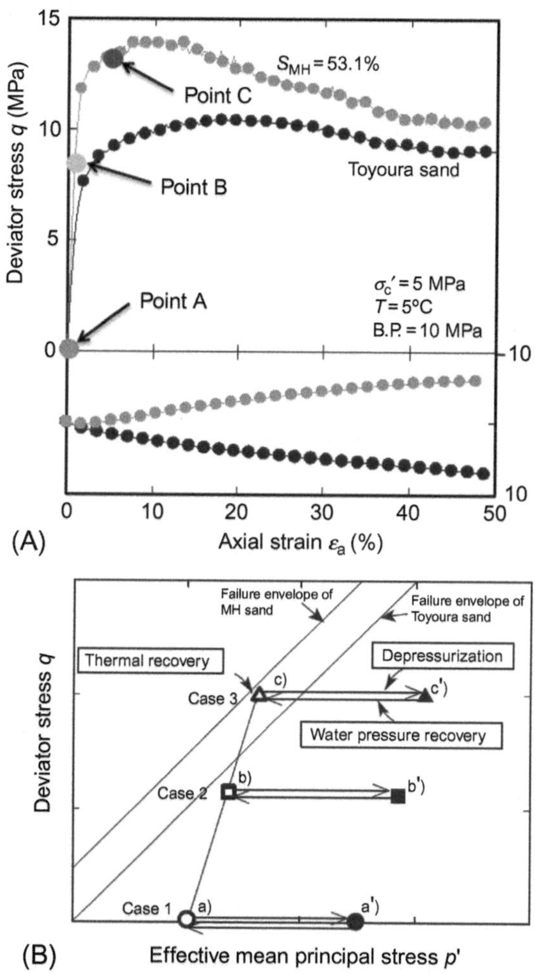

Fig. 7.15 Initial stress before MH dissociation started and image of MH dissociation [18].

7.4.2 Depressurization method

The variations of back pressure, MH saturation, volumetric and axial strains during the depressurization and repressurization stages are illustrated in Figs. 7.16 and 7.17, respectively. In each case there was an initial compressive volumetric strain because of the increase in effective stress. Further volumetric strain occurred as the MH dissipated and bonds dissolved. Dissociation occurred when the back pressure dropped to 4.3 MPa. Dissociation continued for approximately 2–3 h from initiation, with axial strains still increasing up to 3 h after the start of the test. After 5 h the back pressure was ramped back up to 10 MPa over 1 h to represent a restoration of postdissociation equilibrium conditions being restored in situ. In Cases 1 and 2 there was an elastic recovery of axial and volumetric strains due to a decrease in effective stress. However, in Case 3 collapses occurred as the current nonbonded soil moved outside the failure envelope.

Fig. 7.18 shows the stress ratio, axial strain, and volumetric strain relations during the tests. It can be seen that the axial strain increased dramatically when the stress strain curve reached the strength of pure sand as the water pressure recovered. In contrast to the thermal method no deformations were observed despite the increase in effective stresses. After dissociation during back pressure recovery large deformations may occur in the metastable zone.

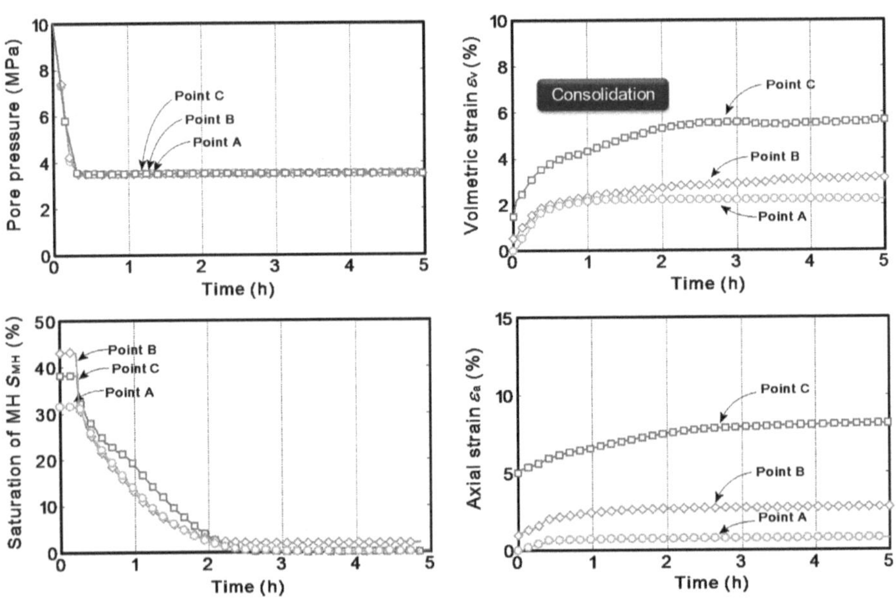

Fig. 7.16 Depressurization method [18].

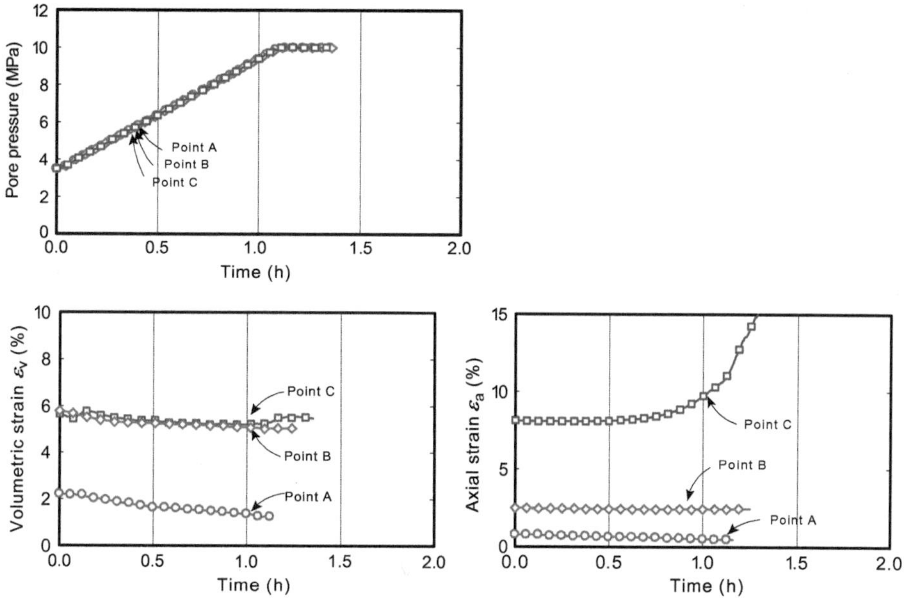

Fig. 7.17 Repressurization stage [18].

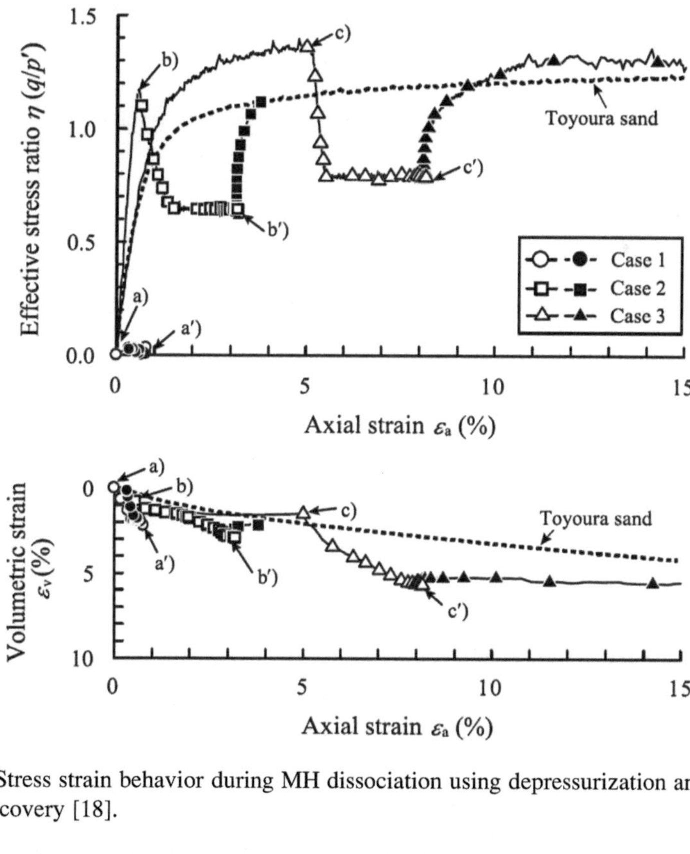

Fig. 7.18 Stress strain behavior during MH dissociation using depressurization and water pressure recovery [18].

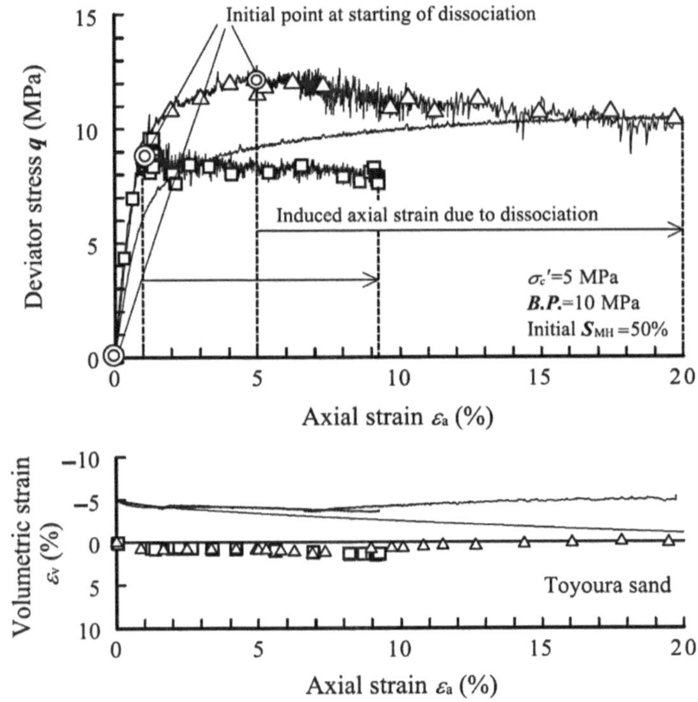

Fig. 7.19 Stress strain behavior during MH dissociation using thermal recovery [18].

7.4.3 Thermal recovery

The temperature was increased to cause dissociation by circulating heated oil through a double walled annulus surrounding the inner cell. The axial load was kept constant after the initial shear stress had been applied (Fig. 7.19). From Fig. 7.20, it is observed for Case 1 that no deformation was occurring during the dissociation of MH. Next, in Case 2, the axial strain increased to axial strain = 9%; however, the failure conditions were not reached. Finally, in Case 3, it was observed that the sediment failed due to dissociation of MH. That is, the sediment collapsed when the methane hydrate dissociated into the state between the failures envelopes for pure sand and the MH-bearing sand. Besides, the deviator stress decreased during thermal recovery in Cases 2 and 3, because the axial load was not kept completely constant due to the deformation occurred very quickly (Figs. 7.19 and 7.20).

7.4.4 Experimental findings

Experimental tests have been carried out on the mechanical and dissociation properties of sands containing MH using innovative high pressure apparatus which allowed control of the back pressure and the temperature. The following conclusions were drawn:

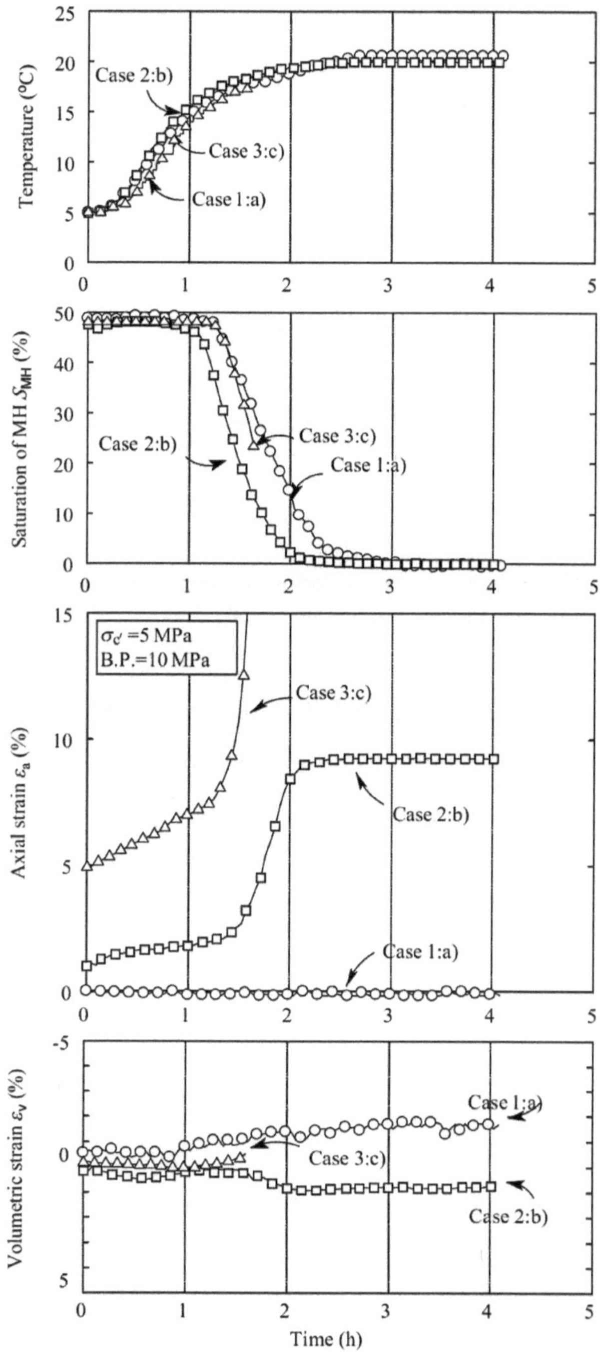

Fig. 7.20 Variation of temperature, MH saturation, and volumetric and axial strains with elapsed time [18].

1. The strength of MH sand increased with MH saturation due to particle bonding and supporting.
2. MHBS sample will be damaged after exploitation (no matter if thermal recovery or depressurization process was used) when the deviator stress applied is larger than the compression strength of pure host sand.
3. In the case of dissociation by depressurization, axial strains were generated by increasing effective stress until a stable equilibrium was reached. However, repressurization led to the failure in the metastable zone.

7.5 DEM simulation of MH dissociation process

Many challenges were encountered in experiments, such as difficulties in replicating samples with the same physical properties or visualizing the changes of microscopic structure, which is believed to underlie the macroscopic behaviors. To achieve these objectives, numerical simulations are required.

The discrete element method (DEM) simulates granular materials (soils) as assemblies of individual particles. It was developed on the basis of simple interparticle contact laws and has the advantage of visualizing the evolution of microscopic particle interactions with good sample repeatability and also of low cost.

7.5.1 Reproduction of MH dissociation process using DEM

The laboratorial thermal recovery and depressurization process was reproduced by Jiang et al. [19] using DEM simulations. As shown in Fig. 7.21, to simulate the MHBS exploitation, each microscale MH bond at contact is idealized as a pure MH sample subjected to a minitriaxial test, where the pore pressure (back pressure) in the MHBS sample is idealized as the "confining pressure" of the pure MH

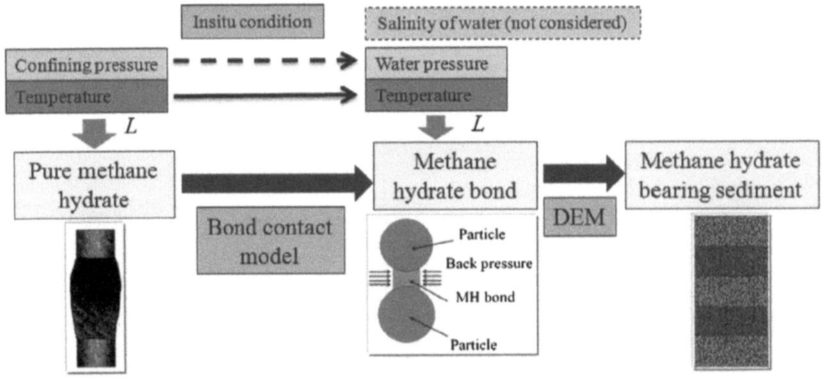

Fig. 7.21 Schematic diagram of connection between lab test of methane hydrate and DEM bond model for simulations of MHBS [19].

sample, equivalent to an experimental test. The controlling parameters are given in Fig. 7.22.

The DEM simulations of thermal recovery are shown in Fig. 7.23, while the DEM simulations of depressurization process are illustrated in Fig. 7.24. By comparing Figs. 7.16 and 7.19 with Figs. 7.23 and 7.24, it is evident that the DEM simulations of thermal recovery and depressurization process capture the sample responses as observed in laboratory tests. The most important advantage of DEM simulation is that it can visualize the micromechanism in a granular material. Some good examples are given in the following section.

7.5.2 Micromechanism associated with MH dissociation

There are two types of contacts in MHBS, i.e., the bonded contact and the unbonded contact. In order to study the effects of temperature and applied deviator stress during thermal recovery, the distributions of the total, bonded, and unbonded contacts in

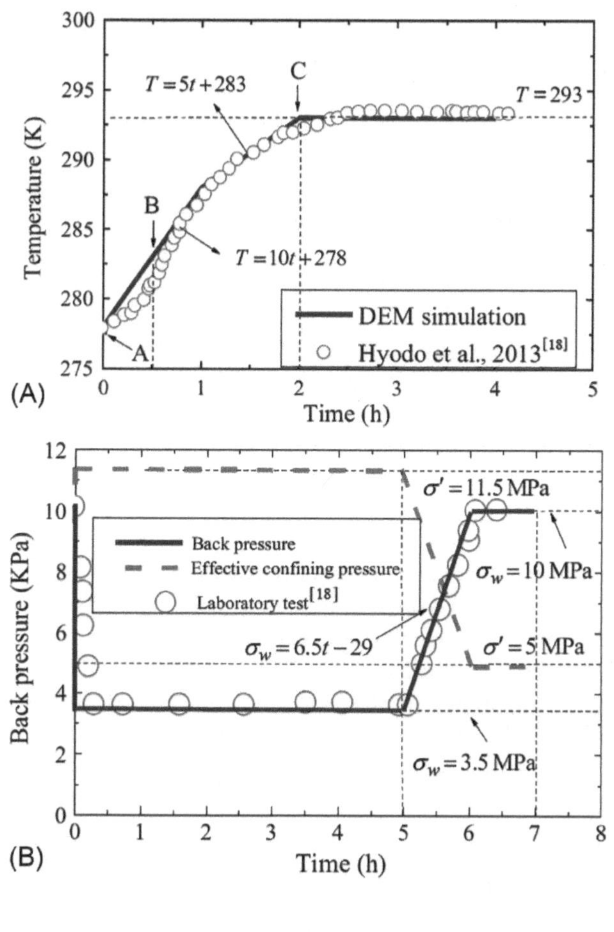

Fig. 7.22 Temperature and confining pressure vs time in laboratory test and DEM simulation (A) thermal recovery and (B) depressurization [19].

Fig. 7.23 DEM simulation results of thermal recovery [19].

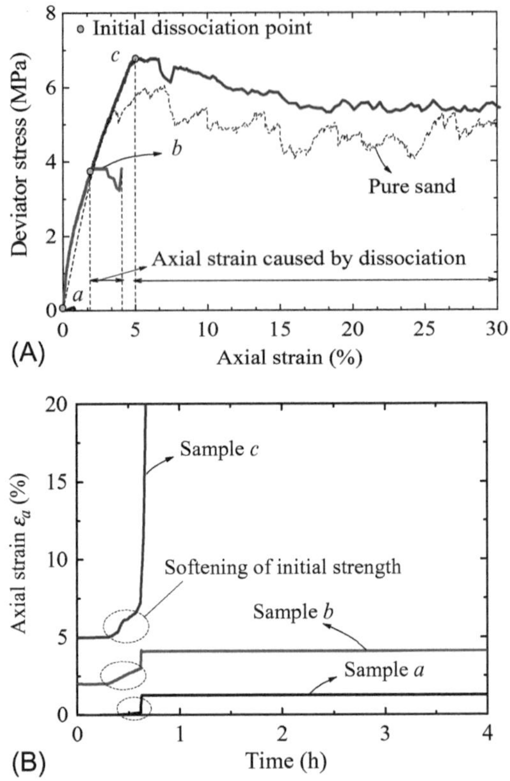

samples *a*, *b*, and *c* at characteristic moment *B* ($t=0.5$ h in Fig. 7.22A) were analyzed and shown in Fig. 7.25. It can be seen that with increasing deviator stress, the total contact distribution changes from slight anisotropy to apparent anisotropy. The portion of unbonded contacts increases with increasing deviator stress as the enclosed area of unbonded contact distribution increases. The distribution of unbonded contact remains anisotropic with a vertical major principal direction, which is consistent with the direction of major principal stress since more particle contacts are required to resist the higher vertical stress. The portion of bonded contacts decreases and its distribution remains anisotropic with a horizontal major principal direction, which is perpendicular to the major principal direction of unbonded contact distribution. For sample *a* with zero deviator stress (i.e., only confining pressure applied), very few bonds were damaged despite the increasing of temperature for 0.5 h (Fig. 7.25A). The contact distribution was similar to the initial condition ($t=0$ h), when the bonds were formed at $\sigma' = 0.2$ MPa. It is noted that the initial total contacts in sample *a* shows slight anisotropy (similar to Fig. 7.25A) caused by specimen compaction. This may reflect the actual anisotropic condition of natural MHBS.

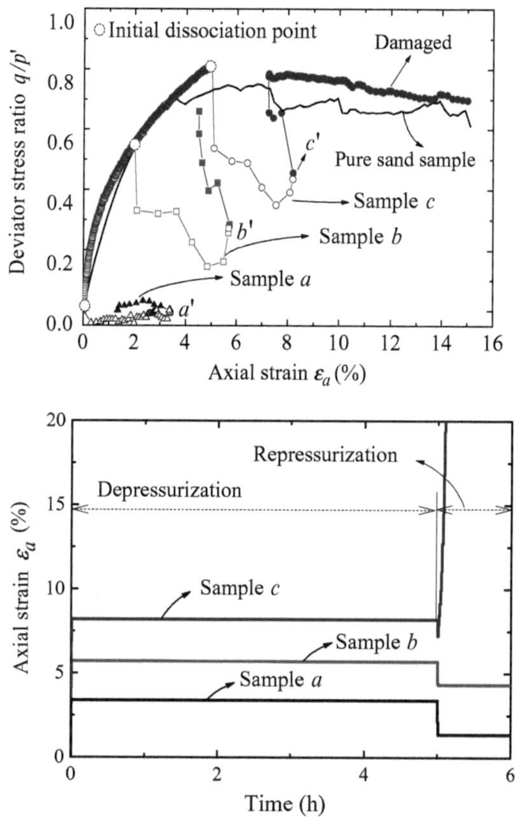

Fig. 7.24 DEM simulation results of depressurization process [19].

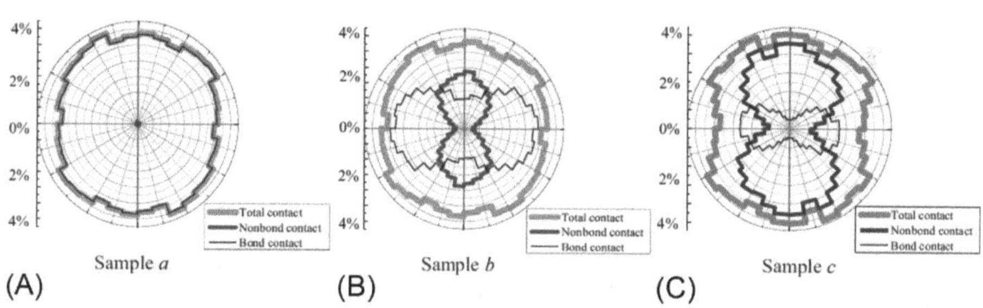

Fig. 7.25 Contact distribution of different samples at $t = 0.5$ h during thermal recovery [19].

However, for sample *b* with same bond strength as sample *a*, MH bonds were apparently damaged (Fig. 7.25B). This difference was caused by the application of a deviator stress. As the deviator stress of 3.8 MPa was applied, forces transmitted in sample *b* were larger than in sample *a*, thus many vertical bonded contact were damaged and bonded contact distribution showed anisotropy with a horizontal major

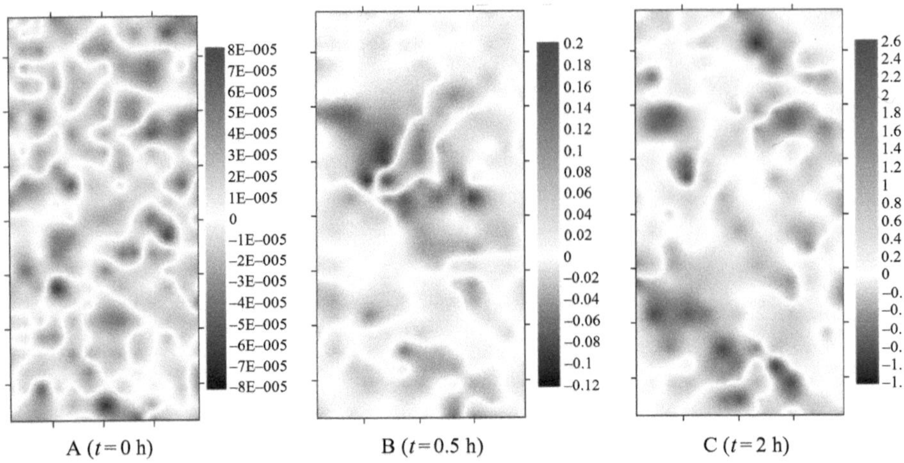

Fig. 7.26 APR distributions of samples *c* at different time during thermal recovery [19].

principal direction. By comparing sample *b* and sample *c*, it is seen that larger force causes more MH bond damages as the area of bonded contact distribution decreases, and the anisotropy of MHBS is more evident (Fig. 7.25C).

On consideration of energy dissipation, general contact displacement can be decomposed into two components: rolling and sliding, which have relationships with particle translational movement, rotation, and particle size. The effects of particle rotation and particle size can be illustrated by the averaged pure rotation rate (APR), which can be expressed as [20]:

$$\omega = \frac{1}{N}\sum_{k=1}^{N}\left[\frac{1}{\overline{R}^{k}}\left(\dot{\theta}_1^k R_1^k + \dot{\theta}_2^k R_2^k\right)\right]$$

Fig. 7.26 shows the APR distributions of sample *c* at points A ($t=0.5$ h), B ($t=0.5$ h) and C ($t=0.5$ h), corresponding to the initial condition, during thermal recovery (still in the stable zone) and after thermal recovery (MH dissociates completely), respectively. Fig. 7.26 shows that with increasing temperature, the maximum absolute value of APR increases, and the distribution of positive and negative values of APR becomes increasingly localized, which indicates the development of shear bands.

Fig. 7.27 depicts the bond ratio distribution of sample *a*, *b*, and *c* at $t=0.5$ h. It shows that the value of bond ratio decreases with increasing deviator stress, which implies that there are more bonds damaged under higher deviator stress.

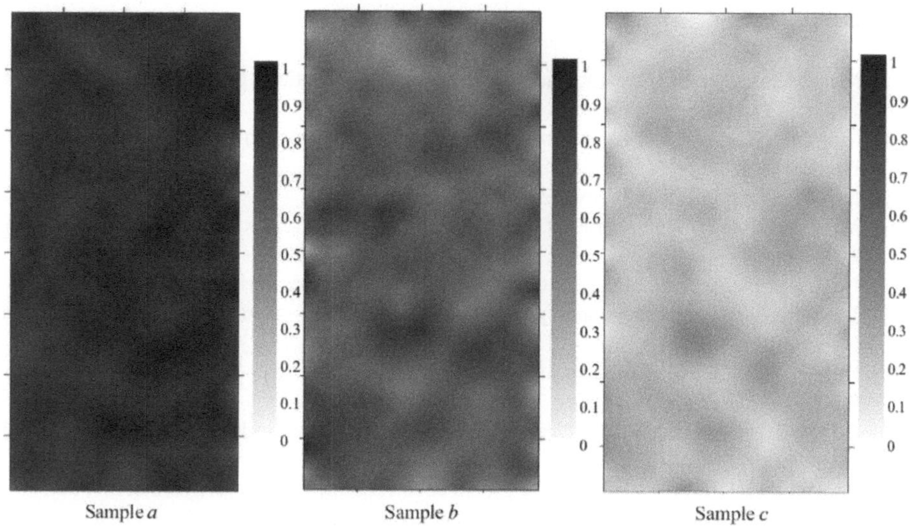

Fig. 7.27 Distribution of the bond ratio in sample *c* during thermal recovery [19].

7.6 Conclusions

Based on the experimental investigations and DEM simulations of the MH exploitation processes using artificial MHBS specimens, the following main conclusions are drawn:

1. The strength of MH sand increased with MH saturation due to particle bonding and supporting.
2. MHBS sample will be damaged after exploitation (no matter thermal recovery or depressurization) when the deviator stress applied is larger than the compression strength of pure host sand.
3. During the depressurization exploitation, although MH dissociates totally, sample with higher deviator stress than the compression strength is not damaged before repressurization and is only damaged after the reinjection of pore water pressure. However, during the thermal recovery exploitation, MHBS is damaged as soon as MH dissociates. This indicates that the depressurization method is safer than the thermal recovery method.
4. During the thermal recovery, the bond contact distribution changes apparently with increasing temperature or deviator stress with a horizontal major principal direction. MH bond contacts dissociate completely at the end of the process. The maximum absolute value of APR increases as the temperature or deviator stress increases.
5. During the depressurization exploitation, the degree of anisotropy of contact distribution only changes slightly due to increase of deviator stress. The maximum absolute value of APR increases as the deviator stress increases.

References

[1] Max M, Lowrie A. Oceanic methane hydrates: a "frontier" gas resource. J Pet Geol 1996;19(1):41–56.

[2] Max MD. Natural gas hydrate: In oceanic and permafrost environments. London (Boston, Dordrecht: Kluwer Academic Publishers; 2003. p. 61–76.

[3] http://sbi.oregonstate.edu/news/201112.html.

[4] Macdonald GT. The future of methane as an energy resource. Annu Rev Energy 1990;15:53–83.

[5] USGS. https://woodshole.er.usgs.gov/project-pages/hydrates/primer.html.

[6] Johnson AH. Global resource potential of gas hydrate—a new calculation. Methane Hydrate Newsl 2011;11(2).

[7] Milkov AV. Global estimates of hydrate-bound gas in marine sediments: how much is really out there? Earth Sci Rev 2004;66(3–4):183–97.

[8] Collett TS, Kuuskraa VA. Hydrates contain vast store of world gas resources. Oil Gas J 1998;96(19):90–5.

[9] Kvenvolden KA, Lorenson TD. The global occurrence of natural gas hydrate. In: Paull CK, Dillon WP, editors. Natural gas hydrates: occurrence, distribution, and detection, vol. 124. American Geophysical Union, Geophysical Monograph Series; 2001. p. 3–18.

[10] Joo YL, Byung JR, Tae SY, et al. Review on the gas hydrate development and production as a new energy resource. KSCE J Civ Eng 2011;15(4):689–96.

[11] Moridis GJ, Collett TS. Gas production from Class 1 hydrate accumulations. In: Advances in the study of gas hydrates. Springer; 2004. p. 83–97.

[12] Liang HF, Song YC, Chen YJ. Numerical simulation for laboratory-scale methane hydrate dissociation by depressurization. Energy Convers Manag 2010;51:1883–90.

[13] Hyodo M, Nakata Y, Yoshimoto N, Ebinuma T. Basic research on the mechanical behavior of methane hydrate-sediments mixture. Soils Found 2005;45:75–85.

[14] Guess M. Japan, China have extracted methane hydrate from the seafloor, https://arstechnica.com/science/2017/05/energy-dense-methane-hydrate-extracted-by-japanese-chinese-researchers/.

[15] Riedel M, Spence G, Chapman N, Hyndman R. Seismic investigations of a vent field associated with gas hydrates, offshore Vancouver Island. J Geophys Res Solid Earth 2002;107 (B9):5–16.

[16] Yuan J, Edwards R. The assessment of marine gas hydrates through electrical remote sounding: hydrate without a BSR. Geophys Res Lett 2000;27(16):2397–400.

[17] Sheng D, Changho L, Santamarina JC. Formation history and physical properties of sediments from the Mount Elbert gas hydrate stratigraphic test well, Alaska North Slope. Mar Pet Geol 2011;28(2):427–38.

[18] Hyodo M, Yoneda J, Yoshimoto N, Nakata Y. Mechanical and dissociation properties of methane hydrate-bearing sand in deep seabed. Soils Found 2013;53(2):299–314.

[18a] Suzuki K, Ebinuma T, Narita H. Features of methane hydrate-bearing sandy-sediments of the forearc basin along the Nankai trough: effect on methane hydrate-accumulating mechanism in turbidite. J Geogr 2009;118(5):899–912 (in Japanese).

[19] Jiang MJ, Fu C, Cui L, Shen ZF. DEM simulations of methane hydrate dissociation by thermal recovery. Comput Geotech 2016;80:410–26.

[20] Jiang MJ, Yu HS, Harris D. A novel discrete model for granular material incorporating rolling resistance. Comput Geotech 2005;32(5):340–57.

Hydropower

8

Ånund Killingtveit[1]
Dept of Civil and Environmental Engineering, Norwegian University of Science and Technology (NTNU), Trondheim, Norway
[1]Corresponding author: anund.killingtveit@ntnu.no

Chapter Outline

Managing Global Warming. https://doi.org/10.1016/B978-0-12-814104-5.00008-9

8.1 Introduction

Hydropower is a renewable energy source where electrical energy is derived from the potential energy of water moving from high to lower elevation. Hydropower is a mature technology and widely used; in 2016, a total of 159 countries in the world reported to have developed hydropower. Hydropower is among the most efficient technologies for production of renewable electrical energy, with a typical efficiency of 90% or better for, "water-to-wire." Hydropower is cost competitive, and is today the only renewable technology that can produce electricity at equal or lower cost, compared to thermal energy sources like coal, oil, or gas, typically in the range of US2–5 c $(kWh)^{-1}$ (2–5 US cents per kilowatt hour).

Hydropower has a high potential for carbon emissions reductions in the global electricity system due to low greenhouse gas emissions (GHG), and a large potential for further capacity increase with low generation cost, compared to both other renewables and to thermal power plants. In total, generation by hydropower contributed nearly 17% of the worldwide electricity supply in 2016, implying that >1 billion people covered their electricity demand from hydropower. Hydropower is the third largest source of electricity generation, behind coal (39%) and natural gas (22%) but far ahead of nuclear (10%) (Fig. 8.1).

Hydropower is the largest source of renewable energy in the electricity sector with a share of 68% of the renewables, more than twice all other renewables combined (32%). The technical potential for increased hydropower generation is large enough to meet substantial further deployment both in the medium (2030) and long term (2050). A realistic scenario is to double the annual generation (4102 TWh in 2016) to over 8000 TWh by 2050.

Adding to its contribution to energy generation, hydropower has a vital role in supporting grid stability, security of supply, energy storage, and balancing unregulated generation from other renewables like wind and solar power. In addition to energy and grid services, hydro reservoirs can deliver other important services such as water supply, irrigation, flood control, navigation, and recreation. Environmental and social concerns, unless carefully managed, represent challenges for further deployment of hydropower.

The first known hydropower plant (HPP) was installed in a house in Cragside, Rothbury, England, in 1870. Industrial use of hydropower started in 1880 in Grand Rapids, Michigan, when a dynamo driven by a water turbine was used to provide lighting. A major breakthrough came when an electric generator was coupled to the water turbine and created the world's first hydroelectric power station (of 12.5-kW capacity)

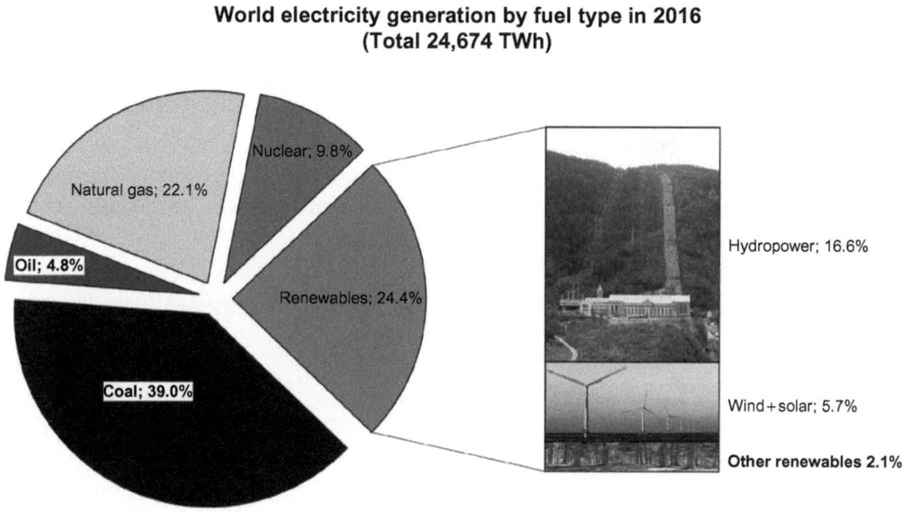

Fig. 8.1 Share of renewables in the global electricity system 2016 [2–4].

which was commissioned on September 30, 1882 on Fox River at the Vulcan Street Plant, Appleton, Wisconsin, United States, lighting two paper mills and a residence [1]. Early hydropower plants were much more reliable and efficient than the fossil-fuel-fired power plants, resulting in a proliferation of small- to medium-sized hydropower stations distributed wherever there was an adequate supply of water and a need for electricity. By 1886, there were 45 hydroelectric power plants in operation in United States. Many other countries followed suit, such as Norway, where the first hydropower plant began operation in 1885 in the town Skien. Only a few years later, in 1890, the town of Hammerfest, north of the Arctic circle in Norway, had electric street lights supplied from a municipal hydropower system. By the end of the century, 14 towns in Norway had electricity supplied from hydropower [5].

8.2 Hydropower generation—Theory

Hydropower is "fueled" by water moving in the hydrological cycle (water cycle). The water cycle is powered by solar energy: Solar radiation reaching the surface of the earth is converted into heat, which, in turn, evaporates water. Nearly 50% of all solar energy input to the earth is used for evaporation and converted into latent energy in the form of water vapor. Some of the water vapor (22%) in brought in over land areas where it later condenses into clouds and precipitation. Some of the precipitation falling on land surfaces generates runoff and flows back toward the sea, under the influence of gravity. Since the water cycle is powered by solar energy, it will continue as long as the sun shines, and makes hydropower a renewable and energy sustainable resource.

The runoff regime (amount and variability of flow) depends on the local climate; first of all on precipitation, air temperature, and potential evaporation, but also on the properties of the catchment; soil type, topography, vegetation, and land use. In most rivers, runoff shows considerable variability in time, both in the short range (hourly and daily), seasonally (summer and winter), and from year to year (dry and wet years). This variability in runoff could lead to variable generation, and problems in meeting demand. The traditional solution to this is to build reservoirs and store water during periods of surplus to be used in periods of drought.

The theoretical output of electrical power (P') for a particular site depends on three main factors, flow (Q), head (H), and efficiency in machinery (η) as illustrated here:

$$P' = \rho \times g \times Q \times H \times \eta \tag{8.1}$$

Here P' is the electrical power output ($\mathrm{J\,s^{-1}} = \mathrm{W}$); ρ is the density of water ($1000\,\mathrm{kg\,m^{-3}}$); g is the acceleration of gravity ($9.81\,\mathrm{m\,s^{-2}}$); Q, the water flow per unit time ($\mathrm{m^3\,s^{-1}}$); H, the elevation drop, usually called head (m); η is the efficiency in the conversion process from water to wire:

$$\text{Potential energy} \rightarrow \text{kinetic energy} \rightarrow \text{mechanical energy} \rightarrow \text{electrical energy}$$

Assuming typical values for density and acceleration of gravity as given here, the equation can be simplified to give power output, P, directly in units of kilowatt:

$$P = 9.81 \times Q \times H \times \eta \tag{8.2}$$

The amount of energy produced depends on the duration of the flow. Assuming the Δt is given in hours (h) the amount of energy, E (in units of kilowatt hours), produced can be computed from:

$$E = P \times \Delta t \tag{8.3}$$

For comparison to other renewable and thermal energy sources, the energy unit Exajoule (where $1\,\mathrm{EJ} = 10^{18}\,\mathrm{J}$) is often used. Here $1.0\,\mathrm{EJ} = 277.78\,\mathrm{TWh}$ and $1.0\,\mathrm{TWh} = 0.0036\,\mathrm{EJ}$. The global hydropower production in 2016 was 4102 TWh, which is equivalent to 14.76 EJ.

Another useful equation gives the energy output in kilowatt hours per cubic meter of water as a function of H and η, and is often called the Energy equivalent of water (EEKV), and is calculated as:

$$\mathrm{EEKV} = P \times 1\,\mathrm{h}/(Q \times 3600\,\mathrm{s}) = 9.81 \times H \times \eta/3600 \tag{8.4}$$

In addition to kilowatt hours (kWh), some other commonly used units of electrical energy are megawatt hour (MWh), gigawatt hour (GWh), and terawatt hour (TWh):

$$1\,\text{MWh} = 1000\,\text{kWh}\,(10^6\,\text{Wh})$$

$$1\,\text{GWh} = 1000\,\text{MWh}\,(10^9\,\text{Wh})$$

$$1\,\text{TWh} = 1000\,\text{GWh}\,(10^{12}\,\text{Wh})$$

The power output, P, in Eq. 8.2 is limited to P_{Max} (nameplate capacity) at the maximum flow capacity of the turbine, Q_{Max}. If the inflow becomes larger than this, water will be lost (spilled) unless there is a reservoir for storage. In most cases, a hydropower plant will be designed with capacity considerably below the maximum possible inflow, and consequently normally spill some water. At periods where the flow is much less than the capacity, the power plant may have to run at reduced capacity and efficiency or even have to stop and bypass water to avoid damage to the turbines.

The *capacity factor* of a power plant, C_f, is the ratio of the actual output of a power plant over a period of time (typically 1 year) compared to its potential output if it had operated at full capacity over the entire period (year). Typical capacity factors for hydropower plants are in the range of 0.4–0.6 [1], but lower and higher values can also be found. World average C_f is 0.45 for hydropower.

8.3 Technology

Hydropower is a renewable energy source derived from the potential or kinetic energy of water, by moving water from higher to lower elevations. It is a proven, mature, predictable, highly efficient and price-competitive technology. Hydropower has among the best conversion efficiencies of all known energy sources, for modern power plants this is typically >90% efficiency, "water to wire."

A hydropower plant typically includes civil structures to collect, store, and transport water, and mechanical and electrical components to convert energy to mechanical and electrical energy. During planning and operation, it is very important to include information about hydrology, hydraulics, and environmental impacts together with information of social and political issues, in order to find an optimal design.

A hydropower plant typically consists of an intake, a "head race" consisting of tunnels and/or pipes, the power station with electrical and mechanical equipment ("Elmek"), and finally a "tail race" consisting of tunnels and/or pipes to the outlet. It may or may not include a dam and a reservoir for water storage. The three main components of Elmek equipment are: turbine(s), generator(s), and transformer(s). In addition, there will be many other components such as gates and valves, electronic equipment for controlling the operation of the station, power cables, switchyards, and grid connections.

A hydropower plant is nearly always tailored to utilize the available water and head, and many different types of turbines have been developed; the most common is the Pelton and Francis turbine for high and medium head situations, and Kaplan and Bulb turbines for lower head and large flow systems. Examples of typical

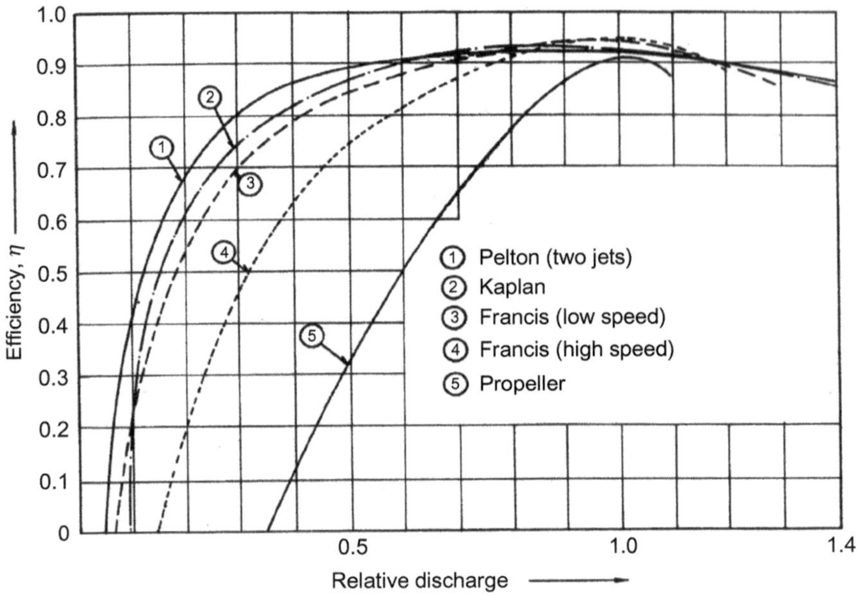

Fig. 8.2 Turbine efficiency versus relative discharge for five different turbine types [6].

efficiency curves for these turbines are shown in Fig. 8.2. The efficiency varies very much with relative discharge; therefore, it is very important to be able to run at or as close as possible to the "best point," for most of the time. For a run-of-river plant with a large variation in inflow, this may be difficult if only one turbine is installed. Increasing the number of turbines will allow a more efficient operation, but will also mean an increase in cost.

With increasing share of wind and solar power plants, there will also be an increasing pressure on hydropower plants to take more responsibility for load balancing, and operation outside the best efficiency range. Today, it is an increasingly important research topic to design turbines that will allow operation over a wider range of flow, while still keeping efficiency at an acceptable level.

8.3.1 Hydropower project classification

Hydropower projects are usually classified into four major types:

- Run of river (RoR)
- Storage (reservoir) based
- Pumped storage
- Instream technologies (Hydrokinetic)

Hydropower projects can also be classified according to size (pico, micro, small, large) and head (low, medium, and high), but there is no clear consensus on classification by project size (installed capacity, MW) or head due to varying policies in different countries. Classification according to size is administratively simple, but is—to some degree—arbitrary: concepts like "small" or "large hydro" are not technically or scientifically rigorous indicators of impacts, economics, or other characteristics [1]. Hydropower projects cover a continuum in scale from units $<1\,kW$ up to megaprojects like China's Three Gorges with 22.5-GW installed capacity (22,500,000 kW) and an annual generation of 93.5 TWh [7].

8.3.2 Run-of-river hydropower plants

An RoR hydropower plant is a plant where little or no water storage is provided; it generates electricity from the available flow of the river. Such plants may sometimes include a short-term storage or "pondage," with some storage capacity ranging from a few hours up to daily flexibility in adapting generation to the demand profile. The generation profile will mostly be determined by the natural river flow conditions or by the release profile from upstream storage if it is part of a cascade. In the absence of any pondage or upstream reservoirs, generation depends entirely on flow and typically has substantial daily, monthly, seasonal, and annual variations.

8.3.3 Storage hydropower plants

Storage hydropower plants include a dam and a reservoir to impound water, which is stored and released later when needed. Water stored in reservoirs provides flexibility to generate electricity on demand and reduces dependence on the variability of inflow. Very large reservoirs can store inflow for months or even years, but they are usually designed for seasonal storage, to supply water during dry seasons. Storage hydropower plants are more flexible than RoR plants, and can be operated to provide baseload power, as well as peakload through its ability to be shut down and start up again at short notice, depending on the demand in the power system. Given their ability to control water flows, storage reservoirs are often built as multipurpose systems, providing additional benefits such as flood control, water supply, irrigation, navigation, and recreation. The primary advantage of hydropower plants with storage is their ability to store large volumes of energy and respond to variable load requirements, from short term (daily peaking) to weekly and seasonal variability. Such reservoirs are becoming increasingly important and valuable also for storing energy from other renewables, such as wind and solar power.

8.3.4 Pumped-storage hydropower plants

In a pumped-storage hydropower (PSH), water is pumped from a lower reservoir into an upper reservoir when electricity generation exceeds demand and released back from the upper reservoir through turbines to generate electricity when demand is higher than the supply. Both reservoir HPPs and PSPs store potential energy as elevated water for generating on demand. The difference is that PSPs can take energy from the grid to pump the water up to a higher level, then return the energy back to the grid later when needed. This cycle can happen several times a day. The round-trip efficiency (pumping-generation) is usually high, from 75% to 85%. Pumped storage currently represents 99% of "on-grid" electricity storage [8]. Some PSH projects can also have natural inflow to the upper reservoir, which will increase the generation. Energy stored in a PSH is directly proportional to the water volume stored in the upper reservoir and the elevation difference between reservoirs. Traditionally, PSH have been used to supply peaking power during the day and allow large fossil and nuclear plants to run at near-constant load.

Another major advantage of PSHs is their ability to interact with other variable renewables such as wind and solar power. PSH installations can store excess energy during periods of high wind or high insolation, and provide back-up reserve, which is immediately dispatchable during periods when the other variable power sources are unavailable.

8.3.5 In-stream (hydrokinetic) hydropower plants

In-stream energy can be derived from the movement (kinetic energy) of water in rivers, streams, and canals, and also from tidal flow and ocean currents. This technology differs from traditional hydropower plants, which rely on the elevation difference (head) between the intake and outlet. Hydrokinetic devices are placed directly in the stream of flowing water and energy is extracted with turbines similar to those used in tidal plants or in the ocean. The major difference to ocean currents is that the river current is unidirectional. The kinetic energy in the flowing water is converted to mechanical energy that drives a generator to produce electricity. Because it is powered by kinetic energy instead of potential energy, it is also known as a "zero-head" turbine. As such, no dams and/or head differential is necessary for the operation of this device; the course of a river remains in its natural state. So far, very few of these power plants have been put in operational use in rivers, but they represent an interesting alternative for harnessing energy from the large number of streams and canals where traditional low-head HPPs cannot be used.

8.4 Classification according to size—Small and large hydro

It has become popular to classify hydropower plants either as "small" or "large," depending on installed capacity. There seem to be a belief that "small is beautiful" and that small hydro is more environmental friendly. In some countries, small hydro schemes are accepted for development and receive subsidies, while larger hydro schemes are not accepted. Sometimes one can even hear the argument that "Large hydro" is not renewable, and should not be included in the group of renewable technologies. This is of course completely wrong and it is difficult to see the logic behind such arguments. All hydropower (except PSH) is powered by the hydrological cycle, which again is powered by solar energy, and completely renewable. What can be discussed is whether all types of hydropower is sustainable or not. In order to answer this it is necessary to study environmental and social impacts, in addition to economical parameters. This will be discussed later in Section 8.10.

If one considers the impact per kilowatt-hour, the picture becomes more complicated. A large hydropower plant could easily have less impact (per kilowatt-hour) than many small plants of the same capacity combined. This was discussed thoroughly in reference [1] where the conclusion was that the use of classification according to size should be avoided, both because there is no clear connection between size and impact, and also because the definition of what is "small" and "large" varies widely from one country to another, as shown in Table 8.1.

8.5 Cutting-edge technology

Though hydropower is a mature technology, there is still a need for and room for technology improvements. Since hydropower is so closely linked to water management, new requirements and environmental policy changes can have a big impact on hydropower projects, and lead to changes for example in flow restrictions and reservoir operation rules, and require new methods for operation and optimization. One such example is the European Water Framework Directive, which has a profound effect on hydropower in many European countries.

With hydropower plants' long lifetime, typically 80 years or more, it will usually be necessary to upgrade machinery and control equipment several times during a plant's lifetime. This gives the possibility to introduce new and improved technology and increase efficiency. Even civil structures, in particular, dams may need refurbishment and major upgrades during such a long lifetime, for example, due to new guidelines for dam safety or climate change impacts.

Cutting-edge technology is not limited to better "hardware" like turbines and generators, it can also be improved planning and operation tools, and of course improvements in dams, tunnels, and other civil structures. Hydropower generation can be increased by optimizing different aspects of plant operation, both for individual units and by better coordination in cascades with many interlinked reservoirs and power

Table 8.1 **Definition of "Small hydro" by installed capacity in some countries [1]**

Country	Small hydro defined by installed capacity (MW)
Brazil	≤30
Canada	<50
China	≤50
EU	≤20
India	≤25
Norway	≤10
Sweden	≤1.5
United States	5–100

plants. Improved hydrological forecasts combined with advanced optimization models are likely to improve operation and at the same time respect new environmental constraints and multipurpose use of the water. New improved optimization methods and software will need to be further developed for better coordination with other renewables, where hydropower could have a key role in the balancing and grid stabilization as the percent of variable generation capacity in wind and solar plants is increasing.

A few interesting areas where new technology is under development can be mentioned. This selection is mostly based on the chapter "Prospects for technology improvement and innovation" in reference [1] and on the Hydropower Development book series [5,6,9,10].

8.5.1 Extending operational regime for turbines

Hydropower turbines, running at fixed speed, have traditionally been optimized for operation at a "best point" defined by speed, head, and discharge. Operating outside this best point (different head and/or discharge) will usually lead to considerable reduction in efficiency and reduced power output. Large hydropower turbines are now close to the theoretical limit for efficiency in a small region near the "best point," but may still be improved for more flexible operation outside this region. This is illustrated in Fig. 8.2, which shows typical efficiency curves for different types of hydro turbines [6]. The application of "Variable-speed" turbines offers advantages in the form of greater flexibility in handling such situations, with higher efficiency as a result.

8.5.2 Utilizing low or very low head—Unpowered dams

Most of the existing hydropower projects were developed in times with lower energy price, where projects with low (<15 m) and very low (<5 m) head were not economically feasible. Most such low head projects were excluded from hydropower potential mapping, for the same reason. Therefore, existing data on hydropower potential for low head sites are probably not complete and sometimes completely absent.

As an example, mentioned in reference [1], in Canada a market potential of 5000 MW of low head hydropower plants was recently identified. In many countries, there is a large untapped potential for producing electricity from irrigation dams, provided low head turbines can be developed at low cost.

There is probably also a large potential for increasing generation by refurbishing and upgrading existing HPP. Many of the existing hydropower plants are >40 years old, and there is a significant potential to increase efficiency and capacity. This potential is not included in Table 8.3. A recent study in Norway based on 20 refurbishment projects of 30- to 60-year-old plants revealed that the generation increased by 23% on average due to increased capacity and efficiency [11]. Similar results were found in studies in United States under the Hydropower Vision project [12]. Here, it is stated that it is possible to increase hydropower capacity from 101 GW today to 150 GW in 2050, a large share of this from sites previously not considered useable.

With increasing energy demand and price, and improved technology for such low head projects, one can see a growing interest in technology development. At extremely low head also hydrokinetic technology may open up new possibilities, since it can extract energy directly from moving water, in the same way as wind turbines extracts energy from moving air.

8.5.3 Fish-friendly hydropower plants

Dams and hydropower plants may create obstacles for fish migration and reduce access to important rearing and spawning sites, and reduce or eradicate fish populations. Safe passage for fish is therefore a major concern in environmental analysis of hydropower projects [9,13]. There are two main categories of technology development, for upstream or downstream migration:

Upstream migration: Fish ladders for upstream migration, combined with flow release and adaption in operational regime to minimize problems when migration actually occurs. Other common devices include bypass channels and fish elevators. In some cases, small corrections at culverts and weirs can improve the situation and increase upstream migration.

Downstream migration: The main objective here is to avoid fish entering into intakes and turbines. New technology for such avoidance systems under development is based on screens, acoustic cannons, light (strobe and laser light), and control of flow direction. If it is not possible to avoid fish to enter the turbines, new type of turbines ("fish-friendly turbines") may minimize the risk of injury or death on passing fish. Some turbine manufacturers claim that the turbines may have good efficiency and still allow 90%–100% of the fish to pass safely.

8.5.4 Tunneling and underground power plants

Tunnels and rock caverns are important construction elements in many large-scale hydropower projects: headrace/tailrace tunnels, access tunnels, powerhouse, surge shafts, power cables and ventilation shafts. In Norway, nearly all large-scale hydropower plants have been built underground since 1960. In 2002 there were 500 underground hydropower plants worldwide, about 40% of these were then found in Norway [10]. An example showing an underground powerhouse is shown in Fig. 8.3. Typical layout of tunnel system and rock caverns is shown in Fig. 8.4. Tunneling technology has evolved from traditional drill-and-blast technology to the use of full-face tunnel boring machines (TBM), increasing speed of tunneling and lowering cost.

Today, there is also an interesting technology development for "microtunnels" that can be used to replace pipes and penstocks for small hydropower plants (<1.5 m diameter), avoiding completely overground work and disturbance in the nature. This technology has been being developed and tested in Norway and is now used successfully in an increasing number of small hydropower plants [14]. Another potentially important use of such microtunnels could be for transmission power cables, avoiding power lines through environmentally sensitive areas.

The use of underground power stations combined with tunnels to transport water gives high flexibility in locating power plants and makes it possible to build efficient systems, superior to traditional systems with surface powerhouse and open-air

Fig. 8.3 Underground hydropower plant Brattset in Orkla river, operational from 1981.

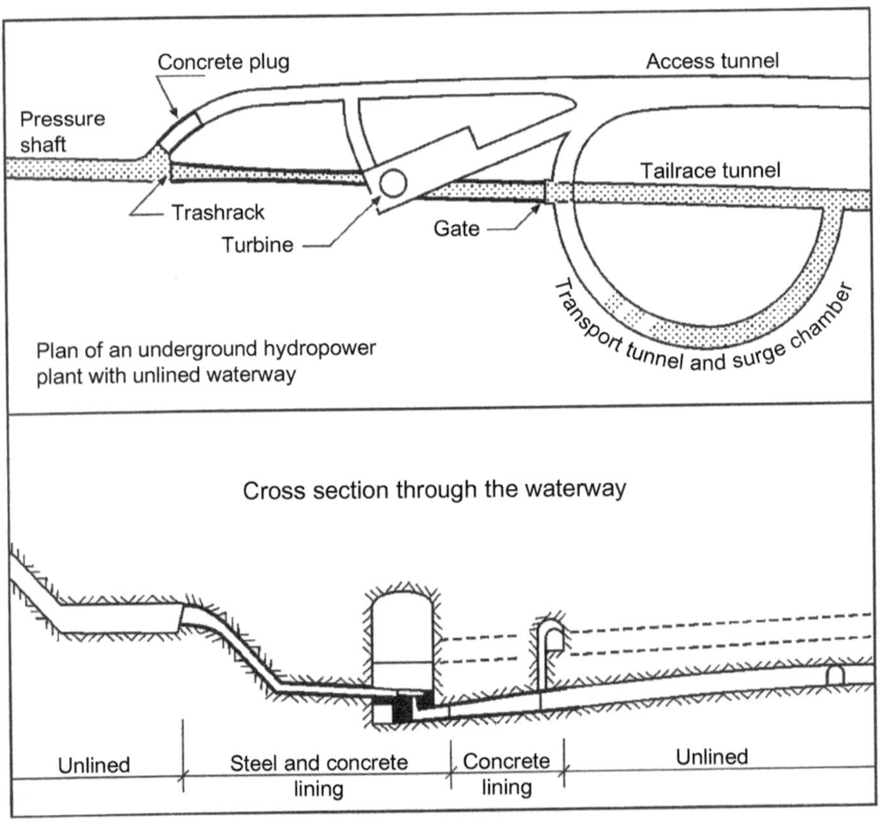

Fig. 8.4 An underground hydropower plant—Typical layout of tunnel system [10].

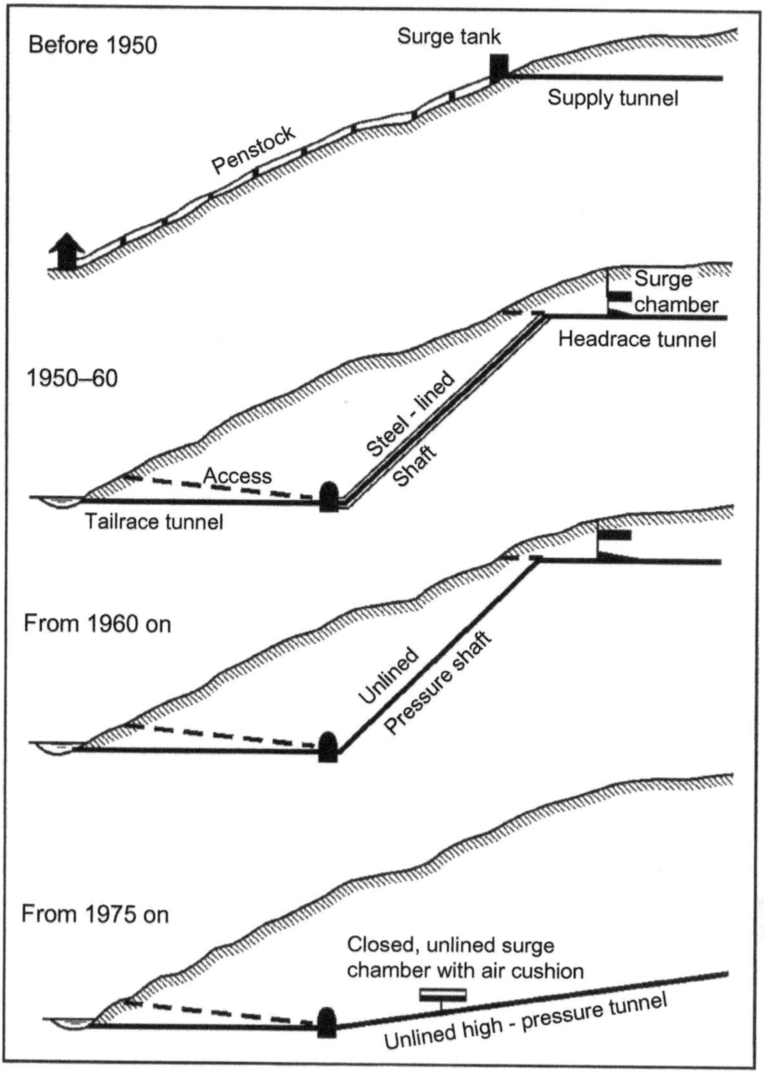

Fig. 8.5 Development of general layout for high-head hydropower plants in Norway [10].

penstocks. Fig. 8.5 show how the layout of high-head hydropower plants have changed gradually, mainly due to advances in tunneling technology. An example of one such system in Norway, the Ulla-Førre power complex, is shown in Fig. 8.6. Here, tunnels and underground power stations are combined with reservoirs and intakes in order to capture and store water from a wide area with many small watersheds that would have been very difficult to develop individually. Most of the water is collected at an elevation of 600 m a.s.l. (600 m above sea level) and pumped up to a very large reservoir (Blåsjø) at 1000 m a.s.l. where it was possible to build dams and create a huge reservoir. The Blåsjø reservoir is the largest in Norway and can store nearly 8 TWh of energy.

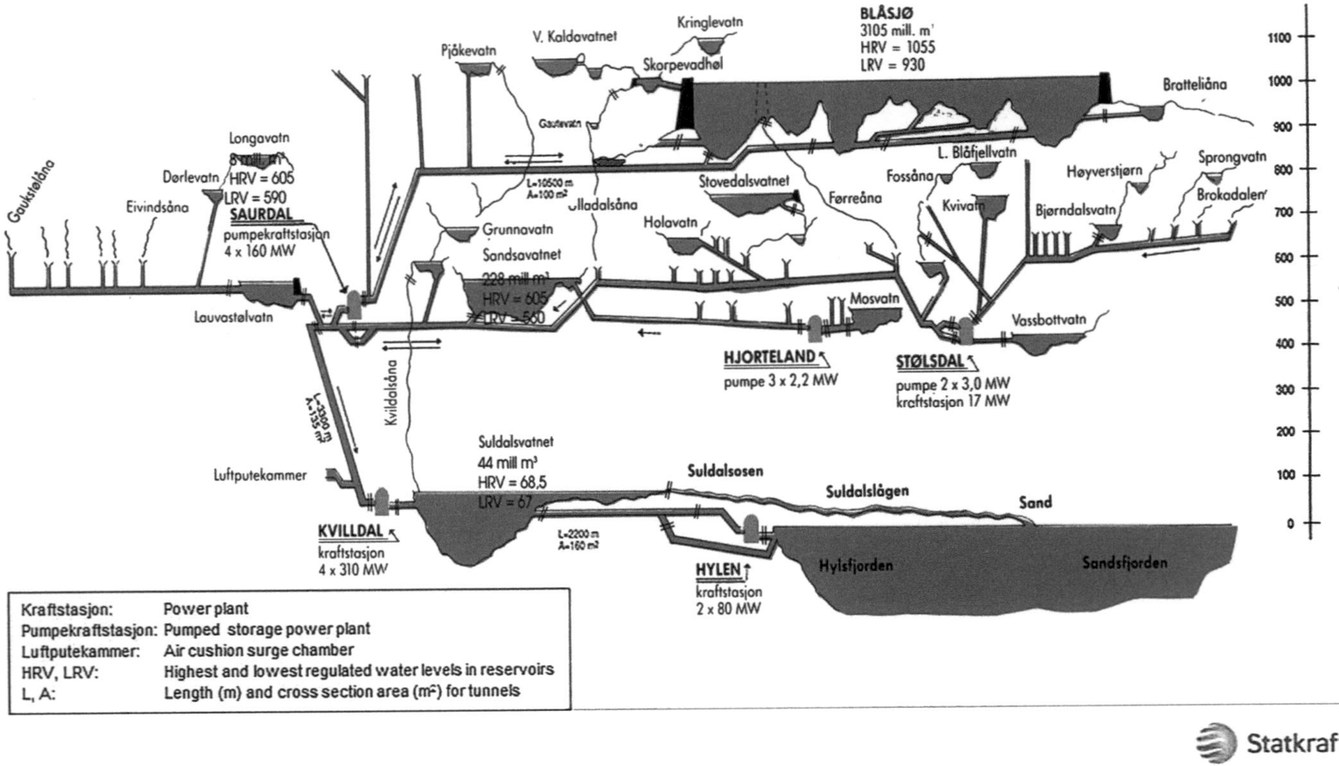

Fig. 8.6 The Ulla-Førre hydropower complex in Norway. Figure from Statkraft, Norway

8.5.5 Surge tanks in hydropower plants

Surge tanks are applied in hydropower plants with long water conduits to reduce pressure forces during the acceleration of the large water masses. They are constructed as intermittent water reservoirs close to the turbines, either with open access to atmospheric air or as a closed volume filled with pressurized air. The surge tanks reduce the length of the water column to be accelerated, and thus the resulting pressure forces. This reduction is necessary to limit the design pressure for structural components and to enable "speed governing" of the turbines.

Closed surge tanks for hydropower plants may be constructed as rock caverns or steel tanks filled with pressurized air ("Air-cushion chambers"), and are applied where the topography or other factors render them more feasible than open surge tanks connected to atmospheric air. The three upper examples in Fig. 8.5 show traditional design with open surge tanks, the lower (From 1975 on) shows the typical layout for a hydropower plant with a closed surge tank. There are two main advantages in using a closed surge tank compared to the open type:

- More flexibility in locating the powerhouse and shorter tunnel system
- Shorter distance from surge tank to turbine—more efficient regulation

A few such air-cushion chambers have been built in Norway, the first ones in Driva HPP (1973), Jukla HPP (1974), and Brattset HPP (1981). Later seven more have been built, with volumes ranging from 2000 to 110,000 m^3.

The transition from fossil to renewable power increases the need for flexible hydropower plants in order to balance the highly variable production from v-RES such as wind and solar. When the size of the power plants increases, the surge tank design becomes more important owing to larger water masses and pressure forces. More frequent start-stop operations further increase the importance of the surge tank, and more surge tank research is considered vital to prepare for future challenges.

Recent research at NTNU has provided better understanding and improved mathematical models of the complex hydrodynamic and thermodynamic processes in closed surge tanks. Improved computer models combined with physical-scale models that can be used to design and operate these optimally have been developed and will be further improved [15].

8.6 Hydropower resources—Potential

8.6.1 Definition of potential

The potential assessment process for hydropower is different from those used for other renewables (wind, solar, bio, …) since hydropower is always site specific, and determined by the combination of available water flow and usable head at each location. All estimates for hydropower potential are therefore based on data from known sites where flow and head is measured and technical design optimized to fit these.

In order to compute the potential for hydropower within a wider area (catchment, region, country, …), it is necessary to identify all feasible sites where a suitable

Table 8.2 Different estimates of global hydropower potential [16]

Estimation method	Potential (EJ)	Comments
Energy in the water cycle	504,000	(40% of solar radiation at earth's surface)
Theoretical potential in runoff	200	Total mass of runoff $\times g \times H$
Technical potential	140–145	Technical potential at known sites
Economic potential at <20 c (kWh)$^{-1}$	50–60	Portion of technical potential with cost low enough to justify a site assessment
Economic potential at <8 c (kWh)$^{-1}$	30	Potential at sites with cost that compete with large thermal power plants

combination of flow (Q) and head (H) can be found, and where hydropower plants can be sited. The potential at each site is computed using Eqs. (8.1)–(8.3) considering both the volume of water and its variability in time. Results are usually given as potential average annual energy generation in units of GWh year^{-1} or TWh year^{-1}.

It can be argued that this definition of potential is not precise because the selection of "feasible sites" will depend on technology, economic parameters, and social and environmental preferences, and all these can change with time. This has led to several definitions of potential: "Theoretical potential," "Technical potential," and "Economic potential." Some also argue for the use of "Sustainable potential." In reference [16], we find an interesting discussion about the use of different types of definitions, and this has been summarized in Table 8.2.

The average annual global runoff has been estimated as 47,000 km^3, out of which 28,000 km^3 is surface runoff, yielding a theoretical potential for hydropower generation of nearly 42,000 TWh year^{-1} (150 EJ year^{-1}).

8.6.2 Global and regional overview

An overview of existing regional and global hydropower potential can be found in Table 8.3, which is based on data from IJHD, World Atlas, 2017 [7]. This table gives the potential in units of TWh year^{-1} for six main regions, and also the percentage share of total global potential. The three different hydropower potentials are: gross theoretical potential, technical potential, and economic potential. Unfortunately, it is not clearly defined at what limit the economic potential is set, or if the same limits have been used in all regions. Asia has ~50% of all feasible potential, North and South America about 30%, and Europe and Africa around 20%.

For each region also the most recent generation estimate (2016) and the share of feasible hydropower potential, which has already been developed, is given in Table 8.4. The share of technical feasible potential, which has already been developed, varies from 50% in Europe to only 6% in Africa, globally only 26% has been

Table 8.3 **Regional and global hydropower potential according to three different definitions [7]**

Region	Gross theoretical hydropower potential		Technical feasible hydropower potential		Economically feasible hydropower potential	
	TWh $year^{-1}$	%	TWh $year^{-1}$	%	TWh $year^{-1}$	%
North America	7601	18	1891	12	1045	11
South America	7848	19	2859	18	1728	18
Europe	3136	7	1195	8	852	9
Africa	4423	11	1647	10	1124	12
Asia	18,248	44	8000	51	4786	50
Australasia/Oceania	658	2	186	1	89	1
World	41,914	100	15,778	100	9624	100

Table 8.4 Share of feasible technical and economical potential already developed [4,7]

Region	Technical feasible hydropower potential (TWh $year^{-1}$)	Economical feasible hydropower potential (TWh $year^{-1}$)	Actual generation in existing system in 2016 (TWh $year^{-1}$)	% of technical potential developed	% of economical potential developed
North America	1891	1045	702	37	67
South America	2859	1728	709	25	41
Europe	1195	852	595	50	70
Africa	1647	1124	106	6	9
Asia	8000	4786	1950	24	41
Australasia/Oceania	186	89	40	22	45
World	15,778	9624	4102	26	43

Table 8.5 Remaining (undeveloped) hydropower potential by 2016

Region	Remaining (undeveloped) share of technically feasible hydropower		Remaining (undeveloped) share of economic feasible hydropower	
	TWh year^{-1}	%	TWh year^{-1}	%
North America	1189	63	343	33
South America	2150	75	1019	59
Europe	600	50	257	30
Africa	1541	94	1018	91
Asia	6050	76	2836	59
Australasia/Oceania	146	78	49	55
World	11,676	74	5522	57

Data from Table 8.4.

developed. The share of economic potential, which has already been developed, varies from 70% in Europe to 9% in Africa; globally 43% of the economic potential has been developed.

Table 8.5 shows the remaining (undeveloped) share of the feasible potential for the same six regions, and globally. One can see that both Africa and South America still have a large undeveloped potential, each with >1000 TWh of economic feasible hydropower. But Asia has the largest undeveloped potential, nearly 3000 TWh of economic feasible hydropower can be developed; a very large share of this is in China.

The "remaining potential" in Table 8.5 does not take into account social and environmental factors that may reduce the potential. For example, in Norway, the remaining technical potential of 84 TWh is reduced to 34 TWh, since 50 TWh is protected for environmental reasons and cannot be developed. There will certainly be similar restrictions in many other countries, but still there is a significant potential for hydropower development, especially in Asia, Africa and South America.

8.7 Existing generation—Regional and global status

8.7.1 Historical trends in hydropower production

As shown briefly in Fig. 8.1 and Table 8.4, hydropower is a major source of electricity with an annual generation of 4102 TWh in 2016, supplying almost 17% of the global electricity. In the last decade, there has been an increasing focus on renewable energy as a means to reduce carbon emissions to the atmosphere, and hydropower has taken a very significant share in this transition.

Fig. 8.7 shows the increase of the world hydropower generation over the past 35 years, from 1971 to 2015. Three trend lines have been added. From 1971 to 1996, there was a steady and nearly linear increase of 50 TWh year^{-1}, in response to increasing demand. From 1996 and almost 10 years on, the growth slowed down

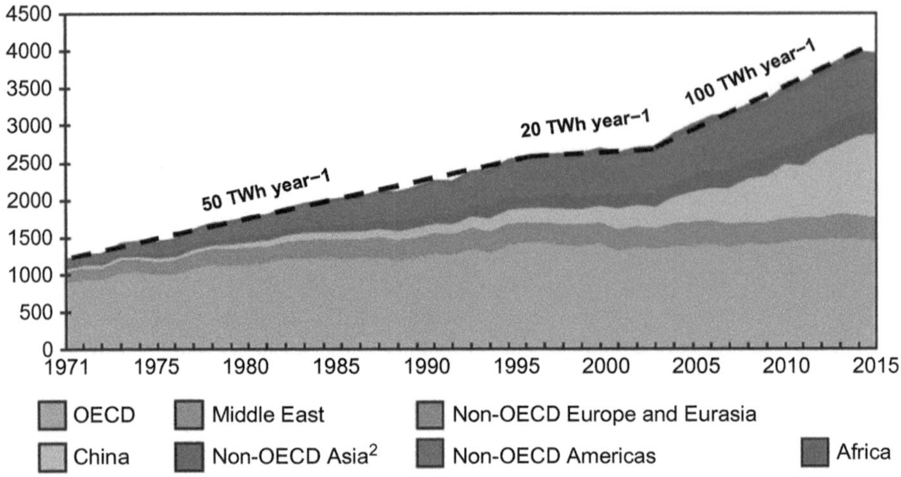

Fig. 8.7 World hydropower production (TWh $year^{-1}$) from 1971 to 2015 by region [17].

to only 20 TWh $year^{-1}$. Finally, from 2005 onwards, the growth increased again, now up to around 100 TWh $year^{-1}$. This rate of growth also continues in 2016. The slowdown in the 10 years around year 2000 has been explained to be a result of an increasingly negative view on hydropower because of its environmental impacts. From 1999 through 2005, hydropower development was largely halted worldwide, reflecting the impact of the World Commission on Dams (WCD) [18]. WCD was convened to review the development effectiveness of large dams and develop guidelines for the development of new dams. The report, published in 2000, challenged existing practices and proposed stringent guidelines for dams, which in turn caused a sharp decrease in investments while the sector and the financial community considered how to respond to new standards and expectations.

After 2005, hydropower development saw a new upswing, possibly due to an increasing concern about climate change, and the strong political determination to move from carbon-emitting technologies to renewable generation. Another important contribution was the intensive efforts led by the International Hydropower Association (IHA) and a multistakeholder range of partners in promoting greater sustainability through the development and use of "sustainability guidelines."

The Hydropower Sustainability Assessment Protocol [19] provides an international common set of methods on how sustainability considerations can be addressed at all phases of a project: early planning, preparation, implementation, and operation.

Fig. 8.7 also very clearly shows the large regional change in hydropower development during this period. At 1971 the largest share, 80% of all hydropower generation, occurred in OECD countries. Further growth up to 2005 mostly took place outside OECD and in particular in "Non-OECD Americas" (South America). From 2005 on, the most striking feature is the increasingly rapid growth in China. In 2016, China alone produced nearly 30% of all hydropower in the world, while all OECD countries combined only contributed by 37% of total generation.

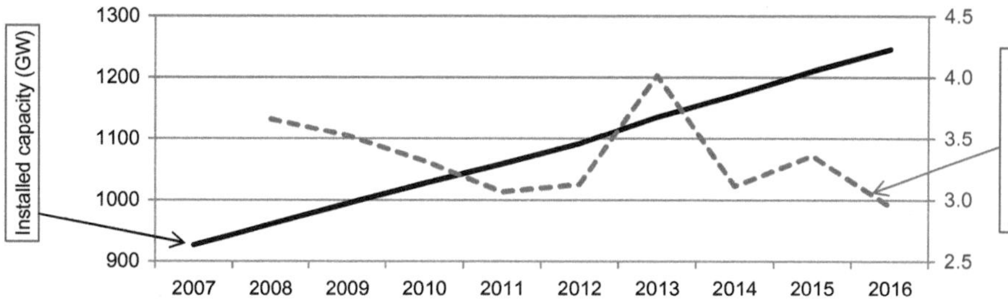

Fig. 8.8 Increase in world hydropower capacity during past 10 years (2007–16).

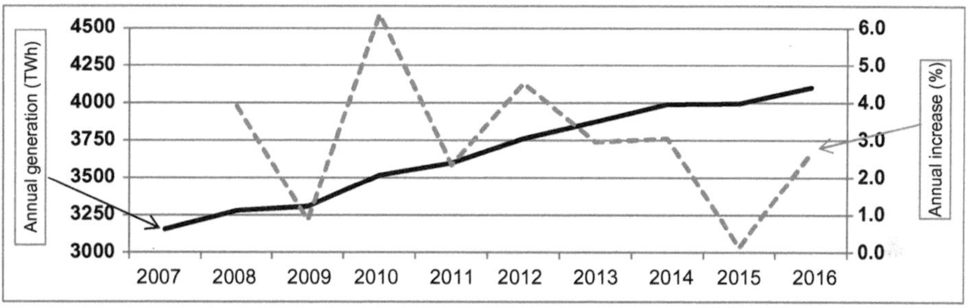

Fig. 8.9 Increase in world hydropower generation during the past 10 years (2007–16).

In the 10 years from 2007 to 2016 the total capacity of hydropower increased by 3%–4% per year, from 926 to 1245 GW; with an average of 32 GW $year^{-1}$, (see Fig. 8.8). The total hydropower generation increased by nearly 1000 TWh, from 3154 TWh in 2007 to 4102 TWh in 2016 (see Fig. 8.9). The annual increase in generation varies from near 0% to >6%. The high variability in generation increase (Fig. 8.9) compared to capacity increase (Fig. 8.8) can be explained by variation in climate and inflow from year to year.

8.7.2 Countries with highest hydropower production

The 10 countries with highest hydropower capacity and production are presented in Tables 8.6 and 8.7. Together, these 10 countries have 66% of all hydropower capacity and 71% of all production in the world. From the tables, one can see China's dominating position with 27% of world total (1114 out of 3995 TWh in generation and 333 out of 1246 GW in capacity). The next three countries are United States, Canada, and Brazil. Here the individual ranks are somewhat different for capacity and generation. Canada is number two in generation, but only number four in capacity. For United States, reverse is true, reflecting the important role of hydropower capacity for power system services, while the energy contribution is of lesser importance.

In Table 8.7 it is interesting to look at the last column, annual hydropower production in kilowatt hours per capita in the different countries. Norway is in front here, with

Table 8.6 The 10 countries with highest hydropower capacity in the world [2,7,17]

Country/ area	Hydropower capacity 2016 (GW)	Hydropower % of all renewables	Total capacity all renewables (GW)
China	334	61.1	546
United States	103	47.8	215
Brazil	98	79.7	123
Canada	81	83.0	98
Russian Fed	51	99.9	51
Japan	50	66.8	75
India	48	52.4	91
Norway	32	96.8	33
Sweden	16	59.1	28
Venezuela	15	99.7	15
World	1246	61.9	2011

nearly 27,000 kWh, while the next country in order, Canada, has less than half of this, 10,496 kWh. Four of the countries have a share of >50% hydropower in their respective electricity systems, Norway (95.9%), Brazil (61.9%), Canada (57.8%), and Venezuela (51.4%). Sweden is also close to 50%. In all these 10 countries combined, hydropower supplies 72% of all renewables electricity, close to the world average.

8.7.3 Share of hydropower in the global energy mix

The share of hydropower in the global energy mix has been quite stable for many years, around 16%–17%, even during a period of strong growth in electricity production. Fig. 8.10 shows the share between main sources of electricity production since 2000. During this period, the world electricity production increased by nearly 9000 TWh; with about 5900 TWh from carbon-based technologies (coal, gas, oil), and the remaining 3100 TWh from renewables. Nuclear power remained nearly constant at 2444 TWh in 2000, increasing to 2661 TWh in 2006 but decreasing again to 2476 TWh in 2016. During this period, hydropower alone contributed an increase of 1350 TWh (44%) of renewable energy. All of the other renewable energies, combined, contributed an increase of 1740 TWh (56%). Wind and solar-based electricity together contributed an increase of about 1300 TWh, slightly less than hydropower.

Fig. 8.11 shows how the share of renewables in electricity generation has changed during the same period. Hydropower has kept a share of 16%–17% during these 16 years, while increasing from 2746 TWh in 2000 to 4102 TWh in 2016. Renewables in total have increased their share from 19% in 2000 to nearly 25% in 2016, and hydropower has given the largest contribution to this increase, followed by wind, solar, and bioenergy.

Table 8.7 **The 10 countries with highest hydropower production in the world [2,7,17]**

	Electricity generation by hydropower in 2015						
Country/ area	**Hydropower generation 2015 (TWh)**	**Hydropower generation— % of all renewables**	**Hydropower generation— % of all electricity**	**Renewable electricity generation 2015 (TWh)**	**Total electricity generation 2015 (TWh)**	**Population million**	**Hydropower production (kWh per capital)**
China	1114	81.2	19.6	1372	5696	1379	808
Canada	381	90.1	57.8	423	659	36.3	10,496
Brazil	360	83.7	61.9	430	582	208	1731
United States	271	48.2	6.3	562	4325	323	839
Russian Fed	170	100.0	15.9	170	1066	144	1180
Norway	139	98.6	95.9	141	145	5.2	26,731
India	150	71.4	11.2	210	1345	1324	113
Japan	91	54.5	9.0	167	1015	127	717
Venezuela	76	100.0	51.4	76	148	31.6	2405
Sweden	75	73.5	46.3	102	162	9.9	7576
World	3995	72.2	16.5	5537	24,255	7442	537

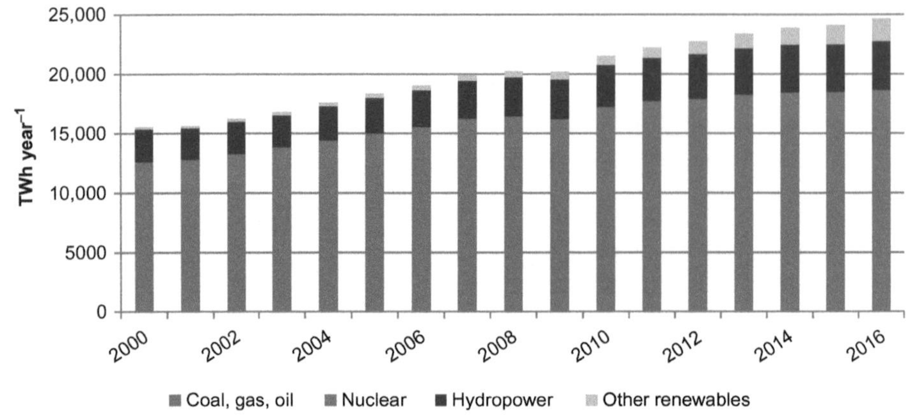

Fig. 8.10 Share of hydropower in the global power system 2000–16 [3].

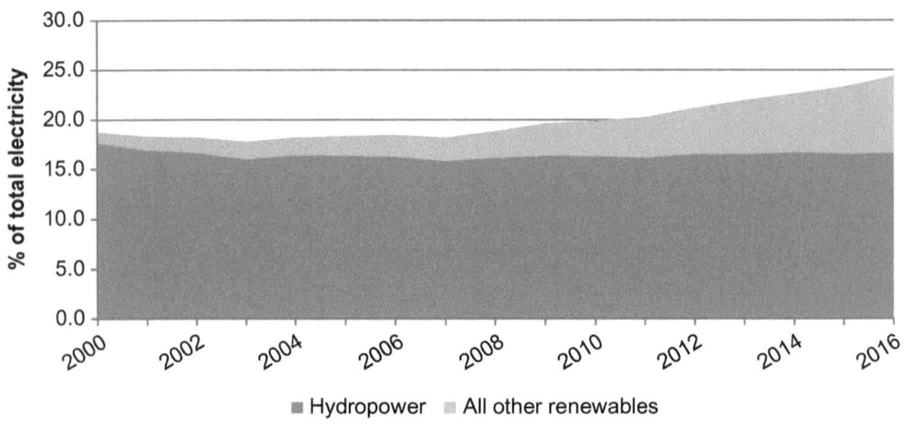

Fig. 8.11 Share of hydropower and other renewables in the global power system 2000–16 [3].

8.7.4 Share of hydropower from small and large hydro

It may be surprising to many to learn that nine out of the ten largest power plants in the world are hydropower plants, see Fig. 8.12. The Kashiwazaki-Kariwa nuclear facility was shut down in 2011 after the Fukushima disaster, and it is still (2017) not back in operation. The nine hydropower plants together generate 450 TWh $year^{-1}$, nearly 2% of world electricity generation, 7.5% of all renewable generation, and 11% of all hydropower generation in 2016! There are also many large hydropower plants under construction. Five of the largest ones that will be completed by 2021 will add a further 174 TWh hydropower (see Fig. 8.12).

In many countries there has been a policy to support small hydropower plants, rather than large ones. This policy seems to be based on the criticism in WCD [18] and a belief that small hydropower is less controversial and more environment friendly than large hydropower. IRENA keeps detailed statistics about hydropower capacity

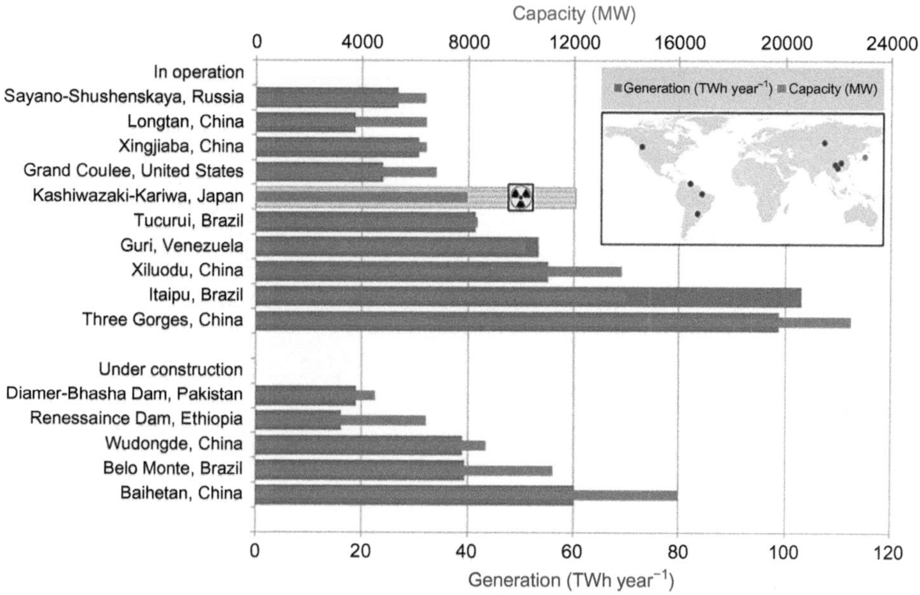

Fig. 8.12 The ten largest power plants in the world in operation (upper) and the five largest hydropower plants under construction in 2016 (lower).
(Data from Wikipedia).

and production from different categories of hydropower, Table 8.8 (Capacity, MW) and Table 8.9 (Generation, TWh). The four categories are based on capacity:

Small hydro	<1 MW
Medium hydro	1–10 MW
Large hydro	>10 MW
Pumped-storage and mixed plants	

From these tables it is clear that the share of small hydro (<1 MW) is very small, and contributes very little to the total capacity and production. Also medium large (1–10 MW) accounts for a small share of total hydropower capacity and production, about 10%. Fig. 8.13 shows the change from 2007 to 2016. Small and medium plants increased their share of production slightly, from 11% to 13% together. Large hydro (>10 MW) accounts for 75% of capacity and 85% of production, with a minor percentage decrease from 2007 to 2016.

8.8 Cost issues

Cost performance for hydropower can best be characterized by using "Levelized Cost of Energy" (LCOE). LCOE usually includes all lifecycle costs for production of energy, but does not include transmission and distribution costs.

Table 8.8 **Share of installed capacity (GW) for different categories of hydropower (2007–16) [3]**

		2007	2008	2009	2010	2011	2012	2013	2014	2015	2016
		Capacity (GW)									
Small hydro	<1 MW	29	29	30	30	30	30	27	28	31	31
Medium hydro	1–10 MW	70	74	78	86	91	96	104	109	117	117
Large hydro	>10 MW	704	731	756	771	791	817	854	882	907	936
Pumped-storage and mixed plants		123	125	130	141	146	149	150	152	155	161
Sum		926	960	994	1027	1059	1092	1135	1171	1210	1246
		Share of total hydropower capacity (%)									
Small hydro	<1 MW	3.2	3.1	3.0	2.9	2.8	2.7	2.4	2.4	2.6	2.5
Medium hydro	1–10 MW	7.5	7.7	7.9	8.4	8.6	8.8	9.2	9.3	9.7	9.4
Large hydro	>10 MW	76.0	76.2	76.1	75.0	74.8	74.8	75.2	75.4	75.0	75.2
Pumped-storage and mixed plants		13.3	13.1	13.1	13.7	13.8	13.6	13.2	13.0	12.8	12.9
Sum		100.0	100.0	100.0	100.0	100.0	100.0	100.0	100.0	100.0	100.0

Table 8.9 Share of generation for different categories of hydropower (2007–16) [3]

		2007	2008	2009	2010	2011	2012	2013	2014	2015	2016
		Annual generation (TWh)									
Small hydro	<1 MW	75	87	88	104	94	110	95	109	112	110
Medium hydro	1–10 MW	260	298	301	351	348	402	418	460	468	420
Large hydro	>10 MW	2723	2801	2826	2965	3064	3158	3266	3325	3318	3475
Pumped-storage and mixed plants		96	92	91	96	93	91	93	96	97	97
Sum		3154	3278	3306	3516	3599	3761	3872	3990	3995	4102
		Share of total annual generation (%)									
Small hydro	<1 MW	2.4	2.7	2.7	3.0	2.6	2.9	2.5	2.7	2.8	2.7
Medium hydro	1–10 MW	8.2	9.1	9.1	10.0	9.7	10.7	10.8	11.5	11.7	10.2
Large hydro	>10 MW	86.3	85.5	85.5	84.3	85.1	84.0	84.3	83.3	83.0	84.7
Pumped-storage and mixed plants		3.0	2.8	2.7	2.7	2.6	2.4	2.4	2.4	2.4	2.4
Sum		100.0	100.0	100.0	100.0	100.0	100.0	100.0	100.0	100.0	100.0

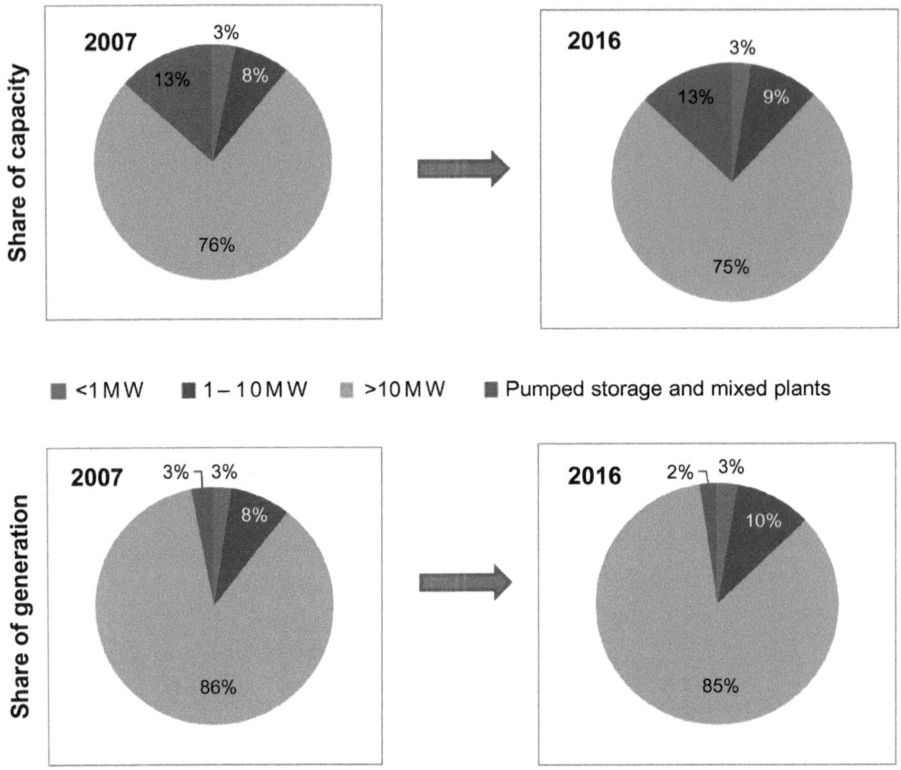

Fig. 8.13 Share of world hydropower capacity and generation for four different categories of Hydropower, change from 2007 to 2016.

Cost for hydropower projects is very site specific, and can therefore vary significantly from one project to another. The main cost components are: (1) investment cost (capital cost), (2) operation and maintenance cost (O&M) and (3) decommissioning cost.

LCOE is defined as the constant unit cost (per kilowatt hour) of a payment stream that has the same present value as the total cost of building and operating a generating plant over its life time. It can be computed as the total costs for the project over its lifetime divided by the total energy output over the same period. The LCOE includes all the cost elements for the entire lifetime of the project, and is usually given as cost in units of US c $(\mathrm{kWh})^{-1}$. LCOE for hydropower depends on the three components mentioned earlier, but also on (4) capacity factor, (5) lifetime of project, and (6) cost of capital (discount rate).

For hydropower, construction cost is usually the largest component in the cost calculation. It is common practice to combine planning and construction costs as Investment costs, and to combine operation, maintenance, and refurbishing as operation and maintenance (O&M)costs. Decommissioning costs are usually not considered, since it is rarely seen that hydropower plants are taken out of operation. In most cases, operating hydropower stations will be refurbished and kept in operation indefinitely. There are no fuel costs, so it is possible to do an LCOE computation for hydropower with the following formula:

$$\mathrm{LCOE} = \frac{\sum_{t=1}^{n} \frac{I_t + \mathrm{OEM}_t}{(1+r)^t}}{\sum_{t=1}^{n} \frac{E_t}{(1+r)^t}} \tag{8.5}$$

where, n, lifetime of project/year; t, time/year;I_t, investment costs during year t; OEM_t, operation and maintenance costs during year t; E_t, electrical energy generation during year t; r, discount rate.

Investment cost (capital cost) includes the cost of civil structures (dams, tunnels, powerhouse, etc.), electrical and mechanical (Elmek) equipment (turbine, generator, transformer, control systems, etc.), access roads, powerlines, cost of planning, and cost of compensating measures (mitigation, resettlement, fish ladders, etc.). The civil engineering cost is usually the largest share for large projects, but for small projects Elmek cost can be the largest [20]. Typical capital cost for large hydro varies from well below \$1000 $(\mathrm{kW})^{-1}$ up to \$3000 $(\mathrm{kW})^{-1}$.For small hydro the cost could be higher, from 1500 to\$6000 $(\mathrm{kW})^{-1}$ or even higher [1], [8,20]. There are also many examples of projects with costs of as low as \$500 $(\mathrm{kW})^{-1}$ for especially good sites.

Lifetime (n) and discount rate (r) are two very important parameters, and have a strong impact on LCOE for a project. Typical lifetime for large hydropower plants is 40–80 years [1], for small hydro it could be lower, 40 years or even less. Hydropower plants are usually refurbished at regular intervals, and actually not taken out of operation at the end of the lifetime. There are many examples of hydropower plants that have been in operation for 100 years or more, with regular upgrading of electrical and mechanical systems, but with no major upgrading of the more expensive civil structures (dams, tunnels, etc.). During a life span of 80–100 years, it will nearly always be necessary to replace much of the Elmek equipment at least once; the cost of this is usually included in the typical O&M costs of 2.5% per year.

Table 8.10 shows the computed LCOE for hydropower for some typical combinations of three important parameters: investment cost, discount rate, and lifetime. O&M cost was assumed to be 2.5% per year with a capacity factor fixed at 45%. This gives LCOE in the range of 1.5–9.5 c $(\mathrm{kWh})^{-1}$ most combinations cluster in the range US\$2–6 $(\mathrm{kWh})^{-1}$ or US\$20–60 $(\mathrm{MWh})^{-1}$.

The importance of interest rate for discounting future incomes to present value is striking. By increasing the interest rate from 3% to 10%, the LCOE is typically more than doubled. The long lifetime of hydropower has little effect on LCOA, especially for high interest rates.

These results can be compared to some of the most recent LCOE estimates for renewable power, see Fig. 8.14 from IRENA report "Rethinking Energy 2017" [21]. The computed LCOE is based on interest rate (cost of capital) of 7.5% for OECD and China, and 10% for the rest of the world. The more mature technologies biomass,

Table 8.10 LCOE estimation for parameters typical for current and future hydropower projects

Investment cost (US$ kW $^{-1}$)	Interest rate (%)	O&M cost (% per year)	Capacity factor (%)	Lifetime (years)	LCOE (US$ kWh $^{-1}$)	Lifetime years	LCOE (US$ kWh $^{-1}$)
1000	3	2.5	45	40	0.017	80	0.015
1000	7	2.5	45	40	0.025	80	0.024
1000	10	2.5	45	40	0.032	80	0.032
2000	3	2.5	45	40	0.035	80	0.029
2000	7	2.5	45	40	0.051	80	0.048
2000	10	2.5	45	40	0.065	80	0.063
3000	3	2.5	45	40	0.052	80	0.044
3000	7	2.5	45	40	0.076	80	0.073
3000	10	2.5	45	40	0.097	80	0.095

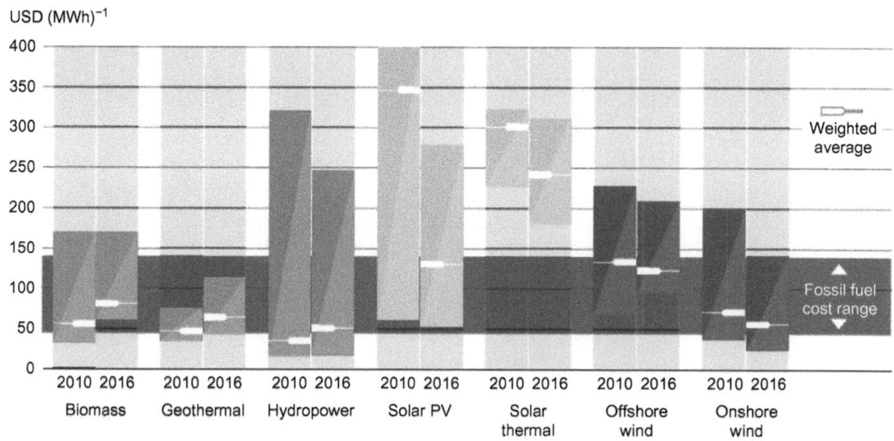

Fig. 8.14 Levelized cost (US$ (MWh)$^{-1}$) of electricity from utility-scale renewable technologies, 2010 and 2016.
From IRENA. REthinking Energy 2017: accelerating the global energy transformation. Abu Dhabi: International Renewable Energy Agency; 2017. ISBN 978-92-95111-06-6 (PDF) http://www.irena.org/publications/2017/Jan/REthinking-Energy-2017-Accelerating-the-global-ene.

geothermal, and hydropower all have seen a slight LCOE increase from 2010 to 2016, while wind has seen a slight reduction. Solar power, both PV and thermal, has seen a dramatic cost reduction in these 5 years. Still, by 2016, hydropower has the lowest average LCOE of all renewables, and is also close to the lowest cost of fossil fuels.

In addition to the low cost of energy, hydropower brings a range of other benefits for the broader energy system, and can support the integration and operation of other renewables, as described more in depth in Section 8.9.

As a comment to Fig. 8.14, we quote some conclusions from another IRENA analysis:

> *Hydropower produces some of the lowest-cost electricity of any generation technology. The LCOE of large-scale hydro projects at excellent sites can be as low as USD 0.02 (kWh)$^{-1}$, while average costs are around USD 0.05 (kWh)$^{-1}$ where untapped economic resources remain. Small-scale hydropower can also be very economic, although typically it has higher costs and is sometimes more suitable as an option for electrification that can provide low-cost electricity to remote communities or for the local grid. There is a clear cost dichotomy for hydropower between regions with remaining economic resources to exploit and those where most of the economic resources have been exploited already. Asia, Africa and South America all experience LCOEs for hydropower projects of on average USD 0.04 to USD 0.05 (kWh)$^{-1}$. In contrast, in regions which have exploited their most economic resources, weighted average LCOE ranges are around USD 0.09 to USD 0.10 (kWh)$^{-1}$ (e.g. in Europe, Eurasia, North America and Oceania). In addition to the higher costs, these regions are also constrained in the amount of economic capacity that still remains to be added. [22].*

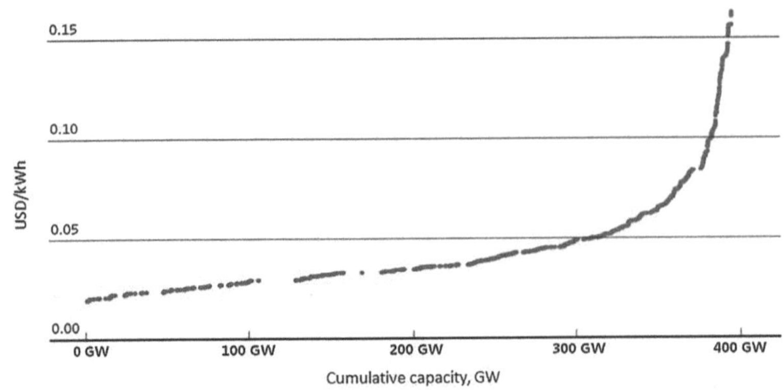

Fig. 8.15 Levelized cost of electricity (LCOE) of some unexploited hydropower resources in the IRENA renewable cost database (US\$ kWh^{-1}) [22].

Fig. 8.15 shows a supply curve for a selection of unexploited hydropower projects up to a limit of US\$0.15 $(kWh)^{-1}$. The projects of >300 GW have LCOE of <US\$0.05 $(kWh)^{-1}$ and the best projects have LCOE as low as US\$0.02 $(kWh)^{-1}$.

8.9 Integration into broader energy system

In addition to providing capacity and energy to meet demand for electricity, hydropower has technical characteristics that enable it to provide other services to make power systems operation more reliable. Hydropower plants can start generating on very short notice and with low start-up costs, manage rapid changes in generation (up/down ramping), and still have a high part-load efficiency. The ability to rapidly change output in response to system needs without suffering large decreases in efficiency makes hydropower plants well suited to providing the balancing services called regulation and load following. Hydropower can also rapidly restore a power station to operation without depending on the electric power transmission network ("black start capability").

Traditionally, power systems have been designed so that generation can follow the load (i.e., demand) pattern closely. This is referred to as the "generation follows demand" paradigm. With increasing amounts of variable and nondispatchable renewables (v-RES) in the power system, the need for extra capacity for load balancing will increase because of higher variability in generation for some technologies, especially for wind and solar PV plants. This variability exists on time scales from minutes up to weeks. Hydropower can manage to change operation very quickly, with startup delay in the order of minutes, and fast ramping time up/down of 40% per minute.

When the share of v-RES increases, it may also be necessary to introduce more storage in the grid in order to save excess generation during periods of high supply and low demand, and return the stored energy during low supply periods. Here, pumped-storage hydropower (PSH) is a very good option, see Sections 8.9.3 and 8.11.2.

In, for example, Denmark and Germany, the high share of variable wind generation is already being balanced by flexible hydropower generation in Norway, connected through subsea power cables. Studies have shown that this balancing capacity can be further increased by constructing new PSH's in Norway. This can be done by using only existing reservoirs; at least 20 GW of such projects has been identified [23]. To utilize them fully, however, grid connection between Norway and Denmark/Germany/United Kingdom needs to be strengthened with an increase of at least 10–15 GW.

8.9.1 Energy management services

Energy management services are needed for safe and efficient operation of the power system. They are needed to maintain the balance between electricity supply and demand, maintain power system stability, and enable electrical energy to be delivered safely, reliably, and economically from generators to consumers.

In addition to these primary functions, grid services also provide various other contributions that make the operation of the power system more efficient and make electricity more affordable for consumers. In principle, grid services and contributions provided by hydropower and PSH plants can be grouped into three main categories [24].

- Bulk power services
  - Energy generation
  - Energy/price arbitrage
  - Generating capacity
- Ancillary and reliability services
  - System inertia
  - Frequency regulation and control (primary, secondary, tertiary)
  - Contingency reserves (spinning, nonspinning, replacement/supplemental)
  - Energy imbalance, ramping, and load following
  - Voltage support
  - Black start service
- Transmission services
  - Transmission congestion relief
  - Transmission investment deferral
  - System-wide effects (portfolio effects)
  - Better integration of variable renewables (e.g., wind and solar)
  - Reduced cycling and ramping of thermal units
  - Reduced overall system operating costs
  - Increased energy security

This is not an exhaustive list but it captures the bulk of the services and contributions typically provided by hydropower and PSH plants to the power system [24].

8.9.2 Energy storage

Power system operators have to manage the electricity feed into the power grid from various sources in order to always balance supply and demand in real time. This requires a considerable level of flexibility from different system resources, in order to manage daily, weekly, and seasonal ramping cycles and minute-to-minute energy balancing. Electricity generation from renewable fluctuates according to resource availability and may not always coincide with the demand profile. Wind power generation varies rapidly with weather conditions, solar PV in addition has daily and seasonal cycles depending on solar radiation.

An example taken from Germany illustrates this challenge, Fig. 8.16. In March 2016, the total installed capacity of wind power in Germany was 39,200 and 38,300 MW of Solar PV. After sunset in the evening March 24, Solar PV stopped producing, and wind calmed down so much that the entire wind power system only produced 550 MW, a capacity factor of only 0.7%!

In order to manage events like this, there must be other power resources that can be started quickly and have enough energy storage to supply power until the variable sources come back in operation. There must be enough of both capacity and energy in storage to handle such real-world situations. For wind power, it is known that calm periods with very low wind can have durations of up to a week or two in Western Europe, for solar power there are both day/night and seasonal cycles and large weather-dependent variability. As long as variable renewable energy sources (v-RES) constitute a small share of the power system, it is necessary to increase generation in thermal plants or import from other countries. A good example is Denmark where wind power is balanced by hydropower from Norway and connecting grid also to Sweden and Germany. But in a future situation where v-RES is dominating, the solution becomes much more difficult. Hydropower, with its ability to ramp up or

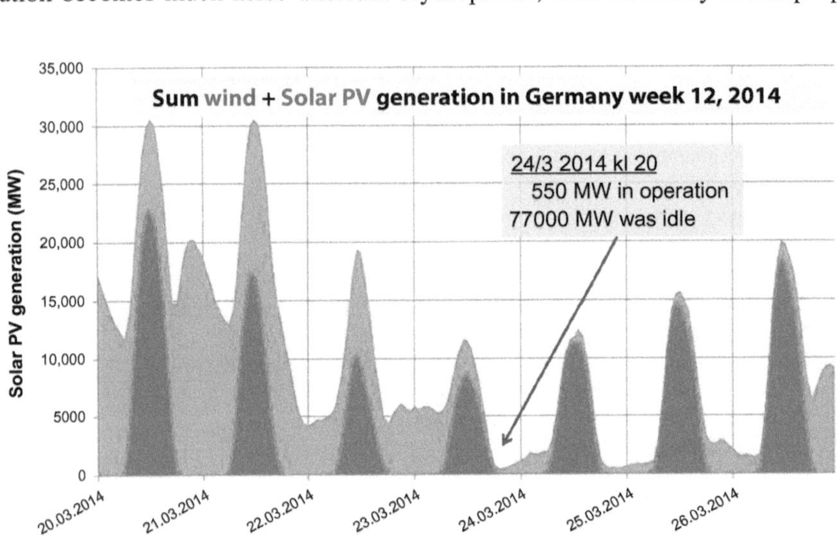

Fig. 8.16 Wind and Solar PV generation in Germany during 1 week in March 2014.

down quickly and produce from energy stored in reservoirs, can contribute very efficiently to the solution.

8.9.3 Pumped-storage hydro

Pumped-storage hydro (PSH) can support other v-RES both by providing storage and capacity. In traditional use of PSH ("time shifting"), pumping mode is utilized to store excess output from thermal plants during periods with low demand, when thermal plants would otherwise operate less efficiently at part load or have to be shut down completely. The PSH then keeps water in reserve in storage in the upper reservoir, ready for generating power during peak period demands. PSH has much the same ability as storage HPPs to provide balancing and regulation services [1].

PSH currently dominates installed grid storage capacity, with 96% of the world's total of 176 GW installed capacity (2016). Other electricity storage technologies in significant use around the world include thermal storage, 3.3 GW (1.9%); batteries, 1.9 GW (1.1%); and other types of mechanical storage, 1.6 GW (0.9%) [4].

More than 75% of PSH capacity is installed in only ten countries, with three of these, China (32.1 GW), Japan (28.5 GW), and United States (24.2 GW), accounting for almost half (48%) of the global PSH capacity.

Several recent reports have concluded that PHS capacity is expected to increase much both in the near (2030) and distant future (2050). But as other new storage technologies like batteries will also be introduced, the share of PSH will probably still be somewhat reduced. In, for example, the Remap study, it was concluded that the PSH share of electrical energy storage will fall from 96% in 2017 to 83%–88% in 2030 [8].

IHA [4] reports that 6.4 GW of PSH was installed in 2016, and that additional 20 GW was under construction. This is seen as an indication of hydropower's increasingly important role in providing flexible support to renewable energy systems. China commissioned 3.7 GW of PSH in 2016, and in its 13th 5-year plan on energy development aims to reach 40 GW by 2020 [4].

PHS systems are evolving to provide additional operating flexibility in order to balance fluctuations in the system. In traditional PHS, power regulation is only available when generating; however, variable-speed PHS systems are being implemented to increase plant efficiency and flexibility by allowing for power regulation in both pumping and turbine mode. Ternary systems, consisting of a motor generator and separate pump and turbine set, can allow for simultaneous pumping and generating, which allows for even finer frequency control [4,8,23].

8.9.4 Role in water management

Water, energy, and climate change are inextricably linked. On the one hand, water availability is crucial for many energy technologies, including hydropower, and on the other hand, energy is needed to secure water supply for agriculture, industries, and households, particularly in water-scarce areas in developing countries. This mutual dependence, the "water-energy nexus," has led to the understanding that water and energy must be addressed in a holistic way, especially regarding climate change and sustainable development.

The challenge to provide energy and water (and food) for sustainable development will require improved regional and global water governance. Since hydropower facilities are often associated with the creation of water storage facilities, hydropower is at the crossroads of these issues and can play an important role in enhancing energy, water, and food security. Therefore, hydropower development is part of water management systems as much as energy management systems, both of which are increasingly becoming climate driven.

In Ref. [24], a number of issues connecting hydropower and water management are discussed. Water management services are defined as both qualitative and quantitative functions in integrated water management, regional development, human development, as well as global and local environmental services:

- Water Quantity Management
  - Flood control
  - Drought control
  - Ground water stabilization
  - Water supply (increased water availability for other uses)
- Water Quality Management
  - Oxygenation
  - Temperature management
  - Cleansing of water (trash removal)
  - Sediment Management
  - Habitat Management (via flow management)
  - Barrier to saline water intrusion
- Regional Development
  - Navigation (transport)
  - Irrigation (agriculture)
  - Leisure and tourism
  - Aquaculture (fisheries and food)
  - Drinking water supply management
  - Industrial water supply management (e.g., cooling water)
  - Improved Infrastructure (roads, access ramps, etc.)
  - Energy-intensive industries
- Human Development
  - Health services
  - Education
  - Sanitation
  - Domestic power
  - Community services
  - Housing
  - Livelihoods
  - Nutrition and food supply
  - Professional competence development
  - New economic activities
- Environmental Services
  - Reduction of GHG emissions

- Reduction of atmospheric emissions
- Creation of wetlands
- Microclimate around reservoirs
- Carbon storage

Hydropower dams and reservoirs are often used for multiple purposes, in addition to electrical energy production [25].

8.10 Sustainability issues

Sustainable development has been defined as "... *development that meets the needs of the present without compromising the ability of future generations to meet their own needs*" [26]. In the process of changing from a fossil to a renewable energy system, it is very important to ensure that the future renewable energy system will be sustainable, considering social, economic, and environmental sustainability. The sustainability issue is of particular importance for hydropower, since it depends on and can have significant impact on an equally important resource, water.

The International Hydropower Association (IHA) has developed "Sustainability Guidelines" and "Sustainability tools" in order to guide the planning, implementation, and operation of hydropower projects [19,27]. The protocol includes between 19 and 23 relevant sustainability topics, depending on the development stage of the project. It is the result of a long process involving many different stakeholders: social and environmental NGOs, governments, commercial organizations, development banks, and the hydropower sector. Some of the topics included are: biodiversity, indigenous people, infrastructure safety, resettlement, water quality, erosion and sedimentation, and downstream flow regimes. The protocol has so far been tested in 16 countries [27].

In order to assess the sustainability of a hydropower project, it is important to study the whole lifecycle of the project, including investigations, planning, construction, operation, maintenance, refurbishing, and eventually decommissioning. The LCOE method discussed in the cost section is based on a lifecycle assessment of economic performance. Besides economic sustainability, which always will have high priority, some of the most important other topics to be investigated and documented are:

- Environmental and social impacts
- Greenhouse gas (GHG) emissions and "carbon footprint"
- Energy payback ratio (EPR)
- Water consumption and "water footprint"
- Sediment issues—Reservoir sedimentation
- Climate change impacts

Hydropower has many advantages compared to other sources of electrical energy: it is renewable, clean, largely carbon-free, and well suited for integration into the grid, often supporting the stability of a grid. But hydropower can also have negative impacts on the aquatic environment, where dams, weirs, diversions, and changes in river flow may create barriers and hinder fish migration, disturb ecosystems, and create other problems [9,13].

8.10.1 Environmental and social impacts

Hydropower schemes often change the flow regime in rivers and the water level regime in lakes and reservoirs, and can therefore have negative consequences for ecosystems and biodiversity. Run-of-river projects usually do not change the flow regime, while storage hydro projects typically lead to changes in the annual flow regime, for example by decreasing summer flows and increasing winter flows. The natural transport of sediments may also be altered, creating problems both in the reservoir and intake pond where sediments are deposited, and downstream of the dam where reduced sediment concentrations may lead to increased bank, riverbed and delta erosion, and decreased stability. Dams, intakes, and other infrastructure may act as barriers for the migration of species, and act as traps for transport of nutrients and sediments. Dams and reservoirs will cause loss of land, and sometimes there is a need for resettlement of people who previously lived in the reservoir area. As part of the Environmental and Social Impact Assessment (ESIA), negative impacts will be analyzed and documented, and possibly mitigated. The ESIA are very central documents to the licensing process while the authorities balance the costs and benefits of the projects, and ultimately define the terms once a license has been granted. In such a process, both monetary and nonmonetary values for the project proposer as well as the society at large are accounted for. Even if impacts cannot be completely avoided, they can be minimized if properly managed during the planning and construction stages. Negative impacts can be reduced by careful design. Sometimes even "win-win situations" for power production and the environment can be created, in particular for refurbishment and extension projects. The concept "Environmental Design" has been widely accepted and is now in use in many hydropower projects as an approach for balancing the power production while optimizing the environmental potential of regulated river basins. The concept is also applicable for other types of renewable energy projects, for example, for wind energy projects [13].

A good overview of social and environmental effects of hydropower development can be found in reference [1] and is summarized in Table 8.11. The planning of hydropower development in most countries has become very complex and time consuming, in order to include all different interests affected by the project. As an example, Fig. 8.17 shows the licensing process for large hydropower plants in Norway.

8.10.2 Greenhouse gas emissions and carbon footprint

Hydropower, like most other renewable energy sources, offers considerable potential for reduction of carbon emissions (GHG) from energy generation by replacing fossil energy generation. Coal-fired power plants typically emit 1000 g of CO_2 for each kilowatt-hour produced, oil-fired plants typically 800 g $(kWh)^{-1}$, and natural gas-fired plants around 4–500 g $(kWh)^{-1}$.

The majority of lifecycle GHG emission estimates for hydropower are between 4 and 14 g(CO_2eq) $(kWh)^{-1}$, but under certain scenarios there is a risk of much larger GHG emissions [1].

A few observations of high GHG emissions from tropical hydropower reservoirs have been raising doubt about the low GHG emission for hydropower, but it is still

Table 8.11 Type of hydropower projects, their main services and distinctive environmental and social characteristics [1]

HPP type	Energy and water management services	Main environmental and social characteristics
All	Renewable electricity generation Increased water management options	Barrier for fish migration and navigation, and sediment transport Physical modification of riverbed and shorelines
Run-of-river	Limited flexibility and increased variability in electricity generation output profile Water quality (but no water quantity) management	Unchanged river flow when powerhouse in dam toe; when localized further downstream reduced flow between intake and powerhouse
Reservoir (storage)	Storage capacity for energy and water flexible electricity generation output Water quantity and quality management; groundwater stabilization; water supply and flood management	Alteration of natural and human environment by impoundment, resulting in impacts on ecosystems and biodiversity and communities. Modification of volume and seasonal patterns of river flow, changes in water temperature and quality, land use change-related QHQ emissions
Multipurpose	As for reservoir HPPs; dependent on water consumption of other uses	As for reservoir HPP; Possible water use conflicts; Driver for regional development
Pumped storage	Storage capacity for energy and water; net consumer of electricity due to pumping No water management options	Impacts confined to a small area; often operated outside the river basin as a separate system that only exchanges the water from a nearby river from time to time

not clear if these observations represent a net emission or emissions that would occur anyway, for example, from a wetland. The GHG footprint of a reservoir is highly dependent on local climate, topography, and human activities in the catchment. Assessment of net carbon emissions from a reservoir needs to take into account the emissions from the catchment before the creation of the reservoir and compare to the emissions after the reservoir has been built.

The International Hydropower Association (IHA) has initiated a project to develop more accurate methodology for assessment of the carbon footprint of a reservoir, and how to mitigate GHG emissions where needed. The methodology will also help to properly allocate GHG footprint to the various services provided by multipurpose reservoirs.

For most hydropower plants, it seems safe to assume that GHG emissions are very low and that hydropower will contribute to reducing GHG emissions by between 500 and 1000 g(CO_2) kWh^{-1}. This amounts to a yearly reduction of $0.5–1 \times 10^6$ t of CO_2

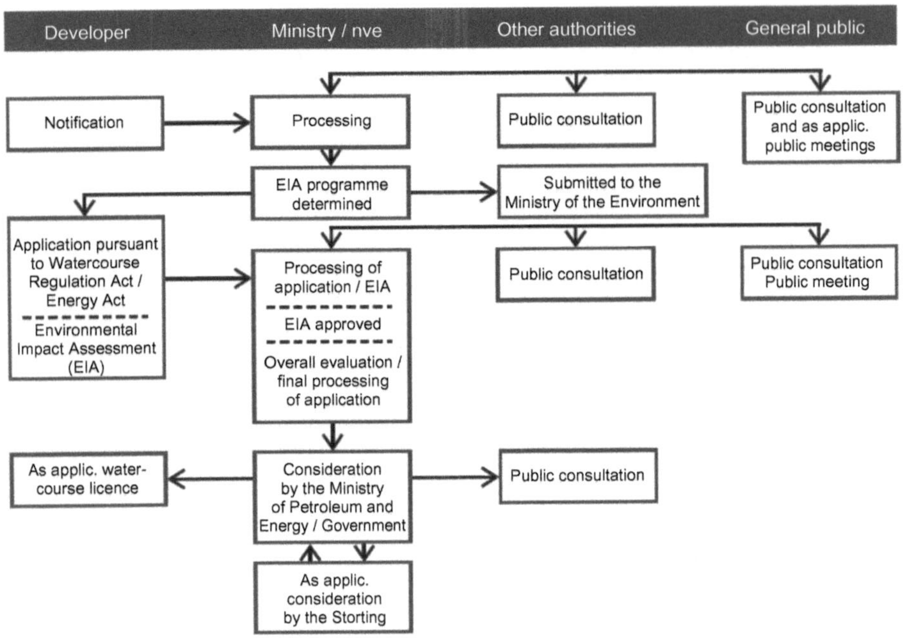

Fig. 8.17 Licensing process for large hydropower (>10 MW) in Norway [37].
From Killingtveit Å. Hydroelectric power. In: Letcher, T. Future energy. Improved, sustainable and clean options for our planet. Elsevier; ISBN 978-0-08-099424-6.

for every 1 TWh. Other renewables also have very low GHG emissions, comparable or slightly higher than hydro, see Fig. 8.18.

8.10.3 Energy payback

The "Energy Payback Ratio" (EPR) or "Energy Return of Investment" (EROI) are among the most commonly used energy indicators and is defined as "*The ratio of total energy produced during a system's normal lifespan, divided by the energy required to build, maintain and fuel it.*" EPR is a lifecycle perspective and shows the net energy contribution from the project during its lifetime compared to the supporting primary energy required to build, maintain, and supply the system.

A high EPR indicates an energy-efficient system, i.e., that the energy output is high compared to the energy required to build, maintain, and supply the system. If a system has EPR close to one, it consumes nearly as much energy during its lifetime as it generates, obviously not a very energy-efficient solution.

Hydropower has the highest EPR of all electricity generation technologies, with EPR up to 267 for run-of-river plants and 205 for storage plants. This can be compared to ratios between 3 and 11 for fossil fuels, 39 for large wind turbines, and 16 for nuclear power plants; see Fig. 8.19.

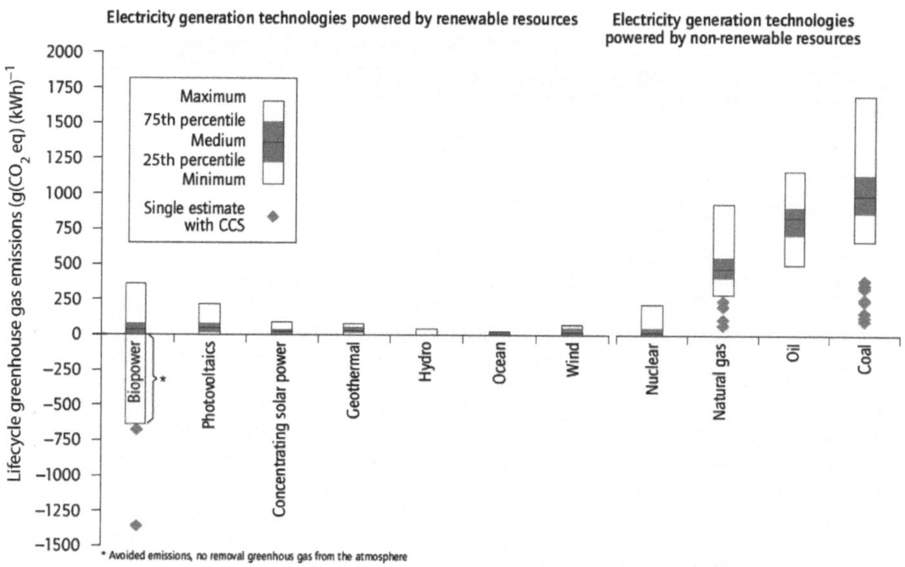

Fig. 8.18 Estimates of lifecycle GHG emissions in electricity generation. [1].
From IPPC. Renewable energy sources and climate change mitigation special report of the intergovernmental panel on climate change. Cambridge University Press; 2012. ISBN 978-1-107-60710-1.

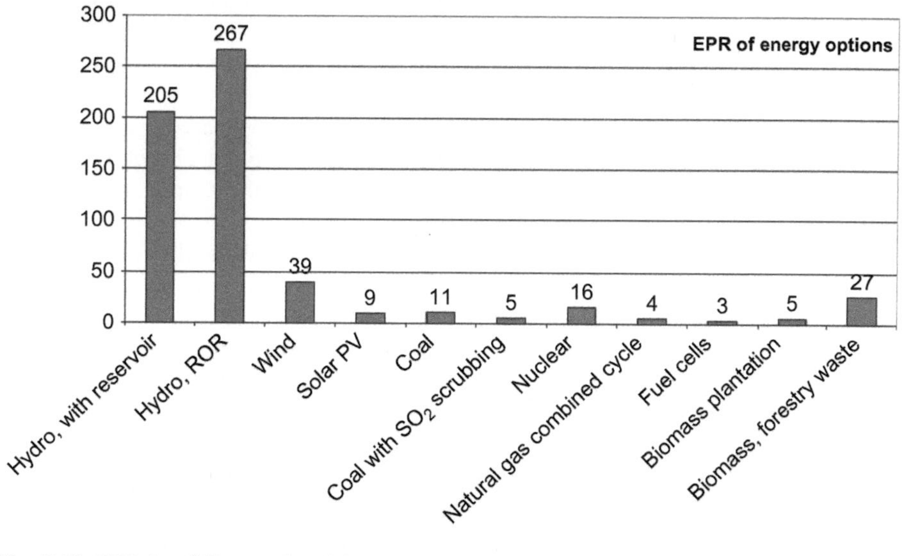

Fig. 8.19 EPR for different electricity generation technologies [1,38].
Based on data from IPPC. Renewable energy sources and climate change mitigation special report of the intergovernmental panel on climate change. Cambridge University Press; 2012. ISBN 978-1-107-60710; Gagnon L. Civilisation and energy payback, Energy Policy 2008; 36:3317–3322.

A detailed study including a number of Norwegian and other European hydropower plants was published in 2012 [29]. Here, 11 reservoir and 7 RoR hydropower plants were studied in a lifecycle perspective. Some of the results are shown in Fig. 8.20. In this study, the reservoir projects achieved the highest EPR values, which ranged from 100 to 500, while RoR projects mostly clustered around 100. One case (not included in the graph) with refurbishing an old power plant gave EPR over 2500 as large parts of the energy investments had already been made.

8.10.4 Water consumption and water footprint

Nearly all electricity generation technologies depend on water in one way or another, and there is a growing understanding of the close relation between water and energy: the "water-energy nexus." Water is a vital resource in the development of energy resources and access to energy is a prerequisite in the provision of water services. It has been estimated that 15% of the world's total water consumption, around $4000 \times 10^9\ m^3$ (4000 billion cubic meters), is withdrawn for energy production [30].

Water consumption is usually defined as "the part of the freshwater, which is not released back to the original watershed; primarily due to evaporation and product integration." Water footprint is defined as "*the volume of water consumption that can be associated with a specific human purpose*." The water footprint of a product (a commodity, good, or service) "*is the total volume of freshwater used to produce the product, summed over the various steps of the production chain*." This definition refers to a lifecycle perspective of the production of a commodity, good, or service. It does not seem to be a clear differentiation between the terms "water footprint" and "water consumption", in the hydropower context they are used synonymously [30].

IPCC Special Report on Renewable Energy Sources [1] revealed potentially very high water consumption rates from hydropower production compared to other renewable technologies, but suffered from few studies and methodological problems. More recent studies have presented new estimates, even higher than those presented by IPCC [25].

The production of hydropower is directly related to the availability of water, and exclusively dependent on locally available water resources as the only possible raw material for the production of electricity. The reduction in the available water will immediately reduce power production. For the hydropower sector, there are at least two important aspects related to the topic "water consumption from energy production." Firstly, high water consumption rates represent a potential reputational risk to the sector. Secondly, water losses from upstream reservoirs and the role of upstream regulations pose a potential financial risk in the case where this results in reduced water availability for power production [30].

For hydropower, it is important to distinguish between water used in the turbines, which is returned back to the river downstream, and water lost by evaporation from reservoirs. If the reservoir is used only for energy generation, the evaporated water can be compared to energy generation, and water consumption (or water footprint) can be calculated, usually in terms of cubic meters of water per megawatt-hour. This can then

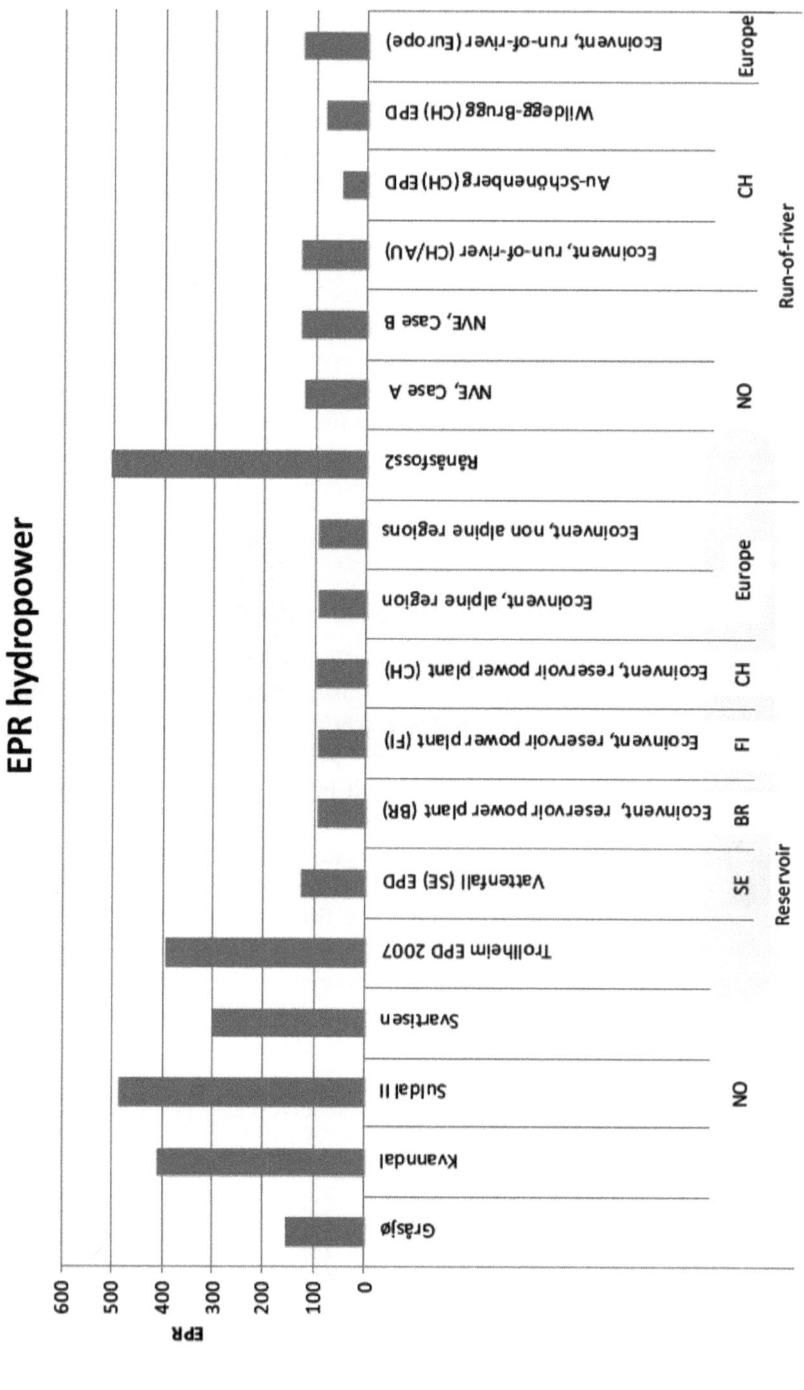

Fig. 8.20 EPR for hydropower plants in Norway and Europe [29].

be compared to water consumption for other renewables or from fossil-based generation.

There are, however, two major complicating factors with hydropower that need to be considered:

- use of observed (gross) or increased (net) evaporation from the reservoir
- allocation of water consumption in multipurpose reservoirs

In most cases, the use of observed (gross) evaporation from a reservoir is misleading, since there would also be some, maybe equally high, evaporation without the hydropower plant. It is therefore important to assess the evaporation losses both before and after the construction of the reservoir, and compute the net evaporation. Many reservoirs used for hydropower are multipurpose reservoirs, providing benefits and also used for water supply, irrigation, flood control, and navigation. In these cases, the fair share of water consumption should be divided between different users. In most published studies, such an allocation between the various functions has not been done and the water losses have been computed from gross evaporation. A discussion and recommendation of methodology and application to reservoirs are given in references [25, 30].

8.10.5 Sediment issues and reservoir sedimentation

Sustainable sediment management, according to the World Bank, "*seeks to maintain long-term reservoir capacity, retarding the rate of storage loss and eventually bringing sediment inflow and discharge into balance while maximizing usable storage capacity, hydropower production, or other benefits*" [4].

For a hydropower plant, sedimentation can result in loss of storage and reduced sustainable (firm) power generation, obstruct intakes and by entrainment of coarse sediment create abrasion in hydromechanical equipment like turbines, gates, and pipes.

A dam and reservoir can dramatically reduce the flow of sediment downstream from the reservoir and impact ecosystems in the downstream river. One example is Lake Nasser in Egypt where nearly all incoming sediments in the Nile are deposited. Outflow from the reservoir is deprived of sediments and this disrupts the sediment balance in the Nile delta, leading to increasing loss of land as coastal erosion removes more sediments than what is supplied by the river.

Available sediment management strategies for a reservoir can be classified into four main groups:

- reducing sediment inflow
- routing sediment through or around the reservoir
- sediment removal
- adapting to sedimentation

Climate change may increase the sediment challenges. Climate change is generally expected to increase sediment yield due to greater frequency of extreme rainfall events, leading to more erosion and sediment transport. In glaciated catchments, the retreat of ice may uncover easily erodible sediments at elevations too high to be stabilized by vegetation. Disappearing glaciers and melting of permafrost at high elevations may also increase frequency of landslides and other geohazards, and thereby increase sediment generation and transport to reservoirs and intakes.

8.10.6 Climate change issues

The hydropower resource potential (Tables 8.3–8.5) is based on observed hydrological conditions, and is therefore valid only for the present climate. In the future, with a changing climate, the hydropower potential could change due to the effects of climate change on the water resources that hydropower depends on:

- changes in river flow volume
- change in river flow variability/seasonality
- change in extreme flow events (floods, droughts)
- change in sediment production and transport

In a recent study [31] the national, regional, and global changes in hydropower generation for the existing hydropower system (per 2012) were investigated, based on a global assessment of climate change and its effect on river flow [32]. At high latitude and in most of the tropics, climate models predict increasing precipitation and runoff, while in the mid latitudes, the prediction is that precipitation and runoff will decrease. Regions with decreasing water resources and decreasing hydropower generation can be found around the Mediterranean, in Southern Africa, in Australia, and in Central and Western America. Regions with increasing water resources and increasing hydropower generation can be found in Northern Europe, in most of Russia, in East Asia, in East Africa, in Canada, and in parts of South America. The study concluded that the total global effect on runoff and hydropower was small, and probably slightly positive.

For future hydropower planning, it will become necessary to carry out climate change analysis in the planning process, and not base all analyses on historical observations alone. The climate is changing, in some regions toward more runoff and more hydropower potential, in other regions toward drier climate and less hydropower potential. The main steps in such analysis are outlined in Fig. 8.21.

Considering as an example the Zambezi river; future climate scenarios indicate a warmer and drier future [33]. This will result in gradually reduced flow and less

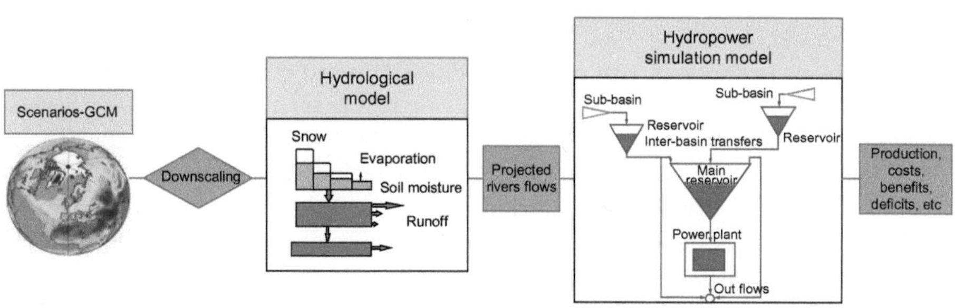

Fig. 8.21 Main steps and methods used in assessing the impacts of climate change on hydropower [33].

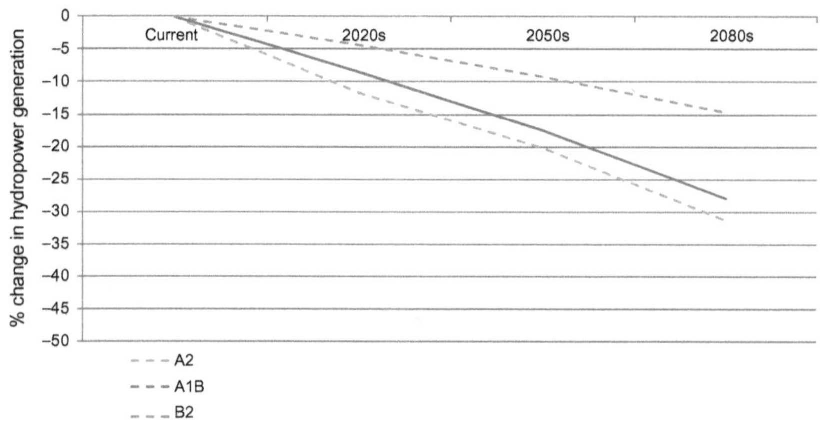

Fig. 8.22 Change in hydropower generation (%) in the Zambezi river for three different emission scenarios (A2, A1B, and B2). Computed values for the 2020s, 2050s, and 2080s compared to current climate (1961–90) [33].

hydropower generation. Scenarios for future hydropower generation, based on modeling approach as illustrated in Fig. 8.21, is shown in Fig. 8.22.

8.11 Hydropower in the future—Potential deployment

8.11.1 Energy production

Hydropower offers significant potential for long-term carbon emissions reductions. The hydropower capacity installed today delivers roughly 17% of world electricity supply and already gives a very significant contribution to reduction in GHG emissions, compared to electricity from fossil fuels.

On a global basis, the potential for further hydropower development is significant. The remaining economic resource potential is large, 4–5000 TWh, and in the near and medium future resource depletion is not likely to restrict hydropower development. Since <25% of the technical potential and <50% of the economic potential has been developed, the prospects for further development are good. If the energy from these projects replaces electricity from coal-fired plants, the total yearly reduction of GHG emissions would amount to 5000×10^6 t of CO_2.

Hydropower is also very cost competitive, with typical cost of US2–5 c $(kWh)^{-1}$. The combination of a large potential and low cost makes it realistic to increase hydropower generation from the present level of about 4100 TWh $year^{-1}$ (2016) by a factor of 2. The IPCC's fourth assessment report (AR4) concluded that hydro could contribute 17% of global electricity supply by 2030, which is equivalent to 5400 TWh $year^{-1}$. The IEA world energy outlook 2010 reference scenario projected 5232 TWh $year^{-1}$ of hydro by the year 2030.

In the most recent projections from US Energy Information Administration (EIA), it is estimated that worldwide hydropower generation will increase from 4100 TWh in

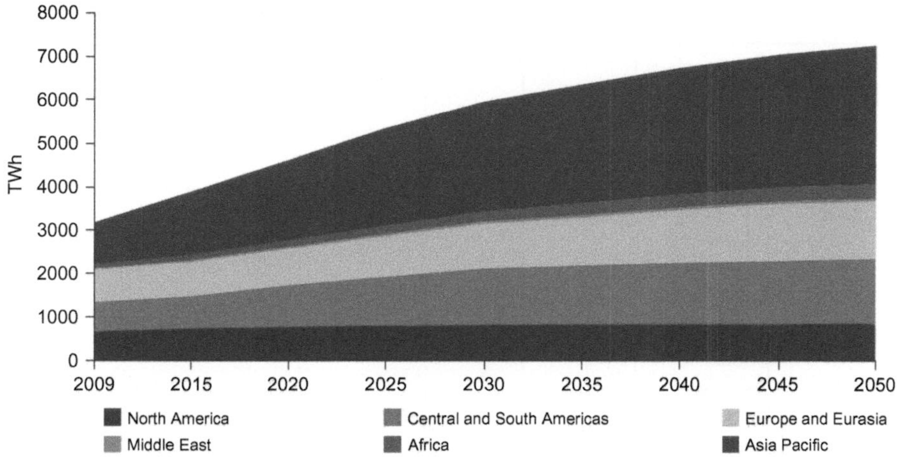

Fig. 8.23 Hydropower generation up to 2050 according to hydropower roadmap vision [20].

2016 to 5678 TWh by 2040, with a share of 16.7% of total electricity generation (34,000 TWh) and 53% of total renewable generation (10,700 TWh) [34]. Some estimates from the industry go even higher, predicting 8700 TWh by 2050 [35], an increase of 4600 TWh compared to today. A recent analysis by IEA [20] shows that hydropower could double its contribution by 2050, reaching 2000 GW of global capacity and over 7000 TWh, Fig. 8.23. The bulk of this growth would come from the large plants in emerging economies and developing countries.

Hydropower also can provide many types of energy and water management services, and help to integrate other variable renewable energy sources in the grid, to further support its deployment. Environmental and social impacts will, however, need to be carefully assessed, managed, and, if possible, mitigated, in order to demonstrate the long-term sustainability.

8.11.2 Pumped-storage hydropower

Interest in pumped-storage hydro is increasing, particularly in regions where solar PV and wind are reaching high levels of penetration or are growing rapidly. The majority of existing pumped-storage capacity is found in Europe, Japan, and the United States. According to data from IHA, this is expected to increase rapidly. There are now (2016) >100 projects in the pipeline around the world, totaling 75 GW of new capacity. This will increase world's total PSH capacity to 250 GW. The majority of the new projects will be operational by 2030 [36].

With continued increasing shares of wind and solar power (v-RES), it is expected that the need for balancing power and storage also will continue to increase. An assessment by IEA for two different scenarios of v-RES [20] shows that PSH capacity may increase to between 412 and 700 GW by 2050, see Table 8.12.

Table 8.12 **Expected PSH capacity by 2050—Analysis by IEA [20]**

		China	USA	Europe	Japan	Rest of the world	Total
IEA low estimate	v-RES % of total energy	21%	24%	41%	18%		
	Hydro % of total energy	14%	6%	13%	12%		
	PSH/ total capacity	4%	4%	6%	11%	2%	
	GW	119	58	91	35	108	412
IEA high estimate (Hi-REN)	v-RES % of total energy	34%	37%	48%	33%		
	Hydro % of total energy	15%	6%	11%	13%		
	PSH/ total capacity	5%	8%	10%	12%	3%	
	GW	179	139	188	39	164	700

8.12 Summary

Hydropower is a renewable energy source where power is derived from the energy of water moving from higher to lower elevations. It is a proven, mature, predictable, and price-competitive technology. Hydropower has among the best conversion efficiencies of all known energy sources (about 90% efficiency, "water to wire"). It requires relatively high initial investment, but has a long lifespan with low operation and maintenance costs. The existing hydropower system has an annual generation of 4100 TWh $year^{-1}$, and contributes to nearly 17% of the world electricity generation.

There is still a large potential for further development, as the total technical potential has been estimated to be nearly 16,000 TWh, of which 9600 TWh has been classified as the economic potential. Globally, 4100 TWh or 26% of technical and 43% of economic potential has already been developed. In Europe, close to 50% of technical (70% of economic) potential, in Asia 24% of technical (41% of economic) potential, and in Africa only 6% (9% of economic) potential has been developed.

Significant potential also exists to use existing water infrastructure that currently lacks generating units (e.g., existing barrages, weirs, dams, canal fall structures, water supply schemes) by adding hydropower facilities. Only 25% of all existing 45,000 large dams are used for hydropower, the other 75% are used exclusively for other purposes (e.g., irrigation, flood control, navigation, and urban water supply schemes).

Hydropower offers significant potential for carbon emissions reductions, since GHG emissions are generally very low, typically $<1\%$ of that from coal power plants. Hydropower is cost competitive, with LCOE typically in the range 2–5 US c $(kWh)^{-1}$, which is comparable to the cost of energy from thermal power plants or even lower. Hydropower has very good efficiency, typically 90% "water-to-wire," and typical energy payback ratio of 100–200, highest of all types of renewables.

Hydropower can provide both energy and water management services and also help to support and integrate other variable renewable energy sources like wind and solar, by providing storage and load balancing services. Environmental and social impacts will need to be carefully managed. The use of new sustainability guidelines and tools for analysis should help in making hydropower sustainable also in the future.

Climate change can be a challenge to hydropower in some regions of the world, but also lead to increasing production in other regions. The future climate and hydrology will change, and it is therefore increasingly important to integrate data and models of climate change impacts in the planning process and design of the future hydropower system.

References

[1] IPPC. Renewable energy sources and climate change mitigation special report of the intergovernmental panel on climate change. Cambridge University Press; 2012, ISBN 978-1-107-60710-1.

[2] Enerdata Global Energy Statistical Yearbook, https://www.enerdata.net/publications/world-energy-statistics-supply-and-demand.html; 2017.

[3] IRENA. Database renewable electricity capacity and generation statistics, http://resourceirena.irena.org/gateway/dashboard/?topic=4&subTopic=17; 2017.

[4] IHA. Hydropower status report 2017, https://www.hydropower.org/2017-hydropower-status-report; 2017.

[5] NTNU. Hydropower development in Norway. Hydropower development, vol. 1. Trondheim: Norwegian University of Science and Technology (NTNU), Department of Hydraulic and Environmental engineering; 1993–2005. This book series consists of 17 volumes published between 1993 and 2005.

[6] NTNU. Mechanical equipment. Hydropower development, vol. 12. Trondheim: Norwegian University of Science and Technology (NTNU), Department of Hydraulic and Environmental engineering; 1993–2005. This book series consists of 17 volumes published between 1993 and 2005.

[7] IJHD. World Atlas & Industry Guide. Int J Hydropower Dams 2017. Wallington, Surrey, UK.

[8] IRENA. Electricity storage and renewables: costs and markets to 2030; 2017. ISBN 978-92-9260-038-9, http://www.irena.org/publications/2017/Oct/Electricity-storage-and-renewables-costs-and-markets.

[9] NTNU. Environmental effects. Hydropower development, vol. 3. Trondheim: Norwegian University of Science and Technology (NTNU), Department of Hydraulic and Environmental engineering; 1993–2005. This book series consists of 17 volumes published between 1993 and 2005.
[10] NTNU. Underground powerhouses and high pressure tunnels. Hydropower development, vol. 14. Trondheim: Norwegian University of Science and Technology (NTNU), Department of Hydraulic and Environmental engineering; 1993–2005. This book series consists of 17 volumes published between 1993 and 2005.
[11] Lia L, Aas MN, Killingtveit Å. Increased generation from upgrading and extension projects. Int J Hydropower Dams 2017;24(4).
[12] U.S. Department of Energy's Office of Energy Efficiency and Renewable Energy: Hydropower Vision Report. https://energy.gov/eere/water/downloads/hydropower-vision-report-full-report.
[13] Forseth, T. and Harby, A.: Handbook for environmental design in regulated salmon rivers. NINA special report 53, 2017, ISBN 978-82-426-2638-7.
[14] Norhard. Norwegian hard rock drilling, In: Pan European networks—Issue 10—May 2014; 2014, https://norhard.no/pan-european-networks-issue-10-may-2014/.
[15] Vereide K. Hydraulics and Thermodynamics of Closed Surge Tanks for Hydropower Plants. (Ph.D. thesis) at NTNU, 2016. p.20, ISBN 978-82-326-1387-820 (electronic version).
[16] Rogner H-H, Aguilera RF, Archer C, Bertani R, Bhattacharya SC, Dusseault MB, Gagnon L, Haberl H, Hoogwijk M, Johnson A, Rogner ML, Wagner H, Yakushev V. Energy resources and potentials. In: Global energy assessment—toward a sustainable future. Cambridge\New York, NY: Cambridge University Press; 2012. p. 423–512. The International Institute for Applied Systems Analysis, Laxenburg, Austria.
[17] International Energy Agency, 2017: Key world energy statistics 2017. https://www.iea.org/publications/freepublications/publication/KeyWorld2017.pdf.
[18] World Commission on Dams. Dams and development: a new framework for decision-making, http://staging.unep.org/dams/WCD/report/WCD_DAMS%20report.pdf; 2000.
[19] International Hydropower Association. Hydropower sustainability assessment protocol November 2010, ISBN 978-0-9566228-1-5, https://www.hydropower.org/hydropower-sustainability-assessment-protocol.
[20] International Energy Agency, 2012: Technology roadmap hydropower 2012. https://www.iea.org/publications/freepublications/publication/2012_Hydropower_Roadmap.pdf.
[21] IRENA. REthinking Energy 2017: accelerating the global energy transformation, Abu Dhabi: International Renewable Energy Agency; 2017, ISBN 978-92-95111-06-6 [PDF], http://www.irena.org/publications/2017/Jan/REthinking-Energy-2017-Accelerating-the-global-energy-transformation.
[22] IRENA. Renewable Power Generation Costs in 2014; ISBN 978-92-95111-53-0, http://www.irena.org/publications/2015/Jan/Renewable-Power-Generation-Costs-in-2014.
[23] Harby A, Sauterleute J, Korpås M, Killingtveit Å, Solvang E, Nielsen TK. Pumped storage hydropower. In: Transition to renewable energy systems. Weinheim: Wiley-VCH; 2013. p. 597–617. ISBN 978-3-527-33239-7.
[24] IEA Hydropower, 2016: IEA hydro annex IX: the economic value of energy and water management services provided by multipurpose hydropower projects (hydropower services) summary report.
[25] Bakken TH, Modahl IS, Raadal HL, Bustos AA, Arnøy S. Allocation recommendations for multipurpose reservoirs—results for water consumption. Water Policy 2016; 18:932–47. https://doi.org/10.2166/wp.2016.009.

[26] WCED. Our common future. Report from the United Nations World Commission on Environment and Development (WCED). Oxford: Oxford University Press; 1987.
[27] International Hydropower Association, http://www.hydrosustainability.org/; 2017.
[28] Modahl IS, Raadal HL, Gagnon L, Bakken TH. How system boundaries affect the energy indicator results for different electricity generation technologies. Energy Policy 2013;63.
[29] Raadal, H.L., Modahl, I.S. and Bakken, T.H.: Energy indicators for electricity production comparing technologies and the nature of the indicators energy payback ratio (EPR), net energy ratio (NER) and cumulative energy demand (CED), 2012, ISBN 978-82-7520-666-2
[30] Bakken TH, Killingtveit Å, Alfredsen K. The water footprint of hydropower production—state of the art and methodological challenges. Glob Challenges 2017;1:1600018, https://doi.org/10.1002/gch2.201600018.
[31] Hamududu B, Killingtveit Å. Assessing of climate change impacts on global hydropower, Energies 2012;5(2):305–22. http://www.mdpi.com/1996-1073/5/2/305.
[32] Milly PCD, Dunne KA, Vecchia AV. Global pattern of trends in streamflow and water availability in a changing climate. Nature 2005;438:347–50.
[33] Hamududu B, Killingtveit Å. Hydropower production in future climate scenarios; the case for the Zambezi river. Energies 2016;9:502. https://doi.org/10.3390/en9070502. http://www.mdpi.com/1996-1073/9/7/502/pdf.
[34] EIA. International energy outlook 2017, https://www.eia.gov/outlooks/ieo/pdf/0484(2017).pdf;2017.
[35] World Energy Council. World energy resources, hydropower 2016, https://www.worldenergy.org/wp-content/uploads/2017/03/WEResources_Hydropower_2016.pdf; 2016.
[36] IHA. Hydropower pumped storage tracking tool, https://www.hydropower.org/hydropower-pumped-storage-tool; 2017.
[37] Killingtveit, Å.: Hydroelectric power. In: Letcher, T.: Future energy. improved, sustainable and clean options for our planet. Elsevier ISBN 978-0-08-099424-6
[38] Gagnon L. Civilisation and energy payback. Energy Policy 2008;36:3317–22.

Solar energy

9

Lee Phillips[1]
Alogus Research Corporation, McLean, VA, United States
[1]Corresponding author: lee.phillips@alogus.com

I'd put my money on the sun and solar energy. What a source of power! I hope we don't have to wait until oil and coal run out before we tackle that.

Thomas Edison

Chapter Outline

9.1 What is solar energy?

Since most things that happen on the Earth can be traced ultimately to energy supplied by the sun, the term *solar energy* is in danger of expanding to become so general that it loses its usefulness. For example, as wind is caused by heating of the atmosphere by the sun, wind power may be considered a form of solar energy. Burning plants or biofuel might equally be counted as a form of solar energy, since plants use visible light as their source of energy and cannot grow without it. It is more useful for us, however, to put these things in different categories. For the purpose of this chapter, we follow the conventional usage, and confine "solar energy" to refer to the exploitation of direct solar radiation, by any means. This therefore includes the conversion of light to electrical power by solid-state *photovoltaic* devices, as well as the direct heating of water or structures by incident infrared energy from the sun.

In fact, the deliberate manipulation of solar infrared radiation by humans is the earliest example of the use of solar energy. It is particularly relevant to dwell on this a bit, in the current context, because the careful husbandry of solar thermal energy in ancient Greece was bound up with the attempt to ameliorate an environmental problem.

In the fifth century BCE, the Greeks found themselves in the midst of an energy crisis [1]. They were running out of fuel: not oil or gas, but wood from trees, which, largely in the form of charcoal, was their source of energy for cooking and heating

Managing Global Warming. https://doi.org/10.1016/B978-0-12-814104-5.00009-0

their houses. The citizenry had burned up most of their usable forests, and, in desperation, and living under a charcoal-rationing regime, were beginning to eye the olive groves.

The authorities addressed this problem by turning to a solution described by Socrates. The great philosopher had written [1] that "in houses that look toward the south, the sun penetrates the portico in winter, while in summer the path of the sun is right over our heads, and above the roof, so that there is shade." He didn't invent the idea, but was reminding his listeners how Greek architecture had traditionally exploited the different paths of the Sun through the sky at different times of year.

The Greeks addressed their energy crisis by arranging the layout of their cities to ensure that each house could take advantage of solar radiation in the way Socrates described. The combination of a proven technology with enlightened government policy worked, and a crisis was averted.

The present volume, and this chapter, exists because we now face a similar crisis at a global scale. The predominant fuels, oil, coal, and natural gas, are not only finite and becoming steadily more scarce, but even in abundance their combustion is altering the composition of the atmosphere and, thereby, causing a significant and measurable effect on the climate. This, in turn, has already begun to harm the quality of life of many populations, human and nonhuman, through the rise in sea level, the increase in intensity of storms and weather volatility, altered migration and predator/prey relationships, and the melting of glaciers and permafrost [2]. It remains to be seen whether an enlightened response on the part of the world's governments will be sufficient and timely enough to mitigate the effects of climate change before the advent of an irreversible disaster. In this chapter, we attempt to inform this process by examining the potential of the adoption of solar energy to reduce greenhouse gas emissions.

In the following sections we look at the past, present, and projected future rates of adoption of solar energy; some recent research in photovoltaics and the resulting improvements in efficiencies and potential for utilization; and, finally, the potential of solar energy to reduce greenhouse gas emissions by replacing a portion of fossil-fuel combustion. First, however, let us complete our answer to the question, "what is solar energy," with a survey of some of its forms.

The ancient Greek practice of designing and situating houses to take the best advantage of sunlight is a technique that is periodically rediscovered in times of fuel scarcity. European colonists in New England borrowed the Greek homebuilding techniques to keep warm in the harsh winters. A resurgence of interest in these architectural practices, under the umbrella term "passive solar," occurred as heating oil and electricity prices began to increase in recent decades.

The next level of sophistication is the construction of systems for storing heat in a fluid medium, usually water, and circulating the fluid through heat exchangers in order to heat living spaces and/or provide hot water. The concept is simple, and the implementation runs the gamut from the black-painted barrels that were sold commercially in the United States in the late 19th century, to more elaborate solar heating systems [3] that pipe water through absorbing and/or focusing panels, with the hot water stored in an insulated tank until needed. In climates where the temperature makes excursions below 0°C, plain water cannot be used, as the expansion of freezing water will damage system components. In this case, a two-fluid system is used, with solar radiation heating a water/antifreeze mixture, or other fluid with a low freezing point that passes through coils embedded in the storage tank, which does double duty as a heat exchanger. Currently, a large variety [4] of sophisticated commercial systems are available for both water and space heating in the home. Thermal solar energy systems have always been deployed throughout the world; lately, the largest installed base per capita can be found [5] in Austria, Cyprus, and Israel.

Paradoxically, perhaps, solar thermal energy can also be used to cool living spaces, with a technique [6] invented by the Romans, and still used in some tropical areas with low winds. A chimney is designed, through situation and surface treatment, to absorb the maximum of incident solar infrared radiation. This heats the air column contained within, and causes it to rise due to buoyancy. The resulting lower pressure within the dwelling can be contrived to draw hot, ambient air through a tube buried beneath the ground; the resulting inflow of geothermally cooled air lowers the air temperature within the house.

Moving up the scale from home-sized installations to utility-scale power generating facilities, we find a form of solar thermal energy used for power generation. Most of these facilities use some variation on the strategy of focusing solar radiation on a working fluid, which is then used to create steam to operate a conventional turbine that generates electricity. There are a handful in the United States, for example, which generate from 100 to 400 MW [7]. These plants come in several different configurations. Linear systems use a field of troughs in the form of parabolic cylinders, with a pipe containing the working fluid mounted at the focus (which is a line, rather than a point). The reflectors rotate along their long axis to track the sun and can create working temperatures of 400°C [7]. Another prominent configuration is that of the *solar power tower*. Here, a field of sun-tracking mirrors combine their effects to focus sunlight upon a vessel mounted atop a tower. The largest solar thermal plant in the United States uses this technology: the Ivanpah Solar Electric Generating System, in the Mojave Desert of California, uses its 347,000 mirrors to generate a peak of 392 MW of electrical power [8].

Many solar thermal plants feature at least a partial solution to the problem of the intermittency of their power source, sunlight. The heat that they produce in the working fluid can be stored, either in the working fluid itself or in a secondary medium, and continue to produce power at night or during sun-occluding weather. This energy storage, which, in photovoltaic systems, must be accomplished with a separate device (electrochemical storage batteries, compressed gas, water pumped to a higher elevation, etc.), can be made an integral part of a solar thermal plant. The working fluid can be water, but in advanced designs it is more likely to be a high temperature molten salt or even a granular material [8].

The other major type of solar energy technology, after solar thermal technology, is photovoltaic. As mentioned earlier, this involves materials that convert incident light into electrical energy, in the familiar form of electron currents. Photovoltaic technology is an indispensable part of the space program. The ubiquitous solar panel arrays can be found on everything from communications satellites to space telescopes; from planetary probes to the Mars rovers.

The first space-bound solar panels date from the early days of photovoltaic technology. In 1958, the Vanguard 1 [9], a small (by today's standards) Earth-orbiting satellite, was launched. Its purpose was to test the concept of launching a satellite into Earth orbit using a three-stage rocket, and to see whether the device could survive launch and the space environment, and also to make geodetic measurements.

Vanguard was only the second satellite launched by the U.S., and the first to employ solar panels. But the panels almost didn't make the trip. Researchers in the U.S. Army Signal Corps had become aware of Bell Telephone's early success with photovoltaic cells and wanted to use them in the satellite [10]. But the Navy, which had been put in charge of the program, felt that the technology was too immature and untested. The Corps, and one scientist in particular, Dr. Hans Ziegler, were so passionate about outfitting Vanguard with solar cells, however, that the Navy finally relented, but under the condition that a battery also be included. As it turned out, the battery died after 19 days [10], (the Sputnik batteries had died after a week) while the solar cells kept the satellite's transmitter alive for 5 years [9]. Vanguard is still in orbit—the oldest artificial satellite orbiting the Earth.

The Vanguard project was one of the gateway events that brought photovoltaic technology to the fore. It not only increased public awareness of the technology but gave greater impetus to research. This resulted in a steady increase in the efficiency of solar cells, along with a steady decline in their cost. Today research is underway in laboratories all around the world into photovoltaic cells, panels, and materials. We'll pick up this topic in a later section.

9.2 Solar energy adoption

The past 50 years have seen the role of solar cells expand from curiosities and toys to a major component of the world's energy systems. When the first photovoltaic solar cell was brought to life [11] by Charles Fritts in 1883, he spoke of his invention eventually competing with Thomas Edison's expanding population of coal-burning power plants. Due to the 1% efficiency [11] of these early cells, however, it would be many decades before this dream started to put down roots in the real world.

By 1954, Bell Labs, in a public demonstration [11], hooked a silicon photovoltaic panel to a toy Ferris wheel and a radio transmitter. The device was a true "panel," with several individual cells connected together to form a solar "battery." This device was 6% efficient: not impressive by today's standards, but an enormous advance for the times. Although it was still too expensive to manufacture for commercial adoption, the *New York Times* was duly impressed [12], gushing that the new invention, made from a "sand ingredient," "may mark the beginning of a new era, leading eventually to the realization of one of mankind's most cherished dreams—the harnessing of the almost limitless energy of the sun for the uses of civilization." The article points out that the energy from the sun is greater than that "of all the reserves of coal, oil, natural gas, and uranium in the Earth's crust", and that the scientists' claim of 6% efficiency compares favorably with steam and gasoline engines.

Bell Labs' new solar technology would find its niche in applications such as satellites (including the Vanguard discussed above) and remote, low-power devices, especially in environments, for example those of high humidity, where electrochemical batteries would degrade quickly.

Solar panels would remain in their niches for quite some time. Even during the energy crisis of the 1970s, when there was an urgent search for alternatives to oil, the efficiency of commonly available solar cells was still in the neighborhood of only 10%, and they were too expensive to be competitive even with the increasingly expensive fossil fuels.

However, due to continuing research throughout the second half of the 20th century, photovoltaic conversion efficiency (the ratio of incident radiative power to electrical power produced, under a standard set of measurement conditions) increased, and manufacturing costs decreased, to the point where photovoltaic panels began to be seen as a viable option by both individual households and electrical utility companies. Using the United States as an example, the consumption of electrical power from solar energy remained fairly stagnant until the last decade, when the combination of lower cost and higher efficiency contributed to a tipping point, after which we can observe exponential growth. Fig. 9.1 shows solar energy consumption in units of 10^{12} BTU (British Thermal Units) *versus* year, from 1989 to 2016. Note that 1 BTU = 1.055 kJ. The exponential fit, shown by the dashed line, to the data beginning in 2010 is very close, with a doubling time of 2.18 years.

In isolation, the data in Fig. 9.1 do not indicate an increasing relative adoption rate for solar energy, because the decade of exponential growth corresponds to a rapid increase in energy consumption overall. This is shown in Fig. 9.2, which plots the U.S. energy consumption from all sources in units of 10^{15} BTUs (solid line). The thick dashed line shows the percentage of this total that is due to solar energy, with the last decade's data fit by an exponential function shown by a thin dashed line. We can see that the relative contribution of solar energy rises exponentially even against the background of rapidly increasing total energy consumption, although with a somewhat longer doubling time of 2.26 years.

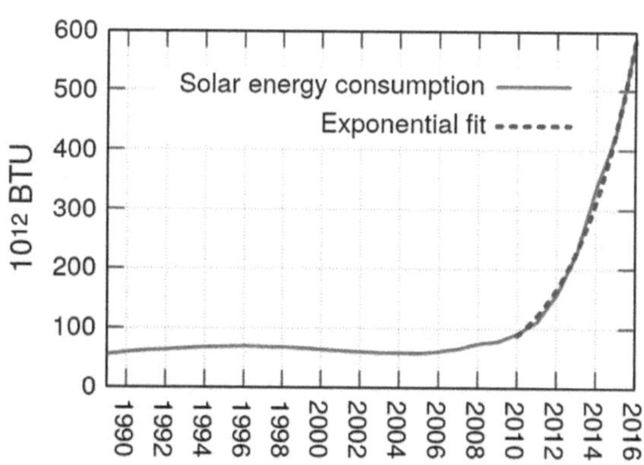

Fig. 9.1 Yearly consumption of solar power in the U.S., with an exponential fit beginning in 2010. Here 1 BTU = 1.055 kJ. Data from the U.S. Energy Information Administration.

There is a fundamental difference between solar power generation and most other forms. People generally do not have personal coal-fired power plants, or backyard natural-gas-powered electrical generating stations, with the exception of small emergency backup generators. However, electricity production, at the scale of individual households, from rooftop or similar-scale solar panel installations, is a significant and growing phenomenon. In order to compare solar energy consumption with that from other sources, statistics are typically compiled that are relevant to utility-scale power generation, and that is the case with the data that we have graphed in Figs. 9.1 and 9.2. The data for these plots are taken from tables compiled by the U.S. Energy Information Administration, which defines utility-scale power generation as employing a facility that generates a nominal power in excess of 1 MW. This naturally excludes household rooftop solar panel installations, and similar facilities. Fig. 9.3 compares solar energy consumption (which is essentially equal to solar energy generation, unlike the case of fossil fuels) at utility scale with "distributed" solar energy, which includes installations with a nominal power under 1 MW. We can see that distributed solar energy consumption was actually somewhat greater than consumption of utility-scale solar energy for several recent years until 2013, when the construction of large solar plants caused utility-scale solar energy to rapidly outstrip the power consumed

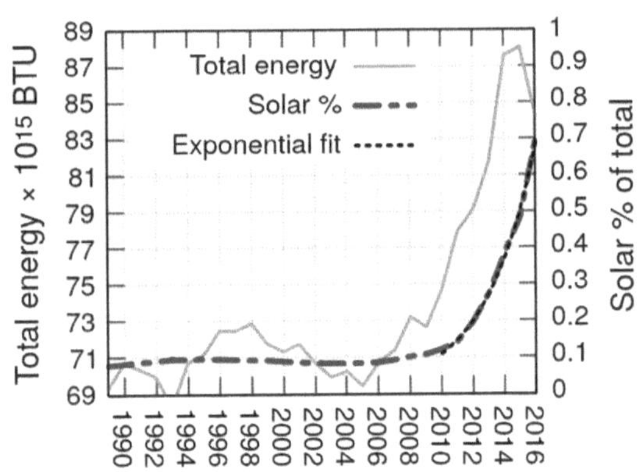

Fig. 9.2 Total consumption of energy in the U.S. compared with solar energy consumption, with an exponential fit beginning in 2010. Here 1 BTU = 1.055 kJ.
Data from the U.S. Energy Information Administration.

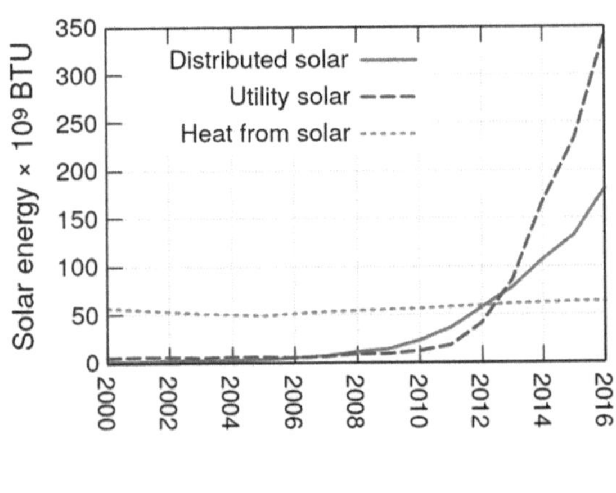

Fig. 9.3 Distributed and utility-scale solar energy. Here 1 BTU = 1.055 kJ.

from rooftop solar panels and the like. These two curves represent data [13] on electrical generation only. Plotted with them are data on solar heat generation through methods other than electricity, including solar water heating, space heating, etc. Interestingly, this value holds steady at about 6×10^{13} kJ (60 billion BTUs) while electrical conversion of sunlight increases dramatically.

The upshot of this is that in comparisons such as the one illustrated in Fig. 9.2 the percentage of solar energy should be considered a lower limit, and, up to about 2013, may be approximately doubled. If the trend of utility-scale solar energy outpacing distributed solar continues to increase in future years, then direct comparisons among different forms of electrical power utilities will become more meaningful.

The growing adoption of solar energy is due not only to the cost and efficiency factors mentioned earlier, but is doubtless also driven in part by the various tax incentives and other programs available at the local, state or provincial, and federal level. For the individual home owner, these incentives amount to lowering the effective cost of a rooftop solar panel installation. In one U.S. county alone, a targeted discount program resulted in 47 people installing rooftop solar panels, with each household saving an estimated average $1250 per year on their electric bills, and a total effective reduction of 206 metric tons in carbon dioxide emissions [14].

In the case of the United States at the federal level, there are considerable incentives built into the tax code that make residential solar power more affordable. Current law applies a 30% tax credit [15] for the installation of new solar energy systems. The credit remains steady until 2019, after which it deceases, reaching 22% in 2021, when the provision expires. Because this is a credit, rather than a deduction from income, it amounts to a net 30% discount in the cost of installation, with no upper limit, creating a significant incentive. It applies to solar water heaters, with some conditions, and to any photovoltaic panel system [16]. The tax incentive has already [17] had a large stimulating effect on solar panel research and development, and to residential adoption. Historically, tax incentives that have been passed with expiration dates have been extended [17]; the declining credit schedule in current law will necessarily be revisited by Congress every year.

In addition to the substantial federal credit, incentives are written into the tax codes of many U.S. states. It is impossible to summarize the situation, as each state has independent tax laws, and the solar energy incentives vary widely. For example, as of 2017, New York State [18] has a 25% tax credit, while New Jersey has no tax credit. However, New Jersey has a sales tax exemption for solar hardware, as does New York. Other incentives that vary by state, and sometimes by locality, and can include: property tax exemptions that exclude the increase in the assessed value of your newly solar-outfitted house, due to its new solar panels, from the consequent increase in property taxes; performance payments, which are a direct, mandated payment from the utility company proportional to the power generated by your panels (called "net metering"); and rebates on the prices of solar equipment and installation, usually paid by the utility as well.

In most countries, regulations mandating net metering are at the national, rather than the provincial level. This incentive is written into the regulations of many countries, including [19] Greece, Italy, Japan, the Netherlands, South Korea, Spain,

Uruguay, Argentina, Brazil, Mexico, and the Philippines. Although such governmental incentives doubtless play a significant role in encouraging the adoption of solar energy for individual households, at the level of utility-scale investment, they may no longer be necessary. Photovoltaics have become the largest market [19] for new investments around the world, due mainly to declining manufacturing costs, despite still being a relatively minor producer of electrical power compared with other renewables, such as hydropower. These declining costs have made solar panels cost competitive, in a "natural" market without subsidies, with fossil fuels in many areas around the world. In fact, solar power's unprecedented low cost per unit energy delivered is driving a renewed intensity of investment and deployment; the world record for the cheapest electricity anywhere of any kind is 1.77¢ $(kWh)^{-1}$ from photovoltaic power [20], contracted by an Italian firm with Mexico and due to supply power to the Mexican grid by 2020. It is predicted [20] by industry experts that electricity from solar plants will continue to decline in price, reaching 1¢ $(kWh)^{-1}$ by 2019.

9.3 Barriers to solar energy adoption

At the level of individual homeowner decision making, there are several barriers to the adoption of solar energy, even in cases where the local climate, utility electrical prices, and the regulatory environment may make the installation of a rooftop panel system a rational choice. The first, obvious potential obstacle is the significant capital cost, which typically is in the neighborhood of US$20 000. While this outlay may be subsidized through several avenues or financed attractively, it is still a major barrier to more widespread adoption of distributed solar energy.

Even if this financial barrier can be overcome, there remain other potential obstacles. The homeowner may lack the information or the ability to perform the financial calculations that would otherwise lead to a rational decision to install solar panels, thereby possibly saving money in the long run: the obstacle here is simple lack of information, possibly combined with innumeracy [21].

Even when a solid financial case cannot be made for a homeowner to install solar panels, he or she may choose to do so for several other reasons. There are anecdotal reports [22] of homeowners deriving psychological benefits from the sensation of achieving independence from the public utility for power. Many homeowners also report that their decision to install was influenced by the desire to contribute to the amelioration of the global climate change problem. Here we come to another set of barriers to solar energy adoption, however. In the United States, in particular, there is a resurgence of hostility and suspicion, in some quarters, directed against the message of scientists about anthropogenic climate change, or simply a counterfactual, largely politically motivated, denial that a scientific consensus exists. There is also simple ignorance: a recent survey revealed [23] that, while greater numbers of Americans accept the reality of anthropogenic climate change than before, only 13% know that "nearly all climate scientists (more than 90%) are convinced that human-caused global warming is happening." It seems likely that at least some of these Americans, were they better informed, would become part of the population that reports concern over global warming as a factor in their decision to install a solar energy system; therefore, their lack of knowledge in the aggregate can be considered a barrier to more widespread adoption of solar energy.

Solar power, along will all other forms of energy that do not consume fossil fuels, competes directly with the economic interests of corporations that profit from the extraction and sale of oil, coal, and natural gas. Every joule of energy that is consumed from a renewable resource replaces a joule of energy that would otherwise need to be made available through chemical combustion, usually of a fossil fuel, or nuclear energy. It may not be surprising, therefore, to learn that the fossil-fuel industry has invested in targeted lobbying campaigns [24] for the purpose of suppressing the adoption of renewable energy, including solar energy. In the United States, this takes the form of direct lobbying of state legislators with the goal of reversing existing incentives in state law that have the effect of encouraging installations of solar panels. It also takes the form of propaganda campaigns [24] designed to inject the idea that net metering discriminates financially against homeowners who are not equipped with solar panels. The recent decisions of the state legislatures in Hawaii, Arizona, Maine, and Indiana to eliminate those states' net metering mandates have been attributed [24] to these lobbying efforts. Even in the absence of a legislative victory, the doubt about the future of net-metering programs raised by lobbying and propaganda efforts are sufficient to dissuade many homeowners from making the substantial investment in rooftop panels, as their financial calculations may assume that the net-metering mandate will remain in force for a substantial fraction of the panels' lifetime.

The situation at the federal level in the United States may be even more dire for the future of solar energy adoption. The current leadership of the Department of Energy shows every sign of being unfriendly to renewable energy in general: the chief of staff of the Energy Secretary is a former high-level executive at a leading fossil-fuel lobbying firm [24], and the Secretary himself has been forthright about his concern for the possible harm that renewable energy could do to the fossil-fuel business.

9.4 Research in solar devices

In the foregoing discussion, we have repeatedly referred to the increasing efficiency and decreasing cost of photovoltaic cells, panels, and materials. But what drives this ever greater efficiency and lower cost? It is entirely due to applied research at laboratories all over the world. In this section, we give a brief survey of some of the most promising recent research [25] into photovoltaics and other solar energy technologies. In the interests of space, we focus on a few key technologies of interest, while acknowledging the important research underway in all areas of solar technology, including concentrated solar power [26].

The photovoltaic effect, the creation of an electron current in a material by the action of light, was first discovered in 1839 by French physicist Alexandre-Edmond Becquerel (Fig. 9.4). Becquerel was a member of a physicist family: his father, Antoine Cesar Becquerel, was a physicist, and his son, Henri Becquerel, was to become famous for discovering radioactivity, and win the Nobel prize. The modern

Fig. 9.4 Alexandre-Edmond Becquerel.
From Figuier, Louis, Les Merveilles de la Science, Vol. 3, p. 73, 1867 (public domain).

photovoltaic cell, made from silicon, was invented in 1954 at Bell Telephone Laboratories, and launched the modern era of solar power.

Much of the research since the Bell Telephone invention has been aimed at increasing the efficiency of silicon cells, simplifying their manufacture, or searching for new materials that also exhibit useful photovoltaic properties. The silicon solar cell is a semiconductor device, similar to a solid-state diode. It can absorb a fixed amount of energy from a photon and use it to move an electron into the material's conduction band; this energy is in the visible light region of the electromagnetic spectrum. Photons with smaller energy (infrared, for example) are reflected or become heat, and larger energy photons (ultraviolet, etc.) have some of their energy wasted. The result is that even for a perfectly manufactured cell operating under ideal conditions, the theoretical limit to photovoltaic efficiency is about 33%. This is exceeded, for example in the comparatively exotic solar panels used in spacecraft, by manufacturing a "multi-junction" cell [27] that can exploit a wider energy band; such cells can achieve efficiencies of 40%.

Some of the most fruitful avenues of research into photovoltaic materials have involved the search for alternatives to the classic silicon cell. One of the most promising alternatives is a class of materials called *perovskites*. Perovskites are synthetic compounds that have an orthorhombic crystal structure identical to the naturally occurring mineral with the same name and that share a structurally similar chemical formula. Depending on the particular elements composing the material, perovskites can feature [28] a wide variety of useful properties, such as superconductivity and photovoltaic activity. Their use in photovoltaic cells is widely considered to be a promising avenue of research, as they have shown an unprecedented increase in efficiency [29] from 3.8% to 20.1% in just 7 years of development. Further gains seem likely, as recently the factors limiting efficiency have been identified [29]. These materials are not expected to exceed the efficiency of silicon cells, and indeed have a lower maximum theoretical efficiency of 31%, but are being investigated because of their potential to be easier and cheaper to manufacture, as well as their ability to be incorporated into novel materials and applications. One example of this is the recent development of a photovoltaic window [30] that could be used in office buildings. Normally nearly transparent (68% transmittance in the visible), exposure to sunlight causes the window to become nearly opaque (less than 3% transmittance). While in the opaque phase, the window generates solar power with an 11% efficiency. Innovative materials research such as this can only accelerate the adoption of solar energy, as they widen its scope of application even into areas where power generation can be seen as a beneficial side effect of a device's primary utility.

Ease and convenience of application is also a motivating factor in the search for new photovoltaic materials. There is great attraction in the idea of solar power collectors in the form of paint-like coatings or transparent films. If such materials become practical, this will aid the adoption of solar energy, as they might be used in situations where rigid and unwieldy solar panels are impractical.

An example of a solar paint is a substance developed [31] at the University of Notre Dame in the United States, made from semiconductor nanoparticles. Although it only exhibits 1% efficiency, it can be applied like ordinary paint, which means it should have a wide variety of applications.

Earlier we divided solar energy devices into two classes: photovoltaics and solar thermal energy systems. However, there is another class of device, which may need its own classification. The *solar thermal fuel* stores solar energy internally and stably over a long period of time, and can release it as heat when exposed to a catalyst or triggering heat [32]. It works through nonchemical transformations of its molecules: in response to solar radiation, the molecules transition into *photoisomers*. The photoisomeric form has the same chemical formula, but a different structure, in which the light energy is stored as added energy in the intramolecular bonds, creating a higher-energy state of the original molecule. When triggered by the catalyst, the molecules transform back to their original shape, releasing the stored bond energy as heat. This heat can be used directly, or converted to electrical power with a thermoelectric generator or heat engine. Since energy storage is integral to the system, no batteries are required, and the fuel can be transported as needed, allowing the stored energy to be used elsewhere.

The research into solar thermal fuels has recently led to the development of a solar thermal fuel in the form of a transparent film [33]; it can be applied, for example, to car windshields, where it would be able to melt ice at night using solar energy absorbed during the day. The impressive rate of progress in of solar thermal fuel research inspires confidence that they will soon make the leap from the laboratory to become a quotidian technology.

Photosynthesis in plants and bacteria involves the absorption of light and atmospheric carbon dioxide. Scientists working at several laboratories have managed to create artificial forms of photosynthesis, which have the advantage of not only replacing greenhouse-gas-emitting combustion, but actually removing CO_2 from the atmosphere as an integral part of their operation. The resulting devices are sometimes called "artificial leaves." One version [34] stores light energy not in carbohydrates, as do plants, but in a fuel called "syngas," which is a mixture of hydrogen gas and carbon monoxide; syngas can be burned directly or turned into more conventional fuels.

9.5 The potential of solar energy to reduce greenhouse gas emissions

Solar power, despite its recent rapid growth in adoption, and the enthusiastic support it receives both from the scientific establishment and energy investors, remains a very small proportion of total energy consumption (see Fig. 9.2). Its potential to drive a significant reduction in the emissions of greenhouse gasses lies in a potential future. If the exponential trend of the last decade continues, solar power will replace a sufficient quantity of fossil-fuel combustion to play a large role in reaching humanity's climate goals [35].

In fact, if the doubling time in the proportion of solar energy to total energy consumption of 2.26 years (see Fig. 9.2) remains the same, then solar energy will reach 100% of the total in approximately the year 2032, as illustrated in Fig. 9.5, where we have adopted the use of a log scale, where the exponential function appears as a straight line. The figure simply extends the exponential fit to the data into the future, assuming a constant growth rate.

Obviously, in this case, the effect of solar energy on the global climate problem would be profound, as it would have completely replaced greenhouse-gas-emitting combustion of fossil fuels in another 15 years. Although such a naive extrapolation obviously needs to be tempered by a serious encounter with reality, such thinking appears to be taken seriously in some quarters [36]. The first obvious problem with this projection is that it predicts that solar energy will not only have replaced fossil fuels, but other forms of renewable energy as well—and there is no reason to expect solar power to supplant all sources of hydropower, wind power, etc.

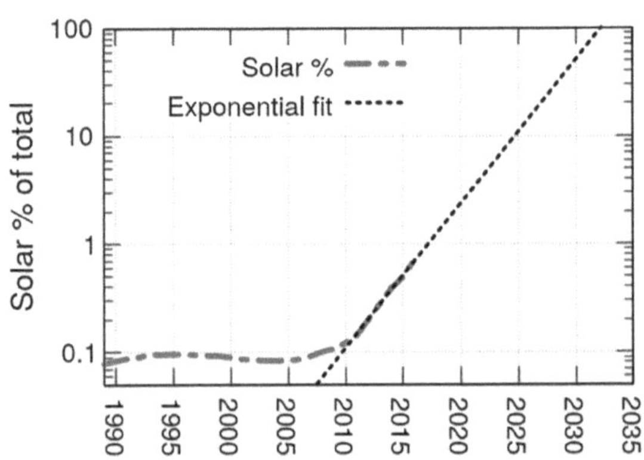

Fig. 9.5 Naive projection of solar energy adoption.

The observation of exponential growth leads naturally to the supposition of a model that can be expressed by the differential equation

$$\frac{dS}{dt} = bS$$

where, in this case, S is the solar energy fraction and b is a growth rate. The meaning of this equation is that the time rate of change of S is proportional to S itself. This has an obvious application in the modeling of populations, where the number of new organisms created every generation is proportional to the number of reproducing organisms at any time. If an equation of this form tells us anything about the mechanism underlying the observed exponential growth of solar adoption, it suggests that we should look for reasons that the rate of growth should be proportional to the installed base at any particular time. Since solar panels do not reproduce themselves, we need, if we are to proceed along this line of speculation, to surmise the cause for dS/dt to be proportional to S. One plausible mechanism may be a kind of reassurance created by the appearance of rooftop solar installations in a neighborhood: a homeowner may not seriously consider investing in solar panels until he is encouraged by their appearance on his neighbors' houses, and motivated by their recounting of a good experience with them.

No matter the underlying reasons for the exponential growth, however, it is clear that it cannot continue indefinitely. As in biological populations, there will be factors that limit growth. In the case of solar energy, these factors, in addition to the political pressure applied by industry lobbying groups mentioned earlier, include rooftop market saturation, where the limited amount of rooftop area situated in desirable weather regions becomes already occupied by panels; cheap, "green" energy supplied to the grid from other renewable sources; reduction in retail electricity prices due to a temporary high supply of natural gas or other fossil fuel, diluting the motivation for investing in solar panels; and any number of unforeseeable factors.

In order to include growth-limiting factors, the differential equation may be modified with an additional term:

$$\frac{dS}{dt} = \frac{b(L-S)}{L} S$$

The new factor L represents the finite capacity of the environment to support (in this case) solar power installations. Note that when S is small, during the early stages of growth, then $L - S \approx L$ and the increase is exponential, but the growth rate decreases and goes to zero as S approaches L.

Although it is certain that L is finite, because of the multitudinous, unforeseeable factors that contribute to its effective value, it is impossible to predict for how long the recently established exponential growth of solar energy adoption will continue, or at what value it will level off. In other words, the future of solar energy adoption is likely to resemble Fig. 9.6, which plots the solution to our second differential equation (sometimes called the *logistic equation*) with L arbitrarily chosen to result in

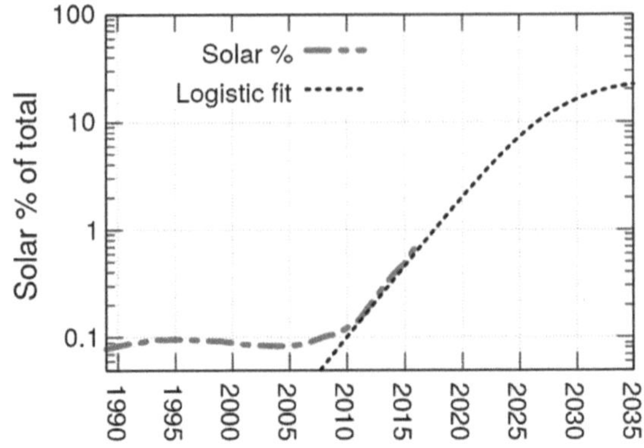

Fig. 9.6 Projection of solar energy adoption based on the logistic equation, with an assumed saturation level of 25%.

saturation at about 25%. Once we enter the phase of saturation, or slowing down of solar energy adoption, it will be possible to estimate L and make an informed prediction of the future; but as we are, according to the data, in the early, exponential phase of solar energy growth, where S is much smaller than L, it is at present impossible to predict the level at which saturation will occur. Consequently, since the equilibrium level of solar energy deployment is impossible to estimate, a quantitative estimate of the amount of greenhouse gas emissions that will ultimately by replaced by solar energy is also not possible, at this still early phase of adoption. It seems clear, however, that solar energy will be a significant portion of the mix of renewables in the future, even if naive predictions [36] of complete dominance in 15 years can be discounted.

Partly because of the suppression efforts by the fossil-fuel industry mentioned, partly due to market saturation in certain regions, and also partly due to the faltering of several manufacturers of solar panels, the exponential growth in solar power consumption may already be showing some initial signs of slowing down [24]. Nevertheless, in 2016 the United States added 15 GW of solar capacity, an increase greater than that from any other source of energy, including fossil fuels.

According to at least one study [37], the world is already on an inexorable path to complete replacement of fossil and nuclear fuels by renewable energy. Solar energy, in its various forms, has a guaranteed place at this table. Solar energy technology has matured to the point where humankind can, if it recognizes the need, extract any part or all of its energy requirements from our star, because, as it is written [38] by the scientists of the Los Alamos National Laboratory, "Sunlight is abundant beyond the energy needs of the entire human race and completely free."

References

[1] Butti K, Perlin J. A golden thread: 2500 years of solar architecture and technology. Cheshire Books 1980.

[2] Haeberli W, Beniston M. Climate change and its impacts on glaciers and permafrost in the alps. Ambio 1998;27:258–65.

[3] Solar water heaters | department of energy. http://energy.gov/energysaver/solar-water-heaters (Accessed 10 November 2017).
[4] Active solar heating | department of energy. http://energy.gov/energysaver/active-solar-heating (Accessed 10 November 2017).
[5] International Energy Agency. Solar heat worldwide markets and contribution to the energy supply. http://www.iea-shc.org/solar-heat-worldwide (Accessed 6 November 2017).
[6] Nugroho AM, Ahmad MH. In: Possibility to use solar induced ventilation strategies in tropical conditions by computational fluid dynamic simulation. Presented at the 6th Sustainable Environmental Architecture Conference, Institut Teknologi Bandung, Bandung, Indonesia; 2005.
[7] U.S. Energy Information Administration. 2016. Solar thermal power plants. https://www.eia.gov/energyexplained/?page=solar_thermal_power_plants (Accessed 10 November 2017).
[8] U.S. D.O.E. Office of Energy Efficiency and Renewable Energy. 2013. Power tower system concentrating solar power basics. https://www.energy.gov/eere/energybasics/articles/power-tower-system-concentrating-solar-power-basics (Accessed 15 November 2017).
[9] NASA. 2017. Vanguard 1. https://nssdc.gsfc.nasa.gov/nmc/spacecraftDisplay.do?id=1958-002B (Accessed 16 November 2017).
[10] Perlin J. From space to earth: the story of solar electricity. Routledge; 1999.
[11] This month in physics history. APS News 2009, 18.
[12] Vast power of the sun is tapped by battery using sand ingredient. New York Times 1954; April 26: p. 1 and 11.
[13] United States Energy Information Administration. 2017. Monthly Energy Review.
[14] Solarize Program Saves County Homeowners $1,250 a Year on Electricity Bills, 2017. http://www.fairfaxcounty.gov/news2/solarize-program-saves-county-homeowners-1250-a-year-on-electricity-bills/ (Accessed 27 November 2017).
[15] Is there a tax credit for solar panels? https://energystar.zendesk.com/hc/en-us/articles/216375477-Is-there-a-tax-credit-for-solar-panels (Accessed 21 November 2107).
[16] Federal tax credits: solar energy systems. https://www.energystar.gov/about/federal_tax_credits/solar_energy_systems (Accessed 21 November 2107).
[17] Martin R. 2015. Tax credit extension gives solar industry a new boom. MIT Technology Review. https://www.technologyreview.com/s/544981/tax-credit-extension-gives-solar-industry-a-new-boom/ (Accessed 21 November 2017).
[18] 2017 United States solar power rankings. Solar Power Rocks. https://solarpowerrocks.com/2016-state-solar-power-rankings/ (Accessed 21 November 2017).
[19] Taxes and incentives for renewable energy. KPMG International, 2015.
[20] Weaver JF. 2017. Cheapest electricity on the planet is mexican solar power at 1.77¢/kWh—record 1¢/kWh coming in 2019, sooner. Electrek. https://electrek.co/2017/11/16/cheapest-electricity-on-the-planet-mexican-solar-power/ (Accessed 22 November 2017).
[21] Paulos JA. 2001. Innumeracy: mathematical illiteracy and its consequences. Holt McDougal.
[22] Phillips L. 2017. The state of residential solar power. Ars Technica https://arstechnica.com/science/2017/02/solar-power-is-it-for-you/ (Accessed 5 February 2017).
[23] Leiserowitz A, Maibach E, Roser-Renouf C, Rosenthal S, Cutler M. 2017. Climate change in the American mind. Yale Program on Climate Change Communication. http://climatecommunication.yale.edu/publications/climate-change-american-mind-may-2017/ (Accessed 22 November 2017).
[24] Tabuchi H. 2017. Rooftop solar dims under pressure from utility lobbyists. New York Times, July 8.

[25] Phillips L. 2017. The future of solar power technology is bright. Ars Technica https://arstechnica.com/science/2017/02/for-a-brighter-future-science-looks-to-re-energize-the-common-solar-cell/ (Accessed 16 February 2017).
[26] Temple J. 2017. To make solar really work, turn up the heat. MIT Technology Review. https://www.technologyreview.com/s/609509/to-make-solar-really-work-turn-up-the-heat/ (Accessed 27 November 2017).
[27] Hou H, Reinhardt KC. 1999. High-efficiency solar cell and method for fabrication. US Patent 5,944,913.
[28] Perovskites and perovskite solar cells: an introduction. https://www.ossila.com/pages/perovskites-and-perovskite-solar-cells-an-introduction (Accessed 27 November 2017).
[29] Miller DW, Eperon GE, Roe ET, Warren CW, Snaith HJ, Lonergan MV. Defect states in perovskite solar cells associated with hysteresis and performance. Appl Phys Lett 2016;109:153902. https://doi.org/10.1063/1.4963760.
[30] Wheeler LM, Moore DT, Ihly R, Stanton NJ, Miller EM, Tenent RC, Blackburn JL, Neale NR. Switchable photovoltaic windows enabled by reversible photothermal complex dissociation from methylammonium lead iodide. Nat Commun 2017;8. https://doi.org/10.1038/s41467-017-01842-4.
[31] Genovese MP, Lightcap IV, Kamat PV. Sun-believable solar paint. A transformative one-step approach for designing nanocrystalline solar cells. ACS Nano 2012;6:865–72. https://doi.org/10.1021/nn204381g.
[32] Kolpak AM, Grossman JC. Azobenzene-functionalized carbon nanotubes as high-energy density solar thermal fuels. Nano Lett 2011;3156–62. https://doi.org/10.1021/nl201357n.
[33] Zhitomirsky D, Cho E, Grossman JC. Solid-state solar thermal fuels for heat release applications. Adv Energy Mater 2015;1502006. https://doi.org/10.1002/aenm.201502006.
[34] Asadi M, Kim K, Liuet C, et al. Nanostructured transition metal dichalcogenide electrocatalysts for CO_2 reduction in ionic liquid. Science 2016;467–70. https://doi.org/10.1126/science.aaf4767.
[35] The Paris agreement. http://unfccc.int/paris_agreement/items/9485.php (Accessed 24 November 2017).
[36] Bushong S. 2016. Futurist Ray Kurzweil predicts solar industry dominance in 12 years. https://www.solarpowerworldonline.com/2016/03/futurist-ray-kurzweil-predicts-solar-industry-dominance-12-years/ (Accessed 26 November 2017).
[37] Kirby A. 2017. 100% renewable electricity in reach by 2050. Climate News Network. https://www.ecowatch.com/100-renewable-electricity-2507753265.html (Accessed 27 November 2017).
[38] Los Alamos National Laboratory. 2015. Perovskite power. http://www.lanl.gov/discover/publications/1663/2015-may/perovskite-power.php (Accessed 19 May 2016).

Wind power: A sustainable way to limit climate change

10

Georgios Nikitas, Subhamoy Bhattacharya*,[1], Nathan Vimalan†, Hasan Emre Demirci*, Nikolaos Nikitas‡, Prashant Kumar**
*University of Surrey, Guildford, United Kingdom, †VJ Tech Ltd, Reading, United Kingdom, ‡University of Leeds, Leeds, United Kingdom
[1]Corresponding author: s.bhattacharya@surrey.ac.uk

Chapter Outline

10.1 Wind among the renewables

During the industrial revolution in the 1800s, fossil fuels (for example, coal in early days and subsequently oil and natural gas) prevailed as the main source of energy by replacing wood and wind. The result was an economic prosperity that led to an exponential increase in population. Furthermore, with the discovery of electricity and the continuous technological advancements, the energy demands have also skyrocketed. Until now fossil fuels have remained as the main source of power with a share of

Managing Global Warming. https://doi.org/10.1016/B978-0-12-814104-5.00010-7

78.4% to the global energy demands [1]. The use of fossil fuels, however, releases large amounts of gases (also known as greenhouse gases) that are associated with climate change. Out of these gases, the largest contributor to climate change is considered to be carbon dioxide (CO_2), the annual emissions of which was calculated by Olivier et al. to be about 36 billion tons for 2015 [2]. Fig. 10.1 shows a time history of growth in population together with annual energy demands and CO_2 emissions worldwide since record began in 1800. It may be observed that the growth of these aspects follows a pattern, which strongly suggests a connection between the CO_2 emissions and the human activities.

The oil crisis in the Middle East during the 1970s together with the major concern of climate change in 1990s and 2000s has sparked a major interest in alternative sources of energy. This is in order to disengage from fossil fuels and decarbonize the economy. The reduction of CO_2 emissions can be achieved by switching to low-carbon energy sources, such as the hydropower, wind, and solar energy, also known as renewable sources of energy. Just to get an idea of how much lower emissions the renewables produce, Fig. 10.2 graphically presents the lifecycle greenhouse emissions range for all various sources of energy. A comparison shows that the emissions produced by renewables are insignificant compared to the nonrenewables and they are mostly associated with the development stage of their infrastructure.

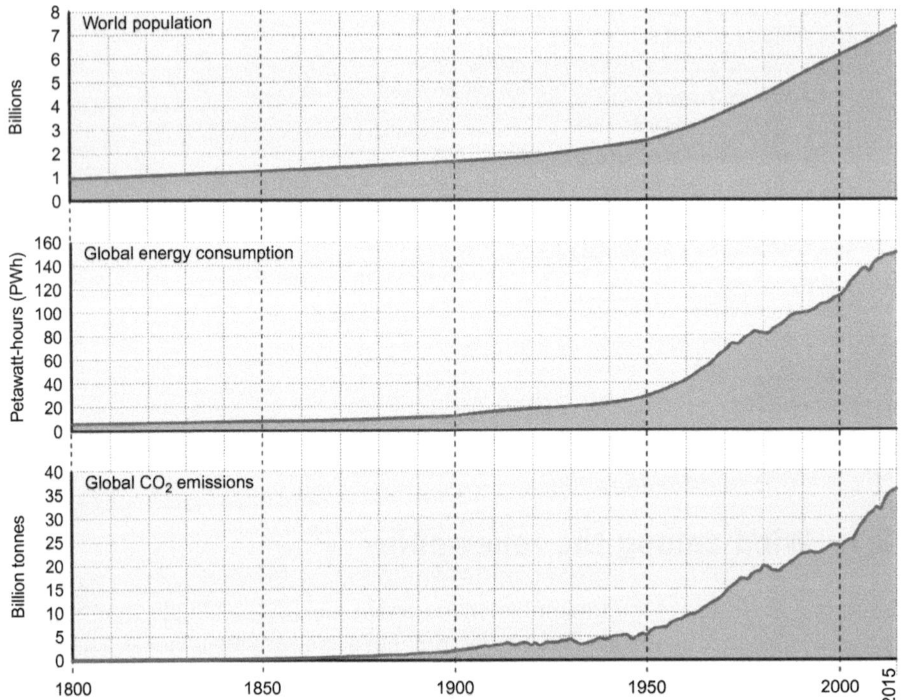

Fig. 10.1 Evolution of population, energy consumption, and CO_2 emissions worldwide (1800–2015).

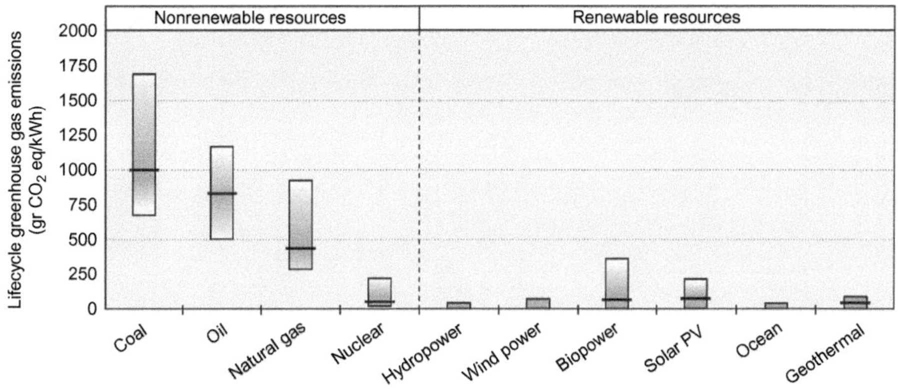

Fig. 10.2 Lifecycle greenhouse emissions range for all the various sources of energy. *Source*: NREL. Opportunities for synergy between natural gas and renewable energy in the electric power and transportation sectors, 2012.

Since 1980s, many countries worldwide have started to invest in greener ways to produce energy by utilizing renewable resources. However, it was not until early 2000s that the share of renewables started to increase significantly. This is not surprising that today the share of renewables in the total energy demand is approximately 20%. However, renewables did not increase equally within the three energy sectors: electricity, transportation, and heating. Wind energy grew faster in the electricity sector by providing almost 25% of the total electricity needs as can be seen in Fig. 10.3, which presents the share of renewable energy to the global electricity production for 2016. Fig. 10.3 also shows the distribution among the different sources.

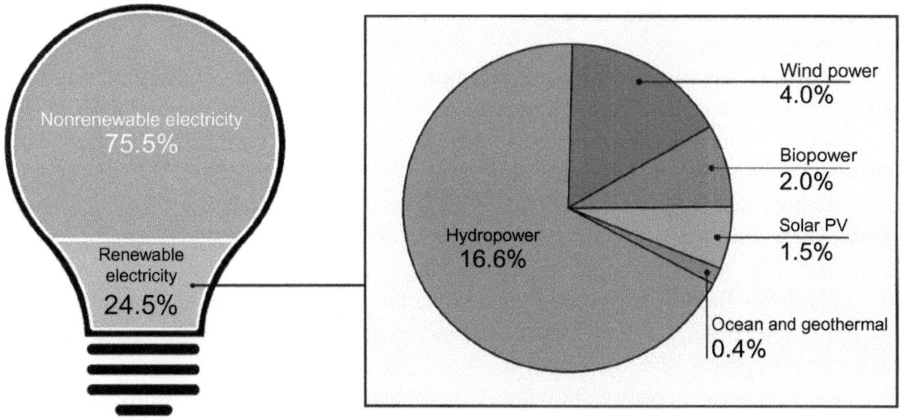

Fig. 10.3 Renewable energy share of global electricity production for 2016. *Source*: REN21. Renewables 2017 global status report. Paris: REN21 Secretariat; 2017.

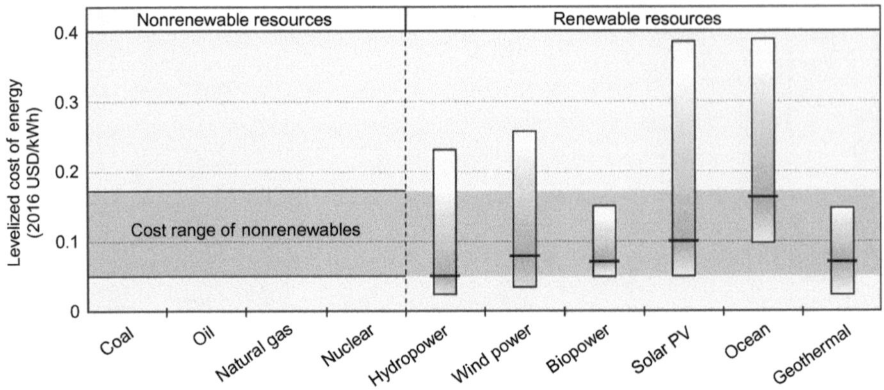

Fig. 10.4 Levelized cost of energy for all the various sources of energy.
Source: IRENA. Renewable power generation costs in 2017. Abu Dhabi: International Renewable Energy Agency; 2018.

The most important parameter that governs the level of investment in renewables is the levelized cost of energy (LCOE), which is the total cost associated with the whole lifecycle of a power plant over the total energy produced. Until the early 2000s, the cost of renewable power plants was considered high and that hindered investment. However, this cost has been dropping and in recent years has been the primary reason behind the improvement in technology and their deployment. Fig. 10.4 presents the LCOE in 2017 for all the various renewable resources compared to the nonrenewable ones. It can be seen that the weighted average cost of all the renewables is now competing with the cost of nonrenewables, but the cost variance is still large. By 2020 these costs are expected to drop below that of the nonrenewables (IRENA, 2018).

10.2 Wind power data

Wind power is one of the major renewable resources alongside hydropower and the most promising one. The power capacity of wind has increased exponentially in the last 20 years, from 6.1 GW in 1996 to 539.6 GW in 2017, taking into account both onshore and offshore installations. Fig. 10.5 shows the cumulative global capacity of wind power for the period 1997–2017. This number might seem insignificant compared to the total energy needs. However, significant technological improvements have been achieved in the recent past, resulting in a reduction of the investment costs for wind energy thereby improving the prospect of massive future investments. Based on future projections, the contribution of wind power to the total global electricity production could reach 18% by 2050 [3].

The wind power capacity shown in Fig. 10.5 is not distributed evenly across the world. Asia has the largest regional capacity with 228.5 GW which accounts for the 42.2% of the global capacity, followed by Europe with 32.9% and North America

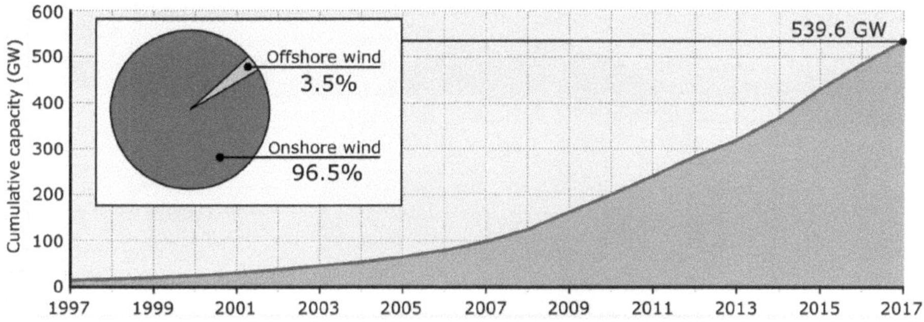

Fig. 10.5 Cumulative global capacity of wind power for 1997–2017.
Source: GWEC. Global wind statistics 2017, Brussels, Belgium; 2017.

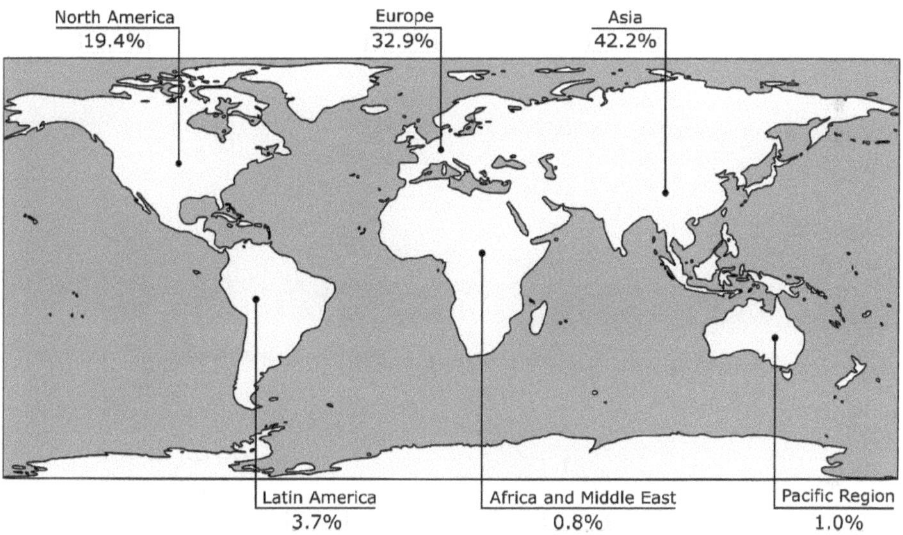

Fig. 10.6 Wind power contribution per region for 2017.
Source: GWEC. Global wind statistics 2017, Brussels, Belgium; 2017.

with 19.4%. These three regions combined make up 94.5% of the wind energy market. Fig. 10.6 graphically presents the share of wind power by region for 2017. If the spatial distribution is made countrywise, it is interesting to note that only 10 countries are responsible for 84% of the global power. As shown in Fig. 10.7, the largest contributor by far is China which alone is responsible for 35% of the global wind power and the second place is held by United States which contributes by 17%, followed by Germany, India, Spain, United Kingdom, France, Brazil, Canada, and Italy [3]. It also happens that these countries are also the most industrialized counties, with the highest energy consumption in the world.

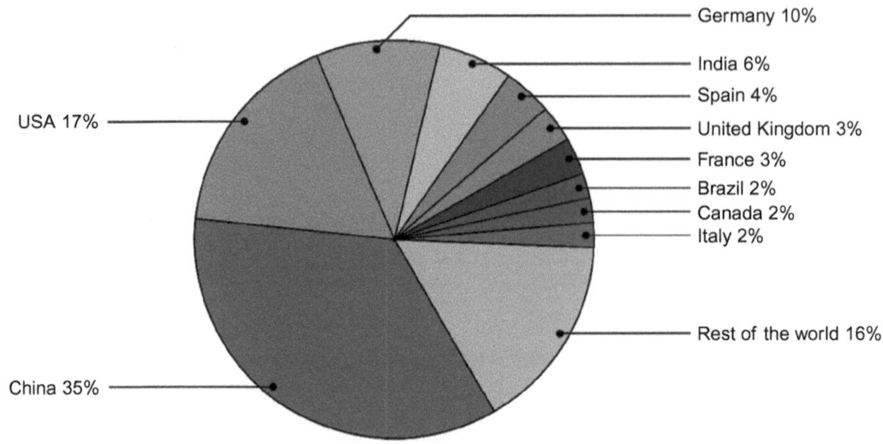

Fig. 10.7 Top 10 countries of wind power generation for 2017.
Source: GWEC. Global wind statistics 2017, Brussels, Belgium; 2017.

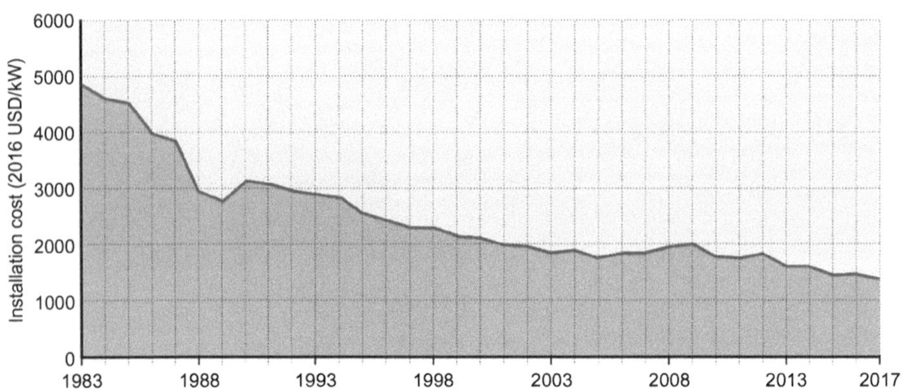

Fig. 10.8 Total cost of onshore wind farm.
Source: IRENA. Renewable power generation costs in 2017. Abu Dhabi: International Renewable Energy Agency; 2018.

More countries are expected to grow as the cost of wind power reduces. The cost reduction however depends mainly on the technology improvements, related to the efficiency of wind turbines, the installation, and maintenance. For the past 30 years, the onshore wind power costs have dropped significantly. Fig. 10.8 shows the average cost of the construction of onshore wind projects from 1983 until 2017 [4]. By comparing the values from Fig. 10.8, it may be noted that for this period, the costs have been reduced by 70%. Even though the cost trend is showing an overall reduction, there is a fluctuation in the cost and is mostly associated with the economic conditions of that time. This decrease in costs can also be seen in terms of LCOE. Fig. 10.9 shows the LCOE for both onshore and offshore wind power generation for 2010 and 2017.

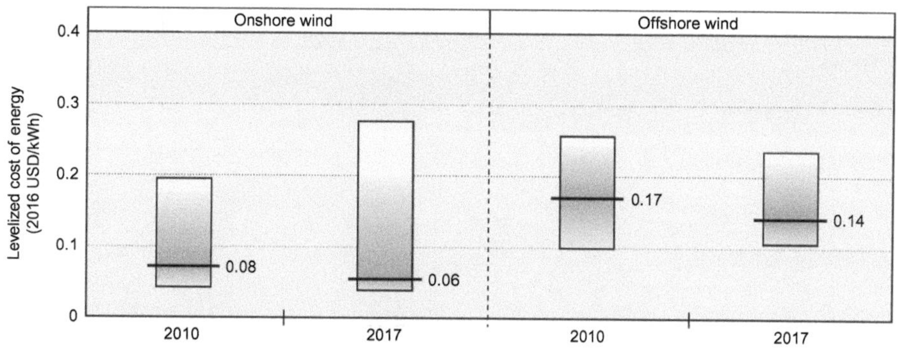

Fig. 10.9 LCOE of onshore and offshore wind power.
Source: IRENA. Renewable power generation costs in 2017. Abu Dhabi: International Renewable Energy Agency; 2018.

It can be seen that there is a significant decrease in the cost of both onshore and offshore, making wind energy a major competitor to the fossil fuels. Another comparison of the costs involved shows the difference between offshore and onshore wind power. While offshore wind power is more expensive at the moment, these values are expected to decrease further more as offshore wind is more scalable than onshore, making it more attractive. It is increasingly common to have offshore wind farms with a capacity of 1 GW.

10.3 Wind energy in a nutshell

10.3.1 What is wind?

The ultimate source of energy responsible for the creation of wind is the sun. Sun bombards the surface of the earth with radiation energy, leading to a continuous warming. Of this energy, a large amount is sent back to space and what remains is transformed into heat energy. Due to the shape of the earth, its surface is not heated evenly resulting in the equator getting more energy than the poles. As a result, there is a continuous heat transfer from the equator to the poles.

The atmospheric air consists of nitrogen and oxygen (totaling 99%) and like all gases, it expands when heated and contracts when cooled. When solar radiation hits the surface of the earth, it causes the air to get warmer, which means that the air gets lighter and less dense. This also results in the rise of warm air in higher altitudes and the creation of areas of lower atmospheric pressure, where the air is warmer. Due to the variance in pressure, the air will move from a high pressure to a lower pressure area, in an attempt to reach equilibrium. Wind can be defined as the movement of atmospheric air masses from a region of high atmospheric pressure to a region of low pressure.

However, the wind circulation system is even more complex. At 30° and 60° latitude, there is a major change in atmospheric pressure creating zones of high and low

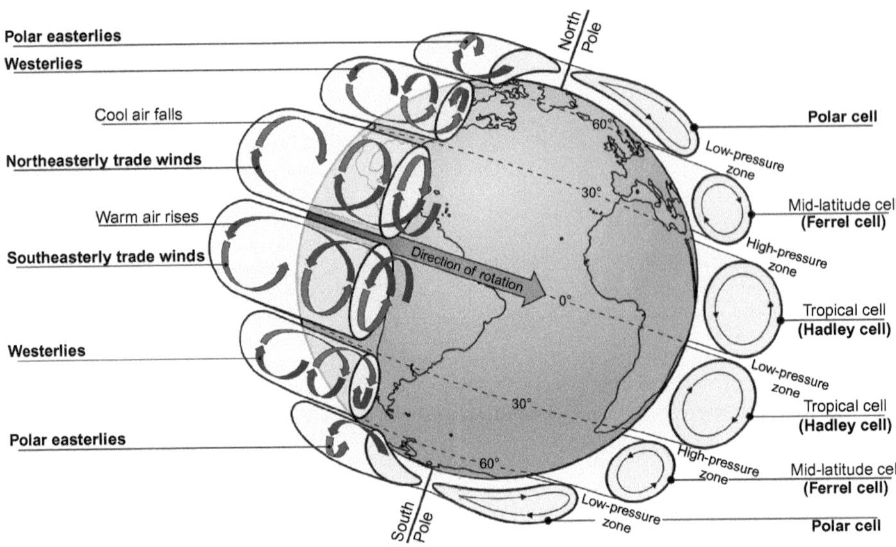

Fig. 10.10 Global wind circulation mechanism.

pressure, respectively. The air circulation within these zones is known as cells and the major winds created as a result are known as trade winds. Due to the diurnal motion of the earth, the winds deflect to the right in the Northern hemisphere and to the left in the Southern hemisphere one, leading to a spiral movement of the air mass. This is due to a phenomenon known as Coriolis effect. Fig. 10.10 describes the mechanism of the global wind system. In addition to the global wind system, there are localized influences as well that are mostly related to the terrain of an area and its proximity to water bodies.

Wind characteristics, such as the speed and direction, are not regular and can have a spatial and temporal variance. More specifically, wind speed is lower around the equator and it increases closer to the poles. The highest wind speeds are developed within the region of 30° and 60° latitude, mostly because of a global wind circulation mechanism. However, the Southern hemisphere has less land obstructions than the Northern one, resulting in higher wind speeds. Therefore, the distribution of land has a major influence on the wind speed as it can obstruct the free movement of air. In other words, wind speed can be higher in the open sea than on land. Also, the wind speed is dependent on the time of the year, i.e., in the Northern hemisphere, the mean wind speed is higher during winter. Further details on wind energy can be found in Refs. [5–8].

In order to use wind as a reliable source of energy, it is essential to have the wind speed and direction variance data. Such a tool is a wind map, showing how the mean wind speed is distributed in an area. Fig. 10.11 shows the global wind map for a period between 1979 and 2012, at an altitude of 100m. The World Bank in conjunction with Technical University of Denmark (DTU) has developed such a global wind map.

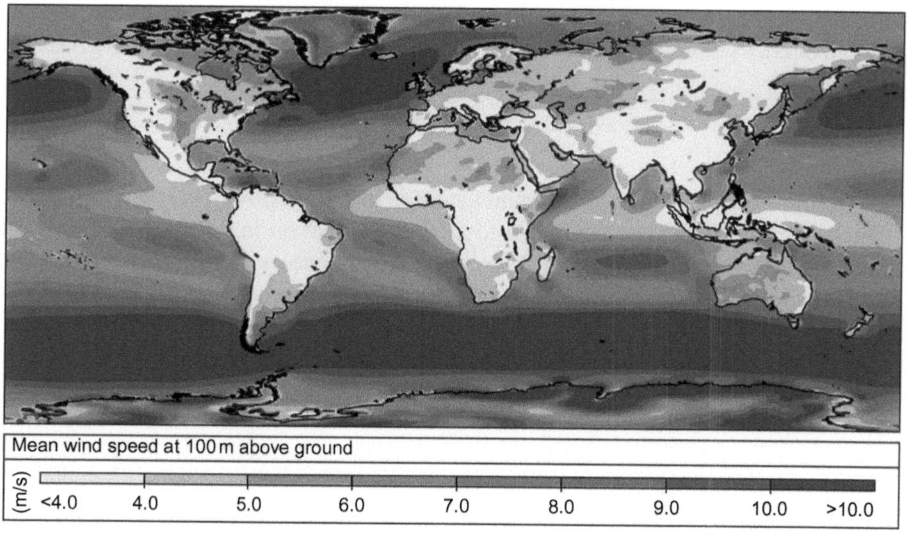

Fig. 10.11 Global mean wind speeds for 1979–2012.

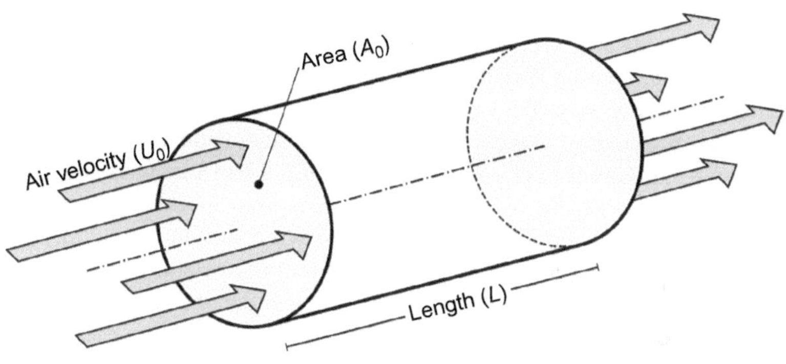

Fig. 10.12 Air flow through a cylindrical pipe of known dimensions.

10.3.2 How much power is there in wind?

The energy contained within the wind is the kinetic energy (E_k) of a mass of air (m) moving at a certain speed (U):

$$E_k = 0.5 \times m \times U^2 \tag{10.1}$$

Assume that a laminar flow of air passes through a cylinder of known dimensions, as shown in Fig. 10.12. The mass (m) passing through the cylinder during each unit of time (dt) will be given by multiplying the density of air (ρ) by the volumetric flow rate (Q). Therefore:

$$\frac{dm}{dt} = \rho \times Q = \rho \times A_0 \times U_0 \tag{10.2}$$

Wind power (P_{wind}) is the flow of kinetic energy through an area and it is defined as the wind energy per unit of time. Using Eqs. (10.1) and (10.2) wind power is given by:

$$P_{wind} = 0.5 \times \frac{dm}{dt} \times U_0^2 = 0.5 \times \rho \times A_0 \times U_0^3 \quad (10.3)$$

Eq. (10.3) shows that the main parameter affecting wind power is the wind speed. A small fluctuation in speed can lead to a large fluctuation in power, i.e., there will be an eightfold rise in wind power if the wind speed doubles. Air density and area of air flow have a smaller effect on the power as they are linear relationship with wind power.

However, not all the available wind power can be transformed into electricity, details of which are given below.

10.4 Wind turbines

10.4.1 History of wind turbine

Modern wind turbines have evolved from windmills. In 1887, the first wind turbine use to produce energy was created by James Blyth in Scotland. It was a cloth-sailed vertical axis windmill with a 10 m diameter rotor and was used to charge accumulators (early batteries) and to power a cottage. During the same year, Charles Brush, the inventor of brush dynamo, designed and built the first wind turbine in the United States. It was a horizontal axis wind turbine with a 17-m diameter rotor including 144 blades driving a 12 kW dynamo and it remained operational until 1909. In 1891, a Danish scientist, Poul la Cour, develops wind turbines to supply electrical power to rural areas of Denmark. He is also known as the person who discovered that wind turbines with fewer blades are more efficient than the ones with many blades. By 1908, 72 wind turbines (ranging from 5 to 25 kW) were supplying electricity across Denmark. The development of the wind turbine technology continued during the next decades and in 1941 the first megawatt-scale turbine (with a power capacity of 1.25 MW) was built in the United States [7].

It was in 1975 that the research on wind turbine technologies increased in intensive. Due to the major oil crisis in the Middle East, US Government funded NASA to develop modern technologies for wind turbines. Many concepts that are used until now, such as steel tube towers, variable-speed generators, composite blade materials, and partial-span pitch control, were introduced through this program. In 1980 the first wind farm was built in the United States and it contained 20 wind turbines, with a capacity of 20 kW each. However, the development of wind turbines slowed down as the oil price dropped, making wind an uneconomical source of energy. Later in 1991, the first offshore wind farm was built in Denmark consisting of 11 turbines.

The concern of excessive CO_2 emissions from the combustion of fossil fuels during the 1990s was the turning point and the interest in wind power began to grow. Since then the development of larger wind turbines combined with lower costs has continued

Fig. 10.13 The turbine created by Charles Brush in 1887 *(left)*; a modern wind turbine *(right)*.

and nowadays wind is considered as one of the most important alternatives to fossil fuels. Just as an indication of the technology improvement through the past century, Fig. 10.13 presents the turbine created by Charles Brush together with a modern wind turbine.

10.4.2 The anatomy of a modern wind turbine

Wind turbines vary in sizes, designs, and power output. Currently, the most common configuration of a wind turbine system consists of a three-blade rotor driving a horizontally mounted generator. The basic elements of a wind turbine are: the rotor, the nacelle, the tower, the transition piece, and the foundation. The different components of a wind turbine are shown in Fig. 10.14. More details on the different components are described in the next section.

The rotor of a wind turbine contains the blades and the hub and is crucial to the efficiency of power output. The blades capture the wind energy from the air passing through and transform it to rotational energy. The shape, dimensions, and number of the blades vary and they are designed in such a way so that they can withstand the forces acting on them and achieve the best possible aerodynamic performance for the required energy efficiency. Most wind turbines have three blades or (less commonly) two blades, which rotate around a central hub on a horizontal axis. There are also wind turbines with the axis of rotation vertical to the ground. These turbines are independent of the wind direction and easier to maintain because the power generator is mounted at ground level. However, they are less efficient at transforming wind energy into electricity and they are only being used in small-scale projects.

The energy captured from the blades is transferred to the main shaft as rotational motion. The main shaft is connected to the gearbox which will change the low speed of rotation of the blades to a higher speed, which feeds the generator and transforms the

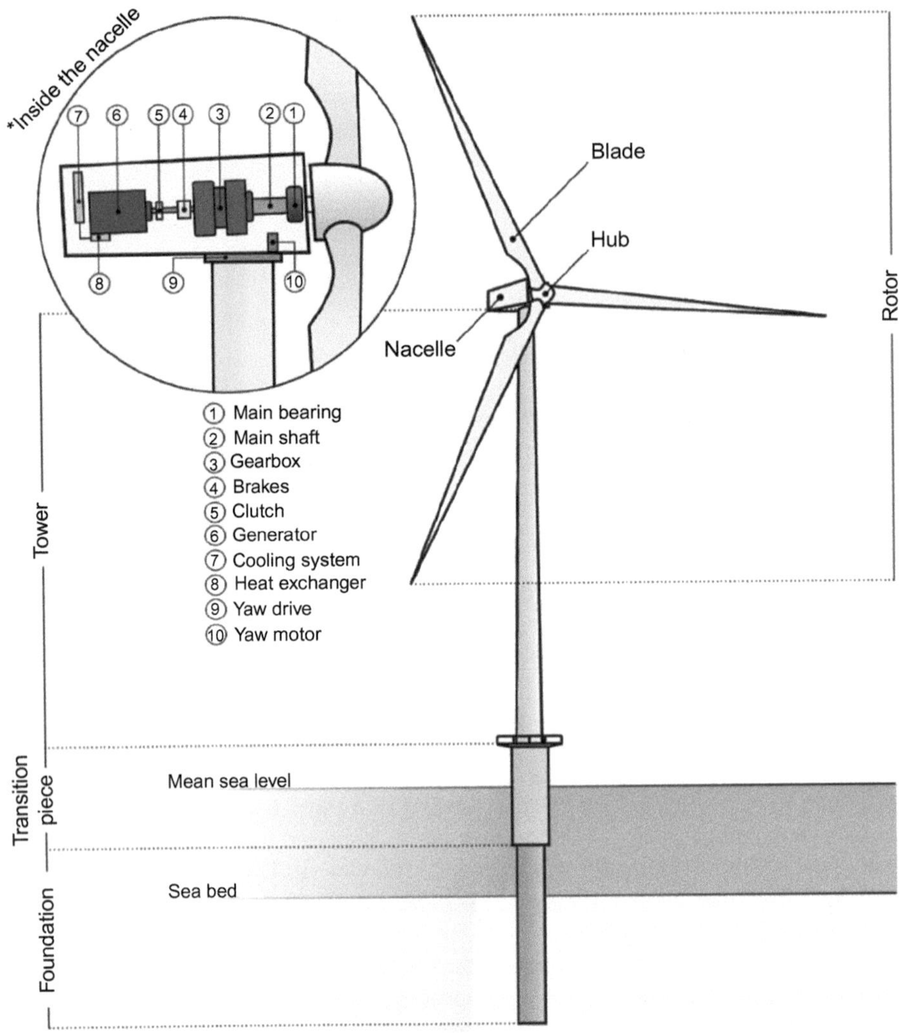

Fig. 10.14 Main components of a wind turbine.

rotational energy to electricity via electromagnetic induction. The main shaft, the brake, the gearbox, and the generator are housed inside the nacelle. The nacelle can rotate horizontally, through the yaw system, in order to align the rotor with the direction of the wind for optimum power performance. All these components are supported by the tower, which is constructed from tubular steel. The design of the tower is mainly influenced by the top mass and the characteristics of the blades.

The wind turbines can either be sited on the mainland (onshore) or in the open sea (offshore). The main difference between onshore and offshore turbines is the

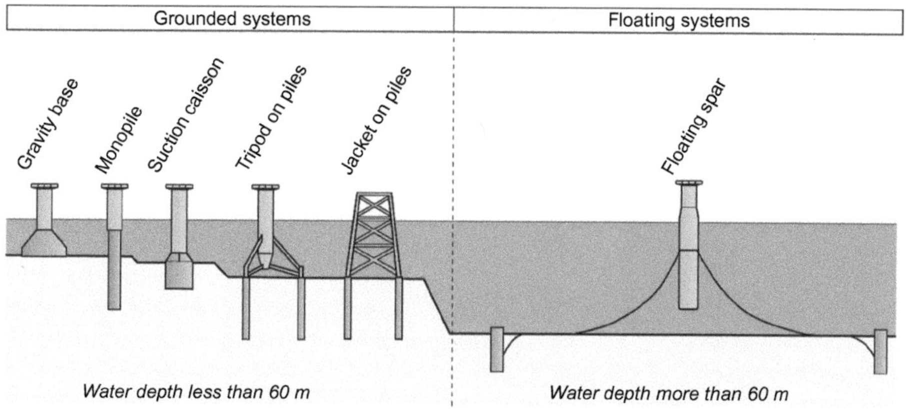

Fig. 10.15 Common types of foundations used to support offshore turbines.

foundation. Foundations constitute the most important design consideration and often determine the financial viability of the project. Many aspects must be considered while choosing and designing the foundation for a particular site. The foundations for onshore turbines are relatively straightforward as they are constructed on stable ground. On the other hand, offshore turbines rest on foundations depending on the site condition which includes water depth and ground profile. For up to 60m of water depth, the foundations are placed directly on the seabed and are of a fixed type. However, for deeper waters, the wind turbines are placed on top of floating structures known as floaters. Fig. 10.15 presents the main types of foundations commonly used today at different depths of water.

The choice of foundation will depend on the following: site conditions (wind, wave, current, seabed condition, ground profile, water depth, etc.), available fabrication and installation expertise, operation and maintenance, decommissioning laws of the land, and finally economics. In 2017 Bhattacharya [9] provided a description of an ideal foundation to support offshore wind turbines:

(1) A foundation is capacity or "*rated power*" specific (i.e., 5 or 8 MW rated power) but not turbine manufacturer specific. In other words, a foundation designed to support a 5-MW turbine can support turbines of any make. There are other advantages in the sense that turbines can be easily be replaced even if a particular manufacturer stops manufacturing them.
(2) Installation of a foundation is not weather sensitive, i.e., not dependent of having a calm sea or a particular wind condition. The installation of the first offshore wind farm in the United States took more time than was expected as a result of unsuitable weather conditions.
(3) Low maintenance and operational costs, i.e., needs least amount of inspection. For example, a jacket-type foundation needs inspection at the weld joints.

It is economical to have a large number of turbines in a wind farm to have the economy of scale and therefore it also requires a large area. If the continental shelf is very steep, grounded (fixed) turbines are not economically viable and a floating system is desirable.

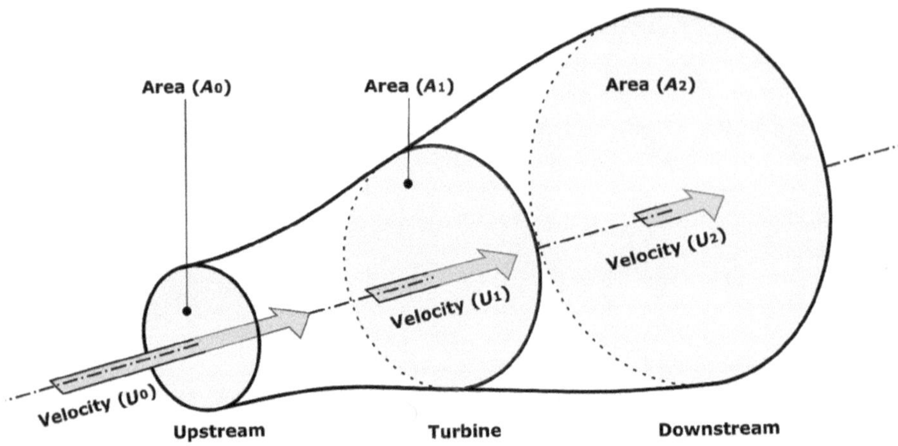

Fig. 10.16 Air flow through a turbine.

10.4.3 How much power can a turbine generate?

As it was mentioned before, not all the wind power can be utilized by a wind turbine. To calculate the power that can be utilized by a turbine, assume that a laminar flow of air passes across the swept area of the turbine (A_1) and there is a change in pressure while the air passes by with a consequent decrease of wind speed as shown in Fig. 10.16. The power output of the turbine ($P_{turbine}$) can be given by:

$$P_{turbine} = \rho \times A_1 \times U_1^2 \times (U_0 - U_2) \tag{10.4}$$

The efficiency of the system can be quantified by the power coefficient (C_p), which is the ratio of the power output of the turbine ($P_{turbine}$) to the wind power (P_{wind})

$$C_p = {}^{P_{turbine}}/{}_{P_{wind}} \tag{10.5}$$

The power output of a wind turbine can be also expressed by:

$$P_{turbine} = 0.5 \times \rho \times A \times U_0^3 \times C_p \tag{10.6}$$

According to Betz law, the maximum power coefficient cannot be larger than 16/27. In other words, the maximum energy that can be used is 59.3% of the wind's energy. The power output of a wind turbine varies with wind speed and this variance is presented in the form of a power curve. Power curve shows how much power a wind turbine produces for different wind speeds. Every turbine has a different curve and the shape of the curve can be influenced by the rotor swept area, the number and shape of blades, and the cut-in, rated, and shut-down wind speeds. A typical example for a 3.6-MW turbine is given in Fig. 10.17.

Wind turbine blades are getting larger and as a result the supporting tower is getting taller. Fig. 10.18 shows the technological progress made in turbine technology with

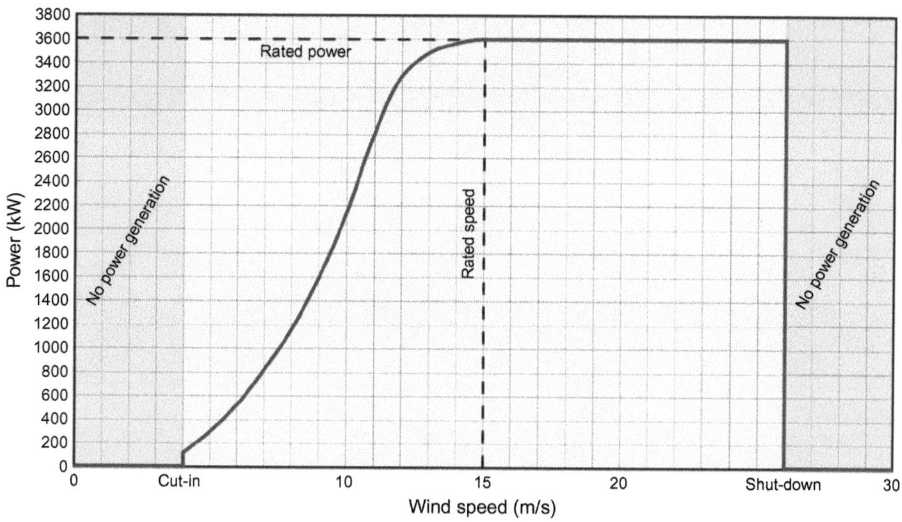

Fig. 10.17 Power performance curve of Siemens 3.6 MW.

Year	1980	1985	1990	1995	2000	2005	2010	2015	2020
Rotor diameter (m)	15	20	40	50	112	126	150	178	252
Rated power (MW)	0.01	0.10	0.50	0.60	2.40	5.00	7.50	10.00	20.00
Top mass (ton)				80	250	340	360	550	880

Fig. 10.18 Growth of sizes of wind turbine and the length of blades.

100-kW machine to 20-MW machine. The top mass is typical for such rated turbines and is also knows as RNA (Rotor-Nacelle-Assembly) mass. Fig. 10.19 shows the location of wind farms either in planning stage or in construction phase along the coast of China and further details can be found in Ref. [9]. It may be noted that modern-day wind farm can produce energy in gigawatt capacity range. On the other hand, Fig. 10.20 shows the location of wind farm in Europe. It is very clear that the future of wind energy is along the coastal region and there needs to be a national grid along the coast.

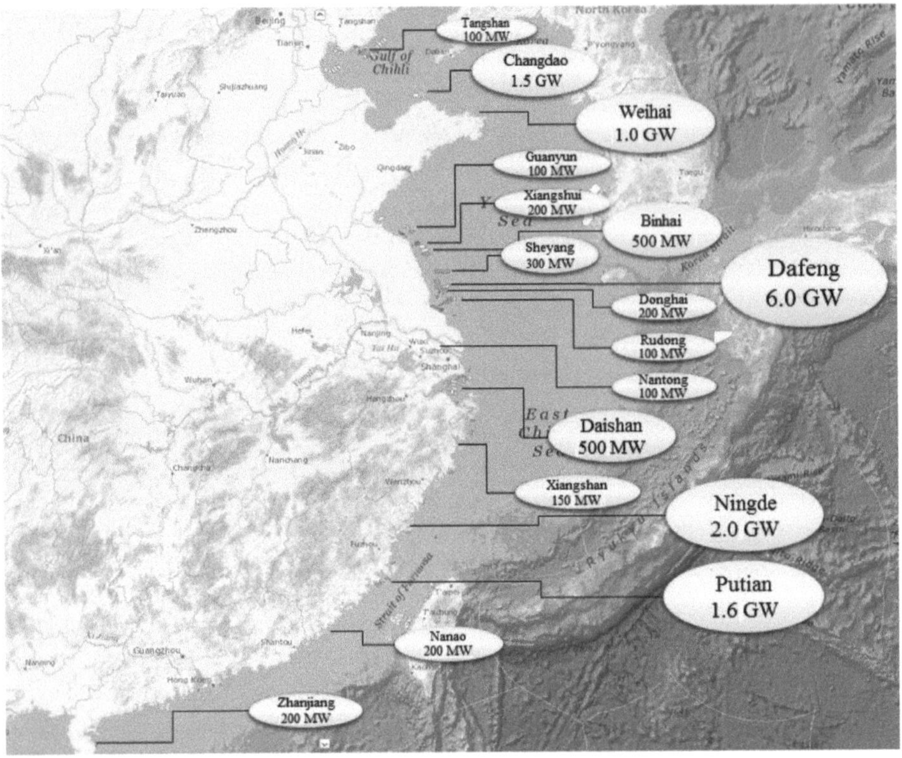

Fig. 10.19 Wind farm location along the coasts of China.

10.4.4 Offshore wind farms

As a way to reduce further planning, construction, and maintenance costs, nowadays many wind turbines are erected at the same time in one location. These power generating facilities (power plant) consisting of many turbines are called wind farms. The main components of a wind farm are the wind turbines, the transmission cables, and the substations. Fig. 10.21 shows a typical layout of an offshore wind farm.

The wind turbines in a wind farm are spaced in such a way that the interference between each other is at a minimum. If the turbines are built too close together, they take wind away from each other, and their overall power efficiency will be reduced. Fig. 10.22 shows the aerial photo of wake turbulence behind individual wind turbines that can be seen in the fog of the Horns Rev wind farm off the Western coast of Denmark. Therefore, they should be sited as far as possible from each other, but at the same time as the distance between the turbines increases, the cost of the cables connecting them increases. The spacing of the turbines must be designed in such a way that the cost will not increase significantly. The geometric layout of a wind farm can be a single line array, or in a square or a rectangle configuration. Typically, the spacing between turbines is equivalent to 3–10 times the rotor diameter and depends on the prevailing wind direction (see Fig. 10.23).

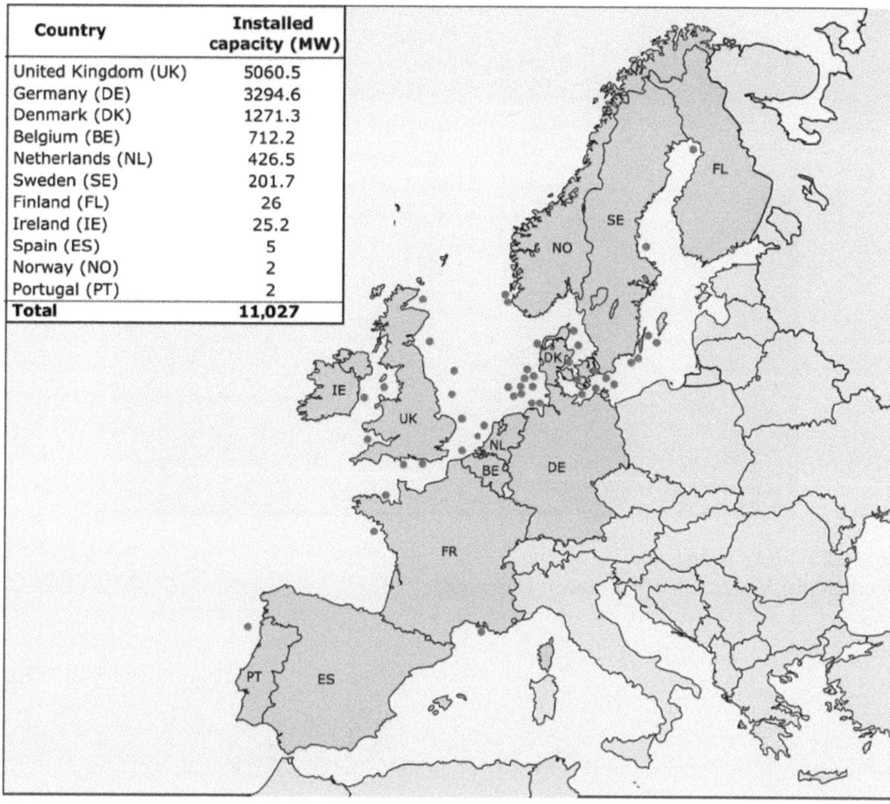

Country	Installed capacity (MW)
United Kingdom (UK)	5060.5
Germany (DE)	3294.6
Denmark (DK)	1271.3
Belgium (BE)	712.2
Netherlands (NL)	426.5
Sweden (SE)	201.7
Finland (FL)	26
Ireland (IE)	25.2
Spain (ES)	5
Norway (NO)	2
Portugal (PT)	2
Total	**11,027**

Fig. 10.20 Wind farm location around Europe.

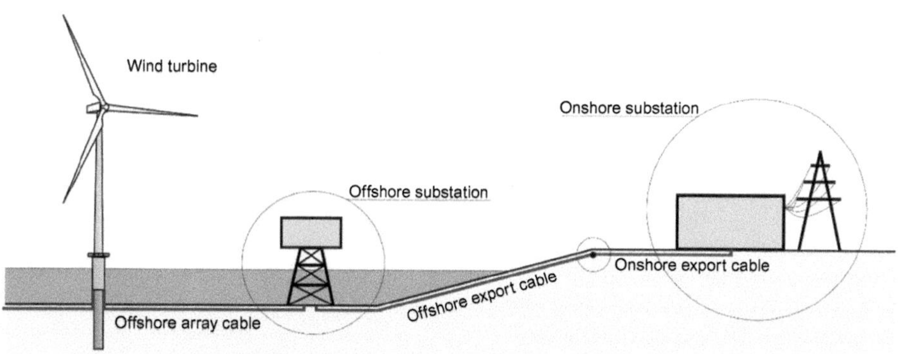

Fig. 10.21 Layout of a typical offshore wind farm.

Fig. 10.22 Wake turbulence.
Photo credit: Vattenfall Wind Power, Denmark.

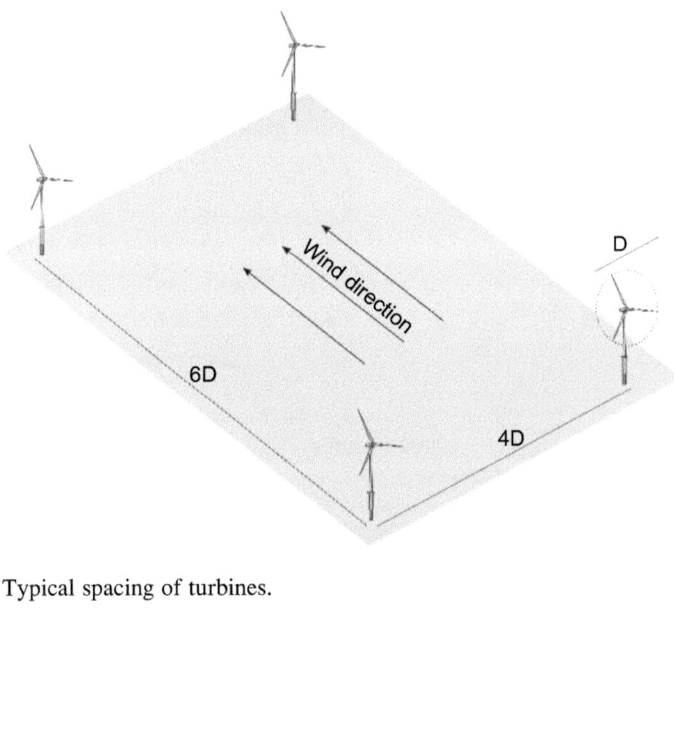

Fig. 10.23 Typical spacing of turbines.

10.5 Offshore wind farm site selection

This section details the considerations for choosing a particular site. Offshore wind farming is considered to be one of the most reliable ways to produce clean green energy; the main reasons are:

(a) The available energy offshore is much greater than onshore. The wind speeds are higher over the sea because there are no mountains or hills to interrupt wind's flow and is thus less turbulent and more consistent.
(b) Larger power capacity can be easily installed offshore in comparison to an equivalent onshore wind farm. The reasons are heavy wind turbine generators (WTG) or towers can be easily transported and installed using sea routes. In contrast, transporting these large and heavy structures/components onshore disrupts the road traffic and is expensive.
(c) Offshore wind farms produce less noise and vibrations and their impact on human beings is small, due to their distance from land. Furthermore, it is easier to gain planning permission for an offshore farm than it is an onshore farm.

The first offshore wind farm was built in Denmark in 1991, but it was only in the 2000s that the offshore wind capacity really developed. Within the last decade, offshore wind power has gained a considerable share in the total wind power, with a total of 14.4 MW, which corresponds to approximately the 3% of the total wind capacity. Currently, United Kingdom is leading in offshore wind power production globally with approximately 5.1 MW. China at the moment has the third largest production and is rapidly expanding its capacity. Fig. 10.24 presents the offshore capacity by country.

The main considerations for choosing an offshore wind farm location are:

(a) *Wind resources*: A thorough knowledge of wind resources in an area is fundamental as it allows an estimation to be made on the wind farms productivity and therefore the financial

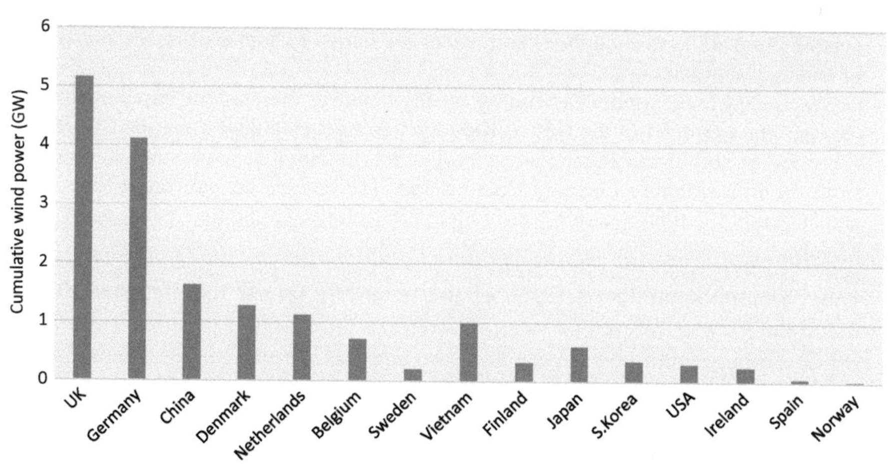

Fig. 10.24 Offshore wind power capacity by country in 2016.

viability of the project. As a thumb rule, a project is not financially viable if the average wind speed at the hub height is below 6 m/s and it is considered a safe investment if the average wind speed at the hub is greater than 8.5 m/s.

(b) *Marine aspects*: Marine aspects would include water depth, wave spectrum at the site (wave height, wave period), current and tide data, exposure to waves and sediment transport, identification of scour-related issues, and if scour protection is needed. Often installation of foundations creates obstacles in the local flow pattern of water which may create turbulence leading to a scouring effect.

(c) *Environmental impact*: For all wind farms, an Environmental Impact Assessment must be completed as a part of the planning process; it covers the physical, biological, and human environment. This involves collecting all types of existing environmental data and assessing the potential impacts that could arise due to the construction and operation of the wind farm. The impacts can range from favorable to less favorable to detrimental. Potential aspects on the biological environment include marine mammals, sea birds that use the area on a regular basis, birds from nearby areas that pass through the area during flight, fish, etc. Other aspects include effect of flora and fauna during the construction (for example, noise due to piling or operating noise) and electromagnetic field generated by subsea cable. Human environment includes a change of landscape. Marine archeology aspects such as ship wreck are also taken in consideration. To carry out the assessment, samples of seabed may be collected and analyzed for worms, barnacles, or other species.

(d) *Power export/grid connection*: One of the important deciding factors is the location of onshore grid connections. Other aspects include the length of submarine cable required which is dependent on the turbine layout, substation location, identify export cable routing (landfall), risk assessment of buried cables, and the transformer options—AC or DC.

(e) *Economics*: Modeling of capital costs and the levelized cost of offshore wind (LCOE) is a function involving many parameters: depth of water, distance from shore, wind speed at the site, port and harbor facilities near the site, socioeconomic condition and access to skilled labor, location of national grid, and hinterland for the proposed development.

(f) *Navigation*: This navigation risk assessment survey investigates whether or not there is a need for exclusion zone due to fishing or navigation or military operations. Cables connecting the wind farms and the export cables are buried to depths of 2–3 m to avoid risk of entanglement with nets.

(g) *Consents and legislations*: Depending on the country, the consent requirements may change. For example, in the United Kingdom, any development more than 100 MW is classified as significant infrastructure project and requires development consent order from the Infrastructure Planning Commission. These rules are subject to amendment and currently the final decision rests with the Secretary of State for Energy and Climate Change.

Recently, Denmark produced 120% of the country's power requirements through wind and the excess 20% power was exported to the neighboring countries. China, started developing wind power very late compared to Europe, made remarkable progress and the ambition is to generate 30 GW of power by 2020 from offshore wind. It is important to note that the growth of offshore wind farm accelerated, in part, due to the Fukushima Nuclear Power Plant (NPP) accident. It is therefore pertinent to discuss this event and its relation to offshore wind farms.

10.6 Case study: Performance of nearshore wind farm during 2012 Tohoku earthquake

A devastating earthquake of moment magnitude M_w9.0 struck the Tohoku and Kanto regions of Japan on 12th March at 2:46 PM which also triggered a tsunami (see Fig. 10.25 for the location of the earthquake and the operating wind farms). The earthquake and the associated effects such as liquefaction and tsunami caused great economic loss, loss of life, and tremendous damage to structures and national infrastructures but very little damage to the wind farms. Extensive damage was also caused by the massive tsunami in many cities and towns along the coast. Fig. 10.26A shows

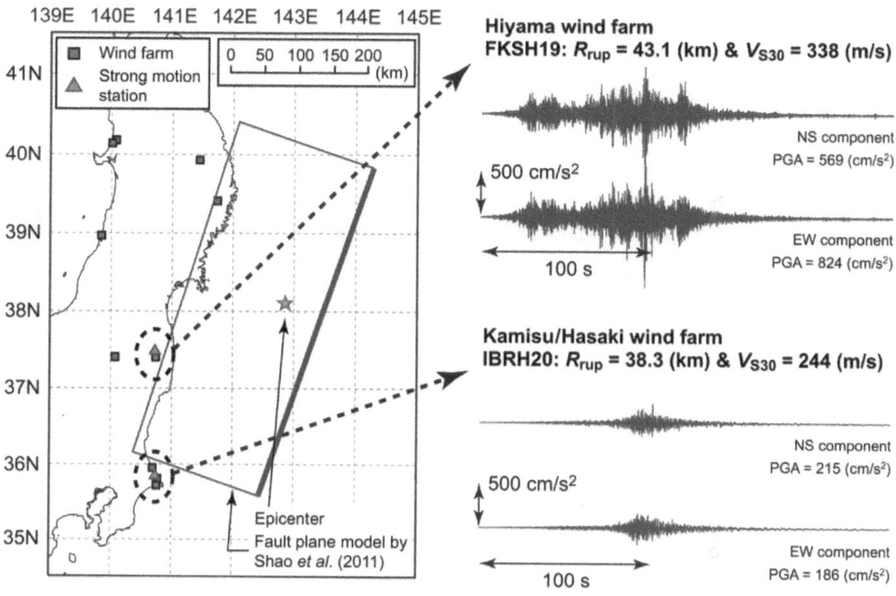

Fig. 10.25 Details of the 2012 Tohoku earthquake and locations of the wind farms.

Fig. 10.26 (A) Photograph of the Kamisu (Hasaki) wind farm following the 2012 Tohoku earthquake; (B) collapse of the pile-supported building following the same earthquake.

photographs of a wind farm at Kamisu (Hasaki) after the earthquake and Fig. 10.26B shows the collapse of pile-supported building at Onagawa. At many locations (e.g., Natori, Ofunato, and Onagawa), tsunami heights exceeded 10m, and sea walls and other coastal defense systems failed to prevent the disaster.

The earthquake and its associated effects (i.e., tsunami) also initiated the crisis of the Fukushima Daiichi NPPs. The tsunami which arrived around 50min following the initial earthquake was 14m high which overwhelmed the 10m high plant sea walls flooding the emergency generator rooms causing the power failure of active cooling system. Limited emergency battery power ran out on 12th March and subsequently led to the reactor heating up and the subsequent meltdown, leading to the release of harmful radioactive material to the atmosphere. Power failure also meant that many of the safety control systems were not operational. The release of radioactive materials caused a large-scale evacuation of over 300,000 people and the clean-up costs are expected to be of the order of tens of billions of dollars. On the other hand, following/during the earthquake the wind turbines were automatically shut down (like all escalators or lifts) and following an inspection—they were restarted.

10.6.1 Why did the wind farm stand up?

Recorded ground acceleration time-series data in two directions (North-South and East-West) at Kamisu and Hiyama wind farms [FKSH 19 and IBRH20] are presented in Fig. 10.27 in frequency domain. The dominant period ranges, of the recorded

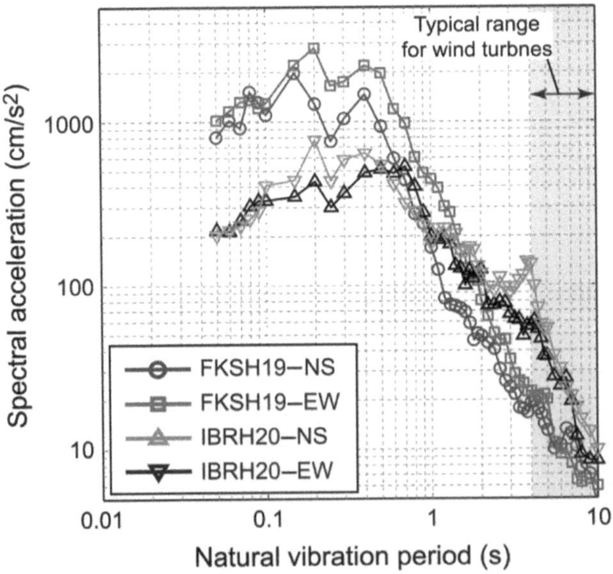

Fig. 10.27 Power spectra of the earthquake and natural frequency of wind turbines.

Fig. 10.28 Semi-submersible foundation for Offshore Fukushima (Japan).

ground motions at the wind farm sites, were around 0.07–1 s and on the other hand the periods of the offshore wind turbine systems are in the range of 3 s. Due to nonoverlapping of the vibration periods, these structures were not "tuned in" and as a result they were relatively insensitive to the earthquake shaking. However, earthquake-induced effects such as liquefaction may cause some damages. Further details can be found in Ref. [10]. Further details on the dynamics of wind turbine structures can be found in Refs. [11–29].

Aside: One may argue, had there been a number of offshore wind turbines operating, the disaster may have been averted or the scale of damages could have certainly been reduced. The wind turbines could have run the emergency cooling system and prevented the reactor meltdown. In this context, it is interesting to note that there are plans to replace the Fukushima NPP by a floating wind farm. The project is at an advanced stage involving a 2 MW semi-subfloating turbine (see Fig. 10.28). An innovative 7-MW oil pressure drive-type wind turbine on a three-column semi-subfloater has also recently been tested. Fig. 10.29 shows a map showing the seismic regions of the world together with the areas where investment in offshore wind is being made.

10.7 Future of offshore wind farm and sustainability

It is increasingly becoming apparent that floating offshore wind farms will be the norm for deeper water developments. Hywind Scotland Pilot Park is the first floating wind farm with rated capacity of 30 MW involving five turbines and located 30 km off the coast of Scotland (see Fig. 10.30). In October 2017, Hywind started delivering

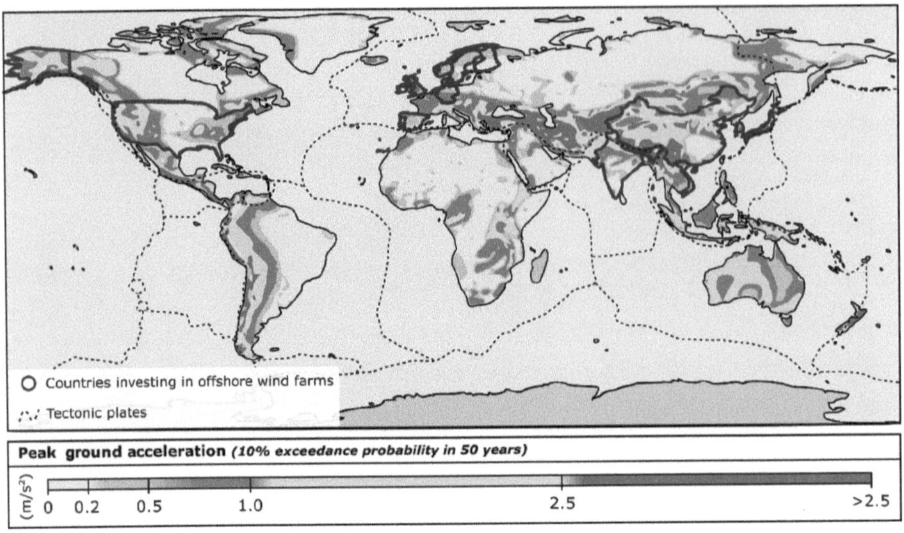

Fig. 10.29 Countries investing in offshore wind farm and seismic hazard map.

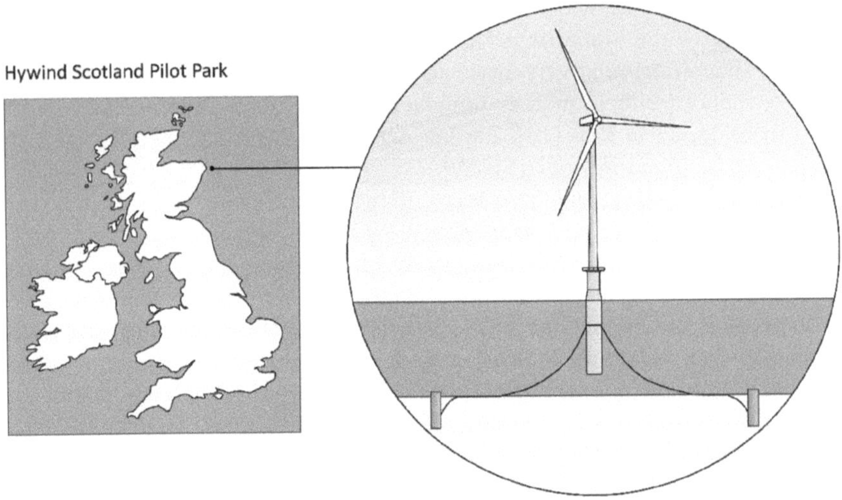

Fig. 10.30 Case study of Hywind Wind Park (location and schematic).

power to Scottish electricity grid. The water depth is 95–120 m and the mean wave height is 1.8 m. The average wind speed at 100 m height is 10.1 m/s and a spar type of concept is being used. The turbine structure is anchored using a conventional three-line mooring system (see Fig. 10.31). The anchor piles have a dimension of 7 m diameter and 16 m long. The control system that is used to dampen out motions

Fig. 10.31 Installation of the turbine: (A) the substructure being pulled to the location; (B) lowering of the substructure; (C) anchored to the seabed with chains; and (D) the RNA together with the tower is being transported for connecting to the substructure.
Photo courtesy: Statoil.

is the blade pitch control. The whole system is full-scale tested through Hywind Demo 2.3-MW turbine installed in 2009 having 5 years of operational data. Fig. 10.31 shows some of the installation details of the floating system.

10.7.1 Sustainability

It is important to discuss the important aspect of sustainability. Following Brundtland report in 1992 to World Commission on Environment and Development, the definition of sustainable development was given as "that development that meets the needs of the present without compromising the ability of future generations to meet their own needs." As a result, an important aspect in offshore wind technology is the decommissioning of whole energy infrastructure including the foundation. This section of the chapter provides a case study of the decommissioning of Lely Wind Farm (Netherlands) where the monopile foundation was also removed. All the self-explanatory photos are provided. In this context, it is necessary to provide an overview of the current legislation on offshore installations.

For the United Kingdom, decommissioning of offshore structures is regulated by UK law and by the OSPAR Decision 98/3 on the Disposal of Disused Offshore Installations (OSPAR Commission, 1998). However, the removal of offshore piles is not required under legislation. Typically, the piles are cut below the seabed, at a suitable distance appropriate and often this distance is around 3 m below the natural seabed.

10.7.1.1 Decommissioning of Lely wind farm

Lely wind farm (four 2 MW turbines in the Ijsselmeer, located approximately 800 m offshore) built in 1992 generated electricity for 22 years and was fully decommissioned in 2017. Crane barges and tugs were used to dismantle the turbines and tower sections (see Fig. 10.32A–F). The monopiles were extracted using vibratory hammer (see Fig. 10.32G–K).

Fig. 10.32 (A) Offshore vessel and crane positioned near the WTG installation. (B) Lift of nacelle and rotor.

(Continued)

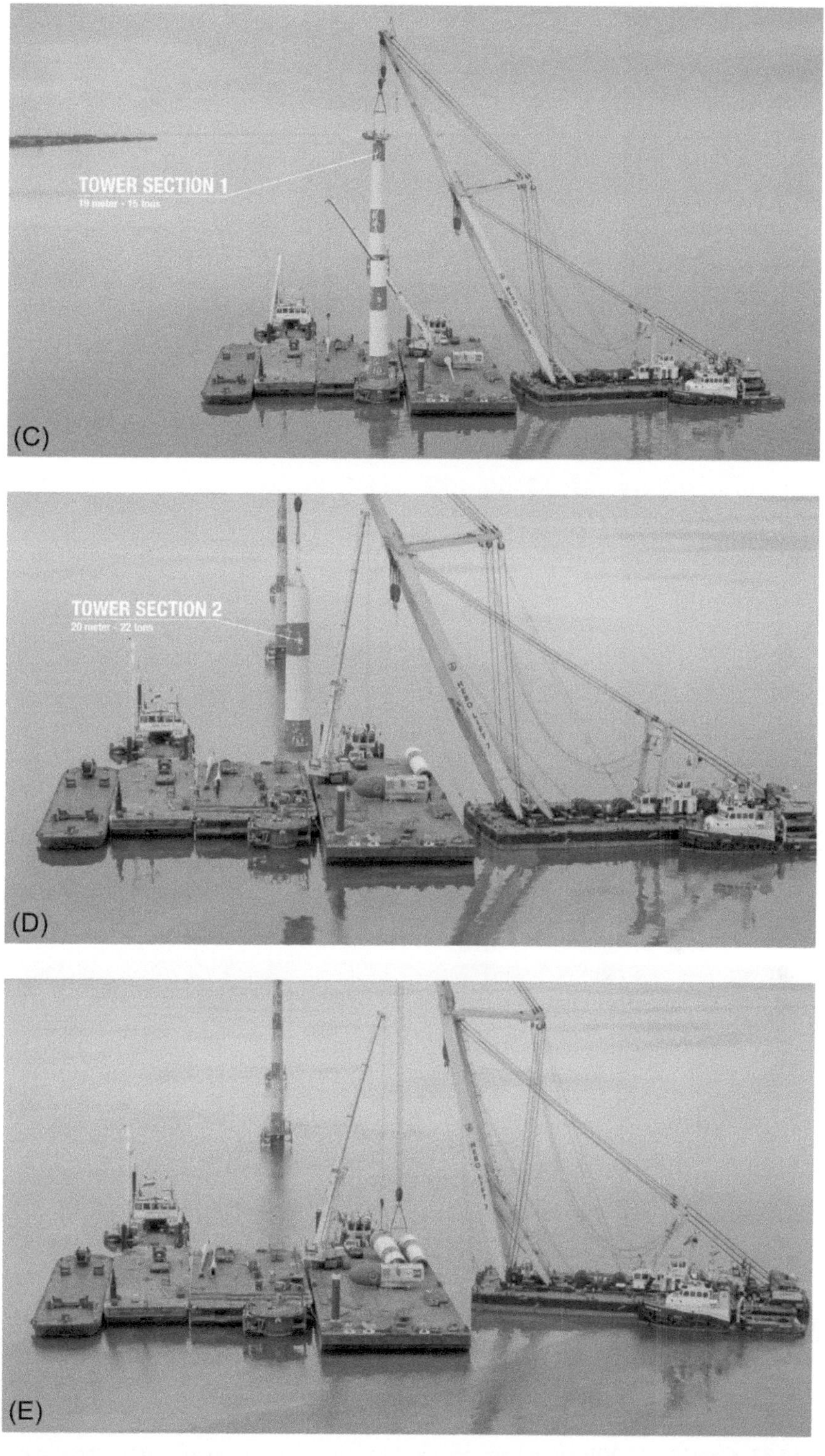

Fig. 10.32, Cont'd (C) Lift of tower section 1. (D) Lift of tower section 2. (E) Superstructure on the vessel.

(Continued)

Fig. 10.32, Cont'd (F) Vibratory hammer used to remove the monopile from the ground. (G) Power pack used along with the vibratory hammer. (H) Extracted monopile being taken off the ground.

(Continued)

Fig. 10.32, Cont'd (I) Lift of the monopile on the vessel. (J) Monopile on the vessel.

10.7.2 ASIDE: Banning of petrol and diesel cars from 2040 and offshore wind energy

In a recent policy and legislation discussion, there is a commitment to ban petrol and diesel vehicles from UK road by 2040 and replace them with electric vehicles. Other countries are also considering this given the consensus reached on climate change agreement by the nations of the world at the 21st meeting of the Conferences of the Parties (COP 21). It has been estimated that based on the available technology, there will be an additional 30 GW electricity requirement and two options are considered here: 10 large-sized NPPs costing circa 25 billion USD each and taking

Fig. 10.33 Relative construction material for NPP and OWT.

approximately 20 years to construct one of them. On the other hand, 30 GW electricity can be produced using 5000 offshore turbines of 6 MW and typical cost of each turbine is about 6–8 million USD. To put the cost aspect in perspective, the strike price, i.e., negotiated price between UK Government and EDF energy for Hinkley NPP development is £92.50 per MWh and in a recent offshore wind auction, the negotiated price for offshore wind is £57.50 per MWh. Furthermore, it takes about 6–9 months to construct a wind farm. The decommissioning of existing NPP is still not very clear in terms of high-level waste. Fig. 10.33 shows comparative photographs of a NPP and OWT, highlighting the steel and concrete necessary to produce electricity.

10.8 Summary

Offshore wind farms are a recent innovation and hold promise in tackling the global challenges of combating climate change. These new technologies are scalable with each modern wind farm capable of generating over 1.2 GW of power. This chapter provides statistics of the installed wind farm around the world and highlights the advantages of offshore wind farms. Current developments are also discussed.

References

[1] REN21. Renewables 2017 global status report. Paris: REN21 Secretariat; 2017.

[2] Olivier JGJ, Janssens-Maenhout G, Muntean M, Peters JAHW. Trends in global CO2 emissions: 2017 report. The Hague/Ispra: PBL Netherlands Environmental Assessment Agency/European Commission, Joint Research Centre; 2016.

[3] GWEC. Global wind statistics 2017; 2017. Brussels, Belgium.

[4] IRENA. Renewable power generation costs in 2017. Abu Dhabi: International Renewable Energy Agency; 2018.

[5] Boyle G. Renewable energy. Oxford: Oxford University Press; 2004.

[6] Quaschning V. Renewable energy and climate change. John Wiley & Sons Ltd; 2010.

[7] Breeze P. Wind power generation. Academic Press; 2016. https://doi.org/10.1016/B978-0-12-804038-6.00014-1.

[8] Letcher T. Why wind energy? [chapter 1]. In: Wind energy engineering: a handbook for onshore and offshore wind turbines hardcover. Elsevier; 2017. ISBN 9780128094518.
[9] Bhattacharya S. Civil engineering aspects of a wind farm and wind turbine structures. [chapter 12]. In: Wind energy engineering: a handbook for onshore and offshore wind turbines hardcover. Elsevier; 2017. ISBN 9780128094518.
[10] Bhattacharya S, Goda K. Use of offshore wind farms to increase seismic resilience of nuclear power plants. Soil Dyn Earthq Eng 2016;80:65–8. https://doi.org/10.1016/j.soildyn.2015.10.001.
[11] Arany L, Bhattacharya S, Macdonald J, Hogan SJ. Design of monopiles for offshore wind turbines in 10 steps. Soil Dyn Earthq Eng 2017;92:126–52.
[12] Arany L, Bhattacharya S, Macdonald JHG, Hogan SJ. Closed form solution of Eigen frequency of monopile supported offshore wind turbines in deeper waters incorporating stiffness of substructure and SSI. Soil Dyn Earthq Eng 2016;83:18–32. https://doi.org/10.1016/j.soildyn.2015.12.011.
[13] Arany L, Bhattacharya S, Adhikari S, Hogan SJ, Macdonald JHG. An analytical model to predict the natural frequency of offshore wind turbines on three-spring flexible foundations using two different beam models. Soil Dyn Earthq Eng 2015;74:40–5. https://doi.org/10.1016/j.soildyn.2015.03.007.
[14] Arany L, Bhattacharya S, Macdonald J, Hogan SJ. Simplified critical mudline bending moment spectra of offshore wind turbine support structures. Wind Energy 2015;18:2171–97.
[15] Bhattacharya S. Challenges in the design of offshore wind turbine foundations. Eng Technol Ref IET 2014; 9, https://doi.org/1049/etr.2014.0041, ISSN 2056-4007.
[16] Bhattacharya S, Cox J, Lombardi D, Muir Wood D. Dynamics of offshore wind turbines supported on two foundations. Geotech Eng Proc ICE 2013;166(2):159–69.
[17] Bhattacharya S, Nikitas N, Garnsey J, Alexander NA, Cox J, Lombardi D, Muir Wood D, Nash DFT. Observed dynamic soil–structure interaction in scale testing of offshore wind turbine foundations. Soil Dyn Earthq Eng 2013;54:47–60.
[18] Lombardi D, Bhattacharya S, Muir Wood D. Dynamic soil-structure interaction of monopile supported wind turbines in cohesive soil. Soil Dyn Earthq Eng 2013;49:165–80.
[19] Nikitas G, Vimalan NJ, Bhattacharya S. An innovative cyclic loading device to study long term performance of offshore wind turbines. Soil Dyn Earthq Eng 2016;82:154–60.
[20] Nikitas G, Arany L, Aingaran S, Vimalan J, Bhattacharya S. Predicting long term performance of offshore wind turbines using cyclic simple shear apparatus. Soil Dyn Earthq Eng 2017;92:678–83.
[21] Yu L, Wang L, Guo Z, Bhattacharya S, Nikitas G, Li L, Xing Y. Long-term dynamic behavior of monopile supported offshore wind turbines in sand. Theor Appl Mech Lett 2015;5(2):80–4.
[22] Guo Z, Yu L, Wang L, Bhattacharya S, Nikitas G, Xing Y. Model tests on the long-term dynamic performance of offshore wind turbines founded on monopiles in sand. ASME J Offshore Mech Arctic Eng 2015;137(4). https://doi.org/10.1115/1.4030682.
[23] Bhattacharya S, Wang L, Liu J, Hong Y. Civil engineering challenges associated with design of offshore wind turbines with special reference to China. [chapter 13]. In: Wind energy engineering: a handbook for onshore and offshore wind turbines. Academic Press (Elsevier); 2017. p. 243–73.
[24] Bhattacharya S, Lombardi D, Wood DM. Similitude relationships for physical modelling of monopile-supported offshore wind turbines. Int J Phys Model Geotech 2011; 11(2):58–68.
[25] Adhikari S, Bhattacharya S. Dynamic analysis of wind turbine towers on flexible foundations. Shock Vib 2012;19:37–56.

[26] Adhikari S, Bhattacharya S. Vibrations of wind-turbines considering soil-structure interaction. Wind Struct Int J 2012;14:85–122.
[27] Bhattacharya S, Adhikari S. Experimental validation of soil–structure interaction of offshore wind turbines. Soil Dyn Earthq Eng 2012;31(5–6):805–16.
[28] Bhattacharya S, Lombardi D, Muir Wood DM. Similitude relationships for physical modelling of monopile-supported offshore wind turbines. Int J Phys Model Geotech 2012;12(2):58–68.
[29] De Risi R, Bhattacharya S, Goda K. Seismic performance assessment of monopile-supported offshore wind turbines using unscaled natural earthquake records. Soil Dyn Earthq Eng 2018;109:154–72.

Storing electrical energy

11

Trevor M Letcher[*,†,1]
[*]Laurel House, FosseWay, Stratton on the Fosse, Somerset, United Kingdom,
[†]University of KwaZulu-Natal, Durban, South Africa
[1]Corresponding author: trevor@letcher.eclipse.co.uk

Chapter Outline

11.1 Introduction

A reliable supply of electricity free from power outages is vital for the stability and economic development of any country. The generation of electricity nowadays relies on an increasingly complex set of electricity suppliers. These include: fossil-fueled power stations burning coal or oil or natural gas, wood- and biomass-burning power stations, nuclear power stations, and an increasing number of renewable energy resources, which include hydroelectric power, wind turbines, and solar photovoltaic panels. The difficulty comes when the grid operators have to balance exactly the electricity produced against the demand by the customers. With the increasing use of intermittent renewable energy, this becomes progressively more and more difficult, as it is sometimes impossible to forecast the exact amount of energy from solar and wind sources.

As a result of climate change and global warming, it is imperative that we continue developing and using renewable forms of energy, free of carbon. The renewable industry is still in its infancy, as judged by the fraction of electricity (24.5%) presently produced by renewable energy worldwide in 2016 [1]. Most of present global renewable energy is pumped hydroelectricity, which makes up 68% of all renewable forms

Managing Global Warming. https://doi.org/10.1016/B978-0-12-814104-5.00011-9

of energy and 16.6% of all electricity worldwide [1]. The only other renewable forms of energy producing any significant amounts of electricity are solar and wind energy, both of which produce intermittent amounts of electricity, with a total of about 5.5% of all worldwide production [1]. The development and implementation of both solar PV and wind energy (onshore and offshore) have advanced enormously over the past decade, but their production is very much outweighed by hydropower. Unfortunately, solar and wind energy often go to waste as they come on stream when there is no demand for electricity; for example, at night when the wind blows and the demand for electricity is low, or during the middle of the day when the sun shines, and again when the demand for electricity is low. Renewable and intermittent forms of energy, such as wind and solar photovoltaic, need storage facilities to be truly effective. A stockpile of stored energy would make the balancing of the national grid a much easier task.

The advances made in developing solar and wind energy have not been supported by similar advances in storing energy. The main reason for this is that electricity is not easy to store. Many of our present energy sources are indeed stored forms of energy: solid coal, liquid oil, or gaseous methane. As such these carbon-based energy sources are very convenient sources of energy and this makes the development of new and difficult technologies less attractive to persuade investors. We all prefer the *status quo* to investigating, researching, and finally implementing new ways of doing things.

If the level of back-up renewable energy is small compared to the conventional generation capacity, the implementation of renewable energy can be forecast with reasonable accuracy. However if the level of intermittent renewable energy is large, in relation to base load electricity generation, the task of balancing the demand becomes very much more complex, difficult, and uncertain [2]. For renewable energy to play a major part in supplying electricity to the national grid, it would have to be backed up with large storage facilities.

11.2 Electricity energy storage

With the increased complexity of electricity generation, electrical energy storage (EES) will be needed more and more to balance the generation and use of electricity in order to ensure grid reliability. EES offers a time dimension in providing electricity when it is needed. The high capital costs of managing peak demands and the investments needed to ensure grid reliability are good reasons for investing in electrical storage technologies [3]. As is often the case, it is money that will finally dictate the future path that grid operators take.

In spite of all the benefits that EES offers, it is still a very small part of the overall grid infrastructure. In September 2017, it was reported that worldwide electrical energy storage amounted to 176 GW, which is less than 2% of the world's electric power production capacity. In the US, 24 GW of electrical power was obtained from stored electricity (about 2.2% of U.S. production capacity and most of that came from pumped hydroelectric storage (PHS)) [4]. In Europe and in Japan, there is a greater reliance on EES with levels of 10% and 15%, respectively. The European Commission is aiming for 27 % of renewable Energy Storage (RES) by 2030 [5].

If we do not develop EES, there are a few options that could be considered, and some are currently being deployed to reduce small surges in the national grids:

- Linking national grids with grids from countries, which have storing facilities or have abundance of electricity from pumped hydro, solar, or wind systems.
- Changing domestic electricity usage so that water heaters and possibly refrigerators are switched off at peak periods and switched on at off-peak times. In this way, peak demands will be reduced and low demands increased. Millions of water heaters have been operated in France for decades; they provide a massive benefit in reducing peak demand, by shifting 5%, i.e., 20 TWh from peak periods to low-demand periods [6].
- Introduce auxiliary electricity-generating units (possible diesel- or gas-burning generators) that can be activated at a moment's notice to maintain the demand level of electricity. All of these are unfortunately expensive and also environmentally unfriendly; developing new EES facilities seems to be the way forward.

The concept of energy storage is not new, though, until very recently, development has been mainly restricted to pumped storage hydroelectricity, which involves the conversion of electrical energy into mechanical and potential energy by pumping water uphill into reservoirs so that when electricity is required the water can be gravity fed through turbines, to produce electricity. This is now a well-established system and has been used throughout the world to store electrical energy during off-peak times.

There are other forms of renewable energy that are more amenable to storage such as concentrated solar energy and thermal energy systems, where the heat energy can be stored as sensible heat or in phase-changing systems such as molten salts. However, these are not electricity-storing technologies, which are the subject of this chapter.

There is some urgency in developing techniques for the storing of electricity: without them global warming takes hold of the planet, CO_2 level's continue to increase unabated, and coal power stations continue to be built and deployed. A major push in developing and deploying EES systems would also encourage more investment in renewable energy sources as investors will see that renewable energy is not wasted. At the moment, with renewable energy schemes expanding, very often electricity from these projects is wasted, electricity that could have been captured and stored. The UK has recently commissioned more offshore wind turbines and in 2017 the UK built more than half the offshore wind turbines installed in Europe. Today, in Europe, there is a cumulative capacity of almost 16 GW of offshore wind and this is predicted to reach 25 GW by 2020. The UK is first with 43% of all grid-connected turbines, followed by Germany (28%), Denmark (12%), the Netherlands (9%), and Belgium (6%), representing the top five markets. Combined, the top five countries represent 98% of all grid-connected turbines in Europe [7]. Without storage facilities, much of the generated electricity will go to waste.

The need for storing electricity is so great that it does appear as though a massive new industry for the storing of intermittent electrical energy is essential. However, there are many problems to solve and difficulties to overcome before this can happen. These include:

- developing ways of integrating intermittent electrical energy into national grids;
- finding out which storing technique is best suited for each country,

- considering what level of domestic in-house storing capacity will be the norm in the future and balancing that with national grid storage;
- deciding on whether we should develop a variety of intermittent renewable options such as solar, wind tide, waves, concentrated solar power, so that when one type of energy falls off, other types can carry the load;
- recognizing that the relative cost of electrical storage might make nuclear power plants more viable, as opposed to more solar and wind energy installations.

Currently, pumped hydro storage accounts for most of global installed energy storage capacity [8]. Compressed air energy storage (CAES) is another form of EES that is being developed and deployed [9]. There are a number of other technologies that could one day be important. These include:

- rechargeable solid batteries (e.g., lead-acid, Li-ion, sodium-sulfur) [10];
- flow redox batteries (e.g., vanadium) [11];
- liquid air electrical storage [12];
- superconducting magnetic storage (SMES) [13];
- chemical energy storage—the production of H_2 or CH_4 [14,15];
- vehicle to grid [16];
- supercapacitors [17];
- hydroelectric storage, e.g., piston-in-cylinder storage [18];
- pumped heat electrical storage [19];
- high-speed flywheels [20];
- rail energy storage [21].

Each storage technique has its own properties and advantages. The power rating is important for large grid storage; times of discharge and response times are important when electricity demand is suddenly increased; life time is a factor when considering systems that have very high initial costs; and the number of cycles that a system can be put through is particularly import for storage systems such as flywheels, batteries, supercritical magnets. A list of attributes and specifications of a number of EES systems is given in Table 11.1.

11.3 Pumped hydropower

Pumped hydroelectricity storage (PHS) is the oldest kind of large-scale energy storage and works on a very simple principle—two reservoirs at different altitudes are required and when the water is released from the upper reservoir to the lower reservoir, energy is created by the downflow, which is directed through a turbine and generator to create electricity. The water is then pumped back to the upper reservoir. The loss of energy in transforming energy from one form to another and back again is of the order of 80% [22]. Presently 95% of all the worldwide large-scale electric storage capacity is provided by PHS facilities [4]. In spite of its usefulness and the fact that it is a well-tried and researched technology, it does have its limitations. The reservoirs, for example, depend on the inflow of water and in many cases the inflow is seasonal; for example, glacial melt and also rainy seasons. Furthermore, it does take time to

Table 11.1 EES specifications

System	Power rating (MW)	Time of discharging	Response time	Cycles or lifetime
PHS	100–3000	4–16h	10s–minute	60 years
CAES	10–1000	2–30h	Minutes	100 years
Flywheel	0.001–20	1s–minutes	<s	10^5 cycles
Li-ion	0.05–200	1min to 8h	<s	10^4 cycles
Lead acid	0.001–100	1min to 8h	<s	6 years
NaS	10–100	1min to 8h	<s	5×10^3 cycles
Redox flow	0.1–200	Hours	<s	10^4 cycles
SC magnet	0.1–1	ms–seconds	<s	10^5 cycles
S capacitor	0.01–1	ms–seconds	<s	10^5 cycles
H_2	0.01–100	Minutes–week	s	30 years
CH_4	1–100	Hours–week	s	30 years

Many of the data are taken from https://www.worldenergy.org/wp-content/uploads/2017/03/WEResources_E-storage_2016.pdf, 14 p (Accessed 11 February 2018).

pump water from the lower reservoir to the top reservoir, and this can only be done in times of excess renewable energy, so there is not a continuous stream of electricity on tap. However, until new storing technologies have been developed, PHS will remain the most important of the storage technologies.

Under suitable conditions, pumped hydro storage does provide a dynamic response and offer critical back-up during periods of excess demand by maintaining grid stability. Its main advantages are its flexibility and the fact that it is the most developed large-scale energy storage technology currently available. Pumped hydro storage is a net user of power—it uses electricity to pump the water back up to the top of the reservoir; in an ideal situation it can resolve intermittency by working in conjunction with other forms of renewable energy.

The largest pumped hydro grid power storage systems in the world, at the moment, is in Bath County, Virginia, USA, and has a generating capacity of 3GW. It is known as the world's biggest battery [23]. A larger PHS facility, which will be the world's largest, is being built in China. It is the Fengning plant in Hebei Province, China, and when completed in 2021 and will have a storage capacity of 3.6GW [24].

In the UK there are four PHS facilities, which amount to over 2.8GW of total capacity and which have an energy storage capacity of about 26.7GWh. The plants with their storage capacity in parentheses are Dinorwig (9.1GWh) and Festiniog (1.3GWh) in Snowdonia, Wales; and, Cruachan (10GWh and Foyers (6.3GWh) in Scotland. The last PHS scheme was commissioned 34 years ago and since then there have been no new developments. However, plans are underway for the extension of existing hydroelectric schemes to incorporate storage facilities and also the building of new relatively small facilities [25]. At the moment, the total storage capacity is only about 14% of the storage capacity required by the UK grid system.

Worldwide PHS facilities are summarized in Table 11.2.

Table 11.2 Worldwide PHS facilities by country

Country	Power capacity (GW)
China	32
Japan	28.3
USA	22.6
Norway	20
Spain	8
Italy	7.1
India	6.4
Switzerland	5.8
France	5.8

Many of the data have been taken from www.irena.org/.../2017/Oct/Electricity-storage-and-renewables-costs-and-markets, 30 p, Table 2 (Accessed 17 February 2018).

It has been mooted that more use should be made of the PHS infrastructure in Norway, for example, to assist other European countries with electricity when back-up falls below the demand. Unfortunately, the reservoirs in Norway and in other countries such as Switzerland are dependent on glacial melt and seasonal rains for keeping the water levels high. Furthermore, the transmission capacity between Norway and the rest of Europe would have to be substantially improved [26].

11.4 Compressed air energy storage

CAES involves the conversion of electrical energy into high-pressure compressed air that can be released at a later time to drive a turbine generator to produce electricity. This makes it ideally suited to work along intermittent energy sources such as solar PV and wind energy. There are a number of storage options, but the best option is to store the compressed air in existing geographical formations such as disused salt mines. Salt caverns are usually free of cracks and fissures as any ingress of water through cracks will dissolve salt, which then crystallizes and creates air-tight seals. CAES is the second biggest form of energy storage currently behind PHS. One advantage CAES has over PHS is that CAES is cheaper to develop.

The first CAES plant to be built was at Huntorf, Germany, in 1978 and is still in operation. The plant consists of two underground salt caverns. The filling process takes 8 h and when electricity is required the compressed air is released and heated by combusting natural gas to get the air to expand. This drives a turbine and can produce electricity at a rate of 290 MW for 2 h [27]. There is a CAES facility in McIntosh, Alabama, USA, with a total storage capacity of 400 MW, and the UK is considering investing in CAES [9].

There are two major problems associated with CAES. The first is that when air is compressed, it heats up. Unfortunately, the warmer the air, the less will be the amount of air that can be stored. The second problem is that on releasing the compressed air, the pressure in the cavern is slowly reduced and this affects the amount of electricity produced by the turbine. The first problem can be dealt with in three ways: adiabatically, by storing the heat and reusing it when the air is expanded to produce power; isothermally, with the aid of heat exchangers; and diabatically, by dissipating the heat to the atmosphere. The second problem can be dealt with by controlling the rate at which the air is discharged and thus creating a constant supply of electricity. Another solution being researched by Seamus Garvey at Nottingham University is to store the air in large energy bags deep in lakes or in the sea. In this way, the pressure of the air leaving the bags remains constant, being the hydrostatic pressure [28]. A good overview of CAES technology can be found in references in [29].

11.5 Battery energy storage systems

Advances in battery technologies have greatly increased their reliability and are important in grid management today. Batteries have the following advantages: able to rapidly respond to frequency regulation signals on the grid; be deployed to manage variable renewable energies; provide energy during peak electricity demands; and condition the flow of power so that it can be used by sensitive electronic equipment. Finally, batteries are a modular solution to energy storage and thus can be mobilized and relocated if they are needed elsewhere.

Many new types of industrial batteries are now replacing the familiar lead-acid battery, which has been a mainstay of the battery energy storage systems (BESS) since its discovery in 1859 by the Frenchman Gaston Planté. One of the first applications was to power lights in stationary train carriages in stations. The lead-acid battery has undergone many improvements and is still being used in many grid-supporting roles such as frequency control and load leveling and is now being used to store intermittent renewable energies. These systems are usually in the 1–40 MW range. Examples include: Berlin (Bewag) 8.5 MW; Shetland Islands, 11 MW; California (Chino) 10 MW; Puerto Rico (PREPA) 20 MW; Alaska (Metlakatla) 1 MW; Hawaii (Kahuku wind farm) 15 MW; and USA (Notrees wind energy project) 36 MW [30].

The lithium-ion battery is the "new boy on the block" and was first used commercially in the 1980s. Its main advantages are; high density, high cycle and long life, fast and efficient charging with little energy wasted, low self-charging rate, light weight compared to the lead acid battery, and there is no need to hold the battery upright as in the case of the lead acid battery [31].

The Li-ion battery, too, has undergone many improvements over the past few decades and is now used for security of supply to national and regional electricity grid networks. A few such BESS schemes using Li-ion batteries are: USA (New York), 16 MW; USA (Laurel Mountains) 32 MW; Japan (Tohoka) 40 MW; Australia (Adelaide) 100 MW [10,31,32]. In the UK there are plans to deploy lithium-ion batteries with a combined capacity of 200 MW, under a new scheme to help grid balance supply and demand within seconds [33]. Projects include a 10-MW storage battery at

Blackburn Meadows and Centrica (parent company of British Gas), is building a 49-MW facility in Barrow-in-Furness, Cumbria, while EDF is working on one of the same size at its West Burton gas power station in Nottinghamshire [34].

The use of battery storage in private houses linked to solar PV installations is becoming very popular. In Germany alone, in conjunction with PV installations, there are around 25,000 domestic installations with total capacity of 160 MWh [35]. The total battery capacity in electric vehicles is growing rapidly and it is possible to use a large number of car batteries as grid energy storage using a technology known as "vehicle to grid" [36].

Sodium-sulfur batteries are also being developed and industrialized especially in Japan and demonstration projects have appeared in the US. These batteries operate at elevated temperatures of 300–350°C and are best suited for grid electricity rather than for vehicular transport [37].

Recently, it was reported that 60% of all the stationary battery projects in the world (441 MW) were Na-S batteries [38,39].

Flow batteries and the vanadium redox flow battery in particular have the potential to be important for grid-level electricity storage. Flow batteries consist of two tanks of electrolytes, which are pumped into a reactor where they generate a charge. The capacity of the system is a function of the size of the tanks holding their respective liquids. The battery system is bulky and has the potential to store large amounts of energy for long periods of time and hence is most suited for grid storage. Furthermore, compared to other grid-scale systems, flow batteries are economical. The working principles of the vanadium redox flow battery, VRFB, can be found in reference [11].

Flow batteries are beginning to be deployed in various parts of the world. The UK commissioned a small system in 2017 with a storage capacity of 1 MWh, and a maximum output of 90 kW, and a lifespan of at least 25 years. In China a 200-MW, 800-MWh plant was commissioned in 2017 (Rongke Power). It is the world's largest VRFB installation. In South Africa, a 450-kWh VRFB system will begin operation in 2018 [40]. In the USA, a 2-MW/8-kWh VRFB system was commissioned in 2017. It looks as though this technology is slowly being accepted by those responsible for national grids [41,42].

Other rechargeable battery systems include: nickel-cadmium, sodium nickel chloride, nickel hydrogen, nickel-zinc, and lithium-cobalt chloride batteries [43]. Most of these are not yet deployed in storing grid electricity.

11.6 Liquid air energy storage

Liquid Air Energy Storage (LAES) uses electricity (wind or solar or excess electricity at low demand times) to cool air (using liquid air or liquid nitrogen at a temperature of −196°C) until it liquefies, stores the liquid air in a well-insulated tank, brings the liquid air back to a gaseous state (waste heat from an industrial process), and uses that pressurized gas to turn a turbine to generate electricity. It uses off-the-shelf components, so can be considered a low-risk technology. Details of the technology can be found in reference. Feasibility and demonstration plants have been erected in Japan, the US, and in the UK at Manchester (5-MW plant) [12].

11.7 Superconducting magnetic storage

When a material is cooled below its critical temperature T_c, the material exhibits zero electrical resistance. Under these conditions, an electric current in a loop made of the material will persist indefinitely with no power source. The electricity is effectively stored in a magnetic field created by the flow of electricity (DC current). In this way, electrical energy can be stored as magnetic energy. When the electrical energy is required, it can be released back into the network by discharging the coil. An SMES system includes three parts: a cryogenically cooled refrigerator; a superconducting coil; and a power-conditioning system. The latter includes an inverter/rectifier to transform A power to DC power and to convert DC power back into AC power. The great advantage of SMES is that the energy loss from AC to DC and back amounts to only about 5%. Of all the electrical storage technologies, SMES loses the least amount of electricity. Another advantage is that it can in principle store large quantities of electricity, which can be discharged almost instantaneously. The fact that it can release large amounts of electricity within a fraction of a cycle means that it is suitable to replace sudden losses or dips in grid power. As a result, SMES is ideally suited for maintaining grid reliability in this modern age with the high penetration of renewable energy sources. Although SMES is ideally suited for grid electricity storage, it has yet to be fully deployed as a result of the high cost of cooling the coils to below T_c (using liquid helium at between 4 K and 20 K or liquid hydrogen at between 20 K and 77 K or liquid nitrogen at 77 K and above) and the high cost of the superconducting wire. The superconductor material is a key issue for the development and deployment of SMES. The superconducting magnetic materials include MgB_2 ($T_c = 39$ K) and a bismuth, strontium, calcium, copper, oxygen complex (T_c ranging from 33 K to 100 K depending on the molar ratio of the elements). The higher the value of T_c, the more cost effective will be the SMES. SMES is currently used for short-duration energy storage and for improving power quality. SMES is a technology that is still very much at the development and research stage. Much of the development work is being done in Japan (a 10-MW, 7.3-MW, and 5-MW plants), in the USA (3-MW and a 1-MJ system) and in Europe in Germany (5-MJ, 2-MJ plants) [13].

11.8 Chemical Storage (H_2 and CH_4)

Intermittent electricity from renewable sources can be used to electrolyze water to produce hydrogen, which can be stored and used directly, especially in motor vehicles or indeed mixed with natural gas (CH_4) and used for heating. However, there is a limit (in the UK it is 0.1% by mole), to the amount of hydrogen that can be mixed with natural gas. Currently, the largest energy storage projects using electrolysis to produce H_2 are in Germany; both Grapzow (1 MW) and Falkenhagen (2 MW) facilities are coupled to wind turbines. An earlier plant, also in Germany (Nuenberg), ran for 8 years, until 1999 as a demonstration plant; it was coupled to solar PV panels. These pilot plants have spawned many demonstration projects and bode well for future deployment [14,44]. This process of using excess electrical energy from renewable sources to electrolyze water to produce hydrogen and indeed to produce methane is known as "power to gas" [45].

Hydrogen together with CO_2 can be used to produce CH_4 using suitable catalysts (Sabatier process). The CH_4 can then be fed into national gas grids. This work is still very much in the development stage [45]. More details can be found in references [14, 15, 45].

11.9 Vehicle-to-grid systems

This is a novel system, which looks into a future when electric cars will be the norm. By drawing power from the grid during the midday hours when solar output is greatest, electric vehicles can soak up the sun-generated power and in turn reduce the evening ramp-up by returning power to the grid. And furthermore you could supply electricity to your home with your own solar power. This is the vision behind the high-tech vehicle-to-grid (V2G) and vehicle-to-home (V2H) energy solutions for the future. These schemes are being researched in many parts of the world including Europe and in the UK where the Government is funding research with almost £30 million. Parliamentary Under-*Secretary* for Transport, Jesse Norman, has said: "As the number of electric vehicles grows and their battery capabilities increase, there is a huge opportunity for them to make a significant contribution to a smart grid" [15].

11.10 Other methods of storing electrical energy

One of the exciting aspects of this field of storing electrical energy is the wide range of options available. We have discussed, very briefly, seven of these possibilities. Many others are being investigated and they include the following five technologies: high-speed flywheels, supercapacitors, pumped heat electrical storage, advanced rail and piston-in-cylinder storage. Many of these have yet to be demonstrated, while others have working pilot projects with the aim of proving their viability.

11.11 Conclusion

Climate change, global warming, extreme weather conditions, all exacerbated by the rapid rise in atmospheric CO_2 are driving the use of renewable energy sources for electricity production. This has led to a rapid rise in the deployment of both solar and wind energy systems. The intermittent nature of solar and wind energy is making it difficult to balance national electricity grids and also making it difficult to use all the energy available. In this chapter, the necessity for developing EES systems has been made together with the need to do so urgently. In addition, options for the storage of intermittent electrical energy have been discussed.

References

[1] Letcher TM. Why solar energy? In: Letcher TM, Fthenakis V, editors. A comprehensive guide to solar energy systems: with a special focus on photovoltaic systems. Oxford: Elsevier; 2018, ISBN: 978-0-12-811479-7. Chapter 1, Figure 1.3.

[2] Letcher TM, Stein G. Integration of PV generated electricity into National Grids. In: Letcher TM, Fthenakis VM, editors. A Comprehensive Guide to Solar Energy Systems: with special focus on photovoltaic systems. Oxford: Elsevier; 2018, ISBN: 978-0-12-811479-7. Chapter 17.

[3] Dunn B, Kamath H, Tarascon J-M. Electrical Storage for the Grid: a battery of choices. Science 2011;334:928–35. https://doi.org/10.1126/science.1212741.

[4] https://www.c2es.org/content/electric-energy-storage/, 3 p (Accesses 12 February 2018).

[5] European Commission, Energy storage—the role of Electricity, Brussels, 2017, 2 p or https://ec.europa.eu/energy/sites/ener/files/documents/swd2017_61_document_travail_service_part1_v6.pdf, 2 p (Accessed 11 February 2018).

[6] https://www.worldenergy.org/wp-content/uploads/2017/03/WEResources_E-storage_2016.pdf, 7 p (Accessed 11 February 2018).

[7] https://www.offshorewind.biz/2018/02/06/eu-goes-from-12-6gw-to-15-8gw-of-offshore-wind-in-one-year/, pp. 1–2 (Accessed 11 February 2018).

[8] Yang C-J. Pumped hydroelectric storage. In: Letcher TM, editor. Storing energy: with special reference to renewable energy sources. Oxford: Elsevier; 2016. p. 25–38, ISBN: 978-0-12-803440-8. Chapter 2.

[9] Donadei S, Schneider G-S. Compressed air energy storage in underground formations. In: Letcher TM, editor. Storing energy: with special reference to renewable energy sources. Oxford: Elsevier; 2016. p. 113–34, ISBN: 978-0-12-803440-8. Chapter 6.

[10] Vetter M, Lux S. Rechargeable Batteries with Special reference to Lithium-ion batteries. In: Letcher TM, editor. Storing Energy: with special reference to renewable Energy Sources. Oxford: Elsevier; 2016. p. 205–26, ISBN: 978-0-12-803440-8. Chapter 11.

[11] Doetsch C, Burfeind J. Vanadium redox flow batteries. In: Letcher TM, editor. Storing Energy: with special reference to renewable energy sources. Oxford: Elsevier; 2016. p. 227–46, ISBN: 978-0-12-803440-8. Chapter 12.

[12] Ding Y, Tong L, Zhang P, Li Y, Radcliffe J, Wang L. Liquid air energy storage. In: Letcher TM, editor. Storing energy: with special reference to renewable energy sources. Oxford: Elsevier; 2016. p. 167–81, ISBN: 978-0-12-803440-8. Chapter 9.

[13] Luo X, Wang J, Dooner M, Clarke J. Overview of current development in electrical energy storage technologies and the application potential in power system operation. Appl Energy 2015;137:511–36.

[14] Chisholm G, Cronin L. Hydrogen from water electrolysis. In: Letcher TM, editor. Storing energy: with special reference to renewable energy sources. Oxford: Elsevier; 2016. p. 315–43, ISBN: 978-0-12-803440-8. Chapter 16.

[15] Wagner L. Overview of energy storage technologies. In: Letcher TM, editor. Future Energy: Improved Sustainable and Clean Options for our Planet. Waltham MA, USA: Elsevier; 2014. p. 627–8, ISBN: 978-0-08-099424-6. Chapter 27.

[16] Zhou Y, Li X, Vehicle to grid technology: a review, 34th Chinese Control Conference 2015, 28–30 July, Hangzhou, China. https://doi.org/10.1109/chiCC.2015.7261068, 2015. ISMB: 978-9-8815-6389-7.

[17] Wagner L, Overview of energy storage technologies in Future energy: improved sustainable and clean options for our planet, Letcher, TM editor, Elsevier, Waltham, MA, USA, ISBN; 978-0-08-099424-6, Chapter 27, 613-632.

[18] Escombe F. Novel hydroelectric storage concepts. In: Letcher TM, editor. Storing energy: with special reference to renewable energy sources. Oxford: Elsevier; 2016. p. 39–67, ISBN: 978-0-12-803440-8. Chapter 3.
[19] http://ease-storage.eu/wp-content/uploads/2016/07/EASE_TD_Mechanical_PHES.pdf (Accessed 17 February 2018).
[20] Bender D. Flywheels. In: Letcher TM, editor. Storing energy: with special reference to renewable energy sources. Oxford: Elsevier; 2016. p. 183–204, ISBN: 978-0-12-803440-8. Chapter 10.
[21] Cava F, Kelly J, Peitzke W, Brown M, Sullivan S. Advanced rail energy storage. In: Letcher TM, editor. Storing energy: with special reference to renewable energy sources. Oxford: Elsevier; 2016. p. 69–86, ISBN: 978-0-12-803440-8. Chapter 4.
[22] Dodds PE, Garvey SD. The role of energy storage in low-carbon energy systems. In: Letcher TM, editor. Storing energy: with special reference to renewable energy sources. Oxford: Elsevier; 2016. p. 9, ISBN: 978-0-12-803440-8. Chapter 1.
[23] https://www.drax.com/technology/every-electricity-storage-technology-need-know/, pp. 1–2 (Accessed 11 February 2018).
[24] http://www.hydroworld.com/articles/2017/09/new-chinese-pumped-storage-hydro-plant-to-be-world-s-largest-when-complete-in-2021.htm, 1 p (Accessed 11 February 2018).
[25] http://www.british-hydro.org/downloads/TypesofHydro/Pumped%20Storage%20report.pdf, 15 p, Table 5-1 (Accessed 11 February 2018).
[26] Hendena AL, Doormanb G, Helsetha A. Economic analysis of large-scale pumped storage plants in Norway. Energy Procedia 2016;87:116–23.
[27] https://www.thegreenage.co.uk/tech/compressed-air-energy-storage/, pp. 1–4 (Accessed 12 February 2018).
[28] Pimm A, Garvey SD. Underwater Compressed Air Energy Bags. In: Letcher TM, editor. Storing Energy: with special reference to renewable Energy Sources. Oxford: Elsevier; 2016. p. 135–54, ISBN: 978-0-12-803440-8. Chapter 7.
[29] Garvey SD and Pimm A, Compressed air energy storage in Storing energy: with special reference to renewable energy sources, TM Letcher ed, Elsevier, ISBN: 978-0-12-803440-8, Oxford, 2016 Chapter 5, pp 87 and https://www.eera-set.eu/wp-content/uploads/Overview-of-Current-Development-on-Compressed-Air-Energy-Storage_EERA-report-2013.pdf, pp. 1–39 (Accessed 12 February 2018).
[30] May GJ, Davidson A, Monahov B. Lead batteries for utility energy storage: a review. J Energy Storage 2018;15:145–57.
[31] Nitta N, Wu F, Lee JT, Yushin G, Li-ion battery materials: present and future, Materials today 2015;18(5) 252-264.
[32] https://www.theverge.com/2017/12/1/16723186/elon-musk-battery-launched-south-australia, pp. 1–2 (Accessed 11 February 2018).
[33] Private communication with Stein G of UK's National Grid.
[34] Guardian Newspaper, 9 October 2017 or https://www.theguardian.com/environment/2017/oct/09/uk-first-mega-battery-plant-come-online-sheffield-eon-renewable-energy, 1 p (Accessed 11 February 2018).
[35] https://www.worldenergy.org/.../World-Energy-Resources-Full-report-2016.10.03.pdf, 865 p (Accessed 11 February 2018).
[36] Wade NS, Taylor PC, Lang PD, Jones PR. Evaluating the Benefits of an electrical energy storage system in a future smart energy grid. Energ Policy 2010;38:7180–8.
[37] Neves D, Silva CA, Connors S. Design and implementation of hybrid renewable energy systems on micro-communities: a review on case studies. Renew Sust Energ Rev 2014;31:935–46.

[38] https://www.worldenergy.org/wp-content/uploads/2017/03/WEResources_E-storage_2016.pdf, 10 p (Accessed 11 February 2018).
[39] Hueso KB, Armand M, Rojo T. High temperature sodium batteries: status, challenges and future trends. Energy Environ Sci 2013;6:734–49.
[40] https://markets.ft.com/data/announce/detail?dockey=1323-13438151-2A3LH9JIC26USHVGBVKF9PJ759 (Accessed 17February 2018).
[41] Chen H, Xu Y, Liu C, He F, Hu S. Storing energy in China—an overview. In: Letcher TM, editor. Storing energy: with special reference to renewable energy sources. Oxford: Elsevier; 2016. p. 522, ISBN: 978-0-12-803440-8. Chapter 24.
[42] https://spectrum.ieee.org/green-tech/fuel-cells/its-big-and-longlived-and-it-wont-catch-fire-the-vanadium-redoxflow-battery (Accessed 17 February 2018).
[43] Omar J, Posada G, Rennie AJR, PerezVillar S, Martins VL, Marinaccio J, Barnes A, Glover CF, Worsley DA, Hall PJ. Aqueous batteries as grid scale energy storage solutions. Renew Sust Energ Rev 2017;68(2):1174–82.
[44] Carroquino J, Roda V, Mustata R, Yago J, Valiño L, Lozano A, Barreras F. Combined production of electricity and hydrogen from solar energy and its use in the wine sector. Renew Energy 2018;122:251–63.
[45] Tichler R, Bauer S. Power to gas. In: Letcher TM, editor. Storing energy: with special reference to renewable energy sources. Oxford: Elsevier; 2016. p. 373–90, ISBN: 978-0-12-803440-8. Chapter 18.

Further reading

[46] https://www.worldenergy.org/wp-content/uploads/2017/03/WEResources_E-storage_2016.pdf, 14 p (Accessed 11 February 2018).
[47] www.irena.org/.../2017/Oct/Electricity-storage-and-renewables-costs-and-markets, 30 p, Table 2 (Accessed 17 February 2018).

Bioenergy

12

Mirjam Röder[1], *Andrew Welfle*
Supergen Bioenergy Hub, Tyndall Centre for Climate Change Research, School for Mechanical, Aerospace and Civil Engineering, The University of Manchester, Manchester, United Kingdom
[1]Corresponding author: m.roeder@aston.ac.uk

Chapter Outline

12.1 The role of bioenergy

Bioenergy is energy generated from organic matter of plant and animal origin such as agricultural and forest residues, energy crops, wood, or organic wastes. Bioenergy is considered a low carbon form of renewable energy, as the natural process of photosynthesis within plants locks atmospheric CO_2 into organic matter, which when combusted however many years later or in whatever form, releases the CO_2 back into

Managing Global Warming. https://doi.org/10.1016/B978-0-12-814104-5.00012-0

the atmosphere. This means there is a removal of carbon at the point of plant growth and a release of this carbon at the point of energy conversion [1].

Bioenergy is the major renewable energy form worldwide, making up about 70% of all primary renewable energy sources [2]. However, it is important to consider how bioenergy is generated and used. Most of the globally used bioenergy is used for traditional domestic use in form of fuelwood, charcoal, agricultural waste, and animal dung for cooking and heating in the global south. This way of using biomass is often highly inefficient and unsustainable as the widespread collection of traditional fuelwood and charcoal production can lead to significant deforestation. Use of bioenergy within traditional cooking systems is also often highly inefficient, and the incomplete combustion of the fuels can lead to significant emissions (in particular particulates) with serious health implications [3]. Worldwide about 2.5 billion people, rely on traditional biomass use and about 1.3 million people, in particular women and children die prematurely every year due to indoor pollution and related respiratory illnesses from traditional biomass use [3,4].

Developing modern bioenergy system in the global south therefore has a large potential for providing sustainable, low-carbon energy facilitating a range of key sustainable development goals (SDG). The United Nations Environment Programme recommend that wherever biomass resources exist with the potential for energy generation, strategies and activities should be developed that target bioenergy pathways that may enhance rural development and alleviate poverty while safeguarding ecosystems [5]. Implemented correctly, bioenergy can provide both urban and rural communities with access to renewable energy and in the process can help: reduce greenhouse gas (GHG) emissions; provide economic and infrastructure development opportunities; while providing potential training, employment, and revenue generation pathways for the estimated 2.5 billion people who are dependent on the land and agriculture [6].

There are many modern bioenergy technologies that can generate energy that replaces fossil fuels and significantly reduces carbon emissions supporting national and global mitigation targets. Therefore, the focus of this chapter will be on modern uses and application of biomass and bioenergy.

12.2 Advantages of bioenergy

Bioenergy is a highly attractive energy option for countries at all stages of development, due to its high flexibility and ability for integration into wide-ranging energy systems [7]. One of the big advantages of bioenergy is its flexibility compared to other renewable energies, such as wind, solar, hydro, or tidal. Bioenergy is highly versatile as it can be generate electricity, heat or transport fuels, and solid, liquid and gaseous energy carriers can be derived from a vast array of biomass feedstocks using different bioenergy pretreatment and conversion technologies. As Fig. 12.1 documents, bioenergy provides multiple options of: thermal (combustion), thermochemical (gasification, pyrolysis, torrefaction), biochemical (fermentation, anaerobic digestion), and chemical (transesterification) conversion.

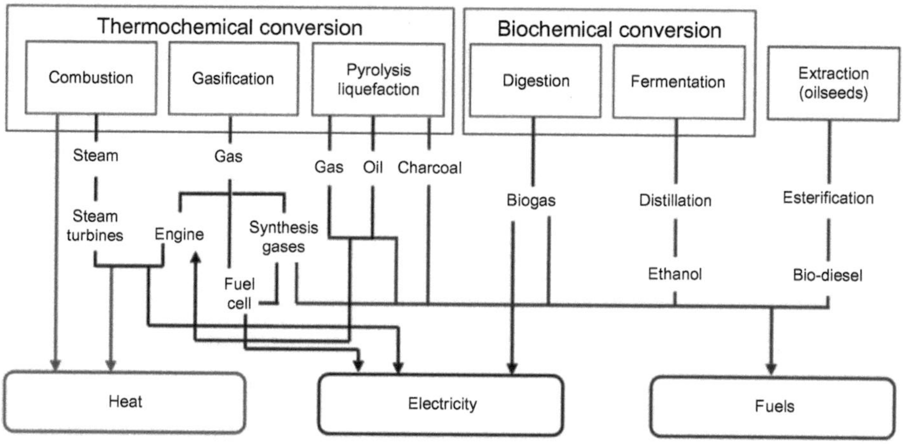

Fig. 12.1 Summary of the core bioenergy conversion pathways & resulting energy vectors [8].

Bioenergy is also easy to store in form of feedstock or energy carrier (solid, liquid, or gashouse fuel) and then can be converted when actually needed. Bioenergy can therefore be considered as a dispatchable renewable energy option that can offer valuable baseload energy compared to the other renewables. Bioenergy system can also complement existing energy infrastructures, technologies, and applications. For example, wood-based biomass in form of wood pellets can be 5 % to 15% cofired in coal power stations without converting the boilers [9]. Some large-scale power stations have converted or installed boilers that can be fired with 100% biomass, leading to significant carbon reductions compared to coal [10–12]. Another example for bioenergy fitting in with existing infrastructures and technologies is the blending of bioethanol and biodiesel with fossil fuels. Biofuels are increasingly being blended with conventional transport fuels, thus reducing the carbon intensity [13]. Biomethane is another example of replacing a fossil fuel (natural gas); it can be used for heating and transport and can be injected directly into an existing gas grid.

Bioenergy can offer various other services apart from providing energy. Organic and wood-based waste materials are valuable bioenergy feedstocks and as such bioenergy is often seen as a waste management option. Using anaerobic digestion for treating and managing livestock manures and slurry and organic (mainly food) waste is a long-established application across the world. Capturing landfill gas and gas from wastewater and sewage sludge are both very common and are main sources of biogas in some countries [14]. The use of agricultural and forest residues for bioenergy is another example of managing waste as residues often get burned in the fields, causing significant CO_2 and particle emissions [11,15,16]. Using such residues for energy can reduce these direct emissions from in-field burning and replace fossil fuels. In general, organic wastes and agricultural and forest residues are valuable bioenergy feedstock, which can provide energy and manage waste at the same time [10,17,18].

12.3 Emission reductions and carbon balance

Bioenergy can make valuable contributions to the transition to a low-carbon future. Replacing fossil-fuel-based energy (electricity, heat, and transport) with bioenergy can lead to significant emission reductions in line with climate change emission targets of many countries. Replacing electricity from fossil fuels and particularly coal with residues from forests and timber processing or waste wood can reduce emissions by up to 90% [10–12]. However, such numbers are specific to selected supply chains and applications and cannot be considered as generic numbers for all types of bioenergy supply chains. Emission reductions of 80% to 90% are typically comparisons of bioenergy applications to high carbon-intense fossil fuels, such as coal and oil. When comparing bioenergy applications to natural gas, emission reductions achieved are much lower due to the comparative lower carbon intensity of natural gas [10–12,19–21]. Therefore, the GHG benefits of bioenergy systems compared to conventional energy systems are highly dependent on the GHG intensity of the alternative fuels.

The GHG performance of different bioenergy systems can be highly variable as there are typically many variations and uncertainties along bioenergy supply chains. This makes the assessment of the carbon intensity and carbon emissions of different bioenergy systems highly challenging. Upstream emissions of supply chains are emissions related to material and energy inputs during biomass production, sourcing, transport and processing, including how biomass is produced, managed, processed, and converted. These emissions vary depending on how biomass is produced and sourced, with potentially great variations between geographic locations where biomass resources may be grown, transported, and processed. As natural conditions also change from year to year, inputs like fertilizer, water, plant protection as well as biomass yields and availability can vary, creating further variations for considerations and with it increased uncertainty of supply-chain emissions [1]. All these variations and uncertainties are important to consider when assessing the overall GHG performance of bioenergy systems and when evaluating their emission reduction potential. Such assessments offer a snapshot of emissions of the specific supply chain.

This becomes even more important when considering bioenergy systems in terms of emission budgets and cumulative emissions. The latter describes the amount of carbon emissions released and accumulated in the atmosphere. The timeframe and amount of the released emissions matters as it determines the amount of emissions that may be released in the context of climate change targets. In principle, the more emissions released in the short term, the less have to be the emissions released in the future as a result of more radical emission reduction efforts [1,22,23]. This is very relevant when considering bioenergy, in particular system where biomass grows slowly in long harvest cycles.

The timeframes and length of growth and harvest cycles of plants used within bioenergy systems will determine the sequestration timeframes when carbon is taken up from the atmosphere and then eventually released through bioenergy combustion. If the utilized biomass is regrown annually (e.g., residues from cereals, grass from pasture), the carbon released to the atmosphere is resequestered within a very short timeframe of months. If biomass is used from plants that have been growing and sequestering for a very long time (e.g., residues from trees, which have been growing for several decades), it will take several years before this carbon is resequestered from the atmosphere and is therefore taking up "space" within the carbon budget. This timeframe is very important in terms of calculating overall emission budgets, and demonstrates the importance of evaluating the true overall GHG performance of bioenergy systems considering all upstream emissions from supply-chain processes and activities, as well as the release of biogenic carbon that has been sequestered during plant growth and is released during bioenergy generation.

Evaluating supply-chain emissions and temporal carbon balances of bioenergy systems allows us to understand how we can reduce emissions to meet our emission targets. This also helps us to assess practices along the supply chain to maximize benefits, emission reductions and carbon savings, and minimize negative impacts.

12.4 Sustainability

Bioenergy differs from all other renewable and conventional energy pathways in that it is directly tied to the farms, forests, and ecosystems from which biomass resources and feedstocks are produced and extracted. This close association within bioenergy systems and supply chains creates the potential for wide-ranging environmental and social impacts that can be both positive and negative. As bioenergy pathways are being assessed in many national energy strategies globally [24], increased mobilization of biomass resource will be required to meet inevitable increases in demand. As a consequence, there are likely to be many limitations and notable sustainability implications relating to this increased global demand of biomass resource.

The EU has developed a series of nonlegally binding sustainability criteria (European Commission, 2010) that in part set guidelines for the levels of GHG savings and sustainability performance that should be achieved through the generation of bioenergy from a given biomass resource. There is also a proposed framework for these guideline sustainability criteria to be made progressively more stringent as the sustainability credentials and potential GHG impact of utilizing increasing levels of biomass resources has become a vital area of discussion with growing focus.

The emission reduction potential of bioenergy for decarbonizing the energy sector is often the focus of bioenergy strategies; however, it is important to see bioenergy beyond carbon as biomass production can compete with other sectors and aspects of society that also rely on these resources. To justify the use of bioenergy as renewable, it has to be conforming to sustainability implications beyond carbon.

Land use is one of the most discussed aspects of the sustainability of bioenergy. In the context of bioethanol and biodiesel production from cereals, corn, rapeseed, and soya, there have been many discussions about a competition between energy crop and food and increasing food prices [25]. Sustainability concerns are even bigger if there is a risk that energy crop production could lead to the clearance of forest and or fuelwood collection to deforestation [26]. With lessons learned during the last few decades, many countries established sustainability guidelines and certifications schemes for bioenergy, in particular biomass sourcing to reduce negative sustainability impacts. However, biomass production can also offer beneficial services to land and land use. Many perennial energy crops can be grown on low-quality and contaminated land, not suitable for food or livestock production and can thus be brought back into production and remediated [27–29]. There is also an increasing use of perennial energy crops for flood prevention and soil protection during floods [27]. In some geographical regions, the growth of biomass as intermediate crops reduces and avoids fallow periods, which in turn improves soil quality and reduces erosion [18].

Additional to environmental and conservational implications, bioenergy can have positive socioeconomic affects and improve livelihoods. Bioenergy is increasingly being introduced as part of existing agricultural and forest systems. For example, instead of burning harvest residues in-field, the residues are utilized for energy purpose. This reduces environmental impact of *in situ* burning and provides farmers and foresters with an additional income and improved energy access, assuming that the energy is locally generated [16]. Hence, bioenergy does not mean that existing systems are replaced but it can become part of it.

It is important that bioenergy is considered as a component of another system, which is often a complex agricultural or forestry system, with various aspects that can be impacted with possible synergies and tradeoffs between these aspects. Maximizing the specific benefits in one aspect is likely to have an impact on another. For example, a feedstock/technology achieves high carbon savings and is cost effective, but affects land access, food production, and biodiversity; a bioenergy system includes many upstream small-scale businesses but is high in cost. This will become ever more apparent in a bioeconomy, with increased competition for resources and services [30,31].

It is therefore important to consider what needs to be achieved when introducing bioenergy into existing systems: is it about low-carbon energy generation and replacing fossil fuels, or providing additional income, or creating jobs, or supporting environmental and conversational benefits, or improving energy security, etc. [18]. For this, it is essential to understand the synergies and tradeoffs of the existing system as well to avoid any unintended consequences.

Encouraging and informing practitioners about best practices and synergies and tradeoff of bioenergy systems can be supported by sustainable measures. The bioenergy sustainability requirements (standards and regulations) will only truly work if all actors and stakeholders along the supply chain and the wider agricultural and forestry system are considered. For example, the EU might incentivize bioenergy systems in relation to a specific emission threshold, but the biomass grower or trader might not be necessarily be driven by that but might follow a sustainable standard of crop/wood production applicable in their location, which also enables the grower to also sell the (usually main) products. This again shows that bioenergy cannot be seen and evaluated outside the context of the actual system in which the biomass is produced and bioenergy used. To achieve high sustainable standards, the different drivers and tradeoffs for the different stakeholders and sectors have to be considered, and this can vary dependent on such issues as: the actual agricultural or forest system; the biomass production/procurement and biomass utilization region (which are often in different geographical locations); the market conditions; and the different target groups for the different products produced.

12.5 Bioenergy case study: Generating low carbon energy from agricultural & food wastes

Policies such as the EU's Waste Framework Directive [32] mean that countries are increasingly applying a waste hierarchy, where management options for different waste streams are ordered in preference to environmental and waste reduction performances. As such, the recovery of energy from waste is the appropriate waste management option for all wastes that cannot be prevented, reused, or recycled with a lower greenhouse gas impact. Wastes with no other uses that may otherwise be diverted to landfill are thus resources highly suitable for energy recovery processes.

Anaerobic digestion (AD) represents a key pathway for the generation of energy from wet waste resources, such as agricultural and food wastes [33]. Biomass resources within this category may include food/crop wastes, animal manures, litter, and slurries. AD is a natural biochemical process in which micro-organisms work in the absence of oxygen to break down organic matter such as that within waste resources. The products of AD are: (i) a renewable fuel—biogas (CH_4 and CO_2 in an equal mix), and (ii) a nitrogen rich fertilizer—digestate [33].

Biogas can be combusted directly within gas boilers to produce heat, or within combined heat and power (CHP) systems to produce both heat and electricity. Biogas may also be "scrubbed" to remove the CO_2 and other substances to produce biomethane, that can in turn be utilized in place of natural gas, as a transport fuel or injected into the national gas grid [34].

12.5.1 Anaerobic digestion of agriculture manures and slurries

Two of the most predominant forms of animal agricultural wastes are manures and slurries. Manures are the direct waste resources generated by the livestock and slurries are a fluid of manures mixed with water. The extent of time that livestock are housed is the key factor that may determine the quantity of these wastes that may be collected. There are several common designs of housing for different livestock, each with different methodologies for waste collection [35].

Manures and slurries produced are typically managed and utilized on the farms in which they are generated where recycling of their nutrients takes place through spreading on the land. Where these waste resources are not managed appropriately, large GHG emissions can result from each step of the resource's life cycle. Potentially large GHG emissions are released through the storage and subsequent spreading of these resources on the land. The specific levels of GHG emissions generated may be highly variable, depending on a broad range of influences, such as the type of storage or spreading systems deployed.

An alternative pathway for managing these waste resources could be through AD systems, where the wastes are digested and biogas is generated to produce renewable fuels. A further product generated following the AD of resources such as manures and slurries is digestate. This digestate typically has greater stability than the raw manures or slurries in that it is less likely to degrade when applied to the land as a fertilizer, therefore providing greater potential benefit for crop growth [36].

12.5.2 Anaerobic digestion of food wastes

The AD of food wastes also produces renewable fuels in the form of biogas, and digestate that may be used as biofertilizers, which in turn can mitigate further emissions by reducing the need for synthetic fertilizers. Significant quantities of food wastes are produced around the world, and much is disposed of through landfill sites.

The diversion of food wastes away from landfill to AD systems may therefore result in large economic benefits by the avoidance of landfill taxes as well as benefits gained through potential generation of bioenergy. AD technology pathways for food wastes are best suited where "wet wastes" (typically <15% solid material) can be collected separately at source and utilized as either single or codigestates within centralized systems [36].

12.5.3 Anaerobic digestion of energy crops as codigestates

The practice of growing energy crops purposely as feedstocks for anaerobic digesters is increasing. AD systems focused on food or agricultural waste feedstocks may also require additional codigestates such as energy crops to increase the dry matter of the mix and the proportion of easily digestible sugars. The additional use of energy crops increases the chemical efficiency of the AD processes, and thus the biogas yields.

Energy crops produced specifically to be utilized as codigestates within AD systems may be grown on lands unsuitable for the production of food crops or as support of normal agricultural crop rotations [18]. Through appropriate crop management and periodic rotation strategies, the production of crops for codigestion may also result in the enhancement of subsequent food crop rotation yields as they can replenish nutrients and condition soils. The most common crop currently utilized within AD systems is maize, which as a food crop may potentially utilize lands that would otherwise be dedicated for food production, and if not produced within appropriately managed systems can result in soil degradation. It is therefore vital to be mindful of any potential direct or indirect unintentional consequences [35].

From a life cycle and GHG impact perspective, using any land currently used for food crop production may result in intensification of production elsewhere and/or indirect land use change (ILUC). There are difficulties in identifying lands that may be classed as "surplus" to requirements for food crop production, and there is a lack of a standard methodology for addressing problems relating to production intensification and indirect land use change - thus, potential impacts may be easily overlooked [37].

12.6 Bioenergy case study: Generating low carbon energy from straws & agricultural residues

Currently across the EU, wood-based resources are the primary bioenergy feedstock of choice for the generation of heat. However, agricultural residues are becoming an increasingly attractive alternative, owing to their widespread availability and low cost [38]. Straw resources represent the most abundant variety of agricultural residues across Europe [39], and reflect a flexible fuel option for the bioenergy sector that can be utilized in its raw form with little pretreatment, or can be processed to provide enhanced fuels such as pellets or briquettes [40].

Straw resources are typically either left on the land, where they may or may not be incorporated into the soils to act as a source of nutrient, or are removed from the land to be utilized within a range of uses or products such as for animal bedding. In most cases, straw harvest strategies both extract and leave straw on the land in varying proportions. The overall availability of straw produced each year typically outweighs the demands for straw products/uses, and the requirements of the land to balance soil nutrient and organic carbon levels [41]. This "surplus" is often returned/left on the land, and it is this potential surplus in straw resource that is attracting particular interest as a potential bioenergy source [42].

Any developed strategies aimed at increasing the utilization of straw resource for bioenergy generation will have to consider and balance any potential negative impacts resulting from greater straw use. The strategies will also have to address potential impacts that increasing demand for straw may have on the existing markets that compete for the resource [43,44].

In terms of GHG impact, the use of straw to fuel bioenergy pathways causes direct biogenic CO_2 emissions, as the carbon content of the straw is instantly emitted through combustion of the fuel [45]. In contrast where straw is either used as/within products or spread/left on the land, the carbon content of the straw will either be locked up over the lifespan of product, or stored within soils—but likewise will be released over time when the straw finally breaks down and decays. Thus, the overall GHG impact of using straw as a bioenergy fuel is similar to the GHG impact of the alternative uses of straw, albeit over different emission timeframes [46]. As such the majority of existing research studies find overall life cycle GHG benefits associated with utilizing straw for energy in situations where the use of fossil fuels are mitigated [47].

It is also important to acknowledge that there may be further direct and indirect upstream GHG emissions associated with straw resources—resulting from activities such as the harvesting of the straw from the land, and/or the processing of the resource. Thus, it is important to carry out life cycle assessments and GHG impact analyses that reflect the varying characteristics of different scenarios, in order to gauge the overall impact of a given bioenergy pathway.

12.6.1 Straw life cycle processes: Harvest, transport, and pretreatment

Straw represents an attractive feedstock option, as it is a largely abundant resource and is very low in cost at the point of source. However the specific characteristics of straw resources mean that steps may need to be taken to enhance its performance as a fuel. For example, in comparison to most wood-based bioenergy fuels, straw typically has low calorific values, low bulk density, high ash content, and the potential to contain corrosive and polluting chemicals [38]. However, many of these negative attributions may be reduced or mitigated through application of pretreatment processes.

The relative low energy density of straw resources is a consequence of the low bulk density of the raw material. This attribute can make the transportation of raw straw highly uneconomical without processing to increase its bulk density. Transport economics ultimately determine the distances and delivery radius to which straw may be delivered to sites for storage, pretreatment, or bioenergy generation [48]. To improve the economics, straw may be processed into bales prior to its removal from the land [49]. Baling not only increases the bulk density of the resource, but also creates uniform shapes that enhance the storage (stacking) potential.

Once removed from the land, straw bales are often further pretreated to remove contaminants, decrease moisture content, and can be processed to produce enhanced forms of fuel such as pellets. Chemical contaminants within straws, such as potassium and chlorine, can be removed largely through leaving the straw on the land to allow rainfall to naturally "wash" the material. Alternatively, pretreatment processes may include resource cleansing steps, where water is sprayed at moderate temperatures (~60°C) to remove contaminants. However, these pretreatment processes will inevitably increase the moisture content of the resource [38]. As such, straw feedstocks will often require drying processes prior to bioenergy conversion, or before further processing such as pelletizing. The specific drying processes applied are typically influenced by the economic considerations and operational requirements of bioenergy plant [50]. Air-drying straw bales during storage (reducing the moisture contents to typically 15%–20%) is the most common and economical drying option. If the straw is to be pelletized, further mechanical drying may be required to decrease the moisture content (typically to 7%–12%) [49].

Straw bales can be utilized directly as fuels within bioenergy plant, or may require further processing, such as chopping, dependent of the specific requirement of the bioenergy systems. Alternatively, once removed from the land, straw bales can be debaled, chopped, and milled with the aim of producing straw pellets. Within this process the straw is ground into fine particles, mixed with binding agents such as molasses, and compressed into pellets. Thus, the pelletization of straw requires a series of additional pretreatment stages that will each incur energy demands, but will result in enhanced fuels with a greater density, better transportation economics, and better storage and fuel qualities compared to the primary resource [49].

12.6.2 Straw bioenergy pathways

A wide range of bioenergy technologies can be applied to combust straw resource. These straw bioenergy systems range widely in their potential scale from small boiler systems suited for domestic energy systems to large commercial or district scale plants. Straw resources may also be gasified to generate syngas, which has high flexibility in its potential application [51]. The choice of conversion technology utilized depends on the characteristics of the feedstock itself and the forms and scales of energy required [52].

The European Commission recognize that generating renewable energy from straws through CHP technologies represents a pathway that shows a potential worth pursuing [53]. Thermal and electricity producing efficiencies may be conservatively assumed to be 60% and 25%, respectively [46,54–57].

Dedicated straw bioenergy systems have been continuously developing since early commercial systems emerged in the late 1970s. Over this timeframe, the conversion efficiencies of straw systems have doubled and their generation of harmful emissions has been much reduced. While a number of years ago large-scale bioenergy heating systems typically favored wood-based fuels, there is now no longer a substantial difference in the energy or economic performance between using straw or wood-based fuels. District-scale systems located strategically in regions with surplus straw resource have been proven to provide cheap renewable heat to local developments and communities [58].

The gasification of straw represents a further bioenergy conversion pathway where high conversion efficiencies may be realized, with systems operating at efficiencies of up to 98-99% carbon conversion [59]. This is achieved as a result of the exergy losses due to internal thermal energy exchange being lower during gasification compared to direct combustion [60]. The resulting syngas produced during gasification may be combusted directly within bioenergy plant or upgraded for potential utilization within the conventional gas grid [61].

The generation of ash following straw combustion can present a challenge for straw bioenergy systems. The excessive build up or mismanagement of straw ash can result in the slagging or fouling of even large-scale bioenergy plant [62]. Landfill is the favored disposal pathway for straw ash; although, provided that any heavy metal elements are recycled, straw ash may also be spread on land as a fertilizer [61]. Straw ash typically contains only limited quantities of nitrogen, but the levels of phosphorus and potassium can be close to 100% of that contained within unburnt straw. Thus, developing ash management strategies that include land spreading can potentially enhance the economics and sustainability of straw bioenergy generation, through mitigating the economic and GHG impacts of utilizing synthetic fertilizers [61].

12.7 Bioenergy case study: Generating low carbon energy from energy crops

Energy crops are a versatile fuel source that when utilized in substitution of fossil fossils can reduce the GHG impact of energy generation. Two of the key forms of energy crop currently grown and utilized for bioenergy are short rotation coppice (SRC) species such as willow and poplar, and energy grass species such as miscanthus [63].

The majority of energy crops grown specifically for the bioenergy sector are produced using conventional farming techniques. The crops may be dried, chopped, chipped, pelletized, or baled to produce fuels that can be directly combusted within small-scale stove or boiler systems, within dedicated biomass energy plants, or even cofired with coal within conventional power plants [64].

12.7.1 Energy crops: Short rotation coppices

Predominant SRC energy crop species include willow and poplar. These are well suited for growth on lands with a wide range of soil classifications and can also be successfully produced on marginal/reclaimed lands. The resulting energy crop fuels may be utilized within a wide range of bioenergy systems to generate heat or power, from domestic to industrial scale [65].

SRC crops are typically planted during the spring season where they are then left to grow over the first year. During the first winter, after planting, the crop stems are then cut back to ground level to encourage the growth of multiple stems (the crop is coppiced) [63]. SRC energy crops may be harvested on a two to five year cycle, with three year cycles being most common. SRC are harvested using conventional farm machinery, producing SRC rods, chips, or billets where they are then further chipped and/or dried as required to prevent decomposition of the material. The roots of SRCs remain in the ground after harvest, where multiple new shoots will emerge with each progressive growth cycle. An SRC plantation will be potentially be viable for up to 30 years before replanting may be necessary [65].

12.7.2 Energy crops: Miscanthus

Miscanthus is a fast-growing perennial C4 rhizomatous grass originating from Asia. It represents a key energy crop as it can be produced on a wide range of soil types, with typical yields of 12–15 t (oven dried) biomass per hectare. Highest yields are achieved on moisture-retentive soils that quickly warm up during the spring season enabling long growing seasons [66]. Miscanthus biomass fuels may also be utilized within a wide range of bioenergy systems ranging from domestic to district scale [67].

After planting, miscanthus stems in the first year may achieve up to 1 to 2 m height, where they typically remain uncut. In successive years miscanthus crops can reach 3.5 m, with the crops having productive life spans of up to 20 years. With the onset of autumn and the cooler season in each annual production cycle, senescence is triggered within miscanthus crops where the nutrients are translocated below ground. By February within production cycles only leafless canes will remain, with any fallen leaf material providing a mulch layer on the soils. Harvesting the crops from February to April therefore results in a low-moisture crop that will be typically baled. The lower the moisture content of the crop, the higher will be the fuel energy yield, as a result of reduced drying prior to chipping or pelletizing [65].

12.7.3 Energy crops, land carbon stocks, and land use change GHG emissions

The carbon stock of a land system may be defined as the total carbon stored within all living plant and organic matter both above and below ground. Carbon stock levels are quantified as a function of the characteristics and area of the land, and the extent of vegetation [68].

Carbon is locked within land systems as plants draw CO_2 from the atmosphere through photosynthesis enabling the growth of vegetation and the development of soil organic matter. As vegetation grows and ecosystems develops toward a climax state, there will be corresponding gradual increases in uptake of CO_2 followed by the stabilization of carbon levels stored within a given land system [69]. At the same time CO_2 is returned to the atmosphere through the respiration of the vegetation, and as a result of the breakdown and decay of organic matter within the soils. CO_2 is also released through "nonrespiratory fluxes" such as fires or the ultraviolet oxidation of organic matter [70]. The transfer of CO_2 between land systems and the atmosphere is substantial, with as much as a seventh of total atmospheric CO_2 passing through vegetation each year. In the absence of any disturbances to stable land systems, these large fluxes of CO_2 will be in equilibrium [69].

Apart from CO_2 emissions, smaller natural GHG emission fluxes of CH_4 and N_2O also take place. Due to global warming, the emissions of these from the land to the atmosphere are expected to increase, thus increasing the overall GHG impact of a land system [71].

Disturbances to land systems such as those driven by land use changes may result in the development of imbalances within the natural GHG fluxes. For example, clearing land for agriculture and the subsequent decay or burning of the cleared organic matter will result in large losses of carbon stock stored within the land to the atmosphere. The land system will then have to develop new balances of photosynthesis/respiration GHG flux [69].

This has strong implications for the development of energy crop plantations; within scenarios where land systems with large existing carbon stocks such as forests are converted to energy crop plantations, there will likely be net fluxes of GHG emissions from the land to the atmosphere. In contrast, within scenarios where abandoned arable lands are converted to energy crop plantations, there would likely be net fluxes of GHG emissions from the atmosphere to the land as the energy crops grow [72].

Attempting to quantify potential levels of GHG impact following land disturbances/land use change is difficult. This complex science has many variables and interactions specific to each different land systems, including the range of human activities interacting with lands and natural systems [69]. It is however possible to estimate carbon fluxes by developing inventories of carbon stocks [73], or through developing methods and datasets to characterize typical carbon stock changes [74,75].

12.7.4 Energy crop cultivation, management, & harvesting

The cultivation, management, and harvesting of energy crops, like with any other crops, can be an energy-intensive process resulting in potentially large GHG emissions [76].

GHG emissions are released as a result of energy and material inputs required by agricultural activities. Fossil fuels may be consumed during planting, cultivation, and harvesting processes. Furthermore, indirect energy and emissions result from the production of fuels and process inputs used in the farming activities [77,78].

Energy and emissions data for different agricultural processes and the contribution due to the production of fertilizers and other chemicals relevant to specific agricultural practice are often hard to define as a result of the large range of process technologies and inputs involved [79]. Thus, emission factors provide key tools for LCAs when evaluating the GHG impact of crop production life cycles.

12.7.5 Resource transportation

The transportation of energy crop resources is a further life cycle process that impacts on the overall GHG. The resulting GHG emissions are dependent on the forms of transport and the distances applicable to a given supply chain of a resource and may indeed outweigh or diminish any emission savings resulting from the utilization of biomass for energy.

12.7.6 Energy crop pretreatment processes

Pretreatment processes may be applied to energy crops to remove impurities, to enhance the resource's storage characteristics, the transportation and utilization flexibility, as well as to improve the energy performance of the resource as a fuel. Like with many other biomass resources, drying processes are those most commonly applied to energy crops. Removing moisture increases the usable energy within the resource, but where mechanical drying systems have been applied there are GHG emissions associated with energy use [80].

The production of energy crop pellets through pelletization is a processing pathway commonly applied to produce enhanced fuels for bioenergy generation. The processes may include drying, chipping, grinding, and milling, followed by compression of the resource and application of lignin as a binding agent [81]. The resulting energy crop pellets have higher energy density compared to the primary resource, making them attractive fuels. However, the pelletization processes can be expensive to undertake and will also result in process GHG emissions that will reduce the GHG performance of bioenergy generated [80].

12.7.7 Energy crop bioenergy pathways

The leading bioenergy conversion technologies for energy crops are pellet boilers and chip burners that produce hot water or steam. These range in scale from domestic to large district systems that generate either dedicated heat or CHP [82–84].

At the domestics scale, biomass boilers and even simple wood stoves fuelled by pellets or chips are becoming increasingly popular as secondary heating systems [33]. Biomass boilers are also the most commonly utilized renewable technology for CHP schemes. However, due to the relative inefficiency of small-scale biomass CHP systems, they are relatively uncommon. At the medium to large scale, in particular over 3 MWe in output, biomass CHP systems can be highly effective renewable energy technologies [85]. District heating systems fuelled by biomass have also been successfully installed in many countries of the world.

12.8 Conclusion

Bioenergy can offer renewable, low-carbon energy systems, sequestering atmospheric carbon as well as offer numerous environmental and socioeconomic benefits and therefore supporting global climate change targets and wider environmental, social, economic, and sustainable targets. There is scientific evidence of the benefits of bioenergy, but results are often subject to variation and uncertainty. Additionally, it is important to consider various sustainable aspects of bioenergy systems beyond carbon. Treating bioenergy only as part of the energy sector will fail to ensure: sustainable biomass production and sourcing, clean applications with low health impacts, and fair and affordable energy vectors. Ensuring that bioenergy offers the required holistic emission reduction, context, specific and long-term approaches are necessary to understand synergies and tradeoff of the bioenergy and related agricultural and forestry systems. Assessing the environmental and wider sustainable impacts of bioenergy, full supply chains as well as direct and indirect stakeholders, their drivers, benefits and challenges needs to be considered. With these, we have to assess and evaluate bioenergy and its impacts in the context of the specific system it is part off and its direct and wider impacts on environment, economy, and society.

References

[1] Röder M, Thornley P. Bioenergy as climate change mitigation option within a 2°C target—uncertainties and temporal challenges of bioenergy systems. Energy Sustain Soc 2016;6(1):1–7. https://doi.org/10.1186/s13705-016-0070-3.

[2] IEA (International Energy Agency). 2016. World energy model. Available from: https://www.iea.org/weo/weomodel/.

[3] WHO. 2016. Household air pollution and health. Geneva Fact sheet N°292. Available from: http://www.who.int/mediacentre/factsheets/fs292/en/.

[4] Gasparatos A, von Maltitz GP, Johnson FX, et al. Biofuels in sub-Sahara Africa: drivers, impacts and priority policy areas. Renew Sust Energ Rev 2015;45:879–901. https://doi.org/10.1016/j.rser.2015.02.006.

[5] Ackom, E., M.B. Pedersen, and J.M. Christensen. 2011. Bioenergy: the potential for rural development and poverty alleviation. Nairobi, Kenya. GNESD-SPM-BET-11/2011. Available from: http://orbit.dtu.dk/files/7590946/Bioenergy.pdf.

[6] FAO (The Food and Agriculture Organization of the United Nations). 2009. How to feed the World 2050. Available from: http://www.fao.org/fileadmin/templates/wsfs/docs/expert_paper/How_to_Feed_the_World_in_2050.pdf.
[7] Ladanai, S. and J. Vinterback. 2009. Global potential of sustainable biomass for energy. Uppsala. Report 013. Available from: https://pub.epsilon.slu.se/4523/1/ladanai_et_al_100211.pdf.
[8] Welfle, A.J. 2014. Biomass resource analyses & future bioenergy scenarios. Available from: https://www.escholar.manchester.ac.uk/api/datastream?publicationPid=uk-ac-man-scw:259603&datastreamId=FULL-TEXT.PDF.
[9] NREL (National Renewable Energy Laboratory). 2009. Biomass cofiring: a renewable alternative for utilities. Golden, Colorado. DOE/GO-102000-1055. Available from: https://www.nrel.gov/docs/fy00osti/28009.pdf.
[10] Röder M, Thornley P. Waste wood as bioenergy feedstock. Climate change impacts and related emission uncertainties from waste wood based energy systems in the UK. Waste Manag 2017; https://doi.org/10.1016/j.wasman.2017.11.042.
[11] Röder M, Whittaker C, Thornley P. How certain are greenhouse gas reductions from bioenergy? Life cycle assessment and uncertainty analysis of wood pellet-to-electricity supply chains from forest residues. Biomass Bioenergy 2015;79:50–63. https://doi.org/10.1016/j.biombioe.2015.03.030.
[12] Thornley P, Gilbert P, Shackley S, et al. Maximizing the greenhouse gas reductions from biomass: the role of life cycle assessment. Biomass Bioenergy 2015;81:35–43. https://doi.org/10.1016/j.biombioe.2015.05.002.
[13] DoE (U.S. Department of Energy). 2018. Biodiesel blends. Washington DC. Available from: https://www.afdc.energy.gov/fuels/biodiesel_blends.html.
[14] BEIS (Department for Business, Energy & Industrial Strategy). 2016. Digest of United Kingdom energy statistics. Available from: https://www.gov.uk/government/collections/digest-of-uk-energy-statistics-dukes.
[15] Nguyen VH, Topno S, Balingbing C, et al. Generating a positive energy balance from using rice straw for anaerobic digestion. Energy Rep 2016;2:117–22. https://doi.org/10.1016/j.egyr.2016.05.005.
[16] Röder M, Stolz N, Thornley P. Sweet energy: bioenergy integration pathways for sugarcane residues. A case study of Nkomazi, District of Mpumalanga, South Africa. Renew Energy 2017;113:1302–10. https://doi.org/10.1016/j.renene.2017.06.093.
[17] Welfle A, Gilbert P, Thornley P. Securing a bioenergy future without imports. Energ Policy 2014;68:1–14. https://doi.org/10.1016/j.enpol.2013.11.079.
[18] Röder M. More than food or fuel. Stakeholder perceptions of anaerobic digestion and land use; a case study from the United Kingdom. Energ Policy 2016;97:73–81. https://doi.org/10.1016/j.enpol.2016.07.003.
[19] Rowe R, Chapman J, Whittaker J, et al. A systematic review of life cycle assessments of bioenergy chains for heat, power and liquid transportation fuel. Comp Biochem Physiol A Mol Integr Physiol 2008;150(3 Suppl):S182. https://doi.org/10.1016/j.cbpa.2008.04.486.
[20] Whittaker C, McManus MC, Hammond GP. Greenhouse gas reporting for biofuels: a comparison between the RED, RTFO and PAS2050 methodologies. Energ Policy 2011;39(10):5950–60. https://doi.org/10.1016/j.enpol.2011.06.054.
[21] Whittaker C, Mortimer N, Murphy R, et al. Energy and greenhouse gas balance of the use of forest residues for bioenergy production in the UK. Biomass Bioenergy 2011;35(11):4581–94. https://doi.org/10.1016/j.biombioe.2011.07.001.

[22] Anderson K, Bows A. Reframing the climate change challenge in light of post-2000 emission trends. Math Phys Eng Sci 2008;366(1882):3863–82. https://doi.org/10.1098/rsta.2008.0138.
[23] Friedlingstein P, Andrew RM, Rogelj J, et al. Persistent growth of CO2 emissions and implications for reaching climate targets. Nat Geosci 2014;7(10):709–15. https://doi.org/10.1038/ngeo2248.
[24] Eurelectric. 2011. Biomass 2020: opportunities, challenges & solutions. D/2011/12.105/5. Available from: http://www.eurelectric.org/media/26720/resap_biomass_2020_8-11-11_prefinal-2011-113-0004-01-e.pdf.
[25] Tomei J, Helliwell R. Food versus fuel? Going beyond biofuels. Land Use Policy 2015; https://doi.org/10.1016/j.landusepol.2015.11.015.
[26] Rathmann R, Szklo A, Schaeffer R. Land use competition for production of food and liquid biofuels: an analysis of the arguments in the current debate. Renew Energy 2010;35(1):14–22. https://doi.org/10.1016/j.renene.2009.02.025.
[27] EEA. 2015. Biomass production as a provisioning service of soil. Available from: https://www.eea.europa.eu/soer.
[28] EEA. 2014. Progress in management of contaminated sites: European Environment Agency. Available from: https://www.eea.europa.eu/data-and-maps/indicators/progress-in-management-of-contaminated-sites/progress-in-management-of-contaminated-1.
[29] WHO Regional Office for Europe. 2013. Contaminated sites and health. Report of two WHO workshops, 106 p. Available from: http://www.euro.who.int/__data/assets/pdf_file/0003/186240/e96843e.pdf?ua=1.
[30] Thornley P, Upham P, Tomei J. Sustainability constraints on UK bioenergy development. Energ Policy 2009;37(12):5623–35. https://doi.org/10.1016/j.enpol.2009.08.028.
[31] Thornley P, Gilbert P. Biofuels: balancing risks and rewards. Interface Focus 2013;3(1). https://doi.org/10.1098/rsfs.2012.0040.
[32] European Commission. 2008. Directive 2008/98/EC of the European Parliament and of the Council. Brussels. Available from: http://ec.europa.eu/environment/waste/framework/.
[33] DECC (Department of Energy & Climate Change). 2013. Generating energy from waste, including anaerobic digestion. Available from: https://www.gov.uk/guidance/generating-energy-from-waste-including-anaerobic-digestion.
[34] Holm-Nielsen JB, Al Seadi T, Oleskowicz-Popiel P. The future of anaerobic digestion and biogas utilization. Bioresour Technol 2009;100(22):5478–84. https://doi.org/10.1016/j.biortech.2008.12.046.
[35] Steffen, R., O. Szolar, and R. Braun. 1998. Feedstocks for anaerobic digestion. Vienna. Available from: http://www.agrienvarchive.ca/bioenergy/download/feedstocks_AD.pdf.
[36] Defra (Department of Environment, Food and Rural Affairs). 2011. Anaerobic digestion strategy and action plan. A commitment to increasing energy from waste through Anaerobic Digestion. London. 56 p. PB 13541. Available from: https://www.gov.uk/government/uploads/system/uploads/attachment_data/file/69400/anaerobic-digestion-strat-action-plan.pdf.
[37] Dauber J, Brown C, Fernando AL, et al. Bioenergy from "surplus" land: environmental and socio-economic implications. BioRisk 2012;7:5–50. https://doi.org/10.3897/biorisk.7.3036.
[38] Kavalov, B. and S. Peteves. 2004. Bioheat applications in the European Union: analysis and perspective for 2010. Petten, The Netherlands. EUR 21401 EN. Available from: https://ec.europa.eu/jrc/en/publication/eur-scientific-and-technical-research-reports/bioheat-applications-european-union-analysis-and-perspective-2010.

[39] Monforti F, Bódis K, Scarlat N, et al. The possible contribution of agricultural crop residues to renewable energy targets in Europe: a spatially explicit study. Renew Sust Energ Rev 2013;19:666–77. https://doi.org/10.1016/j.rser.2012.11.060.
[40] Pastre, O. 2002. Analysis of the technical obstacles related to the production and utilisation of fuel pellets made from agricultural residues. Available from: http://citeseerx.ist.psu.edu/viewdoc/download?doi=10.1.1.473.8636&rep=rep1&type=pdf.
[41] Nicholson, F., D. Kindred, A. Bhogal, et al. 2014. Straw incorporation review. Kenilworth. Research review no. 81. Available from: https://cereals.ahdb.org.uk/publications/2014/july/30/straw-incorporation-review.aspx.
[42] Kretschmer, B., B. Allen, and K. Hart. 2012. Mobilising cereal straw in the EU to feed advanced biofuel production. London. Available from: https://ieep.eu/publications/mobilising-cereal-straw-in-the-eu-to-feed-advanced-biofuel-production.
[43] Panoutsou C, Eleftheriadis J, Nikolaou A. Biomass supply in EU27 from 2010 to 2030. Energ Policy 2009;37(12):5675–86. https://doi.org/10.1016/j.enpol.2009.08.032.
[44] Edwards, R., M. Šúri, T. Huld, et al. 2005. GIS based assessment of cereal straw energy resource in the European Union. Paris. Available from: http://iet.jrc.ec.europa.eu/remea/gis-based-assessment-cereal-straw-energy-resource-european-union.
[45] Petersen, J. and J. Berntsen. 2003. The turnover of soil organic matter on different farm types. Available from: http://orgprints.org/4665/1/4665.PDF.
[46] Parajuli R, Løkke S, Østergaard PA, et al. Life cycle assessment of district heat production in a straw fired CHP plant. Biomass Bioenergy 2014;68:115–34. https://doi.org/10.1016/j.biombioe.2014.06.005.
[47] Gabrielle B, Gagnaire N. Life-cycle assessment of straw use in bio-ethanol production: a case study based on biophysical modelling. Biomass Bioenergy 2008;32(5):431–41. https://doi.org/10.1016/j.biombioe.2007.10.017.
[48] Porter, P., J. Barry, R. Samson, et al. 2008. Growing Wisconsin Energy—a native grass pellet bio-heat roadmap for Wisconsin. Madison, Wisconsin. Available from: https://www.iatp.org/sites/default/files/258_2_104098.pdf.
[49] Zeng, T., M. Kallio, and H. Ovarainen. 2012. Critical review on the pelletizing technology. Metallimiehenkuja, Finland. Available from: https://www.dbfz.de/fileadmin/MixBioPells/publications/D31_Critical_Review_about_pelletising_and_combustion_technology_FINAL.pdf.
[50] Fagernäs L, Brammer J, Wilén C, et al. Drying of biomass for second generation synfuel production. Biomass Bioenergy 2010;34(9):1267–77. https://doi.org/10.1016/j.biombioe.2010.04.005.
[51] Repic B, Dakic D, Djurovic D, et al. Development of a boiler for small straw bales combustion. In: Nathwani J, Ng A, editors. Paths to sustainable energy. 2010. p. 647–64.
[52] Obernberger I. Decentralized biomass combustion: state of the art and future development. Paper to the keynote lecture of the session "Processes for decentralized heat and power production based on combustion" at the 9th European Bioenergy Conference, June 1996, Copenhagen, Denmark, Biomass Bioenergy 1998;14(1):33–56. https://doi.org/10.1016/s0961-9534(97)00034-2.
[53] European Commission and I.f.E.a. Sustainability. 2007. Cereals Straw Resources for bioenergy in the European Union. EUR 22626 EN. Available from: http://publications.jrc.ec.europa.eu/repository/bitstream/JRC35679/PUBSY%205679%20-%20EUR%2022626%20-%20docnum.pdf.
[54] Skøtt, T. and M. Hansen. 2000. Danish biomass solutions—reliable and efficient. Copenhagen. Available from: http://www.videncenter.dk/exportcat/policy_and_market.pdf.

[55] Agency, D.E. 2003. Renewable energy danish solutions: background–technology–projects. Copenhagen.
[56] Preto, F. 2011. Biomass energy background paper. Available from: http://yukonenergy.ca/media/site_documents/charrette/docs/papers/BIOMASS_YEC_Background_Paper.pdf.
[57] Aabakken, J. 2006. Power technologies energy data book—4th ed. AC36-99-GO10337. Available from: https://doi.org/10.2172/891585.
[58] INBIOM. 2011. Straw to energy. Status, technologies and innovation in Denmark. Tjele, Denmark. Available from: http://www.inbiom.dk/en/publications/straw-to-energy.
[59] Stevens, D. 2001. Hot gas conditioning: recent progress with larger-scale biomass gasification systems. Golden, Colorado. Available from: https://www.osti.gov/scitech/biblio/786288.
[60] Prins MJ, Ptasinski KJ. Energy and exergy analyses of the oxidation and gasification of carbon. Energy 2005;30(7):982–1002. https://doi.org/10.1016/j.energy.2004.08.010.
[61] Nguyen TLT, Hermansen JE, Nielsen RG. Environmental assessment of gasification technology for biomass conversion to energy in comparison with other alternatives: the case of wheat straw. J Clean Prod 2013;53:138–48. https://doi.org/10.1016/j.jclepro.2013.04.004.
[62] Tortosa-Masiá A, Ahnert F, Spliethoff H, et al. Slagging and fouling in biomass co-combustion. Therm Sci 2005;9(3):85–98. https://doi.org/10.2298/tsci0503085t.
[63] Defra (Department of Environment, Food and Rural Affairs). 2004. Growing short rotation coppice. Best practice guidelines for applicants to Defra's energy crops scheme. London. Available from: https://www.forestry.gov.uk/pdf/Growing_Short_Rotation_Coppice_tcm6_2004.pdf/$FILE/Growing_Short_Rotation_Coppice_tcm6_2004.pdf.
[64] Defra (Department of Environment, Food and Rural Affairs). 2012. Industrial energy and non-food crops: business opportunities for farmers. London. Available from: https://www.gov.uk/guidance/industrial-energy-and-non-food-crops-business-opportunities-for-farmers.
[65] NNFCC. 2010. Short rotation coppice willow (SRC). York. Available from: http://www.nnfcc.co.uk/publications-store.
[66] NNFCC. 2009. Miscanthus. York. Available from: http://www.nnfcc.co.uk/publications-store.
[67] Defra (Department of Environment, Food and Rural Affairs). 2016. Area of crops grown for bioenergy in England and the UK: 2008–2015. London. Available from: https://www.gov.uk/government/uploads/system/uploads/attachment_data/file/578845/nonfood-statsnotice2015i-19dec16.pdf.
[68] Forestry Commission. 2014 Carbon in live woodland trees in Britain. Available from: https://www.forestry.gov.uk/pdf/FCNFI113.pdf/$FILE/FCNFI113.pdf.
[69] IPCC. 2014. Climate change 2014: Mitigation of climate change. Contribution of working group III to the fifth assessment report of the Intergovernmental Panel on Climate Change. Available from: http://www.ipcc.ch/report/ar5/wg3/.
[70] Chapin FS, Woodwell GM, Randerson JT, et al. Reconciling carbon-cycle concepts, terminology, and methods. Ecosystems 2006;9(7):1041–50. https://doi.org/10.1007/s10021-005-0105-7.
[71] Anderson-Teixeira KJ, DeLucia EH. The greenhouse gas value of ecosystems. Glob Chang Biol 2010;17(1):425–38. https://doi.org/10.1111/j.1365-2486.2010.02220.x.
[72] Zan CS, Fyles JW, Girouard P, et al. Carbon sequestration in perennial bioenergy, annual corn and uncultivated systems in southern Quebec. Agric Ecosyst Environ 2001;86(2): 135–44. https://doi.org/10.1016/s0167-8809(00)00273-5.
[73] UNECE/FAO. 2000. Forest resources of Europe, CIS, North America, Australia, Japan and New Zealand (industrialized temperate/boreal countries) UN-ECE/FAO contribution to the Global Forest Resources Assessment 2000. New York, Geneva. Available from: http://www.unece.org/index.php?id=15729.

[74] Houghton RA. The annual net flux of carbon to the atmosphere from changes in land use 1850-1990*. Tellus B 1999;51(2):298–313. https://doi.org/10.1034/j.1600-0889.1999.00013.x.

[75] Farquhar, G., M. Fasham, M. Goulden, et al. 2001. Chapter 3. The carbon cycle and atmospheric carbon dioxide content. Geneva. Available from: http://www.ipcc.ch/ipccreports/tar/wg1/index.php?idp=95.

[76] EEA (European Environment Agency). 2007. Estimating the environmentally compatible bioenergy potential from agriculture. Copenhagen. Available from: https://www.eea.europa.eu/publications/technical_report_2007_12.

[77] Cherubini F, Ulgiati S. Crop residues as raw materials for biorefinery systems—a LCA case study. Appl Energy 2010;87(1):47–57. https://doi.org/10.1016/j.apenergy.2009.08.024.

[78] Goglio P, Bonari E, Mazzoncini M. LCA of cropping systems with different external input levels for energetic purposes. Biomass Bioenergy 2012;42:33–42. https://doi.org/10.1016/j.biombioe.2012.03.021.

[79] Patyk, A. and A. Reinhardt. 1996. Energy and material flow analysis of fertilizer production and supply. Brussels.

[80] Ireland, D., A. Hall, and D. Jones. 2004. Woodfuel information pack. Farnham. Available from: https://www.forestry.gov.uk/fr/woodfuelinformationpack.

[81] Renewable_Energy_World. 2016. How to make wood pellets—wood pellet manufacture steps. Athens. Available from: http://www.renewableenergyworld.com/ugc/articles/2016/05/how-to-make-wood-pellets–wood-pellet-manufacture-steps.html.

[82] EEA (European Environment Agency). 2008. Maximising the environmental benefits of Europe's bioenergy potential. Copenhagen. Available from: https://www.eea.europa.eu/publications/technical_report_2008_10.

[83] Peterson, D., S. Haase. 2009. Market assessment of biomass gasification and combustion technology for small- and medium-scale applications. Golden, Colorado. NREL/TP-7A2-46190. Available from: https://www.nrel.gov/docs/fy09osti/46190.pdf.

[84] Gerssen-Gondelach SJ, Saygin D, Wicke B, et al. Competing uses of biomass: assessment and comparison of the performance of bio-based heat, power, fuels and materials. Renew Sust Energ Rev 2014;40:964–98. https://doi.org/10.1016/j.rser.2014.07.197.

[85] Ricardo-AEA. 2013. Projections of CHP capacity and use to 2030. Didcot. Available from: https://www.gov.uk/government/uploads/system/uploads/attachment_data/file/191543/Projections_of_CHP_capacity___use_to_2030_2204.pdf.

Quantifying the climate effects of forest-based bioenergy

13

Annette L. Cowie,†,1, Miguel Brandão†,‡, Sampo Soimakallio†,§*
*NSW Department of Primary Industries/School of Environmental and Rural Science, University of New England, Armidale, NSW, Australia, †International Energy Agency (IEA) Bioenergy Task 38, ‡KTH—Royal Institute of Technology, Stockholm, Sweden, §Finnish Environment Institute (SYKE), Helsinki, Finland
[1]Corresponding author: Annette.cowie@dpi.nsw.gov.au

Chapter Outline

13.1 Introduction

Bioenergy has been promoted as a component of climate change and renewable energy policies in many countries. However, over recent years, the climate benefits of bioenergy systems have been increasingly questioned. Concern over the

Managing Global Warming. https://doi.org/10.1016/B978-0-12-814104-5.00013-2

effectiveness of *agricultural* bioenergy for climate change mitigation is mainly related to indirect land-use change (ILUC), the use of agrochemicals and loss of soil organic matter (e.g., [1,2]), while forest-based bioenergy has been challenged with claims that carbon losses from removal of biomass for energy from existing forests create a "carbon debt" (e.g., [3,4]). Some studies have suggested that the magnitude of the carbon loss may, in some circumstances, be so large that the payback period lasts several decades (e.g., [5,6]). Other studies demonstrate cases in which the initial carbon loss, if any, may be paid back after a few years of displacing fossil fuels (e.g., [7,8]).

These widely divergent results present a confusing picture for policy makers and the energy industry. This chapter discusses the key issues in quantifying the climate effects of forest-based bioenergy and reasons for divergent results. We discuss the sensitivity of results to the selection of spatial and temporal boundaries, and the handling of coproduction of wood products and bioenergy. We consider how to include the timing of greenhouse gas (GHG) emissions and carbon sequestration in the assessment of the climate change impacts of forest-based bioenergy. Furthermore, the role of case-specific factors, non-GHG climate forcers, and the choice of metrics to measure climate impacts are considered. Finally, we provide recommendations to guide climate impact assessment of forest-based bioenergy. Our emphasis is on forest biomass used for combustion to produce electricity and/or heat, but the same principles apply to biomass used for other energy applications (liquid biofuels, gaseous fuels).

13.2 The forest carbon cycle

Sustainable bioenergy systems are commonly said to be carbon neutral because the carbon that is released during combustion had previously been sequestered from the atmosphere and will be sequestered again as the biomass is regrown. Forest biomass systems are effectively carbon neutral over time (leaving aside supply-chain emissions) if the forest sequesters the same amount of carbon in the following rotation as was released, so that the long-term average carbon stock remains constant. If the cycle of growth and harvest is sustained, biomass combustion simply returns to the atmosphere the CO_2 that was absorbed as the plants grew, and there is no net release of CO_2 from the forest.

This is fundamentally different from the use of fossil fuels: burning fossil fuels releases CO_2 that has been locked up for millions of years. It takes carbon that was securely stored in geological formations and adds it to the atmosphere. During human-relevant time scales, there is no significant geological sequestration of carbon from the atmosphere through natural processes.

13.3 The bioenergy life cycle

To understand the climate impacts of bioenergy, the full life cycle of the bioenergy system should be considered: the biomass procurement (e.g., growth and harvest in the forest), as well as the conversion to energy products, supply-chain emissions that arise from transport and processing [9]. Construction and demolition of facilities should also be included where applicable, e.g., establishment of forest roads and power distribution infrastructure. Fossil carbon emissions from forest-based bioenergy supply chains are typically a small percentage of the fossil GHG emissions displaced because supply-chain energy use is low [7,10,11], even where international transport is involved [12–14].

The climate impacts of bioenergy systems are mainly related to how the forest carbon cycle is affected by bioenergy, how efficiently biomass is used to create energy products, and the GHG intensity of the energy product displaced. These factors are discussed as follows.

13.4 Forest bioenergy as a coproduct of forestry

It is important to recognize that industrial forest bioenergy is commonly a coproduct of the forest industry, associated with the production of wood products (sawn timber, composite products, and paper). A large share of that energy, e.g., black liquor is used by the forest industry in wood processing. Additional biomass for energy may be derived from harvest slash (e.g., branches, tops, stumps) that would otherwise have decayed in the forest, or it may be obtained from surplus mill residues (saw dust, shavings, etc.) or construction waste. End-of-life wood products are also a possible source of biomass for energy, after using wood products to displace GHG-intensive building materials. Biomass for energy tends to be a low-value product compared with other forest products, and thus has traditionally not been the primary driver in determining forest management and harvest scheduling. Therefore, it is important to consider bioenergy from managed forests within the context of the forest products markets. On the other hand, ambitious targets and mandates to increase the use of bioenergy and various types of subsidies could make the direct use of low-quality roundwood for energy more common.

When considered from a single-stand perspective, forest carbon stocks range widely between young and mature stages (Fig. 13.1A). However, a forest estate generally comprises a series of stands of different ages, harvested at different times, to produce a constant supply of logs. Therefore, in an ideal "normal" forest, when viewed across the estate, the carbon losses at harvest are balanced by gains in the growing stands, and the variations in C stock observed at stand level are smoothed out, so carbon stock of the whole forest is stable (Fig. 13.1B).

The average carbon stock across the estate reflects the net effects of forest growth (influenced by climate and soil), forest management (that is, site preparation, planting, fertilizing, thinning, pruning, and harvesting), and natural disturbances such as fire, windthrow, and insect outbreaks.

A key aspect governing the climate effect of bioenergy is the change (if any) in average carbon stock across the whole forest landscape resulting from forest bioenergy production. When a new harvest regime is introduced, this will be imposed sequentially as each stand is harvested. If the average carbon stock is different under the new regime, a new equilibrium will be reached after one rotation period. While a new forest management regime that extracts more biomass for bioenergy could reduce the carbon stock in the forest, this "GHG cost" (see Fig. 13.1) can be minimized by

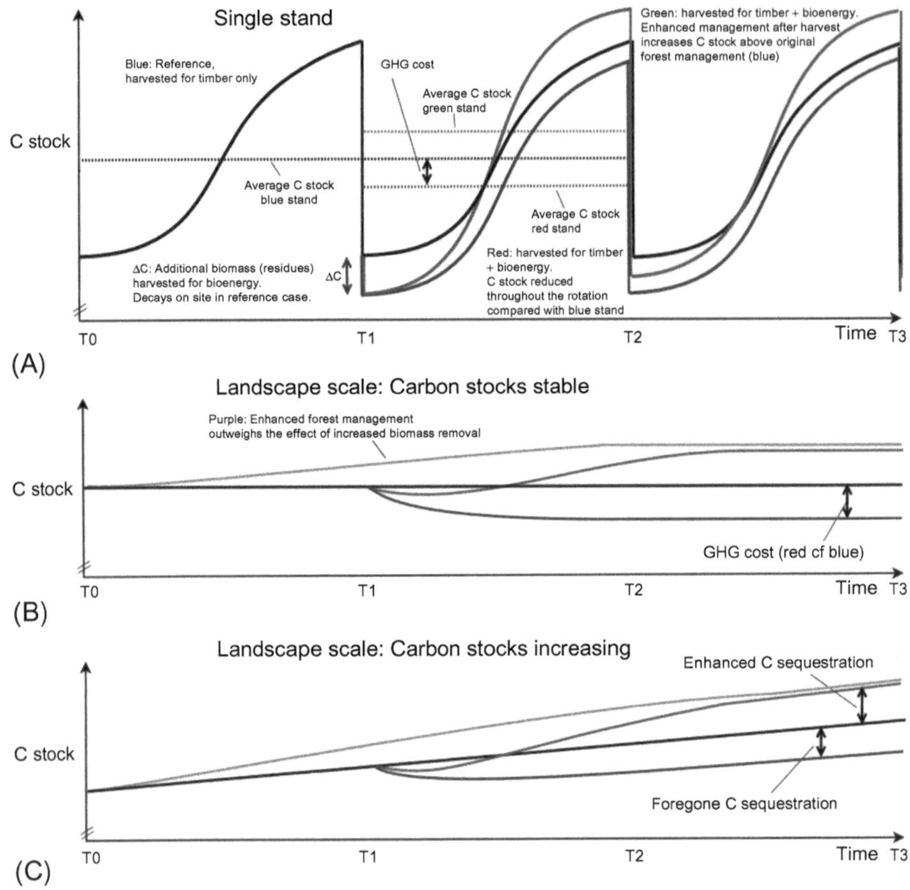

Fig. 13.1 These graphs shows simplified representations of the carbon stocks in a managed forest. They do not show changes in rotation period nor do they show the carbon stock fluctuations around these simplified curves caused by climate variation and forest operations such as thinning. (A) shows the carbon stock (sum of carbon in trees, soil, and litter) of an individual stand, over successive rotations. The blue curve shows the reference scenario, a forest harvested for timber only. The other curves show two alternative scenarios, in which harvest residues (branches and tops), usually left in the forest, are removed for bioenergy at harvest, at time T1 and each successive harvest. The concept of "GHG cost" is illustrated in the red curve: the average carbon stocks are lower compared with the blue stand, due to removal of harvest residues, and, possibly, flow-on effects on soil carbon stocks and forest growth rate. In practice, the GHG cost also includes emissions from any fossil fuel used in the feedstock production, transport, and processing. The green curve illustrates how enhanced forest management can reduce the GHG cost. (B) and (C) show the total carbon stocks summed across a landscape of multiple stands at different stages in the rotation cycle, assuming that all stands follow either the blue, red, or green curves from (A). In reality, the forest carbon stock on the landscape level will reflect a mix of different management approaches applied to different stands, which may include adjustment to the rotation period. An additional curve, in purple, shows a scenario where changes in forest management across the forest landscape outweigh the effect of increased biomass removal for bioenergy, so that the forest carbon stock increases on landscape level.

management practices that enhance growth and thus accelerate C uptake, such as improved site preparation, superior genetic material, and forest fertilization (illustrated by the purple curves in Fig. 13.1B and C). Where there is an increased demand for bioenergy and forest products, forest managers may choose to invest in intensified forest management to enhance growth rates and modify harvest schedules to produce more biomass, which may increase or decrease total forest carbon stocks compared to the case without bioenergy demand. Therefore, a key issue influencing the climate impact of bioenergy is management changes introduced to provide biomass for bioenergy in addition to other forest products, and the consequential impacts on forest carbon stock.

13.5 Factors to consider in quantifying the climate effects of forest bioenergy

Estimating the climate effects of bioenergy systems robustly is fraught with difficulties. To quantify the climate effects of bioenergy, it is critical to apply an approach that considers the whole bioenergy system including multiple life cycle stages, and all relevant climate forcers; is normalized to a functional unit that can be compared with other systems delivering and equivalent quantity of the same function; applies appropriate reference systems and system boundary, consistent with the goal and scope of the study; handles coproducts, and appropriate impact assessment metrics, as detailed here.

13.5.1 *Life cycle perspective*

The assessment should consider the whole life cycle of the bioenergy system, using life cycle assessment (LCA) over the temporal scope determined (see Box 13.1). Unlike national GHG emission reporting and accounting (Box 13.2), which focuses on annual emissions and removals, LCA, as indicated by the name, considers the entire life cycle of a system studied (a product or service). The assessment should

(C) shows a situation where the carbon stocks across the landscape are increasing, such as where the national estate is dominated by young stands; over time, the total carbon stocks increase as these stands mature. Although the total stocks continue to increase in all scenarios in (C), biomass removal can lead to "forgone sequestration" (red curve), though this can be reduced or avoided through enhanced forest management (green and purple curves). Note that the net GHG-mitigation potential of associated bioenergy systems also depends on the GHG displacement efficiency; i.e., a bioenergy system that is associated with declining forest carbon stocks (red curve) can deliver higher GHG mitigation than another bioenergy system that is associated with increasing forest carbon stocks (green or purple curves) if the latter has much lower GHG displacement efficiency. (For interpretation of the references to color in this figure legend, the reader is referred to the web version of this article.)
Source: Cowie A, Berndes G, Smith CT. On the timing of greenhouse gas mitigation benefits of forest-based bioenergy, IEA Bioenergy Executive Committee 2013. Available from http://www.ieabioenergy.com/wp-content/uploads/2013/10/On-the-Timing-of-Greenhouse-Gas-Mitigation-Benefits-of-Forest-Based-Bioenergy.pdf.

Box 13.1 Life cycle assessment

Life cycle assessment (LCA) is a framework for assessing the environmental impacts of product systems and decisions. The steps in LCA are (1) goal and scope definition, (2) life cycle inventory analysis (LCI), (3) life cycle impact assessment (LCIA), and (4) interpretation of the results. The results of LCA are expressed per functional unit (e.g., one MJ of bioenergy; 1 km driven by standard passenger vehicle under standard conditions). Due to the flexibility of the framework, LCA is suitable for small and large-scale product systems, and can be used to aid micro- and macro-level decisions [15]. LCA has been categorized into two main modeling techniques, namely, attributional LCA (ALCA) and consequential LCA (CLCA).

Several definitions for ALCA and CLCA have been proposed and both methods have been applied differently between different studies [16,17]. The "Shonan Guidance Principles" [18] (pp. 132–133) state that ALCA is a "system modelling approach in which inputs and outputs are attributed to the functional unit of a product system by linking and/or partitioning the unit processes of the system according to a normative rule." Conversely, CLCA is a "system modelling approach in which activities in a product system are linked so that activities are included in the product system to the extent that they are expected to change as a consequence of a change in demand for the functional unit" [18]. ALCA focuses on describing all the environmentally-relevant physical flows to and from a product system over its life cycle [19,20]. In contrast, CLCA aims to describe how environmentally relevant physical flows would respond (or would have responded) to a change [19], e.g. a decision to increase production of forest bioenergy. Essentially, ALCA aims to describe the environmental impacts of a system as a component of the total impact of human activities, whereas CLCA aims to answer the question "what happens if?". Typically, average data are applied in ALCA, whereas marginal or incremental data are applied in CLCA [19].

Where a product is derived from a system with several outputs or functions, a method must be used to share the impacts among the products. Allocation based on an attribute such as energy content, mass, or value is used in ALCA as the basis for sharing impacts, whereas allocation is avoided in CLCA by giving credit to the product for products displaced by the co-product. Regardless of the modeling approach chosen, LCA fundamentally aims to describe the environmental impacts of a studied system. However, the system boundaries and other methodological choices, such as allocation and metrics applied in LCIA, vary depending on the goal and scope of the study.

Box 13.2 Bioenergy in national greenhouse gas inventories

The Intergovernmental Panel on Climate Change (IPCC) develops the methods that are used by countries for their reporting under the UNFCCC and accounting under the Kyoto Protocol. The scope of these reports is annual greenhouse gas emissions and removals at national level. According to the IPCC guidelines for national GHG emission reporting, CO_2 emissions from the combustion of biomass are counted as zero in the Energy sector [28]. This is to avoid double counting, because CO_2 emissions from the harvest of forest biomass for energy are included in the Agriculture, Forestry and Other Land-Use (AFOLU) sector. (Formerly the emissions and removals from forestry were reported in the Land Use Land Use Change and Forestry (LULUCF) sector, which was separate from the Agriculture sector. These two sectors are now combined in the AFOLU sector.) Consequently, if all countries follow the IPCC guidelines and report to the UNFCCC, all emissions from the use of biomass for energy will be estimated and reported [29]. However, under the Kyoto Protocol to the UNFCCC only developed ("Annex I") countries have commitments, and are required to account for their emissions against agreed targets. Developing countries do not have commitments, and any decline in forest carbon associated with harvest for biomass that is exported to Annex 1 countries from non-Annex 1 countries is excluded from accounting. Furthermore, reporting changes in forest carbon stock was optional for Annex I countries in the first commitment period (2008–12), so forest carbon losses from biomass harvest in Annex I countries were not accounted for by most countries. There was limited incentive to enhance forest carbon stocks due to national caps on forest sinks. These deficiencies have been partly addressed in the second commitment period of the Kyoto Protocol (2013–20), as accounting for "forest management" is now mandatory for Annex 1 countries. However, forest management emissions and removals are accounted relative to country-specific, projected reference levels representing the "business as usual" baseline. Consequently, harvesting of biomass may or may not create a debit, depending on the overall development of C stock and the agreed "forest management reference level." It should be noted that not all developed countries are committed to the Kyoto Protocol and that the commitments have been defined only until 2020. The Paris Agreement, negotiated in 2015, will commence in 2020, and will cover developing and well as developed countries, so the loopholes noted here associated with incomplete coverage will be largely overcome. However, the rules for forests and bioenergy under the Paris Agreement have yet to be determined. The approach selected will influence the contribution of bioenergy to future targets for emissions reduction, and the incentives for bioenergy and forest sinks.

include, as applicable, GHG emissions and removals related to raw material production, harvesting, transportation, biomass storage and conversion, and final fuel storage, distribution and use. Different amounts and types of fossil fuels and other inputs, such as fertilizers and processing chemicals, are used in different bioenergy systems, depending on the biomass source, energy conversion process, and product produced (i.e., heat, electricity, liquid biofuel). In addition, biomass harvesting influences forest carbon stocks, surface albedo properties, and possibly cloud formation and reflectivity. Biomass storage may cause process emissions.

13.5.2 Reference system

To determine the climate change effects of bioenergy, the bioenergy system must be compared with a reference system. The choice of the "without bioenergy" reference scenario (or counterfactual), against which the bioenergy scenario is evaluated, is critical to the results. For forest-based bioenergy, the reference scenario should describe the alternative use of the forest, as well as the alternative energy source that is displaced by the bioenergy system.

13.5.3 Land-use reference

Comparison with the reference land use allows the inclusion of impacts of change in land use or land management on C stock in biomass and soil. As with the bioenergy system, it is important to consider both management and natural factors in describing the forest carbon stocks in the reference system. This reference scenario may include forest management for a different mix of products and services, or preserving the forest for conservation. In assessing forgone sequestration, it should be recognized that sequestration rate slows as forests approach maturity, and that forests managed for conservation alone may have increased risk of disturbance, such as wildfire [21]. Due to uncertainties, especially those related to future systems, it may be relevant to consider several alternative reference land-use scenarios. The relevant reference land use will be determined by the purpose of the analysis. If the goal is to quantify the contribution of bioenergy as a component of all human activities (attributional approach—see Box 13.1), natural regeneration may be the appropriate reference, whereas if the goal is to estimate the effect of increasing production of bioenergy (consequential approach), the "most likely alternative" land use may be most suitable [22].

13.5.4 Energy source displaced

The reference energy scenario should describe the energy source serving the equal function in the absence of bioenergy. Coal has a higher GHG intensity than natural gas, so displacing coal achieves greater GHG savings than displacing the same energy content of natural gas. In the attributional approach, the GHG intensity of the average electricity mix is typically used (e.g., [23]), but in the consequential approach the relevant reference energy source is the marginal supply that would be used if the bioenergy was not produced, that is, the source(s) that would respond to increases in demand for electricity. In some cases, the marginal supply could be another renewable energy source.

13.5.5 Displacement value

The mitigation value of displacement will depend on the relative efficiency of conversion to energy product in the bioenergy and reference system, and the relative GHG emitted per MJ of energy product. The displacement value is often less than one because more CO_2 is emitted per MJ from biomass compared with coal or gas, due to the difference in chemical composition. This does not necessarily mean that biomass is worse than coal, as some have claimed (e.g., [24]). The difference in climate change effect of bioenergy and fossil energy can only be determined by comparing the bioenergy system with the reference system, considering all the aspects listed here.

The conversion efficiency of biomass to bioenergy greatly affects the climate effect. For example, conversion of biomass to bioliquids, while increasing the energy density and improving ease of transport and storage, is often less efficient than direct combustion of biomass for heat or power [25,26]. However, the actual displacement effect is determined by market-mediated forces - see indirect effects, later.

13.5.6 Handling coproducts

As discussed earlier, bioenergy is often just one product from a managed forest. Biomass for bioenergy is typically a residue, while the high value sawlogs and pulp wood are the main drivers of management decisions, such as rotation length, stocking rate, and fertilizer regime. Coproducts from the multioutput system may displace other products, generating additional mitigation compared to a situation in which wood was not harvested [27]. For example, when biomass is pyrolyzed or gasified, a solid coproduct (biochar) is generated in parallel, which may be used as a soil amendment, thereby reducing fertilizer requirements. The avoided emissions from fertilizer manufacture can be credited to the bioenergy system.

13.5.7 Spatial boundary

The spatial boundary is defined by the goal and scope, and can affect the results dramatically. In the case of long-rotation forestry, the regrowth phase may take many decades. The asynchrony between carbon emissions and sequestration, when considered on the basis of a single stand, has led to concerns that bioenergy does not necessarily deliver climate benefits in the short term: during the period between combustion and regrowth there is additional CO_2 in the atmosphere, causing a warming effect. However, this single-stand perspective does not necessarily provide adequate understanding of the climate effects of a new policy to promote forest bioenergy.

The appropriate boundaries of the assessment are determined by the purpose of the investigation. For example, a policy maker may be concerned with the impacts at a regional or national scale, so assessments to inform policy development should take place at that broad scale. On the other hand, an individual forest owner may be interested in knowing the effects at the estate level. When the assessment is undertaken for the purpose of product labeling, the scale of analysis applies to the system that produces that product.

13.5.8 GHGs and other climate forcers

All climate forcers, such as albedo effects and emissions and removals of greenhouse gases, including CO_2 and non-CO_2 (e.g., CH_4 from stored biomass, N_2O from soil), should be included across the whole life cycle.

13.5.9 Non-GHG climate forcers

Change in albedo (reflectance) from the land surface, which impacts energy balance, should be included in the assessment. In addition to the impact from emissions and sequestration of GHGs, bioenergy systems can affect climate through additional forcing processes, including direct impact on albedo (e.g., [30,31]). Harvest of forests in high latitudes or altitudes with snow cover can increase albedo, reducing global warming [32]. In some circumstances this effect is substantial, even counteracting negative impacts of a reduction in forest carbon stock [33]. However, according to [34], boreal forests double regional cloud condensation nuclei concentrations through emission of organic vapors and the resulting condensational growth of newly formed particles, having a significant cooling impact. Thus, harvesting of boreal forests reduces this cooling impact (thereby increasing warming), which compensates partly or completely the cooling impact of increased surface albedo. It should be noted that there are substantial knowledge gaps and uncertainties related to the complex interrelations between biogeochemical and physical climate forcings of forestry [34–36] . Bioenergy systems may also influence climate through emissions of aerosols, or black carbon, in different ways, depending on the technologies and scenarios considered [37].

13.5.10 Timing of emissions and removals

Conventionally, in LCA, the timing of emissions and removals is not considered: the carbon footprint of a product is calculated by summing the emissions over the entire life cycle (e.g., [23]). The implication is that timing of emissions and removals has no impact on climate change outcomes. In contrast, many current climate change policies provide incentives to delay emissions or sequester carbon temporarily (such as for 100 years). Such policies could "buy time" that could allow society to develop and deploy low-carbon energy systems, and they could assist in avoiding "climate tipping points." However, the timing and irreversibility of such tipping points are uncertain [38], and the climate impacts from delayed emissions may be problematic at later point in time. Some protocols acknowledge that time is significant, and therefore exclude emissions occurring after more than 100 years, or quantify such long-term emissions in a separate category (e.g., [39]).

Furthermore, to achieve equilibrium temperature targets, the exact timing of CO_2 emissions and removals may not be the most important consideration; rather, to mitigate climate change the most important action is to restrict the cumulative total GHG emissions in the longer term [38]. The more ambitious is the stabilization target (e.g., 1.5°C temperature increase), the deeper emission cuts are required [38]. Higher CO_2 emissions in earlier decades require lower CO_2 emissions later [38]. However, in the real world, the short- and long-term impacts are not typically independent from each other, due to relatively long economic lifetime of energy system structures [40]. Consequently, it is likely that achieving a 2°C target in practice will require CO_2 emissions to peak very soon, with deep cuts in emissions in the following decades [38,40]. Furthermore, as the climate sensitivity to the increasing GHG concentrations is not well known, the hedging of climate sensitivity risk calls for deeper early reductions instead of postponing reduction measures [41].

When assessing the effectiveness of bioenergy in climate change mitigation, the timing of the CO2 emissions and C sequestration matters, depending on the given target and the resulting emission reduction window. For example, a forest bioenergy system causing more emissions than a fossil reference fuel in the beginning should possibly provide negative net emissions within the end of this century, in order to efficiently work in achieving the 2°C target (see Fig. 12.46a in [38]). Berndes et al. [42] proposed "GHG emissions space" as a concept to encourage consideration of emissions management in the context of longer-term temperature targets. Focusing on the accumulated emissions up to a given year, society may decide to invest a portion of the emission space, allowed within the GHG target, on the establishment of renewable energy systems. Short-term emissions resulting from the establishment of bioenergy systems may be justified as investment in creating a low-carbon energy system. However, it should be noted that if a bioenergy system does not provide enough emission reductions within the given temporal window, which depends on the given climate change mitigation target, the "invested emissions" will need to be counteracted by other means. This may be difficult, especially considering the ambitious stabilization targets, which require that emissions from the whole society must be cut sharply within the upcoming decades.

13.5.11 Metrics for climate change impact assessment

When a pulse of CO_2 is emitted to the atmosphere a fraction of the CO_2 is taken up by the biosphere, some is dissolved in the ocean, and a fraction (15%–40%) remains in the atmosphere for up to 2000 years [43]. The climate effect of a pulse emission is quantified as the radiative forcing due to the perturbation. The commonly applied metric Global Warming Potential (GWP) quantifies the radiative forcing (extra energy retained) of a GHG pulse emission over a specified time period (commonly 100 years) in comparison with that of a pulse of CO_2 emitted at time zero.

Usually in LCA the GWP (100 years) is applied to calculate the CO_2 equivalent (CO_2-eq.) of non-CO_2 GHGs, and the climate effect is quantified as the total of GHG emissions and removals over the entire life cycle (e.g., [23]) or over the first 100 years of the life cycle of a product (e.g., [39]). Several authors have proposed alternative methods for estimating potential climate impacts of GHG emissions, for application in LCA that do account for the timing of GHG emissions and removals [44]. Clift and Brandão [45] developed the method adopted in the first version of the UK's carbon footprint guideline Publicly Available Specification for the assessment of the life cycle greenhouse gas emissions of goods and services "PAS2050" [46]. Two subsequently published approaches, Levasseur et al. [47] and Cherubini et al. [48], are essentially equivalent to that of Clift and Brandão [45]. Levasseur et al. [47] developed the "dynamic LCA" approach, which quantifies the radiative forcing resulting from an emission according to when it occurs within a defined assessment period, and assigns a reduced impact if emissions are delayed within this period. Cherubini et al. [48] proposed and Guest et al. [49] demonstrated a method, which quantifies the radiative forcing over the assessment period due to the combined effects of a pulse emission from combustion of biomass, followed by CO_2 removal due to the presumed regrowth. They apply a modified characterization factor "GWP_{bio}" that reflects this temporal profile of radiative forcing in comparison with a pulse emission of fossil CO_2, and varies with rotation length of the forest system. It should be noted that this metric does not aim to capture the forest carbon stock change between a bioenergy production scenario compared to a forest reference scenario without the studied bioenergy production, but rather it quantifies the climate effects of actual carbon emissions and the assumed subsequent sequestration. Thus, Pingoud et al. [50] extended the "GWP_{bio}" concept to produce a metric that assesses the climate impacts in comparison to a reference system that includes the alternative use of the forest.

An alternative metric to GWP now gaining traction is the global temperature potential (GTP), which quantifies the effect of GHG emissions on the global temperature at a specified time [51]. GTP is thus more closely related than GWP to the impact of climate change on human and natural systems. Cherubini et al. [52] assessed bioenergy systems and demonstrated that long-rotation forest systems show greater climate change mitigation potential when assessed by GTP than by GWP.

Furthermore, we should bear in mind that the impact pathway from GHG emissions to climate impacts goes through emissions, increased atmospheric concentrations, radiative forcing, temperature increase, and other impacts (such as storms, droughts). Choosing the appropriate metrics depends finally on the goal and scope of the study. There is no one metric that gives the complete picture; thus, it may be relevant to apply several metrics to improve understanding of the factors that influence the outcomes, and the sensitivity of results to alternative metrics. GWP (100 years) is still widely used in policy frameworks, but it is worth recognizing that the preferred metrics might change.

13.5.12 Temporal boundary of assessment

Typically, in LCA, the assessment starts at the "cradle," which commonly includes raw material extraction. In relation to bioenergy, it is debated whether the time period of assessment should commence when the forest was planted, or at the time of harvest.

Fundamentally, this depends on the question posed. If the biomass is produced from reforestation undertaken to meet later bioenergy demand, then the initial removal of CO_2 from the atmosphere as the plants grow before first harvest may be included as part of the bioenergy life cycle. In such a case, a decision to afforest can be combined with a decision to harvest. However, what was considered optimal at some point in time (i.e., afforestation followed by harvest) is not necessarily optimal in the future (i.e., although originally planted for future harvest for bioenergy, it may be decided at the point of harvest that retaining the forest gives greater climate and other benefits). Thus, the impacts of the decision to harvest may also be of interest, independent of the prior decision to afforest and subsequently harvest. These two temporal boundaries can result in very different conclusions. If the biomass is extracted from existing forests, choosing the appropriate time horizon for the assessment is challenging: should the assessment start at the time of harvest, or after harvest, for example? If forest management is changed in advance of the first harvest as a consequence of bioenergy demand, then it may be appropriate to start the accounting clock at the time management was changed. An example of such a change is when forest owners choose to skip precommercial thinning in order to produce a larger bioenergy harvest in later thinning operations. On the other hand, it may be difficult to verify for what purpose the forests have previously been managed. Consequently, for existing forests, both the retrospective and prospective perspectives in accounting may be relevant, as for biomass derived from afforestation or reforestation. The relevant temporal boundary depends on the purpose of the assessment.

The avoided emissions from fossil fuel substitution increase over time as more biomass displaces more fossil fuel. At the same time, carbon cost resulting from increased harvest of forests typically becomes lower over time. Therefore, the time frame chosen to assess the climate impacts has a significant impact on the results and could even turn around the conclusions drawn.

13.5.13 Indirect effects

Comprehensive assessment of the short- and long-term climate effects from expansion of bioenergy systems requires a consequential modeling approach that considers the land, forest products, energy sectors, as well as socioeconomic and biogeophysical effects. This is necessary in the development of policy to inform decisions on appropriate scales of expansion of bioenergy systems and optimal use of biomass for competing energy and material products.

As bioenergy and wood products are interrelated industries, a full understanding of the impacts related to forest bioenergy requires understanding also of the impacts related to industrial wood use. A strategic and rationalized cascading use of biomass, first for materials and subsequently for energy, allows for higher GHG benefits through multiple substitutions compared with direct use of newly harvested wood for energy [53,54] (Pingoud et al. 2010, Gustavsson et al. 2006). In addition, GHG savings from substituting other materials by wood materials may be significantly larger than substituting fossil fuels by bioenergy [55,56] . However, the various possibilities of wood use, as well as the complexity of identifying the displaced construction material, make it difficult to generalize the GHG savings from substitution by wood products.

If land use on a particular parcel of land changes from food production to production of biomass, including wood for bioenergy, land use elsewhere might change to meet the demand for food. This land-use change elsewhere is known as indirect land-use change, and should be included in the assessment of the impacts of bioenergy. However, estimating ILUC is another methodological issue for which consensus remains elusive among scientists, partly because ILUC, by definition, cannot be verified nor ascribed directly to a bioenergy system. Several methods exist; those based on general or partial equilibrium modeling remain the most popular in a policy-making context, while biophysical models also exist for the same purpose (e.g., Schmidt et al. [57]).

The critical question for policymakers should be: will bioenergy incentives facilitate or hinder capacity to stabilize the global climate, preventing warming in excess of 2°C (or 1.5°C)? The answer will depend on how bioenergy incentives affect forest carbon stocks (at the landscape scale), and how they influence fossil-fuel emissions. The impact on forest carbon is location specific, influenced by environmental and socioeconomic factors including forest type, climate, forest ownership, and the character and product portfolio of the associated forest industry. Furthermore, the forest carbon stock response to changes in forest management and harvesting in turn depends on the characteristics of the forest ecosystem. The impact on fossil-fuel emissions will depend on how bioenergy availability impacts on use of fossil fuels. The phenomenon of rebound should be acknowledged: due to impacts of bioenergy on prices of fossil fuels, bioenergy does not necessarily displace the same amount of fossil-fuel energy products [58]. More importantly for long-term climate stabilization, bioenergy incentives may influence investment in fossil-fuel technology and infrastructure. Bioenergy could play an important role in transition to low-carbon energy systems, by providing dispatchable power that can stabilize electricity grids, enabling expansion of other intermittent renewables (solar and wind power).

When considering the climate effects of single bioenergy systems in the prevailing or assumed economic conditions, independent from market responses, an attributional modeling perspective may be more appropriate than consequential modeling. In this case, market-mediated effects are excluded (see Box 13.1). Examples of such cases might be the calculation of the carbon footprint of a bioenergy product to assess compliance within a scheme, or to label a product.

The results of the assessment can be expressed in different ways (Cherubini et al., 2013). For comparison between energy products, the result is expressed as GHG intensity per functional unit (ISO, 2013). It is often relevant also to consider the change in emissions per unit of biomass resource consumed, or per unit land area.

The mitigation value is determined by the net effect of avoided emissions and the "GHG cost." GHG costs arise from emissions from fossil-fuel use along the supply chain, and any decline in the C stock of the forest due to biomass harvest. If the GHG cost is lower than the fossil-fuel emissions avoided by the bioenergy system, there will be an immediate benefit. This could be the case when forest residues are harvested and effectively used to substitute fossil fuels (e.g., [59,60]). If the forest carbon stock is reduced, there will be a delay until the savings from avoided fossil-fuel emissions lead to a net reduction in atmospheric CO_2 (e.g., [3]); and the temporary increase in atmospheric CO_2 will cause global warming. This is typically the situation when growing trees are harvested for energy, as harvesting results in the immediate release of C previously stored in trees, and also in forgone C sequestration if trees would have continued to grow [6,59,60]. The overall forest carbon stock may still increase in the bioenergy scenario, but at a slower rate compared to the absence of harvesting for bioenergy; bioenergy is in this case also associated with forgone carbon sequestration, which should be taken into account when evaluating the net GHG effect (e.g., [61,62]).

13.6 Summary of recommendations

It is proposed that, in order to understand the climate change effects of forest-based bioenergy systems, the following approach should be applied:

- define precisely the question to be answered;
- consider bioenergy production in comparison to a reference system in which the assessed bioenergy is not produced;
- accurately define the reference forest management and energy systems with which bioenergy is compared and consider alternative reference scenarios whenever appropriate (typically there are uncertainties that need to be considered);
- consider whether stand or landscape-level analysis is appropriate, and choose the spatial boundary accordingly;
- recognize that individual harvest decisions are made at stand level while at landscape, regional, or national levels may be more appropriate in order to consider the impacts of the forest economy, as management responds to market forces;
- assess the impact of timing of emissions and sequestration;
- consider whether a retrospective or prospective approach is relevant, and choose the temporal boundary accordingly, acknowledging that as the short-term and long-term climate impacts can be significantly different, the time horizon adopted can substantially influence the conclusions drawn;
- recognize that biomass for bioenergy is typically only one component of a range of products harvested from a managed forest;
- consider all factors that influence forest carbon stocks, including forest management (silviculture, harvest), also natural biotic and abiotic forces (e.g., edaphic, climate, disturbance events);
- commence accounting at the time that management changes in response to bioenergy demand, while acknowledging that the chosen perspective, i.e., retrospective or prospective, might impact on this choice;
- recognize that the earth climate system is altered not only by CO_2, but also by changes in the atmospheric concentration of other gases and aerosols, in solar radiation and in land surface albedo, so the effects of all climate forcers influenced by forest cover and forest management should ideally be included;
- recognize that the climate effect of bioenergy is influenced by market-mediated forces, that impact land use, energy use, forest management, and the wood products sector;
- recognize that a comprehensive analysis of climate impacts of bioenergy is complex and likely to be subject to significant uncertainties and sensitivities; and
- note that the result is specific to each situation.

There are some knowledge gaps that remain to be filled, to enable greater understanding of the climate impacts of bioenergy. These include the need for (i) studies clarifying how the energy and forestry sectors, including forest management, respond to changing forest product markets; (ii) empirical data on forest product supply and demand and land use at resolution that enables comprehensive analyses of alternative scenarios; (iii) studies integrating forest/bioenergy systems modeling with Earth systems/climate science/energy system/integrated assessment modeling; and multidisciplinary research into the interpretation and translation of insights from scenario modeling into policy guidance for management of land use and energy system [63].

13.7 Conclusions

Quantifying the climate effects of forest-based bioenergy is a complex endeavor and is fraught with challenges. Methodological choices influence the results and the related conclusions dramatically. These choices depend on the goal and scope of the study, and are related to the definition of spatial and temporal system boundaries and climate metrics. A comprehensive understanding of the climate effects of bioenergy systems requires a combination of biophysical, climate, and socioeconomic models, including effects on parallel industries (e.g., wood products, agriculture, and energy), in order to robustly inform policy development.

The timing of emissions and sequestration is important when assessing the effectiveness of bioenergy in climate change mitigation, in particular concerning ambitious stabilization targets (e.g., 1.5°C temperature increase) requiring rapid and deep cuts in GHG emissions and increase in carbon sinks [64]. When planning policies, it is very important to recognize that short- and long-term climate impacts of bioenergy might be very different. A shift to a sustainable bioenergy production system resulting in significant climate benefits in the long-term might also result in less beneficial or even harmful short-term impacts. This can be considered as an investment, acknowledging that the related short-term climate impacts should be reduced by other means, particularly if aiming to achieve ambitious climate targets in the short term.

Acknowledgments

We thank Göran Berndes and Tat Smith for their comments on earlier versions of this paper, and IEA Bioenergy for permission to utilize material from [65].

References

[1] Fargione J, Hill J, Tilman D, Polasky S, Hawthorne P. Land clearing and the biofuel carbon debt. Science 2008;319:1235–8.

[2] Searchinger T, Heimlich R, Houghton RA, Dong FX, Elobeid A, Fabiosa J, Tokgoz S, Hayes D, Yu TH. Use of U.S. croplands for biofuels increases greenhouse gases through emissions from land-use change. Science 2008;319:1238–40.

[3] Hudiburg TW, Law BE, Wirth C, Luyssaert S. Regional carbon dioxide implications of forest bioenergy production. Nat Clim Chang 2011;1(8):419–23.

[4] Schulze E-D, Körner C, Law BE, Haberl H, Luyssaert S. Large-scale bioenergy from additional harvest of forest biomass is neither sustainable nor greenhouse gas neutral. GCB Bioenergy 2012;4:611–6.
[5] Walker T, Cardellichio P, Colnes A, Gunn J, Kittler B, Perschel B. Forest Guild, In: - Recchia C, Saah D, editors. Natural Capital Initiative, Biomass sustainability and carbon policy study. Manomet Center for Conservation Sciences; 2010. Available at http://www.biomassthermal.org/programs/documents/20_BiomassSustainabilityCarbon.pdf.
[6] Pingoud K, Ekholm T, Soimakallio S, Helin T. Carbon balance indicator for forest bioenergy scenarios. GCB Bioenergy 2016;8(1):171–82.
[7] Lamers P, Junginger M. The 'debt' is in the detail: a synthesis of recent temporal forest carbon analyses on woody biomass forenergy. Biofuels Bioprod Biorefin 2013;7:373–85.
[8] Gustavsson L, Haus S, Ortiz CA, Sathre R, Le Truong N. Climate effects of bioenergy from forest residues in comparison to fossil energy. Appl Energy 2015;138:36–50.
[9] Ter-Mikaelian MT, Colombo SJ, Chen J. The burning question: Does forest bioenergy reduce carbon emissions? A review of common misconceptions about forest carbon accounting. J Forestry 2015;113:57–68.
[10] Eriksson L, Gustavsson L. Costs, CO2- and primary energy balances of forest-fuel recovery systems at different forest productivity. Biomass Bioenergy 2010;34:610–9.
[11] Lindholm EL, Berg S, Hansson PA. Energy efficiency and the environmental impact of harvesting stumps and logging residues. Eur J For Res 2010;129:1223–35.
[12] Hamelinck CN, Suurs RA, Faaij AP. International bioenergy transport costs and energy balance. Biomass Bioenergy 2005;29:114–34.
[13] Gustavsson L, Eriksson L, Sathre R. Costs and CO_2 benefits of recovering, refining and transporting logging residues for fossil fuel replacement. Appl Energy 2011;88:192–7.
[14] Hoefnagels R, Resch G, Junginger M, Faaij A. International and domestic uses of solid biofuels under different renewable energy support scenarios in the European Union. Appl Energy 2014;131:139–57.
[15] EC-JRC-IES, European Commission—Joint Research Centre—Institute for Environment and Sustainability. International reference life cycle data system (ILCD) handbook—general guide for life cycle assessment—detailed guidance. First edition. EUR 24708 EN. Luxembourg. In: Publications Office of the European Union. 2010.
[16] Soimakallio S, Cowie A, Brandão M, Finnveden G, Ekvall T, Erlandsson M, Koponen K, Karlsson PE. Attributional life cycle assessment: Is a land-use baseline necessary? Int J Life Cycle Assess 2015;20(10):1364–75.
[17] Zamagni A, Guinée J, Heijungs R, Masoni P, Raggi A. Lights and shadows in consequential LCA. Int J Life Cycle Assess 2012;17:904–18.
[18] UNEP/SETAC Life Cycle Initiative, Global guidance principles for life cycle assessment databases (Shonan Guidance Principles), 2011.
[19] Finnveden G, Hauschild MZ, Ekvall T, Guinée J, Heijungs R, Hellweg S, Koehler A, Pennington D, Suh S. Recent developments in life cycle assessment. J Environ Manag 2009;91:1–21.
[20] Curran MA, Mann M, Norris G. The international workshop on electricity data for life cycle inventories. J Clean Prod 2005;13(8):853–62.
[21] Ximenes FA, George BH, Cowie A, Williams J, Kelly G. Greenhouse gas balance of native forests in New South Wales, Australia. Forests 2012;3:653–83.
[22] Koponen K, Soimakallio S, Kline KL, Cowie A, Brandão M. Quantifying the climate effects of bioenergy—choice of reference system. Renew Sust Energ Rev 2018;81:2271–80.
[23] ISO. ISO/TS 14067:2013 greenhouse gases—C footprint of products—requirements and guidelines for quantification and communication. Switzerland: International Organization for Standardization; 2013.

[24] Upton J. What's worse than burning coal? Burning wood 2014. http://grist.org/climate-energy/whats-worse-than-burning-coal-burning-wood/.
[25] Soimakallio S, Mäkinen T, Ekholm T, Pahkala K, Mikkola H, Paappanen T. Greenhouse gas balances of transportation biofuels, electricity and heat generation in Finland—dealing with the uncertainties. Energy Policy 2009;37(1):80–90.
[26] Soimakallio S. Toward a more comprehensive greenhouse gas emissions assessment of biofuels: the case of forest-based Fischer–Tropsch diesel production in Finland. Environ Sci Technol 2014;48(5):3031–8.
[27] Soimakallio S, Saikku L, Valsta LT, Pingoud K. Climate change mitigation challenge for wood utilization—the case of Finland. Environ Sci Technol 2016;50:5127 – 5134.
[28] Houghton J, Meira Filho L, Lim B, Treanton K. Mamaty I. In: Revised 1996 IPCC guidelines for national greenhouse gas inventories. 1997.
[29] IPCC 2014a. Frequently asked questions. http://www.ipcc-nggip.iges.or.jp/faq/faq.html
[30] Georgescu M, Lobell D, Field C. Direct climate effects of perennial bioenergy crops in the United States. Proc Natl Acad Sci 2011;108:4307–12.
[31] Loarie S, Lobel l, Asner G, Mu Q, Fiel C. Direct impacts on local climate of sugar-cane expansion in Brazil. Nat Clim Chang 2011;1:105–9.
[32] Bright R, Astrup R, Strømman A. Empirical models of monthly and annual albedo in managed boreal forests of interior Norway. Clim Chang 2013;120(1–2):183–96.
[33] Bright R, Strømman A, Peters G. Radiative forcing impacts of boreal Forest biofuels: A scenario study for Norway in light of albedo. Environ Sci Technol 2011;45:7570–80.
[34] Spracklen D, Bonn B, Carslaw K. Boreal forests, aerosols and the impacts on clouds and climate. Philos Trans R Soc A Math Phys Eng Sci 2008;366(1885):4613–26.
[35] Bonan G. Forests and climate change: forcings, feedbacks and the climate benefits of forests. Science 2008;320:1444–9.
[36] Anderson R, Canadell J, Randerson J, et al. Biophysical considerations in forestry for climate protection. Front Ecol Environ 2010;9:174–82.
[37] Kupiainen K, Klimont Z. Primary emissions of fine carbonaceous particles in Europe. Atmos Environ 2007;41(10):2156–70.
[38] Collins, M., R. Knutti, J. Arblaster, J.-L. et al. (2013). "Long-term climate change: Projections, commitments and irreversibility," Climate Change 2013: The Physical Science Basis. Contribution of Working Group I to the Fifth Assessment Report of the Intergovernmental Panel on Climate Change [Stocker, T.F., D. Qin, G.-K. Plattner, et al. (eds.)]. Cambridge University Press, Cambridge, United Kingdom and New York, NY, USA. pp 1029–1136.
[39] BSI. (2011). PAS 2050:2011 specification for the assessment of the life cycle greenhouse gas emissions of goods and services (PAS 2050). British Standards Institution, London.
[40] IPCC, 2014b: Climate Change 2014: Mitigation of Climate Change. Contribution of Working Group III to the Fifth Assessment Report of the Intergovernmental Panel on Climate Change [Edenhofer, O., R. Pichs-Madruga, Y. Sokona, E. Farahani, S. Kadner, K. Seyboth, A. Adler, I. Baum, S. Brunner, P. Eickemeier, B. Kriemann, J. Savolainen, S. Schlömer, C. von Stechow, T. Zwickel and J.C. Minx (eds.)]. Cambridge University Press, Cambridge, United Kingdom and New York, NY, USA.
[41] Ekholm T. Hedging the climate sensitivity risks of a temperature target. Clim Chang 2014;127(2):153–67.
[42] Berndes G, Bird N, Cowie A. Bioenergy, land use change and climate change mitigation: Report for policy advisors and policy makers, IEA Bioenergy: ExCo 2010;2010:03. http://www.ieabioenergy.com/publications/bioenergy-land-use-change-and-climate-change-mitigation/.

[43] Ciais, P., C. Sabine, G. Bala, L. et al. (2013). "Carbon and other biogeochemical cycles," Climate. Change 2013: The Physical Science Basis. Contribution of Working Group I to the Fifth Assessment Report of the Intergovernmental Panel on Climate Change [Stocker, T.F., D. Qin, G.-K. Plattner, M. et al. (eds.)]. Cambridge University Press, Cambridge, United Kingdom and New York, NY, USA. pp 465–570.

[44] Brandão M, Levasseur A, Kirschbaum MUF, et al. Key issues and options in accounting for carbon sequestration and temporary storage in life cycle assessment and carbon footprinting. Int J Life Cycle Assess 2013;18:230–40.

[45] Clift R, Brandão M. Carbon storage and timing of emissions—a note by Roland Clift and Miguel Brandão. Guildford, UK: University of Surrey, Centre for Environmental Strategy Working Paper 02/08; 2008.

[46] BSI. PAS 2050:2011 specification for the assessment of the life cycle greenhouse gas emissions of goods and services (PAS 2050). London: British Standards Institution; 2008.

[47] Levasseur A, Lesage P, Margni M, et al. Considering time in LCA: Dynamic LCA and its application to global warming impact assessments. Environ Sci Technol 2010;44 (8):3169–74.

[48] Cherubini F, Peters GB, Bernsten T, et al. CO_2 emissions from biomass combustion for bioenergy: atmospheric decay and contribution to global warming. GCB Bioenergy 2011;3(5):413–26.

[49] Guest G, Cherubini F, Strømman AH. Global warming potential of carbon dioxide emissions from biomass stored in the Anthroposphere and used for bioenergy at end of life. J Ind Ecol 2013;17(1):20–30.

[50] Pingoud K, Ekholm T, Savolainen I. Global warming potential factors and warming payback time as climate indicators of forest biomass use. Mitig Adapt Strateg Glob Chang 2012;17(4):369–86.

[51] Myhre G, Shindell D, Bréon F-M, et al. Anthropogenic and natural radiative forcing. In: - Stocker TF, Qin D, Plattner G-K, et al., Climate Change 2013: The Physical Science Basis. Contribution of Working Group I to the Fifth Assessment Report of the Intergovernmental Panel on Climate Change. Cambridge, United Kingdom and New York, NY, USA: Cambridge University Press; 2013.

[52] Cherubini F, Bright RM, Strømman AH. Global climate impacts of forest bioenergy: What, when and how to measure? Environ Res Lett 2013;8(1)014049.

[53] Pingoud K, Pohjola J, Valsta L. Assessing the integrated climatic impacts of forestry and wood products. Silva Fennica 2010;44(1):155–75.

[54] Gustavsson L, Pingoud K, Sathre R. Carbon dioxide balance of wood substitution: Comparing concrete- and wood-framed buildings. Mitig Adapt Strateg Glob Chang 2006;11 (3):667–91.

[55] Sathre R, O'Connor J. Meta-analysis of greenhouse gas displacement factors of wood product substitution. Environ Sci Pol 2010;13:104–14.

[56] Kurz W, Smyth C, Lemprière T. Climate change mitigation through forest sector activities: principles, potential and priorities 1. Unasylva 2016;67:61.

[57] Schmidt JH, Weidema BP, Brandão M. A framework for modelling indirect land use changes in life cycle assessment. J Clean Prod 2015;99:230–8.

[58] Smeets E, Tabeau A, van Berkum S, Moorad J, van Meijl H, Woltjer G. The impact of the rebound effect of the use of first generation biofuels in the EU on greenhouse gas emissions: A critical review. Renew Sust Energ Rev 2014;38:393–403.

[59] Agostini A, Giuntoli J, Boulamanti A, Marelli L, editors. Carbon accounting of forest bioenergy. Conclusions and recommendations from a critical literature review. JRC technical reports. Report EUR 25354 EN. European Commission: Joint Research Centre; 2013.

[60] Matthews, R., Sokka, L., Soimakallio, S., Mortimer, N., Rix, J., Schelhaas, M.-J., Jenkins, T., Hogan, G., Mackie, E., Morris, A. and Randle, T. 2014. Review of literature on biogenic carbon and life cycle assessment of forest bioenergy. Final Task 1 report, DG ENER project, 'Carbon impacts of biomass consumed in the EU. The Research Agency of the Forest Commission. May 2014.

[61] Kallio AMI, Salminen O, Sievänen R. Sequester or substitute—consequences of increased production of wood based energy on the carbon balance in Finland. J For Econ 2013;19 (4):402–15.

[62] Sievänen R, Salminen O, Lehtonen A, Ojanen P, Liski J, Ruosteenoja K, Tuomi M. Carbon stock changes of forest land in Finland under different levels of wood use and climate change. Ann For Sci 2014;71(2):255–65.

[63] Cowie, A. L., Brandão, M., Fritsche, U.R., Berndes, G., Junginger M. and Smith C.T. 2017 Forests, bioenergy and climate change mitigation. Report of a workshop held Vols. 19–20 May 2014, Copenhagen: IEA Bioenergy. Available from http://task38.ieabioenergy.com/wp-content/uploads/2017/01/Copenhagen_Workshop_Report.pdf

[64] Rockström J, Gaffney O, Rogelj J, Meinshausen M, Nakicenovic N, Schellnhuber HJ. A roadmap for rapid decarbonization. Science 2017;355(6331):1269–71.

[65] Cowie A, Berndes G, Smith CT, On the timing of greenhouse gas mitigation benefits of forest-based bioenergy, IEA Bioenergy Executive Committee 2013. Available from http://www.ieabioenergy.com/wp-content/uploads/2013/10/On-the-Timing-of-Greenhouse-Gas-Mitigation-Benefits-of-Forest-Based-Bioenergy.pdf

Hydrogen fuel, fuel cells, and methane

14

Johannes Lindorfer[1], *Gerda Reiter, Robert Tichler, Horst Steinmüller*
Energy Institute at the Johannes Kepler University Linz, Linz, Austria
[1]Corresponding author: lindorfer@energieinstitut-linz.at

Chapter Outline

14.1 Introduction, characterizing "green gas" options

Development of cutting-edge, environment-friendly, and energy-efficient technologies could reduce the release of anthropological greenhouse gases responsible for global warming, by reducing the use of fossil fuels. Natural gas, a gas mixture that consists primarily of methane (CH_4), is currently fulfilling a large part of the energy requirements in Europe: about 25% of the primary energy demand in the EU is delivered by natural gas [1] in many different applications: it can be used to generate electricity, for heating, as a fuel for cars, and also as a feedstock for industry. Biomethane is the same as methane – and therefore also has the same applications – but it is made from renewable biomass such as energy crops, organic waste, and sewage.

Table 14.1 lists basic production pathways for biomethane. The first three technologies produce raw biogas by microbial anaerobic conversion of biodegradable material. Biogas primarily consists of CH_4 (55–60 vol.%) and CO_2 (40–45 vol.%), and

Managing Global Warming. https://doi.org/10.1016/B978-0-12-814104-5.00014-4

Table 14.1 Renewable production pathways for methane

	Production pathway	Description
a	Anaerobic digestion of biomass	Biogas production followed by upgrading to biomethane
b	Waste water treatment plants	Sewage gas production followed by upgrading to biomethane
c	Landfills	Landfill gas production followed by upgrading to biomethane
d	Gasification of biomass	Production of syngas from biomass (wood, residuals...) followed by methanation
e	Methanation of green hydrogen	Renewable produced hydrogen combined with CO_2 in methanation

Source: own representation.

some trace gases dependent on the fermentation process and on the converted initial substrates [2]. In conventional anaerobic digestions process, biogas is produced in one single digester at ambient pressure and at operational pressure level of 20–57°C dependent on the species of methanogens in the digester. In one single digester, pH ranges between 6.5 and 8 and is one of the hardly adjustable process variables. In vertical digesters, the substrates are moved by gravity from the top of the fermenter to the bottom. By mixing fresh substrate with liquid fermentation residues from the bottom, approximately constant pH value can be reached. In many reactor systems, pH value is one of the important process control parameters although changes on that value show inhibitions often too late. The amount of volatile fatty acids is a precise method for process stability, but online measurements via liquid chromatography are too slow and economically not feasible [3]. The most common continuously operated reactor is named continuously stirred tank reactor (CSTR) and is often used for wet substrates with varied reactor volumes between 1000 and 4000 m^3. If dry substrates (>25% dry matter) are applied, horizontal or vertical plug flow reactors (PFR) with volumes of up to 700 m^3 are applied [4,5].

Biogas is predominantly used for combined heat and power (CHP) generation due to its low calorific value [6]. Instead of using biogas in CHP units, the purification of biogas to biomethane and its subsequent injection into the natural gas grids presents an alternative method of ensuring high-energy utilization, especially for biogas plants located in rural areas without attached heat sinks [7]. By applying upgrading techniques, the gas production and its utilization can be decoupled in terms of time and space [8]. Prior to injection into the gas grid, raw biogas must be purified and its heating value adapted to the natural gas quality [7,9,10]. In this step, the CO_2 content and the typical trace amounts of the raw biogas such as water (H_2O, 5%–10%), oxygen (O_2, 0%–1%), nitrogen (N_2, 0%–2%), hydrogen sulfide (H_2S, 0.005%–2%), ammonia (NH_3,<1%), and siloxane (0%–0.02%) have to be reduced [11]. Commonly used upgrading technologies for reducing the CO_2 content in the raw gas are water and amine scrubbing as well as gas separation membranes and pressure swing adsorption (PSA) [2].

Although currently the predominant proportion of biomethane is produced from biomass with anaerobic digestion and landfills, sewage, it can also be produced by two alternative methods called biomass gasification and power-to-gas. With power-to-gas (renewable), electricity is used to split water into hydrogen (H_2) and oxygen (O_2). In gasification, solid biomass such as woody biomass is converted at high temperatures under influence of O_2 into a mixture of mainly carbon monoxide (CO), H_2 and carbon dioxide (CO_2), called syngas. Syngas is a fuel that can be used directly but it can also be converted into hydrogen, methanol, or methane. The four production methods a-c in Table 14.1 use biomass as an input for biomethane production. In the fifth method, renewable produced H_2 is converted into biomethane by combining it with an external source of CO_2. The volumes of biogas and biomethane have remained modest so far, at least as percentage of natural gas consumption, but an upward trend is discernible. The two alternative biomethane production methods are currently still in their pilot phase but have the potential to significantly increase the volumes of renewable hydrogen and synthetic methane production in the near future [12].

Although biomethane is often simply called green gas, carbon-neutral produced hydrogen can also be named a green gas. Just as methane, hydrogen is a gaseous energy carrier that could be used in many different applications such as power generation and transportation. Fig. 14.1 shows global hydrogen production by process. Natural gas reforming accounts for the largest share with 48%. Only a very small fraction (~4%) is being produced by water electrolysis as the cost and energy consumption of this approach are comparatively high. Every year 140×10^6 t (140 million metric tonnes) of hydrogen are produced worldwide predominantly from fossil sources. Hydrogen generation and consumption generally takes place at the same site to limit safety hazard issues arising from transport and storage. Due to this on-site production of hydrogen, it is difficult to estimate total volumes of global hydrogen production and reported numbers are uncertain [13].

Hydrogen is currently primarily used in refineries (for hydrocracking of crude oil) and in the chemical industry, e.g., as a feedstock for ammonia or methanol production. Ammonia is used to produce fertilizers and its production consumes over half of all produced hydrogen in the EU [13,14]. A gradual shift from our current natural

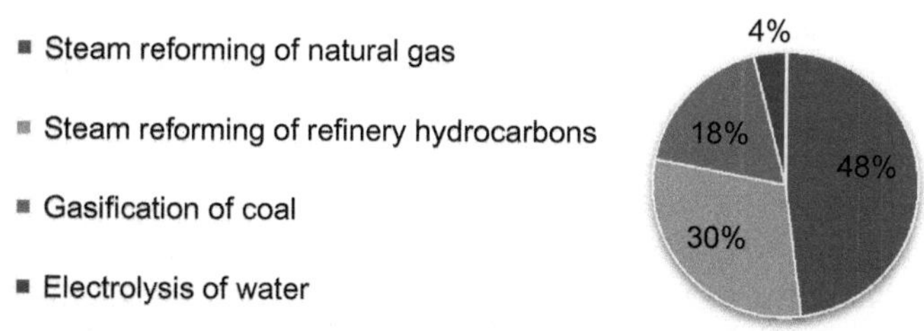

Fig. 14.1 Estimated distribution of hydrogen production by technology.
Source: own representation based on Decourt, Lajoie, Debarre, Soupa. Hydrogen based energy conversion, more than storage: system flexibility. SBC Energy Institute, 2014.

Table 14.2 **Renewable production pathways for hydrogen**

	Technology	Description
A	Power-to-gas	Electrolysis of water with renewable electricity
B	Gasification of biomass	Production of syngas from biomass (wood, residuals...) upgraded to hydrogen
C	SMR of biomethane	Steam methane reforming of biomethane
D	SMR combined with CCS	Conventional steam methane reforming combined with CCS

Source: own representation.

gas-based consumption to a so-called hydrogen economy in which hydrogen becomes the main energy carrier, taking over the role of methane, is discussed in many different studies, articles, and reports [15,16]. .Table 14.2 lists four technologies for producing green hydrogen. These four production methods were identified to emit <60% greenhouse gases compared to the most conventional hydrogen production method (steam methane reforming of natural gas) [17]. Power-to-gas technology, intended to store excess electricity produced when demand is lower than supply, is increasingly discussed as a long-term solution to the intermittency of current renewable energy sources [18–21]. Aspects of how to integrate power-to-gas into the energy system are, for instance, presented in references [22–27]. Most of these evaluations are presented in German because there is currently a strong focus on power-to-gas technology in Germany [28]. With the goal of generating 80% of its electricity from renewable power sources by the year 2050 [29], Germany will have a large demand for energy storage in the near future. Comparable future prospects are projected for the United States [30].

Also by, e.g., photocatalytic methods hydrogen can be generated. Based on the fundamental research character (TRL < 5) of these production pathways, they are not further covered in this chapter.

The focus of international and national actions is predominantly addressing the electricity system, but the necessity of applying interconnections between the energy grids and sectors more and more emerges, especially in the transport sector, where decarbonization has not yet extensively taken place. Table 14.3 shows five alternative fuel that are stimulated in many countries and their application in different mobility modes and ranges [31].

Based on Table 14.3 only biofuels can substitute conventional fuels in all applications. Methane, however, is also very widely applicable as air transport is the only mode in which it cannot be used. Hydrogen is considered to be suitable for most transport modes, except for long-range road freight, air and sea transport. Worldwide countries are stimulating the introduction and use of alternative fuels by regulating and enabling availability and by setting common technical specifications. Renewable power sources that harness solar and wind energy have great potential in this field but are also accompanied by several problems due to their fluctuating and intermittent nature.

Table 14.3 **Coverage of transport modes and travel range by the main alternative fuels**

Fuel	Mode	Road passenger			Road freight			Air	Rail	Water		
	Range	Short	Medium	Long	Short	Medium	Long			Inland	Sea	Maritime
LPG		+	+	+	+	+	+	−	−	+	+	−
Methane	LNG	+	+	+	+	+	+	−	+	+	+	+
	CNG	+	+	+	+	+	−	−	−	−	−	−
Electricity		+	−	−	+	−	−	−	+	−	−	−
Liquid biofuels		+	+	+	+	+	+	+	+	+	+	+
Hydrogen		+	+	+	+	+	−	−	+	+	−	−

+, Adequate; −, inadequate; LPG, liquefied petroleum gas; LNG, liquefied natural gas; CNG, compressed natural gas.
Source: Based on European Commission 2013. Clean power for Transport: a European alternative fuels strategy.

14.1.1 Demand for energy storage and renewable products

According to the IEA, global electricity generation is currently covered by 22.8% of renewable power sources, with water power being the most important (16.6%) [32]. Variable or fluctuating renewable power sources such as wind and photovoltaics only account for 3.1% and 0.9%, respectively. However, the installed capacity of wind power and photovoltaics has grown very rapidly over the past decade. Currently, 370 GW of wind power and 177 GW of photovoltaics are installed globally. Fig. 14.2 shows the expected future capacity of fluctuating renewable power sources for selected countries and the rest of the world.

This strong growth of fluctuating renewable power sources (photovoltaics and wind power) will also lead to an increased demand for energy balancing and storage options. Detailed information on energy system flexibility measures can be for instance gathered from [33]. Available storage technologies are illustrated in Fig. 14.3. Whereas technologies such as flywheels or batteries are better suited for short-term energy storage, energy storage via H_2 or synthetic CH_4 (power-to-gas) is more a long-term energy storage technology. In contrast to the other storage technologies in Fig. 14.3, the output product of power-to-gas is not necessarily electricity. The produced H_2 or CH_4 can also be utilized as transport fuels, for heating purposes or as raw materials in industry. Power-to-gas technology could therefore lead to a hybridization of the energy system, where surplus from electricity generation is utilized to provide renewable products. Surplus electricity could be specified as the electricity that cannot be fed into the public electricity grid or be utilized otherwise. Reasons for that could be a lower electricity demand than the actual generation or that in local grids the electricity network may be too weak to transport peak production from renewables.

However, if power-to-gas is utilized to produce renewable fuels from surplus electricity, the energy carriers are not available for reconversion into electricity in times of low demand. Power-to-gas is thus not necessarily a typical electricity storage

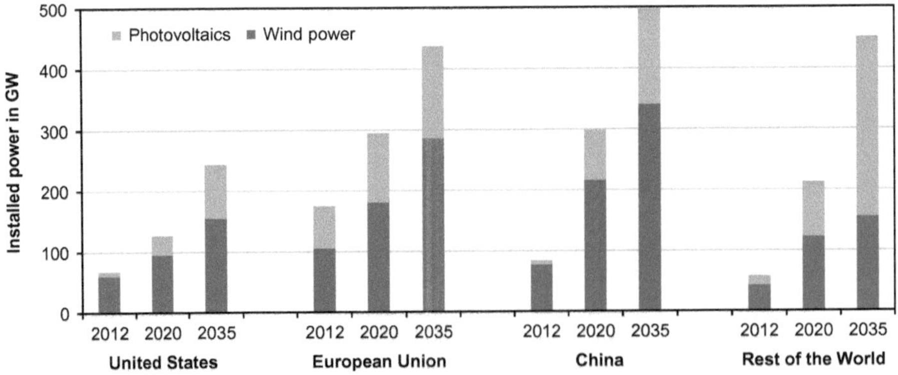

Fig. 14.2 Prognosis of future installed power of fluctuating renewable power sources. *Source*: Data based on IEA (2013) World Energy Outlook 2013—Renewable Energy Outlook. http://www.worldenergyoutlook.org/media/weowebsite/2013/WEO2013_Ch06_Renewables.pdf [Accessed 13 August 2015].

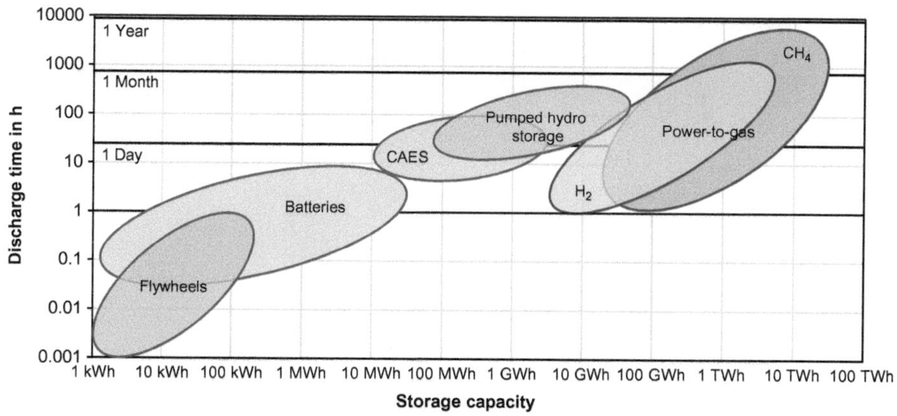

Fig. 14.3 Comparison of different storage technologies by capacity and storage time. *Source*: based on Specht M, Brellochs J, Frick V, Stürmer B, Zuberbühler U, Sterner M, Waldstein G. Storage of renewable energy in the natural gas grid. Erdöl Erdgas Kohle 2010;126:342–346.

technology and could provide renewable gases for transportation or heating purposes. Electricity for times with low power generation but high demand would then have to be provided in other ways – either by typical electricity storage technologies, back-up capacities, or an increased installation of renewable power sources. Increased installation of renewable power sources could on the one hand offer more electricity in times of high demand and would produce more "surplus" for production of renewable fuels via power-to-gas. Power-to-gas technology converts excess electricity from renewable sources to hydrogen or methane gas and could increase the percentage of renewable energy used in the electricity, heating, and transportation sectors. Additionally, the synthesis of methane requires carbon dioxide and, although it is released again when synthetic methane is burned, the cycle can be repeated and the process is deemed carbon neutral for carbon neutral CO_2 separation and electricity supply [34].

In this chapter, various theoretical applications of power-to-gas technology are discussed, and based on reported energy consumption of the main process steps, overall efficiency of each pathway is calculated. Future costs for the various applications of hydrogen and methane from power-to-gas are evaluated from mid-term and long-term perspectives according to specific costs from peer-reviewed literature and component manufacturers. Besides economic benchmarking ecologic analysis of hydrogen and methane fuel from power-to-gas applications is performed considering additional aspect of CO_2 sequestration and utilization.

14.2 Potential pathways of renewable gas technology

With power-to-gas, electricity from fluctuating renewable power sources is utilized for splitting water into hydrogen and oxygen in an electrolyzer. The hydrogen thereby produced has several potential applications, which are illustrated in Fig. 14.4 and are

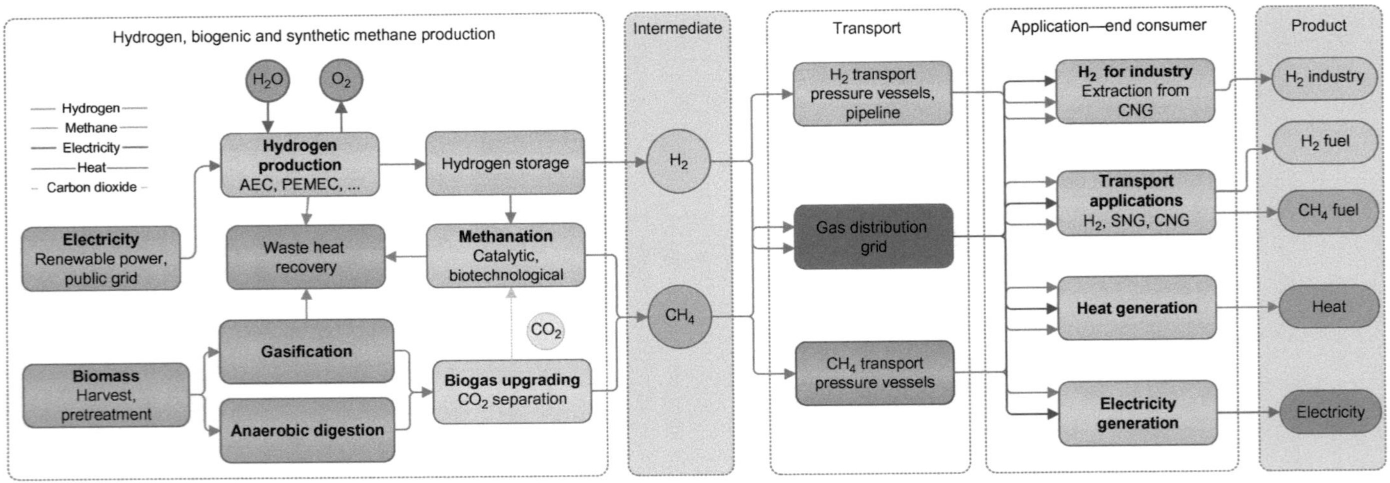

Fig. 14.4 Application pathways for renewable hydrogen, biogenic, and synthetic methane production.
Source: own representation.

described in the following section. Furthermore, the maturity of the main components is discussed, the overall efficiency of the various application pathways is evaluated, and an economic analysis is used to provide cost estimation based on currently available data of renewable gas technology for different applications.

14.2.1 Power-to-gas technology for different types of application

As illustrated in Fig. 14.4, the hydrogen produced in a power-to-gas system can be either directly utilized or further converted to synthetic methane in the CO_2 methanation (Sabatier process) [35]. The synthesis of hydrogen into methane implements additional processes and therefore results in a lower overall conversion efficiency compared to production of hydrogen alone. Nevertheless, one notable advantage of synthetic methane is that it can be easily incorporated in the existing gas infrastructure, comparable to biomethane from biomass. The carbon dioxide needed to complete the conversion to methane could be sequestered from fossil or renewable sources such as flue gas from power plants or biogas upgrading, respectively [36,37]. The type of carbon dioxide source would determine whether the synthetically produced methane would be considered renewable or not. Open questions regarding allocation of carbon dioxide emissions have to be discussed.

In addition to methane, hydrogen can be converted to other hydrocarbons such as methanol, ethanol, dimethyl ether, or formic acid. These liquid hydrocarbons can be utilized in industry or for transportation purposes as diesel or gasoline substitutes. There are some companies that are testing the synthesis of hydrocarbons in laboratory plants, such as Carbon Recycling [38] in Iceland and Sunfire [39] in Germany. Because this hydrogen application is at an early stage of development, information about the existing pilot plants is scarce. Synthesis of hydrocarbons other than methane is consequently not evaluated in this chapter.

14.2.1.1 Feeding hydrogen into the gas distribution system

The hydrogen produced could be directly fed into gas distribution systems, up to certain limits, regarding the volumetric fraction [40]. Excesses from renewable power sources could therefore be shifted to the gas distribution system. In this case, the stored energy could serve for heating, electricity generation, or as fuel in transportation applications. Restrictions on the fraction of hydrogen allowable in the gas distribution systems limit the production capacity of power-to-gas plants. Additionally, storage tanks are required for buffering the flow of produced hydrogen and control systems are necessary for injecting up to the allowed fractional limit. Determining the optimum hydrogen storage volume is one of the remaining challenges. Power-to-gas plants that feed in hydrogen produced from excess electricity have been realized in pilot scale in Germany, e.g., the pilot power-to-gas plant owned by EON in Falkenhagen, which began operating in 2013 [41].

14.2.1.2 Feeding synthetic methane into the gas distribution system

The synthesis of methane out of hydrogen has a lower overall conversion efficiency when compared to hydrogen. However, the clear advantage of synthetic methane is that there are no restrictions on the amount of methane fed into the gas distribution system. As CO_2-methanation is not a state-of-the-art technology, some important parameters must be considered; for instance, the required purity of carbon dioxide or the quality of the resulting methane. Local quality requirements are defined in several standards, such as the ISO 13868 for quality designation of natural gas [42]. If the quality of the produced synthetic methane is too low, an additional gas cleaning process is required. Power-to-gas plants that feed in synthetic methane produced from excess electricity and carbon dioxide have been realized at precommercial scale in Germany, including the power-to-gas plant of the AUDI company in Werlte, which started operating in 2013 [43].

14.2.1.3 Hydrogen as fuel for transport or for industrial applications

At the moment, hydrogen is mostly utilized as a raw material in industrial processes such as materials processing and chemical manufacturing, and in many other applications. At refueling stations, hydrogen could be provided as fuel for transportation. In this scenario, the hydrogen would have to be compressed and stored in high-pressure tanks. The current state of development of hydrogen fuel cells is established, there are already several vehicles (cars, buses) that are already operated in this way. Table 14.4 summarizes data from current fuel cell vehicles [44].

The development of fuel cell vehicles is currently intensified by all major car manufacturers. As standard, 70 MPa (700-bar) pressure tanks are used on cars and 35 MPa (350-bar) pressure tanks on buses. Which pressure tank is used has a significant influence on the range of the vehicles. By using a fuel cell in the vehicle, hydrogen and oxygen are converted into energy, which is available to the electric

Table 14.4 Typical tank sizes and ranges of fuel cell vehicles Here 1 bar = 10^5 Pa

Type	Vehicle	Storage pressure (bar)	Range (km)	Fuel consumption/ kg per 100 km
Car	Daimler B-Klasse F-Cell	700	380	0,97
Car	Toyota FHCV-adv	700	830	0,72
Car	Honda FCX Clarity	50	460	0,87
Bus	Daimler Citaro Fuel Cell Hybrid	350	>250	10–14

Source: Vgl. Frauenhofer-Institut für Solare Energiesysteme ISE (März 2013): S.7.

motor as an energy source. At the same time, there is also an electrochemical conversion into water vapor. The fuel consumption depends on the efficiency of the fuel cell. A fuel cell with an electric motor with an average efficiency of 60% has a 30% higher efficiency than a conventional combustion engine with about 30%. The reason for the increase in efficiency lies in the chemical conversion, since there is no further conversion into mechanical force. Due to these properties, an additional reduction in noise and particulate matter pollution can be recorded [45]. Several hydrogen refueling stations have been built worldwide in the last decade [46]. The future success for hydrogen as fuel for transportation depends on a number of variables, such as the availability of fuel cell vehicles and the prevalence of hydrogen refueling stations. Like electric vehicles, hydrogen-powered vehicles equipped with fuel cells are quiet and have good acceleration and ease of operation.

14.2.1.4 Synthetic methane as fuel for transport

Synthetic methane can be utilized as fuel for mobility and transport where natural gas is currently used, since natural gas is itself composed almost entirely of methane. If the gas quality resulting from the methanation process does not meet the requirements of CNG (compressed natural gas) vehicles, additional gas cleaning is required. CNG refueling stations are state of the art, with high growth rates. At this time, synthetic methane from power-to-gas is among others provided as fuel by the pilot plant run by ZSW Stuttgart and AUDI in Werlte, Germany [43,47]. The future success of using synthetic methane as transport fuel depends on certain variables, such as the availability of CNG vehicles and refueling stations with CNG storage capability.

14.2.1.5 Hydrogen for electricity generation

Hydrogen produced from renewable energy can be transformed back into electricity with a fuel cell or an internal combustion engine, and thus is considered an energy storage medium. Energy storage is required in times of overproduction, when demand for renewable energy is lower than production from these sources. There are several pilot plants that feed in electricity produced from hydrogen into the public electricity grid, such as the four facilities in Germany (hybrid power plant of Enertrag, the EON WindGas facility in Falkenhagen, the RWE demonstration plant in Ibbenbüren, the Thüga-group demonstration facility in Frankfurt/Main in Germany [48]), the MYRTE plant [49] in France, and the DTE Energy Hydrogen Technology Park [40] in the United States.

14.2.2 Maturity of the main components of a power-to-gas system

Depending on the type of application of power-to-gas systems, different components are available with various advantages, disadvantages, and technological maturity. The main components are described here in detail and the challenges of each are addressed.

14.2.2.1 Water electrolyzer

Electrolyzers use an electrical current to split water into hydrogen and oxygen. The various types of electrolytic cells (ECs) available can be characterized as alkaline (AECs), proton exchange membrane (PEMECs), and solid oxide (SOECs) depending on the electrolyte applied for the electrolysis process. Electrolyzer technologies for hydrogen production are discussed in more detail in references [50,51].

The SOECs are operated with high thermal energy input from an external heat source, which leads to accelerated reaction kinetics due to the high temperature. The required amount of electrical energy is therefore reduced and the efficiency of the process increased. Expensive catalysts can be avoided because of the high reaction temperatures [50].

SOEC is the least developed electrolyzer technology and is not included in further evaluations, here. AECs and PEMECs are widely available commercially and therefore are included in this evaluation. Table 14.5 summarizes the main parameters of AECs and PEMECs from mid-term and long-term perspectives. A mid-term perspective represents a time frame of up to ten years, whereas long-term perspective represents a period of more than ten years ahead.

Table 14.5 Main characteristics of AECs and PEMECs from mid-term and long-term perspectives

	Alkaline electrolysis cell (AEC)		Proton exchange membrane electrolysis cell (PEMEC)	
	Mid-term	**Long-term**	**Mid-term**	**Long-term**
Capacity (stack) ($Nm^3\ h^{-1}$)	<760	<1500	<30	< 500
Operating pressure (bar)	Atmospheric or 25–30	<60	<85	60–100
Operating temperature (°C)	65–100	60–90	<80	60–90
Energy demand ($kWh_e\ (Nm^3H_2)^{-1}$)	4.3–7.5	4.1–5.7	5.4–7.4	3.9–4.8
Power range [a] (%)	20–100	10–100	0–100	0–100
Lifetime[a]	15–25	30	6–15	30
Feed water requirements ($\mu S\ cm^{-1}$)	<5	–	<1	–
Hydrogen purity (%)	99,90	–	>99.99	–

[a]Percent of nominal power.

Source: Based on Ursua A, Gandia LM, Sanchis P. Hydrogen production from water electrolysis: current status and future trends. Proc IEEE 2012; Vol. 100(2). p. 410–26. DOI 10.1109/JPROC.2011.2156750; Smolinka T, Günther M, Garche J. Stand und Entwicklungspotenzial der Wasserelektrolyse zur Herstellung von Wasserstoff aus regenerativen Energien. Kurzfassung NOW-Studie. Fraunhofer ISE, FCBAT: 2011. https://www.now-gmbh.de/content/5-service/4-publikationen/4-nip-wasserstoff-und-brennstoffzellentechnologie/now-studie-wasserelektrolyse-2011.pdf [Accessed 26 July 2016]; Maclay JD, Brouwer J, Samuelsen GS. Experimental results for hybrid energy storage systems coupled to photovoltaic generation in residential applications. Int J Hydrogen Energy 2011;36(19):12130–12140. DOI: 10.1016/j.ijhydene.2011.06.089.

AECs employ an aqueous alkaline electrolyte and represent the most highly developed and cheapest electrolyzer technology [52]. Alkaline electrolyzers are available at high capacities and show good performance if operated continuously. Challenges arise in dynamic operation as the auxiliary equipment limits the flexibility of an alkaline electrolyzer and the start-up behavior is slow due to the high thermal capacity of the system. The power range of AECs is narrower than that of PEMECs and the gas quality is low under partial loads [51].

The PEMECs have a much simpler design. They utilize a polymer electrolyte membrane and therefore can tolerate load transients and exhibit faster start-up behavior [51]. Disadvantages are the limited lifetime of the membrane, the small available capacities, and the high costs due to use of noble metal catalysts like Pt. [53] At the moment, PEMECs are in a transitional commercial stage and have lower efficiency and lifetime than AECs, but from a long-term perspective, their efficiency is expected to exceed that of AECs and comparable lifetimes are expected to be reached.

Current challenges that must be met for both types of electrolyzers are to increase efficiency and reliability with fluctuating renewable power sources, increase durability, and lower the high initial investment costs.

The addition of a battery system to the electrolyzer set-up could allow for short-term energy storage to smooth the power output of renewable power sources [54]. On the other hand, an additional battery system makes the whole power-to-gas plant more complex and increases initial investment costs.

14.2.2.2 Methanation reactor

Synthetic methane can be produced from hydrogen in a catalytic methanation process by utilizing CO or CO_2 in the synthesis reactor. The synthesis of methane from hydrogen and CO or CO_2 is called the Sabatier process [22].

Whereas CO methanation is a proven technology in coal gasification processes, CO_2 methanation is currently under development. In the evaluated power-to-gas pathways, a CO_2 methanation reactor is considered for producing synthetic methane due to the additional environmental benefit of the reuse of CO_2. Although the documented efficiencies of the reactor concept are as high as for CO methanation, other challenges arise. Because the methanation process is strongly exothermic, heat management must be addressed. Furthermore, the catalyst exhibits problems with long-term stability and poisoning [55]. Another consideration is the required purity of carbon dioxide, which is strongly affected by the source of origin and determines the overall efficiency of CO_2 methanation [26]. Table 14.6 lists the main parameters of a methanation reactor that operates with carbon dioxide.

As CO_2 methanation is still in the development phase, better efficiencies and lifetimes are expected in the future. Among the research institutes working on the development of CO_2 methanation are, for instance, the ZSW Stuttgart [58], the KIT Karlsruhe in Germany [59], and the Desert Research Institute in the United States [60].

Carbon dioxide sequestration is an important process step of power-to-gas systems that produce synthetic methane. Carbon dioxide is produced in various processes such as burning coal in coal-fired power plants, in industrial processes such as the cement or lime production, in the fermentation process of biogas plants, and in biomass gasification. Theoretically, carbon dioxide could even be extracted from the ambient air, but the efficiency of such absorption processes is very low [36]. There are several CO_2 capture technologies available, many still in development stage. These technologies are not described here but can be found in references [61–63].

14.2.2.3 *Fuel cells*

Fuel cells generate electricity out of various fuels like hydrogen or methane and can be divided into different types depending on the applied electrolyte and resulting in different operating temperatures and pressures. The various fuel cell technologies are compared in detail in [64] and the current cutting-edge advancements are summarized there. In this evaluation, only PEM fuel cells are considered as this is the most developed fuel cell technology operating with hydrogen. Table 14.7 shows the main parameter of a PEM fuel cell.

Advantages of PEM fuel cells are similar to those of the PEMECs described in Section 14.2.2.1 in that they employ a polymer electrolyte membrane; therefore, corrosion and electrolyte management problems are reduced. They operate at low temperatures and exhibit quick start-up behavior. Challenges for PEM fuel cells are the small available capacities, the expensive catalysts, and the sensitivity to fuel impurities [66]. The efficiencies and lifetimes stated in literature vary widely and there is significant potential for improvement.

Table 14.6 **Main characteristics of the reactor for CO_2 methanation (Sabatier). Here 1 bar = 10^5 Pa**

Characteristics	Values	Sources
Operating pressure (bar)	8	[35]
Operating temperature (°C)	180–350	[36]
Energy efficiency (%)	83	[56]
Power range [a] (%)	60–100	[57]

[a]Percent of nominal power.

Table 14.7 **Main characteristics of a PEM fuel cell**

Characteristics	Values	Sources
Max. stack size (kW)	250	[64]
Operating pressure (bar)	1–5	[65]
Operating temperature (°C)	50–100	[64,66]
Electrical efficiency (%)	35–60	[66]
Lifetime	2–10	[67,68]

14.2.3 Efficiency calculation of various hydrogen and methane fuel pathways from power-to-gas technology

Depending on the process steps included and maturity of the components employed, different overall efficiencies can be reached for each individual pathway. The overall efficiency is determined as the ratio of output energy to input energy. The input energy is always the electrical energy from renewable power sources or the public grid, whereas the output energy depends on the pathway. If hydrogen or methane is the final product of a power-to-gas pathway, the lower heating value (LHV) is taken into consideration. If electricity is the final product, the efficiency is related to the electricity output of the power-to-gas system. The comparison of efficiencies is therefore restricted by whether the final product is fuel or electricity. Because efficiency losses occur at each additional step in the conversion process, direct utilization of electricity is preferred. However, storing and making available for use energy that would otherwise be lost, is where power-to-gas technology can provide the most significant impact.

Table 14.8 shows the average efficiencies for the main process steps that were considered in the evaluation of overall pathway efficiencies, all based on the LHV of hydrogen or methane. Overall efficiencies for the various applications of power-to-gas plants were calculated for mid-term and long-term perspectives, as some of the components are not fully developed and still have potential for improvement. Improving these technologies would have the highest impact on the overall process

Table 14.8 Considered efficiencies of the main process steps

Main process step	Efficiency based on LHV/%	Lifetime[a]	Source
Electrolyzer mid-term	55.0	15	[50,51]
Electrolyzer long-term	70.0	30	[51]
H_2 storage low-pressure	99.0	20	[68–70]
Methanation reactor mid-term[a]	72.0	15	
Methanation reactor long-term[a]	85.0	20	
H_2 feed-in[b]	98.5	20	[71]
CH_4 feed-in[b]	98.5	20	[71]
H_2 refueling station 350 bar[c]	93.0	20	[69]
CH_4 refueling station 250 bar[d]	95.0	20	
PEM fuel cell mid-term	48.0	10	[68,72]
PEM fuel cell long-term[e]	70.0	20	

[a]Assumptions according to information in [73].
[b]Including compressor, storage, and feed-in.
[c]Including compressor, storage, and fuel dispensing.
[d]Including compressor, storage, fuel dispensing—assumption according to [74].
[e]Assumption according to electrolyzer efficiency potential.

efficiencies; therefore, the main focus of research and development should be on the electrolyzers, the methanation reactors, and the fuel cell components.

Implementation of battery systems, utilization of released heat, and transport of hydrogen or synthetic methane to refueling stations are not included in efficiency calculations as these parameters vary for each power-to-gas site individually. Distribution of hydrogen via a pipeline has a very high efficiency of 99% [69] and therefore would not penalize the overall efficiency significantly. The addition of heat recovery would actually improve the overall energy efficiency.

Table 14.9 indicates that the overall efficiencies of the pathways with hydrogen as energy vector are higher than when synthesizing methane in methanation reactors because the additional components needed, lower the overall efficiency. Direct utilization of hydrogen or methane for feeding into a gas distribution system or for transportation fuel has higher energy efficiency than if the produced hydrogen is reconverted into electricity in a fuel cell.

14.2.4 Economic analysis of hydrogen and methane fuel from power-to-gas applications

In addition to the efficiency calculation, the costs of the different application pathways for power-to-gas technology and hydrogen or methane fuel products have been evaluated based on the specific costs of hydrogen and methane, which are calculated

Table 14.9 Overall efficiencies of the application pathways for mid-term and long-term perspectives

Application pathway	Final energy carrier	Main components	Overall efficiency (%)	
			Mid-term	Long-term
Feeding hydrogen into the gas distribution grid	H_2	Electrolyzer, H_2 storage, H_2 feed-in	54	68
Feeding in methane into the gas distribution grid	CH_4	Electrolyzer, H_2 storage, CO_2 methanation, CH_4 feed-in	39	58
Hydrogen as fuel for transport or other applications	H_2	Electrolyzer, H_2 storage, H_2 refueling station	51	65
Methane as fuel for transport or other applications	CH_4	Electrolyzer, H_2 storage, CO_2 methanation, CH_4 refueling station	37	56
Hydrogen for electricity generation	Electricity	Electrolyzer, H_2 storage, PEM fuel cell	26	49

Source: own representation.

considering the total annual cost. This is related to the annual amount of energy generated, which is comparable to the levelized cost calculation of electricity (LCOE) [75,76]. The total annual costs are calculated using the so-called annuity method [77], whereas the total annual cost consists of investment related A_K, energy/material related A_V, operational A_B, and additional cost A_S. These deducted from the specific revenue A_E gives the annuity of the total annual payments A (see Eq. 14.1).

$$A = A_E - (A_K + A_V + A_B + A_S) \tag{14.1}$$

A rate of interest of 5% was expected, and the efficiencies and lifetimes of the main components stated in Table 14.4 were used. The investment costs of the various components depend on several parameters, including such as efficiency, manufacturer, and size of the component. Therefore, a trend is provided for the mid- to long-term perspective. As electrolyzers, methanation reactors, and fuel cells are further developed, major reductions in costs are expected over the long term. For the power-to-gas system, operational and maintenance costs between 2% and 4% were assumed [78]. Estimates also included expenditures for feed water for the electrolyzer and carbon dioxide for the methanation reactor.

In Fig. 14.5, the specific investment costs of alkaline electrolyzers are analyzed according to literature data and price inquiries. They are sorted by the year of installation and additionally the rated power of the electrolyzer is given on the right y-axis.

The rated power of the analyzed alkaline electrolyzers shows a wide range from 0.35 MW_{el} to 10 MW_{el} and in most cases the system boundary is not defined exactly. For this reason also, the specific investment costs have a wide range. For alkaline electrolyzers, the actual (year 2015–2018) investment costs are on average about

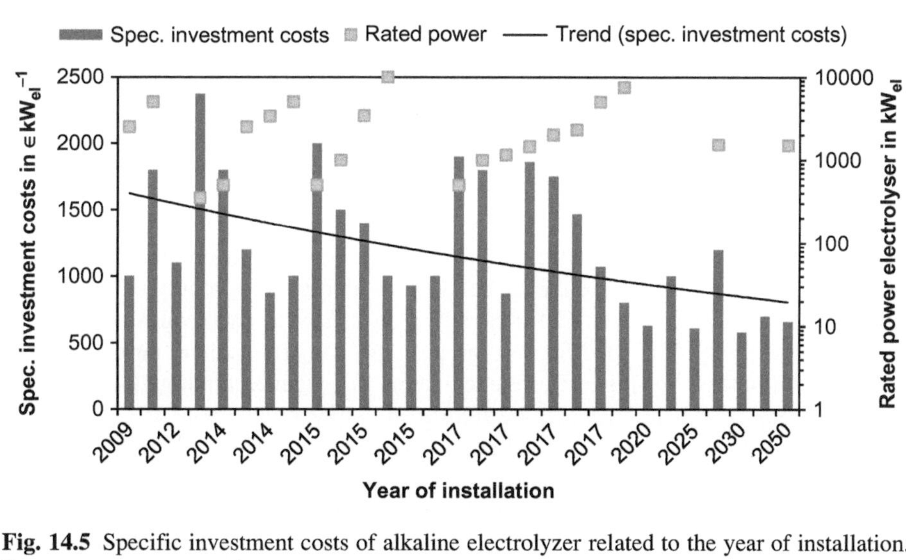

Fig. 14.5 Specific investment costs of alkaline electrolyzer related to the year of installation. *Source*: own representation.

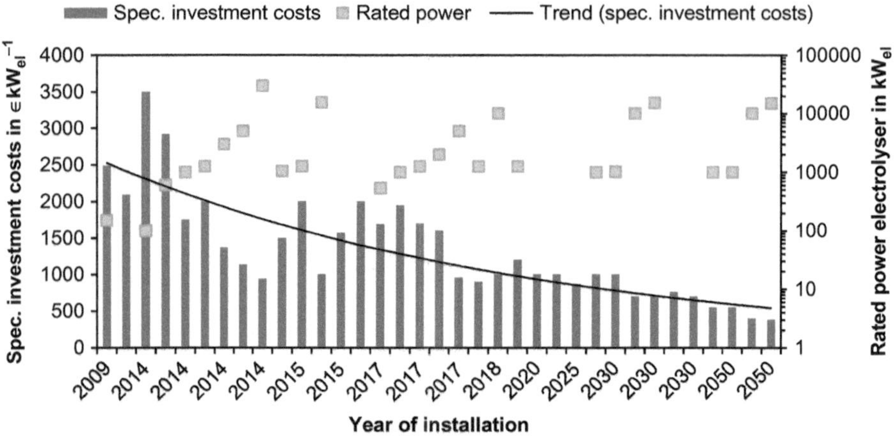

Fig. 14.6 Specific investment costs of PEM electrolyzer related to the year of installation. *Source*: own representation.

1500 € $(kW_{el})^{-1}$. According to the literature data, the costs drop to about 650 € $(kW_{el})^{-1}$ in the year 2050.

A similar screening on specific investment costs was done for PEMEC, see Fig. 14.6.

The actual (year 2015–2018) average investment costs for PEM electrolyzer are also in a range about 1500 € $(kW_el)^{-1}$. But, according to the literature, the costs drop to <500 € $(kW_{el})^{-1}$ in the year 2050.

For the methanation units, there are only limited cost data available, which have a high interval and partly a deficient data basis. The actual (year 2006–2015) average costs according to literature data for a catalytic methanation unit are about 295 € $(kW_{SNG})^{-1}$ (only methanation plant, electrolyzer is not included). The analyzed methanation units differ in capacities from 5 MW_{SNG} to 1000 MW_{SNG}. For the biological methanation units, the actual average costs are about 660 € $(kW_{SNG})^{-1}$.

Defining normalized specific investment costs for carbon capture and sequestration units is not easy, as costs strongly depend on the CO_2 source as well as on the concentration in the source stream. Therefore, it is hard to relate the particular investment costs to a single reference value/unit to provide comparability. More practical for the methanation process is to value the needed CO_2 as an operating supply and therefore represent its costs as € per ton CO_2 depending on its source and sequestration technology, respectively. The specific average costs for CO_2 from direct air capture are about 350 € $(t\ CO_2)^{-1}$ as highest reference value. For other technologies like biogas and wastewater treatment, the costs are significantly lower with about 45 € $(t\ CO_2)^{-1}$.

The electricity costs depend on the integrated energy infrastructure and on the operation mode. For the economic evaluation, three different cases were considered. In the first case it was assumed that only electricity from overproduction is utilized, resulting in low electricity costs but also low potential operating hours, with an estimation of 1500 full load hours per year $(h\ a^{-1})$. It has been stated in [18,22] that negative electricity prices are possible in times of overproduction, but they were not evaluated. The second option is to utilize electricity from the public grid with higher conventional energy prices to reach higher full load hours, assumed to be 7500 h a^{-1}. The electrolyzer could also be connected directly to a renewable power source, such as a large wind farm. Typical full load hours of wind farms are between 2000 and 4000 h a^{-1} depending on the site. If the power of the electrolyzer is smaller than the nominal power of the renewable power source, and a battery is applied for buffering the electricity, higher full load hours could be reached. Therefore, the third option was estimated at 6000 h a^{-1}.

Fig. 14.7 shows that the average cost of electricity input rises sharply with increasing full-load hours. This increase in energy costs, however, has little effect on the specific costs of H_2 and CH_4 fuel products, because the specific investment costs decrease with increasing hours of full load. At higher full-load hours more H_2 or CH_4 could be produced; thus also, more fossil fuels could be replaced. However, the additional electricity purchases needed to increase the hours of full load lead to an increased demand for electricity in the system, which could also lead to an increased load on the power grid.

The consolidated results of the economic calculations are illustrated in Fig. 14.8 and show a range of costs for the various applications for the mid-term, as well as for the long-term perspective.

The lowest costs are achieved for hydrogen that is fed into the gas distribution system or is utilized as fuel and for other applications. Due to the additional investment costs for the methanation reactor, the overall costs for synthetic methane are higher, as can be seen in Fig. 14.8. Nevertheless, the utilization of synthetic methane has other advantages such as feeding-in without restrictions or use in fully developed CNG fueling stations and CNG cars. If hydrogen or methane has to be transported to the site of application, additional costs have to be considered, depending on the transport

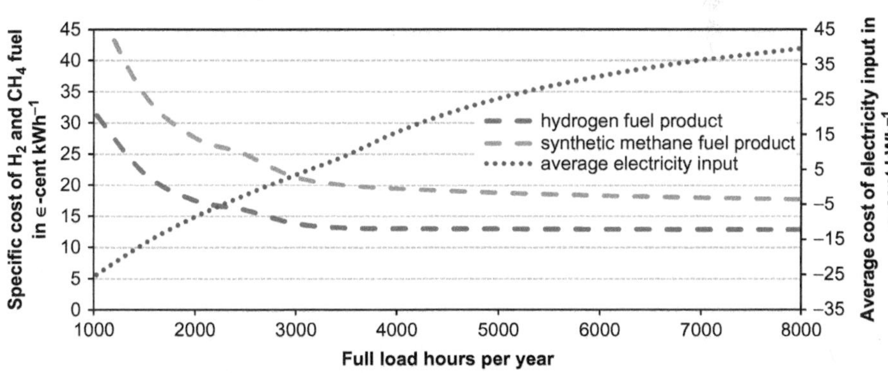

Fig. 14.7 Sensitivity of the costs of H_2 and CH_4 fuel product from power to gas: All values without VAT. The values refer to the lower heating value (LHV) of hydrogen and methane. Included also are the costs for the necessary refueling infrastructure and the end-user grid charges in Austria.
Source: own representation.

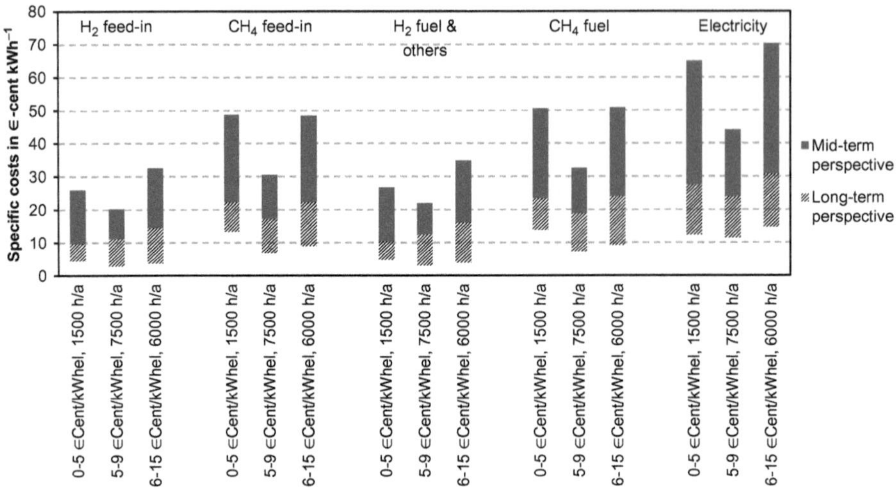

Fig. 14.8 Range of product costs of various power-to-gas applications.
Source: own representation.

distance. The costs for electricity production using power-to-gas as intermediate storage are significantly higher because the overall efficiency of the process decreases with each additional process step. Moreover, investment costs for fuel cells are currently very high. However, if the technology is fully developed and there is high overproduction from renewable power sources and therefore low electricity costs, power-to-gas could be an interesting alternative for storing energy.

It is noted that a lower range of costs results from the higher amount of operating hours described in the second option. Other important influential parameters, besides operating hours, are the electricity costs and the achievable initial investment costs of the main components.

The actual market prices of maximum 10 €-cent $(kWh_{LHV})^{-1}$ for hydrogen {30 €-cent $(kWh_{LHV})^{-1}$ at the fueling station} and maximum 5 €-cent $(kWh_{LHV})^{-1}$ for CNG [79], are (especially for CNG) significantly lower than the calculated average costs of power-to-gas applications for mid-term perspective. Based on the roll-out situation, questions of regulatory framework and dues and taxes on the produced hydrogen are not covered in all countries. Nevertheless, if the main components are improved and efficiency increases, some pathways will be competitive, especially for the long-term perspective. In addition, it has to be regarded that the stated market prices are valid for conventional hydrogen and CNG from fossil sources, and that power-to-gas pathways provide renewable alternatives.

14.2.4.1 Comparison to economic benchmarks

Due to several potential applications of power-to-gas, there is a variety of possible benchmark technologies that differ for each case. If power-to-gas is utilized to store electricity, other electricity storage technologies could be the benchmark technologies. If power-to-gas is utilized for energy transport and thus substitutes power lines, the expansion or building of these power lines is the benchmark. For utilization of H_2 or CH_4 as transport fuels, diesel, gasoline, natural gas, bioethanol or biodiesel are the relevant benchmarks. For the utilization of H_2 in industrial processes, the benchmark technology is steam reforming of natural gas. Other potential benchmarks and comparison to power-to-gas could be found in references [80–82].

The exemplary reference case for power-to-gas and some future costs are compared here to the direct benchmarks natural gas, biomethane, and H_2 from natural gas or biomass gasification. The informations on specific costs have been taken from references [83–85]. Figs. 14.9 and 14.10 show the specific production costs of H_2 and CH_4, respectively. The specific production costs of H_2 and CH_4 from power-to-gas are given for the reference case with a 1-MW PEM electrolyzer, full load hours of 4000 h a^{-1}, and electricity costs of € 30 per MWh. Additionally, investment cost reduction through learning effects (factor 50) have been regarded and one version with higher full load hours of 7000 h a^{-1} and electricity costs of € 40 per MWh has been added.

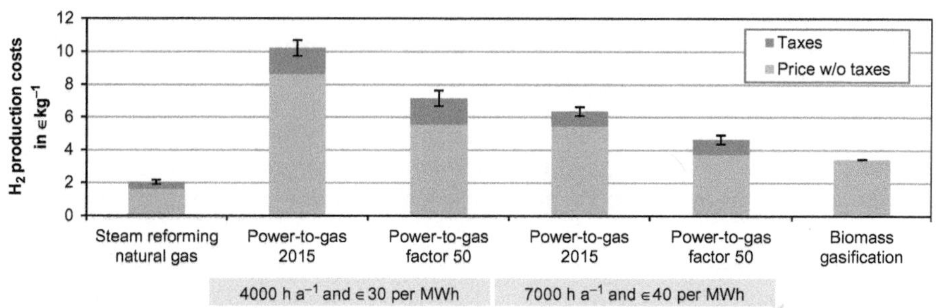

Fig. 14.9 H_2 production costs from power-to-gas in comparison to benchmarks. *Source*: own representation.

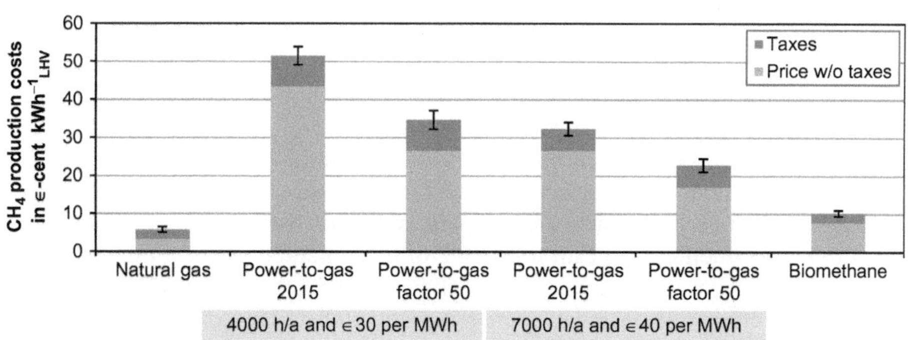

Fig. 14.10 CH_4 production costs from power-to-gas in comparison to benchmarks. *Source*: own representation.

The comparison with benchmark technologies for H_2 production shows that the specific production costs of H_2 from power-to-gas are higher in any of the presented cases. The presented values are specific production costs and the real market prices of H_2 could be much higher in many cases. At Austrian fueling stations, for instance, H_2 currently costs € 9 kg^{-1}. As of January 2017, the total number of hydrogen refueling stations in operation worldwide is 273 [86]. If industrial processes have no on-site production of H_2 due to smaller required amounts, their H_2 costs could easily reach € 30 kg^{-1}. This has also to be considered when comparing specific H_2 production costs. It has also been shown that taxes contribute significantly to specific costs. In the case of H_2 production via power-to-gas, these are the taxes that have to be paid for utilization of electricity.

The comparison of CH_4 from power-to-gas in Fig. 14.10 shows that the specific costs are significantly higher than the costs of the benchmarks natural gas and biomethane. Even with assumed cost reduction through learning effects and higher full load hours, the production costs of CH_4 are still twice as high as that of biomethane production from biogas. However, biogas production is confronted with issues of resource availability and competition with food production (see references [87,88]).

14.2.5 Ecologic analysis of hydrogen and methane fuel from power-to-gas applications

The hydrogen from electrolysis and methane produced by methanation can serve as a substitute for fossil resources, and thus is able to reduce greenhouse gas emissions, especially in the energy sector. The focus of science has recently increased concerning the aspect of CO_2 sequestration in coupling of anaerobic digestion and methanation whereas in both cases, fossil raw materials are replaced, which can contribute to a reduction of CO_2 emissions. However, as long as the captured CO_2 comes from burning of fossil fuels and fossil carbon is the primary energy source with "delayed" release of CO_2 into the atmosphere, the acuate CO_2 savings achieved requires detailed assessment. A real circuit closure in terms of CO_2-neutral energy source is only achieved when the electrical input power and absorbed CO_2 originates from renewable sources. In the following section, a lifecycle analysis for the technological approaches and systems of power-to-gas is exemplary performed in the following section. The power-to-gas system (including all components), the current supply, and the respective application of the product (gaseous hydrogen in industry or mobility) are defined as the technical system boundary, see Fig. 14.11.

Based on the chosen system limits, as well as material and energy balances, the hydrogen production was modeled using the LCA software GaBi to evaluate the ecological performance. The results are compared with the fossil reference technologies of hydrogen production by steam reforming of oil or natural gas, see Fig. 14.12.

Due to the comparatively lower energy efficiency along the entire process chain, the total primary energy demand for hydrogen production in the power-to-gas plant is higher than the demand for steam reforming natural gas or oil. However, when predominantly using renewable electricity, there is a considerable improvement in primary energy input from nonrenewable sources compared to fossil reference technologies.

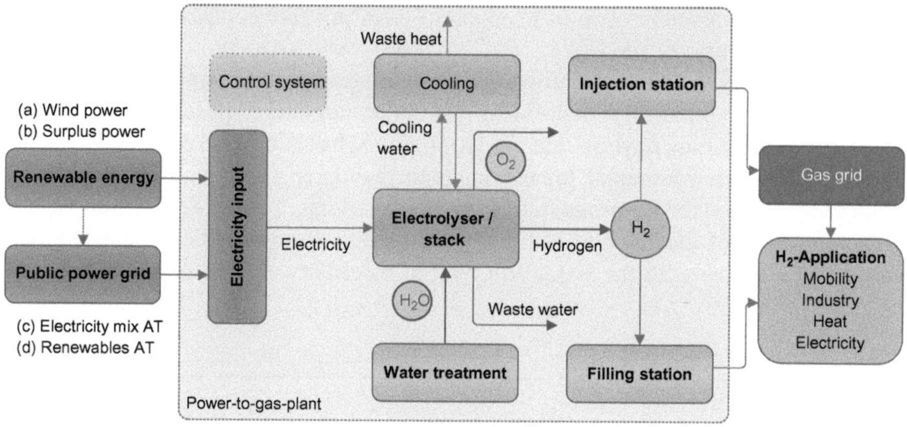

Fig. 14.11 System boundary for the ecological assessment.
Source: own representation

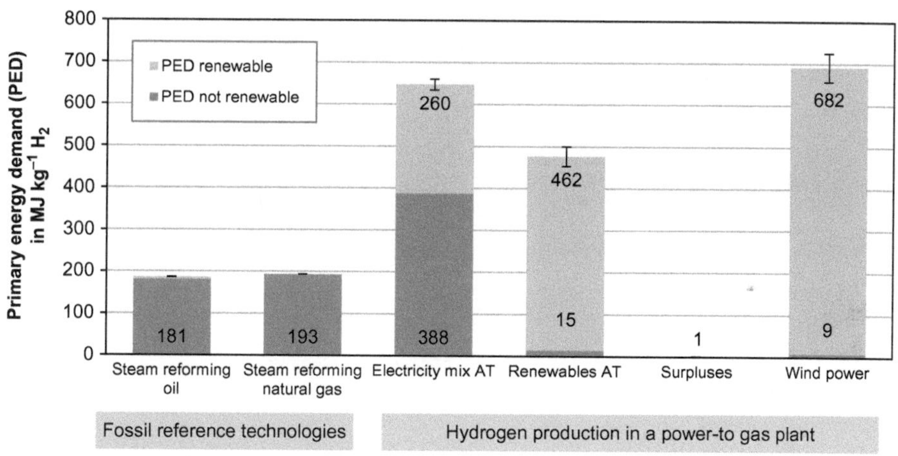

Fig. 14.12 Primary energy demand for hydrogen production via power-to-gas with different power inputs (electrical energy) compared to fossil reference technologies.
Source: own representation.

Fig. 14.13 shows the global warming potential (GWP) of hydrogen production in a power-to-gas system for different origin of electricity input compared to fossil reference technologies.

When using renewable electricity, the GWP of hydrogen is much lower than that of fossil reference technologies. However, if the typical electricity mix in Austria is used, the greenhouse gas emissions per kilogram of hydrogen produced are higher than in the steam reforming of oil and natural gas. When using surplus electricity, the GWP is particularly low compared to the other technologies. The use of electricity from predominantly renewable sources is essential for a positive ecological performance concerning reduction of the GWP.

Based on the GWP of steam reforming of natural gas, the environmental breakeven point can be calculated. In other words, this is the maximum specific GWP that the utilized electricity may have so that the H_2 produced has a lower GWP than the fossil reference. The environmental breakeven point has been calculated in [82] and is 190 g(CO_2) kWh^{-1} for H_2 production from power-to-gas. Table 14.10 shows typical GHG emissions of electricity generation technologies and electricity mixes. Comparing these values with the required GWP of electricity, it gets obvious that only

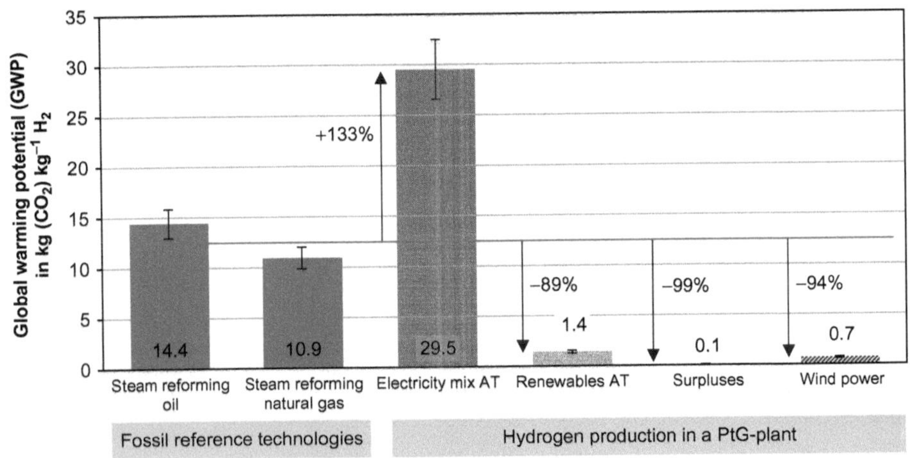

Fig. 14.13 Global Warming Potential (GWP) of hydrogen production via power-to-gas plants with different electricity inputs.
Source: own representation.

Table 14.10 Greenhouse gas emissions of various electricity generation technologies and electricity mixes

	GHG emissions/g(CO_2) kWh^{-1}
Coal power plants	750–1200
Electricity mix EU-27	565
Natural gas power plants	400–550
Electricity mix Austria	406
Photovoltaics	50–100
Wind power, water power	10–40

Source: data from PE International, 2013. GaBi Software with built-in Database (DB), Version 5. Available from http://www.gabi-software.com; Wagner HJ, Koch MK, Burkhardt J, Große Böckmann T, Feck N, Kruse P. CO2-Emissionen der Stromerzeugung. BWK Bd. 59 Nr. 10, 2007. https://www.vdi.de/fileadmin/vdi_de/redakteur_dateien/geu_dateien/FB4-Internetseiten/CO2-Emissionen%20der%20Stromerzeugung_01.pdf [Accessed at 14 August 2015].

renewable power sources or an electricity mix with a majority of renewable generation is suited for H_2 production via power-to-gas.

For reaching economic competitiveness, high full load hours of about 4000 h a^{-1} will be required for power-to-gas plants. Thus, utilization of electricity from the public power grid or from renewable power sources will be necessary in addition to the utilization of surplus electricity. As long as the implementation of power-to-gas plants is low and thus no additional power capacity is needed in the energy system, the ecological performance depends on the direct electricity input (see Fig. 14.13). However, if the implementation of power-to-gas plants is increasing, there will be an additional power demand in the energy system, especially if power-to-gas is utilized to produce renewable transport fuels or H_2 for the industry. For the ecological performance it is relevant, which generation technology provides the additional power that is needed by power-to-gas plants. This could be considered with a marginal electricity supply or a future electricity mix, as suggested by Del Duce et al. [89].

If the produced hydrogen is used in mobility, further reference technologies are relevant for the ecological assessment. These are above all the conventional fuels, i.e., gasoline, diesel, and natural gas. For comparison, a fuel cell car with an electric drive is selected as the vehicle technology. Because of the different vehicle technologies and fuel-dependent consumption, a well-to-wheel analysis is carried out here. The entire production chain is taken into account from the extraction of the raw materials via the processing and transport as well as the operation of the car, see Fig. 14.14.

In addition, the material and energy demand for the production of the vehicle is taken into account.

The Global Warming Potential (GWP) of the well-to-wheel analysis for a hydrogen fuel cell vehicle compared to the fossil reference technologies for vehicles with combustion engines and conventional fuel is illustrated in Fig. 14.15.

The greenhouse gas emissions caused by the production (internal combustion engine, fuel cell) are at the same order of magnitude. Overall, the GWP from the vehicle production part accounts for a significant share of total GWP. The emissions from the fuel for a driving performance of 100 km are particularly low in the case of hydrogen production with renewable electricity.

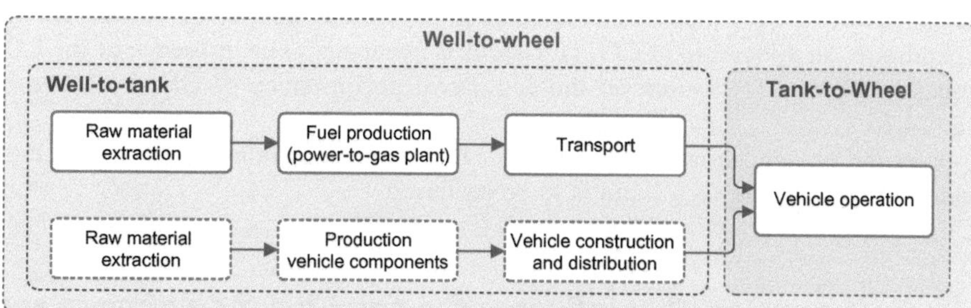

Fig. 14.14 Overview system boundary of LCA Well-to-Wheel analysis. (Source: own representation)

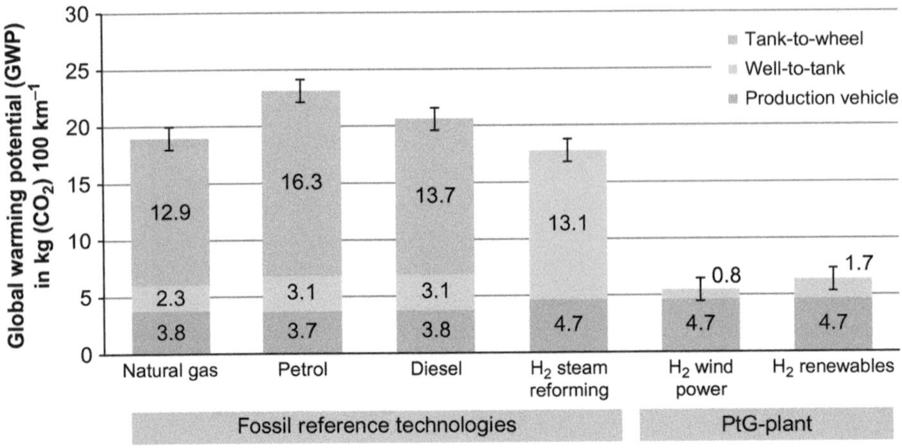

Fig. 14.15 Global Warming Potential (GWP) of different fuels in a well-to-wheel analysis. (Source: own representation)

While no additional emissions (except water) are generated during the operation of hydrogen cars, fossil technologies produce CO_2 emissions during combustion. These direct emissions account for most of the emissions along the entire process chain.

Through the production of hydrogen from renewable electricity and the use in fuel cell vehicles, a considerable reduction of GWP per 100 km of driving can be achieved in the overall comparison (well-to-wheel). The use of hydrogen from natural gas also leads to an improvement in the greenhouse gas balance since the efficiency of the fuel cell vehicle exceeds the efficiency of conventional combustion engines. Moreover, the use of hydrogen does not lead to direct emissions during operation, which improves air quality (especially in agglomeration areas).

If the hydrogen is not used directly, but is further processed to SNG, the necessary CO_2 for this production step must also be taken into account in the assessment of the GWP. Since not only biogenic but also fossil CO_2 sources are suitable for the power-to-gas process and the separation is associated with a certain energy expenditure, an allocation of CO_2 emissions is necessary. The influence of the CO_2 sequestration or CO_2 source on the ecological performance of CH_4 is shown in Fig. 14.16.

From the ecological point of view, an early implementation at sites with a high quality of the existing CO_2 input is to be preferred.

14.2.6 Parameters that influence the potential for hydrogen and methane from power-to-gas systems

The potential of the previously described pathways is influenced by various parameters, the main factors among which are described here.

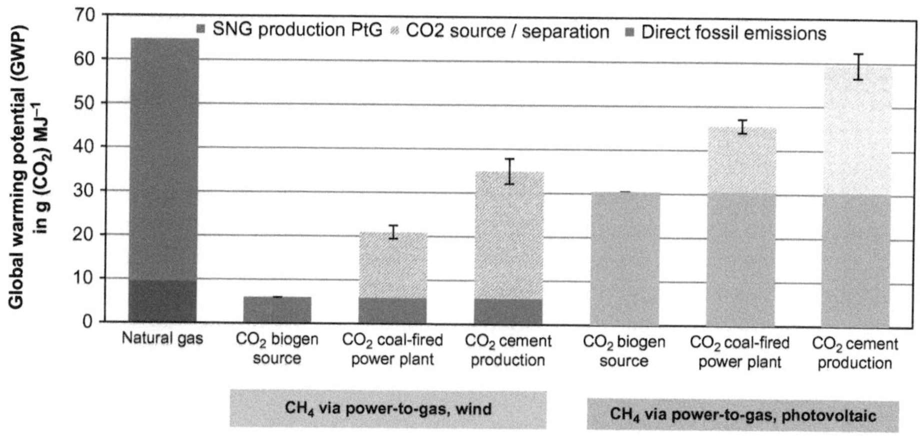

Fig. 14.16 Global Warming Potential of 1 MJ CH_4 from a Power-to-Gas plant compared with the fossil reference energy carrier natural gas.
(Source: own representation)

14.2.6.1 Percentage of renewable energy for electricity, transport, and industry sectors

The demand for power-to-gas technology basically depends on the amount of energy desired from renewable sources for electricity, heat, and transport purposes. With increasing percentage of fluctuating renewable power sources for electricity generation, the need for load leveling and energy storage increases. In the year 2015, the renewable energy share of global electricity production was estimated to be 23.7%, including all types of renewable power sources; whereas the greatest amount (16.6%) was provided by hydropower. Wind and solar photovoltaic actually account for a smaller fraction, but their growth rates are very high, reportedly being 22% and 25%, respectively, over 2014 [90]. Locally, the fraction of wind or solar power could be much higher and load leveling is already required in these regions.

The trend of renewable share in the transport sector is another parameter that influences the potential of power-to-gas. Hydrogen and synthetic methane could serve as alternative renewable fuels and replace conventional fossil fuels. Whereas CNG cars and fueling stations are already widespread, the infrastructure for hydrogen refueling stations and fuel cell cars are at the beginning of their market penetration. At the moment, liquid biofuels account for the largest global share of renewable fuels (4% of all fuels in 2015) [90]. The potential for hydrogen as raw material for chemicals in the industry sector depends on the restriction of industrial greenhouse gas emissions and the desired percentage of renewable products.

14.2.6.2 Availability of energy infrastructure

The potential of the different pathways for power-to-gas technology strongly depends on the available energy infrastructure as already described for the various energy application options in Section 14.2.1.

14.2.6.3 Quality of power lines

The quality of the public electricity (power) network has strong influence on the potential of power-to-gas as weak grids have a strong need for power balancing. Weak public electricity grids near large renewable power sources often cannot absorb the total amount of electricity generated, and therefore have a great need for electricity storage. Expansion of the public electricity grid by building new power lines is often accompanied by strong resistance from local communities and therefore takes a lot of time. Depending on the distance of remote areas from existing public electricity grids, installing new power lines could be cost prohibitive. Therefore, alternatives such as energy transport via the gas distribution system might prove more economical (and less problematic).

14.2.6.4 Availability of carbon dioxide sources

Providing methane via power-to-gas technology requires a source of carbon dioxide for the synthesis. Theoretically, carbon dioxide could be absorbed from the ambient air. However, the separation of carbon dioxide from ambient air is very costly, so alternative sources should be chosen.

14.2.6.5 Alternative systems—Benchmarks

The potential for renewable hydrogen or synthetic methane from power-to-gas plants for application in industry, or for transport purposes, also depends on the availability of alternative renewable production methods. Other technologies for producing renewable hydrogen or methane therefore set the benchmark for power-to-gas systems regarding efficiency, greenhouse gas emissions, and costs. Because power-to-gas systems can be applied for long-term electricity storage, they always have to be compared to alternative energy storage technologies such as batteries, flywheels, compressed air storage, and pumped hydro power.

14.3 Conclusions

The power-to-gas technology could be utilized for a wide variety of applications, particularly in the electrical power, transportation, or industrial sectors. The overall efficiency and economic performance are mainly influenced by the electrolyzer, the methanation reactor, and the fuel cell used; whereas each conversion step imposes additional losses. Further improvement in steady supply of electricity, higher efficiency, and lower costs of initial investment is required for each of these main components. The highest overall efficiency and the lowest costs are achieved for hydrogen production. Further synthesis with CO_2 to CH_4 in a methanation reactor lowers the efficiency and results in higher production costs per kWh. The production costs in the different pathways primarily depend on the amount of operating hours per year, the electricity costs, and the achievable initial investment costs for the main components.

When evaluating the pathways for integration of power-to-gas technology into the available energy infrastructure, different application purposes appear. The functions that can be fulfilled by power-to-gas depend on the grid situation and require comparisons to current benchmarks for the application. A list of the functions and their benchmarks is given as follows.

- Production of a green product (renewable H_2 or CH_4): other renewable hydrogen or methane production technologies as benchmark.
- Utilization of excess electricity: benchmark could be other energy storage technologies or lower operating hours of renewables (e.g., shutdown of wind farms in times of overproduction).
- Transporting energy from remote communities or areas with weak electricity grids: construction of new power lines as alternative.
- Self-supply for standalone systems in remote communities or self-supply by choice: benchmarks are other storage technologies.

The environmental performance of H_2 production via electrolysis is strongly dependent on the electricity input, which should have a global warming potential of $<190\,g(CO_2)\,kWh^{-1}$. In other words, a reduction of greenhouse gas emissions compared to fossil reference technologies can only be achieved if electricity is produced mainly from renewable power sources. Utilization of the greenhouse gas CO_2 for CH_4 synthesis via power-to-gas would substitute natural gas and is thus beneficial from an ecological point of view. Nevertheless, the CO_2 separation from industrial processes or power plants is accompanied by a certain additional energy demand related with additional greenhouse gas emissions. These emissions are characterized by the CO_2 penalty in $g(CO_2)$ additionally emitted per kilogram CO_2 captured. If the CO_2 originates from biogenic sources (no CO_2 penalty), the environmental breakeven for CH_4 synthesis is $113\,g(CO_2)\,kWh^{-1}$ of utilized electricity. When CO_2 is separated for instance from cement production, the additional primary energy demand is comparably high and the environmental breakeven decreases to $63\,g(CO_2)\,kWh^{-1}$ electricity input [82].

The economic performance of power-to-gas is mainly influenced by reached full load hours and total efficiency of the process. Due to the early stage of development, learning and scaling effects could bring significant investment cost reductions. Electricity costs are strongly depending on the type of application, which is again influencing the achievable full load hours. At 4000 full load hours, doubling of electricity costs leads to an increase in specific production costs of +35%. Additional proceeds for oxygen and heat utilization as well as costs of CO_2 have hardly any influence on the specific production costs. Provision of negative balancing power could lead to a significant decrease in specific production costs due to a high achievable proceed for the utilization of electricity. Nevertheless, the participation of power-to-gas plants at the control energy market is connected with a considerable risk, as average prices are fluctuating significantly and cannot be predicted reliably. In comparison to the direct economic benchmarks, the specific H_2 and CH_4 production costs of the presented exemplary power-to-gas plants are higher, even when considering higher full load hours and a potential future investment cost reduction. However, power-to-gas has other benefits such as the possibility of long-term energy storage, the hybridization of the energy system leading to a higher flexibility and the reduction of greenhouse gas emissions when renewable electricity is applied.

The theoretical potential for H_2 and CH_4 as renewable products for mobility or industrial purposes is huge, but the available amount of electricity from renewable power sources may limit this potential. As production costs of H_2 and CH_4 from power-to-gas are higher than for conventional production from fossil resources, the real future demand for green H_2 and CH_4 will strongly depend on the demand for green products in general and on political targets for share of renewables in all energy sectors.

One main issue to be considered for further research is the additional power demand in the energy system induced by H_2 and CH_4 production for transport and industry applications with high full load hours [82]. This could increase the burden on the power grid and the additionally required power generation technologies would influence the ecological performance. For improvement of the economic viability of power-to-gas systems, future research should focus on the improvement of the applied technologies and systems in terms of lower costs and higher system efficiency, especially in part load and dynamic operation.

Acknowledgments

The support of this work by the association Energy Institute at the Johannes Kepler University is gratefully acknowledged.

References

[1] European Commission. EU Reference Scenario 2016 [European Commission].

[2] FNR, editor. Einspeisung von biogas in das Erdgasnetz [injection of biogas into the natural gas network]. Leipzig: Fachagentur Nachwachsende Rohstoffe e.V; 2006.

[3] Drosg, B (2013) Process monitoring in biogas plants, IEA Bioenergy, ISBN 978-1-910154-03-8 https://www.iea-biogas.net/files/daten-redaktion/download/Technical%20Brochures/Technical%20Brochure%20process_montoring.pdf.

[4] De Baere L, Mattheeuws B. State-of-the-art 2008—anaerobic digestion of solid waste. Waste Manage World 2008;9:1–8.

[5] Weiland P, Verstraete W, van Haandel A (2009) Biomass digestion to methane in agriculture: a successful pathway for the energy production and waste treatment worldwide. In: Soetaert W, Vandamme E J (Eds), Biofuels, Wiley, pp 171–195.

[6] FNR, editor. Leitfaden Biogas—Von der Gewinnung zur Nutzung [Biogas guide—From production to utilisation]. Leipzig: Fachagentur Nachwachsende Rohstoffe e.V; 2016.

[7] Niesner J, Jecha D, Stehlik P. Biogas upgrading technologies: state of art review in European region. Chem Eng Trans 2013;35:517–22.

[8] Bajohr S, Graf F, editors. Biogas: Erzeugung, Aufbereitung, Einspeisung [biogas: Production, purification, grid injection]. Munchen: Oldenbourg Industrieverlag; 2013.

[9] EN 16723–1:2016 Natural gas and biomethane for use in transport and biomethane for injection in the natural gas network—Part 1: Specifications for biomethane for injection in the natural gas network.

[10] EN 16723–2:2017 Natural gas and biomethane for use in transport and biomethane for injection in the natural gas network—Part 2: Automotive fuels specification.

[11] Ryckebosch E, Drouillon M, Vervaeren H. Techniques for transformation of biogas to biomethane. Biomass Bioenergy 2011;35:1633–45.
[12] CE Delft, Eclareon, Wageningen Research. Optimal use of biogas from waste streams; 2017 [European Commission].
[13] Stolten D, Emonts B. Hydrogen science and engineering. Jülich, Germany: Wiley-VCH; 2016.
[14] Certify, 2015. Overview of the market segmentation for hydrogen across potential customer groups, Based on key application areas.
[15] Moliner R, Lazaro MJ, Suelves I. Analysis of the strategies for bridging the gap towards the Hydrogen Economy. Int J Hydrog Energy 2016;41:19500–8.
[16] Scipioni A, Manzardo A, Ren J, editors. Hydrogen economy—supply chain, life cycle analysis and energy transition for sustainability; 2017. ISBN [9780128111321].
[17] Certifhy, 2016. Roadmap for the establishment of a wellfunctioning EU hydrogen GO system.
[18] Grueger F, Möhrke F, Robinius M, Stolten M. Early power to gas applications: reducing wind farm forecast errors and providing secondary control reserve. Appl Energy, Available online 12 July 2016, ISSN 0306-2619, https://doi.org/10.1016/j.apenergy.2016.06.131.
[19] Estermann T, Newborough M, Sterner M. Power-to-gas systems for absorbing excess solar power in electricity distribution networks. Int J Hydrog Energy, Available online 16 June 2016, ISSN 0360-3199, https://doi.org/10.1016/j.ijhydene.2016.05.278.
[20] Götz M, Lefebvre J, Mörs M, McDaniel Koch A, Graf F, Bajohr S, et al. Renewable power-to-gas: a technological and economic review, Renew Energy 2016;85:1371–90. ISSN 0960-1481, https://doi.org/10.1016/j.renene.2015.07.066.
[21] Guandalini G, Campanari S, Romano MC. Power-to-gas plants and gas turbines for improved wind energy dispatchability: energy and economic assessment. Appl Energy 2015;147:117–30. ISSN 0306-2619, https://doi.org/10.1016/j.apenergy.2015.02.055.
[22] Sterner M, Jentsch M, Holzhammer U. Energiewirtschaftliche und ökologische Bewertung eines Windgas-Angebotes. Fraunhofer IWES 2011. https://www.greenpeace-energy.de/fileadmin/docs/sonstiges/Greenpeace_Energy_Gutachten_Windgas_Fraunhofer_Sterner.pdf [Accessed 26 July 2016].
[23] Sterner M. Bioenergy and renewable power methane in integrated 100% renewable energy systems. Limiting global warming by transforming energy systems. Dissertation; 2009. http://www.uni-kassel.de/upress/online/frei/978-3-89958-798-2.volltext.frei.pdf [Accessed 25 July 2016].
[24] Specht M, Zuberbühler U, Baumgart F, Feigl B, Frick V, Stürmer B, et al. Storing renewable energy in the natural gas grid. Methane via power-to-gas (P2G): A renewable fuel for mobility. In: Proceedings of the 6th conference "gas powered vehicles—the real economical CO2-alternative", Stuttgart 26–27th of October 2011; 2011. p. 98.
[25] Müller-Syring G, Henel M, Krause H, Rasmusson H, Mlaker H, Köppel W et al. Power-to-gas: Entwicklung von Anlagenkonzepten im Rahmen der DVGW-Innovationsoffensive. Artikel aus gwf-Gas/Erdgas; 2011: 770–777, http://www.dvgw-innovation.de/fileadmin/dvgw/angebote/forschung/innovation/pdf/powertogas2011gwf.pdf [Accessed 26 July 2016].
[26] Bajohr S, Götz M, Graf F, Ortloff F. Speicherung von regenerativ erzeugter elektrischer Energie in der Erdgasinfrastruktur. Artikel in gwf-Gas Erdgas April 2011, S. 200–210. http://www.dvgw-innovation.de/fileadmin/dvgw/angebote/forschung/innovation/pdf/speicherung.pdf [Accessed 26 July 2016].

[27] Vandewalle J, Bruninx K, D'haeseleer W. Effects of large-scale power to gas conversion on the power, gas and carbon sectors and their interactions. Energy Convers Manag 2015;94:28–39. https://doi.org/10.1016/j.enconman.2015.01.038.
[28] Schiebahn S, Grube T, Robinius M, Tietze V, Kumar B, Stolten D. Power to gas: technological overview, systems analysis and economic assessment for a case study in Germany. Int J Hydrog Energy 2015;40(12):4285–94. ISSN 0360-3199, https://doi.org/10.1016/j.ijhydene.2015.01.123.
[29] Bundesregierung Deutschland. Energiekonzept—Erneuerbare Energien. http://www.bundesregierung.de/Webs/Breg/DE/Themen/Energiekonzept/Energieversorgung/ErneuerbareEnergien-Zeitalter/_node.html [Accessed 18 December 2012].
[30] National Renewable Energy Laboratory. (2012). Renewable electricity futures study. Hand, M.M.; Baldwin, S.; DeMeo, E.; Reilly, J.M.; Mai, T.; Arent, D.; Porro, G.; Meshek, M.; Sandor, D. eds. 4 vols. NREL/TP-6A20-52409. Golden, CO: National Renewable Energy Laboratory.
[31] European Commission 2013. Clean power for Transport: a European alternative fuels strategy.
[32] IEA (2013) World Energy Outlook 2013—Renewable Energy Outlook. http://www.worldenergyoutlook.org/media/weowebsite/2013/WEO2013_Ch06_Renewables.pdf [Accessed 13 August 2015].
[33] Lund PD, Lindgren J, Mikkola J, Salpakari J. Review of energy system flexibility measures to enable high levels of variable renewable electricity. Renew Sust Energ Rev 2015;45:785–807.
[34] Reiter G, Lindorfer J. Global warming potential of hydrogen and methane production from renewable electricity via power-to-gas technology. Int J Life Cycle Assess 2015;20(4): 477–89.
[35] Müller B, Müller K, Teichmann D, Arlt W. Energiespeicherung mittels Methan und energietragenden Stoffen—ein thermodynamischer Vergleich. Chemie Ingenieur Technik 2011, 83, No. II, 2002–2013. DOI https://doi.org/10.1002/cite.201100113.
[36] Breyer CH, Rieke S, Sterner M, Schmid J. Hybrid PV-wind-renewable methane power plants. Proceedings European Photovoltaic Solar Energy Conference, Hamburg, Germany; 2011. Pages. 4594–4606. http://www.q-cells.com/uploads/tx_abdownloads/files/6CV.1.31_Breyer2011_HybPV-Wind-RPM-Plants_paper_PVSEC_preprint.pdf.
[37] Reiter G, Lindorfer J. Evaluating CO2 sources for power-to-gas applications – A case study for Austria. J CO2 Utiliz 2015;10:40–9.
[38] Carbon Recycling International, www.carbonrecycling.is [Accessed 26 July 2016].
[39] sunfire GmbH, www.sunfire.de [Accessed 25 July 2016].
[40] Bacyinski R, Whitney B, Gronich S, Garbak J, Bednarz M. DTE Energy Hydrogen Technology Park. Detroit: 2006. http://www.hydrogen.energy.gov/pdfs/progress06/vi_c_1_bacyinski.pdf [Accessed 27 July 2016].
[41] Steiner K. Power2Gas—Praxistest Windstromspeicherung im Erdgasnetz. http://www.eon.com/content/dam/eon-com/download/dwn-news/10737_694/Factsheet_Power2Gas.pdf [Accessed 26 July 2016].
[42] ISO 13686:2013 Natural gas—Quality designation. International Organization for Standardization, Geneva, Switzerland.
[43] Specht M, Jentsch M, Rieke S. Power-to-Gas Technologie. Erneuerbares Methan aus Ökostrom. ZSW Baden-Württemberg, Fraunhofer IWES, Solar Fuel GmbH: 2013. https://www.zsw-bw.de/uploads/media/Broschuere_Erneuerbares_Methan_aus.pdf [Accessed 26 July 2016].

[44] Vgl. e-mobil BW GmbH Landesagentur für Elektromobilität und Brennstoffzellentechnologie [Hrsg.] (2013) Wasserstoff, Infrastruktur für eine nachhaltige Mobilität, S.
[45] Vgl. Leschus, Vöpel (2008), Wasserstoff im Verkehr-Anwendungen, Perspektiven und Handlungsoptionen, S. 14ff.
[46] Ludwig Bölkow Systemtechnik GmbH. Hydrogen Filling Stations Worldwide. http://www.h2stations.org; [Accessed 26 July 2016]
[47] http://www.powertogas.info/power-to-gas/pilotprojekte-im-ueberblick/?no_cache=1 [Accessed 13 July 2016].
[48] Enertrag—Hybridkraftwerk. Dauerthal: 2009. https://www.enertrag.com/90_hybridkraftwerk.html [Accessed 26 July 2016].
[49] Plateforme Myrte. Universitá di Corsica. http://myrte.univ-corse.fr/downloads/ [Accessed 26 July 2016].
[50] Ursua A, Gandia LM, Sanchis P. Hydrogen production from water electrolysis: Current status and future trends. Proc IEEE 2012;100(2):410–26. https://doi.org/10.1109/JPROC.2011.2156750.
[51] Smolinka T, Günther M, Garche J. Stand und Entwicklungspotenzial der Wasserelektrolyse zur Herstellung von Wasserstoff aus regenerativen Energien. Kurzfassung NOW-Studie. Fraunhofer ISE, FCBAT: 2011. https://www.now-gmbh.de/content/5-service/4-publikationen/4-nip-wasserstoff-und-brennstoffzellentechnologie/now-studie-wasserelektrolyse-2011.pdf [Accessed 26 July 2016].
[52] Holladay JD, Hu J, King DL, Wang Y. An overview of hydrogen production technologies. Catal Today 2009;139:244–60. https://doi.org/10.1016/j.cattod.2008.08.039.
[53] Graves C, Ebbesen SD, Mogensen M, Lackner KS. Sustainable hydrocarbon fuels by recycling CO_2 and H_2O with renewable or nuclear energy. Renew Sust Energ Rev 2011;15:1–23. https://doi.org/10.1016/j.rser.2010.07.014.
[54] Little M, Thomson M, Infield D. Electrical integration of renewable energy into stand-alone power supplies incorporating hydrogen storage. Int J Hydrog Energy 2007;32:1582–8. https://doi.org/10.1016/j.ijhydene.2006.10.035.
[55] Project Homepage iC4. Integrated Carbon Capture, Conversion and Cycling. http://www.ic4.tum.de/index.php?id=1235 [Accessed 26 July 2016].
[56] Dickinson RR, Battye DL, Linton VM, Ashman PJ, Nathan GJ. Alternative carriers for remote renewable energy sources using existing CNG infrastructure. Int J Hydrog Energy 2012;35(3):1321–9. https://doi.org/10.1016/j.ijhydene.2009.11.052.
[57] Specht M, Personal communication at "forum power-to-gas" in Regensburg, Germany, 8.10.2012.
[58] http://www.zsw-bw.de/themen/brennstoffe-wasserstoff/power-to-gas.html [Accessed 26 July 2016].
[59] http://ceb.ebi.kit.edu [Accessed 26 July 2016].
[60] http://www.dri.edu/rec-projects/214-renewable-energy-production-via-carbon-capture-and-recycling [Accessed 26 July 2016].
[61] IPPC. Carbon dioxide capture and storage. Cambridge University Press; 2005.
[62] Li B, Duan Y, Luebke D, Morreale B. Advances in CO_2 capture technology: a patent review. Appl Energy 2013;102:1439–47. https://doi.org/10.1016/j.apenergy.2012.09.009.
[63] van den Broek M, Berghout N, Rubin ES. The potential of renewables versus natural gas with CO2 capture and storage for power generation under CO_2 constraints. Renew Sust Energ Rev 2015;49:1296–322. ISSN 1364-0321, https://doi.org/10.1016/j.rser.2015.04.089.

[64] Mekhilef S, Saidur R, Safari A. Comparative study of different fuel cell technologies. Renew Sust Energ Rev 2012;16:981–9. https://doi.org/10.1016/j.rser.2011.09.020.
[65] Luo X, Wang J, Dooner M, Clarke J. Overview of current development in electrical energy storage technologies and the application potential in power system operation. Appl Energy 2015;137:511–36. https://doi.org/10.1016/j.apenergy.2014.09.081.
[66] U.S. Department of Energy. Fuel Cell Technologies Program—Comparison of Technologies. EERE Information Center 2011. http://www1.eere.energy.gov/hydrogenandfuelcells/fuelcells/pdfs/fc_comparison_chart.pdf [Accessed 26 July 2016]
[67] Maclay JD, Brouwer J, Samuelsen GS. Experimental results for hybrid energy storage systems coupled to photovoltaic generation in residential applications. Int J Hydrog Energy 2011;36(19):12130–40. https://doi.org/10.1016/j.ijhydene.2011.06.089.
[68] Ulleberg O, Nakken T, Ete A. The wind/hydrogen demonstration system at Utsira in Norway: Evaluation of system performance using operational data and updated hydrogen energy system modeling tools. Int J Hydrog Energy 2010;35:1841–52. https://doi.org/10.1016/j.ijhydene.2009.10.077.
[69] Wietschel M, Ball M, Cremer C, Hasenauer U, Radgen P, Roser A, et al. In: Ball M, Wietschel M, editors. The hydrogen economy: Opportunities and challenges. Cambridge University Press; 2009.
[70] Forsberg P, Karlström M. On optimal investment strategies for a hydrogen refueling station. Int J Hydrog Energy 2007;32:647–60. https://doi.org/10.1016/j.ijhydene.2006.05.018.
[71] Mueller-Syring G, Henel M. Power-to-Gas: Konzepte, Kosten, Potenziale, DBI Fachforum: Energiespeicherkonzepte und Wasserstoff; 2011. http://www.dbi-gti.de/fileadmin/downloads/5_Veroeffentlichungen/Tagungen_Workshops/2011/H2-FF/07_Mueller-Syring_DBI_GUT.pdf [Accessed 17 December 2012].
[72] Edwards PP, Kuznetsov VL, David WIF, Brandon NP. Hydrogen and fuel cells: Towards a sustainable energy future. Energy Policy 2008;36:4356–62. https://doi.org/10.1016/j.enpol.2008.09.036.
[73] Rieke S. Regenerative Vollversorgung—von der Vision zur Praxis. Hannover, 2011. http://www.bee-ev.de/_downloads/bee/2011/HannoverMesse/20110404_HMI_SoarFuel_Rieke_Vollversorgung.pdf [Accessed 17 December 2012].
[74] Lee J-Y, Cha K-H, Lim T-W, Hur T. Eco-efficiency of H_2 and fuel cell buses. Int J Hydrog Energy 2011;36:1754–65. https://doi.org/10.1016/j.ijhydene.2010.10.074.
[75] Nuclear Energy Agency/International Energy Agency/Organization for Economic Cooperation and Development (2015) Projected Costs of Generating Electricity, ISBN 978-92-64-24443-6, https://www.iea.org/bookshop/711-Projected_Costs_of_Generating_Electricity
[76] de Visser E, Held A. (2014) Methodologies for estimating Levelised Cost of Electricity (LCOE), Project number: DESNL13116, http://res-cooperation.eu/images/pdf-reports/ECOFYS_Fraunhofer_Methodologies_for_estimating_LCoE_Final_report.pdf.
[77] Richtlinie VDI 2067—Blatt 1 Wirtschaftlichkeit gebäudetechnischer Anlagen—Grundlagen und Kostenberechnung. Verein Deutscher Ingenieure, 2012.
[78] U.S. Energy Information Administration (2013) Updated capital cost estimates for utility scale electricity generating plants, http://www.eia.gov/forecasts/capitalcost/pdf/updated_capcost.pdf [Accessed 13 July 2016].
[79] U.S. Department of Energy. Clean Cities Alternative Fuel Price Report. 2016, http://www.afdc.energy.gov/uploads/publication/alternative_fuel_price_report_april_2016.pdf [Accessed 13 July 2016].
[80] Steinmüller H, Tichler R, Reiter G, et al. Power to gas—Eine Systemanalyse. Markt- und Technologiescouting und -analyse. In: Final report on behalf of the Austrian BMWFJ; 2014.

[81] G. Gahleitner, J. Lindorfer (2013) Alternative fuels for mobility and transport: Harnessing excess electricity from renewable power sources with power-to-gas. ECEEE 2013 Summer Study, France.
[82] Reiter G, Lindorfer J. Global warming potential of hydrogen and methane production from renewable electricity via power-to-gas technology. Int J Life Cycle Assess 2015;20(4): 477–89. https://doi.org/10.1007/s11367-015-0848-0.
[83] Abbasi T, Abbasi SA. Renewable hydrogen: prospects and challenges. Renew Sust Energ Rev 2011;15(6):3034–40. https://doi.org/10.1016/j.rser.2011.02.026.
[84] Raine D, Strom H, Maisonnier G, Vinot S, Linnemann J, Trümper S. Analysis of the current hydrogen structure. In: Roads2HyCom; 2009.
[85] J. Schiffers, A. Vogel, B. Meyer-Prescher (2009) Beseitigung technischer, rechtliche rund ökonomischer Hemnisse bei der Einspeisung biogener Gase in das Erdgasnetz zur Reduzierung klimarelevanter Emissionen durch Aufbau und Anwendung einer georeferenzierten Datenbank. Final Report for the BMBF project, Biogaseinspeisung, Vol. 5. http://www.biogaseinspeisung.de/download/Abschlussbericht_Biogaseinspeis.ung_Band_5_Gasnetz_AP3_4.pdf [Accessed at 14 January 2014].
[86] Interactive maps and individual analyses are available at http://www.netinform.net/h2/H2Stations/Default.aspx [Accessed 01 December 2017].
[87] Ajanovic. Biofuels versus food production: Does biofuels production increase food prices? Energy 2011;36:2070–6. https://doi.org/10.1016/j.energy.2010.05.019.
[88] Söderberg C, Eckerberg K. Rising policy conflicts in Europe over bioenergy and forestry. Forest Policy Econ 2012; https://doi.org/10.1016/j.forpol.2012.09.015.
[89] A. Del Duce, P. Egede, G. Öhlschläger, T. Dettmer, H.J. Althaus, T. Bütler, E. Szczechowicz (2013) Guidelines for the LCA of electric vehicles projecteu/fileadmin/dokumente/Guideline_versions/eLCAr_guidelinespdf [Accessed 18 August 2015].
[90] REN21. Renewables 2016 Global Status Report. Paris: 2016, REN21 Secretariat. http://www.ren21.net/wp-content/uploads/2016/06/GSR_2016_Full_Report.pdf [Accessed 26 July 2016].

Further reading

[91] Decourt, Lajoie, Debarre, & Soupa. (2014). Hydrogen based energy conversion, more than storage: system flexibility. SBC Energy Institute.
[92] Specht M, Brellochs J, Frick V, Stürmer B, Zuberbühler U, Sterner M, et al. Storage of renewable energy in the natural gas grid. Erdöl Erdgas Kohle 2010;126:342–6.
[93] Vgl. Frauenhofer-Institut für Solare Energiesysteme ISE (März 2013): S.7.
[94] PE International (2013) GaBi Software with built-in Database (DB), Version 5. Available from http://www.gabi-software.com.
[95] H.J. Wagner, M.K. Koch, J. Burkhardt, T. Große Böckmann, N. Feck, P. Kruse (2007) CO_2-Emissionen der Stromerzeugung. BWK Bd. 59 Nr. 10. https://www.vdi.de/fileadmin/vdi_de/redakteur_dateien/geu_dateien/FB4-Internetseiten/CO2-Emissionen%20der%20Stromerzeugung_01.pdf [Accessed at 14 August 2015].

An overview of ground-source heat pump technology

15

Rao Martand Singh,[1], Abubakar Kawuwa Sani*, Tony Amis*[†]
*Department of Civil and Environmental Engineering, Faculty of Engineering and Physical Sciences, University of Surrey, Guildford, United Kingdom,
[†]GI Energy, Chicago, IL, United States
[1]Corresponding author: r.singh@surrey.ac.uk

Chapter Outline

Managing Global Warming. https://doi.org/10.1016/B978-0-12-814104-5.00015-6

15.1 Introduction

One of the major challenges facing the 21st century is the continuous need to mitigate the environmental damage caused by the burning of fossil fuels: oil, coal, and natural gas. Their extraction, refinement, transport, and burning result in the emission of greenhouse gases (GHG) into the atmosphere, which has negative environmental consequences including extreme weather such as longer and hotter droughts, more heat waves, longer wildfire seasons, extreme rainfall, and inundation of low-lying coastal areas [1,2]. To prevent these threats, there is the need to move away from the use of nonrenewable energy sources. This is supported by the fact that there is a direct correlation between the trends in population increase, primary energy use, and carbon dioxide (CO_2) emission, as shown in Fig. 15.1A. The figure predicts that the energy consumption and CO_2 emission will grow significantly by 136% and 110% respectively, between 1990 and 2040, as compared to the 72% linear population increase. This is as a result of the global increase in per capita energy demand.

To meet the ever-increasing energy need and ensure a clean environment, renewable and sustainable sources of energy such as geothermal, solar, wind, and wave energy should be adopted to substitute or lower the dependence on the use of fossil fuels. There are many renewable energy sources available and it is regrettable that they still only contribute a small percent of the world energy supply, as indicated in Fig. 15.1B.

A number of policies implemented by governmental and nongovernmental organizations around the world are now in place to encourage the use of renewable energy sources. Example of such policies includes the Climate Change Act passed by the UK parliament in 2008 [3,4], the California Global Warming Solutions Act (AB 32) of 2006 [5–9], the Climate Change Authority Act, and the Clean Energy Act of 2011 in Australia [10]. These positive and promising trends should be carefully evaluated, sustained, and encouraged across all the energy sectors, i.e., industry, transport, service, and domestic sector. This is to ensure that efficient systems, which consume less energy or rely on renewable energy sources, are being developed.

For example, in the domestic sector, systems such as ground-source heat pump (GSHP) could be utilized to replace the use of gas radiators and air conditioners in meeting the space heating and cooling demand of buildings. This will drastically bring down the amount of CO_2 emission released, as the heating, ventilating, and air conditioning (HVAC) systems presently consume the major share of energy used in the domestic sector. For instance, in a typical UK residential building, space and water heating account for 79% of the total energy consumed (see Fig. 15.2A), out of which 83% of that consumed energy is derived from fossil fuels as shown in Fig. 15.2B. Thus, the development and implementation of sustainable and renewable energy sources could positively impact on mitigating the amount of CO_2 being emitted into the atmosphere.

This chapter is intended to introduce the reader to the concept of ground-source heat pump technology. It describes the GSHP system, its different components and its working principles. Furthermore, the chapter describes the design considerations,

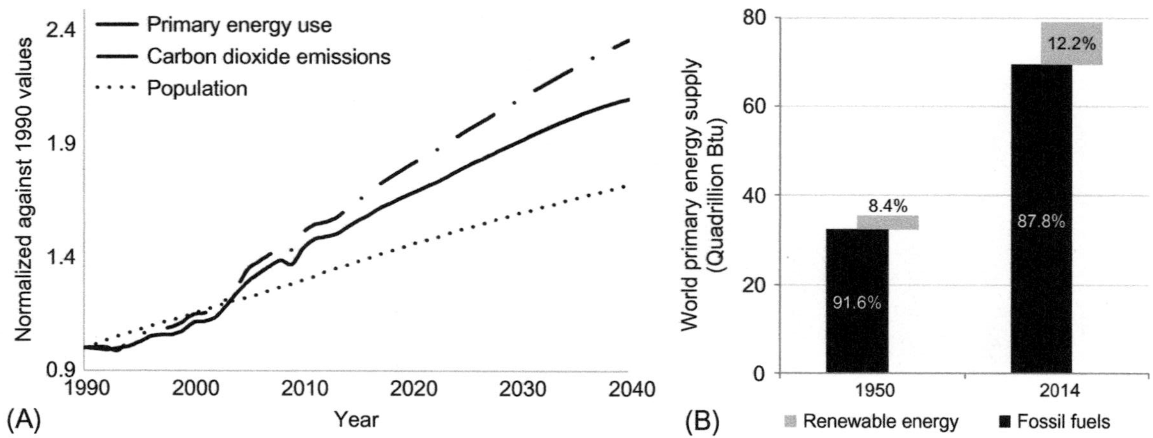

Fig. 15.1 (A) Energy consumption, CO_2 emissions and world population, (B) Energy supply [EIA, 2015].

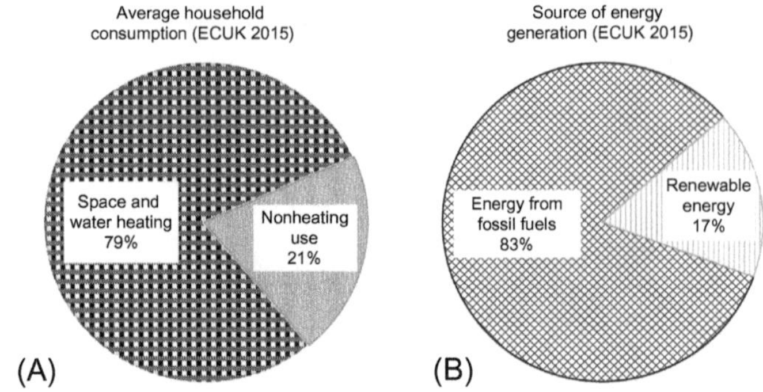

Fig. 15.2 (A) Average household consumption and (B) source of energy generation in the UK [11].

the factors that could enhance or negatively affect the system performance, and also includes a number of case studies of GSHP systems in operation in the UK.

15.2 Ground-source heat pump (GSHP) system

The GSHP system also known as ground-coupled heat exchanger system utilizes the shallow earth surface, 1–100 m, to harness its heat energy in order to meet human thermal comfort needs. The system operates basically by relying on the temperature difference between the ground and the air. For example in winter, the ground temperature is warmer than air temperature, while in summer, the ground temperature is cooler. These differences in temperature allow the transfer of heat energy in achieving building thermal demand throughout the year.

15.3 Components of the GSHP system

The GSHP system, presented in Fig. 15.3, is generally divided into three distinct components by researchers such as Brandl [12]:

- The primary unit or heat exchangers,
- The secondary unit comprises underfloor piping network that delivers heat energy in or out of the building.
- The heat pump unit.

Alternatively, the characteristics of the system can be based on the function of individual units as proposed by Preene and Powrie [13]. Thus, during space heating operation, the heat exchanger becomes the heat-source side, the heat pump, and the load side (secondary pipes). However, to achieve cooling, the source and load side is interchanged.

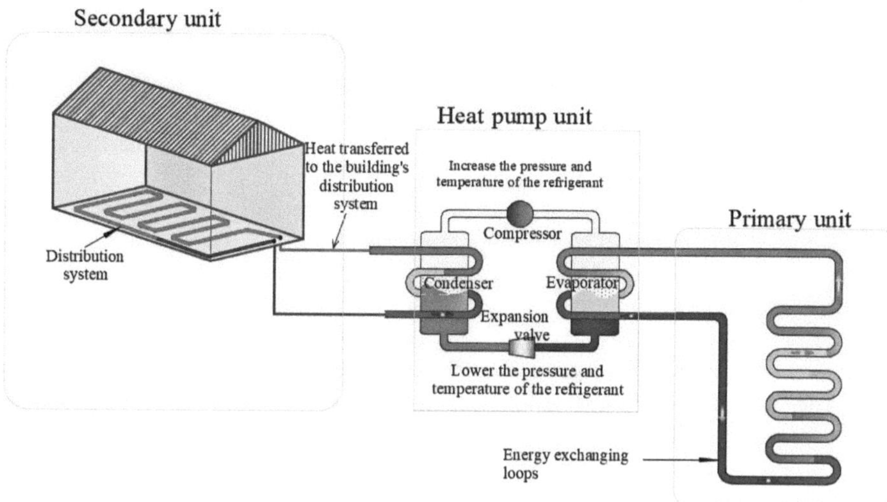

Fig. 15.3 Ground-source heat pump system.

15.3.1 The primary unit

The primary unit also known as the heat exchanger is made up of subsurface elements that aid in exchanging heat energy with the structures above ground surface. These elements, may include boreholes, base slabs, piles, diaphragm walls, tunnels etc., are fitted with energy-exchanging loops or even placed in direct contact with the ground using horizontal trenches.

15.3.2 The secondary unit

The secondary unit commonly consists of a closed network of pipes embedded in floors, walls, ceilings, bridge decks, roads, etc. It delivers heat energy extracted from the ground to the structure to achieve heating, or removes heat from the structure to the ground for heat storage.

15.3.3 The heat pump unit

The first ground heat pump was developed and built by Robert C. Webber in the 1940s [14]. It is a mechanical device that increases or decreases the temperature of the heat energy extracted from the source side. It functions in reverse order to a refrigerator, and is made up of four subcomponents: the evaporator, condenser, compressor, and expansion valve as shown in Fig. 15.3. The heat pump unit operates as a closed system within the closed unit.

For the system to achieve space heating, the cycle starts with the pumping of a constant heat carrier fluid (HCF) through the heat exchanger. The heat extracted by the HCF is retrieved by the refrigerant in the evaporator. The refrigerant begins to boil becoming a low-pressure vapor, which is then fed into the compressor. At the compressor, the low-pressure vapor is compressed and exits at a higher temperature and higher-pressure vapor, which then passes to the condenser. At the condenser, the hot vapor transfers its heat energy to the secondary circuit to be transferred to the building. The vapor (refrigerant) condenses into liquid form, which flows to the expansion valve, which further lowers the temperature and pressure. The refrigerant is then fed into the evaporator for the cycle to start again. Alternatively, to achieve space cooling, the working principle of the system is reversed.

Brandl [12] suggested that temperature in the range of 10–15°C, extracted from the ground, can be elevated to 25–35°C after passing through the GSHP system.

The main parameter used in describing the system is based on ratio of the amount of useful energy that can be obtained to the input power used to run the system. This parameter is commonly referred to as the coefficient of performance (COP) of the system. It is defined as:

$$\text{COP} = \frac{\text{energy output after heat pump/kW}}{\text{energy input for operation/kW}} \tag{15.1}$$

During heating operation, the system should aim for a COP value of 4 and 6 in cooling operation.

For example, let us consider a small household, with a peak heating load of 6 kW, heated by a GSHP system having a COP of 4. Hence, the electrical energy input (E) required to deliver this energy is given as $E = 6\,\text{kW}/4 = 1.5\,\text{kW}$. Furthermore, to calculate the heat energy delivered from the ground, which is $= 6 - 1.5\,\text{kW} = 4.5\,\text{kW}$. Thus, with the GSHP system, we are using 1.5 kW of electrical energy to transfer 4.5 kW of free environmental energy from the ground.

In common practice, heat pumps rely on electricity to drive the system. However, in few instances, heat pumps with internal combustion engine or occasionally absorption systems are used. To prevent environmental pollution, only refrigerants that do not contribute toward ozone layer depletion and with a minimum potential GHG effect should be allowed to drive heat pumps. As a consequence, the use of halogenated fluorinated hydrocarbons should strictly be avoided [12].

In addition to the COP, a seasonal performance factor (SPF) can also be used in characterizing and defining the performance of the system. The SPF of a heat pump system is the ratio of the useful energy output obtained by the system to the input energy used by the system. Therefore, the SPF is dependent not only on the heat pump as the case in COP but also on other energy-consuming units of the system, e.g., the circulation pump. An SPF value ranging between 3.8 and 4.3 can be achieved using standard electrical heat pumps. However, the SPF could be increased by 10%–15% using special devices that allow direct vaporization [12]. The SPF is defined as:

$$\text{SPF} = \frac{\text{usable energy output of the system/kWh}}{\text{energy input for the system/kWh}} \tag{15.2}$$

15.4 Types of ground-source heat pump systems

The heat pump unit in a GSHP system can be connected to the ground heat exchanger in two ways, namely, open-loop and closed-loop ground-source heat exchangers [15].

15.4.1 Open-loop ground-source heat exchangers

An open-loop heat exchanger system operates by circulating underground water, pumped from a water source, through the heat pump to extract heat. After the heat is transferred, the water that exits the heat pump unit is discharged into drains, groundwater, or surface water bodies. The discharged water can be useful for other purposes including irrigation, consumption, etc.

Open-loop heat exchanger systems can be connected via different configurations, depending on the type of water intake and discharge sources. Configurations for three possible cases are given in Fig. 15.4A–C.

In the system shown in Fig. 15.4A, the heat pump is connected to two wells: an intake well and a discharge well. After heat is transferred to the heat pump, the initial water temperature decreases by about 5–7°C, and the water is discharged off into the discharge well. However, in summer, the system can be reversed, providing cooling for the building [16]. During the design of the wells, i.e., intake and discharge wells, the following information is highly required to effectively design the well:

- The design depth of the well will depend on water table level, hydraulic conductivity of the soil, soil type, and the thermal load of the building.
- The diameter of the well is dependent on the diameter of the pump inlet pipe, pump flowrate or the quantity of water to be abstracted by the pump, the thermal load, and the yield of the well, or rate of recharge of the well.

In the second configuration shown in Fig. 15.4B, the heat pump draws water from an intake well, circulates it, and pumps it out to a discharge point, i.e., a surface body of water. In both the cases shown in Fig. 15.4A and B, the wells should be lined with well screens to prevent any loose material or sediments from shutting and filling the well, especially in loose or poor soil lithologies like sands, gravels, and some sandstones [16].

In the third case shown in Fig. 15.4C, water is abstracted from an underground-flooded structure, circulated through the heat pump to extract heat, and discharged to a surface water body. This provides an effective utilization of flooded underground structures such as mines. However, environmental risk assessment should be carried out by an expert to ensure that contaminated water from the aquifer does not cause pollution of the surface water [17].

The installation of an open-loop system is only possible where there are aquifers that can adequately and abundantly supply water to the heat pump. The seasonal groundwater flow and its quality should be studied to ensure an efficient and optimum system design. Similarly, the efficiency of the system is most effective when the underground water temperature is stable; this occurs at several meters below ground surface. Also, it is important to note that the installation of open-loop systems requires

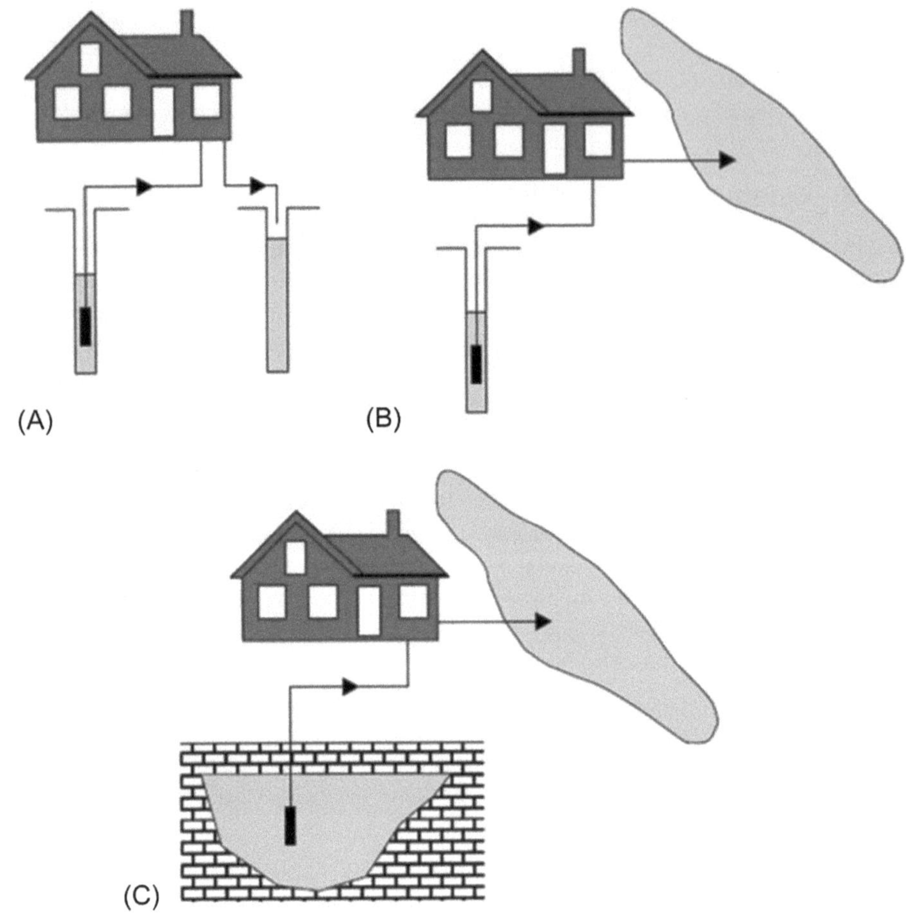

Fig. 15.4 Types of open-loop heat exchanger systems.

legal permission from the relevant government agency. Such permissions are rarely granted without submitting a detailed report on environmental risk assessment carried out and signed by an expert.

Advantages of an open-loop system:

- It is a sustainable and renewable heating and cooling source compared to gas radiators, i.e., it has negligible or zero negative contribution toward GHG emission.
- Suitable for large-scale application, e.g., commercial buildings, with large thermal energy demand.
- Direct exchange of heat between the injected water and the heat pump unit.
- There is negligible or no maintenance required.
- Fairly low operational cost.
- The system payback after few years of operation.
- Possibility of earning from the government through policies such as renewable heat incentive schemes (RHI).

Disadvantages of an open-loop system:

- Not suitable for small residential buildings, e.g., single-family residence.
- Possibility of clogging of the system.
- There is higher likelihood of environmental pollution compared to closed-loop systems.
- It is not feasible in locations with low groundwater flow or lack of an abundant aquifer.
- Require rigorous and tedious legal permissions.

Despite the advantages of the open-loop system, its major setback is the clogging of the system. Clogging may occur, when the system is installed in an environment, which is rich in ocher for example, as a result of the presence of minerals such as iron manganese hydroxides (FeH_4MnO_4), precipitation can occur as a result of changes in temperature [12,16,18]. This is especially common where water is abstracted from flooded and abandoned mines.

However, a recent study by Athresh et al. [17] investigated and proposed a possible way of solving the problem of clogging. The study was carried out on an abandoned coal mine at the National Coal Mining Museum (NCM) for England, on the site of former Caphouse Colliery, located in Overton, near Wakefield, Yorkshire, in the UK at 53.6416°N 1.6251°W with site elevation of about +147 m above sea level. The temperature of the mine water was reported by Banks et al. [19] to be in the range of 13–14°C.

Athresh et al. [17] found that using specially designed filters a mechanical cleaning solution could be used to prevent the clogging associated with open-loop systems. Additionally, they suggested that chemical treatment, via passive treatment through the use of several settling tanks and reed beds, could be employed to prevent system clogging.

15.4.2 Closed-loop ground-source heat exchangers

The closed-loop systems utilize and circulate constant volume of heat carrier fluid (HCF) in a closed plastic pipe network installed in subsurface structures or buried to some depth beneath the ground surface. The exchange of heat between the ground and the superstructure/infrastructure is through an indirect contact in comparison to the open-loop systems. In closed-loop systems, heat is transferred from the ground through the concrete and plastic pipes, via conduction, and transported to the heat pump through HCF circulation, via convection and conduction.

The plastic loops are made from high-density polyethylene/polypropylene (HDPE/HDPP), polyvinyl-chloride (PVC), polybutylene (PB) plastic pipes [20–23], with a nominal internal diameter that ranges between 20 and 44 mm and an appropriate wall thickness of 2.0–2.3 mm [12,24–29].

Once the plastic loops are fitted and secured in place, each of the loops is individually connected to the header or manifold. This allows room for reliability check of the system, and ensures the safety of the primary casing element especially where structural foundation elements are used. Thus, an energy loop with a defect can be independently disconnected at the manifold block to safeguard the system performance.

Enclosed in the energy exchanging loops, is an HCF that serves as a medium of heat exchange between the ground and the heat pump. Typically water, water plus biocide, antifreeze (glycol) or saline solutions are used as the HCF [15,30–33]. However, using an antifreeze-water-based solution increases the HCF viscosity, decreases its freezing point, resulting in more laminar flow behavior of the fluid, thus requiring more energy input to circulate the HCF in the loops.

The heat pump unit can be connected to different types of closed-loop heat exchanger configurations [14,15,31,34,35], as shown in Fig. 15.5A–F. The choice of which particular configuration to adopt should be based on some factors including installation cost, land area availability, thermal demand, building's foundation type, depth of heterothermal influence zone, and whether the installation of the heat pump system is to be carried out prior to or after building construction and occupancy.

In the configuration shown in Fig. 15.5A, HDPE pipes are laid horizontally in the ground beneath the frost influence zone. This usually occurs at a depth ranging between 1.5 and 2.0m. Beyond this depth, daily changes to ground temperature are less significant. This type of configuration should be installed in areas where the ground is in direct exposure of the sun to enhance heat transfer due to solar radiation. Also, infiltration of rain water through the ground also has positive effect on the system performance. The pipes are arranged in such a way that a spacing of about 0.5–0.8m is maintained between the exit and return leg of the pipe network. This reduces the risk of heat exchange between the pipes. In the design of this type of

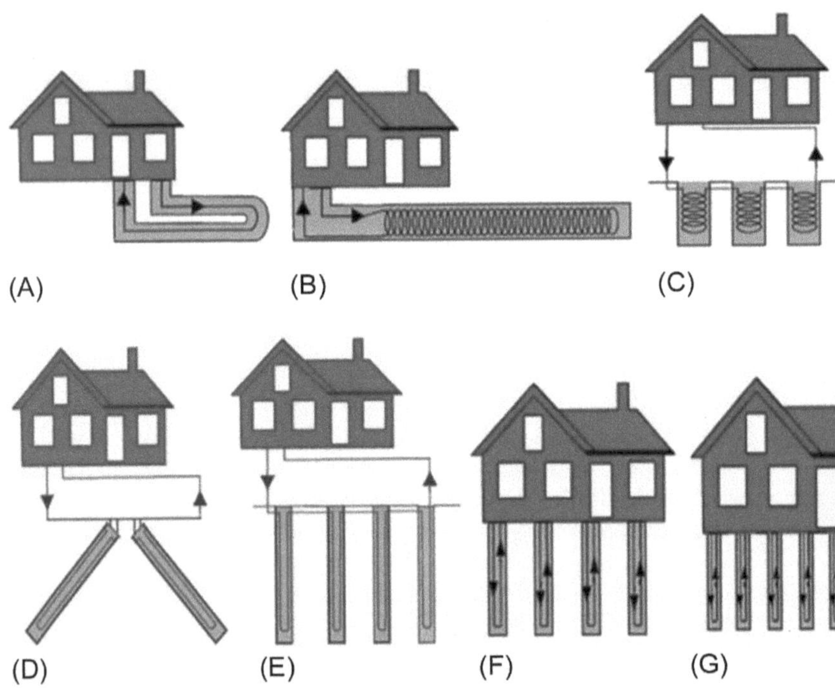

Fig. 15.5 Types of closed-loop heat exchanger systems.

system, it is safe to assume a usable thermal energy output of about 10–40 W per meter length of the loop, depending on soil conditions. The horizontal trench collector requires large area of land to sufficiently supply energy to the building. For example, to heat up a 1 m^2 of living space, a power of about 50 W is needed. Thus, an area of about 1.5–5 times, the area of living space is needed to obtain the required energy to heat the space. This type of configuration is not suitable in densely populated areas. However, in places where land is abundant, e.g., in rural areas, horizontal trench configuration can be used to provide heating for a single-family house. They are relatively simple to implement with very low installation cost [35,36].

Fig. 15.5B shows a slinky pipe configuration. This layout is similar to the horizontal trench installation shown in Fig. 15.5A. However, their only difference is that they allow longer loop length to be incorporated in the horizontal trench. The standard installation procedure recommends the loops to be installed in trenches that are 1.2 m wide. Similarly, in situations where more than one trench is required, the spacing between individual trenches should be at least 5 m apart centers to prevent excessive heat extraction. Additionally, the slinky configuration can also be incorporated vertically, in a narrow trench. The slinky type of configuration could provide 1 kW of energy per 10-m linear length of the pipes [35].

In addition to the configuration in Fig. 15.5B, the loops could also be fitted into a vertical trench in a spiral form as shown in Fig. 15.5C. The spring-like loops are installed to a depth ranging between 2 and 5 m. The distance between each spiral probe to another should be about 3–4 m. This type of configuration is ideally suitable where limited land area is available especially considering it does not require specialized heavy drilling equipment. This is a relatively new approach and could also be used where difficult geological conditions exist and drilling becomes nearly impossible. Under favorable ground conditions, it is possible to yield an output of 100–700 W per helix probe depending on the soil type and its degree of saturation [37–40]. A typical example of this spiral-like probe is the RAUGEO Helix probe manufactured by REHAU unlimited polymer solutions.

Fig. 15.5D shows the diagram of boreholes installed in a slanting form. The construction of this type of configuration occurs by the drilling of boreholes at an angle from a single manifold chamber. This ensures that the tips of the boreholes are placed farther apart from each other to maximize and enhance heat exchange of the system. During the construction stages, a single chamber is excavated, which serves as an access point through which the bores can be drilled and installed. The loops are installed at a sloping angle that ranges between 30° and 65°, within any point along the 360° radial plane [32]. These configurations eliminate the limitations of large-area requirement associated with the horizontally laid configurations, i.e., Fig. 15.5A and B. These allow the slanting boreholes to penetrate beneath existing buildings, gardens, driveways, etc. However, there are legal issues allied with drilling beneath other people's properties and close to government underground structures, e.g., tunnels [34].

The configuration shown in Fig. 15.5E depicts the conventional set-up with the energy loops installed in typical vertical borehole heat exchangers. The bored holes are generally filled with bentonite grout mixed with sand to enhance the borehole thermal conductivity. The borehole diameter and length ranges between 100 and 200 mm, and 20–200 m [18,41–43]. Also, the spacing between neighboring boreholes should be at least 4.5 m apart to ensure effective system performance in the long run [44]. It is estimated that on average, 1 m length of a vertical borehole could provide sufficient energy to heat up 1 m^2 of living space. This type of configuration has gained popularity because it requires a small land area for installation compared to horizontal trench heat exchangers. Furthermore, it provides greater efficiency because it is installed beyond the depth where annual temperature variation occurs. However, its main limitation is the high initial cost, which is attributed to the bore-hole drilling. A single or few boreholes can be installed to provide thermal comfort of a small-to-medium house unit. However, where high-heating loads exist (office buildings, utilities, and others) a group of boreholes, comprising dozens or even hundreds of boreholes, are commonly used; these are often referred to as geothermal fields.

The installation shown in Fig. 15.5F represents the typical case where the energy loops are installed in structural pile foundation elements. They offer similar advantages to the vertical borehole heat exchangers. Also, they are both installed to a depth where the ground annual temperature is relatively constant. Furthermore, the initial drilling cost associated with boreholes is negated because the piles were already needed for structural support. Also, due to the larger diameter of piles, i.e., 300–1500 mm, they offer more flexible and diversified loops arrangement compared to a conventional borehole [21,45]. These elements are generally referred to as energy piles, thermal piles, or geothermal energy piles. It is expected that piles with diameter of 300–500 mm could yield about 40–60 W of heat energy per meter length. Whereas piles with diameter greater than 600 mm are expected to provide about 35 W per square meter of the earth-contact surface area [12,21,45].

In addition, other alternative structural elements such as diaphragm walls, tunnels, and base slabs could be fitted with energy loops to provide the required thermal demand while meeting their primary need of structural support [12,33,46–49].

The configuration shown in Fig. 15.5G shows the scenario where the loops are placed in small diameter bored-holes (up to 300 mm) made using micropiling technology. This new solution offers new opportunities for existing buildings where micropiles are needed to provide additional strength to the existing foundation elements, and as anchors for retaining walls [50].

15.5 Design and installation of ground-source heat pump systems

This section describes the design and installation procedure of the GSHP system and the numerous factors that should be carefully considered during the planning, preliminary and detailed design, and construction process. These factors if carefully and effectively studied and implemented will go a long way to ensure an optimized and efficient system.

The description given in this work is aimed at providing a general information and system specification of a GSHP system. It is not meant to be a detailed design and installation guidelines, training manual, or in any way serve as a substitute to the available design standards. The authors recommend that the design and installation should conform and be carried out in accordance with the available and established standards. For this purpose, readers are referred to ASHRAE handbook [51,52], Verein Deutscher Ingenieure (VDI) 4640 technical guidelines Part 1–4 [53], and the thermal pile design, installation, and materials standards published by the GSHP association in the UK [54], for the design and installation of open-loops and closed-loops GSHP systems.

It is advisable that experts should be involved at the early stages of the project initiation, design, and construction to ensure that the designed and installed GSHP system meet its intended use. Doing this will ultimately result in an effective system design and installation, with the lowest possible implementation cost, and also negligible running cost. This therefore prevents costly refurbishment and renovation of the system as a result of poor and unsatisfactory design process. In most cases, this tends to be much more expensive than the cost of a well-designed system, which has been correctly installed.

15.6 Factors affecting the performance of a GSHP system

There are several factors that directly and indirectly affect the performance of a GSHP system. These factors could be categorized based on the components of the system, i.e., primary, secondary, and heat pump unit, as shown in Fig. 15.3.

Considering the secondary unit or building component, there are factors that should be carefully and thoroughly studied during the planning, design, and construction stages. These factors could vary from one building type to another, and with the behavioral characteristics of the household occupants. These factors have long been recognized in past applications [12,45,55,56], and are itemized as follows:

(a) Early ecological energy planning for the building
(b) Desired hourly heating and cooling load of the building
(c) Possible heat losses in the system should be accounted for and factored into the heating/cooling load calculations.
(d) Building usage
(e) Building size
(f) Building physics, e.g., insulation of roofs, walls, and ceilings, size and quality of windows, etc.
(g) Type of occupancy
(h) Sustainability of the secondary pipes, radiators, and other subcomponents that comprise the secondary unit.

Similarly, the heat pump component as a single entity has some factors that could influence the overall performance of the GSHP system. These factors if carefully examined could maximize the useful energy output that can be obtained from the system. These factors are as follows:

(a) Type of the GSHP system, i.e., either open-loop or closed-loop system
(b) Type of the heat pump unit, i.e., either monodirectional (for heating or cooling only) or bidirectional system (for both heating and cooling)

(c) Heat pump size, which is determined based on the thermal load
(d) Flowrate capacity of the circulation pump
(e) Length of energy loop, diameter, and its thermal properties
(f) Type, mixture, and viscosity of circulation fluid
(g) Heat carrier fluid flow velocity and flow condition (laminar or turbulent)
(h) Coefficient of performance of the heat pump

In addition to these factors, there are other factors that need to be cautiously investigated due to their significant influence on the performance of the primary unit. The importance of a specific or some of these factors could differ based on the type of primary heat exchanger unit of the GSHP system. For example, information about groundwater flowrate could be among the major factors that should be accounted for in open-well GSHP system design, as compared to the closed-loop GSHP system. Also, information about soil strength parameters could be very important in load-bearing elements (e.g., pile foundations, diaphragm walls, tunnels, and base slabs) as compared to their use in non-load-bearing elements (e.g., borehole and trench collectors). Nonetheless, despite the different preferences, they all require the following general information to warrant effective system performance. These factors are:

(a) Hydrogeological ground properties, which includes
  (i) Groundwater table level and its seasonal fluctuation
  (ii) Direction of groundwater flow and its velocity
(b) Geothermal properties of the soil, such as
  (i) Soil thermal conductivity
  (ii) Ground temperature gradient
  (iii) Ground undisturbed temperature
(c) Geotechnical properties of soil, such as
  (i) Soil water content, hydraulic properties
  (ii) Soil density, void ratio
  (iii) Shear strength parameters
(d) Structural properties of the building (applicable to structural elements, e.g., piles), e.g.,
  (i) Bored hole depth and their arrangement/spacing
  (ii) Foundation type and size
  (iii) Method of installation
(e) Soil minerals and chemical composition

Furthermore, in cases where foundation or structural elements are used as heat exchangers, their primarily role of structural support should not be overlooked in the course of meeting the thermal demand. In such situations, the imposed thermal loads should not be excessive to ensure the overall pile capacity is not exceeded. Through using a control system, the temperature limit can be maintained within an acceptable range, which allows for system switch-off, to warrant long-term system performance; thus, assuring the longevity and efficiency of the whole GSHP system [45].

It is common in practice, as part of the design process, to make sure an evaluation of the system is carried out to confirm the amount of heating and cooling it can provide. Equally, the design will need to identify the top-up heating and cooling energy required from other conventional heating and cooling systems, in order to fully understand the amount of energy the GSHP can provide and what is needed from the other conventional system. Similarly, it is a good practice to allow for some margin of error, during the design process, between the energy required and the expected system capacity. A common recommendation is to allow for 10% margin between needed demand and the expected system energy output [57]. This is reviewed accordingly as the project progresses toward completion [45].

Moreover, it is strongly advised that a numerical model and simulation of the whole GSHP system, made up of the primary, secondary, and heat pump unit, should be carried out in large scale projects (i.e., with heating and cooling demand of more than 50 kW) [12,45]. This will help in providing the general concept of the system working principle during service life. And most importantly, it will help in ensuring that any possible issues that may likely develop and were overlooked during the design stages, are recognized and resolved [45].

15.7 Environmental, social, and economic impact of the GSHP system

The main purpose of using renewable and sustainable energy resources is to provide clean energy with no or insignificant negative impact on the environment. This section investigates the various influences that the installation of the GSHP system and its use could potentially have on the environment.

15.7.1 Environmental influence of the GSHP system

The environmental impact of the GSHP system can be analyzed in numerous ways. From a wider perspective, the key benefits of using the GSHP system in achieving sustainable space heating and cooling are that of reducing the amount of greenhouse gas emissions, by minimizing the use of fossil fuels. For example in the UK, the domestic sector consumes the third largest share of the total energy generated in the country [11]. Next to transportation and the industrial sector, it contributes to about 106.3×10^6 t of carbon-dioxide-equivalent emissions into the atmosphere [58].

Similarly, in an average household in the UK, the heating, ventilating, and air conditioning (HVAC) systems used about 79% of its total energy as shown in Fig. 15.2A. This clearly highlights how significantly beneficial it would be on the environment to move away from using fossil fuels for achieving space heating and cooling.

However, it can also be argued that the use of heat pumps could indirectly contribute toward the emission of GHG into the atmosphere, because heat pumps require electricity as an energy input to operate, and this is generally produced in a fossil-fuel-burning power plant. Thus, a working heat pump, which utilizes electricity from such power plants to operate, contributes its own share toward GHG emission. And the volume of the GHG produced at the power plant directly depends on the type of fossil fuel used during energy generation and on the size of the turbine and its associated energy input requirement. A concept, which evaluates the impacts of the gases released by power plants on global warming, was developed at Oak Ridge National Laboratory (ORNL) in the early nineties. The concept was sponsored by the US Department of Energy and is known as total equivalent warming impact (TEWI). It is a measure of the combined direct and indirect effects of GHG emissions into the atmosphere over the lifetime of any power plant [59]. It is quantified in kilograms of carbon dioxide equivalent per kilowatt hour of electricity generated.

Alternatively, where hydroelectric, nuclear, solar, wind power, or any other renewable sources of energy are used to produce the electricity used by the heat pump, the GSHP system becomes fully environmentally friendly with zero emission.

In addition, where structural elements are used as heat exchangers, the influence of increase in temperature on the mechanical behavior of such structural elements should be analyzed in detail. Numerous researchers have shown that the cyclic heating and cooling process results in the expansion and contraction of the structural foundation elements, e.g., piles; thus, superimposing additional thermal stresses and strains on the foundation element. However, those thermally imposed changes are thermoelastic in nature and are reversible [14,24,26,28,57,60–74]. Nevertheless, the thermomechanical behavior of foundation elements should be carefully studied especially where the piles are to be installed in problematic soils such as swelling clays.

15.7.2 Social impact of the GSHP system

The social influence of the GSHP system can be evaluated from different perspective, considering myriads of factors that could influence its development or serve as a barrier against its acceptability. Within this section, factors that contribute either positively or negatively on the development of the system are analyzed. These factors include a psychological influence, information availability, and possible educational barriers.

15.7.2.1 Psychological perception of the GSHP system

The development of the heat pump system concept took place in the late 18th century, while the integrating and coupling of heat pumps with different heat exchange sources occurred in the early and mid-19th century [45]. Despite the availability of this technology for over a century, its acceptability and development has been slow. This is most likely due to the fear of adopting new technologies and the concern about maintenance and risk of possible system failure due to lack of adequate system awareness. Furthermore, the high installation cost of the system in comparison to other classical and cheaper solutions, which rely upon fossil fuels, could have been among the possible reasons of slow GSHP system acceptance. However, today, most countries are investing and encouraging the development of sustainable energy solutions, which harbor no or negligible negative impact on the environment. As a consequence, ensuring adequate information about all possible sustainable energy solutions is readily and easily available to both investors and end users. These encourage the high uptake of renewable systems such as pile heat exchanger (PHE) closed-loop systems.

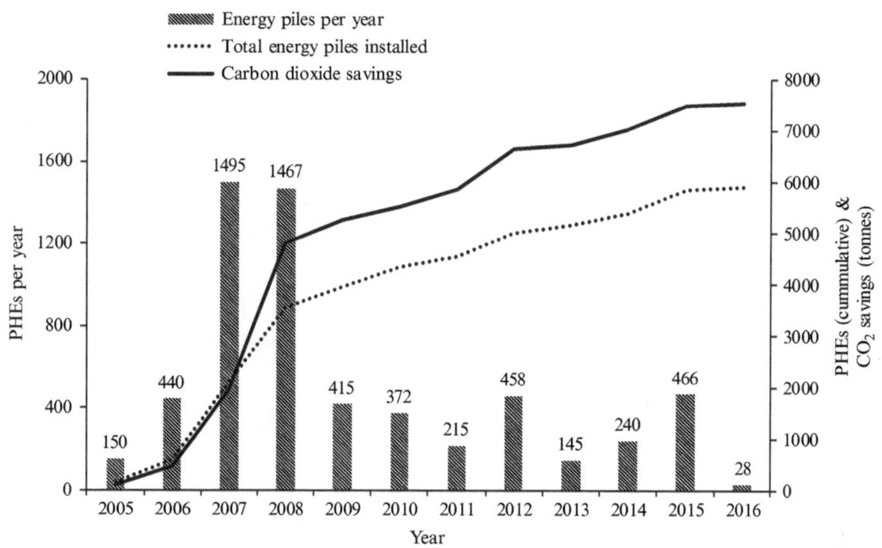

Fig. 15.6 PHEs Installed in the UK and their resultant annual CO_2 Savings as of December 2016 [45].

For example, in the UK, there were about 150 PHEs installed by 2005. However, by 2016, the number has drastically increased to about 5891; this provides an annual CO_2 savings of about 7545 t as shown in Fig. 15.6.

15.7.2.2 Information availability and educational barrier

The manner and ways in which information about the GSHP system is communicated could influence the public perception of GSHP, making it more difficult to make the correct decisions. The availability of many communication channels, such as the internet, makes information more readily and easily available. However, it comes with a negative implication too. With information scattered all over the internet, finding the right and correct information becomes even trickier and difficult; this is particularly true when considering the increase in the number of unreliable sites and advert.

However, recently, nongovernmental bodies, such as the Ground-Source Heat Pump Association (GSHPA) in the UK, formed by experts on the GSHP system technology, are now bridging the gap. In 2012, the association published a very well-detailed document, given in readily accessible form, covering professional and objective information to the potential GSHP investors and users [54]. The report covers areas such as:

- Preliminary design stage specifying the various roles and responsibilities each party has toward achieving a successful project design and implementation.
- Similarly, it has also covered both the thermal and mechanical analyses and design of the pile heat exchanger.

In addition to the information availability, well-structured educational programs such as workshops, seminars on GSHP systems should be organized and encouraged. The workshops should be aimed different stakeholders including engineers, designers, architects, energy sector experts, financial investors, and decision makers. The workshops should present clear information about the technology and its associated positive significance of sustainable and economical ways of achieving space heating and cooling process, ranging from properly designed systems and installation details from inception to completion.

Furthermore, training and certifications for the design, installation, and construction of the system should be encouraged and enforced. This ensures that those involved in the design and construction process of the system are well qualified.

15.7.3 Economic analysis of the GSHP system

The economic significance of the GSHP system can be viewed by analyzing the initial investment cost of the system, its associated running cost, and its efficiency. In comparison to other conventional heating and cooling systems, the cost of installing the GSHP is high and depends on the type of heat exchanger unit chosen and the size of the system. Equally, where retrofitting of the building is required, the investment cost may be significantly higher.

However, with advancement in technology, more modern and mechanized approaches are now available for installing the heat exchanger units making GSHP systems more economically viable and attractive. Furthermore, the increase in number of manufacturers of heat pump and its components gave rise to an open and more competitive market; this is expected to keep the overall price down.

Another factor that significantly contributes toward the economic benefit of the system is its low running cost. A GSHP system that is appropriately designed and correctly installed should have a low maintenance requirement involving regular servicing to ensure the effective functioning of the system. For example, the servicing of the system could be carried out after a year or after a few years after installation to check the heat pump speed, possible leakages in the secondary unit, and other subcomponent to ensure that optimum designed output capacity is maintained …

In addition, the incentive schemes being offered by various governments to encourage the public toward investing in renewable energy systems such as GSHP is yielding a positive result. One such scheme is the renewable heat incentive (RHI) scheme in the UK. As of 2016, the RHI tariff for a conventional GSHP system was 19.33 pence per kilowatt hour. However, as of September of 2017, the RHI has been increased to 19.86 p $(kWh)^{-1}$, which is an increase of about 3%. The RHI for a GSHP is far higher than the incentive for other renewable sources such as air-source heat pump (ASHP) and biomass, which have RHIs of 10.18 and 6.54 p $(kWh)^{-1}$, respectively [75].

Furthermore, the overall efficiency of the system could be attributed to the performance of each subcomponent. Ensuring that the building is well insulated, does go a long way toward guaranteeing maximum efficiency, assuming that the heat pump system working at optimum capacity. Also, improving the heat exchanging properties of the heat exchanger does make for a more economical GSHP system [76] by improving the yield output capacity of the system, while utilizing lower energy input. This can be directly linked to the COP of the GSHP system. An open- or closed-loop GSHP system that is designed to utilize 25% of its input energy to conjure 75% free renewable energy from the ground provides an overall savings in comparison to sourcing the total energy from nonrenewable ways. Thus, aiming for higher system performance combined with RHI schemes will ensure quick recovery of the initial investment cost, which is between 2 and 10 years, depending on the system size, climatic and geological conditions.

15.8 Case studies of GSHP system

In this section, different case studies of UK installed GSHP systems are discussed with the aim of illustrating to potential stake holders that the technology can be implemented in both existing and proposed structures and can be carried out on small-scale or large-scale projects in order to achieve sustainable space heating and cooling.

15.8.1 Gloucester police headquarters, Quedgeley, Gloucester, UK

The first major GSHP project scheme to be installed in the UK was the Gloucester police headquarters, at Gloucester, in the year 2005. The three-storey building, shown in Fig. 15.7, has a floor area of 8500 m^2. It was equipped with 150 vertical closed-loop borehole heat exchangers installed at an average depth of 98 m. Nine reversible heat pump units, connected to the boreholes at the manifold, allow the circulation of HCF to achieve space heating and cooling operations, via the use of underfloor secondary heat energy exchange elements. The GSHP scheme currently provides 860 kW of cooling and 765 kW of heating energy. An active CO_2 management system was

(a) (b)

Fig. 15.7 (A) Gloucester Police Headquarters, Quedgeley, Gloucester, UK and (B) during construction [77].

incorporated into the GSHP to monitor and limit the carbon emission of the building. In the first year of operation, 8% of fossil fuel (gas) was used in the building, thus providing about 600 MWh of energy in savings. This is equivalent to an annual savings on fuel of £60,000 [77].

15.8.2 Robert Gordon University, Garthdee Campus, Aberdeen, UK

Another example of a GSHP scheme installed in the UK was that constructed at Robert Gordon University, at the Garthdee campus in October of 2013. The system comprised 66 vertical closed-loop borehole heat exchangers drilled through granite rock to a depth of 200m. Despite the challenges associated with the drilling, granite rock had its own advantages. This includes using it for heat storage owing to its higher volumetric heat capacity. Eight reversible heat pumps were connected to the boreholes, which delivered 900kW of heating and 900kW of cooling to the School of Art and Architecture, situated next to river Dee as shown in Fig. 15.8.

Each of the eight heat pumps was equipped with a computerized control system to allow each pump to be individually switched between heating or cooling mode, depending on the internal and external air and ground temperatures. The installed system was the largest commercial GSHP system in Scotland. Modeling results indicate an anticipated COP of 5 for heating and 6 for cooling [78].

15.8.3 Kingsmill Hospital, Mansfield, Nottinghamshire, UK

A new state-of-the-art hospital (Kingsmill hospital) was built in Nottinghamshire, over a land area of 21-ha site, and completed in January of 2011. The building (shown in Fig. 15.9) had a 140,000 m^2 total useable area, spaciously designed to house 920 beds. With the aim of reducing the overall energy consumption of the hospital estate,

Fig. 15.8 Robert Gordon University, Garthdee Campus, Aberdeen, UK [78].

Fig. 15.9 Kingsmill Hospital, Mansfield, Nottinghamshire, UK [79].

various heating schemes including onsite electricity generation using mines gas (Methane), combined heat and power (CHP), geothermal energy piles, photovoltaics, and a closed-loop lake GSHP system were considered during the feasibility studies. The closed-loop lake system was chosen and built. It comprised 140 stainless steel heat exchangers submerged under the surface of the Kings Mill reservoir, which is used for water supply and recreation, located to the southeast of the building. Forty-two EKW130 water-source heat pump units deliver a total of 10 MW of energy (5.4 MW of cooling and 5 MW of heating) to the building via the use of stainless steel radiators (called slim Jims). It was the largest geothermal loop lake system installed in Europe. The system has a COP of 4 for heating and 7 for cooling. It provides an annual savings of 9600 MWh of electricity and gas, which is equivalent to about £120,000. In terms of emission, it saves and prevents the release of 1700 t of CO_2 into the atmosphere. This is equal to taking 600 cars off the road [79].

15.8.4 Plas Newydd mansion, Anglesey, Wales, UK

A 300-year-old, grade 1 listed, 16th century country house (Fig. 15.10A) located next to Menai Strait in Anglesey, Wales, was fitted with a GSHP system. There were three floors to the building with a total floor area of 5000 m^2. Prior to the installation of the

(a) (b)

Fig. 15.10 Plas Newydd mansion, Anglesey, Wales, UK [80].

GSHP system, the building relied upon oil boilers to achieve space heating. The boilers used about 1500l of oil per day in winter, enough to heat a normal house for 10 months. In the year 2011, the 7th Marquess of Anglesey came across an article in a magazine regarding GSHP system installed at Howard castle. He thought this surely could work. After series of consultations and preliminary studies, a water-source heat pump (WSHP) system was chosen because of the close proximity of Menai Strait. Four units of submersible heat pumps, WPF 66 type (Fig. 15.10B), manufactured by Stiebel Eltron were installed. The heat pumps have a 300-kW capacity and connected via open-loop configuration [80]. Pipes of 200 mm in diameter, covered by concrete caissons and natural stones, convey water from the sea (53 m away) to extract heat and return it back via the outlet pipe. The system has a COP of 4.08 and SPF of 2.82. The project was completed and became operational by May of 2014. The total cost of the project was £600,000. It currently provides an annual savings of £40,000 in fuel [81].

15.8.5 Kingston Heights, Kingston upon Thames, Surrey, UK

Kingston Heights (shown in Fig. 15.11A) is a mixed-use development, built at a cost of about £70 million. The building is situated next to river Thames, in Kingston upon Thames, Surrey, UK. It is made up of 56 affordable homes, 81 luxury apartments and 145 rooms occupied by Hilton Doubletree Hotel. The building utilize an open-loop WSHP system, which abstracts water from the river and passes it through a stainless steel filter (located at 2.5 m below the water surface) fitted with automated backwash system (shown in Fig. 15.11B). The filter also prevents any marine life from entering into the system. In addition, a second filter system removes any silt that might have passed through the first filter.

A total of thirty-nine heat pumps were installed to fully supply the 2.3-MW space-heating demand of the building. Each heat pump deliver heat energy to about 3–5 flats to achieve space heating, via the use of underfloor heating, and also provide hot water.

(a) (b)

Fig. 15.11 Kingston Heights, Kingston upon Thames, Surrey, UK [82].

About 13 million liters of water is circulated through the heat pump per day, to extract heat energy, and pumped back into the river. This is equivalent to the volume of water that can fill up five standard Olympic-size swimming pools. The system ensures that the temperature of the water that returns back to the river is maintained within ±3°C of the initial river temperature. This safeguards the marine habitat in the river. The system provides an annual savings of 500t of CO_2 [82].

15.8.6 The Crystal, Docklands, London, UK

The Crystal building, shown in Fig. 15.12, located at the Docklands in London, serves as the headquarters of Siemens. Completed in 2012, it is the first building to reach both the BREEAM® Outstanding and LEED® Platinum status in the world. Much of this success is due to its sophisticated and integrated active and passive energy design elements making it one of the most sustainable buildings that exist today.

Designed with the aim of improving our knowledge of urban sustainability, the building has an area of 3500 m^2 and is 18 m high. Photovoltaic panels, connected to Siemens inverters, cover two-third of the roof of the building, which is 88 m long and 45 m wide. The solar panels supply 20% of the building's electrical demand. The remaining 80% is provided via green electricity grid [83].

The space heating and cooling demand of the building is achieved through the use of a closed-loop GSHP system. One-hundred-and-ninety-nine pile foundation elements, reaching a depth of 150 m and with a total length of 17 km, were fitted with energy exchanging loops. The system achieves space cooling by extracting heat from the building, reinjecting, and storing it into the ground during summer for subsequent use in winter, thus achieving a reversible heating and cooling process. Two reversible heat pumps are used to deliver the 614 kW of cooling and 614 kW of heating requirement of the building. The cooling process is achieved by passing chilled water through secondary pipe units mounted on the floor beams. The space heating is attained by passing hot water via underfloor secondary heating elements [84].

Fig. 15.12 The Crystal, Docklands, London, UK [83].

15.8.7 NEO, Bankside, London, UK

The iconic NEO Bankside project is made up of four apartment towers, located next to the Tate Modern on the South Bank of river Thames. Completed in 2011, the buildings (shown in Fig. 15.13) consist of 217 residential units with retail stores at ground level. As the building were designed with supporting pile foundations, a vertical closed-loop GSHP scheme was chosen to meet its heating and cooling requirement. This allows the use of the pile foundation elements for the dual purpose of structural support and for extracting or rejecting thermal energy from/to the ground. One-hundred-and-thirty-two piles, with a depth reaching up to 52 m, were equipped with energy loops to allow the circulation of water-based fluid to transfer heat in or out of the building. Bentonite slurry was used to fill in the piles—a technique that has been successfully employed on other projects after its first implementation on the NEO Bankside project [85].

Seven heat pumps (NSKW type), each having a capacity of 130 kW [86], deliver 650 kW of cooling and 760 kW of heating energy to the building. The system has been in use since 2011; however, no data are publicly available regarding its carbon emission savings.

15.8.8 Bvlgari hotel, Knightsbridge, London, UK

The 10-storey Bvlgari hotel and apartment complex with six levels of basement is located on the south side of Knightsbridge, in London, UK (shown in Fig. 15.14). With the basement extending to a depth of over 24 m below platform level, it is used to house a double-height ballroom, dining facilities, swimming pool, plant rooms, and a GSHP system. The pile foundations were used to support the building and were fitted with energy exchanging loops. In addition to that, a diaphragm wall was required to maintain structural integrity of the basement due to the requirement of its retained height and also prevent groundwater ingress [87]. Similarly, the diaphragm wall

Fig. 15.13 NEO Bankside apartment buildings, London, UK [85].

Fig. 15.14 The Bvlgari hotel, Knightsbridge, London, UK [87].

was equipped with energy loops, thus making it the first energy diaphragm wall project in the UK. A hybrid GSHP scheme, made up of the diaphragm wall and 50 energy piles, delivers 150 kW of peak heating and 150 kW of peak cooling to the hotel. The whole project was completed and commissioned in 2012.

15.8.9 Embassy court, St John's Wood, London, UK

Another example of incorporating GSHP scheme into the development of luxury apartments is the Embassy court, located on Wellington road, in London. The construction of the building was completed in 2009 and is shown in Fig. 15.15A. The GSHP system installed in the building is made up of a hybrid system of 10 vertical boreholes with a depth to 100 m, (located 5 m from London underground), and 100 energy piles. Four heat pumps, each having a capacity of 130 kW, were connected to the heat exchangers to supply the thermal load requirement of the building. In summer, the GSHP system provides 370 kW of cooling energy to the building and 200 kW of heat energy to achieve space heating. The building has energy efficiency and CO_2 environmental impact ratings of 86% and 85%, respectively [88]. This is classed as "class B" under the EU Directive 2002/91/EC as shown in Fig. 15.15B.

15.8.10 One New Change, London, UK

The One New Change retail and office development, located next to St. Paul's cathedral, London, is another innovative building equipped with a hybrid GSHP system. The construction of the building was completed in 2010 and is shown in Fig. 15.16. The hybrid heat exchangers is made up of one pair of open-loop wells and 192 energy piles, with some reaching to a depth of 45 m and 2.4 m in diameter. The system supplies about 2300 kW of cooling and 2400 kW of heating energy to the building. Currently, the system prevents the emission of 170,820 kg of

(A)

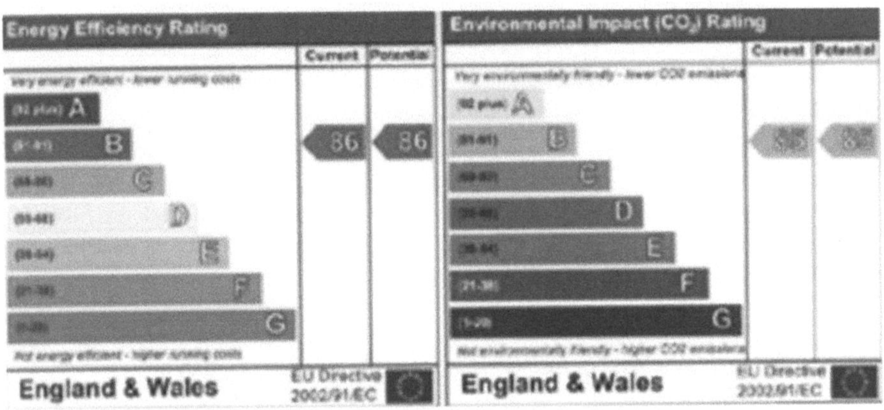

(B)

Fig. 15.15 The Embassy court, St John's Wood, London, UK [88].

Fig. 15.16 One New Change building, London, UK [89].

CO_2 per year. In addition, it provides an annual savings of £27,000 on the running cost. It also benefits from the RHI scheme, with an annual contribution of about £60,000 from the government toward encouraging the adoption of renewable energy [89].

15.9 Conclusion

This chapter presents a general overview of the GSHP technology with information on the different components of the system, their working principle, and the function of each subcomponent. In addition, it highlights the design and installation guide including the various factors that positively impact on the overall system performance. Furthermore, the environmental, social, and economic significance of the system have been discussed together with a number of case studies describing the implementation of GSHP schemes in the UK.

References

[1] Council Climate. Key Issues for the New Climate Agreement 2015;9.
[2] UNFCCC. Adoption of the Paris Agreement. Proposal by the President. Paris Climate Change Conference—November 2015, COP 21 2015. FCCC/CP/2015/L.9/Rev.1.
[3] Bowen A, Rydge J. Climate change policy in the United Kingdom. Grantham Research Institute on Climate Change and the Environment 2011.
[4] Fankhauser S, Kennedy D, Skea J. The UK's carbon targets for 2020 and the role of the committee on climate change. Building a Low-Carbon Future: The Politics of Climate Change 2008:99–110.
[5] Holst DR. California Climate Policy to 2050, 2015.
[6] Yeh S, Yang C, Gibbs M, Roland-Holst D, Greenblatt J, Mahone A, et al. A modeling comparison of deep greenhouse gas emissions reduction scenarios by 2030 in California. Energ Strat Rev 2016;13–14:169–80. https://doi.org/10.1016/j.esr.2016.10.001.
[7] Salkin PE. Sustainability and Land Use Planning: Greening State and Local Land Use Plans and Regulations to Address Climate Change Challenges and Preserve resources for future generations 2009; 121.
[8] Brown EG, Nichols MD, Corey RW. Proposed first update to the climate change scoping plan: Building on the framework February 2014 table of contents 2014.
[9] Franco G, Cayan D, Luers A, Hanemann M, Croes B. Linking climate change science with policy in California. Clim Change 2008;87:7–20. https://doi.org/10.1007/s10584-007-9359-8.
[10] Kember O, Jackson E, Chandra M. GHG mitigation in Australia: an overview of the current policy landscape. World Resources Institute Working Paper 2014:1–40.
[11] ECUK. Energy consumption in the UK. 2015. http://webarchive.nationalarchives.gov.uk/20160317154000/https://www.gov.uk/government/statistics/energy-consumption-in-the-uk (Accessed 1 September 2016).
[12] Brandl H. Energy foundations and other thermo-active ground structures. Géotechnique 2006;56:81–122. https://doi.org/10.1680/geot.2006.56.2.81.
[13] Preene M, Powrie W. Ground energy systems: from analysis to geotechnical design. Géotechnique 2009;59:261–71.
[14] Akrouch GA. Energy piles in cooling dominated climates. Texas A&M University; 2014.

[15] Suryatriyastuti ME, Mroueh H, Burlon S. Understanding the temperature-induced mechanical behaviour of energy pile foundations. Renew Sustain Energy Rev 2012;16:3344–54. https://doi.org/10.1016/j.rser.2012.02.062.
[16] Banks D. Introduction to thermogeology: ground source heating and cooling. 2nd ed. John Wiley & Sons; 2012.
[17] Athresh AP, Al-Habaibeh A, Parker K. The design and evaluation of an open loop ground source heat pump operating in an ochre-rich coal mine water environment. Int J Coal Geol 2016;164:69–76. https://doi.org/10.1016/j.coal.2016.04.015.
[18] Zeng HYY, Diao NRR, Fang ZHH. A finite line-source model for boreholes in geothermal heat exchangers. Heat Transf Asian Res 2002;31:558–67. https://doi.org/10.1002/htj.10057.
[19] Banks D, Skarphagen H, Wiltshire R, Jessop C. Heat pumps as a tool for energy recovery from mining wastes. Geol Soc Lond Spec Publ 2004;236:499 LP–513.
[20] Akrouch GA, Sánchez M, Briaud J-L. Thermo-mechanical behavior of energy piles in high plasticity clays. Acta Geotechnica 2014;9:399–412. https://doi.org/10.1007/s11440-014-0312-5.
[21] Loveridge F. The thermal performance of foundation piles used as heat exchangers in ground energy systems 2012:206.
[22] Hamada Y, Saitoh H, Nakamura M, Kubota H, Ochifuji K. Field performance of an energy pile system for space heating. Energ Buildings 2007;39:517–24. https://doi.org/10.1016/j.enbuild.2006.09.006.
[23] Park H, Lee SR, Yoon S, Choi JC. Evaluation of thermal response and performance of PHC energy pile: Field experiments and numerical simulation. Appl Energy 2013;103:12–24. https://doi.org/10.1016/j.apenergy.2012.10.012.
[24] Bourne-Webb PJ, Amatya B, Soga K, Amis T, Davidson C, Payne P. Energy pile test at Lambeth college, London: Geotechnical and thermodynamic aspects of pile response to heat cycles. Géotechnique 2009;59:237–48. https://doi.org/10.1680/geot.2009.59.3.237.
[25] Bozis D, Papakostas K, Kyriakis N. On the evaluation of design parameters effects on the heat transfer efficiency of energy piles. Energ Buildings 2011;43:1020–9. https://doi.org/10.1016/j.enbuild.2010.12.028.
[26] Singh RM, Bouazza A, Wang B, Haberfield CH, Baycan S, Carden Y. Thermal and thermo-mechanical response of a geothermal energy pile. World Geothermal Congress 2015;2015:7.
[27] Wang B, Haberfield C, Baycan S. Field investigation of a geothermal energy pile: initial observations. Geotech-FrOrg 2013;3415–8. https://doi.org/10.1179/1937525514Y.0000000002.
[28] Mccartney JS, Murphy KD. Strain distributions in full-scale energy foundations. J Deep Found Inst 2012;6:26–38. https://doi.org/10.1179/dfi.2012.008.
[29] Jeong S, Lim H, Lee JK, Kim J. Thermally induced mechanical response of energy piles in axially loaded pile groups. Appl Therm Eng 2014;71:608–15. https://doi.org/10.1016/j.applthermaleng.2014.07.007.
[30] Gao J, Zhang X, Liu J, Li KS, Yang J. Thermal performance and ground temperature of vertical pile-foundation heat exchangers: a case study. Appl Therm Eng 2008;28:2295–304. https://doi.org/10.1016/j.applthermaleng.2008.01.013.
[31] Zagorscak R, Thomas HR. A review on performance of energy piles and effects on surrounding ground. Inženjerstvo Okoliša (Environ Eng) 2016;3:33–45.
[32] Kurevija T, Strpić K. Hydraulic and thermogeological design differences between two-loop vertical and inclined coaxial borehole heat exchangers. Renew Energy 2018;117:314–23. https://doi.org/10.1016/j.renene.2017.10.077.

[33] Brandl H, Adam D, Markiewicz R. Ground-sourced energy wells for heating and cooling of buildings. Acta Geotech Slov 2006;1:5–27.
[34] Omer AM. Ground-source heat pumps systems and applications. Renew Sustain Energy Rev 2008;12:344–71. https://doi.org/10.1016/j.rser.2006.10.003.
[35] Kovačević MS, Bačić M, Arapov I. Possibilities of underground engineering for the use of shallow geothermal energy. Gradevinar 2013;64:1019–28.
[36] Rees SJ. An introduction to ground-source heat pump technology. Elsevier Ltd; 2016. https://doi.org/10.1016/B978-0-08-100311-4.00001-7.
[37] REHAU unlimited polymer solutions. RAUGEO HELIX PROBE PE-Xa: The alternative to probes and horizontal collector n.d. https://www.gshp.org.uk/documents/REsearchseminar2010/Ralf%20Winterling%20Spiral%20Probes.pdf.
[38] Equipment Energy. REHAU RAUGEO Helix Probes n.d. http://www.edenenergy.com/rehau/raugeo-helix-loop/ (accessed December 18, 2017).
[39] REHAU unlimited polymer solutions. RAUGEO HELIX® PE-XA n.d. https://www.rehau.com/download/790542/raugeo-helix-sales-brochure.pdf.
[40] Go Geothermal Ltd. Helix Probe 2017:1–2. https://www.gogeothermal.co.uk/product-detail.asp?c=2&p=66.
[41] Loveridge F, Powrie W. Pile heat exchangers: thermal behaviour and interactions. Proc ICE—Geotech Eng 2013;166:178–96. https://doi.org/10.1680/geng.11.00042.
[42] Sliwa T. Analysis of a heat pump system based on borehole heat exchangers for a swimming pool complex in Krynica. S-Poland 1999;357–84.
[43] Bose JE, Parker JD, McQuiston FC. American Society of Heating R and A-CE. Design/data manual for closed-loop ground-coupled heat pump systems. Atlanta, GA: American Society of Heating, Refrigerating, and Air-Conditioning Engineers; 1985.
[44] Gultekin A, Aydin M, Sisman A. Thermal performance analysis of multiple borehole heat exchangers. Energ Conver Manage 2016;122:544–51. https://doi.org/10.1016/j.enconman.2016.05.086.
[45] Sani AK, Singh RM, Amis T, Cavarretta I. Thermal and thermo-mechanical performance of pile heat exchangers: a review. Renew Sustain Energy Rev 2018;.
[46] Brandl H. Energy piles for heating and cooling of buildings. Seventh International Conference & Exhibition on Pilling and Deep Foundations, 1998.
[47] Katzenbach R. New technologies and applications: materials and equipment in near surface geothermal systems. DFI J 2015;https://doi.org/10.1179/1937525514Y.0000000012.
[48] Markiewicz R, Adam D. Energy from earth-coupled structures, foundations, tunnels and sewers. Géotechnique 2009;59:229–36. https://doi.org/10.1680/geot.2009.59.3.229.
[49] Xia C, Sun M, Zhang G, Xiao S, Zou Y. Experimental study on geothermal heat exchangers buried in diaphragm walls. Energ Buildings 2012;52:50–5. https://doi.org/10.1016/j.enbuild.2012.03.054.
[50] Design T. The green micropile the design and analysis of a geothermal micropile system a major qualifying project submitted to the faculty of Worcester n.d.
[51] Ashrae Ashrae Guide and Data Book (Fundamentals). 2002:148.
[52] ASHRAE. ASHRAE Handbook—HVAC Applications. 2011.
[53] Reuss M, Konstantinidou E, Sanner B. 10 years VDI 4640—German guidelines for ground coupled heat pumps, UTES and direct thermal use of the underground. Proceedings of ECOSTOCK, Stockton College, New Jersey, USA 1998:8.
[54] GSHP Association. Thermal pile design, installation and materials standards. Ground Source Heat Pump Association, Milton Keynes 2012;82.

[55] De Moel M, Bach PM, Bouazza A, Singh RM, Sun JO. Technological advances and applications of geothermal energy pile foundations and their feasibility in Australia. Renew Sustain Energy Rev 2010;14:2683–96. https://doi.org/10.1016/j.rser.2010.07.027.
[56] Dincer I, Sanner B. Shallow Geothermal Energy. GHC Bull 2001; June 19–25. https://doi.org/10.1016/S1364-0321(99)00011-8.
[57] Bourne-Webb PJ, Bernard J, Friedemann W, Pralle N, Uotinen VM, Widerin B. Delivery of energy geostructures. In: Energy Geostructures; 2013. p. 229–63.
[58] Department for Business E&IS Final UK greenhouse gas emissions national statistics 1990–2015. 2017.
[59] Fischer SK. Total equivalent warming impact: A measure of the global warming impact of CFC alternatives in refrigerating equipment. Int J Refrig 1993;16:423–8. https://doi.org/10.1016/0140-7007(93)90059-H.
[60] Ng CWW, Shi C, Gunawan A, Laloui L. Centrifuge modelling of energy piles subjected to heating and cooling cycles in clay. Proc Inst Civil Eng: Struct Build 2014;4:310–6. https://doi.org/10.1680/geolett.14.00063.
[61] Ng CWW, Shi C, Gunawan A, Laloui L, Liu HL. Centrifuge modelling of installation effects on helical anchor performance in sand. Can Geotech J 2015;52:1045–57.
[62] Laloui L, Nuth M, Vulliet L. Experimental and numerical investigations of the behaviour of a heat exchanger pile. Int J Numer Anal Meth Geomech 2006;30:763–81. https://doi.org/10.1002/nag.499.
[63] Arson C, Berns E, Akrouch GA, Sanchez M, Briaud J-L. Heat propagation around geothermal piles and implications on energy balance 2013:628–35.
[64] McCartney JS, Rosenberg JE. Impact of heat exchange on side shear in thermo-active foundations. Geo-Frontiers 2011: Advances in Geotechnical Engineering, ASCE; 2011, p. 488–98.
[65] McCartney JS, Rosenberg JE. Impact of heat exchange on side shear in Thermo-active foundations. Adv Geotech Eng 2011;1–10.
[66] Laloui L, Moreni M. In-situ thermo-mechanical load test on a heat exchanger pile. Proceedings of 4th international conference on deep foundation practice 1999:273–9.
[67] Singh RM, Bouazza A, Wang B. Near-field ground thermal response to heating of a geothermal energy pile: Observations from a field test. Soils Found 2015;55:1412–26.
[68] Rotta Loria AF, Gunawan A, Shi C, Laloui L, Ng CWW. Numerical modelling of energy piles in saturated sand subjected to thermo-mechanical loads. Geomech Energy Environ 2015;1:1–15. https://doi.org/10.1016/j.gete.2015.03.002.
[69] Di Donna A, Rotta Loria AF, Laloui L. Numerical study of the response of a group of energy piles under different combinations of thermo-mechanical loads. Comput Geotech 2016;72:126–42. https://doi.org/10.1016/j.compgeo.2015.11.010.
[70] Wang B, Bouazza A, Singh RM, Haberfield C, Barry-macaulay D, Baycan S. Post-temperature effects on shaft capacity of a full-scale geothermal energy pile. J Geotechnol Geoenviron Eng 2014;141:4014125. https://doi.org/10.1061/(ASCE)GT.1943-5606.0001266.
[71] McCartney JS, Murphy KD, Henry KS. Response of an energy foundation to temperature fluctuations. Proceedings of the international foundations conference and equipment exposition (IFCEE 2015) 2015:1691–1700. https://doi.org/10.1061/9780784479087.153.
[72] Stewart MA, McCartney JS. Strain distributions in centrifuge model energy foundations, Oakland, California, USA: 2012, p. 4376–85.
[73] Amatya BL, Soga K, Bourne-Webb PJ, Amis T, Laloui L. Thermo-mechanical behaviour of energy piles. Géotechnique 2012;62:503–19. https://doi.org/10.1680/geot.10.P.116.

[74] Goode III J.C., McCartney JS. Centrifuge modeling of end-restraint effects in energy foundations. J Geotechnol Geoenviron Eng 2015;141:1–13. https://doi.org/10.1061/(ASCE)GT.1943-5606.0001333.
[75] RHI. Renewable Heat Incentive 2018. http://www.energysavingtrust.org.uk/scotland/grants-loans/renewables/renewable-heat-incentive.
[76] Faizal M, Bouazza A, Singh RM. Heat transfer enhancement of geothermal energy piles. Renew Sustain Energy Rev 2016;57:16–33. https://doi.org/10.1016/j.rser.2015.12.065.
[77] Facts K, Cooper M. Gloucester Police Headquarters Quedgeley, Gloucester, UK Consultant: 2005.
[78] Gi-energy. Robert Gordon University Garthdee campus, Aberdeen, UK. 2013.
[79] Skanska. Kings mill hospital. In: UK. 2008.
[80] Stiebel-Eltron. heat pump WPF 66 type specifications. 2008. https://doi.org/10.1128/JB.00375-08.
[81] Newydd BP. 21st-century renewable energy from the Irish Sea for a centuries-old mansion cutting-edge technology meets nature and heritage conservation; 2014.
[82] RICHARD VENGA. Case Study: Kingston Heights. 2016.
[83] Andy P. The crystal, Royal Victoria Docks, London. 2012.
[84] Siemens plc. The crystal: one of the most sustainable buildings in the world. 2016.
[85] Gi-energy. CASE STUDY NEO Bankside. 2011.
[86] Envision. NSKW 130 heat pump Specification. 2012.
[87] Amis T, Robinson CAW, Wong S. Integrating geothermal loops into the diaphragm walls of the Bulgari Hotel Knightsbridge London. 11th DFI/EFFC international conference on geotechnical challenges in Urban Regeneration, London, 2010, p. 1–10.
[88] Gi-energy. Embassy court St John's wood, London, UK. 2009.
[89] Gi-energy. Case study: One New Change, London, UK. 2011.

Further reading

[90] Gao J, Zhang X, Liu J, Li K, Yang J. Numerical and experimental assessment of thermal performance of vertical energy piles: An application. Appl Energy 2008;85(10):901. https://doi.org/10.1016/j.apenergy.2008.02.010.

Geological sequestration of carbon dioxide

16

Mojgan Hadi Mosleh[*,1], Majid Sedighi*[*], Masoud Babaei*[†], Matthew Turner*[‡]
[*]School of Mechanical, Aerospace and Civil Engineering, The University of Manchester, Manchester, United Kingdom, [†]School of Chemical Engineering and Analytical Science, The University of Manchester, Manchester, United Kingdom, [‡]IHS Global Limited, Enterprise House, Tetbury, United Kingdom
[1]Corresponding author: mojgan.hadimosleh@manchester.ac.uk

Chapter Outline

16.1 Introduction

The climate of the Earth is changing as a result of the increasing concentration of greenhouse gases. Carbon dioxide is the main anthropogenic greenhouse gas and its concentration in the atmosphere is continuously increasing. Therefore, many countries have pledged to reduce the emissions of greenhouse gases [1]. One option is to capture and store the carbon dioxide in porous rocks in deep geological formations.

The carbon capture and storage (CCS) technology provides a solution for the decarbonization of energy and manufacturing sectors. According to the International Energy Association [2], it is anticipated that fossil fuels will produce approximately 80% of the total global energy demand in 2030 with a breakdown of coal, oil, and gas

Managing Global Warming. https://doi.org/10.1016/B978-0-12-814104-5.00016-8

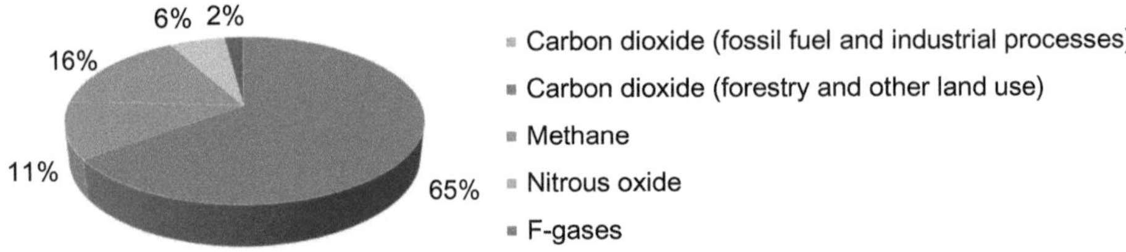

Fig. 16.1 Global emission by gas [3].

to be 26%, 28%, and 26%, respectively. Despite increasing global attentions to the renewable and low carbon energy alternatives and technological developments in this field, it is predicted that fossil fuels will continue to be the major contributor of world energy generation for many years to come.

Fossil fuels and industrial processes are the greatest emitter of carbon dioxide to the atmosphere as shown in Fig. 16.1. Improving the efficiency of energy consumption and implementing carbon storage will have significant impacts on the overall greenhouse gas emission. CCS schemes have been reported to contribute to 20% reduction by 2050 [4].

CCS technology involves capturing the carbon dioxide from power plants and industrial plants (in particular the iron, steel, cement, and bulk chemicals manufacturing industries) and injecting the captured carbon dioxide in a compressed form to suitable deep geological strata where carbon dioxide can be safely stored over a long geological period. Our understanding of carbon storage processes in subsurface environment has been significantly improved over the last decade as a result of substantial research. Several review articles have been published in the literature that provides a comprehensive state-of-the-art review of different aspects of CCS [5–8]. This chapter aims to provide an overview of the CCS technology with a focus of presenting essential learning materials for undergraduate and postgraduate students of interdisciplinary syllabuses of civil engineering and earth science programs including energy and environmental geotechnics.

16.2 Overview and engineering aspects of the technology

CCS is a technology that can capture up to 90% of the carbon dioxide (CO_2) emissions produced from the use of fossil fuels in electricity generation and industrial processes, thus preventing the carbon dioxide from entering the atmosphere. The process of CCS takes place in three separate stages [9]:

1. CO_2 capture and separation: This component of the CSC technology includes the capture and separation of CO_2 from the flue gas or other process stream.
2. CO_2 compression and transportation: The gas is compressed to a physical state that is categorized as supercritical phase. The supercritical gas is transported to the selected geological site by pipeline.
3. CO_2 storage: The gas is injected into a selected reservoir where it is safely stored for a geologically significant period of time.

CO_2 capture and storage projects will require a significant amount of engineering activities in different phases. Once carbon dioxide is captured, the carbon will need to be compressed and transported to the storage site via pipelines. Appropriate number of boreholes will need to be designed and constructed for injecting the compressed carbon dioxide. For technologies that involve recovery of hydrocarbon through the injection process, facilities should be provided for handling the hydrocarbon fluid (liquid or gas). There will be a water treatment plant to clean any water removed during the storage of carbon dioxide prior to disposal.

16.2.1 CO_2 capture technologies

In principal, there are four basic systems of CO_2 capture from fossil fuel power plants, fuel processing plants, and the industrial plants. These systems are:

Precombustion: This is the process of removing the carbon from the fuel prior to combustion. The hydrocarbon fuel is generally converted to a mixture of hydrogen and carbon monoxide (syngas) from a primary fuel which is followed by a shift reaction to convert CO to CO_2.

Postcombustion: This is similar to precombustion apart from the fact that carbon dioxide is removed from the flue gas stream after the fuel has been combusted. The postcombustion is commercially available. In this process, the gas is passed through a solvent followed by either heating or depressurizing in order to release a high-purity CO_2 stream.

Oxy-fuel combustion capture: This is a method by which fossil fuels are burnt in pure oxygen yielding almost pure CO_2. There are a number of demonstration projects utilizing oxy-fuel combustion capture in operation.

Capture from industrial process streams: Purification of natural gas and production of hydrogen-containing synthesis gas for the manufacture of ammonia, alcohols, and synthetic liquid fuels are examples of CO_2 capture from process streams.

16.2.2 Compression and transportation

Once the CO_2 has been separated at the source, it will need to be compressed prior to transportation. The most cost-effective way to transport the CO_2 to the storage site is to transport it in a supercritical state [5]. The density of pure supercritical CO_2 is approximately 500 times higher than that of gaseous CO_2. In other words, approximately 500 m^3 of gaseous CO_2 with atmospheric pressure and temperature conditions can be stored in 1 m^3 at supercritical conditions. The critical point of CO_2 is 31 °C at 73 atm. Gaseous CO_2 can also be transported by pipelines using a compressor to push the gas through the pipeline. Sometimes a pipeline will have intermittent compressors to keep the gas flowing. In order to prevent corrosion to the pipeline, the CO_2 must be clean and free of water, hydrogen sulfide and other impurities.

If transporting under supercritical conditions, the interaction of the gas mixtures with the pipeline will need to be understood and specialized alloys may be needed to mitigate potential corrosion issues [5]. Water will have to be removed from the gas stream because if the CO_2 were to come into contact with water, it would form carbonic acids which will corrode the pipeline. The pipeline will have to be constructed to withstand the pressures well above the supercritical state of approximately 80 bar and may need to be engineered to withstand pressures well over 100 bar.

The solvent properties of supercritical CO_2 are known to be detrimental to the elastomers commonly used in valves, gaskets, coatings, and O-rings, used for sealing purposes. At high pressures supercritical CO_2 diffuses into the elastomers and, when the pressure is reduced, blistering and even explosions can occur as the material decompresses. Many of the elastomeric materials currently used in oil and gas pipelines are therefore not suitable for CO_2 transportation [10].

In the United States CO_2 is transported from existing power plants to an enhanced oil recovery (EOR) fields at high pressures and routed through sparsely populated areas; the consequence of an accident is therefore likely to be relatively small because the CO_2 will dissipate before affecting the local inhabitants. However, if networks are to be developed onshore for the collection of CO_2 from industry sources in densely populated countries such as the United Kingdom, this will require the routing of pipelines close to population centers. In the UK pipeline design code, the minimum distance between a pipeline and occupied buildings is defined by the substance being transported. However, supercritical CO_2 has not been classified and so, in order for pipeline networks to be designed in the United Kingdom, new regulations need to be put in place for both on- and offshore pipelines.

16.2.3 Drilling and well completion

The drilling will be another important aspect of the operation, ensuring that: the boreholes are drilled into the storage reservoir accurately; drilling is done at a sufficient depth; the borehole layout is well designed. The success of this will be determined largely by the geology of the chosen site and also by the well pattern design which should be as cost-effective way as possible. In order to enhance the permeability, the correct direction of drilling is vital. Two drill teams will be required: the rig operators and the directional tool operators. This will increase the cost of drilling.

16.2.4 CO_2 storage

Geological sequestration of carbon dioxide (CO_2) is defined as the capture of CO_2 directly from anthropogenic sources and disposal into geological formations for geologically significant periods of time. CO_2 can be stored in geologic formations by different processes and mechanisms. At the initial stages of injection, supercritical CO_2 is trapped under a low-permeability geological medium. The CO_2 can further react with the solid and aqueous phases and can become dissolved into the water phase, referred to as solubility trapping. It can also react, either directly or indirectly, with the minerals and organic matter in the geological phases; this is known as mineral trapping (see Ref. [11]). The effectiveness of geological storage depends on a combination of physical and geochemical trapping mechanisms, which will be directly related to the host rock formations. The most effective storage sites are those where CO_2 is immobile because it is trapped permanently under a thick, low-permeability seal or is chemically converted to solid minerals or is adsorbed on the surfaces of coal or mineral micropores or through a combination of physical and chemical trapping mechanisms [5].

The physical properties of CO_2 that is stored depend on pressure and temperature. CO_2 is a gas at normal temperature and pressure, and it can turn into liquid at a moderate pressure of 6.4 MPa at 298 K. It becomes supercritical when the temperature is higher than 304 K and the pressure is greater than 7.38 MPa. Carbon dioxide density increases rapidly at approximately 800 m depth, when the CO_2 reaches a supercritical state.

16.3 CO_2 geological storage options

A geological storage site should have adequate capacity and injectivity, a satisfactory sealing caprock or confining unit, and a sufficiently stable geological environment to avoid compromising the integrity of the storage site. There are mainly three geological settings that have been suggested and studied for the long-term sequestration of CO_2:

- Active and depleted oil and gas fields at which oil and gas recovery can be enhanced, called enhanced oil or gas recovery (EOR and EOG)
- Deep unmineable coal seams that have a capacity to enhance methane recovery, termed enhanced coalbed methane (ECBM) recovery
- Deep saline aquifers

Storage options such as EOR, EGR, and ECBM have added benefits of producing commercially useful by-products such as oil and gas that can offset cost. This can have a significant benefit by offsetting the significant capital and operating expenditure incurred, over and above that provided by a carbon trading scheme.

16.3.1 Enhanced oil recovery and enhanced gas recovery

Oil and gas reservoirs are porous rock formations (usually sandstones or carbonates) containing hydrocarbons (crude oil and/or natural gas) that have been physically trapped. There are two main types of physical traps that are expected in carbon sequestration in depleted oil and gas reservoirs: (1) stratigraphic traps, created when changes have occurred in rock types, and (2) structural traps, in which the rocks have been folded or faulted to create a trapping reservoir. Oil and gas reservoirs are ideal geologic storage sites because they have held hydrocarbons for thousands to millions of years and hence are suitable for CO_2 storage. Although EOR and EGR have the lowest capacity of all options for genealogical CO_2 sequestration [12], their architecture and properties are well known as a result of exploration for and production of hydrocarbons. In addition, due to the industrialization of these sites, infrastructure exists for CO_2 transportation and storage [13]. The CO_2 gross retention efficiency defined as the percentage of injected CO_2 that is not being produced and does not need to be reinjected has been reported to be 60% for some of the US and Canadian CO_2 flooding operations [14–16]. The total volume of CO_2 consumed by US CO_2-EOR schemes has been about only 11% of the total US CO_2 emissions from industrial sources alone. Nonetheless, CO_2-EOR could be an enabling catalyst for larger scale sequestration efforts [17].

The most effective storage sites are those where the interplay of the identified trapping mechanisms (structural and stratigraphic, residual, solubility, and mineral trapping) helps contain the injected CO_2 beneath thick and impermeable caprock formations, potentially for millions of years. However, in short-term and immediate applications, residual and solubility trapping mechanisms are the ones that can be capitalized. Therefore, regulatory and monitoring frameworks can be designed to make sure that these two mechanisms are taking place as effectively as possible. For a CO_2-EOR process, the two mechanisms interactively secure CO_2 storage and contribute to incremental hydrocarbon recovery.

Laboratory and field studies have established that CO_2 can be an efficient agent featuring different mechanisms by which it can displace oil from porous media, including oil swelling, interfacial tension, and viscosity reduction, increasing the injectivity index due to solubility of CO_2 in water and subsequent reaction of carbonic acid with minerals. An underlying physical process for these mechanisms of oil recovery is molecular diffusion. Molecular (interphase and intraphase) diffusion is responsible for the mixing of CO_2 into oil at the pore level through a rate-controlling mechanism that governs the gas-oil miscibility. Under these conditions, the first-contact or dynamic miscibility is achieved between oil and CO_2 phases and the oil unswept and by-passed in secondary oil recovery stage by water flooding becomes mobilized. In secondary recovery, the molecular diffusion is responsible for multiple contact miscibility achieved through a vaporizing gas drive mechanism (where gas vaporizes intermediate components from oil and becomes oil-like) and condensing gas drive mechanism (where rich-gas intermediate components condensate into in-place oil and oil becomes gas-like). In tertiary recovery, the molecular diffusion leads to mobilization of waterflood residual oil by swelling of residual oil blobs when CO_2 diffuses through a blocking water phase. For a wide range of conventional and unconventional reservoirs, e.g., heavy oil extraction in VAPEX [18] or oil extraction by solution-gas drive [19], diffusion can act as an important transport process.

In fractured reservoirs, the dispersive and segregated flux through fractures tends to accentuate compositional differences between matrix and fracture hydrocarbons [20] and as a result the incremental oil recovery from CO_2 injection processes in fractured reservoir is more influenced by molecular diffusion. As oil remains in matrix blocks in fractured reservoir after primary recovery, the gravity drainage mechanism provides initial recovery of oil. The density difference between the gas in the fracture and the oil in the matrix results in oil production until gravitational forces are equalized by capillary forces. In low-permeability matrix the dominant mechanism is molecular diffusion of oil and gas. In small size matrix blocks and high capillary pressure, gravity drainage is very low or ineffective. Injection of dry gas causes mass transfer between the gas in the fracture and the gas/oil system saturating the matrix blocks [21]. The process leads to horizontal movement of CO_2 in addition to gravitational drainage. When the rate of oil recovery during secondary or tertiary oil displacement by injection gas is significantly affected by diffusion, multicomponent molecular diffusion coefficients are important parameters to be determined. Compositional variations in CO_2-oil interactions lead to compositionally dependent diffusion coefficients that will result into significantly different prediction of EOR efficiency when compared to the results derived by constant (compositionally invariant) diffusion coefficients [22].

16.3.2 Enhanced coalbed methane recovery

Coal seams that are considered unmineable (typically too deep, too thin, or lacking the internal continuity to be economically mined with technologies) may have the potential for CO_2 storage. Many power plants are located near coal seams, which would reduce the transportation costs [5]. Sequestration of carbon dioxide in coal seams can be considered as an efficient way of safely sequestering carbon dioxide in an adsorbed state that is expected to be stable for geologically significant periods [23]. In addition, the large ratio of surface area to volume in coal seams allows storage of up to seven times as much gas as in other natural reservoirs. The worldwide carbon dioxide sequestration potential has been estimated at 150 Gt of carbon dioxide [24].

Coal preferentially adsorbs CO_2 over CH_4 which is naturally found in coal seams. In other words, coal has higher affinity to carbon dioxide that is approximately 2–13 times that of methane. This property, known as adsorption trapping, is the basis for CO_2 storage in coal seams. Methane gas is typically recovered from coal seams by dewatering and depressurization (coalbed methane recovery or CBM) or by injection of N_2 (N_2-ECBM). However, a significant amount of methane can still remain trapped in coal that can be released by carbon dioxide. The process of injecting and storing CO_2 in unmineable coal seams to enhance methane recovery is called enhanced coalbed methane recovery (Fig. 16.2).

Coal is a sedimentary rock which is a mixture of organic (C_nH_n) and inorganic chemicals (ash/mineral matter) formed from gradual compaction and thermal alteration of plant matter [25]. The degree of alteration (or metamorphism) of coal that occurs as a coal matures from peat to anthracite is referred to as rank. The lowest rank of coal is lignite which is generally high in moisture content and volatile matter. As coal rank increases, the volatile matter and moisture content generally decreases. The carbon content of coal increases from subbituminous, bituminous, semianthracite to anthracite [26]. As the organic matter is gradually altered through coalification process, it is compacted to a great extent so that a 30-m thick sequence of lignite may yield just 1 m of anthracite [26]. During this process methane (CH_4) is produced and accumulates in the coal seam. The trapped methane in the coal seam is then absorbed in micropores on the coal surface or remains within the natural fracture system (cleats).

Several molecular structures of coal have been suggested in the literature. For instance, Marzec [27] suggested a two-phase structure for coal in which a small amount of components with low molecular weight is trapped within a three-dimensional macromolecular network with covalent cross-links. Jones et al. [28] suggested a three-dimensional structure of randomly oriented small aromatic clusters.

The porous structure of coal generally contains micropores and macropores [29]. Micropores as part of the coal matrix have a radius less than 2 nm (2×10^{-9} m) and

Fig. 16.2 Schematic diagram of carbon sequestration in coal seam and methane recovery.

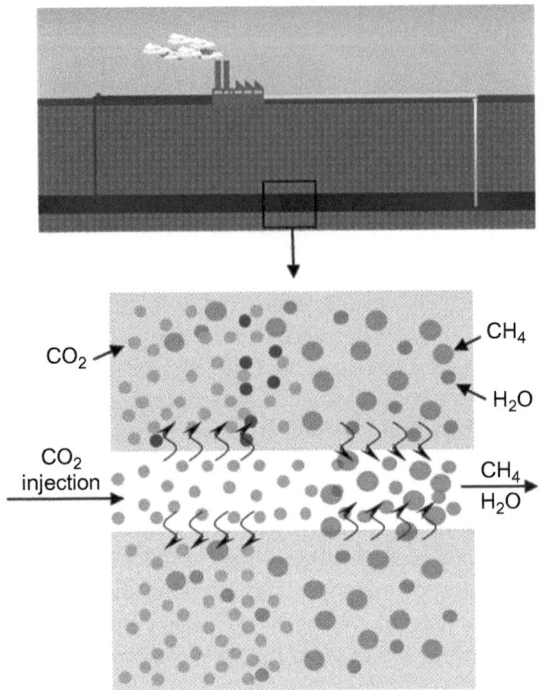

make up approximately 70% of its total porosity. Macropores with a radius greater than 50 nm consist of a naturally occurring network of fractures called the cleat system. The transitional pores or mesopores with radius between 2 and 50 nm do not generally appear in coal. Therefore, due to the presence of micropores and macropores, the coal is generally considered to be a dual porosity rock [29].

The gas transport in the coal seam usually includes the flow of gas through the naturally fractured porous network (cleats), diffusion into the organic coal matrix, and storage within the micropores in an adsorbed state. Nevertheless, there is a lack of understanding of what happens when the carbon dioxide is injected into a coal seam because the transport properties of the CO_2 are highly related to the chemical and physical changes that occur during the adsorption and desorption processes.

Transport in the coal usually includes the flow of gas through the naturally fractured porous network (cleats) and diffusion into the organic coal matrix. The storage includes the adsorption and desorption of gases on and from the coal matrix, as well as the stability of the adsorbed carbon dioxide on the coal. Since the transport properties of the coal are highly related to the chemical and physical changes that occur during the adsorption and desorption processes, there is a lack of understanding of what happens when the carbon dioxide is injected into a coal seam under in situ conditions. More importantly, there is a lack of understanding on the fate of the adsorbed gas, i.e., extent of carbon dioxide release due to pressure depletion in the reservoir [30]. The prediction of the adsorption capacity and the long-term stability of the sequestered gas require knowledge of how the gas is held in place and what factors might induce its release [5,31–33].

16.3.3 Saline formations

Saline formations or deep saline aquifers are layers of sedimentary porous and permeable rocks saturated with salty water (brine). These formations can be found in both onshore and offshore sedimentary basins. The saline formations have potential for the CO_2 storage. Trapping mechanisms include the CO_2 dissolving in the brine (solubility trapping), reacting chemically with the minerals and fluid to form solid carbonates (mineral trapping), or becoming trapped in the pore space (volumetric trapping). Carbon dioxide (density 0.6–0.7 $g\,cm^{-3}$) is lighter than saline water and a layer of carbon dioxide can be formed at the top of the formation layer. It is therefore important that a regionally extensive confining zone (often referred to as caprock or seal) overlies the porous rock layer; as a result the potential leakage is minimized. Saline formations are estimated to have much larger storage potential for CO_2 than oil and gas reservoirs and unmineable coals because they are more extensive and widespread [13].

Initially, the injected supercritical critical carbon dioxide displaces some of the saline pore water within the aquifer. It then begins to interact with water, aqueous chemicals, gas, and rock via a chain of geochemical reactions that leads to formation (precipitation) of new minerals. There are four main mechanism of carbon dioxide storage in saline aquifers: structural/stratigraphic, solubility, residual, and mineral trapping [11]. These mechanisms occur at different timescales. The most dominant storage/trapping mechanism is through structural and stratigraphic process by which the carbon dioxide is accumulated as a free phase below the cap rock. Carbon dioxide can be dissolved in aqueous solution: a solubility mechanism. In a longer geological timescale, geochemical reactions occur between the carbon dioxide, water, and minerals in the host rock that yields precipitation of carbonate phases that is considered as mineral trapping mechanism. In very arid regions, deep saline formations may be considered for future water desalinization. The Sleipner Project in the North Sea is the best available example of a CO_2 storage project in a saline formation.

16.4 Existing worldwide projects

The technology of carbon sequestering is still being developed. At present, there are only few fully integrated, commercial-scale CCS projects in operation and currently over 5 Mt CO_2 per year is stored from these plants [11]. Assuming worldwide CO_2 emissions from anthropogenic sources are approximately 30×10^9 t, this represents just 0.01% of total global emissions. As stated earlier, storage options such as EOR, EGR, and ECBM have the added benefit of producing commercially useful by-products such as oil and gas. This can have a significant benefit offsetting the significant capital and operating expenditure these plants incur over and above that provided by the carbon trading scheme. Following projects are some of the examples for worldwide commercial-scale CCS projects:

- The *Sleipner* project located in Norway: Capturing CO_2 from the natural gas produced at the Sleipner gas field and injecting it into a saline aquifer, located above the gas reservoir. The operation was started in 1996.
- The *Snøhvit (Snow White)* project located in Norway: The CO_2 is separated out from unprocessed gas and then returned via a separate pipeline for storage beneath the seabed.
- The *In Salah* project located in Algeria: Removing CO_2 from gas produced, followed by large-scale reinjection into an underground sandstone formation 2 km below the surface.
- The *Rangely* project located in North America: CO_2 from a natural gas processing plant in Wyoming is injected for EOR and storage at the Rangely field in Colorado.
- The *Weyburn-Midale* project in North America: Involves the capture of CO_2 from a coal-based synthetic fuels plant in North Dakota. The captured CO_2 is compressed and sent via pipeline to an oil field in Canada, where it is also used for EOR as well as sequestration.

16.5 Environmental aspects

The activities of carbon capture and sequestration may cause potential risk and impacts to the surrounding environment. Despite the fact that a key objective of CCS is to reduce the level of carbon dioxide emission to atmosphere, it has been identified that there are potential environmental footprints and impacts associated with it.

The environmental issues associated with the CCS can be summarized as:

(1) Impact on water resources

During the construction phase, potential discharges from pipeline and spillage that can happen accidentally are among the concerns with regard to the potential impact on surface water. Discharges of wastewater from capture plants are also considered as source of contamination. Some of the contamination sources such as the pollution from runoff are similar to those that can be found in other industrial activities. The leakage of carbon dioxide can induce surface water acidification.

The potential contamination of groundwater can be related to events that the spillage and leakage can enter the groundwater. The potential leakage of carbon dioxide from the well, fault, and caprock during the operational phase and after closure are mechanisms that can create contamination risk to the groundwater. Main pathways and mechanisms for carbon dioxide migrations to the groundwater include the cap rock and via capillary and diffusion processes and through the faults and fracture networks due to advection process.

Another source of contamination of water is related to the outflow water. When CO_2 is injected into the formation, the brine water that exists in the target formation of reservoir can be replaced. The outflow water produced can therefore contain chemicals that can be classified as toxic. Similar issue exists in CBM where a significant quantity of water may be produced. Data from the US state that water production can vary between 25 and 400 barrels per day per well [34]. The produced water cannot be immediately discharged or used due to high salinity and potential hazardous chemical and therefore it requires further treatment. The total dissolved solids (TDS) in coal reservoirs vary between 200 and 10,000 $mg\,L^{-1}$ [35]. A TDS level as high as

180,000 mg L^{-1} in coal seams has been reported [5]. Typically water produced from coalbed methane is contaminated with sodium, barium, bicarbonates, and iron. The concentration will vary based on coal seam depth, metabolism process, aquifer recharge, and other geological conditions. CBM-produced water is typically disposed through injection into subsurface aquifers, surface discharge, or surface ponds. Occasionally, CBM-produced water is reused for beneficial purposes such as irrigation and livestock watering. Water designated for reuse must meet requirements under several Federal and State regulations, including the Clean Water Act, the Safe Drinking Water Act, and the Resource Conservation and Recovery Act. Current opportunities for reuse of produced water are dependent on the quality, which varies widely, but often the high salinity precludes its use.

(2) Impact on air, ground, and climate

The local air quality can be affected through the construction and operational phase. This is expected due to production and suspension of dust, fuels used by machineries, and shipping processes. Release of carbon dioxide to soil due to spillage and accidental leakage can reduce soil pH and increase potential mobilization of heavy metals. Gas emission from capture plants can also be considered to have impact on air quality which depends on the technology used. Atmospheric leakage of carbon dioxide is another aspect that can occur after the completion of the injection process and in a longer term due to improper well sealing and plugging. The integrity of wellbore seal and caprock plays an important role in preventing potential leakage of carbon dioxide in long term.

(3) Induced seismicity

Potential short-term seismicity effects and long-term geomechanical impacts due to high volume injection of fluids into the deep subsurface have been discussed in the literature. The knowledge is mainly related to the experience gained in oil and gas industry. The shear failure of a fault is theoretically controlled by the changes in normal stress, pore pressure, and shear cohesion across the fault provided by the Coulomb theory. High volume/pressure fluid injection can affect the stress state of subsurface strata and affect the existing faults as the pore pressure changes. Induced seismicity has been reported in various subsurface activities including mine subsidence, oil and gas field depletion, fluid injection for secondary oil recovery, solution mining, deep drilling programs, wastewater disposal, enhanced geothermal systems, and reservoir impoundments [36]. Foulger et al. [37] have compiled 180 cases where fluid injections have been reported as the source of induced seismicity. There is also added risk associated with the leakage of the injected fluid (e.g., CO_2) due to potential caprock rupture [37]. Water injected for EOR has been reported to cause 38 cases of fluid injection-induced seismicity [37]. Potential seismicity induced by injection of water and carbon dioxide has been discussed in the literature [8]. The lower viscosity and bulk modules of carbon dioxide than water have been described as reasons that lower risk of induced seismicity should be expected compacted to water injection [38]. Foulger et al. [37] describe only two cases of induced seismicity among approximately 100 EOR injection sites where CO_2 has been used.

16.6 Summary

Different geological sequestration options have been suggested that have been subjected to a rigorous research and development. The CCS technologies can contribute as a global solution to the low carbon agenda and agreed reduction of carbon dioxide in various countries. Sequestration of carbon dioxide in depleted oil and gas reservoirs or unmineable coal seams can offer financial offsets through enhanced recovery of hydrocarbons, and the existence of transport and drilling infrastructure in such reservoirs can facilitate the process. The engineering developments of carbon sequestration require in-depth consideration of the complexity of the deep subsurface conditions and environments. As such the expertise and knowledge of applied earth scientists and geotechnical engineers play an important role in developing and implementing sustainable solutions for geological storage of carbon dioxide. In addition, considerations of environmental aspects and concerns related to implementation of carbon dioxide during the entire chain of capture, transport, and storage play an important role in successful development of technology.

Acknowledgment

This chapter is mainly based on the lectures given by Dr. Hadi Mosleh on Energy Geotechnics module for the new MSc in Advanced Geotechnical Engineering at the University of Surrey (United Kingdom).

References

[1] Berlin K, Sussman RM, Hendricks B. Global warming and the future of coal, the path to carbon capture and storage. Center for American Progress; 2007. p. 68.
[2] IEA. World energy outlook. International Energy Agency; 2010. p. 736, ISBN 978-92-64-08624-1.
[3] IPCC. Climate change 2014: mitigation of climate change. UK: Cambridge University Press; 2014.
[4] DECC. CCS roadmap: supporting deployment of carbon capture and storage in the UK. Department of Energy and Climate Change; 2012.
[5] White CM, Smith DH, Jones KL, Goodman AL, Jikich SA, LaCount RB, DuBose SB, Ozdemir E, Morsi BI, Schroeder KT. Sequestration of carbon dioxide in coal with enhanced coalbed methane recovery—a review. Energy Fuel 2005;19:659–724.
[6] Bachu S. Review of CO_2 storage efficiency in deep saline aquifers. Int J Greenhouse Gas Control 2015;40:188–202.
[7] Shukla R, Ranjit P, Haque A, Choi X. A review of studies on CO_2 geological sequestration and caprock integroity. Fuel 2010;89:2651–64.
[8] Aminu MD, Nabavi SA, Rochelle CA, Manovic V. A review of developments in carbon dioxide storage. Appl Energy 2017;208:1389–419.
[9] Hester RE, Harrison RM. Carbon capture; sequestration and storage, issues in environmental science and technology. RSC Publishing; 2010.
[10] Forbes S, Verma P, Curry TE, Friedmann SJ, Wade SM. CCS guidelines: guidelines for carbon dioxide capture, transport and storage. Washington, DC: World Resources Institute (WRI); 2008.

[11] IPCC. In: Metz B, Davidson O, de Coninck H, Loos M, Meyer L, editors. Carbon dioxide capture and storage. UK: Cambridge University Press; 2005. p. 431.
[12] Bachu S, Shaw J, Pearson R. Estimation of oil recovery and CO_2 storage capacity in CO_2 EOR incorporating the effect of underlying aquifers. The SPE/DOE symposium on improved oil recovery; 2004.
[13] Carbon utilisation and storage atlas. 4th ed. US Department of Energy (NETL); 2012.
[14] Hadlow RE. Update of industry experience with CO_2 injection. SPE annual technical conference and exhibition; 1992.
[15] Stephenson DJ, Graham AG, Luhning RW. Mobility control experience in the Joffre Viking miscible CO_2 flood. SPE Reserv Eng 1993;8(03):183–8.
[16] Holt T, Jensen J-I, Lindeberg E. Underground storage of CO_2 in aquifers and oil reservoirs. Energy Convers Manag 1995;36(6):535–8.
[17] Duda JR. Carbon dioxide—enhanced oil recovery, untapped domestic energy supply and long term carbon storage solution. National Energy Technology Lanboratory (NETL); 2010. p. 32.
[18] Yang C, Gu Y. A novel experimental technique for studying solvent mass transfer and oil swelling effect in the vapour extraction (VAPEX) process. Canadian international petroleum conference; 2005.
[19] Li X, Yortsos YC. Theory of multiple bubble growth in porous media by solute diffusion. Chem Eng Sci 1995;50(8):1247–71.
[20] Da Silva FV, Belery P. Molecular diffusion in naturally fractured reservoirs: a decisive recovery mechanism. SPE annual technical conference and exhibition; 1989.
[21] Kazemi A, Jamialahmadi M. The effect of oil and gas molecular diffusion in production of fractured reservoir during gravity drainage mechanism by CO_2 injection. EUROPEC/EAGE conference and exhibition; 2009.
[22] Babaei M, Mu J, Masters AJ. Impact of variation in multicomponent diffusion coefficients and salinity in CO_2-EOR: a numerical study using molecular dynamics simulation. J Pet Sci Eng 2017;162:685–96.
[23] Smith DH, Sams WN, Jikich S, Ertekin T. Simulating carbon dioxide sequestration/ECBM production in coal seams: effects of permeability anisotropies and the diffusion-time constant. SPE annual technical conference and exhibition, Denver; 2002. p. 156–63.
[24] Gale J, Freund P. Coal-bed methane enhancement with CO_2 sequestration worldwide potential. Environ Geosci 2001;8(3):210–7.
[25] Thomas L. Coal geology. Dargo Associates Ltd: Wiley Publishing; 2002.
[26] Speight JG. Handbook of coal analysis. Hoboken, NJ: John Wiley & Sons; 2005.
[27] Marzec A. Macromolecular and molecular model of coal structure. Fuel Process Technol 1986;14(1):39–46.
[28] Jones JM, Pourkashanian M, Rena CD, Williams A. Modelling the relationship of coal structure to char porosity. Fuel 1999;78(4):1737–44.
[29] Flores RM. Coal and coalbed gas: Fueling the future. Waltham: Elsevier; 2014.
[30] Hadi Mosleh M, Turner M, Sedighi M, Vardon PJ. Carbon dioxide flow and interactions in a high rank coal: permeability evolution and reversibility of reactive processes. Int J Greenhouse Gas Control 2018;70(1):57–67.
[31] Ozdemir E, Morsi BI, Schroeder K. Importance of volume effects to adsorption isotherms of carbon dioxide on coals. Langmuir 2003;19(23):9764–73.
[32] Hadi Mosleh M, Sedighi M, Vardon PJ, Turner M. Efficiency of carbon dioxide storage and enhanced methane recovery in a high rank coal. Energy Fuel 2017;31(12):13892–900.
[33] Hadi Mosleh M, Turner M, Sedighi M, Vardon PJ. High pressure gas flow, storage and displacement in fractured rock—experimental setup development and application. Rev Sci Instrum 2017;88(1):1–14. 015108.

[34] Hightower M. Managing coalbed methane produced water for beneficial uses, initially using the San Juan and Raton basins as a model. A presentation by the Sandia National Laboratories, http://wrri.nmsu.edu/conf/forum/CBM.pdf; 2003.
[35] Taulis M, Milke M. Chemical variability of groundwater samples collected from a coal seam gas exploration well, Maramarua, New Zealand. Water Res 2013;47:1021–34.
[36] Davies R, Foulger G, Bindley A, Styles P. Induced seismicity and hydraulic fracturing for the recovery of hydrocarbons. Mar Pet Geol 2013;45:171–85.
[37] Foulger GR, Wilsona MP, Gluyasa JG, Juliana BR, Davies RJ. Global review of human-induced earthquakes. Earth Sci Rev 2018;178:438–514.
[38] Verdon JP. Significance for secure CO_2 storage of earthquakes induced by fluid injection. Environ Res Lett 2014;9:64022.

Section D

Reducing CO_2: Industry, Farming and Improved Efficiency

Polymers from plants: Biomass fixed carbon dioxide as a resource

17

Janet L. Scott[1], *Antoine Buchard*
Department of Chemistry and Centre for Sustainable Chemical Technologies, University of Bath, Bath, United Kingdom
[1]Corresponding author: j.l.scott@bath.ac.uk

Chapter Outline

17.1 Introduction

17.1.1 The CO_2 balance

While there are a large number of greenhouse gases, GHGs, in the Earth's atmosphere, carbon dioxide has gained most attention, largely due to the relatively high concentration and rate of increase that is attributed to human activities. Prior to the rapid industrialization that commenced in the 18th century, CO_2 levels were 280ppm and had not changed significantly for millennia [1]; however, it is predicted that, post 2016, CO_2 levels will not drop below 400 ppm for the foreseeable future [2]. Significant anthropogenically generated CO_2 arises from the burning of fossil fuels

Managing Global Warming. https://doi.org/10.1016/B978-0-12-814104-5.00017-X

[3], but the conversion of fossil carbon into chemicals and materials, which are later allowed to degrade, is another source of "released" carbon, albeit a minor one that "disappears into the noise" of CO_2 measurements. Nonetheless, efforts to move chemical and plastics production away from reliance on fossil carbon resources and toward the use of renewable feedstocks (and energy) continue apace.

17.1.2 Scope

After fuels of various forms, polymer production constitutes the largest sector of the chemical industry that uses fossil carbon as a feedstock [4]. Producing chemicals and polymers from renewable plant-derived feedstocks offers an opportunity to reduce reliance on fossil carbon sources and to mitigate some of the CO_2 released as a consequence. Here we focus on polymers derived from plants: (synthetic) polymers produced from monomers prepared from plant biomass, and (natural) (bio)polymers obtained directly from plants in large volume, such as cellulose. Industrially produced, biomass-derived polymers predated oil-based polymers, so there is a very large body of knowledge stretching back some decades, but here only newer developments with the potential to have a measurable impact on the use of fossil carbon resources, such as oil or gas, have been selected. In addition, CO_2 as a feedstock is considered and monomers produced by the action of micro-organisms, by "white biotechnology," are included.

17.2 Monomers and polymers from plant biomass

17.2.1 The need for sustainable synthetic polymers

In our modern societies, synthetic polymers (the main component of plastic materials) have replaced many traditional materials such as wood, stone, metal, and glass. Today, polymers are used everywhere: they protect food and prevent spoilage, insulate electric cables, save fuel by making cars lighter yet safer, are used as fabric for clothing, in contact lenses, etc. Many promising new technologies also rely on sophisticated polymers, for example solar cells and artificial organs and limbs. In 2016, polymer production exceeded 320×10^6 t (320 million metric tons) worldwide [5], and their demand is steadily increasing. To enable a sustainable society that promotes and maintains our current living standards, polymers are needed. Yet polymers can also create serious environmental challenges. Currently, more than 99% of polymers come from depleting oil resources [6], and by 2050 the plastics industry will account for 20% of the total oil consumed annually [7]. Furthermore, as these polymers are made to be durable in use, they are also durable in waste. This remains a challenge even if more and more plastics are recycled or incinerated to recover some energy (but then releasing CO_2). As a specific example, currently 72% of all plastic packaging waste generated ends up in landfills (40%) or escape the collection system (32%), and will persist for centuries, releasing pollutants and harming wildlife. It has been estimated that, as of 2015, out of the 6300 Mt of plastic waste that have been generated so far by mankind, around 9% have been recycled, 12% have been incinerated, and 79% has

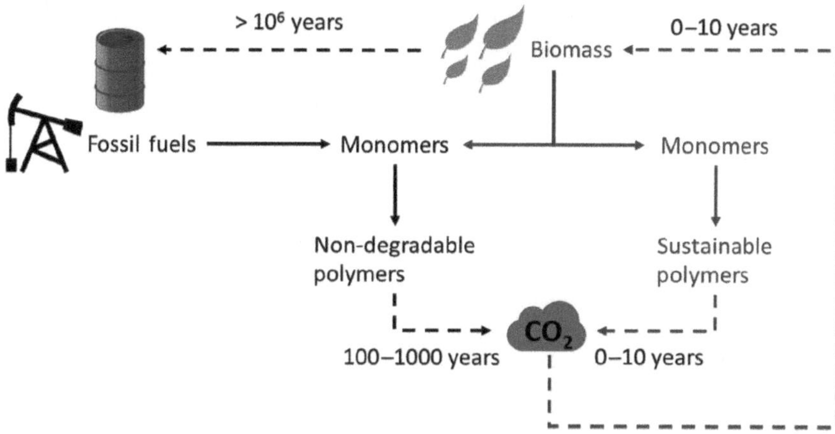

Fig. 17.1 Main strategies for the production of monomers and polymers from biomass (plain lines represent chemical processes, dashed lines represent natural biological processes).

accumulated in landfills or the environment [8]. Moving away from petrochemical feedstocks toward renewable resources, in particular from plants, can address some of these environmental challenges and mitigate the influence of plastics production, use and end of life, on climate change.

From a chemist's perspective, plant-derived macromolecules (cellulose, lignin, and other polysaccharides) and small molecules (sugars, vegetable oils, and terpenes) are an attractive, alternative abundant supply of annually renewable carbon for the preparation of synthetic polymers with desired properties [9]. Two main strategies exist, Fig. 17.1.

The first strategy involves the production from biobased resources of the monomers that currently dominate the commodity petrochemical polymer industry. This approach however does not address the end-of-life requirements of a circular economy. The second strategy involves the conversion of renewable feedstocks into completely novel polymers that exhibit useful properties, including enhanced (bio)-degradability [10–12]. For both strategies, the main challenge is to develop efficient transformations of these abundant resources into useful compounds that are also commercially competitive.

17.2.2 Monomers from vegetable oils

Vegetal oils are important renewable raw materials of the chemical industry, due to their main applications as surfactants, lubricants, and coatings. According to the United States Department of Agriculture, the annual production of major vegetable oils was 188×10^6 t in 2016/2017, which included, in order of importance: palm, soybean, rapeseed, sunflower seed, kernel, peanut, cottonseed, coconut, and olive oils [13]. The main structural components of these oils are triglycerides, triesters of glycerol with fatty acids, classified as saturated, monounsaturated, and polyunsaturated depending on the number of alkene functionalities they contain. Some fatty acid moieties such as ricinoleic acid (from castor oil) and vernolic acids (from vernonia oil) contain additional functional groups (epoxide and alcohol, respectively), but their production remains minor. One challenge for chemistry focused on the valorization of these feedstocks is that the chemical compositions of triglycerides may vary both between crops and within a particular crop.

Historically, triglycerides have been used directly to make materials. For example, linoleum, a resin used in flooring, is traditionally made from linseed oil by crosslinking Diels-Alder reactions at high temperature. Vegetable oils are however generally processed by transesterification to produce fatty esters and glycerol, and both components have been used to produce renewable monomers.

Glycerol (also a major by-product of the biodiesel industry) can be used as a crosslinking agent in resin production, or as a raw material for the production of monomers such as epichlorohydrin [14], lactic acid [15] (then lactide), or dihydroxyacetone cyclic carbonate. The latter resulting semicrystalline polycarbonate is a high-melting thermoplastic with a melting point of 246°C that is structurally very similar to Carilon, a petroleum-derived alternating copolymer of ethylene and carbon monoxide [16].

The transformation of fatty esters into renewable monomers has been reviewed extensively recently [17,18]. Common methods exploit the fatty ester alkene groups in thiol-ene reactions, acyclic diene metathesis, epoxidation, and radical or thermal crosslinking reactions. A strategy with considerable potential and growing interest is the use of catalysis to isomerize the internal alkene group to the chain end to produce long aliphatic chain α,ω-diesters or α,ω-diols by alkoxycarbonylation processes [19]. These monomers can then undergo conventional condensation polymerizations and yield bioderived polyesters with properties that are intermediate between those of nonpolar long-chain polyalkenes and more polar short-chain polyesters [20]. Finally, if α,ω-diamines are used as the comonomers, nylons can be produced. Due to the presence of a secondary alcohol group, nylon 11; nylon 6,10; and nylon 4,10 can also be made from ricinoleic acid from castor oil (which is however twice as expensive as more abundant oils). Strategies for the valorization of vegetable oil triglycerides into monomers for polymer production are summarized in Fig. 17.2.

17.2.3 Monomers from lignin

Lignin is one of the three main components of lignocellulosic feedstock (with hemicellulose and cellulose). Structurally, it consists of a crosslinked polymer network of phenolic rings and aliphatic side chains, which provide rigidity and strength to plant cell walls (particularly in woody biomass). Although the structure of lignin varies depending on the source, it is a highly abundant, and a potential renewable source of aromatic compounds such as vanillin, ferulic acid, eugenol, creosol, and other sinapyl alcohol derivatives, which can be used directly or transformed into monomers. These aromatic structures are desirable in monomer and polymer synthesis to impart rigidity, crystallinity, and thermal resistance [21]. For example, Gatoresin® is a commercial renewable copolymer of hydroferulic and ferulic acids that has been developed by Miller and coworkers [22]. Lignin-based vanillin and acetic anhydride can

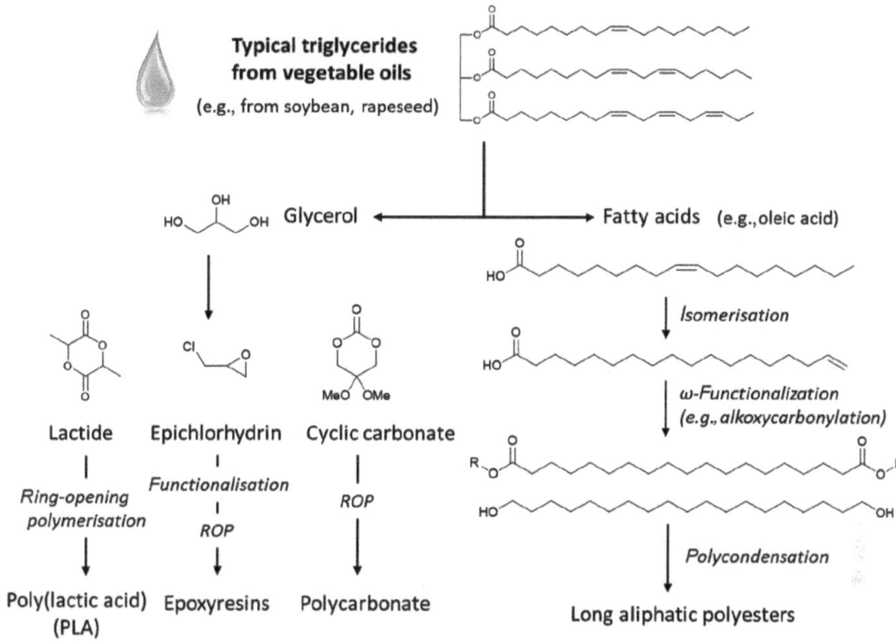

Fig. 17.2 Selected strategies for the valorization of triglycerides from vegetable oils into monomers

be reacted to afford acetylferulic acid, then hydrogenated to produce acetyldihydroferulic acid. Their copolymerization then yields poly(ferulic-*co*-dihydroferulic acid), which exhibits thermal properties and functionality similar to those of polyethylene terephthalate (PET) [23]. Miller's team have also developed polyethylene ferulate and polyethylene coumarate from ferulic acid and *p*-coumaric acid, as well as the related monomers obtained by hydrogenation of the main-chain double bond. By controlling the comonomer feed ratios, materials can be obtained with thermal properties in the range of polystyrene [24]. However, despite numerous lignin valorization processes being developed and studied (Fig. 17.3), a major challenge in the field remains to establish a cost-effective, large-scale process for the production of such monomers and many would argue that direct use of lignin as a component of copolymers is a more viable route for converting plant fixed carbon dioxide into polymer-based materials. This is discussed in greater detail in Section 17.3.2.

17.2.4 Monomers from terpenes

Terpenes are a vast family of naturally occurring hydrocarbon compounds, which are produced by a variety of plants from isoprene building blocks, and which display enormous structural diversity. Terpenes are abundant, inexpensive, and do not directly compete with food sources. Turpentine is for example the volatile fraction isolated from pine resin during the kraft process (paper manufacturing), and is produced on

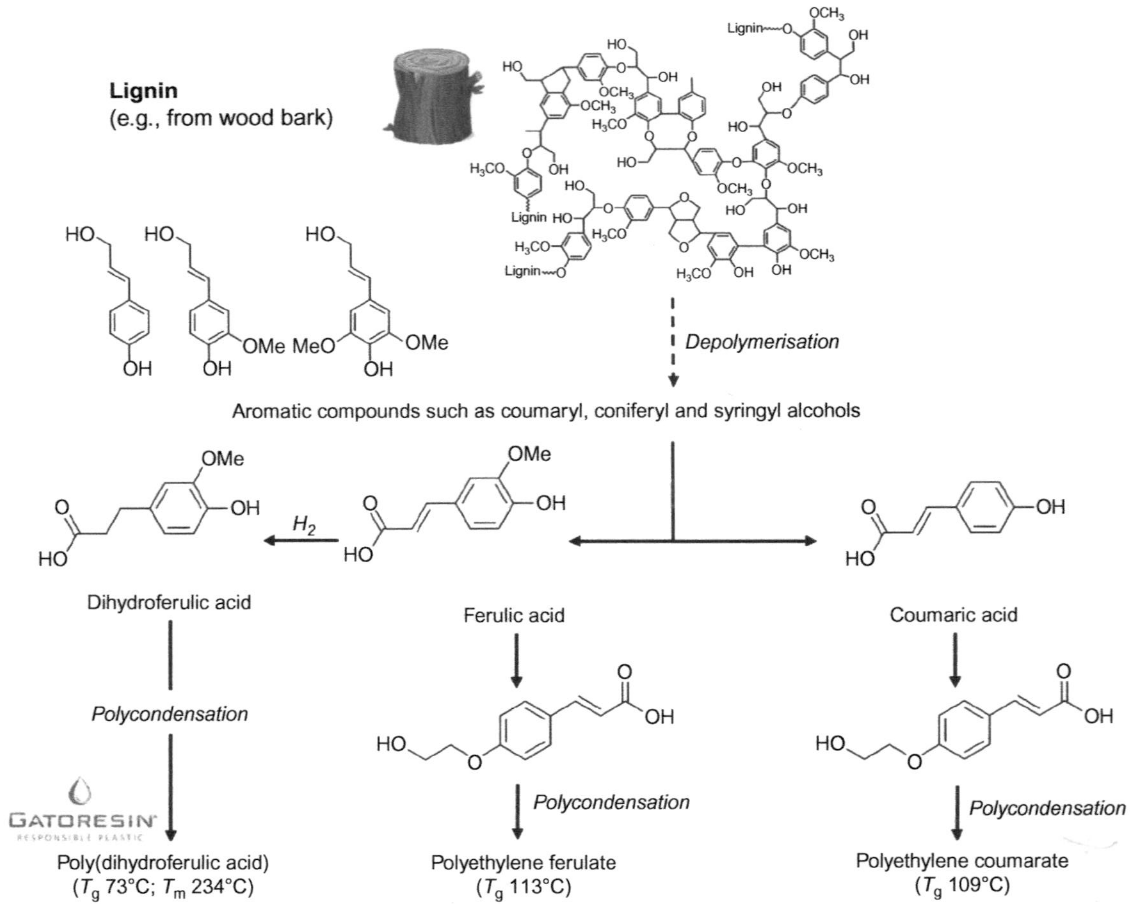

Fig. 17.3 Selected strategies for the valorization of aromatic precursors from lignin into monomers. PDFA exhibits thermal properties functionally similar to those of polyethylene terephthalate (PET), while both polyethylene hydroxycinnamate can be copolymerized with their hydrogenated analogues to match polystyrene thermal properties.

a scale of 330kt per annum. Despite the vast range of terpenes available, α-pinene, β-pinene, myrcene and limonene are the most widely studied as monomers for production of polymers, because of their abundance, low cost, and ease of isolation. Terpenoids, such as menthol or carvone, which feature additional functional groups, have also been used in polymer synthesis [25–27].

Due to the presence of several alkene groups, the first approach has been to use radical or cationic polymerization to produce terpene homo and copolymers (for example with styrene). α-Pinene, despite being the most abundant component of turpentine, has however not been polymerized so far because of its lack of a reactive exocyclic double bond. Conversely, high-molecular-weight homopolymers of β-pinene (which display a glass transition temperature of around 85°C) can be produced, including using commercial metal-based commercial catalyst systems (e.g., Schiff-base nickel catalyst with methylaluminoxane initiator) [28]. The resulting polymers possess a mechanical strength comparable to commercial polyolefins and still behave as a flexible plastic material. The hydrogenation of poly(β-pinene) can further improve thermal resistance, durability, and optical properties [29]. Recently, a study also reported the radical homopolymerization of limonene, albeit in low conversion. The resultant poly(limonene) had a glass transition temperature, T_g, of 116°C [30]. Myrcene, which has three double bonds offering multiple polymerization sites, has only recently been polymerized in a controlled way; using reversible addition-fragmentation chain-transfer (RAFT) polymerization, yielding a polymer at 96% 1,4-myrcene, with T_g of −60°C [31]. Hillmyer and coworkers chose a different strategy to produce a polymer from myrcene, and developed a cyclic conjugated monomer, 3-methylenecyclopentene, from myrcene via ring-closing metathesis, which yielded a 1,4-regiocontrolled polymer via cationic polymerization [32]. In general, the majority of polymers obtained by direct polymerization of terpenes have not exhibited properties that could allow these to compete with current commercially produced plastics.

Another strategy has therefore been to functionalize terpenes (or terpenoids) into monomers that can produce more competitive polymers, Fig. 17.4. For example, the epoxidation of the internal double bond of limonene has attracted a lot of attention to produce renewable linear polycarbonates by copolymerization with carbon dioxide (another abundant renewable feedstock) [33], catalyzed by zinc [34] or aluminum complexes [35]. The resulting material has attractive thermal (glass transition temperature, $T_g = 130$°C) and optical properties (transmission 94%, haze 0.75%) and scale-up has been realized on the kilogram scale [36]. Coates and coworkers have further shown that a stereocomplexed, semicrystalline polymer with enhanced mechanical properties can be prepared by the cocrystallization of the two possible enantiomerically pure polymers [37]. Poly(limonene dicarbonate) has also been developed from commercially available mixture of *cis/trans* (+)-limonene oxide and CO_2, with tuneable T_g values of up to 180°C [38]. Since this exemplary work on limonene oxide(s), a general research trend has been to focus on epoxide starting materials that can be generated from renewable feedstocks. Limonene can also be transformed into commodity monomer terephthalic acid [39], which, when combined with ethylene glycol from bioethanol (*vide infra*), can be used for the synthesis of fully biobased poly(ethylene terephthalate) (PET) [40]. However, the synthetic route

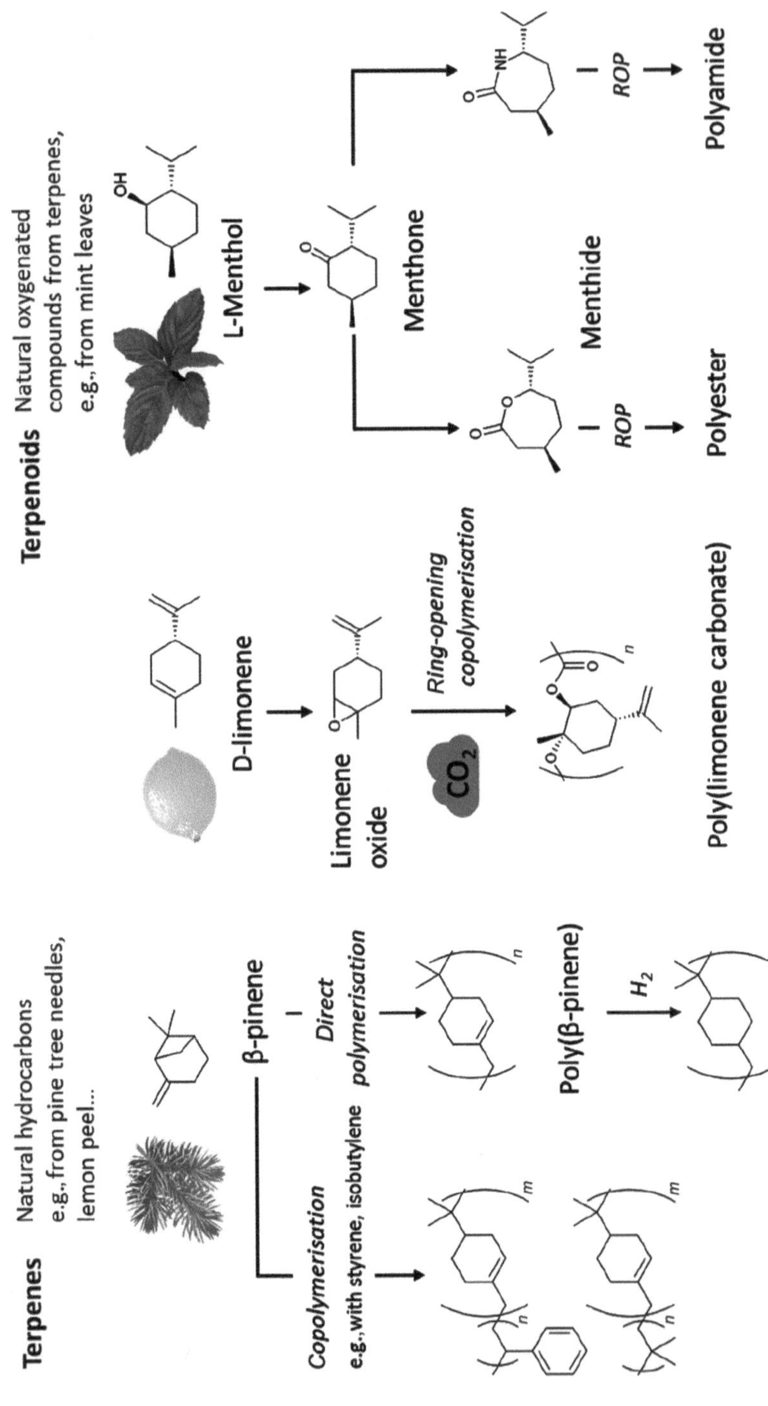

Fig. 17.4 Selected strategies for the valorization of terpenes and terpenoids from plants into monomers and polymers.

reported is not so far environment friendly or commercially viable, nor does it address the end-of-life issues associated with PET. One last example worth highlighting is the work done on the use of menthol, a commercial terpenoid, to produce a renewable aliphatic polyester, via the ring-opening polymerization of a lactone monomer (menthide). The resulting poly(menthide) displays a low glass transition temperature and was used as a midblock for the synthesis of degradable thermoplastic elastomers with outstanding performances [41]. Using a Beckman rearrangement, the lactam analogue of menthide has also been synthesized and the resulting poly(amides), although at a very early stage, exhibit high melting crystalline domains ($T_m = 300°C$), making them promising for biobased versions of Nylon-6 [42]. However, in addition to a costly and polluting Baeyer-Villiger oxidation step, the cost of starting menthol has been prohibitively expensive (about $20 per kilogram) for poly(menthide) to penetrate the commercial elastomer market [10].

17.2.5 Monomers from sugars

Among renewable resources, carbohydrates are particularly promising raw materials as they are widely available (more than 150×10^9 t of polysaccharides are produced naturally per year), environmentally benign, and they are likely to impart biocompatibility and degradability properties to polymers due to their high oxygen content [43,44]. Glucose, the most abundant monosaccharide, is currently obtained by saccharification of starch or natural sucrose hydrolysis, but could also be commercially obtained in the future from lignocellulosic sources of lesser value, without competing with food crops. In general, polysaccharides can be converted using chemical or enzymatic processes into monosaccharides (pentoses or hexoses), and into unsaturated carboxylic acids, polyols, and furan derivatives, which can be manipulated further to afford versatile commodity monomers [45], Fig. 17.5.

As previously mentioned, one popular industrial approach has been the synthesis of traditional monomers and platform chemicals from renewable resources. One main advantage is that these bioreplacement strategies are compatible with current infrastructure and markets, provided similar purity and costs can be achieved. Among these processes, the now well-established fermentation of sugars into bioethanol, and subsequent dehydration into ethylene, is a promising route to polymers derived from plants [46]. Ethylene can be polymerized directly to produce polyethylene or can be used as a precursor for the synthesis of other monomers including ethylene oxide, tetrafluorethylene, ethylene glycol, and vinyl acetate. In particular, bioethylene glycol produced in this fashion is already used industrially in the synthesis of partially biobased PET, popularized as Coca-Cola's PlantBottle® [39].

Glucose can also be readily transformed into other building blocks such as lactic acid or succinic acid, which are polymerized directly or reacted further to produce useful monomers.

The production of succinic acid by microorganisms (up to 1 mole per mole of glucose) is already operated on an industrial scale in Europe by Succinity, a joint venture between Corbion-Purac and BASF established in 2013. Succinic acid can

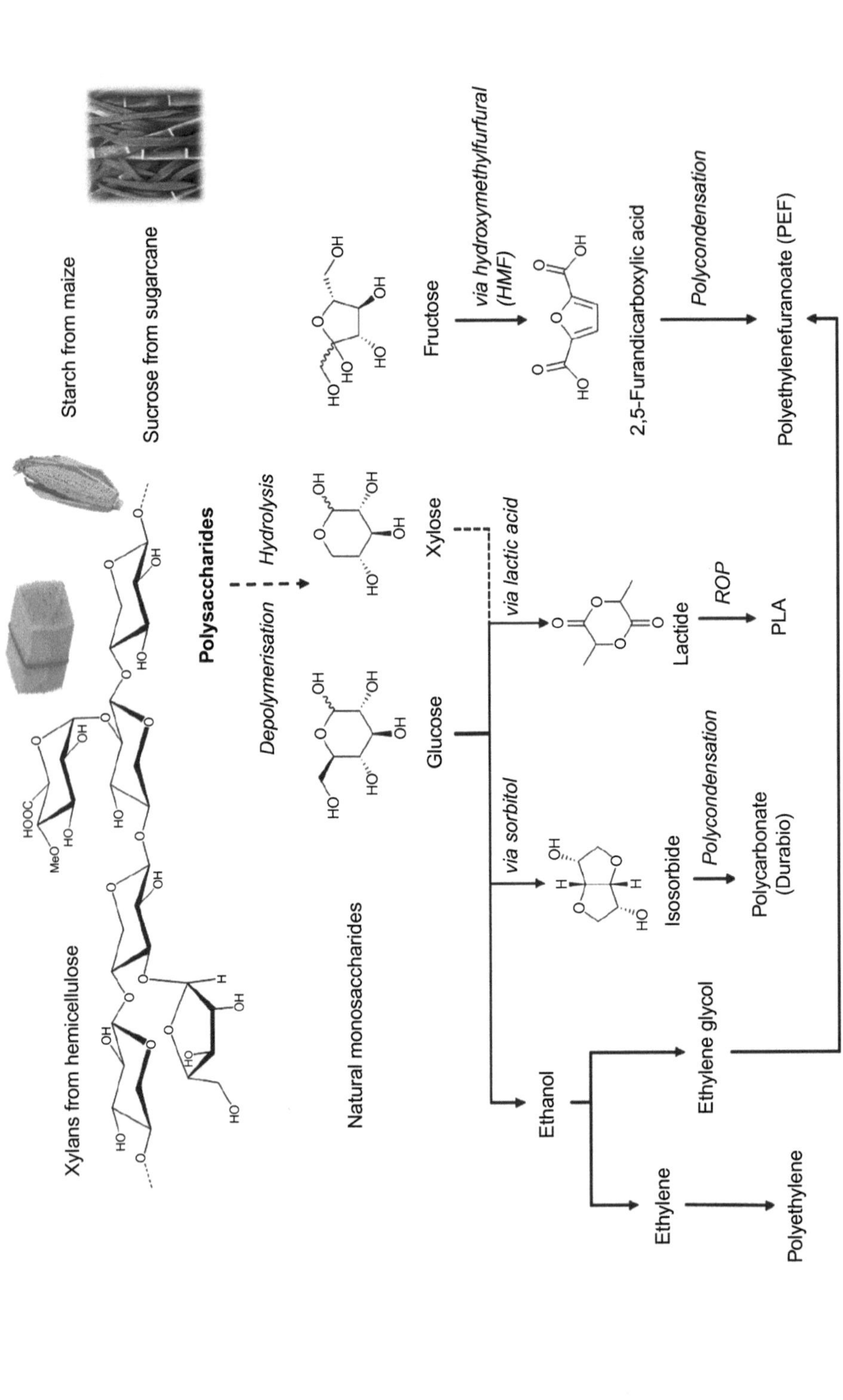

Fig. 17.5 Selected strategies for the valorization of carbohydrates from plants into monomers and polymers.

then react by polycondensation with bioderived 1,4-butanediol (itself obtained by reduction of succinic acid) to produce polybutylene succinate (PBS). PBS is a semicrystalline biodegradable polymer, which is produced commercially on a scale of around 40 kt per year, and used as a barrier in packaging and in blends [47]. Another success story is that of poly(lactic acid) (PLA), which is produced by the ring-opening polymerization of lactide, the dimeric lactone obtained from lactic acid. PLA, despite a modest T_g of 55°C that can limit its application at higher temperatures, can replace petrochemical-derived plastics in some types of packaging and fibers [48,49]. Furthermore, PLA can be mechanically recycled, and is even compostable at high temperatures, degrading back to lactic acid, which can be metabolized naturally. Lactic acid is already produced on a scale of around 400 kt annually, including by companies such as Galactic, Natureworks, and Total-Corbion, from the microbial fermentation of starch-rich crops such as maize, or sugar cane. Recently, the use of cheaper substrates such as glycerol [15] and lignocellulosic feedstocks has been reported and may improve further economic viability, as well as alleviate some of the current environmental impacts of PLA production, notably the use of water and fertilizers and competition with food crops. For example, the Xylex technology allows the transformation of waste streams from the forest products industry into valuable streams rich in C5 sugars (while removing and isolating lignin) [50], which can then be converted them into racemic lactic acid via the Versalac chemical process [51]. Both technologies have been developed by Plaxica and are now propriety of Sappi Limited [52]. 1,4:3, 6-Dianhydrohexitols (isosorbide, isomannide, and isoidide) are bicyclic diols, which have also been investigated as renewable diols for the synthesis of polymer (polyester and polycarbonate) with high T_g [53]. These diols, which can be obtained by dehydration of reduced sugars such as sorbitol or mannitol, are attractive due to their rigidity, chirality, and nontoxicity. Mitsubishi's Durabio™ polycarbonate is thus a commercial renewable alternative to Bis-Phenol A-based Polycarbonate, made from isosorbide [54].

Finally, an area of growing industrial interest is the production of furan derivatives from sugar crops (via a fructose intermediate), which has been the subject of various recent reviews. It is worth highlighting that such processes are currently moving toward industrialization. Avantium has thus developed a catalytic technology, named YXY, to convert fructose into various furanic compounds, in particular 2,5-furandicarboxylic acid (FDCA), which has been working on a pilot plant scale since 2011 [55]. The polycondensation of FDCA with ethylene glycol produces poly(ethylene furanoate), a biobased replacement for commodity PET, which can be recycled in the same way, and have furthermore superior characteristics, including barrier (oxygen, CO_2, and water) and thermal properties [56]. When combined with biobased ethylene glycol, PEF is a fully bioderived polymer and a suitable material for the packaging of drinks, food and nonfood products. In 2016, a joint venture (Synvina) was formed between BASF and Avantium to bring renewable FDCA and PEF and its applications to the market.

17.3 Polymers from plants

Natural polymers produced by plants are a carbon dioxide sink and are biodegradable—these form part of the natural CO_2 cycle, thus "borrowing" nature's polymers for a period to reduce use of polymers derived from fossil carbon (e.g., oil, coal) is a valid carbon dioxide mitigation strategy. As most natural polymers thus employed are returned to the biological nutrient cycle post use, the gains arise from offsetting emissions that would arise from the use of fossil carbon sources in similar biodegradable applications (which points to a conundrum: using fossil carbon resources to prepare biodegradable polymers provides yet another route for converting fixed fossil carbon into emitted CO_2).

17.3.1 Polysaccharides

Natural polymers comprising sugar units, polysaccharides, are produced in Nature as a form of energy storage, e.g., glycogen or starch and as the basis of structural materials, e.g., cellulose Fig. 17.6. The former serve as foodstuffs for human and animals, so converting these into materials to replace fossil-carbon-derived plastics or chemicals can result in competition with food production. The structural polysaccharides, however, may be obtained from nonfood crops, or even "waste" sources. Thus, use of cellulose from agricultural plant residues, or chitin (and chitosan) from crustacean shell waste or fungi [57] represents potentially sustainable sources of polymers for replacement of synthetic polymers in some applications. While crustacean shells are clearly not obtained from plants, fungi are amenable to growth in bioreactors [58], although there is some evidence to suggest that mycorrhizal fungi can contribute to *release* of CO_2 when growth is enhanced in higher CO_2 environments [59].

One of the challenges associated with replacing fossil-oil-derived plastics such as poly(ethylene), poly(propylene), poly(styrene), or poly(ethylene terephthalate), used widely in packaging with plant-derived polysaccharide materials is that the former are thermoplastic and thus amenable to thermoforming processes such as extrusion, injection-molding or blow-molding, while native starch, cellulose, and chitin cannot be readily thus formed. Starch may be rendered moldable by the addition of

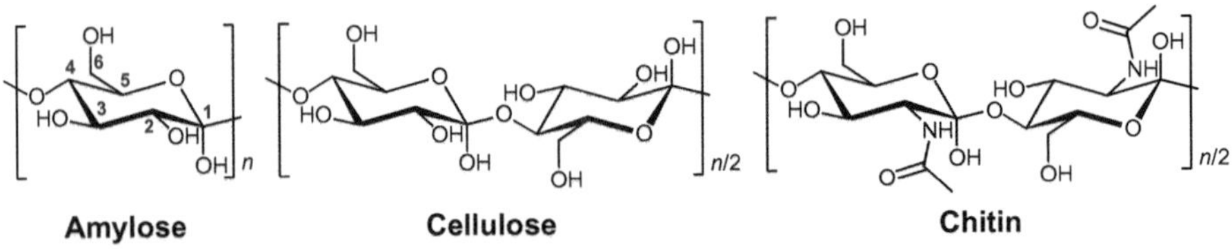

Fig. 17.6 Structures of some natural polysaccharides: amylose from starch, cellulose, and chitin. Starch comprises linear $\alpha(1 \rightarrow 4)$-linked glucose units, while branched amylopectin has additional chains grafted onto the amylose backbone by $\alpha(1 \rightarrow 6)$-glycosidic linkages. Glycogen is similar, but more highly branched. Cellulose and chitin are linear polymers with $\beta(1 \rightarrow 4)$-glycosidic linkages and deacetylation of chitin yields (dilute acid) soluble chitosan.

Fig. 17.7 Foamed regenerated cellulose (*left and middle*) prepared by phase inversion of cellulose and 1-butyl-3-methylimidazolium-2-carboxylate in ionic liquid, 1-butyl-3-methylimidazolium chloride, by the action of dilute hydrochloric acid (*reaction at right*), which results in evolution of carbon dioxide and regeneration of the ionic liquid, facilitating ionic liquid recycling.
Reproduced from Scott JL, Unali G, Perosa A, A "by-productless" cellulose foaming agent for use in imidazolium ionic liquids, Chem Commun 47 (2011) 2970–2972 with permission from the Royal Society of Chemistry.

plasticizers (and water) including urea, glycerol, sorbitol, and glycerin and thus formed "thermoplastic" starch, TPS, has long been mooted as a material suitable for food packaging [60]. Barrier properties of starch films may be enhanced by addition of small lipophilic biomolecules, such as stearic acid, which also reduces water solubility [61].

While the most abundant natural biopolymer, cellulose is less amenable to plasticization or dissolution, recent advances in solution processing of cellulose using ionic liquid mediated processes [62] offer promising routes to forming cellulose materials. Combination of ionic liquids with dipolar aprotic solvents, so-called organic electrolyte solutions, OESs, provides lower-density cellulose dopes and careful screening of OES solvents allows selection of greener alternatives, including biobased solvents, such as δ-butyrolactone or γ-valerolactone [63]. Low-density, high-volume foams are even accessible using a foaming agent that results in regeneration of an ionic liquid (Fig. 17.7), so facilitating recovery and recycling [64].

Although naturally derived polysaccharides represent significant CO_2 sinks, decomposition at the EOL results in a return of this CO_2 to the atmosphere; thus, the imperative for considering EOL options that keep carbon cycling in the manufacturing loops remains, as discussed later. Unmodified polysaccharides remain biodegradable, so these do offer options for pollution mitigation. For example, replacing persistent plastic microbeads, used in some "washaway" personal care products, with cellulose [65] or chitin [66] alternatives removes a source of oceanic microplastic pollution [67].

17.3.2 Lignin

As lignin is the second most abundant biopolymer in terrestrial biomass [68] and also occurs in some seaweeds [69], it is an attractive source of renewable biopolymer formed by carbon dioxide fixation by plants. The very challenges that can frustrate cost-effective and efficient depolymerization of lignin to produce small-molecule phenolic or monolignol fragments for use as monomers (described earlier), including the energy-intensive nature of the fragmentation and purification processes, as well as the tendency for lignin fragments to repolymerize into even more recalcitrant polymers, point to opportunities to use lignin as a polymer in its own right, *exploiting* the highly crosslinked and/or rich functionalization of the biopolymer. As plant lignins can vary significantly in structure, and thus in properties, direct use of unmodified lignin may be restricted to applications such as production of carbon fibers [70]. Recently, it has been suggested by Ragauskas et al. that manipulation of biosynthetic pathways might provide a means to tune lignin compositions [71] providing lignins enriched in, or depleted of, one of the three key "monomer" units, syringyl (S), guaiacyl (G), or hydroxyphenyl (C). Some such monolignol homopolymers are known in nature, Fig. 17.8; thus, the authors claim the "Both the mining of genetic variants in native populations of bioenergy crops and direct genetic manipulation of biosynthesis pathways have produced lignin feedstocks with unique properties for coproduct development" [70].

In 2010, only 1 million of the ca. 50×10^6 t of lignin produced in the pulping processes used to separate cellulose from lignin (and other components of woody biomass) was used in specialty products, with much of the remainder burned for energy recovery [72]. Further, industrially lignins are produced by a range of processes, leading to varying degrees of reduction in degree of polymerization (and thus molecular weight, M_w) and alterations to the chemical functional groups decorating the polymer. Fractionation of lignins, for example into solvent soluble and insoluble fractions, can provide materials with narrower M_w ranges, and differentiated properties. For example, Li and McDonald fractionated three commercially produced lignins into methanol soluble and insoluble fractions [73]; the methanol soluble fractions contained fewer aromatic structures, a greater number of β-O-4 linkages and, not surprisingly, lower M_w lignin polymer, resulting in lower T_g and thermal stability. Careful selection of extraction solvents, which are recyclable (albeit at the cost of further energy use), allowed recovery of low M_w lignin fractions with reasonably well-defined functional group content [74]. Such, lignin "macromers," large lignin polymer fragments that are

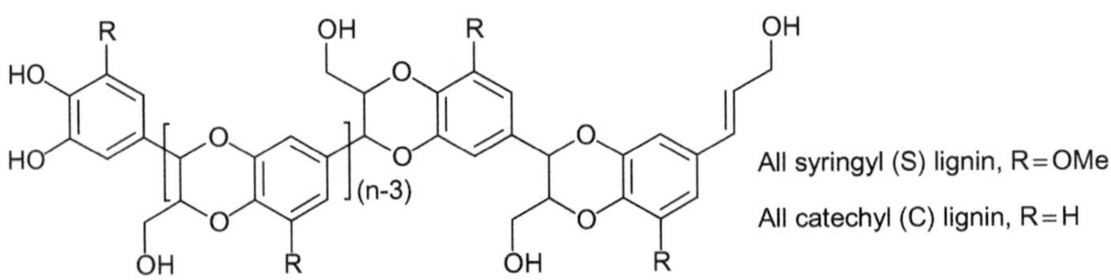

Fig. 17.8 Monolignol homopolymers, composed exclusively of syringyl, S, or catechyl, C, units. All S lignin may be isolated from poplar overexpressing ferrulate-5-hydroxylase, while linear all C lignin is found in the seed coats of some plants [70].

not broken down to substituent monomer units, find utility in the synthesis of resins (both epoxy and phenolic), polyesters, and polyurethanes.

Frequently, the lignin macromer polyols are used directly as crosslinkers in epoxide resins (which nonetheless require use of typical epoxide reagents, such as epichlorhydrin) or included in phenol-formaldehyde resins as partial phenol replacements. "Activation" of lignin by depolymerization and enhancement of reactive functional group density may be required to enhance reactivity toward formaldehyde [75]. More frequently, lignin is further reacted to provide macromers bearing functional groups suitable for production of copolymers, Fig. 17.9.

A wide range of chemistries have been applied to modify lignins, including nitration, amination, alkylation/dealkylation, carboxylation, and halogenation, among others [76]. Demethylation lignins are by-products of production of the popular dipolar aprotic solvent, dimethyl sulfoxide, DMSO. Lignin (as a cheap source of methoxy substituted aromatics) is reacted with molten sulfur to form dimethyl sulfide, which is subsequently oxidized to DMSO. Such dealkylated lignins bear a greater density of phenol groups and may be reacted with formaldehyde in adhesives or resins used in preparation of article board and laminated wood products [75].

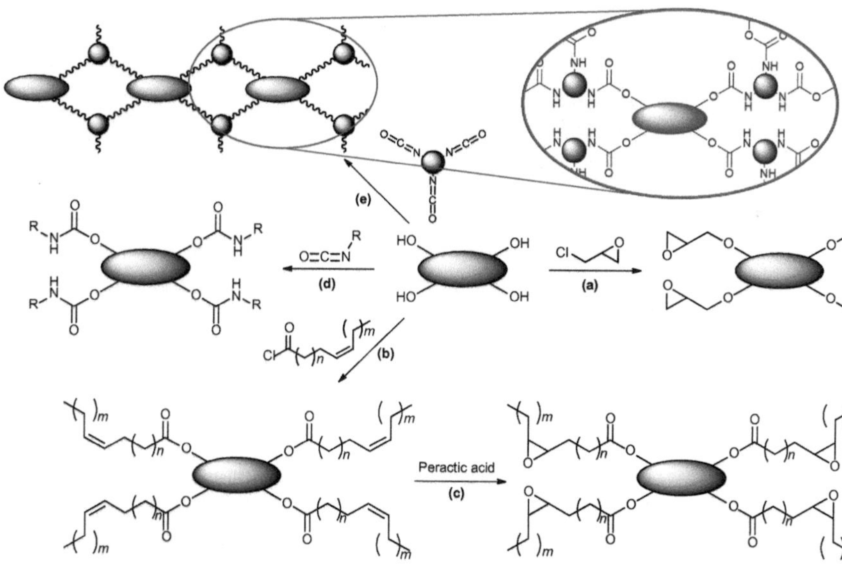

Fig. 17.9 Lignin "macromers" can be incorporated into copolymers by derivatization and reaction of alcohol functionality. For example, by installation on epoxide groups, either directly (a), or following reaction with fatty acid (FA) chlorides (b) and epoxidation of FA alkene moieties (c). Reaction with isocyanate terminated oligomers or polymers yield polyurethanes (d) and, if branched isocyanate terminated blocks are employed (e) highly crosslinked polyurethanes are possible (although not with the regular structure suggested by the schematic representation here).

17.4 Critical considerations for plant-derived chemicals for GHG mitigation

Using plant sources for chemicals, including plastics, allows exploitation of Nature's natural CO_2 fixing mechanisms, but sustainability gains, including CO_2 mitigation, must be measured if decisions are to be made based on numbers, not intuition; mitigation of CO_2 release must be balanced against potential other sources of pollution; and business models must be selected such that holistic management of chemicals across the entire life cycle is achieved and resource efficiency maximized.

17.4.1 Life cycle assessment

Metrics, such as those derived from life cycle assessment (LCA) [77,78], are required for clear and accountable comparison of technologies. While CO_2 equivalents released (or sequestered, i.e., negative CO_2 equiv.) is one critical metric, there are a multitude of other factors that must be considered if a material or technology is to be considered "environment friendly." Energy demand, waste generation, global warming potential (from all GHGs), human and ecotoxicity, eutrophication potential, and acidification potential must all be measured and compared to allow informed technology choices to be made [79,80].

LCA is sensitive to context [81]; thus, it is the life cycle of the *product* that must be compared as end-of-life options can vary depending on use, as well as processes used in manufacture and source of raw materials [82]. Further, full LCA requires extensive datasets, which are not always available for new biobased chemicals and methods applicable early in the research and development cycle can be applied to avoid "dead-end" research. Broeren et al. provide an overview of early-stage methods (applied to biobased chemicals) and conclude that the most widely used indicator is that of climate change, followed by resource indicators for energy, and measures of eutrophication and acidification potential [83]. Only a small percentage of the studies deal with water and land-use indicators, although these are now included in the European standard [84].

LCA-based comparisons of fossil-carbon versus biobased chemicals abound, but some recent examples illustrate the importance of careful comparison. High-density poly(ethylene), produced via dehydration of ethanol from Belgian sugar beet or wheat, compares favorably with the equivalent polymer from fossil oil in respect of climate change potential: −1018 and −1091 kg CO_2 equiv. per ton of sugar-beet- and wheat-derived ethylene, respectively, *versus* 1396 kg CO_2 equiv. per ton for oil-derived C_2H_4. In this case, the bio- and oil-derived monomers are chemically identical and the polymers produced indistinguishable, so only sources of raw material and processes for conversion to the C_2H_4 monomer need be considered [85]. However, biobased ethylene scores significantly less well in *all* other indicators except fossil-fuel depletion—the biobased polymer leads to increased ozone depletion, photochemical oxidation, acidification, and eutrophication compared to the oil-based material [84].

Clearly, comparison of biobased chemicals that are not direct "drop-in" (chemically identical) replacements will present even greater challenges. High-quality data are often not available, particularly for crops where regional differences in agricultural technology and inputs are critical to the analysis [86] and vastly differing end-of-life options for biobased (and biodegradable) polymers can be a defining factor in all LCA categories, but particularly GHGs [87]. Comparison of biobased polymers poly(lactic acid), PLA, thermoplastic starch, TPS, with oil-derived poly(ethylene terephthalate), PET showed that landfilling of the biobased and biodegradable PLA and TPS could lead to significantly *greater* global warming potential due to methane evolution [86], Fig. 17.10.

Thus, end-of-life recycling options that retain material in the manufacturing loop could be as critical for biobased materials as for fossil-carbon-derived plastics if gains from CO_2 sequestration by plants are to be realized. Biodegradable is not always best for this measure.

17.4.2 Keeping carbon in the manufacturing loop: The Circular Economy

Chemicals and polymers are not manufactured and used in isolation, but are some of the tangible materials of an industrial economic system, particularly with regards to the end-of-life fate. For example, consider the industrial economic models described by Stahel [88]:

1. *The Linear Economy* is an open-ended model with "no built-in tendency to recycle, which is reflected by treating the environment as a waste reservoir," Su et al. [89]. Products are discarded at the end of their useful life and growth depends on ever-increasing consumption. In such an economy, CO_2 mitigation is best achieved by reduction in energy use and even in "locking up carbon" in persistent polymers, i.e., using nonbiodegradable polymers to ensure that mineralization to CO_2 and H_2O does *not* occur. However, it is clear that such an approach leads to ever-growing mountains of solid waste and attendant pollution, including in oceans and other sensitive ecosystems, potentially leading to unintended negative impacts [90,91]. Further, materials are lost from the manufacturing cycle and new production required, leading to further CO_2 emissions as a consequence of energy production.
2. *The Circular Economy* is often described as a new concept, but was described by Boulding more than five decades ago, in a piece entitled "The Economics of the Coming Spaceship Earth" [92]. This closed system bears finite resources and must also contain all of the effluvia of human endeavors—as we now note with increasing CO_2 levels ascribed to anthropogenic sources. Stahel identifies two types of models: "those that foster reuse and extend service life through repair, remanufacture, upgrades and retrofits; and those that turn old goods into as-new resources by recycling the materials" Stahel [88]. Here, CO_2 may also be considered to be a chemical included in material and manufacturing loops [93] and thus "leakage" from the system is a loss of circularity. In such a system, biobased chemicals and polymers (utilizing plant-trapped CO_2) must be kept in the cycle and biodegradation should be delayed for as long as possible, while the chemical/material is passed through many loops.
3. *The Performance Economy* no longer relies on ownership of goods by consumers, but on delivery of a service or "unit of function." In most forms of this model, the manufacturer, not the user, owns the goods and thus has an interest in the most efficient use of same.

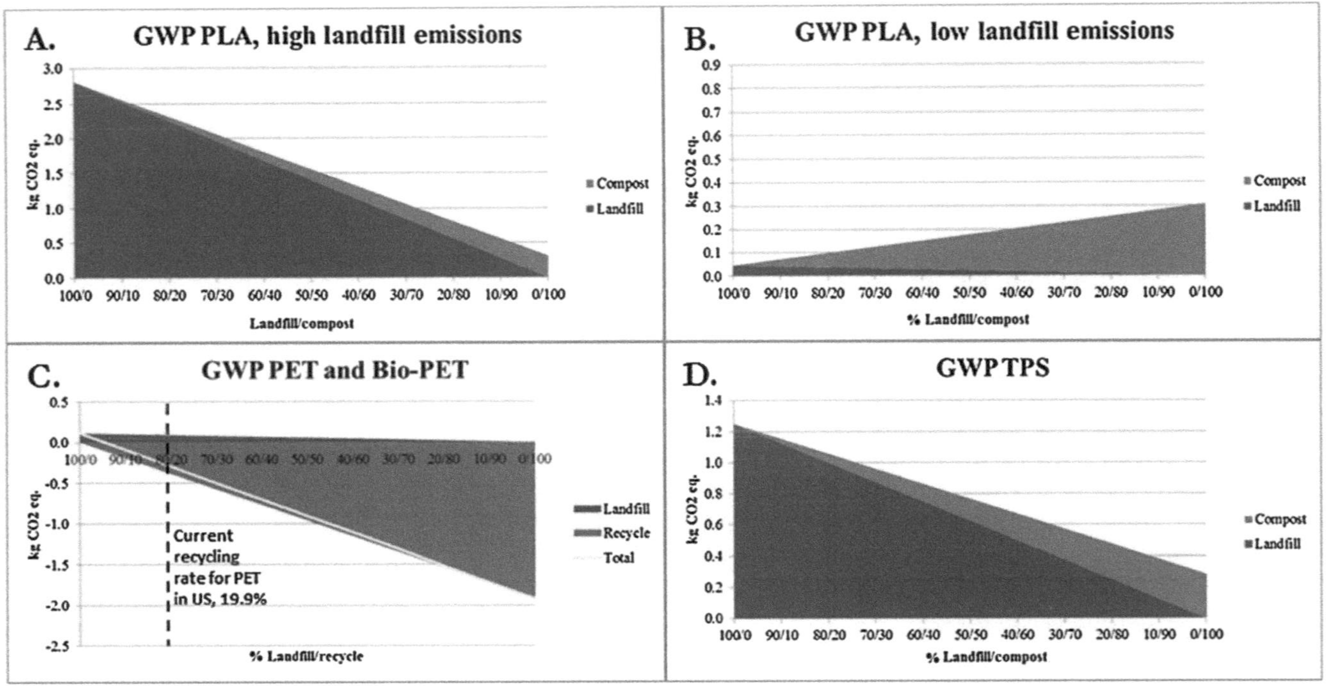

Fig. 17.10 Global warming impacts for different end-of-life scenarios for PLA, PET/BIO-PET, and TPS scaled from 100% landfilling and 0% recycling/composting to 0% landfilling and 100% recycling/composting for high (A) and low (B) emission scenarios for PLA, PET, and Bio-PET (C), and TPS (D).

Reprinted from Hottle TA, Bilec MM, Landis AE. Biopolymer production and end of life comparisons using life cycle assessment. Resour Conserv Recycl 122 (2017) 295–306, Copyright 2017, with permission from Elsevier.

> The manufacturer "Sells goods and molecules as service through rent, lease and share business models," Stahel [94]. Such a chemical leasing model has been espoused by the United Nations Industrial Development Organization (UNIDO) as particularly beneficial in the management of chemicals [95,96], although data supporting successes remain scarce [97].

From the point of view of managing CO_2 emissions, adoption of a Performance Economy could be considered to be a supreme enabler of circularity in chemicals management, including chemicals from plants. In a perfectly Circular Economy waste disappears, becoming a resource [98], and energy- and resource-intensive primary production is replaced by lower-impact secondary production. However, 21st century humans, accustomed to boundless growth economics, can respond counterintuitively, *increasing* consumption, and so negating sustainability gains, in a phenomenon that Zink and Geyer have termed "*Circular Economy Rebound*" [99].

References

[1] Intergovernmental Panel on Climate Change. Climate change 2001: The scientific basis. New York: Cambridge University Press; 2001.

[2] Betts RA, Jones CD, Knight JR, Keeling RF, Kennedy JJ. El Niño and a record CO2 rise. Nat Clim Chang 2016;6:806–10.

[3] Geider RJ, Delucia EH, Falkowski PG, Finzi AC, Grime JP, et al. Primary productivity of planet earth: biological determinants and physical constraints in terrestrial and aquatic habitats. Glob Chang Biol 2001;7:849–82.

[4] Hopewell J, Dvorak R, Kosior E. Plastics recycling: challenges and opportunities. Philos Trans R Soc B 2009;364:2115–26.

[5] Messe Düsseldorf. 2016. Plastics Europe Deutschland e.V. Plastics—the Facts 2016; www.plasticseurope.org (Accessed 5 January 2018).

[6] www.european-bioplastics.org/market/ (Accessed 5 January 2018).

[7] World Economic Forum, Ellen MacArthur Foundation, McKinsey and Company. 2016. The new plastics economy: rethinking the future of plastics, www.ellenmacarthurfoundation.org/publications (Accessed 5 January 2018).

[8] Geyer R, Jambeck JR, Law KL. Production, use, and fate of all plastics ever made. Sci Adv 2017;3:e1700782.

[9] Hillmyer MA. The promise of plastics from plants. Science 2017;358(6365):868–70.

[10] Schneiderman DK, Hillmyer MA. 50th Anniversary perspective: there is a great future in sustainable polymers. Macromolecules 2017;50(10):3733–49.

[11] Zhu Y, Romain C, Williams CK. Sustainable polymers from renewable resources. Nature 2016;540:354–62.

[12] Kristufek SL, Wacker KT, Tsao TY-Y, Lu Su L, Wooley KL. Monomer design strategies to create natural product-based polymer materials. Nat Prod Rep 2017;34:433–59.

[13] United States Department of Agriculture. 2017. Oilseeds: world markets and trade; www.fas.usda.gov/data/oilseeds-world-markets-and-trade (Accessed 13 December 2017).

[14] Santacesaria E, Vitiello R, Tesser R, Russo V, Turco R, Di Serio M. Chemical and technical aspects of the synthesis of chlorohydrins from glycerol. Ind Eng Chem Res 2014;53:8939–62.

[15] Sharninghausen LS, Campos J, Manas MG, Crabtree RH. Efficient selective and atom economic catalytic conversion of glycerol to lactic acid. Nat Commun 2014;5:5084.

[16] Simon J, Olsson JV, Kim H, Tenney IF, Waymouth RM. Semicrystalline dihydroxyacetone copolymers derived from glycerol. Macromolecules 2012;45(23):9275–81.
[17] Meier MAR, Metzger JO, Schubert US. Plant oil renewable resources as green alternatives in polymer science. Chem Soc Rev 2007;36:1788–802.
[18] Maisonneuve L, Lebarbe T, Grau E, Cramail H. Structure-properties relationship of fatty acid-based thermoplastics as synthetic polymer mimics. Polym Chem 2013;4:5472–517.
[19] Goldbach V, Roesle P, Mecking S. Catalytic isomerizing ω-functionalization of fatty acids. ACS Catal 2015;5:5951–72.
[20] Stempfle F, Ortmann P, Mecking S. Long-chain aliphatic polymers to bridge the gap between semicrystalline polyolefins and traditional polycondensates. Chem Rev 2016;116:4597–641.
[21] Nguyen HTH, Short GN, Qi P, Miller SA. Copolymerization of lactones and bioaromatics via concurrent ring-opening polymerization/polycondensation. Green Chem 2017;19:1877–88.
[22] Miller S A, Mialon L. 2015. Poly(dihydroferulic acid): a biorenewable polyethylene terephthalate mimic derived from lignin and acetic acid and copolymers thereof, US Patent 9,080,011 (University of Florida).
[23] Mialon L, Pemba AG, Miller SA. Biorenewable polyethylene terephthalate mimics derived from lignin and acetic acid. Green Chem 2010;12:1704–6.
[24] Nguyen HTH, Reis MH, Qi P, Miller SA. Polyethylene ferulate (PEF) and congeners: polystyrene mimics derived from biorenewable aromatics. Green Chem 2015;17:4512–7.
[25] Thomsett MR, Storr TE, Monaghan OR, Stockman RA, Howdle SM. Progress in the synthesis of sustainable polymers from terpenes and terpenoids. Green Mat 2017;4(3):115–34.
[26] Winnacker M, Rieger B. Recent progress in sustainable polymers obtained from cyclic terpenes: synthesis, properties, and application potential. ChemSusChem 2015;8: 2455–71.
[27] Wilbon PA, Chu F, Tang C. Progress in renewable polymers from natural terpenes, terpenoids, and rosin. Macromol Rapid Commun 2013;34:8–37.
[28] Yu P, Li AL, Liang H, Lu J. Polymerization of b-pinene with Schiff-base nickel complexes catalyst: synthesis of relatively high molecular weight poly(b-pinene) at high temperature with high productivity. J Polym Sci A Polym Chem 2007;45(16):3739–46.
[29] Satoh K, Nakahara A, Mukunoki K, Sugiyama H, Saito H, Kamigaito M. Sustainable cycloolefin polymer from pine tree oil for optoelectronics material: living cationic polymerization of β-pinene and catalytic hydrogenation of high-molecular-weight hydrogenated poly(β-pinene). Polym Chem 2014;5:3222–30.
[30] Singh A, Kamal M. Synthesis and characterization of polylimonene: polymer of an optically active terpene. J Appl Polym Sci 2012;125(2):1456–9.
[31] Hilschmann J, Kali G. Bio-based polymyrcene with highly ordered structure via solvent free controlled radical polymerization. Eur Polym J 2015;73:363–73.
[32] Kobayashi S, Lu C, Hoye TR, Hillmyer MA. Controlled polymerization of a cyclic diene prepared from the ring-closing metathesis of a naturally occurring monoterpene. J Am Chem Soc 2009;131(23):7960–1.
[33] Hauenstein O, Agarwal S, Greiner A. Bio-based polycarbonate as synthetic toolbox. Nat Commun 2016;7:11862.
[34] Byrne CM, Allen SD, Lobkovsky EB, Coates GW. Alternating copolymerization of limonene oxide and carbon dioxide. J Am Chem Soc 2004;126(37):11404–5.
[35] Peña Carrodeguas L, González-Fabra J, Castro-Gómez F, Bo C, Kleij AW. Al^{III}-catalysed formation of poly(limonene)carbonate: DFT analysis of the origin of stereoregularity. Chem Eur J 2015;21:6115–22.

[36] Hauenstein O, Reiter M, Agarwal S, Rieger B, Greiner A. Bio-based polycarbonate from limonene oxide and CO_2 with high molecular weight, excellent thermal resistance, hardness and transparency. Green Chem 2015;18(3):760–70.
[37] Auriemma F, De Rosa C, Di Caprio MR, Di Girolamo R, Ellis WC, Coates GW. Stereocomplexed poly(limonene carbonate): a unique example of the cocrystallization of amorphous enantiomeric polymers. Angew Chem Int Ed 2015;54(4):1215–8.
[38] Kindermann N, Cristòfol A, Kleij AW. Access to biorenewable polycarbonates with unusual glass-transition temperature (T_g) modulation. ACS Catal 2017;7(6):3860–3.
[39] Colonna M, Berti C, Fiorini M, Binassi E, Mazzacurati M, Vanninia M, et al. Synthesis and radiocarbon evidence of terephthalate polyesters completely prepared from renewable resources. Green Chem 2011;13(9):2543–8.
[40] Pang J, Zheng M, Sun R, Wang A, Wang X, Zhang T. Synthesis of ethylene glycol and terephthalic acid from biomass for producing PET. Green Chem 2016;18:342–59.
[41] Hillmyer MA, Tolman WB. Aliphatic polyester block polymers: renewable, degradable, and sustainable. Acc Chem Res 2014;47(8):2390–6.
[42] Winnacker M, Vagin S, Auer V, Rieger B. Synthesis of novel sustainable oligoamides via ring-opening polymerization of lactams based on (−)-menthone. Macromol Chem Phys 2014;215(17):1654–60.
[43] Galbis JA, García-Martín MG, de Paz MV, Galbis E. Synthetic polymers from sugar-based monomers. Chem Rev 2016;116:1600–36.
[44] Gregory GL, López-Vidal EM, Buchard A. Polymers from sugars: cyclic monomer synthesis, ring-opening polymerisation, material properties and applications. Chem Commun 2017;53:2198–217.
[45] Chen GQ, Patel MK. Plastics derived from biological sources: present and future: a technical and environmental review. Chem Rev 2012;112:2082–99.
[46] Zhang M, Yu Y. Dehydration of ethanol to ethylene. Ind Eng Chem Res 2013;52: 9505–14.
[47] Delidovich I, Hausoul PJC, Deng L, Pfützenreuter R, Rose M, Palkovits R. Alternative monomers based on lignocellulose and their use for polymer production. Chem Rev 2015;116:1540–99.
[48] Inkinen S, Hakkarainen M, Albertsson AC, Sodergard A. From lactic acid to poly(lactic acid) (PLA): characterization and analysis of PLA and its precursors. Biomacromolecules 2011;12:523–32.
[49] Auras R, Harte B, Selke S. An overview of polylactides as packaging materials. Macromol Biosci 2004;4:835–64.
[50] Marshall E L, Cooke G D. 2017. Biomass processing method, World Patent 060722 (Plaxica Limited).
[51] Marshall E L, Goodier P J, Cooke G D. 2015. Lactate production process, World Patent 155533 (Plaxica Limited).
[52] www.plaxica.com (Accessed 1 January 2018).
[53] Fenouillot F, Rousseau A, Colomines G, Saint-Loup R, Pascault J-P. Polymers from renewable 1,4:3,6-dianhydrohexitols (isosorbide, isomannide and isoidide): a review. Prog Polym Sci 2010;35(5):578–622.
[54] www.mcpp-global.com/en/asia/products/brand/durabiotm/ (Accessed 1 January 2018).
[55] www.avantium.com/yxy/yxy-technology/ (Accessed 1 January 2018).
[56] Burgess SK, et al. Chain mobility, thermal, and mechanical properties of poly(ethylene furanoate) compared to poly(ethylene terephthalate). Macromolecules 2014;47:1383–91.
[57] Ghormade V, Pathan EK, Deshpande MV. Can fungi compete with marine sources for chitosan production? Int J Biol Macromol 2017;104:1415–21.

[58] Satari B, Karimi K, Taherzadeh MJ, Zamani A. Co-production of fungal biomass derived constituents and ethanol from citrus wastes free sugars without auxiliary nutrients in airlift bioreactor. Int J Mol Sci 2016;17:302.
[59] Cheng L, Booker FL, Tu C, Burkey KO, Zhou L-S, Shew HD, et al. Arbuscular mycorrhizal fungi increase organic carbon decomposition under elevated CO_2. Science 2012;337:1084–7.
[60] Khan B, Niazi MBK, Samin G, Jahan Z. Thermoplastic starch: a possible biodegradable food packaging material—a review. J Food Process Eng 2017;40:e12447.
[61] Schmidt VCR, Porto LM, Laurindo JB, Menegalli FC. Water vapor barrier and mechanical properties of starch films containing stearic acid. Ind Crop Prod 2013;41:227–34.
[62] Swatloski RP, Spear SK, Holbrey JD, Rogers RD. Dissolution of cellulose with ionic liquids. J Am Chem Soc 2002;124:4974–5.
[63] Gale E, Wirawan RH, Silveira RL, Pereira CS, Johns MA, Skaf MS, et al. Directed discovery of greener cosolvents: new cosolvents for use in ionic liquid based organic electrolyte solutions for cellulose dissolution. ACS Sustain Chem Eng 2016;4:6200–7.
[64] Scott JL, Unali G, Perosa A. A "by-productless" cellulose foaming agent for use in imidazolium ionic liquids. Chem Commun 2011;47:2970–2.
[65] Coombs OBrien J, Torrente-Murciano L, Mattia D, Scott JL. Continuous production of cellulose microbeads via membrane emulsification. ACS Sustain Chem Eng 2017;5:5931–9.
[66] King CA, Shamshina JL, Zavgorodnya O, Cutfield T, Block LE, Rogers RD. Porous chitin microbeads for more sustainable cosmetics. ACS Sustain Chem Eng 2017;5:11660–7.
[67] Fendall LS, Sewell MA. Contributing to marine pollution by washing your face: microplastics in facial cleansers. Mar Pollut Bull 2009;58:1225–8.
[68] Lewis NG, Lantzy TR. Lignin in adhesive introduction and historical perspective. In: Hemingway RW, Conner AH, Branham SJ, editors. Adhesive from renewable resources. New Orleans: ACS Symposium Series; 1989. p. 13–26.
[69] Martone PT, Estevez JM, Lu FC, Ruel K, Denny MW, Somerville C, et al. Discovery of lignin in seaweed reveals convergent evolution of cell-wall architecture. Curr Biol 2009;19:169–75.
[70] Frank E, Steudle LM, Ingildeev D, Spörl JM, Buchmeiser MR. Carbon fibers: precursor systems, processing, structure, and properties. Angew Chem Int Ed 2014;53:5262–98.
[71] Ragauskas AJ, Beckham GT, Biddy MJ, Chandra R, Chen F, Davis MF, et al. Lignin valorization: improving lignin processing in the biorefinery. Science 2014;344:1246843.
[72] Stewart D. Lignin as a base material for materials applications: chemistry, applications and economics. Ind Crop Prod 2008;27:202 – 207.
[73] Li H, McDonald AG. Fractionation and characterization of industrial lignins. Ind Crop Prod 2014;62:67–76.
[74] Boeriua CG, Fiţigău FI, Gosselink RJA, Frissen AE, Stoutjesdijk J, Peter F. Fractionation of five technical lignins by selective extraction in green solvents and characterisation of isolated fractions. Ind Crop Prod 2014;62:481–90.
[75] Lorente E, Torras C, Berrueco C, Salvadó J. Transformation of lignin from bioethanol production for phenol substitution in resins. Wood Sci Technol 2017;51:1209–25.
[76] Laurichesse S, Avérous L. Chemical modification of lignins: towards biobased polymers. Prog Polym Sci 2014;39:1266–90.
[77] ISO. Environmental management—life cycle assessment—principles and framework. ISO No. 14040, Geneva, Switzerland: ISO; 2006.
[78] ISO. Environmental management—life cycle assessment—requirements and guidelines. ISO No. 14044, Geneva, Switzerland: ISO; 2006.

[79] Hunt RG, Franklin WE. LCA—how it came about. Int J Life Cycle Assess 1996;1:4–7.
[80] Klöpffer W. Life cycle assessment. Environ Sci Pollut Res 1997;4:223–8.
[81] Clavreul J, Guyonnet D, Christensen TH. Quantifying uncertainty in LCA-modelling of waste management systems. Waste Manag 2012;32:2482–95.
[82] Mittal V K, Bailin S C, Gonzalez M A, Meyer D E, Barrett W M, Smith R L. Toward automated inventory modeling in life cycle assessment: the utility of semantic data modeling to predict real-world chemical production, ACS Sustain Chem Eng Article ASAP, DOI: https://doi.org/10.1021/acssuschemeng.7b03379.
[83] Broeren MLM, Zijp MC, Waaijers-van der Loop SL, Heugens EHW, Posthuma L, Worrell E, et al. Environmental assessment of bio-based chemicals in early-stage development: a review of methods and indicators. Biofuels Bioprod Biorefin 2017;11:701–18.
[84] CEN EN 16760:2015. Bio-based products—life cycle assessment. European Commission Standard. Brussels, Belgium: CEN; 2015.
[85] Belboom S, Léonard A. Does biobased polymer achieve better environmental impacts than fossil polymer? Comparison of fossil HDPE and biobased HDPE produced from sugar beet and wheat. Biomass Bioenergy 2016;85:159–67.
[86] Grabowski A, Selke SEM, Auras R, Patel MK, Narayan R. Life cycle inventory data quality issues for bioplastics feedstocks. Int J Life Cycle Assess 2015;20:584–96.
[87] Hottle TA, Bilec MM, Landis AE. Biopolymer production and end of life comparisons using life cycle assessment. Resour Conserv Recycl 2017;122:295–306.
[88] Stahel WR. Circular economy. Nature 2016;531:435–8.
[89] Su B, Heshmati A, Geng Y, Yu X. A review of the circular economy in China: moving from rhetoric to implementation. J Clean Prod 2013;42:215–27.
[90] Gall SC, Thompson RC. The impact of debris on marine life. Mar Pollut Bull 2015;92:170–9.
[91] Santos RG, Andrades R, Boldrini MA, Martins AS. Debris ingestion by juvenile marine turtles: an underestimated problem. Mar Pollut Bull 2015;93:37–43.
[92] Boulding KE. The Economics of the Coming Spaceship Earth. In: Jarrett H, editor. Environmental quality in a growing economy. Baltimore, MD: Resources for the Future/Johns Hopkins University Press; 1966. p. 3–14.
[93] Budzianowski WM. Implementing carbon capture, utilisation and storage in the circular economy. Int J Global Warm 2017;12(2):272–96.
[94] Stahel WR. Analysis of the structure and values of the European Commission's Circular Economy Package. Waste Resource Manage 2017;170:41–4.
[95] Schwager P, Decker N, Kaltenegger I. Exploring green chemistry, sustainable chemistry and innovative business models such as chemical leasing in the context of international policy discussions. Curr Opin Green Sustain Chem 2016;1:18–21.
[96] United Nations Industrial Development Organization case studies on Chemical Leasing, 2016. Available at: http://chemicalleasing-toolkit.org/node/8 (Accessed 26 December 2017).
[97] Schwager P, Dunjic B, Kaltenegger I. Success and failure of the chemical leasing model in addressing sustainability challenges: evidence from practice. Curr Opin Green Sustain Chem 2017;8:14–7.
[98] European Commission, 2014. Towards a circular economy: a zero waste programme for Europe. Available at: http://ec.europa.eu/environment/circular-economy/pdf/circular-economy-communication.pdf (Accessed 26 December 2017).
[99] Zink T, Geyer R. Circular economy rebound. J Ind Ecol 2017;21(3):593–602.

Carbon Dioxide Utilization as a Mitigation Tool☆

18

Peter Styring[1]
UK Centre for Carbon Dioxide Utilisation, Department of Chemistry, Dainton Building, Brook Hill, The University of Sheffield, Sheffield, United Kingdom, Department of Chemical & Biological Engineering, Sir Robert Hadfield Building, The University of Sheffield, Sheffield, United Kingdom
[1]Corresponding author: p.styring@sheffield.ac.uk

Chapter Outline

18.1 Introduction

As has been discussed previously, carbon dioxide (CO_2) levels in the atmosphere are increasing at an alarming rate. For millennia, the Earth has maintained an efficient carbon cycle where emissions are balanced by natural processes such as photosynthesis and oceanic dissolution. However, since the advent of the industrial revolution, emissions have by far exceeded natural consumption. We have three realistic options by which to halt and then reverse this catastrophic increase. The Lansink ladder or hierarchy (Fig. 18.1) was proposed in the Dutch Parliament in 1979 by Ad Lansink [1]. Priority should always be afforded to methodologies to avoid waste or emissions in the first instance. Moving down the hierarchy, reuse, recycling, and energy

☆Dedicated to Professor Peter M Maitlis FRS on the occasion of his 85th birthday.

Managing Global Warming. https://doi.org/10.1016/B978-0-12-814104-5.00018-1

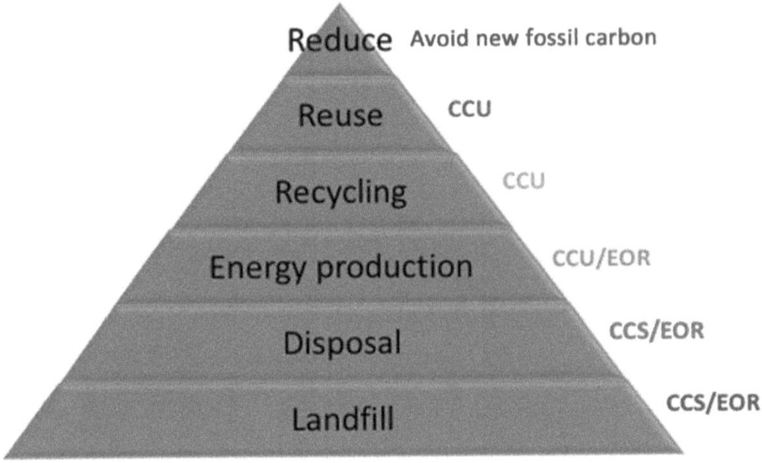

Fig. 18.1 The Lansink ladder or hierarchy for priorities in waste management.

conversion are attempted with decreasing priorities. In the worst-case scenario, the waste is disposed of; in this scenario it is landfill. However, it should be noted that no single technology will be appropriate in reality and scenarios need to be considered where each becomes a part of an overall toolkit for mitigation.

Using the Lansink Hierarchy approach to CO_2 mitigation, the first approach would be to avoid emissions in the first instance. As the majority of anthropogenic CO_2 emissions come from combustion processes, this would mean a radical transition away from fossil fuels to weather-dependent or renewable power technologies such as solar, wind, tidal, or hydro. While this is progressing, it will be decades before a fully decarbonized energy system can be achieved. This is because as well as power generation, which is being effectively decarbonized, space heating is still very much dependent of natural gas technologies. On a positive note, combusting methane to carbon dioxide sees a reduction in relative greenhouse gas potentials from 24 to 1, respectively: but this is only a sensible route if the methane comes from nonfossil or synthetic sources.

Furthermore, we need to look at the sectors in which these emissions could be avoided, being primarily power and transportation. While efforts are being made to address this transition, there are difficulties, including long-haul road and air transportation where high-energy density hydrocarbon fuels are going to remain important for many years to come. The second realistic approach, using Lansink, is to reuse and recycle the CO_2 [2–7]. This will be the main focus of this chapter but we will also aim to put this in a global context. So, what defines CO_2 utilization? In a true utilization process, carbon-oxygen bonds are broken and new bonds are made. The chemical entering the process is carbon dioxide; however, the chemical leaving the process is completely different [8].

While carbon capture and storage (CCS) in geological formations offers a long-term waste disposal strategy for this key greenhouse gas [9], carbon dioxide utilization

(CDU) adds value by treating the gas as a commodity chemical rather than a waste [2]. Furthermore, CCS is the worst-case scenario in the Lansink Hierarchy. As there will be no single solution to CO_2 emissions mitigation, it is clear that these two technologies represent different but complementary approaches to dealing with this problem emission. While CDU adds value to the process it should not be considered an alternative to CCS, but rather a key component of the complete mitigation strategy toolbox. So, will CDU be the solution to rising emissions mitigation? Almost certainly not if one considers the vast quantity of gas that needs to be abated. Figures for 2017 put CO_2 emissions at ~36 Gt year^{-1} [10]. That equates to 9.87 Gt year^{-1} of carbon as the elemental mass of C is 12.01 Da and the relative molecular mass of CO_2 is 44.01 Da. To put that into perspective, it is necessary to consider the bulk uses of carbon in society. If we consider the top five bulk chemicals and their annual production, this puts the problem into perspective. These chemicals are shown in Table 18.1 [11].

In total, this is ~3% (315 Mt) of global annual emissions. Of course, the permanency of the long-term storage can be questioned as the fate of the products is not considered in this case. In the case of urea, this is certainly only temporary storage as its major application is in agricultural fertilizers where the majority of the carbon dioxide is re-emitted during its use. Aside from commodity chemicals, it is necessary to consider synthetic fuels as a potential use of CO_2. This is an interesting scenario because the vast majority of CO_2 is emitted through the combustion of fuels to release energy. Consequently, to regenerate fuels, energy must be supplied in order to satisfy the First Law of Thermodynamics that energy cannot be created nor destroyed but can be transferred from one form to another. Fuels production accounts for ~2.3 Gt of carbon [11], so it can readily be seen that this is now a considerable part of the emissions portfolio (23%). Combining fuels and the top 5 chemicals represents 26% of the total. So, is the production of CO_2 fuels really a scenario that can be applied to mitigation strategies? This is an area of considerable debate on many levels. True synthetic fuels involve the making of carbon-carbon (C—C) and carbon-hydrogen (C—H) bonds, while at the same time breaking energetically favorable carbon-oxygen (C—O) bonds. Another approach in the suite of technologies is CO_2-enhanced oil recovery

Table 18.1 Annual global production quantities of the top five commodity chemicals

Chemical	Annual production (Mt)	Production as a percentage of annual carbon emissions (%)
Ethene (ethylene)	128.3	1.3
Propene (propylene)	97.8	1.0
Benzene	42.4	0.4
Urea	35.8	0.4
Butadiene	10.5	0.1

(CO_2-EOR) where the CO_2 is used as a working fluid and is unchanged chemically at the end of the process, although some is stored in the treated oil field. We will look at this comparison and the accounting problems that result later in this chapter.

Estimates of carbon utilization in 2015 were 250 Mt(CO_2) year^{-1} [12] or 68 Mt(C) year^{-1} (EC, 2015) with urea production being the major product. By comparison, CCS technologies in 2016 accounted for <30 Mt year^{-1} CO_2 sequestration or 8 Mt year^{-1} of carbon. Of this amount, only 2.15 Mt(CO_2) year^{-1} were conventional sequestration with the remainder being CO_2-EOR technologies. Predicted growth in CCS projects is 45 Mt of CO_2 by 2024, again with the majority being EOR projects. If a scenario analysis to 2050 is carried out, then relative to the reduction in emissions set under the Paris Agreement from COP21 of ~32 Gt(CO_2) year^{-1}, a best-case analysis based on current deployment projections estimates only 125 Mt(CO_2) year^{-1} can be achieved. This is only 0.4% of the target reduction. Therefore, unless there is a major disruptive technology change in CCS, additional mitigation approaches are required. The coupling of renewable energy storage to CO_2 fuels production therefore represents one such potential disruptive approach. This is because if captured CO_2 is converted into a synthetic hydrocarbon, then that same quantity of fossil hydrocarbon does not need to enter the supply chain. This is an example of fossil carbon avoidance and so equates to a higher level of the Lansink ladder. Of course, for this approach to be viable, a secure supply of cheap CO_2 is required for the reprocessing and so another disruptive technology that would be required is a means of producing CO_2 of an acceptable quality at least 50% cheaper than is possible using current capture processes.

Anastas and Warner developed the 12 Principles of Green Chemistry [13] that have been used to demonstrate how chemical syntheses should be developed to produce a sustainable approach. These principles are:

1. Prevent waste
2. Atom economy
3. Less hazardous chemical synthesis
4. Design safer chemicals
5. Safer solvents and auxiliaries
6. Design for energy efficiency
7. Use renewable feedstocks
8. Reduce derivatives
9. Use catalysis
10. Design for degradation
11. Real-time pollution potential
12. Safer chemistry for accident prevention

It is clear that these can be mapped on to a Lansink approach. Avoidance of waste should always be a priority. Atom economy reactions are the nadir of a green chemistry approach; however, in reality this is a goal that is very difficult to achieve as most chemical reactions produce some kind of by-product. Even a by-product as benign as water brings with it an environmental cost as energy is required to produce the hydrogen, which is a part of that waste. So which aspects of these Principle can realistically be achieved in CDU?

Table 18.2 The proposed 12 principles of CO_2 chemistry [14]

C	Catalysis is crucial
O	Origin of the CO_2?
2	Tomorrow's world may be different
C	Cleaner than existing process?
H	High volume of high-value chemicals?
E	*E*-factor must be low
M	Maximize integration
I	Innovative process technology
S	Sustainability is essential
T	Thermodynamics cannot be beaten
R	Renewable (and reasonable) energy input
Y	Your enthusiasm is not enough

At the 2015 Faraday discussion on carbon dioxide utilization, Poliakoff proposed 12 principles of carbon dioxide chemistry [14], analogous to the 12 Principles of Green Chemistry. Using CO2CHEMISTRY as the acronym, the proposal took the form shown in Table 18.2.

The methodology again realizes the importance of catalysis as an underpinning technology, while relating this to the thermodynamic challenges associated with CO_2 conversion. It also recognizes that chemistry alone will not solve the problems but rather that integration with other technologies is essential. Feedstocks and energy supply are important and of course sustainability must be a driver. The final point is interesting and open to interpretation. It is important that we do not "overegg" or "greenwash" the subject. Enthusiasm needs to be backed up with a realization that while technology can be produced to overcome a huge number of situations, it must be achievable economically. If a technology is so expensive that even going to scale it cannot be reduced, then industry and society will not adopt it. While initial costs will be high (think of the cost of computers and televisions over time as new technologies evolve), societal uptake may reduce the costs through economy of scale. It is also likely that some tax mechanism must be in place to make costs equitable. While production costs need to reduce, the penalty for emissions must increase, so it is not economically preferable to emit to waste. The current cost of carbon capture is approximately €100 $[t(CO_2)]^{-1}$ while the ETS trading price is only €6 $[t(CO_2)]^{-1}$.

Using this approach, we can consider the possible products that could be produced. Fig. 18.2 shows a snapshot of the chemical possibilities. This does not consider the energy or coreactants needed to affect the change, simply the chemistry to produce them.

While the figure does not show the full range of products that can be produced from CO_2 it does contain the key materials that represent the greatest potential contribution toward mitigation: mineralization and synthetic fuels. These will be considered later; however, at this stage we need to consider the second of Poliakoff's principles. Where

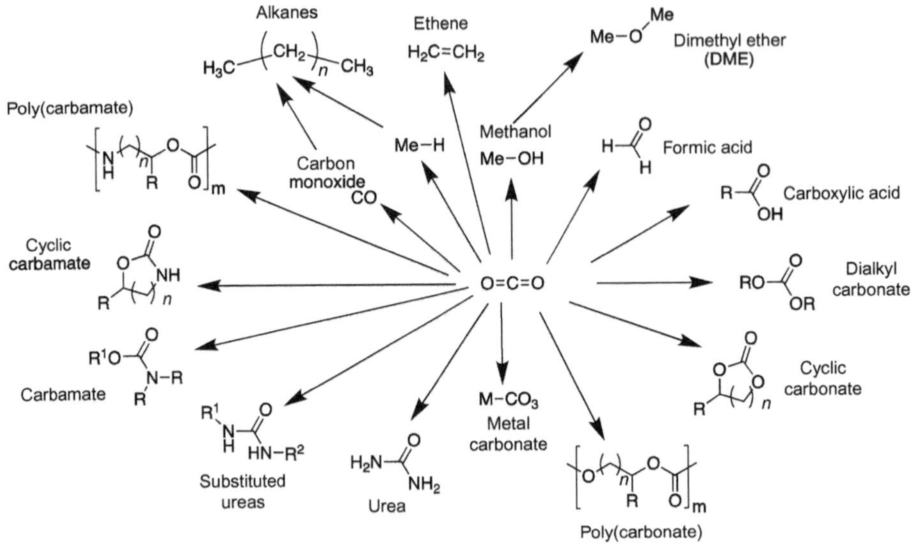

Fig. 18.2 Some of the products that can be produced using carbon dioxide utilization chemistries.

does the CO_2 come from and how pure does it need to be? Whether we are considering storage or capture, the common denominator is carbon capture. Or precisely how to capture CO_2 at low financial and environmental penalty?

18.2 Carbon capture

The terms carbon capture and storage and carbon dioxide capture and utilization have a commonality. They both emphasize the need for carbon capture. Current state-of-the-art capture technologies are largely restricted to amine absorption. So why do we need capture? The waste CO_2 is typically emitted in power and industrial processes. Typical concentrations of CO_2 in power sector flue gases are around 12%. There is a perceived necessity to selectively capture the CO_2 and purify it from the other flue gases. This is a very large, energy-intensive process. As has been stated, typical capture costs are of the order of €100 per tonne of CO_2 captured. However, the carbon price, either a carbon tax or emission levy, is typically around €6 per tonne. Therefore, to make this process viable, either the carbon price must be increased through taxation, or the cost of capture must be reduced significantly, or both. Another approach is to eliminate the need for CO_2 purification completely. This poses an interesting challenge, but one that can potentially be solved by careful selection of the chemistries used [11].

It has been said that CO_2 utilization is not feasible on thermodynamic grounds because CO_2 is unreactive. Instead, it is said that efforts should be directed to purifying CO_2 for other uses (such as EOR) and sequestration. But what are the energetics of carbon capture versus utilization? CO_2 adsorption by MEA is approximately $+73\,kJ\,mol^{-1}$ for a 30% aqueous solution at 40°C. Regeneration is approximately $+69\,kJ\,mol^{-1}$ at the same CO_2 concentrations and at 130°C [15]. This is a round-trip process of $+142\,kJ\,mol^{-1}$ to obtain pure CO_2. This means that for a power generation process there is a 30% parasitic loss in energy production to drive the process. The situation is not significantly different for other so-called advanced amines, although there is a slight reduction in enthalpies and associated energy losses due to the process. By contrast, the catalytic conversion of CO_2 into methanol and hydrocarbons can be exothermic, as shown in Fig. 18.3 [16]. If amines can purify the CO_2, then surely the carbon dioxide utilization process should be able to perform the same task. After all, it is not just amines that will selectively react with carbon dioxide. The major impurity in flue gas streams is nitrogen, an inert molecule. Synthetic reactions in the laboratory are often carried out with a nitrogen atmosphere, particularly if a catalyst is used; so why separate out an inert gas? If the other acid gases, such as NOx and SOx, can be removed and the waste stream dehumidified, then there is a good chance of selective reaction with carbon dioxide under significantly milder conditions.

Studies carried out by the Global CO_2 Initiative (GCI) have suggested that the mitigation potential of CO_2 utilization is $\sim 7\,Gt\,year^{-1}$ [9]. The report concludes that there are two major processes that contribute to the mitigation potential: accelerated mineralization and synthetic fuel production. Furthermore, these technologies also contribute the greatest economic potential. At significantly lower potential is the production of methanol and polymers. Methanol is interesting as it is a key intermediate in the synthesis of dimethyl ether, a diesel drop-in fuel, while specialist polymers although low on mitigation potential offer an economic incentive to technology development. Fig. 18.4 shows potential product pathways for organic-based products. Mineralization is not included as a complex series of products is possible, from simple low-cost aggregate materials to concrete and pure metal carbonates.

In order to focus on mitigation potential, we will now focus on synthetic CO_2-derived fuels and accelerated mineralization.

Dry methane reformation

$$CO_2(g) + CH_4(g) \rightleftharpoons 2\,CO(g) + 2\,H_2(g) \qquad \Delta H = +247.3\ kJ\ mol^{-1}$$

Fischer-Tropsch reaction

$$CO_2(g) + 3\,H_2(g) \rightleftharpoons [\text{-}CH_2\text{-}](g) + 2\,H_2O(g) \qquad \Delta H = -110.8\ kJ\ mol^{-1}$$

Methanol synthesis

$$CO_2(g) + 3\,H_2(g) \rightleftharpoons CH_3OH(g) + H_2O(g) \qquad \Delta H = -49.7\ kJ\ mol^{-1}$$

Methanol to gasoline (MTG) process

$$2\,CH_3OH(g) \rightleftharpoons [\text{-}CH_2\text{-}CH_2\text{-}](g) + 2\,H_2O(g) \qquad \Delta H = -89.5\ kJ\ mol^{-1}$$

Fig. 18.3 Conversion of carbon dioxide to fuels showing exothermic reactions.

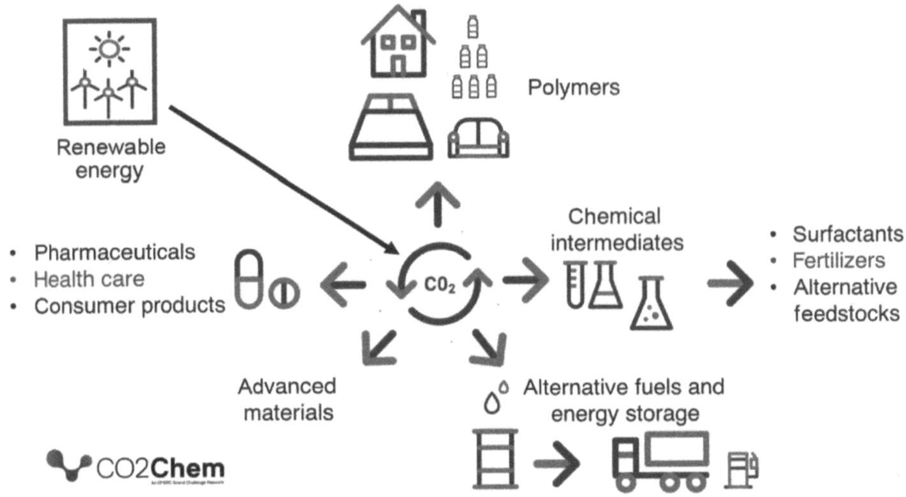

Fig. 18.4 A product-driven view of carbon dioxide utilization for organic materials.

18.3 CO_2-derived synthetic fuels and CO_2-enhanced oil recovery

The notion that CO_2 can be converted back to a viable fuel is often disputed based on grounds of both energy conversion and CO_2 mitigation. The problem is often a blurring or even disregard of boundaries for such processes and little appreciation of a full supply-chain analysis. We need hydrocarbon fuels. That is clear. Long-haul road transportation of heavy goods and aviation transport require fuels with high-energy density. Synthetic fuels offer a potential solution to a transition to a low-carbon transportation fuel model while retaining the current engine technologies and logistics chains for supply and delivery. It is unlikely that aircraft will operate on battery technologies for the foreseeable future, although it is acknowledged that Siemens are beginning to make significant advances in electric aircraft technologies [17] for short-haul flights. Likewise, Tesla have announced an electric truck/semiunit [18], but the range is only 480km (300miles), or 805km (500miles) for the two battery configurations. How can synthetic CO_2 fuels help in mitigating emissions? In a business as usual (BAU) scenario, transportation fuel use will remain constant and no mitigation is achieved. Indeed, emissions will increase cumulatively. However, there is the possibility that fuels will need to display diversity. Local and regional transport solutions can be transitioned to electric vehicles (EVs) if reliable renewable sources can be guaranteed, while long-haul transport relies on fossil-based hydrocarbon fuels. This is a good example of partial avoidance. If the EVs are powered on weather-dependent ("renewable") energy alone, then emissions are avoided by not using internal combustion vehicles (ICVs). While no CO_2 is actually captured, emissions are nevertheless reduced through avoidance. Now let us extend this to synthetic fuels derived from CO_2. We can capture CO_2 from a point source emitter in the short-term,

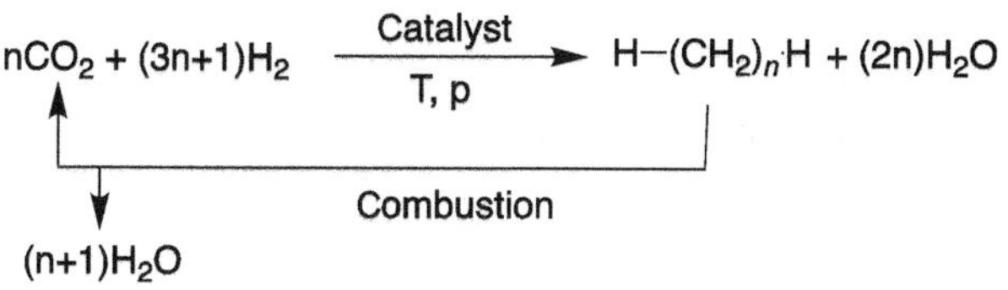

Fig. 18.5 An idealized carbon fuel cycle in which the fuel produced is reoxidized on combustion.

with the possibility of direct air capture (DAC) from the atmosphere in the long-term once emissions reductions are achieved. We can convert that CO_2 into a fuel using renewable energy and hydrogen, preferably obtained through water electrolysis using renewable electricity. On combustion, that fuel again is converted to CO_2. We, therefore, have a CO_2 cycle. If weather-dependent power provides the energy to drive the process, then we have a closed carbon cycle. This can be seen as an acceleration of what happened naturally in the conversion of CO_2 into plants through photosynthesis, which then decayed in the fossilization process to give the oil and gas we use today.

If we look at an ideal synthetic fuel mass balance (Fig. 18.5), we see:

For the time being, let us just say that the carbon is in a balanced cycle between reduction and oxidation of the carbon. We will consider the hydrogen shortly. This cycle does not account for the energy conversion in the system or the emissions associated with the synthesis of the fuel. If we consider a fossil fuel, such as diesel, then combustion will release the same amount of CO_2 as a synthetic diesel fraction. However, the availability of the diesel is often neglected when making comparisons. When considering fossil diesel, we need to also be aware of the economic and environmental costs of its production: exploration, wellhead recovery, transportation to a refinery and subsequent refining. We also need to appreciate that the when the fossil diesel (on average $C_{18}H_{38}$) is combusted, it will release 18 molecules of new carbon into the environment. So, while synthetic diesel can be regarded as carbon neutral in the simple mass balance, fossil diesel is certainly not. The carbon neutrality of synthetic diesel does not imply that the same carbon that is incorporated into the fuel is reused, rather that there is another carbon captured from a concentrated source that now enters the cycle while the original carbon leaves.

The question is where does CO_2-EOR fit into this philosophy? In many ways, CO_2-EOR may be regarded as a last-resort technology for exploiting oil fields. Once conventional extraction has recovered most of the oil, fields are flooded with water to recover stranded oil deposits (H_2O-EOR). Once complete, a switch to CO_2-EOR recovers the last remaining, and generally lower quality, oil and gas. Some, but not all, of the CO_2 is sequestered in the oil field [19]. The exact amount depends on the geology and topology of the site but is rarely 100% sequestration. The oil that is then recovered needs to be separated from the entrained CO_2, with some of the CO_2 recycled in the process. The oil is then separated from the water that is expelled from the pores and the oil subjected to the usual cracking and refining to give the fuel. So how much CO_2 is actually sequestered and how much oil is produced? This is very much dependent on the nature of the well and the geology of the rocks. Blunt [20] has suggested a break-even point for the production of crude oil where the amount of CO_2 permanently sequestered is balanced by the same amount emitted by its ultimate combustion. Where the boundary is set in the CO_2-EOR process is very much open to debate and controversy. If we are to compare like with like, we need to be very transparent on setting boundary conditions. Here are a few scenarios.

1. The CO_2-EOR process sets the output boundary at the well head. Oil is produced but its ultimate fate is not accounted for. No CO_2 is emitted once the fuel is produced.
2. The CO_2-EOR process sets the boundary at the end of life of the fuel so that the amount of CO_2 released on combustion is accounted for.
3. The CO_2-fuels process sets the boundary at the end of process so that fuels is produced and stored but not combusted. No CO_2 is emitted once the fuel is produced.
4. The CO_2-fuels process sets the boundary at the end of life of the fuel so that the amount of CO_2 released on combustion is accounted for.

Each scenario can be viewed in isolation, but this has little value. In order for true comparisons to be made, only two scenario combinations are valid. The emissions and costs associated with fuel production are compared (scenarios 1and 3), or those associate with end of life of the fuel are compared (scenarios 2 and 4). In reality, only the final combination is really valid as the properties of the two fuels (natural and synthetic) are very much different. Synthetic fuels are typically free from aromatics and sulfur compounds, and so particulate and SOx emissions are lower than fossil fuels. Therefore, only all emissions associated with end of life can be truly compared.

Blunt and coworkers [20] have described the CO_2-EOR process and the parameters that govern its production in the recovery of stranded hydrocarbons. There are many considerations that must be addressed. The sequestered CO_2 is not gaseous, but in the supercritical state. It therefore has a density of $\sim$450 kg m^{-3}, although this depends on the exact temperature and pressure of the injected fluid. Crude oil on the other hand is considerably denser, typically 870 kg m^{-3} for heavy Brent crude. So, 1 m^3 of $scCO_2$ and heavy crude will have masses of 450 and 870 kg, respectively. If the amount of CO_2 sequestered is considered over a number of scenarios from full oil-CO_2 immiscibility to complete miscibility, then a range of outcomes result. In general, if the boundary conditions are set once the recovered oil is combusted, then more CO_2 is released than was originally sequestered [8]. This suggests that while CO_2-EOR may be used to generate income for the carbon capture process, it is unlikely to be a mitigation tool as it introduces more new fossil carbon into the supply chain than is realistically sequestered.

18.4 Magic hydrogen (r-H_2)

To convert CO_2 back to a valuable fuel, energy is required, which must be renewable, low-carbon energy. Furthermore, dihydrogen will also be needed to deoxygenate, or reduce, the CO_2. Where this hydrogen comes from is key to the sustainability of the fuel. In many scenarios for renewable fuels or other chemical products, dihydrogen is often assumed to be available but no rationale is given as to where it is sourced. Because of this, it has been referred to as "Magic Hydrogen" as it appears from nowhere [21]. The current source of dihydrogen is primarily through the steam reformation of methane (SMR).

$$CH_4 + 2H_2O \rightarrow CO_2 + 4H_2 \tag{18.1}$$

While dihydrogen is produced, so is CO_2 so there is a conundrum. Another way to look at this is that the process converts methane, with a GHG potential of 24 into CO_2 with a GHG potential of 1. Therefore, it theoretically reduces the environmental impact: assuming that the methane would otherwise be released. Obviously if the methane is being recovered from gas fields, this would not be the case. The question arises as to whether this is an effective use of methane as it could otherwise be directly combusted to produce heat and the same amount of CO_2. However, due to the high price of CO_2 purification, it currently attracts a higher commercial price than methane. One strand of CCU research is the methanation of CO_2, the reverse of the SMR reaction. In this case, it would seem sensible to use the methane directly for energy applications as the CCU route offers no benefits, and indeed comes at high cost. However, if the dihydrogen is produced using renewable methods (r-H_2), the situation changes. Dihydrogen can be produced by the electrolysis of water:

$$2H_2O \rightarrow 2H_2 + O_2 \tag{18.2}$$

This can be achieved at low temperatures using polymer electrode membrane (PEM) cells [22] technologies or at high temperatures using solid oxide electrolysis calls (SOECs) [23]. If we now consider the synthesis of methane from renewable routes, we have:

$$\begin{aligned} &4H_2O \rightarrow 4H_2 + 2O_2 \\ &CO_2 + 4H_2 \rightarrow CH_4 + 2H_2O \\ &CO_2 + 2H_2O \rightarrow CH_4 + 2O_2 \end{aligned} \tag{18.3}$$

While this may not seem to be an effective process, it does illustrate an interesting concept. Weather-dependent renewable energy sources are by their very nature intermittent. Therefore, there will be times where the electrical energy produced will be in excess. Electrical energy storage systems such as batteries are efficient for short-term, diurnal storage and suffer leakage and capacity deletion over repeated cycles. However, gases and liquids offer easier and more efficient storage capabilities over prolonged periods. For example, energy storage across seasons. The infrastructure is already present to handle gaseous and liquid fuels making this a so-called drop-in technology. So, is this mitigation? Yes, because by using CO_2 as a recycled feedstock, it allows the equivalent quantity of new fossil methane to be prevented from entering the supply chain. Avoidance is the primary route to mitigation as discussed previously in the Lansink model, so is the preferred option.

18.5 The relative importance of synthetic hydrocarbons and oxygenates in a low-C fuel economy

For the foreseeable future, aviation transportation will remain reliant on hydrocarbon fuels. The regulative framework surrounding aviation fuel will mean that alternative fuels will require extensive testing and approval before they can be accepted as

Fig. 18.6 Synthesis of DME through methanol condensation.

replacements for Jet A. Synthetic hydrocarbons therefore represent the best short-term solution to low-carbon aviation fuels, as long as they are derived from renewable sources such as carbon dioxide and weather-dependent energy. By contrast, long-haul road transportation is already embracing alternative low-carbon fuels to replace diesel. These include a range of oxygenates starting with methanol, dimethyl ether (DME, shown in Fig. 18.6) and butanol [24].

Synthetic fuels from carbon dioxide can be produced by a variety of methods. The most common involves the conversion of carbon dioxide and water into syngas; carbon monoxide and hydrogen. This is the reverse water-gas shift (RWGS) reaction, which is followed by conversion of the syngas to higher hydrocarbons using Fisher-Tropsch (FT) Technology [25]. One of the main problems with FT synthesis is the lack of selectivity for particular chain lengths, meaning that either short hydrocarbons such as methane or ethane are produced in the low temperature (200–250°C) process, or long-chain hydrocarbon waxes in the high temperature (320–375°C) process that require a further refinement through hydrocracking [25]. If technologies could be developed that would selectively target the middle distillate fractions, then FT technologies would become cheaper and more widespread. Sunfire, a German spin-out company, has demonstrated an efficient process for the conversion of CO_2 into diesel hydrocarbons, so-called Blue Crude, using a combination of high-temperature electrolysis in solid oxide fuel cells (SOFCs) followed by FT conversion [26]. While the cost of such synthetic diesel is somewhat higher than fossil diesel, scaling of the technology will no doubt bring costs down in the future.

Other routes to hydrocarbons from CO_2 include the methanol to gasoline (MTG) process [27], sometimes known as the methanol to hydrocarbons process. This methodology first converts CO_2 to methanol under catalytic conditions and the methanol produced is further catalytically reduced with chain extension to give a range of hydrocarbons. This has been achieved commercially over a zeolite-based catalyst developed by the Mobil oil company [28]. This was the basis of the Fuels from Air [29] process using air captured CO_2 and energy from wind turbines. Another problem with the production of hydrocarbons from CO_2 is the amount of hydrogen needed to reduce the number of carbons within target molecules. If we assume that petroleum (gasoline), kerosene, and diesel possess on average 8, 12, and 18 carbons, respectively, then $6n+1$ (49, 61, and 109) hydrogen molecules are required in each case. The problem is that for each CO_2 molecule reduced, two molecules of water are produced: very expensive by-products. It would be both economically and environmentally more efficient to develop hydrocarbon replacement fuels, still derived from carbon dioxide, which possess similar or better engine performance while requiring fewer hydrogen molecules in their production.

Methanol, already described as an intermediate used in the MTG process, requires only three equivalents of hydrogen for its production from carbon dioxide. This is because methanol is a partially reduced form of carbon dioxide. Methanol can be used as a blending agent in gasoline and diesel engines, but due to its hydrophilicity and corrosivity, major modifications to the engine are required; otherwise, seals can become damaged. This means that methanol can only be added to gasoline up to around 18% concentration. However, in certain maritime engines, methanol can be used as a direct drop-in fuel as has been demonstrated by the Stena Line in Scandinavia [30]. The situation is improved if methanol is converted through condensation to give dimethyl ether (DME). This has been found to be an effective direct drop-in fuel to replace diesel. Unlike diesel, which on average possesses 38 hydrogen atoms, DME contains only six. The future of DME as a diesel replacement is looking bright. Volvo [31], Ford [32], and Mack [33] trucks have developed alternative diesel vehicles running on DME for use in the North American market. Oberon Fuels in California [34] have developed a commercial process for the production of DME to satisfy the demand generated by DME powered trucks. A direct drop-in fuel for gasoline engines is butanol (C_4H_9OH) [11]. Butanol, unlike methanol, is hydrophobic and consequently noncorrosive. Furthermore, it has the same energy density as gasoline and diesel [35]. Butanol contains only 10 hydrogen atoms so is again less hydrogen demanding than diesel. Table 18.3 shows the energy densities of a number of fuels that can be synthesized from CO_2, and in many cases, these are superior to the hydrocarbon itself.

Dowson and Styring have reported the synthesis of butanol in a stepwise process including the use of CO_2 at key stages [11]. Surprisingly, the overall process is exothermic (Fig. 18.7) with only the Claisen condensation being mildly endothermic. The primary CO_2 utilization step is a reaction between the gas and a Grignard reagent, in this case methyl Grignard to give acetic acid. While this step is noncatalytic, the magnesium can be recovered through electrolysis in a well-known and efficient industrial process [36]. This is an example of chemical looping (Fig. 18.8) where the metal is recovered and regenerated and fits the profile required for an efficient adoption of Lansink principles.

The process works well on a laboratory scale but is economically nonviable. This would improve as the process is scaled and heat recovery is used to harvest energy released in the exothermic processes. A basic technoeconomic analysis (TEA) on the CO_2 to acetic acid process has been reported to show how scaling benefits the process [37].

Table 18.3 Gravimetric and volumetric energy densities for different fuels

Fuel	Specific energy ($MJ\,kg^{-1}$)	Energy density ($MJ\,L^{-1}$)
Heavy fuel oil	41	–
LPG	46.4	26
Diesel	48	35.8
Gasoline	46.4	34.2
Jet-A	42.8	37.4
DME	28.8	19.3
Butanol	35	36

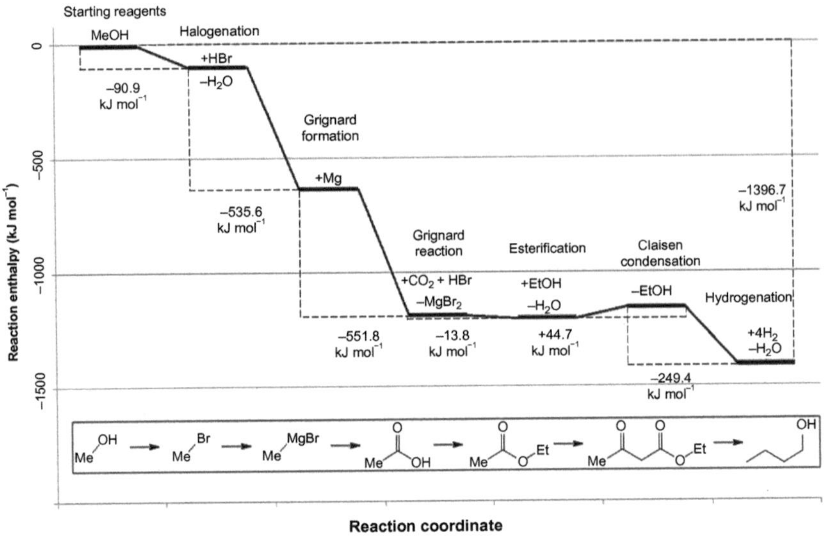

Fig. 18.7 The stepwise conversion of methanol to butanol demonstrating an overall exothermic process.

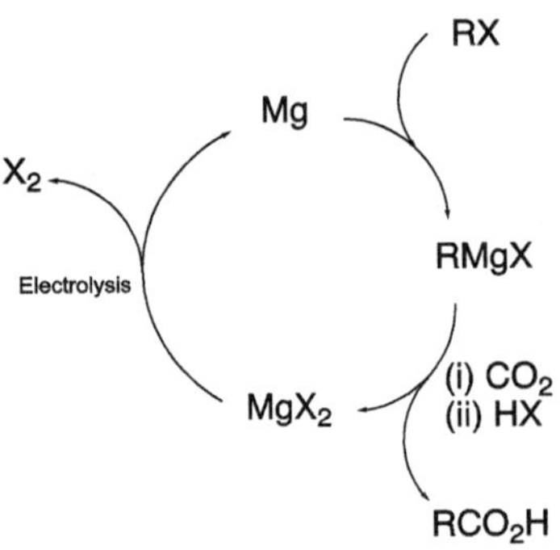

Fig. 18.8 The concept of metal looping in the synthesis of carboxylic acids using Grignard reagents.

It is important to realize that a CCU process may not necessarily be a single synthetic step as seen in Fig. 18.9. It may be just one process in a sequence: there may also be several CCU steps within that sequence. A good example of this is the synthesis of butanol from CO_2.

In the reaction scheme, each fragment containing a CO_2-derived carbon is shown in blue. Firstly, CO_2 is converted to methanol, which is then converted to the methyl halide and then the methyl Grignard reagent. Reaction with a second molecule of CO_2 produces the acetic acid intermediate. A further molecule of CO_2-derived methanol is condensed to give the methyl acetate ester, which undergoes a Claisen condensation with a second molecule of the ester to give methyl acetoacetate and a recovered molecule of methanol. The latter is then reduced to give the isomers of butanol. Overall, six CO_2 molecules are used in the synthesis of butanol. Four are incorporated into the butanol itself, one is recovered as the methanol condensate and the third is converted to methane in the final reduction step. So, while CO_2 is a key component in this process, it is also clear that methanol also plays a crucial role. While it may appear to be low in the GCI assessment, this is likely due to the fact that it plays a much more important role as a key intermediate in the synthesis of other more valuable fuels as shown in Fig. 18.10.

Public acceptance of alternative fuels is essential if they all to make a long-term impact. Studies have shown that both CCS and CCU technologies are regarded as methods to facilitate the continued use of fossil fuels [38]. Therefore, they are regarded less as mitigation tools and more as a means toward economic sustainability by the public. What must be emphasized is that CCU enables the continued use of high-energy density fuels but allows the amount of new fossil carbon entering the supply chain to be reduced. If we do not consider alternative carbon sources, then the future of transportation looks bleak. It is not inconceivable that governments will limit the use of fossil fuels in the future. Indeed, New York City has announced a disinvestment in oil and gas technologies as it moves toward a more sustainable city [39]. Synthetic fuels derived from CO_2 offer a partial solution by which such scenarios can be avoided. It is not an unreasonable to assume that signs in airports showing canceled flights will become increasingly common if fuel use restrictions are imposed. We may be entering an era

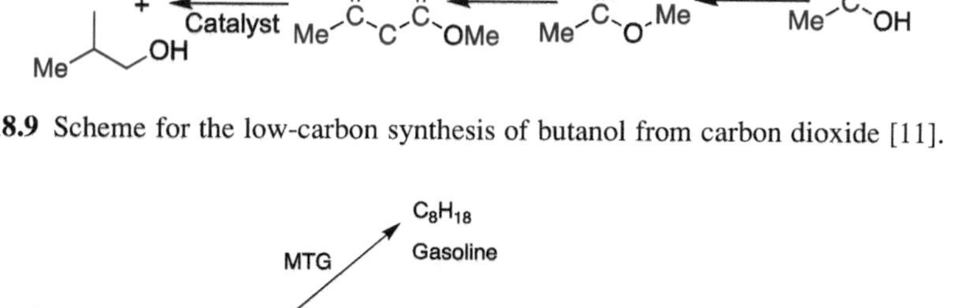

Fig. 18.9 Scheme for the low-carbon synthesis of butanol from carbon dioxide [11].

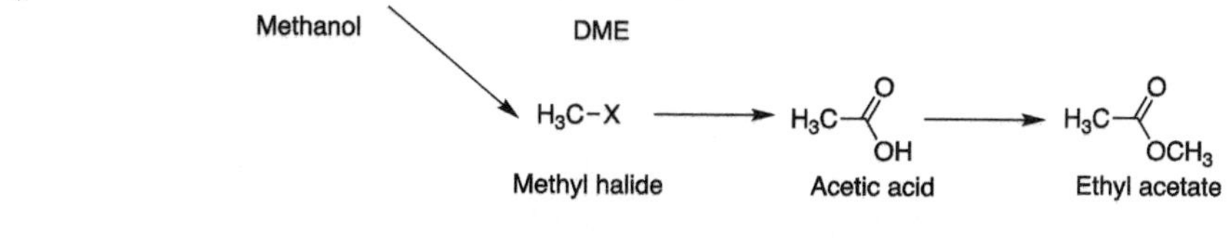

Fig. 18.10 The key role of methanol in the production of synthetic fuels.

where air travel becomes restricted to essential journeys and long-haul leisure air travel becomes the preserve of the super-rich or even curtailed completely.

18.6 Accelerated mineralization

The largest potential mitigation is offered by mineral carbonation. The potential products include low-carbon cement, aggregates, and building materials [16]. These will also lead to permanent storage of CO_2, but in the built environment rather than in geological storage sites. Carbon dioxide reacts with inorganic species, such as calcium and magnesium oxides and silicates to produce the metal carbonates in an exothermic reaction. This carbonation process is a naturally occurring weathering process, but this can also be achieved on an industrial scale. The key here is acceleration. If the conditions are optimized, then the reaction can occur in minutes and seconds rather than years and months. Achieving a particle size that enhances carbonation through the mechanical grinding of rocks can be energetically and commercially unfavorable [40]. However, the use of accelerated mineralization to treat solid inorganic wastes, such as steel slags and air pollution control residues (APCRs), can not only be carried out efficiently but adds considerable financial benefit [41]. The simplified process can be regarded as:

$$MO + CO_2 \rightarrow MCO_3 \quad (18.4)$$

The resulting carbonates can be used in construction materials, a concept clearly demonstrated by the Hills group and the company Carbon8 [42–44]. The materials have structural properties similar to conventional materials. It is essential that such materials have properties that are acceptable to the notoriously conservative building regulators if these are to be accepted by the construction industries. In the case of Carbon8, who produce building materials from air pollution control residues, the main pinch point is the high cost of carbon dioxide. The process is viable because APCRs attract a gate price because of their toxic nature. The accelerated mineralization process means that the otherwise toxic waste becomes safe so producing valuable materials while reducing the need for toxic landfill, which comes at a high environmental and economic burden.

An interesting application of accelerated mineralization in the geological environment has been demonstrated by CarbFixin collaboration with Reykjavik Energy, a geothermal energy company in Iceland [45]. The captured CO_2 is injected at an elevated temperature and under high pressure directly into basalt rock formations. The conditions used, favor an accelerated process and leads to truly permanent geological storage of CO_2. A recent collaboration between the company and Climeworks in Switzerland has seen air captured directly from ambient air being injected in the CarbFix process [46].

The concept of accelerated mineralization has been extended by the Styring Group to investigate mine waste tailings as the cheap source of minerals, looking in particular at particle size dependence in the mineralization process and varying process conditions in

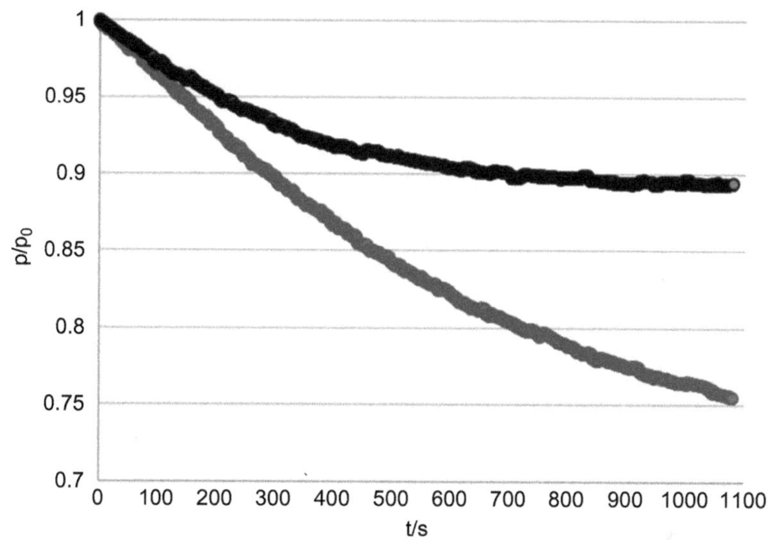

Fig. 18.11 Accelerated mineralization of lead mine tailings using simulated flue gas (black line, 12% CO_2: 88% N_2) and simulated biogas (gray line, 40% CO_2: 60% CH_4).

small-scale reactors. This has been carried out using pure CO_2 streams, CO_2 at flue gas concentrations in nitrogen and in simulated biogas (60% methane, 40% carbon dioxide). The latter is interesting as it allows simultaneous capture of CO_2 in the mineralization process while upgrading the biogas to the more desirable biomethane (>95% methane), which has a significantly higher calorific value and superior flame qualities [47]. In the case of biogas upgrading, the higher CO_2 concentration in the crude gas aids the uptake kinetics. Fig. 18.11 compares the rate of CO_2 reaction against time for synthetic flue gas (black line, 12% CO_2: 88% N_2) and synthetic biogas (gray line, 40% CO_2: 60% CH_4). For flue gas, the reaction results in almost complete consumption of the CO_2 after 1100 s (18 min). For synthetic biogas, the consumption is faster. Over the same time period, 25% of the gas is consumed out of the 40% available: 63% consumption of CO_2. If the reaction is allowed to continue to 30 min, <5% of the CO_2 remains.

The significance of using accelerated mineralization to upgrade biogas is that two products are generated. Mineralized feedstock for use in the production of building materials and biomethane, which has a higher calorific value and hence higher economic value than raw biogas. If this is coupled to the use of inorganic wastes into the process, then even greater value is added: economically and environmentally.

18.7 Polymers

While fuels offer short-term storage and mineralization long-term sequestration in the built environment, the conversion of CO_2 to polymers offers an intermediate solution. That may be both a benefit and an inhibitor to development. While polymers have intermediate lifetimes, this is often related to their application rather than their chemical and mechanical stability. One of the most visible environmental problems has been the accumulation of polymers in the natural environment that pollute because they do not degrade. Even when degradation occurs this is often mechanical rather than chemical, posing increasing risks to ecosystems. Therefore, the priority, while using CO_2 to reduce gaseous emissions, is also to design biodegradability or easy chemical recycling into the polymer design, using sustainable feedstocks wherever possible.

The mitigation potential for polymers from CO_2 was assessed as part of the McKinsey report for the Global CO_2 Network [9]. Due to the limited market the potential was deemed to be low, although there was commercial potential due to their high market value. If we consider the mitigation potential, then those polymers used in construction where there is a reasonable degree of longevity, and therefore storage potential, are considered here. There are several excellent reviews of CO_2-polymer chemistry [48–51], and so only an overview is given here.

Poly(carbonate)s derived from epoxides and CO_2 represent a class of polymers where a high proportion of the final material is derived from the original CO_2 molecule. The polymer formed actually meets the atom efficiency criterion of the 12 Principles of Green Chemistry. That is to say, all the atoms contained in the starting materials are retained in the final product with no material loss through by-product formation. In the case of poly(ethylcarbonate), derived from ethylene oxide, 50% of the molecular weight of the polymer is derived from the CO_2 ($C_2H_4O + CO_2 \rightarrow C_3H_4O_3$), while for poly(propylcarbonate) it is 33%. These materials are useful in construction as a direct replacement for glass due to their high transparency and mechanical strength, including their use in security shields. An example of the synthesis of poly(carbonate)s is shown in Fig. 18.12.

Poly(urethane polyol)s have been widely reported in the so-called Dream Reaction and Dream Process reported by Bayer Materials Science, now Covestro, in Germany [52]. Poly(acrylate)s are an intriguing class of polymers that can be derived from CO_2. Dowson has reported the synthesis of acrylic acid monomer from CO_2 and vinyl Grignard reagent [37]. However, the vinyl fragment typically comes from petrochemical sources and is so less sustainable. Another approach is to directly react ethylene with CO_2 to give acrylic acid, especially if the ethylene could be derived from CO_2 as well [53]. The possible routes to poly(methyl acrylate) are shown in Fig. 18.13. Interestingly, the methanol used for the synthesis of the ester can also be produced from CO_2 as described previously. If the ethylene route can be made to be environmentally and economically viable, then each carbon atom in the final polymer is sourced from CO_2, heightening the environmental credentials.

Again, we should remember that when carbon accounting is applied in the life cycle analysis of the polymers, each carbon atom in the target product that is derived from CO_2 is one less carbon from fossil sources that is entering the supply chain. If the CO_2 is derived from a waste gas stream, we can think of it as a "green carbon," rather than a "black carbon" from fossil oil sources. Often avoided carbon is more important than sequestered carbon in the drive toward greenhouse gas mitigation.

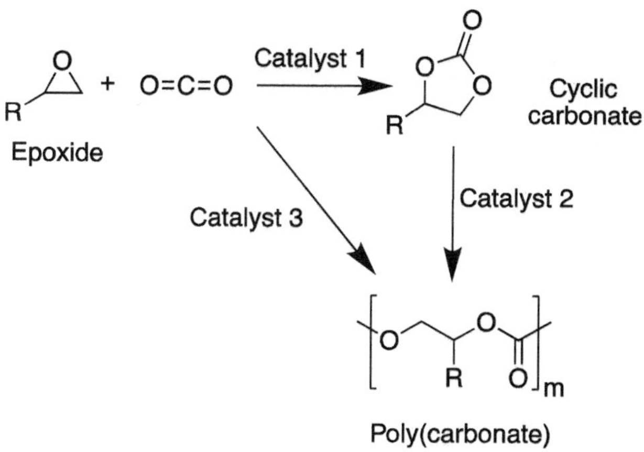

Fig. 18.12 The indirect and direct synthesis of poly(carbonate) from carbon dioxide and an epoxide.

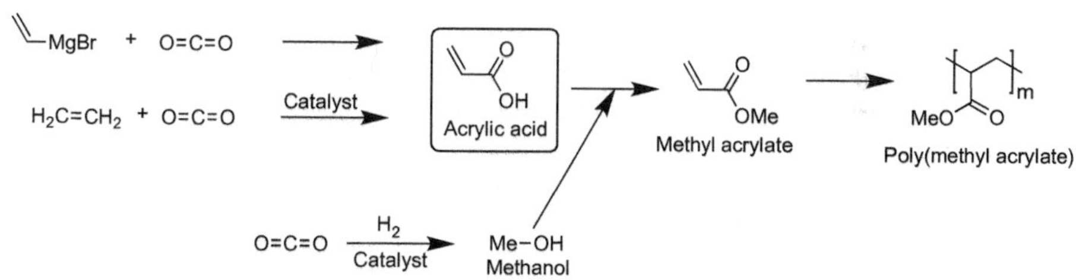

Fig. 18.13 Possible synthetic routes to poly(methyl acrylate) from CO_2.

18.8 Catalysis is key

Around 90% of the world's industrial processes involve some form of catalysis. Most of these are based on traditional thermal methods, perhaps also at elevated pressures. With growing concern surrounding a sustainable future, care must be exercised in the catalyst choice. Noble metal catalysts are expensive and fast approaching depletion. Unless they can be effectively recycled, their use becomes questionable. Pitts [54] has reported the concept of endangered elements and has suggested that we should only be looking to use Earth abundant elements and metals in future manufacturing processes. This is a huge challenge as it involves moving away from traditional processes. These will not always meet the required specifications for individual reactions. Life Cycle Analysis (LCA) and Technoeconomic Analysis (TEA) [55,56] will be required to balance catalyst abundance, sustainability, and efficacy. However, if we don't act now to reinvent the chemicals industry, it may become too late. Carbon dioxide chemistry has been at the forefront of pushing forward this revolution and North in particular has pioneered the concept of resource recovery in organic chemistry using recovered metal waste [57].

Most of the early CDU research focused on traditional thermal methods, including homogeneous and heterogeneous catalysts. In recent years, new approaches have emerged. In particular, electrocatalysis, photoctalysis, and plasmolytic reaction chemistries have begun to emerge at the laboratory scale.

18.9 Where does the future lie?

Carbon dioxide mitigation should not be regarded as a single technology solution. Rather it should be regarded as a suite of technologies working independently and together. While we need to tackle climate change, it must be achieved responsibly and economically. There are obvious questions surrounding the longevity of capture. However, if the product being discussed is a secondary product, and produced from waste and renewable energy, then that is *de facto* mitigating CO_2 emissions through avoidance. As a society, we need to change our approach to the production of chemicals. We currently rely on petrochemicals when we should be looking for more sustainable replacements. In order to achieve this, the cost of carbon dioxide needs to be lower. There needs to be a step change in capture technology. Another approach will be to develop reactions that can tolerate impure CO_2 streams, using the reaction itself to separate out the CO_2. Carbon capture using amines uses reaction chemistry to produce the carbamate products that are then decomposed to release the pure CO_2. Why can't this approach be used to produce products that can be isolated and add value?

The key mitigation targets are transportation fuels and minerals, simply because of the sheer volumes that can be processed. As we have discussed, the fuels are changing to add performance at lower cost through reducing the hydrogen burden. These fuels are cleaner than fossil oil-based fuels, due to lower sulfur and particulate emissions. Mineralization is a relatively simple process; however, we can improve on these processes by developing more intensified and efficient technologies to carry out conversions faster and with lower energy penalties.

New approaches to reaction chemistry and processes are required. Photocatalysis [58] will become increasingly important as we try to replicate nature. Photosynthesis has developed to harness CO_2 and solar energy to provide energy storage for plants. Artificial photosynthesis is being developed to mimic, not copy, nature by producing targeted product using solar energy [59]. While in its early stages, this is set to advance at a rapid rate as investment increases. Other emerging technologies include electrocatalysis [60,61] and plasma-catalyzed reactions [62–64].

Perhaps most importantly, the field of CO_2 utilization needs to enter the public consciousness. Governments and funding agencies need to be aware of the potential, as well as the limitations, of carbon dioxide utilization [7]. We must be careful not to "Greenwash" the public. While the quantities of CO_2 that can be directly mitigated are at the moment small, there is evidence that this will increase as the technology matures. We also need to embrace the concept of carbon avoidance. If we replace fossil carbon with recycled carbon, then we automatically mitigate emissions using the highest of the Lansink hierarchy interventions. As the European Commission SCOT project have said: "Using carbon dioxide as a resource; not as a waste" [65]. Whether or not CDU or CCS are appropriate to any given situation is dependent on numerous factors. Some that need to be addressed are:

- Is there a reliable and easily accessed supply of CO_2 emissions at suitable concentrations?
- Are there storage facilities close to the emission site that could support CCS?
- Is there a local source of renewable energy that can be used in a CDU process?
- Are there suitable local and sustainable sources of coreactants, such as water and hydrogen?
- What are the costs of storage or utilization? Does it make economic sense to invest in either of these technologies?
- Are there any other, more economic options?

In the end, the decision on whether a particular technology will be suitable comes down to Location, Location, Location; politically, the decision will rest on Economics, Economics, Re-election! Any roll out of a new technology needs to be fully justified. Therefore, there will be an increasing demand to carry out life cycle assessment (LCA) and technoeconomic analysis (TEA) that are transparent and adhere to a recognized comparative framework. It is important that we compare like with like in order to adequately evaluate any technology development.

18.10 Conclusions

Carbon dioxide mitigation can be achieved by utilizing the gas. The full suite of available technologies is outside the scope of a single chapter in a book and readers are directed to more extensive reviews. The coming years will help to establish the scale of such mitigation. This will be primary mitigation through the capture and conversion of emissions, and also secondary mitigation through new carbon avoidance. Neither CCS nor CDU should be ruled out at this stage of technology development. Both are unproven at scales appropriate to the task set by the COP21 Paris agreement. It may be that these technologies both mature to adapt to local needs. It may be that a new technology emerges that supersedes both. We should not limit ourselves until there is clear scientific evidence one way or another. Investment is needed to develop new technologies, but we also need to address the low cost of emitting the gas.

Acknowledgments

The author thanks the University of Sheffield for the opportunity to carry out research into carbon dioxide utilization and the Engineering and Physical Sciences Research Council for funding. In particular, funding to the CO2Chem Grand Challenge Network in three awards since 2010 has allowed the Styring Group to consider not only the Science and Engineering of CCU but also the social, economic, and political aspects of this exciting technology. Dr. Susan Handford-Styring is also acknowledged for her dedication in putting up with the author's left-field ideas over 25 years of marriage! The author would also like to acknowledge the support and inspiration of his PhD supervisor many years ago who sowed the seeds of interest in catalysis.

References

[1] Lansink A, 1979, http://www.isonomia.co.uk/?p=2556 [Accessed 18 February 2018].

[2] Styring P, Quadrelli EA, Armstrong K, editors. Carbon capture and utilisation: closing the carbon cycle. Amsterdam: Elsevier; 2014.

[3] Carbon capture and utilisation in the green economy: using CO2 to manufacture fuel, chemicals and materials, Styring P, Jansen D, de Coninck, H, Reith H, Armstrong K, CO2Chem Publishing, Sheffield, 2012. 978-0-9572588-0-8 ISBN 978-0-9572588-1-5 (eBook). Available from: co2chem.co.uk/carbon-capture-and-utilisation-in-the-green-economy.

[4] Artz J, Muller TE, Thenert K, Kleinekorte J, Meys R, Sternberg A, et al. Sustainable conversion of carbon dioxide: an integrated review of catalysis and life cycle assessment. Chem Rev 2018;118:434–504. https://doi.org/10.1021/acs.chemrev.7b00435.

[5] Aresta M, Dibenedetto A, Angelini A. J CO_2 Util 2013;3–4:65–73.

[6] Lanzafame P, Abate S, Ampelli C, Genovese C, Passalacqua R, Centi G, et al. Beyond solar fuels: renewable energy-driven chemistry. ChemSusChem 2017;10:4409–19. https://doi.org/10.1002/cssc.201701507.

[7] The Royal Society. The potential and limitations of using carbon dioxide. Available from: http://royalsociety.org/low-carbon-energy-programme; 2017.

[8] Armstrong K, Styring P. Assessing the potential of utilization and storage strategies for post-combustion CO_2 emissions reduction. Front Energy Res 2015;3. 8https://doi.org/10.3389/fenrg.2015.00008.

[9] A roadmap for the global implementation of carbon utilization technologies: transforming CO2 from a liability to an asset at significant market scale, CO2 Sciences and The Global CO_2 Initiative; 2016. Available from: http://www.globalco2initiative.org/webinar [Accessed 18 February 2018].

[10] Multiple Authors. Carbon capture and storage (Faraday discussions). London: The Royal Society of Chemistry; 2016.

[11] Dowson GRM, Styring P. Demonstration of CO_2 conversion to synthetic transport fuel at flue gas concentrations. Front Energy Res 2017;5:26. https://doi.org/10.3389/fenrg.2017.00026.

[12] Marzano C, Petrov P, Bøwadt S, Transforming CO2 into value for a rejuvenated European economy, Directorate-General, Key Enabling Technologies, European Commission; 2015.

[13] Anastas PT, Warner JC. Green chemistry: theory and practice. New York: Oxford University Press; 1998.

[14] Poliakoff M, Leitner W, Streng ES. The 12 principles of CO_2 chemistry. Faraday Trans 2015;183:9–17.

[15] Kim S-T, Kang J-W, Lee J, Min B-M. Analysis of the heat of reaction and regeneration in alkanolamine-CO2 system. Korean J Chem Eng 2011;28(12):2275–81. https://doi.org/10.1007/s11814-011-0126-1.

[16] France LJ, Edwards PP, Kuznetsov VL, Almegren A. The indirect and direct conversion of CO_2 into higher hydrocarbon fuels. In: Styring P, Quadrelli EA, Armstrong K, editors. Carbon dioxide utilisation: closing the carbon cycle. Amsterdam: Elsevier; 2015. p. 161–82.

[17] http://www.airbus.com/newsroom/press-releases/en/2017/11/airbus--rolls-royce–and-siemens-team-up-for-electric-future-par.html [Accessed 18 February 2018].

[18] http://www.forbes.com/sites/alanohnsman/2017/12/28/orders-pile-up-for-tesla-semi-that-doesnt-yet-exist-as-rival-electric-trucks-get-rolling/#51c816864a0a [Accessed 18 February 2018].

[19] Perera S, Pathegama R, Rathnaweera T, Ranathunga A, Koay A Choi X. A review of CO_2-enhanced oil recovery with a simulated sensitivity analysis. Energies 2016;9:481. https://doi.org/10.3390/en9070481.

[20] Blunt M, Fayers FJ, Orr Jr FM. Carbon dioxide in enhanced oil recovery. Energy Convers Manag 1993;34:9–11.

[21] North M, Abrantes P, Remiezowicz E, Bardow A, Dodson J, Manning T, et al. CO_2 reduction reactions: general discussion. Faraday Discuss 2015;183:261–90. https://doi.org/10.1039/C5FD90080H.

[22] Buttler A, Spliethoff H. Current status of water electrolysis for energy storage, grid balancing and sector coupling via power-to-gas and power-to-liquids: a review. Renew Sust Energ Rev 2018;82:2440–24543. https://doi.org/10.1016/j.rser.2017.09.003.

[23] Zheng Y, Wang JC, Yu B Zhang WQ, Chen J, Qiao JL, Zhang JJ. A review of high temperature co-electrolysis of H_2O and CO_2 to produce sustainable fuels using solid oxide electrolysis cells (SOECs): advanced materials and technology. Chem Soc Rev 2017;46:1427–63. https://doi.org/10.1039/c6cs00403b.

[24] Dowson G, Styring P. Conversion of carbon dioxide to oxygenated organics. In: Styring P, Quadrelli EA, Armstrong K, editors. Carbon dioxide utilisation: closing the carbon cycle. Amsterdam: Elsevier; 2015. p. 141–59.

[25] Maitlis PM. In: Maitlis PM, de Klerk A, editors. Greener Fischer-Tropsch processes. Weinheim: Wiley-VCH; 2013. p. 13.

[26] http://www.sunfire.de/en/applications/fuel.

[27] Chang C, Silvestri A. The conversion of methanol and other O-compounds to hydrocarbons over zeolite catalysts. J Catal 1977;47:249–59.

[28] Norahim N, Yaisanga P Faungnawakij K, Charinpanitkul T, Klaysom C. Recent membrane developments for CO_2 separation and capture. Chem Eng Technol 2018;41:211–23. https://doi.org/10.1002/ceat.201700406.

[29] Fuels from Air, http://www.westbeaconfarm.co.uk/fuelsfromairltd/index.html [Accessed 18 February 2018].

[30] Stena Line Methanol, http://news.stenaline.co.uk/pressreleases/stena-line-launches-the-world-s-first-methanol-ferry-1137516 [Accessed 18 February 2018].

[31] Volvo Trucks, http://www.volvogroup.com/en-en/news/2017/jan/mack-trucks-tests-alternative-fuel-dme.html [Accessed 18 February 2018].

[32] Ford Trucks, http://www.aboutdme.org [Accessed 18 February 2018].

[33] Mack Trucks, http://www.macktrucks.com/community/mack-news/2017/mack-partners-with-new-york-city-department-of-sanitation-oberon-fuels-for-dme-truck-demo/ [Accessed 18 February 2018].

[34] Oberon Fuels DME Process, qa. http://oberonfuels.com/technology/oberon-process/ [Accessed 18 February 2018].

[35] Energy densities, en. http://wikipedia.org/wiki/Energy_density [Accessed 18 February 2018].

[36] Crystal water treatment process, http://www.crystal.com [Accessed 18 February 2018].

[37] Dowson GRM, Dimitriou I, Owen RE, Reed DG, Allen RWK, Styring P. Kinetic and economic analysis of reactive capture of dilute carbon dioxide with Grignard reagents. Faraday Discuss 2015;183:47–65.

[38] Jones CR, Radford RL, Armstrong K, Styring P. What a waste! Assessing public perceptions of carbon dioxide utilisation technology. J CO2 Util 2014;7:51–4.

[39] Office of the Mayer, City of New York, http://www1.nyc.gov/office-of-the-mayor/news/022-18/climate-action-mayor-comptroller-trustees-first-in-the-nation-goal-divest-from#/0 [Accessed 18 February 2018].

[40] Madeddu S, Priestnall M, Godoy E, Kumar RV, Raymahasay S, Evans M, et al. Extraction of mg$(OH)_2$ from mg silicate minerals with NaOH assisted with H_2O: implications for CO_2 capture from exhaust flue gas. Faraday Discuss 2015;183:369–87. https://doi.org/10.1039/c5fd00047e.
[41] Gunning P, Hills CD, Carey PJ. Accelerated carbonation treatment of industrial wastes. Waste Manag 2010;30:1081–90.
[42] Tripathi N, Hills CD, Singh RS, Carey PJ. Construction: use waste for building. Nature 2017;550:457.
[43] Tota-Haharaj K, Hills C, Monrose J. In: Novel permeable pavement systems using carbon-negative aggregates. NexGen Technologies for Mining and Fuel Industries. 1; 2017. p. 683–96.
[44] Chen QY, Tyrer M, Hills CD, Yang XM, Carey P. Immobilisation of heavy metal in cement-based solidification/stabilisation: a review. Waste Manag 2009;29:390–403. https://doi.org/10.1016/j.wasman.2008.01.019.
[45] Matter JM, Stute M, Snæbjörnsdottir SO, Oelkers EH, Gislason SR, Aradottir ES, et al. Rapid carbon mineralization for permanent disposal of anthropogenic carbon dioxide emissions. Science 2016;352:1312–4. https://doi.org/10.1126/science.aad8132.
[46] http://www.climeworks.com/public-update-on-carbfix/ [Accessed 18 February 2018].
[47] Styring P, Reed DG, Sicheng L, Dowson GRM. In: Kinetic analysis of the accelerated mineralization of lead mine tailings. NexGen Technologies for Mining and Fuel Industries. 2; 2017. p. 949–56.
[48] Lankanke J, Wolf A, Peters M. Polymers from CO_2—an industrial perspective. In: -Styring P, Quadrelli EA, Armstrong K, editors. Carbon dioxide utilisation: closing the carbon cycle. Amsterdam: Elsevier; 2015. p. 59–71.
[49] Liu SJ, Wang XH. Polymers from carbon dioxide: polycarbonates, polyurethanes. Curr Opin Green Sustain Chem 2017;3:61–6. https://doi.org/10.1016/j.cogsc.2016.08.003.
[50] Trott G, Saini PK, Williams CK. Catalysts for CO_2/epoxide ring-opening copolymerization. Philos Tran A Math Phy Eng Sci 2016;374(2061)20150085. https://doi.org/10.1098/rsta.2015.0085.
[51] von der Assen N, Bardow A. Life cycle assessment of polyols for polyurethane production using CO_2 as feedstock: insights from an industrial case study. Green Chem 2014;16:3272–80. https://doi.org/10.1039/c4gc00513a.
[52] https://www.co2-dreams.covestro.com/en [Accessed 18 February 2018].
[53] Plessow PN, Schäfer A, Limbach M, Hofmann P. Acrylate formation from CO2 and ethylene mediated by nickel complexes: a theoretical study. Organometallics 2014;33:3657–68.
[54] Pitts M. Let's take better care of our rare earth elements. New Scientist 2011;(2799):26–7 February 15.
[55] Sadhukhan J, Ng KS, Hernandez EM. Life cycle assessment. In: Biorefineries and chemical processes: design, integration and sustainability analysis. Chichester: John Wiley & Sons, Ltd.; 2014. https://doi.org/10.1002/9781118698129.ch4.
[56] Pérez-Fortez M, Tzimas E. Techno-economic and environmental evaluation of CO_2 utilisation for fuel production: Synthesis of methanol and formic acid. Joint Research Centre, Institute for Energy and Transport, report EUR 27629 EN, 2016.
[57] Lamb KJ, Dowsett MR, Chatzipanagis K, Scullion ZW, Kroger R, Lee JD, et al. Capacitance-assisted sustainable electrochemical carbon dioxide mineralisation. ChemSusChem 2018;11:137–48. https://doi.org/10.1002/cssc.201702087.

[58] Wang H, Jia J, Song PF, Wang Q, Li DB, Qian CX, et al. Efficient electrocatalytic reduction of CO_2 by nitrogen-doped nanoporous carbon/carbon nanotube membranes: a step towards the electrochemical CO_2 refinery. Angew Chem Int Ed 2017;56:7847–52. https://doi.org/10.1002/anie.201703720.
[59] Faunce T, Styring S, Wasielewski MR, Brudvig GW, Rutherford AW, Messinger J, et al. Artificial photosynthesis as a frontier technology for energy sustainability. Energy Environ Sci 2013;6:1074–6. https://doi.org/10.1039/c3ee40534f.
[60] Pander JE, Ren D, Huang Y, Loo NWX, Hong SHL, Yeo BS. Understanding the heterogeneous electrocatalytic reduction of carbon dioxide on oxide-derived catalysts. ChemElectroChem 2018;5:219–37. https://doi.org/10.1002/celc.201701100.
[61] Zhang WJ, Hu Y, Ma LB, Zhu GY, Wang YR, Xue XL, et al. Progress and perspective of electrocatalytic CO_2 reduction for renewable carbonaceous fuels and chemicals. Adv Sci 2018;5:1700275. https://doi.org/10.1002/advs.201700275.
[62] Fridman A. Plasma chemistry. New York: Cambridge University Press; 2008.
[63] van Rooij GJ, van den Bekerom DCM, den Harder N, Minea T, Berden G, Bongers WA, et al. Taming microwave plasma to beat thermodynamics in CO_2 dissociation. Faraday Discuss 2015;183:233–48. https://doi.org/10.1039/c5fd00045a.
[64] Moss M, Reed DG, Allen RWK, Styring P. Integrated CO_2 capture and utilization using non-thermal plasmolysis. Front Energy Res 2017;5:20. https://doi.org/10.3389/fenrg.2017.00020.
[65] Wilson G, Travaly Y, Brun T, Knippels H, Armstrong K, Styring P, et al. A vision for smart CO2 transformation in Europe. CO2Chem Media & Publishing Limited; 2015, ISBN 978-0-957258839.

Greener farming: managing carbon and nitrogen cycles to reduce greenhouse gas emissions from agriculture

19

Laurence G. Smith[1], Nicolas H. Lampkin
The Organic Research Centre, Hamstead Marshall, Newbury, United Kingdom
[1]Corresponding author: laurence.s@organicresearchcentre.com

Chapter Outline

Managing Global Warming. https://doi.org/10.1016/B978-0-12-814104-5.00019-3

19.1 Climate change and agriculture

Climate change can be defined as a "problem of many hands," [1] i.e., with so many actors no one person can be reasonably held to account. While this diffusion of responsibility has led to delays in progress toward greenhouse gas (GHG) reduction, there has been an increasing awareness of the important role that agriculture and the food system needs to play in ensuring progress toward global mitigation. Unlike many industrial processes, which rely on fossil energy and material inputs, agriculture, like forestry, is a biological process involving management of carbon (C) and nitrogen (N) cycles where these elements are both fixed (captured) and released. Opportunities for GHG mitigation within farming are therefore related to the better management of C and N cycles.

Improvements in the management of these cycles area are now seen as a critical to ensure that global temperatures stay under 2°C, and recent estimates highlight the scale of the issue (agriculture is estimated to be responsible for around a quarter of global GHGs [2]). At the same time, there is recognition of an urgent need to produce enough safe, nutritious food for future generations, with populations predicted to expand to 9 billion by 2050 [3]. Progress in both areas must be achieved alongside the protection of natural resources, to ensure the adequate provision of ecosystem services (e.g., biodiversity, safe clean air and water). Addressing this multifaceted problem has been highlighted as one of greatest obstacles currently faced by humanity [4].

Despite the scale of the challenge, the opportunities for progress are substantial, in particular as agriculture is one of the main sources of the high-impact GHGs methane (CH_4) and nitrous oxide (N_2O). These gases are considered at least 28 and 265 times worse than CO_2 over 100 years, respectively [2]. Livestock farming, is the primary source, responsible for about 50% and 60% of the human-induced emissions of CH_4 and N_2O, through enteric fermentation (the process by which CH_4 is generated through digestion of fibre in ruminant livestock) and through the nitrogen losses associated with manure and mineral fertilizer use [5,6]. Livestock farming is also responsible for the bulk of the emissions from land-use change, with vast areas of tropical forest in South America being converted to cropland for feed production, in particular following the expansion of soy production in the global south (in Brazil and Argentina, the area planted with soy rose from just over 10 million ha in 1980 to 48 million ha in 2009 [7]). Emissions from deforestation alone accounted for 8% of total anthropogenic emissions in 2010, and although this has reduced from levels of 17% in the 1990s, land-use change for agricultural purposes still represents a major source of anthropogenic emissions that need to be addressed through improved governance, infrastructure, and environmentally efficient production [2].

Given the magnitude of livestock's impact, some have recommended a widespread reduction in meat consumption. At the same time it is recognized that a complete conversion to more vegetarian consumption could have unintended consequences, for example in relation to the supply of by-products from livestock production such as manure and leather [8]. In addition productivity increases may be more effective than adjusting consumer demand patterns from a GHG mitigation perspective [9]. In addition, crops produced for human consumption represent a considerable component of the total GHG contribution from agriculture, with N_2O emissions from N fertilizer application and CH_4 emissions from paddy rice cultivation particularly important sources [2]. It is also recognized that livestock farming can play a useful role in the promotion of greenhouse gas mitigation through soil carbon sequestration in grazing systems, although net gains in this area should exceed increases in CH_4 emissions for benefits to accrue [10].

Management of carbon in cropping systems can also encourage better GHG efficiency through reduced reliance on manufactured inputs (e.g., through the use of composts and green manures) and better soil quality [2]. Managing nitrogen more effectively can also help to promote GHG mitigation through N_2O reduction, and in particular the redistribution of nitrogen from areas, which currently apply too much (such as China) to areas where nitrogen supply is likely to be limited (such as in sub-Saharan Africa) could help to improve productivity while reducing environmental losses [11]. Replacing manufactured nitrogen fertilizers with "natural" sources, such as legumes, which capture N through biological fixation in root nodules and through the effective use of manures, composts, and slurries, can also offer effective mitigation strategies, depending on the application context [12].

Agriculture and the food system therefore represent both a source and a (potential) sink in the context of global GHGs. They also represent sectors in which the entire human race plays a part, either as a consumer, a producer or both. While the scale of the challenge is therefore somewhat daunting, there are also considerable opportunities for improvement through adaptation to farming practice, farm structures, and/or consumer behavior. This chapter will explore these current issues in more detail, and ways in which natural cycles can be harnessed and managed more effectively with particular regard to carbon (C) and nitrogen (N).

19.2 Greener farming systems

Greener, more environmentally friendly, farming systems potentially have a significant contribution to make to reducing GHG emissions. There is a wide range of examples of such systems, including those that improve the *efficiency* of input use, for example through precision farming, and those that reject certain inputs or practices and *substitute* more benign ones. In this chapter, we focus more on examples of approaches that involve *redesigning* systems to optimize the ecological interactions underpinning the systems, variously characterized as functional biodiversity, ecofunctional intensification, agroecosystem management, and agroecology. Examples of such approaches include integrated pest management, integrated farming, conservation agriculture, organic farming, agroforestry, and permaculture [5].

19.3 Improvements in carbon cycling for greenhouse gas mitigation

The carbon cycle is a natural process vital for the development of all life on Earth. Effective management of carbon in plant biomass and agricultural soils is particularly crucial for ensuring the long-term sustainability of the food system. Although the Earth's carbon cycle had been able to self-regulate for millennia, human activity has recently created an imbalance in this natural process, through the burning of fossil fuels, which has released massive amounts of CO_2 into the atmosphere. This has led to raised temperatures and increased occurrence of droughts and flooding across the globe. Farming systems and individual practices can help to redress this by building carbon reserves within above- and below-ground biomass. A recent estimate suggests that increases in Soil Organic Carbon (SOC) alone have the potential to offset around 7% of total global emissions [2]. Above-ground biomass generation can also lead to substantial greenhouse gas offsets, and in a "no-animal products" scenario, it would be theoretically possible to provide enough food for 9 billion people while reducing the impacts of land-derived greenhouse gas emissions to one-third of a reference "business-as-usual" case, in part by regrowing vegetation on freed-up pasture and cropland [13].

There is an increasing recognition of the offset potential of agriculture within international climate policy and there have been recent moves to promote carbon sequestration in countries faced with extreme soil degradation (e.g., in Australia where the Government has introduced a "Carbon Farming Initiative," which allows farmers to generate income by restoring vegetation in agricultural landscapes). The "4 per 1000" initiative agreed at the 2015 United Nations Climate Change Conference also aims to increase global soil carbon stocks as a compensation for global emissions from anthropogenic sources [14] and although the ambitious targets set within this initiative may be scientifically flawed, due to a lack of sufficient available-N [15], increasing soil carbon concentrations is still a worthy aim and increased soil carbon can make a useful contribution to GHG mitigation. A range of system and practice level approaches that can be applied to help achieve such objectives and are explored—see later.

19.3.1 Mixed farming

A focus on productivity and the generation of economies of scale within farming have led to a major increase in specialized production since the mid-20th century and a separation of cropping and livestock production [16]. While this trend has led to increased food outputs per unit of land, it has been accompanied by increased environmental degradation within agricultural areas, in particular as a result of the pollution resulting from the use of agrochemical inputs and soil carbon losses [17]. A more integrated approach can help to reduce the climate-related impacts associated with the production of animal feed by ensuring that a greater proportion of animals' nutritional requirements are met by grassland and other forage crops produced on farm [7]. Such methods can help offset the impacts associated with the recent growth in feedlot-based production, i.e., systems that depend on grain-based feed that is often grown on deforested land in South American countries [18]. The use of mixed-farming also allows for the transfer of manure from livestock to cropping areas, allowing for a recoupling of C and N cycles, improved soil organic carbon concentrations, and reduced manufactured N-fertilizer use [19]. Wilkins [16] also highlights the potential role of "mixed farming at a distance" (i.e., transfer of manures from specialized livestock to cropping farms), although the practicalities of transporting manures and slurries over any distance may be prohibitive.

Whether mixed approaches genuinely offer a more efficient alternative, in terms of total GHG reduction, is still a matter of some debate, as many mixed farms will fail to reach the economies of scale associated with more specialized, intensive modes of production [20,21]. In any case, the flexibility inherent in mixed crop-livestock systems constitutes a real strength that could allow for greater long-term environmental sustainability in livestock production, in the face of increasingly challenging environmental conditions [5].

19.3.2 Integrated farming and conservation agriculture

It has also been suggested that integrated production systems applying organic and nonorganic practices may represent a "sustainable optimum" by obtaining the best of both worlds (i.e., combining lower-input approaches used within the organic sector with targeted manufactured N fertilizer and herbicide application to maximize yields and reduce weed burdens [22]). Such systems can reduce the land-use demands associated with lower-yielding forms of agriculture by improving productivities per hectare or acre [22]. They have also been shown to increase soil organic matter in arable rotations through the adoption of zero tillage systems and cover crops, albeit supported by the use of herbicides and [23,24]. The development of integrated farming has been encouraged within Europe by organizations such as Linking Environment and Farming (LEAF) in the UK who define a certified IFM (Integrated Farm Management) approach for farming as "*a whole farm business approach that delivers sustainable farming*" although studies comparing the performance of integrated, conventional and organic systems are rare, work that does exist in this area has highlighted the potential benefits that the integrated approach provides in terms of fossil-energy efficiency and economic performance [5].

19.3.3 Organic farming

Organic farming places a special emphasis on long-term sustainability and working with ecological principles and practices in the design and management of agroecosystems, in order to protect natural resources [5,25,26]. Development of soil health is a central tenet and the combination of practices applied within these systems has been found to offer real improvements in terms of soil carbon reserves, in particular through the use of ley/arable rotations as in mixed farming, but emphasizing legumes for biological nitrogen fixation and the application of organic fertilizers [27]. Although organic management can offer an effective, carbon-friendly management system in terms of both sequestration and reduced emissions per unit of land area, lower yields on organic farms may lead to a requirement for increased agricultural land areas in the absence of changes in human diets or in the use of human edible proteins for feeding livestock [28].

This could result in increased emissions overall at a global level and most system comparisons to date have missed this important element [29]. Studies comparing the effects of intensive and extensive/semiextensive farming systems have drawn similar conclusions, i.e., that the environmental impacts from lower yielding and "wildlife-friendly farming methods" can be greater than more intensive production, due to the need for more cultivated land [30]. Although such comparisons can miss "real-world" effects (in particular the effect of economic drivers on agricultural expansion

[31]), the conclusions drawn still support the need for more production from fewer resources, as one part of any progress toward "Sustainable Intensification" [5]. Others highlight that changes to the food system (e.g., reducing waste) could allow for the "best of both worlds," i.e., that reducing the demands placed on agriculture would allow for adequate food production and at the same time make the widespread adoption of less intensive approaches such as organic farming feasible [28,32].

Organic farms also place a special emphasis on creating a closed system with respect to carbon management through the integration of plant and livestock production (i.e., the grass/clover from leys and arable crops provides feed for livestock and the livestock provide a mobile source of organic matter for fertilizing grassland and crops through manure and slurry [33]). Ruminants therefore play a vital role in many organic systems, by making productive use of fodder legumes, which also provide the main source of nitrogen to the farm. This approach aligns with European organic standards, which dictate that at least 60% of the diet (on a dry matter basis) for ruminants shall be forage based. Within organic pig and poultry farms, emphasis is also placed on providing access to forage via ranging areas. This approach can reduce the need for imported feed and recent work highlights that foraging areas can provide for 50% of the of the feed energy requirements in organic pig production through crops such as lucerne and other legumes with a high-protein content [34]. A study of six poultry meat farms in central Italy also highlighted the greenhouse gas benefits of the lower input approaches used within organic poultry production, with particular regard to reduced feed inputs. The same study highlighted the need for a greater emphasis on the benefits provided by ranging areas with regard to ration formulation [35]. A lack of suitable breeds hinders developments in this direction however, and improvements in this area are likely to be necessary for environmental and animal welfare improvements to be realized within the organic sector [36].

19.3.4 Agroforestry and permaculture

The agroforestry approach (i.e., integrating perennial trees and shrubs with crop and livestock production) represents another agroecological approach to production that can enhance carbon capture while reducing pressure on existing forests [5]. Permaculture is a form of agroforestry that also considers the design of the farm from an energetics perspective. Such systems have the potential to increase the total production of biomass and carbon capture compared with the production of crops and trees separately, through improved solar energy capture for photosynthesis. The return of leaf litter to the soil also contributes to soil carbon sequestration. The potential for agroforestry systems to sequester carbon depends on a number of factors including system design, tree density per unit area, species composition and age, environmental factors such as climate, management, and the end product. Schroeder estimated average carbon storage by agroforestry systems as 9, 21, 50, and 63 Mg C ha^{-1} in semiarid, subhumid, humid, and temperate regions, with higher rates in temperate regions reflecting longer rotations and longer-term storage [37]. In a trial in Oregon, USA, a silvopastoral system with Douglas-fir (*Pseudotsuga menziesii*)/perennial ryegrass

(*Lolium perenne*)/subclover (*Trifolium subterraneum*) was found to be more efficient in storing C than tree plantations or pasture monocultures [38]. In the 11 years since establishment, the silvopastoral system had accumulated 740 and 520 kg ha^{-1} year^{-1} more C than forests and pastures, respectively. This was thought to be a result of higher biomass production and more active nutrient cycling patterns within the silvopasture system compared to tree and pasture monocultures. Intercropping also allows for multiple crop outputs (e.g., fruit, nuts, livestock products) from the same area of land and improved weight gains from browsing and increased shade provided to animals (i.e., increased shade limits high body temperatures and appetite suppression [39]).

19.3.5 Pasture-based approaches

The adoption of pasture-based livestock rearing also represents an agroecological approach that can be used to promote improved carbon management in agriculture, through improved grassland management. The use of forage-based diets in ruminant livestock farming has seen increased popularity in recent years through the work of organizations such as the Savory Institute in the US and the Pasture Fed Livestock Association in the UK, who promote the use of holistic grazing methods such as rotational and mob grazing (i.e., approaches that consist of long resting periods followed by short grazing periods with high stocking rates). The proponents of such approaches claim that such methods can improve soil carbon contents by encouraging deep rooting through "shocking" plants and by trampling organic matter into the soil [40]. A recent claim suggests that the widespread adoption of such methods could lead to substantial removals of carbon from the atmosphere, i.e., over 500×10^9 t (500 billion tonnes) over 40 years, which would be enough, to "reverse climate change" since about 555×10^9 t carbon (i.e., 2035×10^9 t of CO_2) have been released into the atmosphere since the start of the industrial revolution [40,41]. Recent reports have found that these claims are unsupported by scientific evidence and overestimate the eligible grazing land to which holistic grazing practices could be applied [41,42]. Nevertheless, the adoption of improved grazing methods can offer a partial solution to agricultural GHGs, and improved grassland management is likely to help restore soils on the vast areas of agricultural land that are declining or in a degraded state (i.e., over 30% of the world's agricultural soils according to a recent estimate [43]).

Making efficient use of the grassland areas of the world (i.e., between 20% and 47% of the Earth's area [44]) is also key to the development of food security and grazing livestock (e.g., cattle, sheep) are uniquely placed to make effective use of these areas [45]. Recent studies have tended to ignore this "land-use" question and the non-human-edible component of livestock diets in determining feed conversion efficiencies when positioning ruminant livestock as poor convertors of feed (see Table 19.1), for example Godfray et al. [3] state that the "feed required to produce 1 kilogram of meat" was 8, 4, and 1 "kilogram of cereal per animal" for cattle, pigs, and broiler chickens, respectively. Such assessments, and the recommendations that

Table 19.1 Feed Conversion Efficiencies for livestock products expressed in terms of total feed energy used (in Megajoules—MJ) per unit of food energy produced. When considering the total amount of energy used, ruminant livestock products (milk, beef, lamb) tend to perform worse than monogastric livestock (pig and poultry meat, eggs). When considering the human-edible component of the diet, the situation is reversed, i.e., the mainly grass-fed systems such as upland suckler beef perform better than monogastric production due to the high grassland/ forage component in the diet

	Total energy efficiency (MJ input per MJ of output)	**Human edible energy efficiency (MJ input per MJ of output)**
Milk	4.5	0.47
Upland suckler beef	40	1.9
Lowland suckler beef	37	4.2
18—20 month beef	23.3	3.2
"Cereal" beef	13.2	6.2
Upland lamb	62.5	3.6
Lowland lamb	52.6	2.5
Pig meat	9.3	6.3
Poultry meat	4.5	3.3
Eggs	4.9	3.6

Adapted from Wilkinson J. Re-defining efficiency of feed use by livestock. Animal 2011;5:1014–1022.

build on them, miss the importance of feed resource type in the rearing of livestock, as illustrated in Table 19.1 [17], and that 60% of global agricultural land is pasture.

Despite the advantages that ruminant livestock can provide by making good use of grassland and other non-human-edible feeds, it should be remembered that ruminants do not add carbon to the system, they can only convert it into forms that are able to be used more effectively (e.g., via animal products and manure). At the same time, ruminant agriculture is the main source of CH_4 and emissions from the sector are increasing as livestock numbers expand to keep up with rising demands [41]. In addition, a reliance on grassland/forage within pasture-fed systems can lead to higher CH_4 emissions per unit of product, as a result of reduced feed digestibility and increased dry matter intakes [46] although severe difficulties relating to health and longevity can occur in cereal/concentrate intensive systems (e.g., through increased acidosis), which may, paradoxically, lead to an increase in herd size and greater CH_4 emissions overall [47]. Lower replacement rates on dairy farms feeding forage-based diets can also lead to reduced greenhouse gases if dairy calves produced are used for beef production, with the increased longevity of lower input herds reducing or offsetting the emissions associated with the unproductive rearing of dairy cows and the need for suckler cows [7,48] although this depends on a demand for the milk product as the supply from suckler beef and associated GHGs would increase if the demand for milk falls and the demand for meat increases.

19.3.6 Individual practices for improved carbon management

Individual practices are also being encouraged to improve soil carbon management in agriculture, for example within the French Ministry of Agriculture's action plan for the development of sustainable agroecological production, which seeks to encourage the uptake of measures such as manure and compost application on agricultural soils. Full-greenhouse gas accounting should be applied when considering the impacts of such practices, as soil carbon increases may be offset by increased emissions from other sources (e.g., increased N_2O from manure storage or increased CH_4 emissions from livestock) [49,50]. The incorporation of fertility building leys within crop rotations can also improve soil carbon concentrations by increasing the amount of carbonaceous material incorporated within the soil, and a review of field studies carried out in the US found that legume and manure-based farming systems achieved similar levels of SOC increase, the results suggesting that the ley period alone is more significant than additions of manure, within farming systems applying both methods [51].

The use of cover crops (i.e., crops that replace bare fallow during winter period that are plowed under as green manure before sowing of the next main crop) also has the potential to build levels of soil carbon in a range of cropping situations by adding biomass to the soil. A recent study estimated that a global adoption for cover crops could compensate for 8% of global GHG emissions from the agriculture sector through increased soil carbon concentrations, [52]. Cover crops offer the additional advantage of not requiring additional land compared to other extensification methods [52].

The adoption of reduced or zero tillage, a central focus of Conservation Agriculture, also has the potential to increase soil carbon concentrations within cultivated soils, [53] although zero tillage is difficult in organic systems because the development of weeds associated with this system cannot be controlled with the use of herbicides, only by mechanical weed control [54]. Recent work has also highlighted that the increases of SOC from reduced tillage are often associated with higher concentrations near the soil surface, rather than an increase in SOC stocks, and that the need to combine reduced tillage systems with inversion tillage to control perennial weeds will almost certainly result in any gains being lost through mineralization [23]. Others have highlighted that the larger soil aggregates, low gas diffusivity, and greater water retention near to the soil surface that results from reduced tillage can result in making the soil less aerobic and lead to higher N_2O emissions [55].

The production of biochar (i.e., the solid, charcoal-like product prepared from the pyrolysis of various crop residues and woody materials) within agriculture also has potential to offset GHG emissions by providing a renewable source of energy while simultaneously converting woody and plant biomass to stable forms of C [56]. A recent study estimates that its widespread uptake could offset 12% of current anthropogenic GHG emissions without endangering food security, soil conservation, or habitat provision, although the authors highlight that strict sustainability protocols should be ensured where biochar production is applied, in order to avoid excessively long carbon-payback times, during which net emissions are increased before any net reduction is achieved [57].

19.3.7 Carbon sequestration and saturation limits

Any comparison of systems or practices should also consider that an increase in soil carbon concentrations that results in a change in management is finite and only likely to occur within the first 20 years, as a new steady state is reached. Long-term studies have confirmed this trend, finding that rates of accumulation over time are nonlinear and will differ according the previous land use and the amount of organic material applied in recent years. For instance, the 140-year Broadbalk Experiment at Rothamsted Research, UK, found that on the "farmyard manure" plots the rate of increase was greatest in the early years of application, and reduced over time as the soil approached a new steady state [58]. It is therefore essential that the initial, large gains in soil carbon sequestration, resulting from the addition of organic materials (e.g., livestock manure), are not extrapolated year on year under an assumption that the same increase will be observed indefinitely [58]. Smith et al. [59] support this view, pointing out that terrestrial biomass sources only remove carbon from the atmosphere until the maximum capacity for the ecosystem is reached—which may take 15–33 years, depending on the farming system and/or management practice. This phenomenon is referred to as "sink saturation" and is illustrated in Fig. 19.1.

Soussana et al. [50] contest the concept of a saturation point in agricultural soils through their assessment of the greenhouse gas budget of nine European grassland

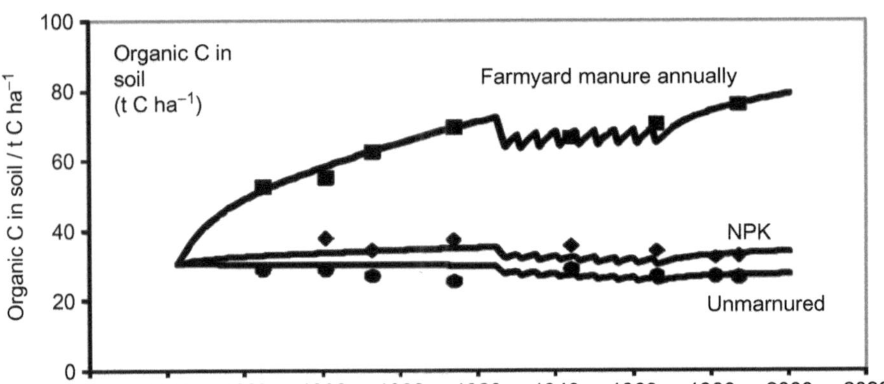

Fig. 19.1 The accumulation of total organic carbon in the topsoil (0–23 cm) of silty clay loam soils at Rothamsted, UK, under three treatments: no fertilizers or manure applied since 1844; P, K,Mg plus 144 kg(N) ha^{-1} since 1852; farmyard manure at 35 t ha^{-1} fresh weight applied since 1885 plus 96 kg(N) ha^{-1} since 1968.
Source: Powlson, DS, Whitmore, AP, Goulding, KWT. Soil carbon sequestration to mitigate climate change: A critical re-examination to identify the true and the false. Eur J Soil Sci 2011;62:42–55; Smith LG, Williams AGW. The effects of organic management on greenhouse gas emissions and energy efficiency in livestock production. In: Vaarst M, Roderick S (editors). Improving organic animal farming. Cambridge: Burleigh Dodds Science Publishing (in press).

sites, which used an eddy-covariance method to measure the Net Ecosystem Exchange (NEE) of carbon at a fine temporal resolution. Losses of carbon as CH_4 were also measured across the sites, and the exports and imports of organic-C as manure, animal body mass, and harvested plant biomass were recorded and accounted for on top of the NEE. The results revealed that established grassland sites showed a large sink potential through a negative carbon balance (i.e., C imports exceeded exports), whereas the newly sown grass-clover mixtures displayed a net loss. The results from this study contradict the widely held view that permanent grassland systems tend to be at the sink-saturation point referred to in Fig. 19.1, and that newly sown grasslands can store more carbon than established pastures [50] although others suggest that a perpetual increase in soil carbon concentrations under long-term/permanent grassland is unjustifiable, given observed results from long-term trials [60].

19.4 Improvements in nitrogen cycling and nitrogen-use efficiency for greenhouse gas mitigation

Managing nitrogen effectively in agricultural systems is also essential for ensuring the continued productivity and sustainability of modern agriculture. Despite this, an exponential increase in the amount of manufactured N fertilizer applied to cropland over the last 70 years has not been accompanied by any substantial increase in nitrogen-use efficiency (NUE), with the amount of N fertilizer taken up by crops remaining at a relatively low level (over 50% of the N applied to cropland being lost to the environment according to a recent estimate [61]). At the same time, NO_x emissions from agroecosystems have been increasing, causing losses of biological diversity, eutrophication, dominance of certain weed species, and increased greenhouse gas emissions [62], with a recent estimate suggesting that CO_2 and N_2O associated with the manufacture and use of N fertilizer used on cropland are responsible for 70% of all agricultural GHG emissions [58].

Substantial room for improvement therefore remains and there is a huge potential to decrease rates of nitrogen application across the globe while maintaining current levels of production [61,63]. A range of farm system and practice-level interventions could be applied to improve performance in this area and the benefits that could accrue from their adoption are outlined in the following sections.

19.4.1 Mixed farming

As highlighted earlier, a better integration of cropping and livestock production can potentially contribute to increasing production efficiencies at the local and global scales [61]. As with carbon cycling, improved nitrogen-use efficiency can be obtained by providing feed for animals from legume and grass mixes within leys on the same farm. This reduces the use of imported fertilizer used for cropping by supplying nitrogen through biological fixation within fertility building leys [19]. At the same time, mixed farming systems can help avoid the stockpiling of nutrients in the form of manure and slurry, as is often seen in specialized livestock systems, leading to substantial increases in N_2O emissions and other environmental problems [54,64]. High rates of net productivity and NUE can also be achieved within mixed farming systems, by making effective use of grassland and cropping areas to maximize outputs per hectare or acre of land [5,22].

19.4.2 Integrated farming

The adoption of integrated farming approaches can also help to address nitrogen-use efficiency in agriculture, and in a comparison of integrated, organic, and conventional farming, the use of reduced tillage, early drilling of autumn-sown crops, and mineral N measurements led to optimal NUEs within an integrated system [65,66]. The results from this study suggested that application of appropriate controls for minimizing pest, disease, and weed burdens combined with farmer knowledge and application of planning tools (nutrient budgeting combined with soil tests) are key elements in attaining improved performance [65]. Groot et al. [67] also found that gradual adjustments to fertilizer application, feed rates and milk production could lead to improved nitrogen efficiencies and better economic performance on dairy farms applying integrated, lower-input approaches to production.

19.4.3 Organic farming

Organic systems, in addition to securing the benefits from integrating crops and livestock described here, can also contribute to improved NUE in agriculture, by avoiding the emissions associated with manufactured N fertilizer referred to earlier (the main source of N within organic farms is biological nitrogen fixation, within the fertility building ley period of the crop rotation [5]). In addition, organic farms can ensure that N is retained in forms that are more stable and thereby help to reduce losses of N_2O to the atmosphere, in particular through the use of legumes and manures [19]. Despite these advantages, it has been highlighted that the N use efficiency of organic fertilizers is low, compared to equivalent quantities of inorganic fertilizer, as the N can often be released too late in the season, or after the growing season has ended [68]. Such effects are often influenced by factors that are beyond the direct control of the farmer, such as the soil moisture, aeration, and temperature, which affect the rate of N mineralization from organic material [47]. These effects have led some authors to conclude that the N_2O emissions associated with the application of manure can be higher than for mineral fertilizer, depending on soil type [69]. The amount of nitrogen supplied by legumes can also be highly uncertain and difficult to predict, with ranges of <100 to $700\,kg(N)\,ha^{-1}\,yr^{-1}$, depending on the legume that is used [70]. Claims have therefore been made that the total N budget on organic farms will be inherently more unreliable than for conventional farms where the total N input (as manufactured fertilizer) is more easily predicted [71].

Improvements in N use efficiency on organic farms are particularly related to improvements in timing of crop cultivations and sowing/planting of crops following legumes [72], and in the effective use of manures as a nutrient resource [47]. This management must take into account the fact that nutrients from manure are available at a slower rate compared to mineral fertilizer, that fresh manure contains more readily available nitrogen than composted material, and that much of the available N in manure may be lost through the release of ammonia (NH_3) during composting [73]. Manure analysis and improved timing may help to improve management in terms of improving the prediction of available N [47]. Nitrogen leaching from manures can also be reduced, through the use of cover and catch crops and green manures, which can also improve soil structure, and enable significantly lower mobile nitrogen concentrations, further reducing N_2O emissions [5]. Improvements in N utilization can also be made by breeding crops to improve their N-use efficiency [74].

19.4.4 Agroforestry systems

Improved nitrogen-use efficiency in crop production can also be encouraged through the use of agroforestry systems, in particular through deposition of leaf litter by tree species that produce high amounts of leaf biomass (e.g., hybrid poplar), which can reduce the need for mineral fertilizer and the associated N_2O emissions [75], as well as the use of nitrogen-fixing tree species. Decreased leaching in agroforestry systems may also have the potential to reduce N_2O from denitrification in surface water [39]. The ammonia (NH_3) abatement potential of agroforestry systems has also been highlighted through a recent project, which illustrated the benefits of enhanced tree cover in agricultural landscapes [76].

19.4.5 Individual practices for improving nitrogen-use efficiency (NUE) and reducing N_2O emissions

Increasing the global supply of nitrogen from legumes via symbiotic N fixation has been highlighted as one of the most promising options for improving NUE and reducing N_2O emissions in agricultural systems. A recent study highlights that that NUE is generally higher (and the N surplus relatively lower) for agricultural systems and individual countries with a higher proportion of N inputs derived from symbiotic N fixation (see Fig. 19.2, [61]).

Increasing the presence of peas, beans, and other legume crops within cropping and livestock systems is therefore likely to contribute to GHG mitigation [77]. Developments in this area will require changes to national consumption patterns and technological developments to allow for greater "marketability" of legume-derived products (unlike cereals, legumes have received little attention in this area [77]). Reforms to ration formulations for livestock and a greater focus on funding streams for farmer-focused education and research for the improvement of rates of N fixation could also help to encourage uptake [78]. Similarly within grazing systems, improving the supply of N from clover and other legumes within forage production can lead to improved performance with respect to N_2O efficiency and productivity through a reduced reliance on manufactured fertilizer, improved control of internal parasites,

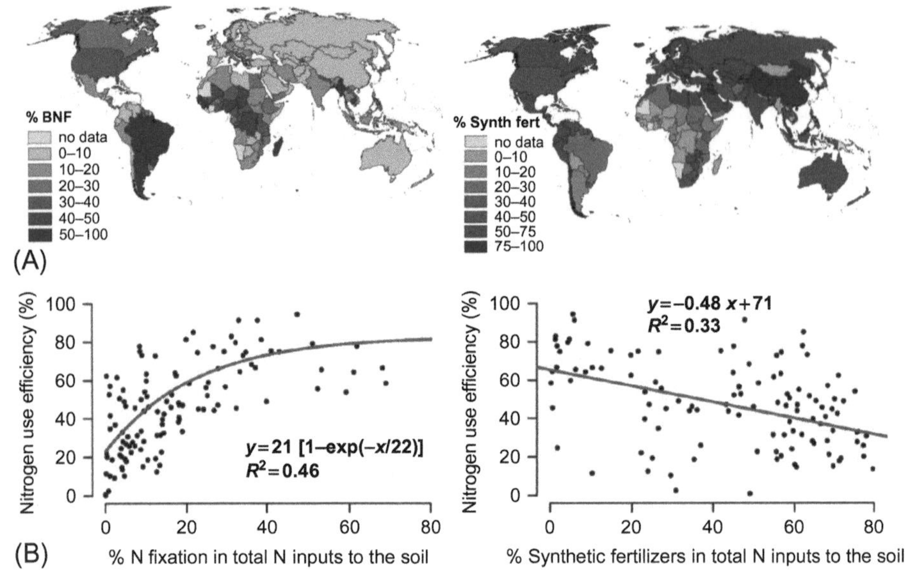

Fig. 19.2 (A) Share of nitrogen derived from Biological N Fixation (BNF) and synthetic N fertilizer as a percentage of total N inputs to soils on croplan in 2000–2009. (B) the relationship between NUE and the proportion of N in soils from biologically fixed nitrogen or synthetic fertilizer in 2000–2009. The results serve to highlight the high N efficiency of legume-based production, and areas where these systems dominate such as soybean-producing areas in South America and countries in Asia and Africa producing significant amounts of rice, groundnuts, and beans.
Source: Lassaletta L, Billen G, Grizzetti B, Anglade J, Garnier J. 50 year trends in nitrogen use efficiency of world cropping systems: the relationship between yield and nitrogen input to cropland. Env Res Lett 2014;9:105011.

and increased levels of production per hectare [5]. The numerous benefits that legumes can provide in terms of reduced N_2O, improved soil carbon sequestration, biodiversity, productivity, and human health have led to claims that they should play a major role in the securing the future sustainability of the food system [77,79].

As within the field of carbon management, cover cropping can also help to improve NUE within a range of agricultural systems. Applying cover cropping alongside targeted manufactured N fertilizer can lead to highly N-efficient production systems by overcoming the N synchronization issues associated with a reliance on organic sources of fertility (see earlier) while reducing N losses [80]. Organic fertilizers with a high N availability (e.g., digestate from anaerobic digesters, poultry manure) could also be used to improve NUEs in cropping systems by supplying N in readily available forms at times of peak demand, although such sources can increase the occurrence of nitrophilous weeds, can lead to a reduction in SOC contents in the case of digestate [81], and may have other impacts on soil and plant health. The use of perennial arable crops can also help to reduce nitrogen losses in a range of systems by keeping the soil covered and improving N synchrony, although lower yields, weed susceptibility, and pest and disease management issues may limit uptake [81].

The application of modern technological solutions can also offer substantial improvement in NUE within a range of farming systems, in particular the use of precision farming technology that can allow for fertilizer application to follow the "4 R's" principles (i.e., the right source, at the right rate, at the right time, in the right place [82]) and through the use of novel sensors (e.g., for remote sensing of spatial patterns in chlorophyll content in plants) to allow for detection of plant stress and application of N at times of peak demand [83]. The use of synthetic additives (i.e., nitrification or urease inhibitors) can also help to reduce N_2O emissions within a range of cropping systems, while the organic amendment of biochar can be used to the same effect (i.e., decreasing N_2O emissions by facilitating the transfer of electrons to soil-denitrifying microorganisms, which promotes the reduction of N_2O to N_2 [4,84]).

19.5 Context specificity and data sources used in greenhouse gas assessments

It is important to note that the relative efficacy of many of the practices (e.g., legumes vs manufactured N fertilizer) and systems (e.g., organic vs nonorganic) identified earlier will depend on the unit of comparison, i.e., there is a general trend for lower GHG impacts per unit of land within low-input systems, whereas the differences are less clear per unit of product, as illustrated in Fig. 19.3.

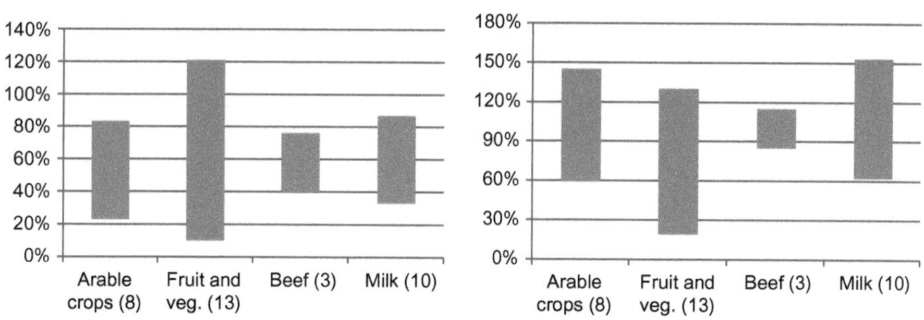

Fig. 19.3 Results from a review of Life Cycle Assessment (LCA) based comparisons of the Global Warming Potential of organic/nonorganic production. The impacts of organic production are expressed as a percentage of nonorganic per unit of area (left) and per unit of per unit of product (right).
Source: Meier MS, Stoessel F, Jungbluth N, Juraske R, Schader C, Stolze, M. Environmental impacts of organic and conventional agricultural products—are the differences captured by life cycle assessment? J Environ Manage 2015;149:193–208; Smith LG, Williams AGW. The effects of organic management on greenhouse gas emissions and energy efficiency in livestock production. In: Vaarst M, Roderick S (editors). Improving organic animal farming. Cambridge: Burleigh Dodds Science Publishing (in press).

There is considerable debate over the most suitable approach, with some suggesting that a unit of product-based assessment of greenhouse gas performance is the most relevant metric, in view of growing populations, increasing demands for food and limited land areas [29] and others suggesting that a land area-based comparison better reflects a farming system's function as a producer of nonmarket goods (e.g., biodiversity [85]). Comparisons based on the amount of product can also overlook the wide disparity that exists in the nutrient and water content of contrasting foodstuffs such as eggs, milk and meat, and such comparisons can hide the increased impacts and land-use associated with imported feed [86]. Additionally, comparisons based on individual product types (e.g., arable crops, beef) may miss the importance of the farming system (e.g., with respect to the amount of land needed for the production of a given amount of food) and the tradeoffs associated with particular modes of production such as organic farming [5].

A number of methods for addressing this issue have been proposed in recent years, for example, the Net System Output (NSO) approach developed by the Organic Research Centre in the UK, which considers the sum total of the energy (GJ) and nitrogen (kg) outputs produced by a farm and converts these to a "wheat equivalent" value (tWeq) for farming systems, using a standard energy and nitrogen value. This allows all commodities and system types, and their associated GHGs, to be assessed on the same basis [5]. Similarly, the Kaya-Porter Identity (KPI) developed by John Porter at the University of Copenhagen applies a system-level assessment to determine the GHG intensity of the energy used for production (kg CO_2 Eq. MJ^{-1}), the energy intensity of the production (MJ kg^{-1} dry matter), an estimate of areal productivity (kg dry matter ha^{-1}), and areal land-based GHG emissions (CO_2-eq. ha^{-1}) within an individual production system, region, or country [87]. By using approaches such as the NSO and KPI, it is possible to deconstruct the impacts of various production systems (as opposed to crops or livestock products) and identify the impacts of their wider adoption.

Data sources and variation in environmental factors and/or typical practices between countries or regions can also lead to wide differences in the estimates of GHGs from livestock. For example, it is important to know whether IPCC Tier 1 (default global or regional emission factors), Tier 2 (regional) or Tier 3 (system specific) approaches are applied in the estimation of global emissions [88]. Such differences can lead to starkly different results in greenhouse gas assessment, in particular as a result of the disparity associated with the calculation of emissions resulting from fertilizer management and application [89]. Further research in this and other areas would allow for the generation of more accurate, system and/or practice specific estimates in determining the total GHG impacts associated with the food system.

19.6 Future research

Many of the interventions and system approaches mentioned here are currently in development and further research is needed to improve the understanding of their impact with regard to the C and N cycles. In particular, an improved understanding of the soil organic carbon dynamics and the sequestration potential associated with a range of agricultural practices would help to ensure progress in this underdeveloped area [90]. Improved understanding of carbon storage in the subsoil is also important to address, and a recent study highlights a major issue relating to the shallow-soil-sampling depth in many studies [27]. With over 50% of SOC residing in the subsoil [91], this is potentially an important omission. Accounting for carbon sequestration in greenhouse gas assessments of products is also a challenge facing the research community and the lack of a consistent basis for accounting for the GHG offsets from SOC increases is hindering developments in this area [92].

Similarly with regard to nitrogen management, more work is needed to reduce uncertainty in the estimation of N_2O emissions that occur on a range of agricultural soils and agriclimatic zones [93]. Research on the efficacy of slow-release fertilizers and nitrification and urease inhibitors within a range of environments could also be applied to allow for the broader uptake of this technology, which can be used to alongside mineral fertilizer application to reduce losses to the environment and save farmers' money [82]. Substantial improvements in nitrogen-use efficiency (NUE) could also be obtained through improved feeding practices, and it has been highlighted that some crops such as lucerne, soybeans, and some cereals may have high NUE as crops, but can also lead to high N losses in manure in livestock systems due to excesses of protein in the diet [94,95]. Balancing the high protein feeds and forages with other feedstuffs that have a higher energy-to-protein ratio (e.g., maize or cereal silages) or providing high-energy supplements (e.g., concentrates and sugar-processing by-products) could help to reduce losses associated with high-protein diets in feeding regimes, although a balance needs to be achieved in order to minimize consumption of human-edible components [94].

Breeding crops and animals for improved NUE could also help to foster improvement, and the use of older animals in dairy farming may help to reduce N excretion per kg of milk, as a greater proportion of the protein consumed is used for milk production, as opposed to maintenance [74,96]. Managing diets for CH_4 reduction also has the potential to address emissions within the livestock sector, and certain feed additives such as oils and nitrates may have the potential to lead to inhibitory effects on methanogenesis within ruminant livestock, although increasing dietary fats in livestock ratios may depress feed intake and consequently reduce animal productivity [97]. The economics of adding vegetable oils to animal diets may also be prohibitive in view of rising prices on the world market and high-oil by-product feeds such as distiller's grains, and meals from the biodiesel industry may provide a more cost-effective alternative [97]. There is also some evidence that livestock diets rich in tannins (e.g., diets with a high clover/legume content) produce less methane than grass-only diets through a suppression of fibre degradation in the rumen [98] although a recent meta-analysis, conducted as part of the UK Greenhouse Gas Platform project, concluded that there is insufficient evidence for high-legume/high-tannin pastures to be considered an effective option for lowering direct CH_4 emissions from ruminant livestock [99] and reduced digestibility and antinutritive factors associated with high tannin diets may also affect livestock productivity [97]. Nevertheless, the nutritional and animal health (anthelmintic, bloat prevention) benefits associated with the use of tannins in livestock diet, combined with the potential for the reduction in N_2O emissions, and the absence of N requirements for plant growth may make these plants attractive from an environmental sustainability standpoint, if applied carefully [97].

More research on the coupling of N and C cycles within agriculture could also lead to improved GHG performance within the sector by reducing N losses to the environment and at the same time ensuring adequate C returns to the soil. Further work in this area could help to ensure optimal stocking densities, which can minimize pressure on ecosystems (e.g., from overgrazing or excess build up of N reserves in the form of manure [19]). More research on the potential benefits of mixed farming systems incorporating crops and livestock could also allow for the more effective transfer of C and N between cropping and animal enterprises [64]. Further work on the combined C sequestration and N_2O emission prevention benefits that biochar provides could also help to encourage its uptake within combined food and energy production systems [56].

A focus on greenhouse gas emissions alone can also eclipse the extent to which farming systems can contribute toward multiobjective and internationally binding sustainability targets (e.g., the UN Sustainable Development Goals [82]) and future research could instead consider how land management can be adapted to promote healthy and sustainable economies using a range of environmental, economic, and social assessment criteria [100]. In this context, it is possible that diverse and low-input farming systems such as organic farming offer greater potential to be successfully woven within agricultural landscapes. This perspective challenges the view that conservation and agriculture are diametrically opposed, instead viewing agricultural systems as part of an integrated whole that can promote the development of healthy and sustainable economies [31,100].

It has also been suggested that future research on agricultural systems requires collaboration and communication between the "science communities" and the "practice sector" to enable the demonstration of economically viable, observable, compatible solutions [101]. Such direct engagement with end users could help to ensure that any solutions identified are applicable to local conditions and socioeconomic contexts. A provision for such "bottom-up" research processes within the European-Commission-funded European Innovation Partnership program will help to encourage progress within this underdeveloped area, although similar publicly funded programs are now required at a global level to ensure that innovation technologies and system-level solutions end up in the hands of farmers [78].

19.7 Conclusions

Agriculture has a major role to play in meeting global greenhouse gas mitigation targets, and managing carbon and nitrogen cycles effectively can help to ensure real progress in this area. Encouraging the uptake of improved practices will require improved application of outreach programs, research funding, and regulatory frameworks [82]. Progress in this area has been slow to date and action is now urgently required to overcome political inertia and to encourage the uptake of innovative technology, practices, and farming systems [102]. At a regulatory level, more concrete, measurable targets with respect to carbon and nitrogen management and other environmental impacts are needed help to foster progress toward the Rio+20 Sustainable Development Goals (SDGs) [82,102]. Encouraging farmer uptake is also likely to require improved approaches to extension that effectively communicates the benefits (in particular, the cost savings) that can accrue from the adoption of greenhouse gas mitigation measures such as nitrification inhibitors and replacing manufactured N with biological fixation by legumes [103].

Simply considering how more food can be produced with less impact can also create more problems than it solves and a broader "food-system" perspective may help to ensure more effective progress toward GHG mitigation [104]. In particular the adaptation of human diets, including reduced meat and processed food consumption, and the reduction of food waste, could allow for the scaling up of forms of agriculture offering greater protection of environmental resources and improved ecosystem services [28,32]. The degree of complementarity between sustainable production and consumption habits is still unclear however and there is a pressing need to account for real-world complexity in the development and application of innovative food systems, particularly in smallholder agriculture, given its current importance in addressing food security [105].

As highlighted earlier, both the scale of the challenge and the opportunities for improvement are considerable. Addressing soil management and carbon sequestration is key given the prevalence of the soil ecosystem, and even if carbon stocks cannot be increased on agricultural land, there is a urgent need to protect what currently remains and to halt the destructive practice of deforestation [41,90]. At the same time, there is an increasing recognition of the complexity of the issues faced within the realm of sustainable land management and agricultural subsidy and a need to develop land use that is appropriate to land type in order to find "least-bad" solutions for feeding growing populations and meeting GHG reduction targets [41].

References

[1] van de Poel I, Fahlquist JN, Doorn N, Zwart S, Royakkers L. The problem of many hands: climate change as an example. Sci Eng Ethics 2012;18:49–67.

[2] Smith P, Clark H, Dong H, Elsiddig E, Haberl H, Harper R, House J, Jafari M, Masera O, Mbow C, Ravindranath N, Rice C, Robledo Abad C, Romanovskaya A, Sperling F, Tubiello F. Agriculture, forestry and other land use (AFOLU). In: Edenhofer O, et al., editors. Climate change 2014: Mitigation of climate change. Contribution of working group III to the fifth assessment report of the intergovernmental panel on climate change. Cambridge, United Kingdom and New York, USA: Cambridge University Press; 2014.

[3] Godfray HCJ, Beddington JR, Crute IR, Haddad L, Lawrence D, Muir JF, Pretty J, Robinson S, Thomas SM, Toulmin C. Food security: the challenge of feeding 9 billion people. Science 2010;327:812–8.

[4] Smith P, Haberl H, Popp A, Erb K, Lauk C, Harper R, Tubiello FN, Siqueira Pinto A, Jafari M, Sohi S. How much land-based greenhouse gas mitigation can be achieved without compromising food security and environmental goals? Glob Change Biol 2013;19:2285–302.

[5] Lampkin N, Pearce B, Leake A, Creissen H, Gerrard C, Girling R, Lloyd S, Padel S, Smith J, Smith L, Vieweger A, Wolfe M. The role of agroecology in sustainable intensification. Report for the Land Use Policy Group. The Organic Research Centre and Game & Wildlife Conservation Trust; 2015.
[6] Harwatt H, Sabaté J, Eshel G, Soret S, Ripple W. Substituting beans for beef as a contribution toward US climate change targets. Clim Change 2017;1–10.
[7] Idel A. Livestock production: A climate change and food security hot spot. In: United Nations conference on trade and development (UNCTAD) trade and environment review. Organisation for Economic Cooperation and Development (OECD); 2013.
[8] Hallström E, Carlsson-Kanyama A, Börjesson P. Environmental impact of dietary change: a systematic review. J Clean Prod 2015;91:1–11.
[9] Havlík P, Valin H, Herrero M, Obersteiner M, Schmid E, Rufino MC, Mosnier A, Thornton PK, Böttcher H, Conant RT. Climate change mitigation through livestock system transitions. Proc Natl Acad Sci U S A 2014;111:3709–14.
[10] Mottet A, Henderson B, Opio C, Falcucci A, Tempio G, Silvestri S, Chesterman S, Gerber PJ. Climate change mitigation and productivity gains in livestock supply chains: insights from regional case studies. Reg Environ Change 2017;17:129–41.
[11] Smith P. Delivering food security without increasing pressure on land. Glob Food Secur 2013;2:18–23.
[12] Niggli U, Fließbach A, Hepperly P, Scialabba N. Low greenhouse gas agriculture: mitigation and adaptation potential of sustainable farming systems. Rome: FAO; 2009.
[13] Stehfest E, Bouwman L, van Vuuren DP, den Elzen MGJ, Eickhout B, Kabat P. Climate benefits of changing diet. Clim Change 2009;95:83–102.
[14] Minasny, B, Malone, BP, McBratney, AB, Angers, DA, Arrouays, D, Chambers, A, Chaplot, V, Chen, Z-S, Cheng, K, Das, BS, Field, DJ, Gimona, A, Hedley, CB, Hong, SY, Mandal, B, Marchant, BP, Martin, M, McConkey, BG, Mulder, VL, O'Rourke, S, Richer-de-Forges, AC, Odeh, I, Padarian, J, Paustian, K, Pan, G, Poggio, L, Savin, I, Stolbovoy, V, Stockmann, U, Sulaeman, Y, Tsui, C-C, Vågen, T-G, van Wesemael, B, and Winowiecki, L. Soil carbon 4 per mille. Geoderma 2017; 292: 59–86.
[15] van Groenigen JW, van Kessel C, Hungate BA, Oenema O, Powlson DS, van Groenigen KJ. Sequestering soil organic carbon: a nitrogen dilemma. Environ Sci Technol 2017;4738–9.
[16] Wilkins R. Eco-efficient approaches to land management: a case for increased integration of crop and animal production systems. Philos Trans Roy Soc B: Biol Sci 2008;363:517–25.
[17] Wilkinson J. Re-defining efficiency of feed use by livestock. Animal 2011;5:1014–22.
[18] Deblitz C. Feedlots: a new tendency in global beef production. Agri Benchmark 2012;.
[19] Soussana J-F, Lemaire G. Coupling carbon and nitrogen cycles for environmentally sustainable intensification of grasslands and crop-livestock systems. Agric Ecosyst Environ 2014;190:9–17.
[20] Ross S, Topp K, Ennos R, Chagunda M. Breeding, feeding and management to reduce the emissions intensity of dairy production. Rural policy Centre research briefing 2014/06. Scotland's Rural College: Aberdeen; 2014.
[21] Peters GM, Rowley HV, Wiedemann S, Tucker R, Short MD, Schulz M. Red meat production in Australia: Life cycle assessment and comparison with overseas studies. Env Sci Tech 2010;44:1327–32.
[22] Tuomisto HL, Hodge ID, Riordan P, Macdonald DW. Comparing energy balances, greenhouse gas balances and biodiversity impacts of contrasting farming systems with alternative land uses. Agric Sys 2012;108:42–9.

[23] Powlson DS, Bhogal A, Chambers BJ, Coleman K, Macdonald AJ, Goulding KWT, Whitmore AP. The potential to increase soil carbon stocks through reduced tillage or organic material additions in England and Wales: A case study. Agric Ecosyst Environ 2012;146:23–33.
[24] Bhogal A, Chambers B, Whitmore A, Powlson D. The effects of reduced tillage practices and organic material additions on the carbon content of arable soils. Summary report for Defra project SP0561. Rothamsted Research/ADAS: Notts/Harpenden; 2006.
[25] Muller A, Bautze L, Meier M, Gattinger A, Gall E, Chatzinikolaou E, Meredith S, Ukas T, Ullmann L. Organic farming, climate change mitigation and beyond: reducing the environmental impacts of EU agriculture. Brussels: IFOAM EU Group; 2016.
[26] Griese S, Bautze L, Trappenberg A-K, Meredith S, Gall E, Wierzbicka J. Viable climate friendly farming: Socio-economic strategies. Brussels: IFOAM EU Group; 2016.
[27] Gattinger A, Muller A, Haeni M, Skinner C, Fliessbach A, Buchmann N, Mäder P, Stolze M, Smith P, Scialabba NEH. Enhanced top soil carbon stocks under organic farming. Proc Natl Acad Sci U S A 2012;109:18226–31.
[28] Muller A, Schader C, Scialabba NE-H, Brüggemann J, Isensee A, Erb K-H, Smith P, Klocke P, Leiber F, Stolze M. Strategies for feeding the world more sustainably with organic agriculture. Nat Commun 2017;8:1290.
[29] Notarnicola B, Sala S, Anton A, McLaren SJ, Saouter E, Sonesson U. The role of life cycle assessment in supporting sustainable agri-food systems: a review of the challenges. J Clean Prod 2017;140(Part 2):399–409.
[30] Green RE, Cornell SJ, Scharlemann JPW, Balmford A. Farming and the fate of wild nature. Science 2005;307:550–5.
[31] Perfecto I, Vandermeer J. The agroecological matrix as alternative to the land-sparing/agriculture intensification model. Proc Natl Acad Sci U S A 2010;107:5786–91.
[32] Reganold JP, Wachter JM. Organic agriculture in the twenty-first century. Nature Plants 2016;2:15221.
[33] Watson CA, Atkinson D, Gosling P, Jackson LR, Rayns FW. Managing soil fertility in organic farming systems. Soil Use Manage 2002;18:239–47.
[34] Crawley, K. Fulfilling 100% organic pig diets: Feeding roughage and foraging from the range: Technical note. Improved contribution of local feed to support 100% organic feed supply to pigs and poultry (ICOPP) project report; 2015.
[35] Castellini C, Boggia A, Cortina C, Dal Bosco A, Paolotti L, Novelli E, Mugnai C. A multicriteria approach for measuring the sustainability of different poultry production systems. J Clean Prod 2012;37:192–201.
[36] van de Weerd H, Keatinge R, Roderick S. A review of key health-related welfare issues in organic poultry production. Worlds Poult Sci J 2009;65:649–84.
[37] Schroeder P. Carbon storage benefits of agroforestry systems. Agroforestry Systems 1994;27:89–97.
[38] Sharrow SH, Ismail S. Carbon and nitrogen storage in agroforests, tree plantations, and pastures in western Oregon, USA. Agroforestry Systems 2004;60:123–30.
[39] Smith J, Pearce BD, Wolfe MS. Reconciling productivity with protection of the environment: Is temperate agroforestry the answer. Renew Agr Food Syst 2012;27:1–13.
[40] Savory A, Butterfield J. Holistic management: a new framework for decision making. Washington: Island Press; 1998. p. 644.
[41] Garnett, T, Cécile, G, Muller, A, Röös, E, Smith, P, de Boer, IJM, Zu Ermgassen, E, Herrero, M, van Middelaar, C, Schader, C, and van Zanten, H. Grazed and confused? Ruminating on cattle, grazing systems, methane, nitrous oxide, the soil carbon sequestration question—and what it all means for greenhouse gas emissions. Food Climate Research

Network: Oxford Martin Programme on the Future of Food & The Environmental Change Institute, University of Oxford; 2017.
[42] Nordborg M. Holistic management—a critical review of Allan Savory's grazing method. SLU/EPOK—Centre for Organic Food & Farming; 2016.
[43] Nations U. Global land outlook. Secretariat of the United Nations Convention to Combat Desertification: Bonn, Germany; 2017.
[44] FAOSTAT FAO. Accessed 13 2017;11:17.
[45] O'Mara FP. The role of grasslands in food security and climate change. Ann Bot 2012;110:1263–70.
[46] Williams AG, Audsley E, Sandars DL. Determining the environmental burdens and resource use in the production of agricultural and horticultural commodities. Main report. Defra research project IS0205. London: Defra; 2006.
[47] Novak SM, Fiorelli JL. Greenhouse gases and ammonia emissions from organic mixed crop-dairy systems: A critical review of mitigation options. Agron Sustain Dev 2009;22.
[48] Flysjö A, Cederberg C, Henriksson M, Ledgard S. The interaction between milk and beef production and emissions from land use change – Critical considerations in life cycle assessment and carbon footprint studies of milk. J Clean Prod 2012;28: 134–42.
[49] Paustian, L, Babcock, B, Hatfield, JL, Lal, R, McCarl, BA, McLaughlin, S, Mosier, A, Rice, C, Roberton, G, and Rosenberg, N. Agricultural mitigation of greenhouse gases: Science and policy options. In first National Conference on carbon sequestration. 2004. Washington DC: Council on Agricultural Science and Technology.
[50] Soussana JF, Allard V, Pilegaard K, Ambus P, Amman C, Campbell C, Ceschia E, Clifton-Brown J, Czobel S, Domingues R, Flechard C, Fuhrer J, Hensen A, Horvath L, Jones M, Kasper G, Martin C, Nagy Z, Neftel A, Raschi A, Baronti S, Rees RM, Skiba U, Stefani P, Manca G, Sutton M, Tubaf Z, Valentini R. Full accounting of the greenhouse gas (CO2, N2O, CH4) budget of nine European grassland sites. Agric Ecosyst Environ 2007;121:121–34.
[51] Marriott EE, Wander MM. Total and labile soil organic matter in organic and conventional farming systems. Soil Sci Soc Am J 2006;70:950–9.
[52] Poeplau C, Don A. Carbon sequestration in agricultural soils via cultivation of cover crops–a meta-analysis. Agric Ecosyst Environ 2015;200:33–41.
[53] King JA, Bradley RI, Harrison R, Carter AD. Carbon sequestration and saving potential associated with changes to the management of agricultural soils in England. Soil Use Manage 2004;20:394–402.
[54] El-Hage Scialabba N, Müller-Lindenlauf M. Organic agriculture and climate change. Renew Agr Food Syst 2010;25:11.
[55] Ball BC, Crichton I, Horgan GW. Dynamics of upward and downward N2O and CO2 fluxes in ploughed or no-tilled soils in relation to water-filled pore space, compaction and crop presence. Soil Tillage Res 2008;101:20–30.
[56] Jeffery S, Bezemer TM, Cornelissen G, Kuyper TW, Lehmann J, Mommer L, Sohi SP, Voorde TF, Wardle DA, Groenigen JW. The way forward in biochar research: Targeting trade-offs between the potential wins. GCB Bioenergy 2015;7:1–13.
[57] Woolf D, Amonette JE, Street-Perrott FA, Lehmann J, Joseph S. Sustainable biochar to mitigate global climate change. Nat Commun 2010;1:56.
[58] Powlson DS, Whitmore AP, Goulding KWT. Soil carbon sequestration to mitigate climate change: A critical re-examination to identify the true and the false. Eur J Soil Sci 2011;62:42–55.

[59] Smith P, Martino D, Cai Z, Gwary D, Janzen H, Kumar P, McCarl B, Ogle S, O'Mara F, Rice C, Scholes B, Sirotenko O, Howden M, McAllister T, Pan G, Romanenkov V, Schneider U, Towprayoon S. Policy and technological constraints to implementation of greenhouse gas mitigation options in agriculture. Agric Ecosyst Environ 2007;118:6–28.

[60] Smith P. Do grasslands act as a perpetual sink for carbon? Glob Change Biol 2014;20:2708–11.

[61] Lassaletta L, Billen G, Grizzetti B, Anglade J, Garnier J. 50 year trends in nitrogen use efficiency of world cropping systems: The relationship between yield and nitrogen input to cropland. Env Res Letters 2014;9:105011.

[62] Tilman D, Cassman KG, Matson PA, Naylor R, Polasky S. Agricultural sustainability and intensive production practices. Nature 2002;418:671–7.

[63] Mueller ND, West PC, Gerber JS, MacDonald GK, Polasky S, Foley JA. A tradeoff frontier for global nitrogen use and cereal production. Env Res Letters 2014;9.

[64] Lemaire G, Franzluebbers A, de Faccio Carvalho PC, Dedieu B. Integrated crop–livestock systems: Strategies to achieve synergy between agricultural production and environmental quality. Agric Ecosyst Environ 2014;190:4–8.

[65] Goulding K, Jarvis S, Whitmore A. Optimizing nutrient management for farm systems. Philosophical Transactions of the Royal Society of London B: Biological Sciences 2008;363:667–80.

[66] Leake A. The effect of cropping sequences and rotational management: An economic comparison of conventional, integrated and organic systems. Aspects of Applied Biology 1996;47:184–95.

[67] Groot JCJ, Rossing WAH, Lantinga EA. Evolution of farm management, nitrogen efficiency and economic performance on Dutch dairy farms reducing external inputs. Livest Sci 2006;100:99–110.

[68] McNeill AM, Eriksen J, Bergström L, Smith KA, Marstorp H, Kirchmann H, Nilsson I. Nitrogen and Sulphur management: Challenges for organic sources in temperate agricultural systems. Soil Use Manag 2005;21:82–93.

[69] Rochette P, Janzen H. Towards a revised coefficient for estimating N2O emissions from legumes. Nutr Cycl Agroecosyst 2005;73:171–9.

[70] Ledgard S. Nitrogen cycling in low input legume-based agriculture, with emphasis on legume/grass pastures. Plant and Soil 2001;14.

[71] Leach K, Roberts D, Swain D. Assessment and improvement of the efficiency of nitrogen use in clover based and fertilizer based dairy systems. 2. Beyond benchmarking to internal system components. Biol Agric Hortic 2002;20:17.

[72] Lampkin N. Organic farming. Ipswich, UK: Old Pond Publishing Ltd.; 2002.

[73] Watson C, Öborn I, Eriksen J, Edwards A. Perspectives on nutrient management in mixed farming systems. Soil Use Manage 2005;21:132–40.

[74] Wolfe M, Baresel J, Desclaux D, Goldringer I, Hoad S, Kovacs G, Löschenberger F, Miedaner T, Østergård H, Lammerts van Bueren E. Developments in breeding cereals for organic agriculture. Euphytica 2008;163:323–46.

[75] Thevathasan N, Gordon A. Ecology of tree intercropping systems in the north temperate region: Experiences from southern Ontario, Canada. Agrofor Syst 2004;61:257–68.

[76] Bealey W, Braban C, Theobald M, Famulari D, Tang Y, Wheat A, Grigorova E, Leeson S, Twigg M, Dragosits U, Dore A, Sutton M, Nemitz E, Loubet B, Robertson A, Quinn A, Williams A, Sandars D, Valatin G, Perks M, Watterson D. Agroforestry systems for ammonia abatement. Final report, project AC0201. London: Defra; 2013.

[77] Voisin A-S, Guéguen J, Huyghe C, Jeuffroy M-H, Magrini M-B, Meynard J-M, Mougel C, Pellerin S, Pelzer E. Legumes for feed, food, biomaterials and bioenergy in Europe: A review. Agron Sustain Dev 2014;34:361–80.
[78] Peoples M, Brockwell J, Herridge D, Rochester I, Alves B, Urquiaga S, Boddey R, Dakora F, Bhattarai S, Maskey S. The contributions of nitrogen-fixing crop legumes to the productivity of agricultural systems. Symbiosis 2009;48:1–17.
[79] Stagnari F, Maggio A, Galieni A, Pisante M. Multiple benefits of legumes for agriculture sustainability: An overview. Chem Biol Technol Agric 2017;4.
[80] Torstensson G, Aronsson H, Bergström L. Nutrient use efficiencies and leaching of organic and conventional cropping systems in Sweden. Agron J 2006;98:603–15.
[81] Smith LG, Tarsitano D, Topp CFE, Jones SK, Gerrard CL, Pearce BD, Williams AG, Watson CA. Predicting the effect of rotation design on N, P, K balances on organic farms using the NDICEA model. Renew Agr Food Syst 2016;31:471–84.
[82] Zhang X, Davidson EA, Mauzerall DL, Searchinger TD, Dumas P, Shen Y. Managing nitrogen for sustainable development. Nature 2015;528:51–9.
[83] Mulla DJ. Twenty five years of remote sensing in precision agriculture: Key advances and remaining knowledge gaps. Biosyst Eng 2013;114:358–71.
[84] Cayuela ML, Sánchez-Monedero MA, Roig A, Hanley K, Enders A, Lehmann J. Biochar and denitrification in soils: When, how much and why does biochar reduce N2O emissions? Sci Rep 2013;3:1732.
[85] Van der Werf HM, Tzilivakis J, Lewis K, Basset-Mens C. Environmental impacts of farm scenarios according to five assessment methods. Agric Ecosyst Environ 2007;118: 327–38.
[86] Salou T, Le Mouël C, van der Werf HM. Environmental impacts of dairy system intensification: the functional unit matters! J Clean Prod 2017;140:445–54.
[87] Bennetzen EH, Smith P, Soussana J-F, Porter JR. Identity-based estimation of greenhouse gas emissions from crop production: case study from Denmark. Eur J Agron 2012;41:66–72.
[88] Herrero M, Henderson B, Havlík P, Thornton PK, Conant RT, Smith P, Wirsenius S, Hristov AN, Gerber P, Gill M. Greenhouse gas mitigation potentials in the livestock sector. Nat Clim Change 2016;6:452–61.
[89] Whittaker C, McManus MC, Smith P. A comparison of carbon accounting tools for arable crops in the United Kingdom. Environ modell softw 2013;46:228–39.
[90] Stockmann U, Adams MA, Crawford JW, Field DJ, Henakaarchchi N, Jenkins M, Minasny B, McBratney AB, Courcelles VR, Singh K, Wheeler I, Abbott L, Angers DA, Baldock J, Bird M, Brookes PC, Chenu C, Jastrow JD, Lal R, Lehmann J, O'Donnell AG, Parton WJ, Whitehead D, Zimmermann M. The knowns, known unknowns and unknowns of sequestration of soil organic carbon. Agric Ecosyst Environ 2013;164:80–99.
[91] Batjes NH. Total carbon and nitrogen in the soils of the world. Eur J Soil Sci 1996;47:151–63.
[92] Goglio P, Smith WN, Grant BB, Desjardins RL, McConkey BG, Campbell CA, Nemecek T. Accounting for soil carbon changes in agricultural life cycle assessment (LCA): A review. J Clean Prod 2015;104:23–39.
[93] Yang Y, Tao M, Suh S. Geographic variability of agriculture requires sector-specific uncertainty characterization. Int J Life Cycle Assess 2017;1–9.
[94] Eckard R, Grainger C, De Klein C. Options for the abatement of methane and nitrous oxide from ruminant production: A review. Livest Sci 2010;130:47–56.

[95] Powell JM, Rotz C. Measures of nitrogen use efficiency and nitrogen loss from dairy production systems. J Environ Qual 2015;44:336–44.
[96] Børsting CF, Kristensen T, Misciattelli L, Hvelplund T, Weisbjerg MR. Reducing nitrogen surplus from dairy farms. Effects of feeding and management. Livest Prod Sci 2003;83:165–78.
[97] Hristov A, Oh J, Lee C, Meinen R, Montes F, Ott T, Firkins J, Rotz A, Dell C, Adesogan A. Mitigation of greenhouse gas emissions in livestock production: a review of technical options for non-CO_2 emissions. In: Gerber P, et al., editors. Animal Production and Health. Rome: FAO; 2013. p. 1–206.
[98] Hess H, Tiemann T, Noto F, Carulla J, Kreuzer M. Strategic use of tannins as means to limit methane emission from ruminant livestock. Int Congr Ser 2006;1293(4).
[99] McBride J, Morrison S, Yan T. Impact of increased dietary starch concentration or the addition of oils, lipids, tannins or saponins in the diet on enteric methane emissions. Northern Ireland: Agri-Food and Biosciences Institute (AFBI), Agriculture Branch; 2012.
[100] McIntyre B, Herren H, Wakhungu J, Watson R. Agriculture at a crossroads: Global report. Washington D.C: Island Press for International Assessment of Agricultural Knowledge, Science, and Technology for Development; 2008.
[101] Robertson M, Llewellyn R, Mandel R, Lawes R, Bramley R, Swift L, Metz N, O'Callaghan C. Adoption of variable rate fertiliser application in the Australian grains industry: status, issues and prospects. Precis Agric 2012;13:181–99.
[102] Spangenberg JH. Hot air or comprehensive progress? A critical assessment of the SDGs. Sustain Develop 2017;25:311–21.
[103] Wilson B. Communicating carbon reduction schemes to farmers; busting preconceptions, driving efficiency and profit. Taunton, UK: Nuffield (UK) farming scholarships trust report; 2017.
[104] Ingram J. Perspective: Look beyond production. Nature 2017;544:S17.
[105] Horlings L, Marsden T. Towards the real green revolution? Exploring the conceptual dimensions of a new ecological modernisation of agriculture that could 'feed the world'. Glob Environ Change 2011;21:441–52.

Section E

Geo-Engineering

Geoengineering: Sunlight reflection methods and negative emissions technologies for greenhouse gas removal

20

Renaud de Richter[*,1], *Sylvain Caillol*[*], *Tingzhen Ming*[†]
[*]Institute Charles Gerhardt, University of Montpellier, Montpellier, France, [†]School of Civil Engineering and Architecture, Wuhan University of Technology, Wuhan, China
[1]Corresponding author: renaud.derichter@gmail.com

Chapter Outline

Managing Global Warming. https://doi.org/10.1016/B978-0-12-814104-5.00020-X

20.1 Why geoengineering?

Scientists have proposed numerous methods to moderate anthropogenic global warming in the event that mitigation and adaptive strategies against climate change are not successful or sufficient. This includes the hypothesis that renewable energy production will not increase rapidly enough to replace and stop fossil fuel use in the next 2 or 3 decades. These alternative methods have been called geoengineering (GE; climate engineering, climate remediation, or climate intervention) because large-scale intentional implementation is necessary to have climatically relevant effects.

According to the Oxford English Dictionary, GE consists of "the deliberate large-scale manipulation of environmental processes that affects the Earth's climate in an attempt to counteract the effects of global warming." Some of these proposals have been discussed by the Intergovernmental Panel on Climate Change (IPCC) [4], and are described in Chapters 6 and 7 of the Contribution of Working Group I of the 5th IPCC assessment report [5]. The US National Research Council of the National Academies released two separate reports on GE [6,7].

20.2 The Earth's energy budget imbalance is due to greenhouse gases

The Earth's energy fluxes are summarized in Fig. 20.1 [8]. Increasing the amounts of GHGs in the atmosphere traps infrared heat; consequently, there is a need to compensate for the increasing imbalance in the Earth's energy budget, characterized by reduced global outgoing longwave (LW) terrestrial radiation remaining trapped by the GHG effect.

GE strategies are almost always categorized into two distinct families of technologies: solar radiation management (SRM), which intends to reduce the incoming energy from the Sun; and carbon dioxide removal (CDR), which aims to trap less outgoing infrared energy from the Earth. However, the second category is rather reductive, as CO_2 is not the only GHG responsible for global warming, and does not include several other technologies allowing more infrared to escape to the outer space.

Consequently, it has been proposed [9] to enlarge the second family of GE methods to earth radiation management (ERM), which considers several other possible longwave LW radiation management strategies, such as enhancing clear-sky radiative cooling, which utilizes the atmospheric transparency window (8–13 μm) [10], or thinning cirrus clouds as less heat is trapped if cirrus clouds are reduced [11]. Other authors have proposed to name this category of GE techniques thermal radiation management [12], but this term is already used for energy applications, such as photovoltaic cells.

The GE methods being suggested are called "radiation management strategies," and they directly influence the Earth's radiation balance and therefore temperature. For instance, sunlight reflection methods consist of reducing incoming shortwave

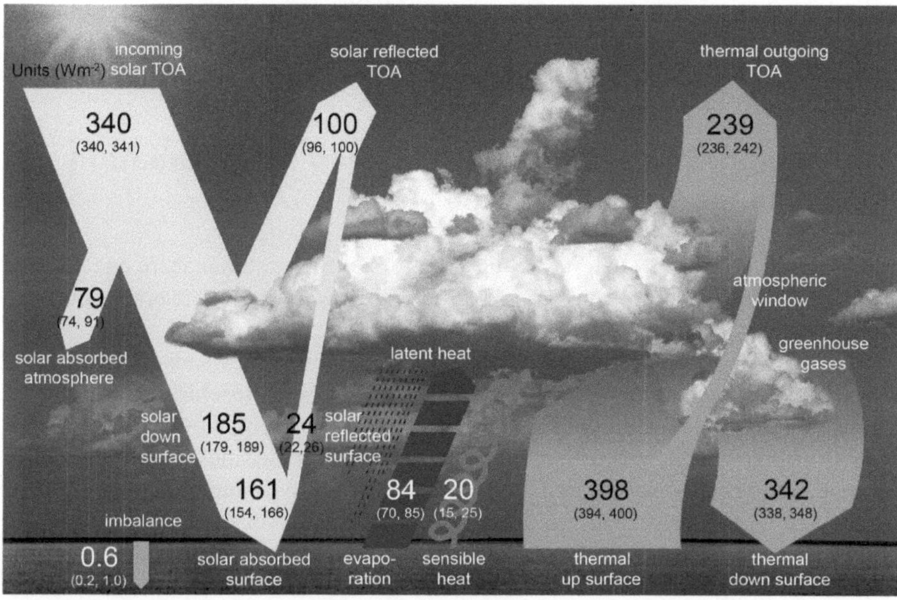

Fig. 20.1 The Earth's mean global energy budget with magnitudes and uncertainty ranges of the individual energy fluxes in $W m^{-2}$.
Reproduced with permission from Wild M, Folini D, Schär C, Loeb N, Dutton EG, König-Langlo G. The global energy balance from a surface perspective. Clim Dyn 40 (2013) 3107–34. doi: 10.1007/s00382-012-1569-8.

(SW) sunlight radiation that reaches the Earth surface, acting like a parasol (the umbrella effect), or removing GHGs and thinning cirrus clouds, which enhance the amount of Earth's LW radiation that can escape into outer space and help rebalance the energy budget.

In simple words, the aim of GE strategies is to cool the planet either by reducing the amount of solar energy received or by enhancing the amount of terrestrial energy discharged into space (ERM). In Fig. 20.1, the energy fluxes in yellow are targeted by SRM strategies, while the ones in orange (plus the two other in red and in dark blue) are targeted by ERM

Nearly two-thirds of the published articles on climate engineering (CE) are related to moral hazards, ethics, law, justice, governance, and other social sciences. The "Asilomar Conference Recommendations," the "Oxford Principles," and the "Code of Conduct for Responsible Geoengineering Research" are some of the governance rules that have already been proposed [13]. The environmental, governance, and societal aspects and impacts, or political and ethical issues of CE are discussed in other chapters in this book. The goal of this chapter is to briefly describe the principal GE methods with an overview of some of their advantages and disadvantages as well as their feasibility and gross estimated costs. Furthermore, we summarize the research conducted on the search for negative emissions technologies (NETs), i.e., industrial methods aimed at removing GHGs from the atmosphere.

20.3 Sunlight reflection methods and solar radiation management

It was shown in the 1990s that providing enough SO_2 to the stratosphere can produce sufficient aerosols to cool the Earth, offset global warming on average, but also severely impact the ozone budget.

On June 15, 1991, Mt. Pinatubo, a volcano on the Philippine Islands, exploded. The plume of gases and some ash penetrated the stratosphere, reaching an altitude of 21 miles. Three weeks later, the Earth was encircled by nearly 30×10^6 tons (30 million metric tons) of sulfate aerosol clouds created by sulfur dioxide and the other sulfur gases ejected, which remained in the stratosphere for nearly 2 years. Throughout 1992 and 1993, the amount of sunlight that reached the surface of the Earth was significantly reduced by these stratospheric sulfate aerosols, which reflected solar rays back to space by acting as millions of small mirrors.

Despite the warming effects of the El Niño event during 1991–1993, the observed climate effects of the Mt. Pinatubo explosion were 0.5–0.6°C surface cooling in the Northern Hemisphere, together with a faster-than-ever-observed ozone depletion rate. Furthermore, the largest recorded ozone hole in the Southern Hemisphere appeared in 1992.

Artificially injecting stratospheric aerosols and marine cloud brightening are examples of SRM techniques that will be described here.

The main advantage of SRM strategies is that a relatively cheap and rapid deployment is theoretically possible to "buy time." Due to the lack of mitigation efforts to avoid potentially catastrophic warming, buying time is supposed to allow humans to get rid of their fossil-fuel addiction and deploy enough decarbonized energies, and massively implement NETs.

The principal drawbacks of SRMs are that they do not address the principal causes of global warming (excess GHGs in the atmosphere), or those of ocean acidification, and propose to reduce SW radiation while the real problem is excess LW radiation. For instance, SRM by stratospheric aerosols could pose serious risks to humans and natural systems with regional consequences, such as to deplete the ozone layer, modify precipitation patterns as well as temperature patterns, with reductions or increases in certain regions; even if the average global expected cooling result is reached.

Other SRM approaches are described in Chapter 7 of IPCC WG1-AR5 pp. 627–635 [5], with an assessment of the principal methods, as well as in a National Academies report dedicated to the principal SRM methods [6]. One of the principal concerns is the possibility of radically altering the monsoon patterns in some regions.

20.3.1 Overview of the principal SRM proposals

The Wikipedia dedicated page and the references cited therein [14] provide a good summary of the numerous SRM methods imagined by scientists.

Some are quite anecdotal and propose to block or disperse sunlight before it reaches the Earth, which look like science fiction, have prohibitive costs, and are very poorly reversible: these include space-based orbital mirrors and sunshades between the Earth and the Sun, using a giant mirror, or a giant Fresnel lens, or millions of small light-weight mirrors and even a crazy idea based on moon dust deliberately produced by explosions.

The other methods are based on increasing the Earth's albedo, which is the ratio of the sunlight reflected by our planet to that received. Earth's overall reflectivity or solar reflectance can be modified by land or ocean-based strategies, as well as by modifying the reflectivity of tropospheric clouds or by creating stratospheric aerosols.

The principal SRM approaches are summarized in Fig. 20.2 (reproduced with permission from Marc Jamous and IPSL/UVSQ).

20.3.2 Principal SRM proposals for increasing surface reflectance

The terrestrial and oceanic approaches consist of whitening surfaces to increase albedo:

- *Cities*: The initial approach would be to paint roofs white. It has been shown [15] that by changing the materials and pigments in which colored roof tiles and coatings are made, an increasing surface reflectance of nearly 30% can be achieved, while also increasing thermal emittance and consequently reducing heat absorbance. The concept of a "cool roof" has been extended to reducing the urban heat island using "cool asphalt" and "cool pavements" [15]. According to an atmospheric model, the authors found that increasing the average albedo of roofs and pavement by 0.1 in all global urban areas could offset the warming caused by 57 Gt of CO_2 [16]. In hot regions, cool roofs also allow energy savings for air conditioning year round.
- *Deserts or mountain glaciers*: Cover thousands of square kilometers with plastic reflective sheeting, except that they have a short lifespan and high cost.
- *Agricultural land, grasslands, and forests*: Select crops, trees, and plant varieties with high albedo, either natural or genetically modified.
- *High latitude forests*: Deforestation has been proposed in order to increase the local albedo but it reduces the natural carbon sink, reduces biodiversity, and has catastrophic ecological consequences and many other drawbacks.
- *Oceans*: Boat-made hydrosols or microbubbles produced by injecting air and using natural tensoactive agents [17] or by foams [18] or by reflecting floating barges in polystyrene-like plastic polymers, or by increasing sea ice at the poles.
- *Poles*: "Saving the Arctic" [19] and preserving downwelling currents has been proposed by using snow cannons in floating barges driven by wind energy during winter periods [20], and also by producing larger ice caps [21] or increasing the thickness of sea ice [22], as illustrated by Graham Murdoch in Fig. 20.3.

Floating ice acts as a good insulator, as the water below cools quite slowly and its thickness does not increase quickly: drilling holes on superficial ice above lakes or

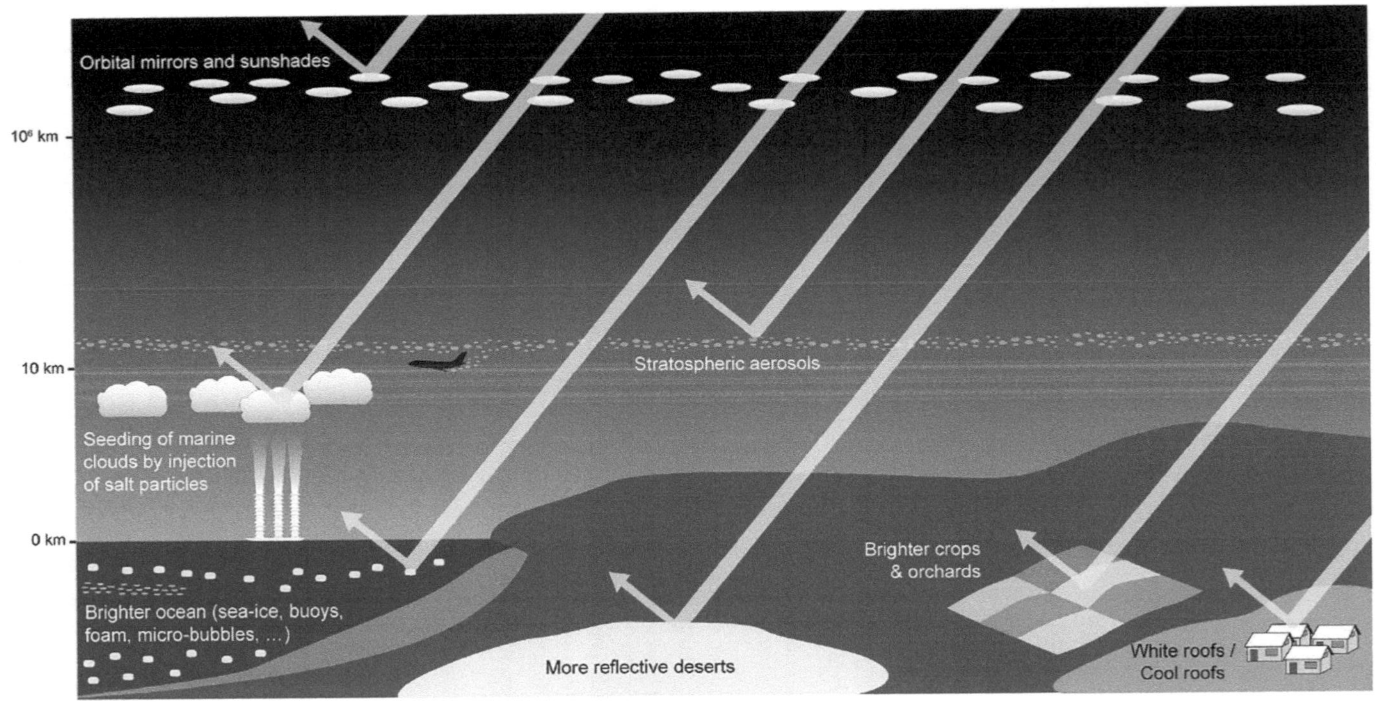

Fig. 20.2 Overview of the principal sunlight reflection methods. Courtesy of Mr. Marc Jamous and IPSL/UVSQ.

Fig. 20.3 Method proposed for preserving downwelling currents [20].
Illustration from Graham Murdoch reproduced with permission, initially in Pacella RM. Duct tape methods to save the earth: re-ice the Arctic. Popular Science, 2007. https://www.popsci.com/scitech/article/2007-06/duct-tape-methods-save-earth-re-ice-arctic (Accessed 16 November 2017).

rivers and pumping water above, or breaking sea ice with ice breakers allows for thicker ice by enhancing contact between the cold atmosphere (negative temperatures) and warmer but almost freezing water, such as breaking the insulating layer and creating thermal bridges.

This last point is one of the most important, as disappearance of the Arctic sea ice and the retreat of ice sheets stops the ice-albedo feedback and allows the open sea to store heat in a possible vicious circle: more LW heat is absorbed by low albedo dark waters resulting in warmer sea surface with less sea ice, resulting in less SW reflected back to space by high albedo sea ice and consequently more LW heat absorbed, even if an increase in outgoing LW radiation is possible by radiative cooling under a clear sky [23]. Increasing surface albedo in cities, landscapes, deserts, and oceans by SRM can help, but much less than that by preserving sea ice from disappearing and preventing glaciers, Greenland, and the Arctic and Antarctic poles from melting.

A similar approach is already widely used in Nordic countries to prevent permafrost from melting during the summer along roads, railways, and pipelines: two-phase thermo-siphons (thermal diodes or heat pipes) are used to transfer heat from underground to the winter freezing air, keeping soils frozen [24].

Breaking the insulating layer and creating thermal bridges is another mitigation strategy: Russian scientists [25] have proposed to slow thawing of the Siberian permafrost by bringing back grazing animals to those regions, as trampling herds of herbivores might remove a thick layer of insulating snow in search of food and in other parts they might compact the snow. This would allow some parts of the soil surface to cool through night sky radiative cooling and other parts with thinner compacted ice and better thermal conductivity to be in contact with very cold air [26]. Both effects lower the permafrost temperature. The methane emissions from ruminants are supposedly smaller than those from melting permafrost or due to organic soil microbial decomposition, and the small albedo change has little èffect during long polar winter nights.

20.3.3 Can GE methods prevent a sea-level rise?

Storing water in some parts of Antarctica has been proposed to delay the sea-level rise, where temperatures stay several tens of degrees below freezing [27]. However, Antarctica alone is already losing about 160×10^9 tons (160 billion tons) of ice per year to the ocean [28] and the future predicted rise will be faster and is accelerating; consequently, the size of the problem, the costs, and the feasibility make this proposal problematic. How will the water pumps and wind farms that provide the energy be maintained? Will the method induce disturbances in the climate system, such as currents, salinity, and temperature? What consequences will the Earth's weight redistribution have from Greenland to the Antarctic?

According to some authors, stabilizing climate temperature by SRM will not prevent the ocean from rising [29], and SRM will not prevent Greenland from melting unless the Earth is cooled back to the lower temperatures seen in the 1900s [30].

20.3.4 Principal SRM proposals for increasing clouds albedo in the troposphere

The tropospheric SRM approaches consist of increasing the number of cloud condensation nuclei, which increases the number of water droplets or atmospheric ice particles and reduces their size, as the amount of humidity available is unchanged, which increases their reflectivity due to the Twomey effect.

- *Marine cloud brightening (MCB)*: This strategy consists of spaying seawater under the oceanic boundary layer to seed marine stratocumulus clouds with numerous submicrometer particles to enhance the concentration of cloud droplets, thereby increasing cloud albedo and probably its longevity [31,32].

  Plans to perform some experiments of "controlled cloud perturbation" were recently announced in the USA [33].
- *Cirrus cloud thinning (CCT)*: Low-altitude clouds (mainly containing water) generally have an overall cooling effect, as they trap lower amounts of LW radiation but are able to reflect larger amounts of SW radiation than higher-altitude clouds, such as cirrus clouds (mainly containing ice), which reflect lower amounts of SW radiation back to space and trap larger amounts of LW radiation between them and the Earth's surface; therefore, a warming effect occurs. Consequently, it has been proposed to reduce, thin, or remove cirrus clouds to trap less infrared heat [34]. However, CCT should not be considered an SRM strategy and in IPCC subchapter 7.7.2.4, the "Cirrus Thinning" approach is erroneously classified as

SRM [5]. One possible CCT method is to seed larger ice crystals to reduce the optical thickness of the clouds, increase crystal sedimentation velocities, and, thus, shorten cirrus cloud lifetime.

20.3.5 Principal SRM proposals for increasing stratospheric albedo by aerosols or by particles

- *Stratospheric sulfate aerosols (SSAs)*: As already described in the introduction to SRM, this proposal is based on the observations made after the 1991 Mt. Pinatubo explosion, which provided a good model for geophysical scientists to understand how to deliberately alter the climate of the Earth. Although this method was suggested well before by several scientists, it was a Nobel prize laureate [35,36] who removed a taboo when he suggested that man-made stratospheric sulfur injections might be used to counteract some of the effects of global warming and should be studied.

  Most of the GE articles published in the last decade relate to SRM by SSAs. Many studies are related to ethics, governance, and other social sciences. Alternatives to SSA have been proposed, with instead of sulfates, numerous other types of chemicals, often solid particles, such as alumina or diamond [37], calcite or limestone [38], black carbon (BC), and [39] titanium [39,40]. Even small photophoretic levitation discs (radius 5 μm and thickness 50 nm) with three different layers (aluminum oxide, metallic aluminum, and barium titanate) [41] have been proposed to increase particle lifetimes while reducing their interaction with ozone chemistry.

The energy fluxes targeted by SRM methods are illustrated in Fig. 20.4 [8].

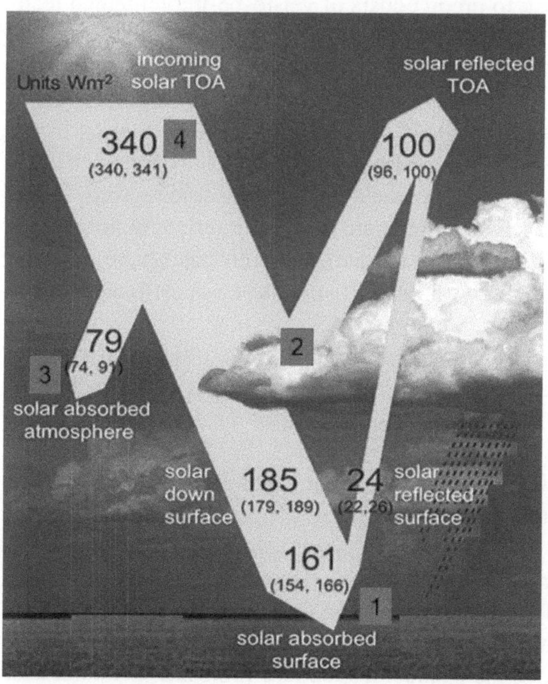

Fig. 20.4 Individual energy fluxes, which are the targets of solar radiation management (SRM) methods (part of Fig. 20.1). Some of the incoming sunlight can be scattered back to space by increasing: (1) surface reflectance (Section 20.3.2); (2) clouds albedo in the troposphere, mainly by marine cloud brightening (Section 20.3.4); (3) stratospheric albedo by aerosols or by particles (Section 20.3.5) and also by (4) orbital mirrors or sunshades that prevent the SW radiation to reach the Earth atmosphere.
Reproduced with permission from Wild M, Folini D, Schär C, Loeb N, Dutton EG, König-Langlo G. The global energy balance from a surface perspective. Clim Dyn 40 (2013) 3107–34. https://doi.org/10.1007/s00382-012-1569-8.

20.3.6 Mixed SRM approaches

The simultaneous application of a "cocktail" of GE methods (SSA+CCT) has been proposed [42]. It is known that the SSA strategy alone can reduce the ice crystal nucleation rate in cirrus clouds, and thus produce optically thinner cirrus clouds allowing more LW to escape into space. The model [43] attributes 60% of overall net GE forcing to this outgoing LW radiation (which means that the effect of reflected incoming SW radiation is smaller). Deploying SSA+CCT methods in concert would decrease the cooling effect by reflecting sunlight, but local substantial differences would remain, with some regions wetter and others drier, some areas hotter and some areas colder.

In other cases, GE methods have several effects, such as ocean surface foams used to inject sea salt in the atmosphere, having some MCB effects [18]; organic gases released by natural processes, such as dimethyl sulfoxide and halogenated compounds released by phytoplankton produce cloud condensation nuclei [44,45], which can be used for MCB and consequently a CDR method, such as ocean iron fertilization (OIF) which will be described later, by enhancing phytoplankton production and release of sulfur and halogenated gases, enhances MCB [46].

20.3.7 Comparative cost estimates, safety, and risks for SRM methods

A limited number of attempts have been made to estimate the direct costs of SRM or CDR technologies. It is very difficult to predict costs of yet-to-be-implemented technologies. It is almost impossible to predict how a massive scale-up would affect costs, as only a few have been tested on a small scale. Would mass production and the learning curve reduce prices? Or would real costs escalate due to the large demand in some commodity (steel or sulfur) in the case of rapid and massive implementation in the event of a climate emergency? Unwanted side effects of SRM technologies are often mentioned (increasing ocean acidification, increased drought in some regions, possible disruption of the Asian and African summer monsoons [47], reducing the precipitation needed by the food supply of billions of people), but their costs have not been quantified; therefore, it is misleading to give apparently precise costs estimates for GE methods.

Here, we only provide a very short overview of possible costs for GE options. The report of the Royal Society of London in 2009 provided a summary of the estimated costs of various GE technologies [48], which was completed by a review of ~300 references made by the Kiel Earth Institute in Germany [49–51] and by other publications [52] that compared cost evaluations of different SRM and CDR options, or estimated the radiative forcing potential of different GE proposals [53].

Cost estimates have been given for SSA delivery systems in the stratosphere using military aircraft, airships, guns, or rockets [54–56]. A fleet of only 43–66 specially designed and modified aircraft (Gulfstream C-37A) costing in total US $\$2.1–3.2 \times 10^9$ (US $2.1–3.2 billion) and performing several rotations per day could inject 1 Mt $year^{-1}$ of sulfur just above the tropical tropopause (18.3 km) at a total cost of US \$2.4–2.7 × 10^9 per year (US \$2.4–2.7 billion per year), including yearly maintenance, depreciation rate, and interest [54]. The amount of sulfur required to cool down the Earth of the desired radiative forcing (~2.3 $W\,m^{-2}$) depends on the type and size of the particles and on several other parameters, but are usually estimated to 5 $Mt\,year^{-1}$ when employing SO_2 and only half with H_2S, so the delivery costs would have to be multiplied accordingly. Only the delivery costs have been presented, and they do not include the costs for the chemicals, which are very toxic and corrosive, requiring very careful handling and rigorous safety precautions [56].

A fleet of only 250 ships, costing each less than US \$3 million and lasting 25 years, is estimated to be sufficient for MCB [57]. Interestingly, some opponents to all GE methods estimate that 5000–30,000 ships would be necessary for MCB, which is 1–2 orders of magnitude larger [58].

Some mistakes seem to exist in the report [48], such as the annual cost given for MCB to achieve a cooling of 1 $W\,m^{-2}$ estimated to US \$200 million per year, whereas MCB proponents give an estimate of less than US \$50 million per year [57]. The same report states "medium affordability" for MCB, while it states "high affordability" for SSA, which had costs estimates in the order of US \$8 billion per year using aircraft [59].

Yearly cost estimates for painting urban surfaces white or for increasing desert surface albedo are estimated to be 3–4 orders of magnitude larger than for SSA or MCB [48] and are not detailed here.

A comparative evaluation of different methods is difficult and can be quite subjective. The Royal Society report [48] quotes "5" (high safety) for the urban surface albedo modification, but quotes only "1" (very low safety) for the desert surface albedo modification, and "2" (low safety) for both the SSA and the MCB. This is curious as the duration of sea salt spayed in marine clouds is in the order of 3 days and only represents about 1% of the existing present transfer of salt from the sea to air and most will fall in the sea [57], while the life expectancy of sulfates in the stratosphere is nearly 2 years. Any new sulfate injection would be rapidly well dispersed and distributed all over the Earth's stratosphere. Consequently, even if human-made sulfate injections are immediately stopped, the excess would remain and may have increased negative effects, particularly on the ozone hole, allowing more dangerous UV radiation to reach the Earth's surface. MCB can be adapted and finetuned by region and season, which is not the case for SSA.

The objective of SRM technologies is to reduce global warming and help maintain an average temperature as stated in the Paris agreement goals; however, if the objective is to alleviate specific consequences of climate change, MCB has a number of advantages, such as specifically reducing the intensity of hurricanes, or of ENSO events, or to reduce Arctic sea-ice melt [60].

The low costs and small number of airplanes or ships needed to perform SRM by SSA or MCB, and the possibility of weather modifications [58] have raised the risks of militarization and weaponization of GE methods, with their possible use by rogue governments or countries [61]: this subject is treated in the GE chapters on governance and ethics in this book.

20.4 Negative emissions technologies

To keep the global temperature rise to 2°C or lower, the Paris climate agreement and the integrated assessment models and scenarios used by the IPCC rely heavily on NETs [62] and particularly on bioenergy production with carbon capture and sequestration (BECCS) [63].

A very good overview of the principal NETs and their capacity, scalability, limiting factors, estimated costs, and possibilities was written in 2011 [64].

Some authors consider that CDR is a synonym for NETs [65], thinking that "CDR is the only geoengineering technique that allows negative emissions and the reduction of anthropogenic carbon in the atmosphere" [66]. This is incorrect on at least three points: first, CDR in not a single technique, but consists of a portfolio of numerous different methods that will be described in this subchapter; second, removing CFC-12 or HCFC-22 not only reduces the amount of anthropogenic carbon in the atmosphere but is also much more efficient on a mole per mole basis than removing CO_2, as their global warming potential (GWP) on a 100-year horizon differs by 3–4 orders of magnitude [67] (10,200, 1760, and only 1, respectively, by definition for CO_2); third, the CO_2 removed by a CDR method needs to be stored and possible leaks must be monitored for centuries, which is not the case when removing other GHGs, which can be transformed into safer compounds for the climate and for the ozone layer.

20.4.1 Carbon dioxide removal

A US National Academies report outlined the principal CDR methods [7]. The intent of CDR measures is to decrease the atmospheric concentration of CO_2, which is the main cause of the GHG effect, by enhancing natural sinks or by using chemicals together with technical and physical engineered processes. This is possible because CO_2 is a weak acid able to form salts with alkaline compounds. The isolated carbonate salts dissociate, often by thermal processes, and pure CO_2 is released and geologically stored in saline aquifers or in depleted fossil-fuel reservoirs.

The natural terrestrial carbon sink can be enhanced by means of forestation, afforestation, and land management but requires fertilizer and water and may compete for fertile agricultural land and food production, as carbon is not stored permanently.

A very slow but almost permanent and quite important natural sink for CO_2 is mineralization by reacting with alkaline or ultramafic rocks: this sink can be enhanced by chemically, thermally, or mechanically accelerated weathering of olivine, serpentine, silicates of MgO, CaO, and Fe-oxides as well as numerous other rock types.

The oceanic CO_2 sink can be enhanced in some oceanic regions that are poor in chlorophyll but rich in P and N, by means of iron fertilization. Iron is an essential micronutrient for all organisms and particularly for phytoplankton. In some other oceanic regions, the CO_2 sink can be enhanced by mechanical upwelling of macronutrients and fertilizers.

In contrast to ocean sinks, terrestrial vegetation sinks were included in the Kyoto Protocol as offsets for anthropogenic GHG emissions.

According to IPCC AR5 [68], "the boundary between CDR and mitigation is not clear and there could be some overlap between the two current definitions." "Some CDR methods fall under the category of geoengineering, though this may not be the case for others, with the distinction being based on the magnitude, scale and impact of the particular CDR activities."

The Potential Effects of CDR Methods and SRM on the Carbon Cycle are discussed in the 5th IPCC report (Chapter 6.5, pp. 546–552) [5].

20.4.2 Most discussed CDR methods [1–3]

The first three CDR methods described here are the more natural methods, based on biomass, forestry, and agriculture and, consequently, they should be less costly and closer to deployment, although they are dependent on irrigation and fertilizers and with land limitations and possible competition for food production. These methods are far from a possible large-scale deployment and vulnerable to reversal by forest fires and drought. Then, more technological methods are described, which require more research and development, cost more in theory, but are less vulnerable to reversal.

- *Forestation and afforestation*: Consists of planting trees on a global scale, but this should not compete with agricultural land, water, or fertilizer use and food production. Preventing deforestation and allowing the growth of secondary forests can remove significant amounts of carbon from the atmosphere [69], but currently tropical forests are net carbon sources [70]. Among the drawbacks: forests have a lower albedo than deserts; planting forests in deserts, such as the Sahara, should not prevent desert dust (from the Bodélé Depression in Chad) to be swept away by winds and to travel over thousands of kilometers, as it is known that its iron dust content fertilizes large parts of the equatorial Atlantic, and its phosphor content fertilizes the Amazon Rainforest [71], so this dust transfer should be preserved.
- *Bioenergy with carbon capture and sequestration (BECCS)*: Requires growing trees, collecting and drying the biomass, and transporting it to thermal power plants to burn the biomass instead of fossil fuels, to produce electricity and heat. To become carbon negative, the technology requires that the CO_2 generated during the process is captured and locked away for thousands of years by carbon capture and sequestration (CCS). The process is summarized in Fig. 20.5. In comparison, leaving fossil fuels, such as coal, unused underground where it is carbon neutral, while extracting and burning coal with efficient CCS is carbon positive because the capture yields are never 100% and because of the emissions produced during extraction and transportation. The Paris climate agreement and numerous IPCC scenarios rely on NETs based on BECCS [62,72,73].

  Many scientists warn that growing enough biomass requires years and that biological, ecological, and economic limitations exist [74,75], as fertilizer use can release N_2O or CH_4 and offset the amount of eq.-CO_2 stored. Simple in principle, BECCS requires a careful evaluation. For instance, industrial processes to extract and transport millions of tons of coal are well established, while processing and collecting similar amounts of already existing biomass is more energy intensive. Two tons of wood chips (with 25% humidity) are required to produce as much energy as using 1 ton of anthracite. Rising bioenergy demand in Europe is

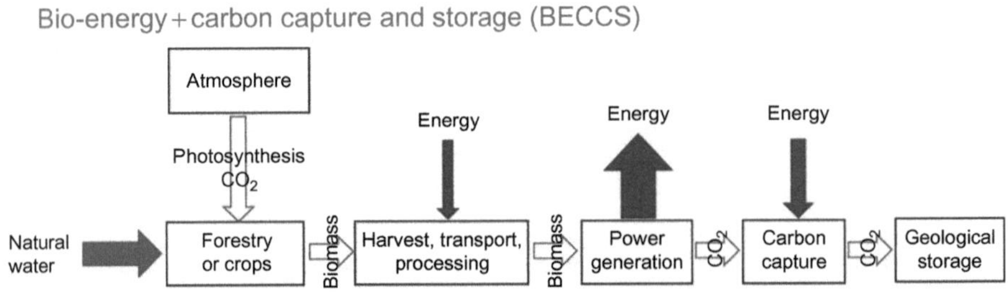

Fig. 20.5 Process of bioenergy production with carbon capture and sequestration (BECCS). Reproduced with permission from Davies P. Can water treatment be a negative emissions technology? 1st international conference on sustainable water processing, Sitges, Spain, 11–14 September, 2016.

currently increasing the export of wood pellets from southeastern US. Sequestering 1 ton of CO_2 by processing switchgrass requires 2.1 times more switchgrass biomass [74], while it only removes nearly 0.5 tons of CO_2 from the atmosphere due to a "rebound" effect, as CO_2 outgasses from the oceans because of the air-water CO_2 equilibrium [76].

- *Fertigation with BECCS*: Experiments have been carried in India to grow short-rotation woody crops (*Melia*, *Salix*, *Poplar*, and *Eucalyptus*) irrigated with domestic waste water or gray water (which is rich in nutrients and organic load) and the amount of biomass produced per unit area was 1.5–3.2 times higher than the same trees grown in adjacent land parcels with typical irrigation water [78]. These experiments also demonstrate that this vegetation can be used as a pollutant filter, as the biological and chemical oxygen demand was reduced by almost half. Similar experiments conducted with eucalyptus enabled an increase in biomass production from 29.7 kg per tree to 34.2 kg per tree [79]. A comparison of the water and land use for producing the same amount of bioenergy using a traditional BECCS system and a fertigation and BECCS system shows that fertigation allows almost a 2.8 times reduction of water $m^3\ t(CO_2)^{-1}$ and of 1.4–8 land $m^2\ t(CO_2)^{-1}$ [77].
- *Biochar*: Consists of charring or burning biomass without oxygen (roasting or pyrolysis) and burying it, so the carbon is locked up in the soil, with the advantage that it acts as a fertilizer substitute and improves crop production. Biochar has many advantages, as it reduces the emissions of GHGs (methane and nitrous oxide) and prevents nutrient loss if mixed with composting materials. The pyrolysis process produces heat and syngas, which can be used for energy production. A bioenergy-biochar system integrated in a process allowing CO_2 capture and sequestration (BEBCS) has better carbon efficiency [80]. Of course collecting, transporting, and drying the biomass, making the char and burying the final product causes CO_2 emissions, but the overall process is a NET, with additional beneficial effects on agriculture. Biochar amendments increase soil total nitrogen, soil organic carbon, and available nutrients, while decrease bioavailable toxic metals like As, Cr, Co, Ni, and Pb [81]. Biochar amendments increase shoot biomass and grain yield and reduce N_2O and CH_4 emissions when applied to crops.
- *Biochar from low-grade biomass and from waste organic material such as sewage sludge*: Solar-driven pyrolysis and gasification of low-grade carbonaceous materials from biomass (wheat straw and *Scenedesmus* algae) and from waste treatment (sewage sludge, SS) is possible [82]. Of course much SS is already used to produce biogas (CH_4), but this only

consumes 30%–50% of the feedstock, and the remaining is usually used as compost, but can be converted into biochar.

An advantage of making biochar from SS is that the profits can be used to help develop sewage systems to reduce contamination of water for drinking due to waste from housing areas. Only a small part of the world's population benefits from toilets connected to waste water treatment plants via sewers and very few other people benefit from good cesspits or pit latrines able to do the same job.

SS may contain heavy metals and other potentially toxic elements, and although the pyrolysis process enriches the heavy metal content in the biochar, it is a proven way to reduce their availability to the soil and plants and consequently to be safer than using SS directly as a soil amendment [83]. Even SS amended with only a portion of biochar enhances immobilization of potentially toxic elements and reduces their bioavailability and mobility in soils amended by the mixture [84]. Serpentine soils often contain toxic heavy metals, which induce phytotoxicity: biochar amendments reduce their bioavailability and improve plant growth [85].

No published study describes pyrolysis of mixtures of alkaline silicates (like olivine) with carbonaceous materials to produce biochar and syngas, but one could expect that thermal activation would enhance mineral weathering, as biochar prepared with aluminum hydroxide, magnesium hydroxide, and iron oxide produces metal oxyhydroxide–biochar composites that capture CO_2 [86,87]. Alkalinized sludge that has been sanitized with lime (containing CaO and MgO) corrects soil acidity and improves the fertilizing power of SS alone, probably by increasing Zn, Mn, Mg, and K availability [88]. No CO_2 capture information has been published about biochar made from these types of sludges.

- *Direct air capture (DAC) with CCS*: Removing CO_2 directly from ambient air where CO_2 is diluted (450 ppm) requires building machines that can move and pass enough air through a collector with an alkali to make an intermediate carbonate salt during the CSS process. Several names have been given to the different variants and methods, such as "artificial trees" [89]. The proposed methods try to reduce their overall costs by several strategies to compensate for the penalty due to the smaller CO_2 concentration in the air than at the stack exhaust of a fossil-fuel power plant FFPP. Obtaining high yields is not important for DAC (while it is from the exhaust of FFPPs with CCS). A weak alkali can be used (CO_2 is a weak acid), reducing the energy needs for its regeneration comparatively to using strong alkalis; also other regeneration methods can be employed, such as using moisture or vacuum swing-driven absorption-desorption cycles over solid polymers. The DAC apparatus can be installed close to the sequestration site, reducing infrastructure costs for CO_2 transportation.

  The first DAC prototypes have been successfully tested [90] and the first plants started operation in 2017. The future costs of carbon capture seem to be decreasing [91] and scalability is not a problem, as there is no need to increase the size of the units, just their number.
- *Enhanced weathering of rocks*: Ultramafic rocks, silicates, and other alkaline minerals, such as serpentine, olivine, and many others, contain oxides (MgO and CaO) that naturally capture atmospheric CO_2 to form stable carbonate minerals (easy to store), but the kinetics are rather slow (hundreds of years). Several physical, chemical, electrochemical, and biological methods exist to enhance the speed of mineral weathering [92]. By reducing the size of a mineral rock to millimeters or even nanometers, its exposed surface to atmospheric air is increased by several orders of magnitude favoring the reactions. The speed of the reactions can also be accelerated by working with aqueous solutions of carbonates, by increasing the temperature and/or the pressure and/or the CO_2 concentration, which requires running the reaction in closed reactors, with filtration and precipitation steps (to remove silica). But batch processes require transporting and moving large quantities of material, which

increases costs. Strong acids can help solubilize the minerals and if an alkali is coproduced with the acid used, this base will be used to capture the CO_2. This can be performed by several electrochemical processes [93]. Many microorganisms, bacteria, mycorrhizal fungi [94], plant roots [95], lichens [96], and gut microbiota of soil worms [97] produce complexing agents, such as oxalate, to extract and transport soluble forms of the micronutrients they need and are thus are able to accelerate carbonate mineralization, reduce soil acidity, and improve biomass production. The mining industries already dispose of huge quantities of rock pieces that have been crushed and ground more or less finely to extract the ore: some propose to become carbon-neutral by carbonating their mine tailings [98] in situ, where it is cheaper [99], than ex situ at FFPPs where the carbonation of daily emissions of a 1-GW plant would require up to ~55,000 tons of mineral per day [100]. Mineral carbonation technologies have been reviewed [101] and include not only natural minerals, but also industrial wastes such as cement kiln dust, waste cement, alkaline paper mill waste, red mud, nickel tailings, asbestos tailings, steel slag, coal fly ash, and oil shale ash. For the latter two ash wastes, the impact for CO_2 sequestration is negligible as only a very small fraction of the emissions made by these industrial processes can be captured.

- *Ocean alkalinity enhancement and ocean liming*: These are variants of the previous method, which was first intended to reduce, oppose, or thwart ocean acidification due to increasing amounts of acidic CO_2 dissolved in the oceans and which threatens coral reefs, shell-forming marine organisms (plankton, echinoderms, and benthic mollusks), calcifying species, and many other forms of marine life [102].

  Some mine tailings from the mining industry that are located close to the sea and contain alkaline mineral wastes and no toxic heavy metals could be transported at low cost to the beach to make a profit of the comings and goings of the waves and tides on the coast, which will provide mechanical energy to reduce their size, favor contact with water saturated in CO_2 and HCO_3^-, with some biogenic organic mono-, di-, or tri-carboxylic acids present in seawater. The alkalinity of the minerals will reduce ocean acidification [103,104].

  Alkaline compounds, such as lime CaO or hydroxides ($Ca(OH)_2$, NaOH), or rock types like MgO, silicates must be ground up, crushed, dispersed, and dissolved in the ocean to increase its pH and enhance ocean alkalinity, but also to increase its ability to store carbon. In the case of industrially prepared alkali compounds like lime, the expression "ocean liming" has been used. The observed result is faster, but the risks (environmental impacts of localized elevated pH or of codissolution of trace metals) and the costs could also rise [105]. This process is represented in Fig. 20.6. As production of lime from limestone generates CO_2 gas, it can be coupled with CCS.

  Olivine minerals also contain various amounts of iron that can increase ocean productivity by transforming mineral carbonates into organic carbon by a fertilization effect (described in the next section) [106], thus reducing acidity even more.

  But current international legislation and treaties do not allow dumping chemicals into the oceans, so ocean liming and other ocean alkalinity enhancement methods will require changes in international regulations.

- *Ocean nourishment, ocean fertilization, and ocean iron fertilization (OIF)*: Consists of adding nutrients to the ocean in selected locations to increase marine phytoplankton growth, which draws down CO_2 from the atmosphere [107,108]. The major problem is that human activities (mainly agricultural runoff) already add too much phosphorus, nitrogen, and other nutrients in the lakes, rivers, and oceans, inducing algal blooms, which deplete the oxygen level and induce hypoxia and eutrophication of freshwater bodies, estuaries, and coastal waters [109] and "dead zones," where no animal life is possible [110].

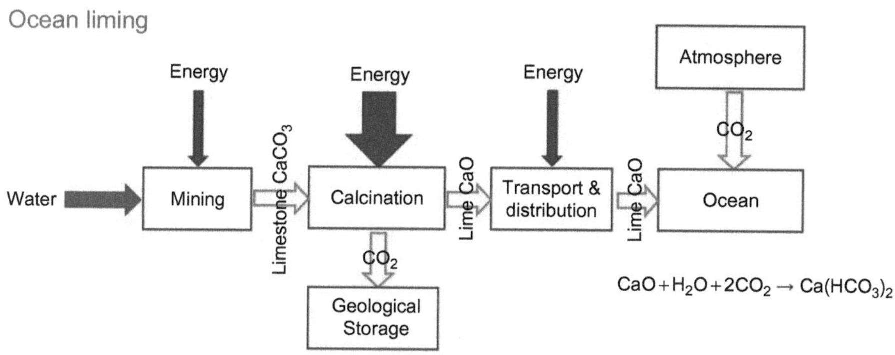

Fig. 20.6 The ocean-liming process.
Reproduced with permission from Davies P. Can water treatment be a negative emissions technology? 1st international conference on sustainable water processing, Sitges, Spain, 11–14 September, 2016.

The most discussed method is OIF: fertilize by adding iron compounds; the rationale is that several large areas of the oceans have very low amounts of chlorophyll but they are rich in phosphorus, nitrogen, and other nutrients, except iron, which is an essential micronutrient for life (microorganisms, plankton, plants, trees, animals, and humans). Iron is the fourth more abundant element on Earth, but it oxidizes readily and the Fe(III) oxides are quite insoluble, so they are not bioavailable. Several OIF experiments have reported phytoplankton blooms and sometimes deep carbon export (>1000 m) of a significant part of the biomass produced [111]. However, after ingestion and excretion by fish, significant amounts of the remaining organic material and detritus will sink to the bottom of the ocean and be stored for long periods, as part of it decomposes and releases CO_2 into the atmosphere, but also N_2O and CH_4, which are two GHGs of higher GWP than that of CO_2. One possible explanation is that diatom growth cannot be enhanced in seawater limited by silicon [112] and that plankton types without a solid skeleton do not sink deep enough; another possible explanation is that circulation by currents and upwelling patterns re-exposes their carbon biomass to the atmosphere [113].

OIF has been widely criticized, as it can change biodiversity and species repartition, might have possible unintended effects and drawbacks and, as already mentioned earlier, is forbidden under current international treaties and legislation, as it is considered dumping chemicals and pollutants into the oceans. Terrestrial vegetation sinks have entered the Kyoto Protocol as offsets for anthropogenic GHGs, but ocean sinks have not, and several scientists suggest OIF experiments should be allowed [114,115].

- *Micronutrient recycling from deeper water layers or artificial upwelling*: This involves bringing nutrient-rich, subeuphotic water up to the surface to feed phytoplankton and stimulate primary production by using a passive wave-powered ocean pipe with a flap valve [116,117]. This approach would enhance fixation of dissolved CO_2 from surface waters into organic carbon, which sinks as organic particulates and can be stored for centuries. Renewable energy production with this type of system has been proposed [118]. Critics of this elegant GE scheme have argued that colder water from the bottom layers contains more dissolved CO_2 and consequently the upwelling will start by releasing CO_2; or that particles will sink into the deep water and change the phytoplankton community composition [119].

Furthermore, coupled carbon-climate computer model simulations have determined that the biomass produced will be "stored on land" (not stored underwater, which is astonishing as it is not explained if it is stored for years and years on tourist beaches). The same model simulations showed that if artificial upwelling was stopped, in few years, the atmospheric CO_2 concentrations would rise quickly and for decades to centuries to levels higher than would have been experienced if this CDR method had not been engaged [120]. The effects of a sudden stop of power production or of Internet can have deleterious consequences in only seconds, but the article qualifies artificial upwelling of playing "the sorcerer's apprentice," without explaining for which reasons the 7 million pipes could possibly be stopped abruptly without replacement.

Other types of climate models have determined that implementing ocean pipe technology would damage the thermal structure of the ocean, with possible disruption of the ocean thermocline, which could exacerbate long-term warming instead of avoiding it [121].

A potential improvement in the method was proposed with a second component of pumping to send down particulates produced into the deep ocean for long-term sequestration [122], but this was promptly criticized as costing too much energetically and with added problems of eutrophication, deoxygenation, and acidification [123]. In conclusion, this very simple and elegant GE method raised many critiques, but the idea has not been abandoned [124].

The principal CDR approaches are summarized in Fig. 20.7, reproduced with permission from Marc Jamous and IPSL/UVSQ.

20.4.3 Carbon capture and sequestration

CCS has been developed to capture CO_2 from flue gas exhaust originating at FFPPs where it is relatively concentrated (~10%). Capturing the CO_2 before it is released into the atmosphere is not a negative emission as it only prevents new emissions. CCS is not considered a GE method or a NET, but many IPCC scenarios rely on CCS [125]; thus, it has been used as an integral part of numerous GE proposals and deserves a short description [126].

Although other media for CO_2 capture are possible, such as metal organic frameworks and ionic liquids, alkaline solutions are often used.

When using an alkaline media (amines or aqueous solutions of KOH, NaOH, ammonia), a salt is formed between the weak acid (CO_2) and the strong alkali, which is often separated by precipitation and then dissociated by a thermal process to recycle the base and obtain almost pure CO_2. The CO_2 is dried, compressed, transported, and injected into saline aquifers or in depleted fossil-fuel geologic reservoirs where it can safely remain stored for thousands of years [127]. In theory, it is cheaper and easier than removing CO_2 from the atmosphere, but CCS yields are not 100%. Alkali regeneration is energy intensive and incomplete, and it uses consumables and requires significant investment costs. Consequently, CCS is expensive and reduces energy production while increasing electricity costs. To give an example of one of the technologies called supercritical pulverized coal: to build a 550-MW plant costs US $\$1.2 \times 10^9$ (US \$1.2 billion) without CCS and US $\$2.1 \times 10^9$ (US \$2.1 billion) with CCS. The consumables (alkali+energy) increase the electricity price from 9.64 US\$ kW^{-1} to 17.84 US\$ kW^{-1}, respectively, without and with CCS.

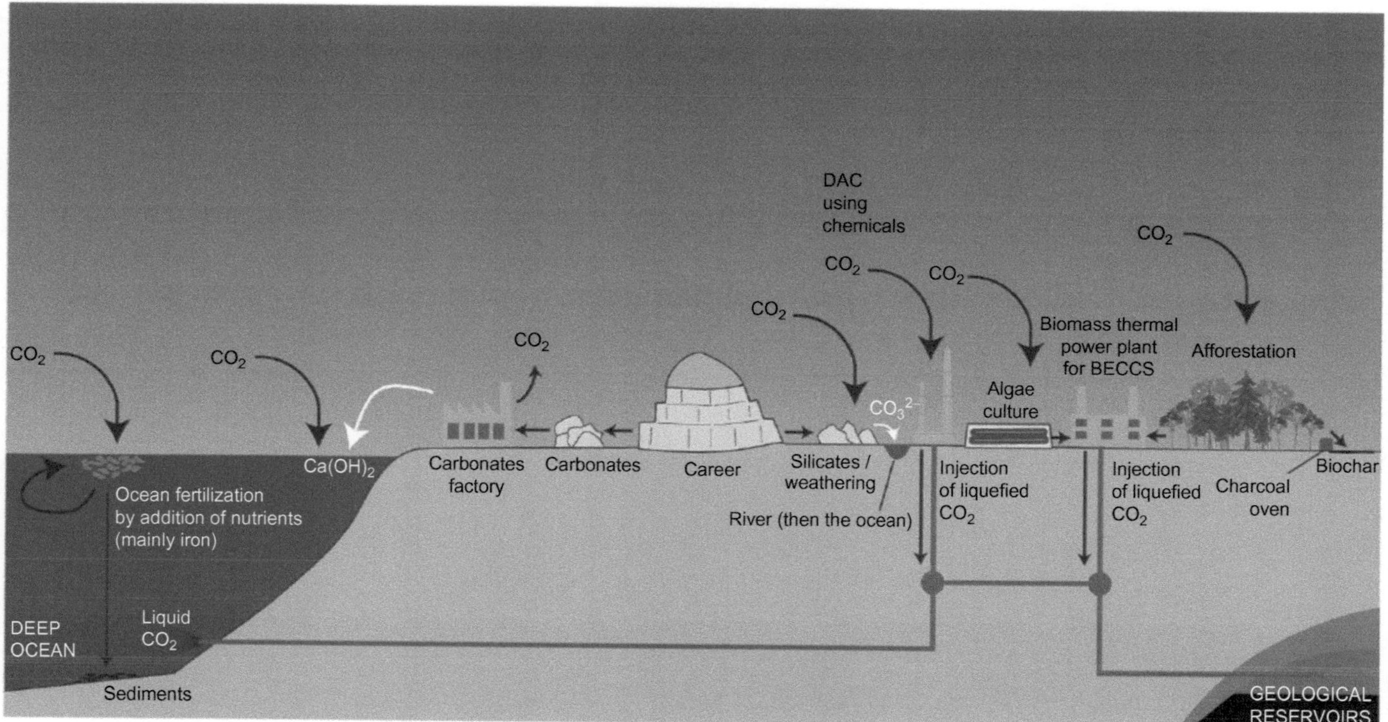

Fig. 20.7 Overview of the principal carbon dioxide removal methods.
Courtesy of Mr. Marc Jamous and IPSL/UVSQ.

As CCS technologies are both costly and highly energy intensive, other combustion technologies producing directly a concentrated CO_2 stream ready for sequestration are actively studied, such as chemical-looping combustion [128,129], which consists in having a different oxygen carrier in the form of combinations of metal oxides (Fe/Mn/Ni/Si/Cu/Ca/Mg). Typically, a chemical-looping combustion system consists of two reactors: a fuel reactor where the fuel reacts with oxygen released from a metal oxide structure at 800–1000°C; and an air reactor where the reduced metal oxide (its metallic form) is oxidized (regenerated) back to its metal oxide form to be reused in another reduction-oxidation cycle. This type of combustion process without NO_x emissions allows recovering pure CO_2 without additional energy penalties for its separation.

Several anecdotic GE proposals have been developed for instance to store CO_2 in the center of Antarctica where temperatures stay and are already tens of degrees Centigrade below zero. To do so, it is proposed seeding the atmosphere with monoethanolamine, a chemical compound commonly used for CO_2 capture in submarines and industrial processes, which makes stable crystals when dry and that would fall just in the local area where the seeding is done [130]. Another proposal consists in installing 20 GW of wind farms to provide electricity to gigantic freezers in the heart of Antarctica to make "CO_2 snow" by cooling the air to −140°C [131]. However, both proposals lack feasibility and are not credible solutions. An unanswered question is how is the CO_2 to be kept stored once frozen? Will the process need sealed chambers resistant to high pressure and running the freezers forever? Can the risk of abrupt termination and very rapid and massive CO_2 release be avoided? What are the quantities and costs of monoethanolamine? How to perform maintenance on the wind farms?

Other more serious CO_2 storage possibilities have been proposed, such as very deep ocean storage (~4–6 km), where it would not mix with (or be dissolved in) the water layer above; thus, no ocean acidification [132] or by formation of CO_2 hydrates or clathrates in deep marine sediments [133]. But the preferred options for CO_2 storage are saline aquifers and depleted oil reservoirs or abandoned coal mines [126].

20.4.4 Current development of CCS and BECCS

Many, if not all, IEA and IPCC scenarios for limiting global warming well below 2°C rely on CCS and BECCS.

No "large-scale" BECCS facility exists yet. It is worth noting that the global portfolio of energy from biomass and of BECCS projects is expanding slowly [134]:

- The largest pure biomass plant (not equipped with CCS) uses wood pellets for power generation (Severn Gorge, UK) and has a capacity of 740 MW (it is a former coal-fired 1000 MW facility converted in 2013). The second (265 MW, paper mill facility) and third largest (210 MW, burns peat as fuel) are in Finland, but neither of them is equipped with CCS. Several others of similar size are starting operation in China, such as the Zhanjiang biomass power plant in Guangdong Province, and some others that cofire coal with some amount of biomass.
- The world's first large-scale bio-CCS facility entered operation in April 2017 in Illinois, USA. It is not a plant producing electricity from burning biomass, but a facility producing

bioethanol from corn and able to store up to 1.1 $Mt(CO_2)$ $year^{-1}$ in a saline aquifer. Half a dozen other smaller bio-CSS facilities producing ethanol have been funded as demonstration projects mainly in Kansas, USA, and in the Netherlands [135].

The global portfolio of "large-scale" CCS projects is also expanding slowly, with the largest coal-fired CCS power plant that started up in January 2017 (Petra Nova project in Texas, USA) capturing up to 1.4 $Mt(CO_2)$ $year^{-1}$.

Much industrial practice and experience has been gained over the years [136], but scaling up from megatons to gigatons for massive CCS deployment remains a big challenge [137]. In addition, CCS cannot deal with diffuse sources (from cars, trucks, ships, airplanes), nor with past emissions, which is why CDR methods are being developed.

At the end of 2017, there were only 38 "large-scale" CCS facilities in operation or in the planning stage around the world [138]. The biggest CCS facilities already in operation are capturing on the order of 2 Mt of CO_2 annually {nearly 40 $Mt(CO_2)$ $year^{-1}$ all together}.

The CCS capacity needs to expand 10-fold in 2025 to be on track to meet the 2°C scenario, and 4000 $Mt(CO_2)$ $year^{-1}$ or 100 times more than today is needed by 2040.

Representative 2°C scenarios used in the IPCC models could require deployment of as much as 24 GW $year^{-1}$ of BECCS by 2040, requiring a cumulative US $\$1.9 \times 10^{12}$ (US $1.9 trillion) in total capital investment for BECCS through 2050 [139].

20.4.5 Comparative cost estimates, safety, and risks for the most common CDR methods

No large changes were found in the literature among the costs estimates proposed in 2011–2012 for some of the most common CDR methods [64,140,141], which are summarized in Table 20.1.

Table 20.1 Principal carbon dioxide removal (CDR) methods and some innovative CDR proposals

NET	Brief description	Storage medium (Estimated abatement cost)	Principal advantages & disadvantages
Avoiding deforestation, agricultural land management, reforestation, and other forestry	Enhance forest CO_2 uptake by planting young trees, removing surface dead organic matter, increase organic carbon in sols	Biomass and soil organic carbon {$0–20 t $(CO_2)^{-1}$} [144]	Wood used for architecture purposes stores carbon Fighting forest fires is expensive Considered as mitigation strategies (not geoengineering) [145]

Continued

Table 20.1 Continued

NET	Brief description	Storage medium (Estimated abatement cost)	Principal advantages & disadvantages
Afforestation	Planting forests on abandoned land or irrigating deserts	Biomass and soil organic carbon {\$20–100 t $(CO_2)^{-1}$} [146]	Decreases land-surface albedo resulting in a global warming effect. Risk of increasing food prices. Irrigation and fertilizers required by arid zones. Local temperature reduction and weak precipitation increase, with rain enhancement at distance [147]. Deserts dust from Sahara has fertilization effects by iron in ocean, and by phosphorus in Amazon, if no more desert dust lowers productivity elsewhere
BECCS Bioenergy with chemical CO_2 Capture and Storage	Biomass combustion (for electricity production and/or heat)+Char production	Compressed CO_2 in geological storages {\$45–250 t $(CO_2)^{-1}$} [148,149]	Improved soils fertility, reduced GHG emissions, reduced nutrients needs Forest biodiversity reduction, risk of competition with food production [141]
Fertigation + BECCS	Same as BECCS with fertilizers in irrigation systems at the precise time and necessary rate	Compressed CO_2 in geological storages (costs expected lower than for BECCS)	Lower amounts of fertilizers used, lower N_2O of CH_4 emissions, reduced water consumption, and possibility to recycle and reuse domestic wastewater [150,151]

Table 20.1 **Continued**

NET	Brief description	Storage medium (Estimated abatement cost)	Principal advantages & disadvantages
BIOCHAR & Bioenergy transportable gases or liquids	Torrefaction or pyrolysis of biomass with production of stable "char" and useful products for energy generation	Soil organic carbon {$0–135 $t(CO_2)^{-1}$} [152]	Allows energy production, soil remediation, carbon sequestration, improve soil health and fertility, reduces some GHG emissions by soil microorganisms, but biochar is hydrophobic and induces local soil water repellency [153]
BIOCHAR from Sewage Sludge	Same as Biochar but using wastes dried by solar energy	(costs expected lower than for Biochar)	Composting, using as organic amendment or incinerating Sewage Sludge emit GHGs with much higher global warming potential than CO_2. Biochar production saves > eq-CO_2 emissions
Mineral weathering of ultramafic rocks and silicates	Natural very-low-speed kinetics of carbonation of ultrabasic silicate rocks such as olivine-rich dunite can be enhanced	Solid metal carbonates and bicarbonates on land [96], dissolved ones in the oceans [154]	Does not interfere with terrestrial land use. If done in coastal environments, it uses wave energy and friction for size grain reduction and counteracts ocean acidification. Possible reuse of some mine waste tailings [155] Might release heavy metals

Continued

Table 20.1 Continued

NET	Brief description	Storage medium (Estimated abatement cost)	Principal advantages & disadvantages
Ocean Liming (with CCS)	Calcination at high temperature of limestone $CaCO_3$ yields lime CaO, which when added to the oceans captures CO_2 and reduces its acidity	Dissolved bicarbonates CO_3^{2-} and hydrogen carbonates HCO_3^- in the oceans {\$70–160 t $(CO_2)^{-1}$} [105,140]	Might be helpful to limit ocean acidification, but less to remove atmospheric CO_2. Transformation of limestone into quicklime releases ~1.4 moles of CO_2 per mole, consequently CCS has to be performed at the first step [156]
DAC, Direct Air Capture	Capturing diluted CO_2 from air by solid or liquid sorbents, usually regenerated by heat while the CO_2 released is purified and compressed	Pure CO_2 compressed for geological storage {\$40–600 t $(CO_2)^{-1}$} [141]	Slightly more energy intensive than conventional CCS at the stack of power plants, as requires large surface contact areas with air. Can compensate for dispersed CO_2 emissions of the transportation sector or for past ones
Desalination brines decomposition by solar energy	Solar thermal decomposition of salts in desalination reject brines yield MgO+HCl	$MgCO_3$ or $Mg(HCO_3)_2$ in the ocean (electrical energy penalty in the range estimated for that of ocean liming) [157]	Reduces ocean acidification. Synergies with the infrastructure of desalination plants [77,158]

We have already seen that the published estimates are broad and variable for SRM technologies from one study to the other. It seems worse for CDR, as the report of the Royal Society [48] provided ranges of costs estimates for SRM (Table 3.6, p. 35) in US\$, but the report speaks of affordability for CDR and only provided rough qualitative estimates of the costs (Table 2.9, p. 20) indicated by words: low (for afforestation), medium (for BECCS), and high (for CDR with chemicals). This has been well decrypted by several articles [142,143].

DAC is a good example to illustrate the difficulty for estimating what the future cost of a technology in early development can be after large-scale deployment: some reports estimate DAC costs as high as 600–1000 US\$ $t(CO_2)^{-1}$ captured from the air [159,160] and were widely criticized by developers of the technology [89,161], basing their arguments on physics, thermodynamics, and entropy differences compared to CO_2 capture at concentrated sources (FFPP stacks). Recent results obtained by experimental pilot plant tests by the companies Global Thermostat, Climeworks, and CarbFix with three slightly different technologies allowed them to announce costs from 100 down to 30 US\$ $t(CO_2)^{-1}$ captured from the air [91], which was in line with what was anticipated for long-term total DAC costs: from 40 to 140 US\$ $t(CO_2)^{-1}$ [162]. However despite the progress that has been made, DAC still has opponents that think it can "at best be a peripheral activity, at worst a distraction" with "marginal" utility [163].

A very interesting article evaluated numerous parameters, such as land intensity, the need for nutrients, energy, and water requirements and costs, as well as the biophysical limits of the most common CDR technologies [75]. Fig. 20.8 illustrates the changes in the CO_2 compartments (land, ocean, atmosphere, and geological) induced by the most common CDR methods described in this article [75].

Other estimate of the costs [164], as well as possible limitations [165] for ocean liming, enhanced weathering, OIF, forestation, BECCS, biochar, soil carbon management, and DAC can be found in those articles, together with other CDRs, such as the construction industry, timber, carbon negative cement, and concrete made using hemp fibers that could be used in negative emissions building materials.

Creating wetlands and burying biomass at sea or on land are further examples of "biological" NETs, but there are limitations that can be illustrated for example by the fact that foresting an area of $2.64 \times 10^6 km^2$ (area larger than the UK) might only store 0.7–1.54 $Gt(CO_2)$ $year^{-1}$

The safety and risks of the most common CDR methods have been mentioned: droughts, heat waves, forest fires, storage vulnerable to reversal, dependence for many on large-scale geological sequestration technologies orders of magnitude larger that already tested, limitations in nutrients and fertilizers, dependence on rainfall or irrigation, competition for agricultural land for food production, reduction of biodiversity, etc.

Several startup companies have been created to develop commercial DAC technologies, including Global Research Technologies, Global Thermostat, Carbon Engineering, Climeworks, Infinitree, Skytree, Soletair, and several others [166].

A useful comparison of the processes and technologies can be found [167], as well as an estimate of the final costs [168], as several pilot plants have been built and one company has achieved costs as low as \$30 $t(CO_2)^{-1}$ [169] thanks to the use of renewable geothermal energy and free heat. Rather than using pure CO_2 compressed as a supercritical fluid, CO_2 in water is being injected underground in a process that converts 95% into carbonated basalt rock on a 1- to 2-year time scale for storage in stone after the reaction, avoiding the necessity to control CO_2 leaks. This last technology is a mix of two CDR technologies: DAC and enhanced weathering of rocks, with no need to extract and grind the rock, which explains lower costs.

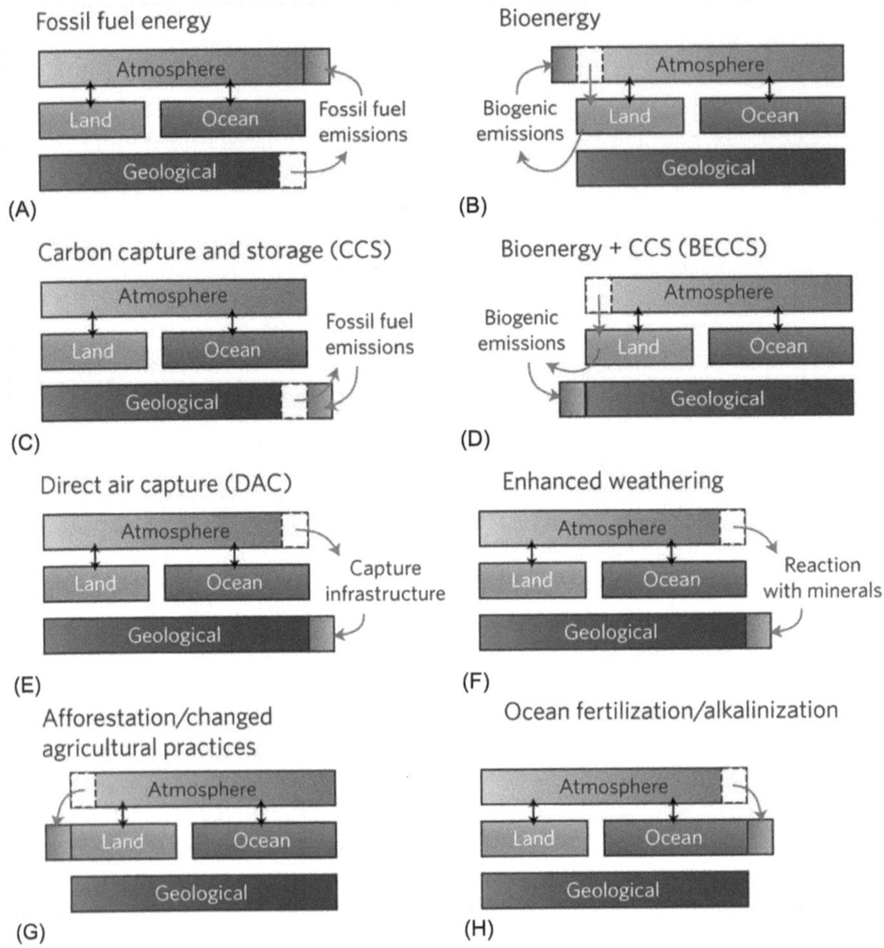

Fig. 20.8 The carbon flows between the different reservoirs are represented: (A) using fossil fuel energies adds geological carbon to the atmosphere; (B) using bioenergy avoids net addition of carbon to the atmosphere when the bioenergy feed stocks re-growing rates matches; (C) carbon capture and storage (CCS) target to avoid emission by capturing the CO_2 resulting from fossil fuels combustion and storing it in geological reservoirs; (D–H) represent some negative emissions technologies (NETs) which remove carbon from the atmosphere: (D,E) through biological (bioenergy + CCS) or industrial processes (direct air capture (DAC) + CCS); (F) through enhanced weathering of minerals; (G,H) through biological uptake by land (afforestation, agriculture), or by the oceans (fertilization), as well as by oceans alkalinization. Reproduced with permission from Smith P, Davis SJ, Creutzig F, Fuss S, Minx J, Gabrielle B, et al. Biophysical and economic limits to negative CO_2 emissions. Nat Clim Chang 6 (2016) 42–50.

20.4.6 Innovative CDR proposals

The CDR approaches described here are the most commonly mentioned and some have been the object of numerous studies and evaluations of their potential, costs, safety, limitations, and even of their public acceptance [64]. New emerging CDR proposals are described now and several deserve the interest of the scientific community with the goal of disposing of a variety of technologies to try to limit global warming below 2°C. For example, enhanced mineral weathering, ocean alkalinity, and ocean liming are related by their terrestrial and oceanic counterparts in the same way as are afforestation and OIF. Curiously, few oceanic counterparts to biochar or BECCS exist, which should be possible, as it is based on biomass production and the oceans are responsible for half of Earth's productivity. Several electrochemical technologies using renewable electricity (from PV, CSP, and wind turbines) propose to electrolyze inexpensive brine and low-cost alkali minerals to generate ocean alkalinity, which will remove atmospheric CO_2. The acidity removed will be neutralized by base minerals. Some processes are associated with production of carbon-negative H_2, others with desalination and high-temperature electrochemical processes are associated with the production of negative-emissions iron, cement, and carbon nanotubes.

Oceanic counterpart to DAC, which can be called DOC for direct ocean capture, proposes the use of electrochemical technologies to generate acidity in order to extract from seawater almost pure CO_2 (to be directed to geological sequestration), reducing ocean acidity and as there is an equilibrium between the partial pressures in the gas and liquid phases, allowing more atmospheric CO_2 to dissolve in the oceans' surface.

- *Ocean Afforestation*: Expanding natural populations of macroalgae to cover up to 9% of the world's ocean surface should make it possible to absorb atmospheric CO_2 by harvesting the macroalgae to produce up to 12,000 Mt year^{-1} of biomethane from marine origins via anaerobic digestion and storing in saline aquifers the bio-CO_2 released from marine biogas production (19,000 Mt year^{-1}) [170]. Coupling CCS to bio-CH_4 combustion could help store an additional 34,000 Mt year^{-1} of CO_2. An algal forest can be maintained or even expanded by recycling the macroalgal nutrients remaining after anaerobic digestion, which would also increase sustainable fish production. This scheme [170] can provide enough bioenergy to replace all of today's fossil fuel needs, while reducing ocean acidification and increasing primary productivity and biodiversity. In a supplement to this article, it was stated that increased N_2O emissions are expected but limited and can be managed. Transforming marine organic residue obtained after anaerobic digestion into biochar and using it for traditional land agricultural purposes can help with this purpose. As increased fish production is anticipated, fewer ruminants will be needed to feed the future 10 billion human beings on Earth, saving CH_4 emissions from cows' guts and N_2O emissions from agricultural land whose production is devoted to feed livestock. In addition, more carbon might be stored if part of the organic detritus from the fish sinks deep enough into the bottom of the oceans.
- *Bioenergy production from microalgae and seaweed*: A similar scheme could be developed either in the sea or in artificial ponds at land-based facilities [171]. Bio-oil, bio-fuel, biodiesel production from algae is being studied over the past 3 decades, as algae has high photosynthetic efficiency and high biomass yields [172,173], but they seem difficult to harvest, filter, and dry in a cost-competitive way. The economic viability of this process can be improved by a wide range of applications, such as algae production for food, cosmetics, and medicines [174]. Among other advantages: there is no competition with agricultural land

for food production, low nutrient requirements, and the possibility to produce biochar from residues [175]. Excess nitrogen, phosphorus, and nutrients removed from agricultural waters, urban wastewater, and from rivers before discharge into estuaries and oceans by microalgae can be used to prevent oceanic dead zones due to anoxia and eutrophication [176,177]. The reverse has been proposed for oceanic regions depleted of N, P, or of both N and P [178]: an additional 0.9 Gt C $year^{-1}$ (i.e., 3.3 Gt(CO_2) $year^{-1}$) sequestration capacity would require doubling current phosphate production.

- *Burying crop residues in the deep ocean*: According to one study [179] soil sequestration is about 14% efficient, cellulosic ethanol production is only 32% efficient, while ocean sequestration of crop residue carbon is 92% efficient (must be weighed down with ballast). Similarly, burying wood underground has been proposed [180], as anaerobic conditions under a sufficiently thick layer of soil would prevent decomposition of the buried wood.
- *Metal extraction with plants*: Nickel mining is energy intensive and polluting but can be performed by nickel hyperaccumulating plants (e.g., *Alyssum corsicum*), which also enhance weathering of olivine and serpentine [181]. After harvesting the plants, they are burned to collect the metal in ash (up to 10% of nickel) and it might be possible to integrate the collection operation into a BECCS process.
- *Enhancing downwelling ocean currents*: This technology was already mentioned among the SRM methods as an idea to artificially make sea ice during winter in the Arctic using snow cannons on floating barges fueled with renewable energy produced by wind turbines on the barges [20]. The initial goal was to prevent collapse of the Gulf Stream. However, as CO_2 is more soluble in cold water and downwelling currents have not reached saturation, the additional cubic kilometer of cold sinking water would trap significant amounts of CO_2 and, thus, become a NET, as these type of oceanic currents stay 600–1000 hundred years in the deep oceans [182].

  Due to global warming, sea-ice extent and volume has decreased in the Arctic Ocean. Scientists [183] estimate that the Arctic Ocean sink for CO_2 has tripled over the last 3 decades due to sea-ice retreat, enhancing air-to-sea CO_2 flux by 28% per decade.

  But this GE proposal was not initially intended either to change the albedo (SRM), or to capture CO_2 (CDR), but to cool the deep oceans by transferring latent heat of freezing from surface seawater to the air (ice cannons), allowing more heat (LW radiation) to escape in outer space.

  This example together with thinning of cirrus clouds illustrates the limitations of SRM and the CDR categorization of GE technologies and the need for a wider family that is briefly presented at the end of this chapter [24].

- *Fuel cells, lime, and CCS*: Fuel cells can be powered by different energy carriers, such as natural gas or methanol. The process produces electricity, heat, and releases pure CO_2 and water. Researchers have proposed valorizing the heat production by coupling the energy-generating process with fuel cells to the production of lime (CaO) from limestone ($CaCO_3$), as this second process also produces pure CO_2 [184].

  Both pure CO_2 fluxes can easily be captured at a reduced cost and geologically stored in deep saline aquifers as an example. The lime obtained can be used to produce the cement and concrete needed by the building industry or to fight ocean acidification by enhancing ocean alkalinity or ocean liming [156]. A member of the same team has patented a variant of this method in which fuel cells are fed with carbon monoxide [185]. The patent describes CO generation by CO_2 reduction with hydrocarbons (e.g., with methane according to the reaction: $CO_2+CH_4 \rightarrow 2CO+H_2$) and by syngas production.

- *Electrochemical acceleration of chemical weathering*: NaOH and HCl can be produced via electrolysis of brine using existing industrial chloralkali processes. HCl can be neutralized via the dissolution of silicate rocks (the kinetics are highly enhanced by very low pHs) and then the dissolved products and NaOH are added to the ocean to increase alkalinity [186], but the process is energy intensive. To reduce costs, the locations where transport of silicate rocks and reagents is minimal, with available cheap electricity and at close proximity to the oceans must be carefully chosen. The problem of diluting the alkali and the speed of addition to the ocean without increasing the pH too much and disturbing the biological balances as little as possible applies here, as well as for ocean liming.

  A similar process consists in the electrochemical splitting of $CaCO_3$, a natural product, which is essentially insoluble in water, but which can be electrochemically dissolved to produce primarily dissolved aqueous calcium bicarbonate. Consequently, a net capture of up to 1 mol of CO_2 for each mole of $CaCO_3$ split can be realized [187]. The author estimates the net process cost to be less than $\$100\,t(CO_2)^{-1}$.
- *Solar thermal decomposition of desalination reject brine*: The overall result is a carbon-negative reverse-osmosis desalination process, even if extra investment and energy are required to treat the concentrated salted brine that is usually rejected [157]. Reject brine is dewatered by nanofiltration and then treated by multiple effect distillation to reduce the energy requirement of the subsequent thermal decomposition of $MgCl_2$, which is split into HCl and MgO and added back into the ocean with NaCl brine. Conversion to carbonates takes up the CO_2. Concentrated solar energy can be used, as the operating temperatures for thermal decomposition are <600°C. The HCl is neutralized separately by reacting with naturally abundant Mg/Ca silicates.
- *Electrochemical transformation of desalination reject brine*: $MgCl_2$ in the brine is split into an HCl and $Mg(OH)_2$ suspension by electrolysis in a cell with a gas diffusion anode. Then, the HCl is neutralized by reacting with Mg/Ca silicates, while the magnesium hydroxide is returned to the sea with the NaCl brine where it captures CO_2 to form magnesium carbonate or hydrogeno-carbonate, which neutralizes ocean acidity. This process is represented in Fig. 20.9. Synergies exist with the infrastructure of desalination plants [77,158] but even if worldwide desalination capacity is increasing {currently up to $30 \times 10^9\,m^3\,year^{-1}$ (30 billion $m^3\,year^{-1}$) of freshwater output}, this CDR can only potentially make a small contribution {estimated to $0.25\,Gt(CO_2)\,year^{-1}$ in 2015} to the required needs of NETs.
- *Concentrated solar power, lime, and CCS*: The idea of producing lime from limestone using solar energy followed by CCS using the pure CO_2 released and then using the lime in ponds or lakes to capture more CO_2 and restart the process was described by the French Administration in charge of the national underground and mines (Annex 1, Fig. A.2 in [188]).
- *Solar-driven electrodialysis process, lime, and CCS*: The same French Administration described earlier proposed a solar-driven electrodialysis process using NaCl salted water to generate both NaOH and HCl, with the alkali being used to capture atmospheric CO_2 and producing pure Na_2CO_3, which is then reacted with the acid to release pure CO_2 for CCS, while leaving NaCl salted water to restart the solar-driven electrodialysis process (Annex 1, Fig. A.1 of [188]).
- *Solar-driven electrodialysis process and mineral carbonate storage*: In a variant of the same process (Annex 1, Fig. A.1 of [188]) using renewable thermal solar energy, the HCl is used to produce $CaCl_2$ and $MgCl_2$ from natural CaO and MgO silicate minerals, which are later reacted with Na_2CO_3 to yield mixed Ca and Mg carbonate precipitates (dolomite) for mineral storage, leaving NaCl salted water to restart the solar-driven electrodialysis process.
- *Artificial photosynthesis* [189] *and sustainable sunlight-to-fuel pathways:* Photocatalysis allows H_2 to be produced from H_2O [190], which can be coupled to photocatalytic

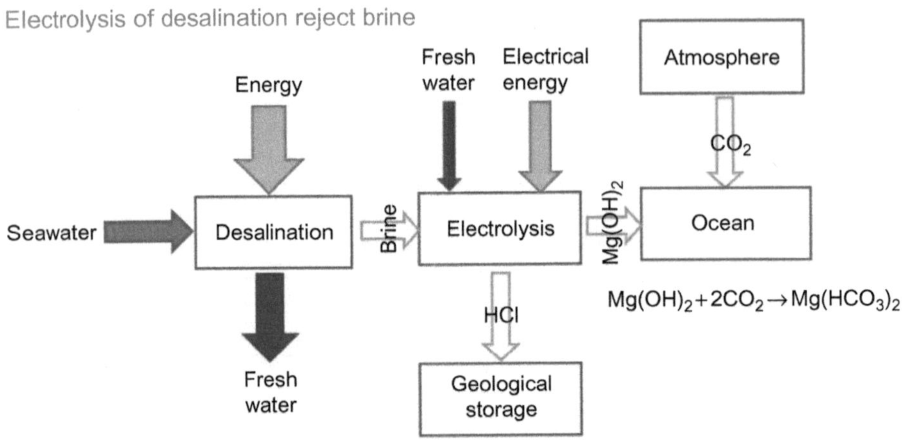

Fig. 20.9 Diagram representing the carbon dioxide removal (CDR) process starting with rejected brine and desalination.
Reproduced with permission from Davies P. Can water treatment be a negative emissions technology? 1st international conference on sustainable water processing, Sitges, Spain, 11–14 September, 2016.

conversion of CO_2 into C_1 fuels and chemicals (CO, CHO, CH_2O, CH_4, and H_3COH) [191,192]. Other processes are possible, such as CO_2 conversion into fuels by photocatalytic and electrocatalytic processes [193]; "solar photo-thermo-chemical alkane reverse combustion," [194] which means thermal photocatalytic conversion of CO_2 into C_nH_{n+2} fuels [195], as well as some others. However, only few researchers from the GE community seem to be involved in such research areas [196,197] and reciprocally researchers in the photocatalysis field poorly mention CDR or NETs in their articles [198].

- *Carbon capture and utilization (CCU) and carbon capture, utilization, and sequestration (CCUS)*: Several uses have been found for CO_2: mainly to inject it into fossil-fuel wells to enhance oil recovery; during the chemical synthesis of some polymers and useful chemicals [199]; as well as for several very short term uses, such as carbonated beverages. However, these are only "niche markets" that utilize three to four orders of magnitude less CO_2 than that is produced yearly [200].

  Fire extinguishers can be filled with compressed CO_2, and it has been proposed to fight forest fires using finely divided serpentine slurries [201] that form a thin cake on the burning material once deposited on burning trees, preventing oxygen from reaching the tree, preventing the escape of inflammable gases and withdrawing large quantities of heat from the fire. After the fire is extinguished, the serpentine calcined material has been activated and will rapidly weather in contact with the atmosphere and capture CO_2. The release of magnesium and increasing soil pH are additional beneficial effects that help trees grow faster. Thermal activation of Mg and Ca-silicates during pyrolysis of biomass to make biochar mixed with powdered alkaline rocks might enhance mineral carbonation, but this aspect does not seem to have yet been well explored [86,87], although the carbonation of coal and oil ash residues, as well as of iron, cement, and numerous other industrial wastes has been studied.
- *Electrolytic dissolution of silicate minerals*: Several experiments involving electrolysis of diluted Na_2SO_4 solutions, using an anode encased in powdered wollastonite, have been

conducted to produce H_2 and enhance CO_2 capture in solution due to the large increase in alkalinity [93]. Similar electrochemical processes were previously patented to dissolve divalent metal silicate minerals to form metal hydroxides that absorb CO_2 [202], allowing for the formation of metal-hydrogeno-carbonates and metal-bicarbonates.

It has been proposed that ocean thermal energy can be converted to generate electricity, with a similar electrochemical process to generate H_2 (or NH_3) as an energy carrier and to store the salt carbonates obtained in the deep ocean in the very large natural marine carbon storage reservoir, which would help mitigate ocean acidification [203,204].

The net reaction is: Mg_2SiO_4 (solid)+4 CO_2 (g)+4 $H_2O \rightarrow 2\ Mg^{2+}$+$SiO_2$ (solid)+4 HCO_3^-+2 H_2 (g)+O_2 (g).

- *Molten carbonate fuel cells (MCFCs)*: MCFCs are high-temperature fuel cells that operate at approximately 650°C (no noble metals required) and use an electrolyte composed of a molten carbonate salt mixture suspended in a porous chemically inert ceramic matrix. Electrons are combined with O_2 from air at the cathode, and with injected CO_2 to form carbonate ions CO_3^-. The negatively charged CO_3^- ions move through the electrolyte (molten carbonate salts) to the anode where they combine with protons (generated by the fuel that is provided, generally H_2) to maintain the charge balance. The byproducts at the anode are CO_2 and H_2O. The almost pure CO_2 is separated and recycled to the cathode side.

  Research and development is being conducted to provide CO_2 to the cathode directly from the flue stack of an FFPP. In this case, the almost pure CO_2 generated at the anode is separated from the water vapor in a cooling system where H_2O is condensed and the CO_2 is sent for sequestration [205].

  Consequently, MCFCs can be placed in the flue gas stream of an FFPP to separate CO_2 by transferring it from the cathode side to the anode side where CO_2 is captured while electricity is simultaneously generated, providing additional power, higher system efficiency, and a reduction in CO_2 emissions [206,207].

  When the fuel used is biogas of natural origin (e.g., from anaerobic digestion of sludge from wastewater treatment plants), negative emissions are possible [208] as long as a sequestration system is used and its corresponding emissions are lower than the amount of CO_2 sequestered.

  Unlike phosphoric acid, alkaline, or polymer electrolytic membranes fuel cells, MCFCs do not require an external reformer to convert H_2 to more energy-dense fuels. The high temperature at which MCFCs operate helps to convert the fuel directly into H_2 within the fuel cell by internal reforming, which reduces costs. As MCFCs are more resistant than other fuel cell types to impurities, scientists hope to sufficiently improve them to resist sulfur compounds and particulates, in order to be directly capable of internal reforming coal into H_2.

  A similar method was first developed at room temperature using fuel-cell concentrators with a membrane sandwiched between two gas diffusion electrodes where CO_2 transport was achieved by an electrochemically pumped concentration gradient [209].
- *Solar thermal electrochemical processes (STEPs)*: STEPs allow ambient CO_2 to be captured at temperatures of 750°C in molten lithiated carbonate electrolytes (Li_2O/Li_2CO_3) leading to the production of carbon nanofibers and carbon nanotubes [210]. Thermochemical and experimental results at an electrolysis voltage as low as <1 V indicate that the principal reaction allows carbon plus oxygen to be obtained according to the following overall reaction: $Li_2O+2CO_2 \rightarrow Li_2CO_3+C+O_2$ [211]. The lithium carbonate electrolyte can be replaced as reaction media by melted calcium and/or barium carbonate for successful carbon deposition.

  Numerous STEP variants have been developed to produce iron (without CO_2 emissions), lime for cement production (without CO_2 emissions), and to generate energy-rich chemicals, such as ammonia, hydrogen, carbon nanotubes, and carbon nanofibers [212–214]. The full

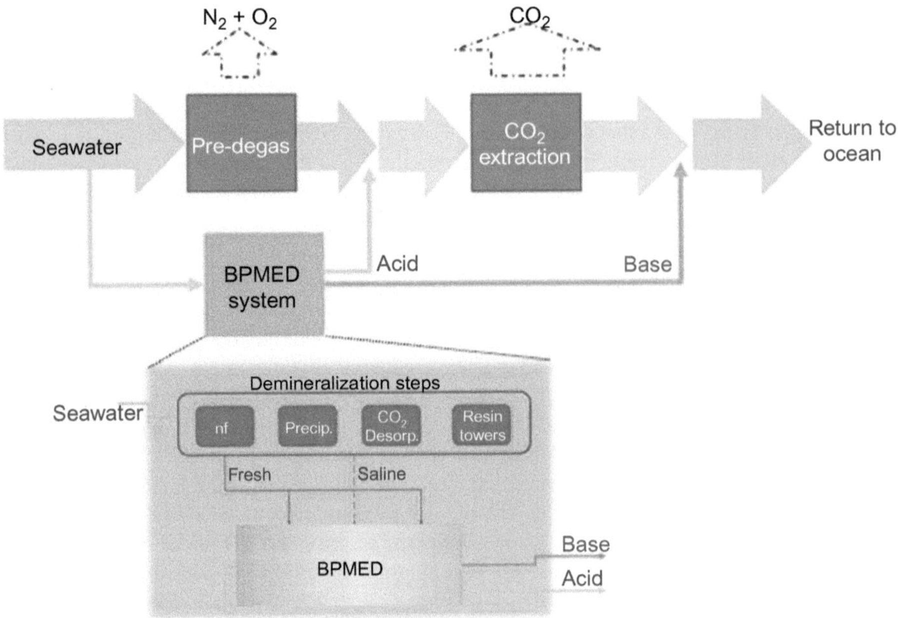

Fig. 20.10 Diagram representing the process of direct oceanic CO_2 removal and capture (DOC) using bipolar membrane electrodialysis (BPMED).
Reproduced with permission from de Lannoy C-F, Eisaman MD, Jose A, Karnitz SD, DeVaul RW, Hannun K, et al. Indirect ocean capture of atmospheric CO_2: part I. Prototype of a negative emissions technology. Int J Greenhouse Gas Control (2018).

sunlight spectrum can be utilized with solar thermal energy to decrease the energy required for carbon capture, while the visible sunlight generates the electronic charge to drive electrolysis.

Using STEPs, CO_2 can be captured as solid carbon and stored with high solar energy efficiency (34%–50%), or transformed into CO feedstock to synthesize together with STEP generated H_2, useful chemicals and synthetic solar fuels for all types of vehicles.

- *Direct ocean capture (DOC)*: involves CO_2 extraction from seawater by electrochemical processes using bipolar membrane electrodialysis (BPMED) represented in Fig. 20.10. This is the oceanic counterpart to the direct CO_2 capture from air (DAC). Seawater contains hydrogenocarbonate and carbonate ions together with some dissolved CO_2 all in close chemical equilibrium. By generating some acidity electrochemically in one of the system compartments some CO_2 outgases removing this weak acid from the ocean (to be collected and sequestred), while the alkalinity generated in the other compartment displaces the atmosphere-seawater equilibrium and allows an equivalent amount of atmospheric CO_2 to be dissolved in seawater [215,216]. Similar processes are able to also generate hydrogen and thus provide an interesting energy carrier [217].

The CDR overview given in this chapter is far from exhaustive [218]. Several other possible CDR methods have been briefly suggested [64,181].

20.5 Greenhouse gases removal

Climate change is due to increased emissions of several GHGs (not only to CO_2) and the removal of other GHGs from the atmosphere will also help to reduce the overall GHG effect. Thus, a larger concept than CDR has been introduced as greenhouse gases removal (GGR), but there are few definitions of this GGR concept and few scientific articles have considered it. A glossary of EPA terms from 2006 gives the following definition: "GHG removal. Absorption or sequestration of GHGs from the atmosphere," which seems to mainly apply to CO_2, as destruction or transformation of CH_4, N_2O, or halogenated gases, such as chlorofluorocarbons, does not enter this definition and yet is possible.

In 2017, several UK scientific research institutions launched a research program on GGR with £8.6 million in grants [219], intended to counteract global warming by investigating ways to remove GHGs from the atmosphere at climatically significant scales, but almost all of them have been awarded to CDR.

Previous strategies dealing with GHGs were mainly based on preventing or mitigating emissions to prevent or reduce the amount of GHG emissions in the agricultural and forestry sectors [220], as well as from the livestock sector [221], because N_2O emissions are mainly due to fertilizer use, and because CH_4 emissions principally due to rice paddies, enteric, and manure fermentation and to the urban, transportation, energy, and industrial sectors [125,222].

20.5.1 *Methane removal and tropospheric (surface) ozone removal by enhancing the halogen natural sink*

Surface or tropospheric O_3 is dangerous to all living beings, while stratospheric O_3 forms the protective "ozone layer," which prevents harmful UV radiation from reaching the Earth's surface. Notably, tropospheric ozone is a short-lived GHG that is neither well mixed nor distributed, but is third in importance due to its radiative forcing.

Some articles have mentioned GE methods that consider atmospheric CH_4, but no concrete technology has been described: the first proposes assessing all suitable technologies for "CH_4 air capture" [223] and the second suggests catalytic aerosols dispersed in the atmosphere, which are able to process nearly one-tenth of the atmosphere each year, but no suitable aerosol material was identified by the authors [224].

Chlorine atoms in the troposphere remove nearly 2.5% of the CH_4 and up to 20% in some locations [225]. It has been demonstrated that sunlight iron photocatalysis by Fe(III)/Fe(II) can regenerate Cl° under acidic conditions [226].

Atmospheric removal of CH_4 by increasing the chlorine atmospheric sink has been proposed [227], together with removal of tropospheric ozone (O_3), as halogen radicals destroy O_3 [228], but it is more efficient if performed by Br or I atoms [229]. Fig. 20.11 illustrates different ways of producing iron salts aerosols to enhance the chlorine sink of CH_4 and tropospheric O_3.

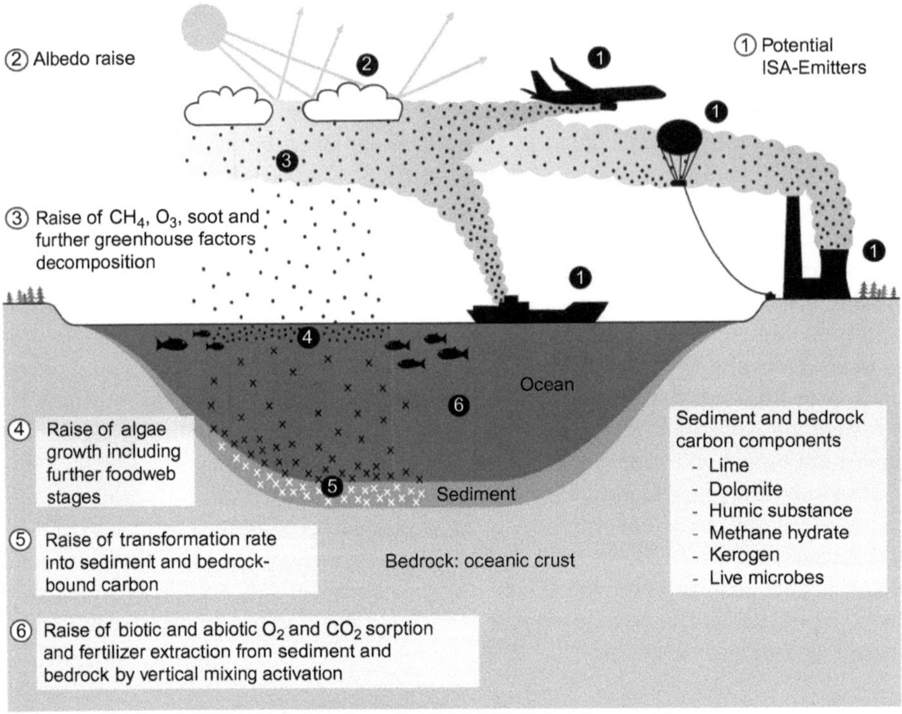

Fig. 20.11 Concept of increasing the chlorine atom sink of methane and tropospheric ozone. Reproduced from Oeste FD, de_Richter R, Ming T, Caillol S. Climate engineering by mimicking natural dust climate control: the iron salt aerosol method. Earth System Dynam 8 (2017) 1–54. doi: 10.5194/esd-8-1-2017.

In the section on OIF, it has been seen that iron is an important micronutrient and is depleted in many oceanic regions. As a matter of fact, an important part of marine productivity is due to anthropogenic iron emissions from FFPPs [230] but only the iron-fertilizing effect on phytoplankton after oceanic deposition has been considered [231].

Sulfates [232] and iron deposition [233] on wetlands, peat lands, rice paddies, and other terrestrial landscapes inhibit CH_4 production or enhance CH_4 consumption by soil microorganisms. Consequently, it has been demonstrated [227] that the effects of Cl°, iron salt aerosols, and sulfate emissions by coal power plants and by the steel industry on tropospheric CH_4 and O_3 concentrations need to be considered in the future.

Last but not least, antipollution regulations and the accelerated decrease in fossil-fuel use will reduce atmospheric acidity and emissions of sulfates, chlorine, iron derivatives, and fertilizers (N, P). This may induce a rapid warming by reducing the amount of cooling aerosols [234] and also decrease marine productivity [235–237], increase atmospheric CH_4 emissions from wetlands [238], and reduce the atmospheric CH_4 sink due to the chlorine radical.

20.5.2 Methane, nitrous oxide, and ozone-depleting compounds removal by photocatalysis

Fig. 20.12 illustrates new large-scale GE methods for GGR that have been recently proposed to:

- *Remove N_2O* [239] applied directly to atmospheric air based on photocatalytic reduction of N_2O to produce nitrogen and oxygen. N_2O reduction occurs even in the presence of O_2 [240].
- *Remove CFCs and other halogenated gases* (included in the Montreal protocol as ozone-layer-depleting compounds), which are powerful GHGs due to their large GWP [67]. The method is based on photocatalyzed oxidoreduction [241,242] and generally yields CO_2 and the corresponding halogenated acids.
- *Remove CH_4* [227,243] by oxidation and transformation into CO_2, which has a much lower 20 year GWP (GWP_{20} is 86 for CH_4 and 1 for CO_2, respectively). Several photocatalysts are possible to perform the complete oxidation of CH_4, one of the them is zinc oxide with 1% silver [244].

Many GGR methods rely on implementing solar updraft towers, which produce renewable electricity and prevent further CO_2 emissions from the FFPP they have replaced. A solar updraft tower consists of a tall chimney in the middle of a large

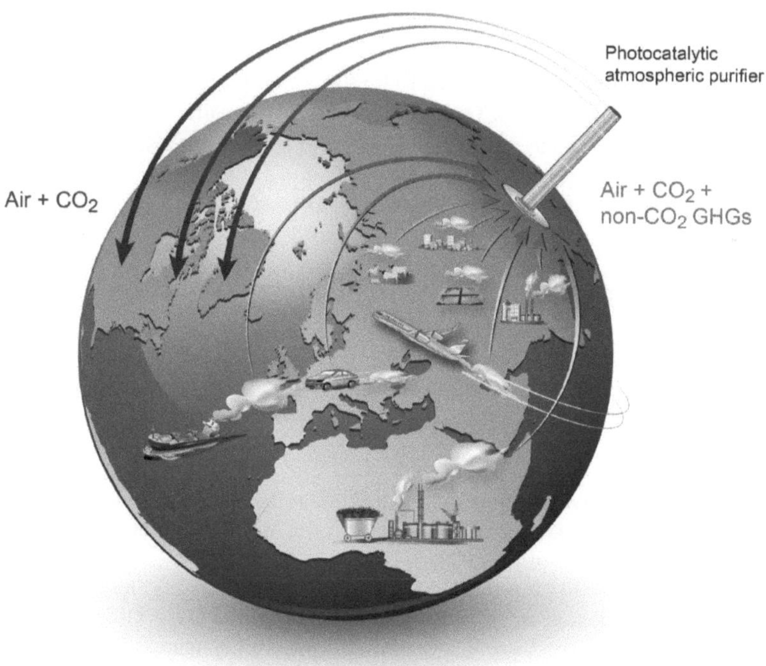

Fig. 20.12 The greenhouse gas removal (GGR) concept.
Reproduced from de_Richter R, Ming T, Davies P, Liu W, Caillol S. Removal of non-CO_2 greenhouse gases by large-scale atmospheric solar photocatalysis. Prog Energy Combust Sci 60 (2017) 68–96.

greenhouse, where the stack effect and the greenhouse effect create artificially hot wind. The air at ambient temperature enters by the periphery of the greenhouse, warms, and gets out at the top of the chimney. The airflow created in hot dry deserts allows electricity to be produced with turbines, day and night, with no intermittency thanks to the thermal inertia of the soil or by thermal energy storage in water bags [245].

20.5.3 *Main advantages of GGR*

The biggest advantage of GGR for non-CO_2 GHGs over CDR/CCS is that there is no need for storage. After CO_2 is captured, it needs to be purified, compressed, and transported to a sequestration site where it will be injected.

The proposal consisting to produce cost-competitive electricity and at the same time removing from the atmosphere N_2O, CFC, and CH_4 by photocatalysis under solar updraft towers seems attractive. This technology can make money, being competitive with existing technologies (solar, wind, and nuclear), while removing GHGs, which is not the case for many CDR methods [191], nor of any SRM proposal that requires important subsides to be implemented and maintained in operation. But in order to have effects at a significant climatic level, it requires replacing numerous FFPPs.

Soot or black carbon (BC) and organic carbon are solid carbonaceous materials mainly produced by burning biomass and other incomplete combustion processes. They are not GHGs, but are considered to be the second largest contributor to anthropogenic radiative forcing after CO_2 [246], being of the same order of magnitude as CH_4. The method proposed to remove CH_4 and O_3 [227] might also reduce the lifetime and the amount of BC. Reducing the amount of tropospheric O_3, carbon black, and/or CH_4 can save millions of lives annually [247,248].

20.6 Earth (or thermal) radiation management

Substantial variability exists between different GE proposals and numerous studies have ignored the wider portfolio of GE options for tackling climate change, often only concentrating on a limited number of SRM methods. As already mentioned in the subchapter about SRM:

- Cirrus clouds thinning works differently than SRM, as the goal is to allow more infrared LW radiation to escape into outer space, while preventing a smaller amount of incoming solar SW radiation to be reflected [34,249].
- The technology proposed for preserving downwelling currents [20] illustrated in Fig. 3 is a SRM technology, as it increases albedo by producing sea ice, but it is also a CDR method, as it allows more CO_2 to be dissolved in seawater (solubility increases when temperature decreases) and to be stored in the deeper layers of the oceans for 600–1000 years, while the real aim of this proposal is to cool surface seawater, to enhance downwelling currents, by transferring heat to air and allowing more outgoing LW radiation.

Materials and methods allowing night-time radiative cooling in a clear sky have been developed [250,251] and may replace wet cooling towers for CSP [252].

As shown in Fig. 20.13, which is part of Fig. 20.1, the outgoing energy flux by the "atmospheric window" consists of a portion of the infrared electromagnetic spectrum that can be transmitted through the atmosphere without absorption and re-emission downwards to the surface by GHGs and which is mainly in the 8–13 μm region. Thus, a possible way to cool the Earth is by developing materials that can absorb LW and/or SW radiation and re-emit LW radiation through an atmospheric window [253]. These types of materials allow daytime radiative cooling [254–256].

All of these methods help compensate for the increasing imbalance in the Earth's energy budget, characterized by a reduction in global outgoing LW terrestrial radiation trapped by the GHG effect: it has been proposed to call them ERM [9,24].

The Earth can be considered as a thermodynamic machine with a heat source at the equator and two heat sinks at the poles. But at any latitude the upper atmosphere layers and the extragalactic space are also heat sinks. The GHGs are basically slowing the heat exchanges between the ground, the high atmosphere and the outer space at 3 K acting as an insulating layer. So, to counteract the increase of the greenhouse effect, rather than equator-poles heat exchanges it could be more useful to enhance vertical heat exchanges by creating thermal bridges. This is the basic idea of many ERM technologies have been proposed for each one of the outgoing energy flues shown in Fig. 20.13 [24]. One of them consists of building very tall two-phase thermo-siphons (also called heat pipes and more rarely thermal diodes), which can transfer sensible heat from the surface to an altitude above the boundary layer where there are less GHGs, allowing LW radiation to escape into the outer space [257]. The proposed devices are intended to produce renewable energy with no-CO_2 emissions, thus to decarbonize the energy sector.

Unfortunately, the GE community has not yet become interested in ERM and very poorly in GGR methods and seems to prefer studying SRM and CDR technologies, sometimes with a commercial objective.

20.7 Geoengineering patents

The area of GE patents is not covered by this chapter, but it is worth mentioning that several patents describe original methods that have not been published in scientific articles and, consequently, were not reviewed here. For example, methods to remove stratosphere ozone-depleting agents and to artificially regenerate ozone are patented under the World international patent application WO/2007/033448.

There are many patents devoted to SRM techniques, such as United States Patent (USP) Application 2012-0117003, which specifically refers to GE methods; USP 4997632, which aims to remove GHGs in the stratosphere by spraying particles.

As examples, USP 9363954 and USP 9422739 relate to an apparatus or a structure for transporting and dispersing solid particles into the Earth's stratosphere; USP application 2008-0203328 describes a sunscreen to deploy in the outer space to reduce global warming; USP application 2010-0127224 provides a method for mitigating global warming by injecting reflective aerosols within the stratosphere.

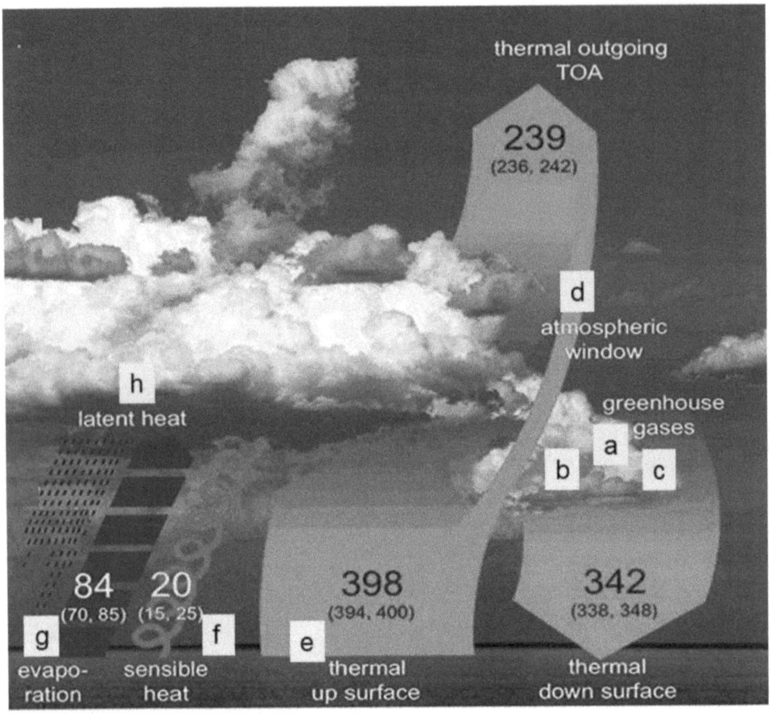

Fig. 20.13 Individual energy fluxes of outgoing longwave radiation targeted by Earth radiation management (ERM) (part of Fig. 20.1). Less heat is absorbed by the atmosphere and re-emitted downwards to the surface if there are less GHGs, which is the aim of: (a) CDR (Section 20.4); (b) GGR (Section 20.5); and (c) cirrus clouds thinning. Methods enhancing clear sky nighttime radiative cooling and new materials for daytime radiative cooling allow more IR radiation to be emitted by the surface (e) in the 8–13 μm longwave region (i.e., in the atmospheric window (d)) and as this wave lengths are not trapped by the atmosphere that enhances the amount of outgoing heat. Several devices have been proposed [24] (Section 20.6), like very tall closed two-phase thermo-siphons that enhance the transfer of sensible heat (f) form surface to high enough above the boundary layer, or which enhance atmospheric convection and evaporation (g) and the transfer of latent heat (h) in altitude where the IR radiation can easily escape to space.
Reproduced with permission from Wild M, Folini D, Schär C, Loeb N, Dutton EG, König-Langlo G. The global energy balance from a surface perspective. Clim Dyn 40 (2013) 3107–34. doi: 10.1007/s00382-012-1569-8.

Some of these patents have been listed on websites [58,258], but the inventories are far from exhaustive.

Numerous patents are related to CDR and describe methods, systems, processes, apparatus, or absorbents for removing CO_2 from air or from industrial exhaust flue gases. These last patents are devoted to mitigating CCS strategies (removing CO_2

at the stack exhaust before it enters the atmosphere), while others might enter among the classification of GE methods.

Of course, numerous GE patents also target ocean-based methods and processes, such as USP application 2012-0011050: spraying inorganic nutrients aimed to increase sea surface albedo, cloud nucleation activity, carbon sequestration, and yields for pelagic aquaculture.

The ethics and governance aspects about GE patents are often debated [259,260].

20.8 What is next?

A mix of technologies is needed to provide all energy needs for the future. In a similar way, no single GE method will solve the global warming problem and a mix of techniques will probably be needed.

After extensively evaluating some of the SRM methods individually, scientists have just started to evaluate them in combination (to be deployed simultaneously) [42]: "We simulate the climate effect of cocktail geoengineering that combines stratospheric sulfate aerosol increase and cirrus cloud thinning. Cocktail geoengineering can offset carbon dioxide-induced changes in global mean temperature and precipitation simultaneously" say the experts in this study. The good news is that the simulations showed that if both methods are deployed in concert, it would decrease warming to preindustrial levels, as desired and, rainfall on a global level would also stay at preindustrial levels. However, the model shows that global average climate might be largely restored, it demonstrates that substantial differences remain locally: with some areas getting much wetter and other areas getting much drier.

Some attempts have been made to consider "SRM as CDR," with the idea that SRM can reduce the possible future atmospheric carbon burden [261], but many advocate that CDR, GGR, and NETs should not be considered SRM GE [262], as GGR share many characteristics with existing mitigation measures [263]. Mitigation is defined as an anthropogenic intervention to reduce the GHGs sources or enhance its sinks. Both CCS and avoiding deforestation reduce GHG emissions; both BECCS and reforestation enhance GHG sinks: CDR and GGR also and consequently should be fully considered as mitigation.

Should humans deploy technological options, with potentially serious negative climatic side effects, in order to respond to the impacts of other human-made technologies, with serious negative climate impacts? Humans are performing a very risky giant worldwide meteorological experiment consisting of adding to the climate system warming (GHGs) and cooling emissions (sulfates, aerosol pollution, and fertilizers: NO_x, P, and Fe). Should humans increase the number of climate interventions and add some more chemicals by deploying stratospheric SRM on a GE scale? Scientists have already found more than 20 reasons not to implement SRM using sulfate aerosols [264]. Are we already caught between Scylla and Charybdis and sure that the risks of not deploying sunshade GE will be clearly higher? Should humans deliberately select a (not so) well-considered risk, as Odysseus did: knowing that "some loss" is inevitable?

Undoubtedly, these questions cannot be answered and much more research is needed to evaluate the different options available.

Very interesting work is being performed under the initiative of the "Model Intercomparison Project (MIP)," with very large international cooperation to compare the climatic models [265]. GeoMIP started studying the possible efficiency, effects and drawbacks of some of the GE proposals [266]. Now, the CDR-MIP will start working on CDR proposals [267] and mention of GGR is made in the project. A new Fe-MIP that considers iron biogeochemical aspects has started [268] and hopefully it will also consider iron fertilization effects on the continents and oceans.

However, this is a very small part of the research urgently needed to try to keep global warming of 1.5–2°C as aimed by the Paris climate agreement. Scientists on the NETs area urgently need research funding in order to assess the different techniques and to be able to provide the best possible guidance to the policymakers to decide of the better methods to rapidly implement to address the global warming problems and, of course, the research needs to be conducted under strict ethics and governance rules [269,270].

20.9 Conclusions

Geoengineering SRM is an ensemble of ideas to cool the planet by reflecting away incoming sunlight without treating the cause of global warming.

NETs (CDR and GGR) regroup an increasing number of credible proposals to remove the planet's GHGs, and are necessary to try reaching the Paris agreement goals [62], but for the moment humans continue adding GHGs into the atmosphere. Fig. 20.14 summarizes the SRM methods and the most discussed CDR proposals (reproduced with permission from Mrs Rita Erven/the Kiel Earth Institute in Germany [49–51]).

ERM is a set of proposals to enhance heat transfer from the surface to outer space to help the Earth equilibrate its energy budget imbalance due to GHGs, but few researchers are working in this area.

None of these approaches can replace cutting GHG emissions, as SRM, NETs, or ERM cannot reverse all harmful effects of GHGs. The evidence is that deep and rapid emissions cuts are imperative and that possibly some GE methods could help alleviate global warming, the melting of ice caps, sea-level rise, ocean acidification, and other major risks.

With 10 billion inhabitants expected on Earth by mid-century, it seems difficult and unlikely that agriculture and industry will produce zero emissions. Consequently, actively removing CO_2 and other GHGs from the atmosphere will probably be necessary.

The objective of this chapter was to briefly review the approaches already on the table. No doubt that numerous other NETs will be proposed and we hope that they will quickly be evaluated for their efficiency, cost, feasibility, security, and acceptability, as rapid deployment will probably be needed to work at the scale to deliver the goal of the Paris agreement. The earlier will be the best, as there are many ecological,

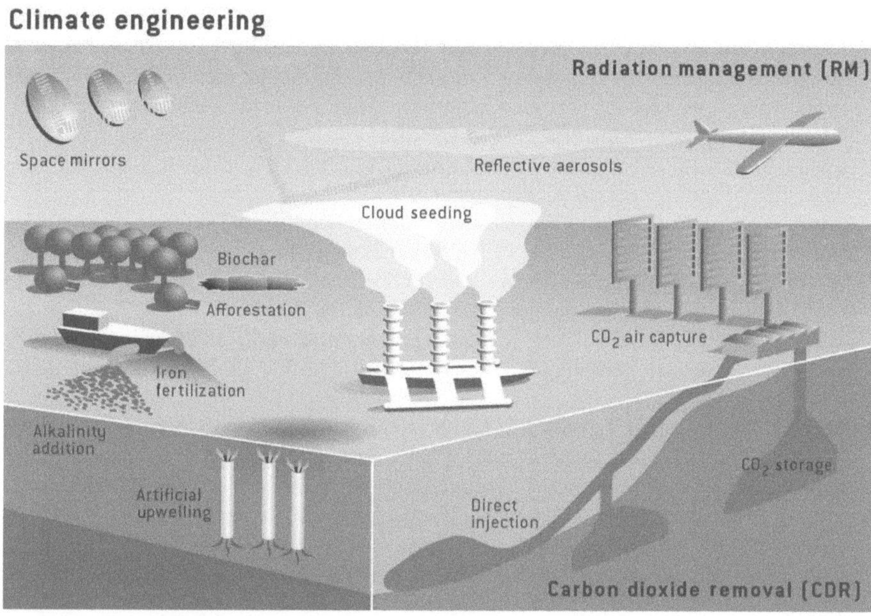

Fig. 20.14 Summary of the principal sunlight reflection methods (*top*) and of the most discussed carbon dioxide removal methods (*bottom*).
Courtesy of Mrs Rita Erven/GEOMAR from Rickels W, Klepper G, Dovern J, Betz G, Brachatzek N, Cacean S, et al. Gezielte Eingriffe in das Klima? Eine Bestandsaufnahme der Debatte zu Climate Engineering. Kiel Earth-Institute; 2011, http://www.kiel-earth-institute.de/sondierungsstudie-climate-engineering.html?file=files/media/downloads/CE_gesamtstudie.pdf (Accessed 23 October 2017).

chemical, or physical limitations to the amount of land-based and ocean-based CDR and GGR that can realistically be performed in case of overshoot of the carbon budget [271] with already this century minimal estimated costs of US $\$89–535 \times 10^{12}$ (US \$89–535 trillion) [272].

References

[1] Vaughan NE, Lenton TM. A review of climate geoengineering proposals. Clim Chang 2011;109:745–90.
[2] Lenton T, Vaughan N. Geoengineering responses to climate change: selected entries from the encyclopedia of sustainability science and technology. Springer Science & Business Media; 2012https://doi.org/10.1007/978-1-4614-5770-1.
[3] Zhang Z, Moore JC, Huisingh D, Zhao Y. Review of geoengineering approaches to mitigating climate change. J Clean Prod 2015;103:898–907.
[4] Edenhofer O, Pichs-Madruga R, Sokona Y, Field C, Barros V, Stocker T, et al. 2011. IPCC expert meeting on geoengineering: Meeting report, 20–22 June, Lima, Peru. Available at: http://www.ipcc-wg2.org/meetings/EMs/EM_GeoE_Meeting_Report_final.pdf. ISBN 978-92-9169-136-4 (Accessed 21 September 2017).

[5] IPCC-AR5-WGI. in: Stocker TF, Qin D, Plattner G-K, Tignor M, Allen SK, Bosc Hung J, Nauels A, Xia Y, Bex V, Midgley PM, editors. Climate change 2013: The physical science basis. Contribution of working group I to the fifth assessment report of the intergovernmental panel on climate change. Cambridge, United Kingdom and New York, NY, USA: Cambridge University Press; 2014. 1535 p, https://doi.org/10.1017/CBO9781107415324http://www.climatechange2013.org/images/report/WG1AR5_ALL_FINAL.pdf. Accessed 21 September 2017.

[6] National Research Council. Climate intervention: reflecting sunlight to cool earth. 234 p, National Academies Press; 2015, ISBN: 978-0-309-31482-4.

[7] National Research Council. Climate intervention: carbon dioxide removal and reliable sequestration. National Academies Press; 2015. ISBN: 978-0-309-30529-7, 155 p.

[8] Wild M, Folini D, Schär C, Loeb N, Dutton EG, König-Langlo G. The global energy balance from a surface perspective. Clim Dyn 2013;40:3107–34. https://doi.org/10.1007/s00382-012-1569-8.

[9] Mitchell DL, Mishra S, Lawson RP. Cirrus clouds and climate engineering: new findings on ice nucleation and theoretical basis. Planet Earth 2011-Global Warming Challenges and Opportunities for Policy and Practice 2011 ISBN: 978-953-307-733-8 Available at: https://cdnintechopencom/pdfs-wm/21120pdf (Accessed 18 September 2017). Croatia: InTech. pp. 257-88.

[10] Chen Z, Zhu L, Raman A, Fan S. Radiative cooling to deep sub-freezing temperatures through a 24-h day–night cycle. Nat Commun 2016;7.

[11] Storelvmo T, Boos W, Herger N. Cirrus cloud seeding: a climate engineering mechanism with reduced side effects? Phil Trans R Soc A 2014;372:20140116.

[12] Muri H, Kristjánsson J, Storelvmo T, Pfeffer M. The climatic effects of modifying cirrus clouds in a climate engineering framework. J Geophys Res-Atmos 2014;119:4174–91.

[13] Heyward C, Rayner S, Savulescu J. 6 Early geoengineering governance: the Oxford principles. Phil Technol Environ 2017;2:103.

[14] Wikipedia. 2017. Solar radiation management. https://en.wikipedia.org/wiki/Solar_radiation_management (Accessed 15 October 2017).

[15] Akbari H, Kolokotsa D. Three decades of urban heat islands and mitigation technologies research. Energy Build 2016;133:834–42.

[16] Menon S, Akbari H, Mahanama S, Sednev I, Levinson R. Radiative forcing and temperature response to changes in urban albedos and associated CO_2 offsets. Environ Res Lett 2010;5:014005.

[17] Seitz R. Bright water: hydrosols, water conservation and climate change. Clim Chang 2011;105:365–81.

[18] Evans J, Stride E, Edirisinghe M, Andrews D, Simons R. Can oceanic foams limit global warming? Clim Res 2010;42:155–60.

[19] Save-The-Arctic. Arctic News http://arctic-news.blogspot.fr/; Arctic Methane Emergency Group http://ameg.me/; Greenpeace campaign "Save the Arctic" http://www.greenpeace.org/usa/arctic/ and https://www.savethearctic.org/; The Climate Restoration Foundation https://www.climate-restoration-foundation.com/ice-cap-thickening (All Accessed 15 October 2017).

[20] Zhou S, Flynn P. Geoengineering downwelling ocean currents: a cost assessment. Clim Chang 2005;71:203–20.

[21] Haqq-Misra J. Should we geoengineer larger ice caps? Futures 2015;72:80–5.

[22] Desch SJ, Smith N, Groppi C, Vargas P, Jackson R, Kalyaan A, et al. Arctic ice management. Earth's Future 2017;5:107–27.

[23] Notz D. The future of ice sheets and sea ice: between reversible retreat and unstoppable loss. Proc Natl Acad Sci 2009;106:20590–5.
[24] Ming T, de_Richter R, Liu W, Caillol S. Fighting global warming by climate engineering: is the Earth radiation management and the solar radiation management any option for fighting climate change? Renew Sust Energ Rev 2014;31:792–834.
[25] Zimov SA, Zimov N, Chapin IIIF. The past and future of the mammoth steppe ecosystem. In: Louys J, editor. Paleontology in ecology and conservation. Springer; 2012. ISBN: 978-3-642-25037-8. p. 193–225.
[26] Gouttevin I, Menegoz M, Dominé F, Krinner G, Koven C, Ciais P, et al. How the insulating properties of snow affect soil carbon distribution in the continental pan-Arctic area. J Geophys Res Biogeosci 2012;117:.
[27] Frieler K, Mengel M, Levermann A. Delaying future sea-level rise by storing water in Antarctica. Earth System Dynam 2016;7:203.
[28] McMillan M, Shepherd A, Sundal A, Briggs K, Muir A, Ridout A, et al. Increased ice losses from Antarctica detected by CryoSat-2. Geophys Res Lett 2014;41:3899–905.
[29] Applegate PJ, Keller K. How effective is albedo modification (solar radiation management geoengineering) in preventing sea-level rise from the Greenland Ice Sheet? Environ Res Lett 2015;10:084018.
[30] McCusker K, Battisti D, Bitz C. Inability of stratospheric sulfate aerosol injections to preserve the West Antarctic Ice Sheet. Geophys Res Lett 2015;42:4989–97.
[31] Latham J, Bower K, Choularton T, Coe H, Connolly P, Cooper G, et al. Marine cloud brightening. Phil Trans R Soc A 2012;370:4217–62.
[32] Salter SH, Stevenson T, Tsiamis A. Engineering ideas for brighter clouds. p 131-161. In "Geoengineering of the climate system" Edited by RM Harrison and RE Hester. Vol. 38. 271 p. Royal Society of Chemistry. ISBN-13: 978-1849739535.2014.
[33] Wood R, Ackerman T, Rasch P, Wanser K. Could geoengineering research help answer one of the biggest questions in climate science? Earth's Future 2017.
[34] Mitchell DL, Finnegan W. Modification of cirrus clouds to reduce global warming. Environ Res Lett 2009;4:045102.
[35] Crutzen PJ. Albedo enhancement by stratospheric sulfur injections: a contribution to resolve a policy dilemma? Clim Chang 2006;77:211–20.
[36] Rasch PJ, Crutzen PJ, Coleman DB. Exploring the geoengineering of climate using stratospheric sulfate aerosols: the role of particle size. Geophys Res Lett 2008;35.
[37] Weisenstein DK, Keith DW, Dykema J. Solar geoengineering using solid aerosol in the stratosphere. Atmos Chem Phys 2015;15:11835–59.
[38] Keith DW, Weisenstein DK, Dykema JA, Keutsch FN. Stratospheric solar geoengineering without ozone loss. Proc Natl Acad Sci 2016;113:14910–4.
[39] Jones AC, Haywood JM, Jones A. Climatic impacts of stratospheric geoengineering with sulfate, black carbon and titania injection. Atmos Chem Phys 2016;16:2843–62.
[40] Ferraro AJ, Charlton-Perez AJ, Highwood EJ. Stratospheric dynamics and midlatitude jets under geoengineering with space mirrors and sulfate and titania aerosols. J Geophys Res-Atmos 2015;120:414–29.
[41] Keith DW. Photophoretic levitation of engineered aerosols for geoengineering. Proc Natl Acad Sci 2010;107:16428–31.
[42] Cao L, Duan L, Bala G, Caldeira K. Simultaneous stabilization of global temperature and precipitation through cocktail geoengineering. Geophys Res Lett 2017;44:7429–37.
[43] Kuebbeler M, Lohmann U, Feichter J. Effects of stratospheric sulfate aerosol geoengineering on cirrus clouds. Geophys Res Lett 2012;39.

[44] Facchini MC, Mircea M, Fuzzi S, Charlson RJ. Cloud albedo enhancement by surface-active organic solutes in growing droplets. Nature 1999;401:257–9.
[45] Topping D, Connolly P, McFiggans G. Cloud droplet number enhanced by co-condensation of organic vapours. Nat Geosci 2013;6:443.
[46] Wingenter OW, Elliot SM, Blake DR. New directions: enhancing the natural sulfur cycle to slow global warming. Atmos Environ 2007;41:7373–5.
[47] Irvine PJ, Ridgwell A, Lunt DJ. Climatic effects of surface albedo geoengineering. J Geophys Res-Atmos 2011;116.
[48] Shepherd JG. Geoengineering the climate: science, governance and uncertainty, Royal Society, 2009. ISBN: 978-0-85403-773-5, https://royalsociety.org/~/media/Royal_Society_Content/policy/publications/2009/8693.pdf. Accessed 22 October 2017.
[49] Klepper G, Rickels W. Climate engineering: economic considerations and research challenges. Oxford University Press; 2014.
[50] Rickels W, Klepper G, Dovern J, Betz G, Brachatzek N, Cacean S, et al. Gezielte Eingriffe in das Klima? Eine Bestandsaufnahme der Debatte zu Climate Engineering. http://www.kiel-earth-institute.de/sondierungsstudie-climate-engineering.html?file=files/media/downloads/CE_gesamtstudie.pdf (Accessed 23 October 2017). Kiel Earth-Institute; 2011.
[51] Klepper G, Rickels W. The real economics of climate engineering. Econ Res Int, Article ID 316564 20 p. 2012.
[52] Caldeira K, Bala G, Cao L. The science of geoengineering. Annu Rev Earth Planet Sci 2013;41:231–56.
[53] Lenton TM, Vaughan NE. The radiative forcing potential of different climate geoengineering options. Atmos Chem Phys 2009;9:5539–61.
[54] McClellan J, Keith DW, Apt J. Cost analysis of stratospheric albedo modification delivery systems. Environ Res Lett 2012;7:034019.
[55] McClellan J, Sisco J, Suarez B, Keogh G. 2011. Geoengineering cost analysis: final report AR10–182 (Prepared by Aurora Flight Sciences for The University of Calgary). http://justinmcclellan.com/GeoengineeringReport.pdf (Accessed 22 October 2017).
[56] Robock A, Marquardt A, Kravitz B, Stenchikov G. Benefits, risks, and costs of stratospheric geoengineering. Geophys Res Lett 2009;36.
[57] Salter S. 2017. Nullius in Verba: thoughts on the Royal Society 2009 geoengineering report relating to Cloud Albedo. https://groups.google.com/forum/#!msg/geoengineering/8MsSdPtxsq8/9HH2axFsBwAJ (Accessed 22 October 2017).
[58] Bronson D, Mooney P, Wetter KJ. Retooling the Planet - Climate Chaos in Geoengineering Age, Swedish Society for Nature Conservation; 2009.. ISBN: 978-91-558-0172-4 A report prepared by ETC Group, http://www.etcgroup.org/sites/www.etcgroup.org/files/publication/pdf_file/Retooling%20the%20Planet%201.2.pdf. Accessed 22 October 2017.
[59] Blackstock JJ, Battisti DS, Caldeira K, Eardley DM, Katz JI, Keith DW, et al. Climate engineering responses to climate emergencies. IOP Conf Ser Earth Environ Sci 2009;6:452015.
[60] MacCracken MC. On the possible use of geoengineering to moderate specific climate change impacts. Environ Res Lett 2009;4:045107.
[61] Cairns R. Climates of suspicion: 'chemtrail'conspiracy narratives and the international politics of geoengineering. Geogr J 2016;182:70–84.
[62] Anderson K. Duality in climate science. Nat Geosci 2015;8:898–900.
[63] Burns W, Nicholson S. Bioenergy and carbon capture with storage (BECCS): the prospects and challenges of an emerging climate policy response. J Environ Stud Sci 2017;1–8.

[64] McLaren DK. Negatonnes—an initial assessment of the potential for negative emission techniques to contribute safely and fairly to meeting carbon budgets in the 21st century. 2011. Friends of the Earth, London http://www.homepages.ed.ac.uk/shs/Climatechange/Carbon%20sequestration/negatonnes.pdf (Assessed 11 November 2017).
[65] Tavoni M, Socolow R. Modeling meets science and technology: an introduction to a special issue on negative emissions. Clim Chang 2013;118:1–14.
[66] Vichi M, Navarra A, Fogli P. Adjustment of the natural ocean carbon cycle to negative emission rates. Clim Chang 2013;118:105–18.
[67] de_Richter R, Ming T, Caillol S, Liu W. Fighting global warming by GHG removal: destroying CFCs and HCFCs in solar-wind power plant hybrids producing renewable energy with no-intermittency. Int J Greenhouse Gas Control 2016;49:449–72.
[68] IPCC. Climate change 2014: synthesis report. Contribution of working groups I, II and III to the fifth assessment report of the Intergovernmental Panel on Climate Change [Core Writing Team, R.K. Pachauri and L.A. Meyer (eds.)]. IPCC, Geneva, Switzerland, 151 p. http://ar5-syr.ipcc.ch/ipcc/ipcc/resources/pdf/IPCC_SynthesisReport.pdf (Accessed 21 September 2017).
[69] Houghton RA, Nassikas AA. Negative emissions from stopping deforestation and forest degradation, globally. Glob Chang Biol 2017;.
[70] Baccini A, Walker W, Carvalho L, Farina M, Sulla-Menashe D, Houghton R. Tropical forests are a net carbon source based on above ground measurements of gain and loss. Science 2017;358:230–4.
[71] Bristow CS, Hudson-Edwards KA, Chappell A. Fertilizing the Amazon and equatorial Atlantic with West African dust. Geophys Res Lett 2010;37.
[72] Larkin A, Kuriakose J, Sharmina M, Anderson K. What if negative emission technologies fail at scale? Implications of the Paris Agreement for big emitting nations. Clim Pol 2017;1–25.
[73] Fuss S, Canadell JG, Peters GP, Tavoni M, Andrew RM, Ciais P, et al. Betting on negative emissions. Nat Clim Chang 2014;4:850–3.
[74] Smith LJ, Torn MS. Ecological limits to terrestrial biological carbon dioxide removal. Clim Chang 2013;118:89–103.
[75] Smith P, Davis SJ, Creutzig F, Fuss S, Minx J, Gabrielle B, et al. Biophysical and economic limits to negative CO_2 emissions. Nat Clim Chang 2016;6:42–50.
[76] Zickfeld K, MacDougall AH, Matthews HD. On the proportionality between global temperature change and cumulative CO_2 emissions during periods of net negative CO_2 emissions. Environ Res Lett 2016;11:055006.
[77] Davies P. Can water treatment be a negative emissions technology?. 1st international conference on sustainable water processing, Sitges, Spain, 11–14 September; 2016.
[78] Pandey A, Singh M, Srivasatava R, Vasudevan P. Pollutant removal potential, growth and nutritional characteristics of short rotation woody crops in grey water vegetation filter system. J Sci Ind Res 2011;70:610–5.
[79] Shankhwar AK, Srivastava R. Biomass production through grey water fertigation in Eucalyptus hybrid and its economic significance. Environ Prog Sustain Energy 2015;34:222–6.
[80] Woolf D, Lehmann J, Lee DR. Optimal bioenergy power generation for climate change mitigation with or without carbon sequestration. Nat Commun 2016;7:13160.
[81] Khan S, Chao C, Waqas M, Arp HPH, Zhu Y-G. Sewage sludge biochar influence upon rice (Oryza sativa L) yield, metal bioaccumulation and greenhouse gas emissions from acidic paddy soil. Environ Sci Technol 2013;47:8624–32.
[82] Arribas L, Arconada N, González-Fernández C, Löhrl C, González-Aguilar J, Kaltschmitt M, et al. Solar-driven pyrolysis and gasification of low-grade carbonaceous materials. Int J Hydrog Energy 2017;42:13598–606.

[83] Yuan H, Lu T, Huang H, Zhao D, Kobayashi N, Chen Y. Influence of pyrolysis temperature on physical and chemical properties of biochar made from sewage sludge. J Anal Appl Pyrolysis 2015;112:284–9.
[84] Bogusz A, Oleszczuk P, Dobrowolski R. Adsorption and desorption of heavy metals by the sewage sludge and biochar-amended soil. Environ Geochem Health 2017;1–12.
[85] Bandara T, Herath I, Kumarathilaka P, Hseu Z-Y, Ok YS, Vithanage M. Efficacy of woody biomass and biochar for alleviating heavy metal bioavailability in serpentine soil. Environ Geochem Health 2017;39:391–401.
[86] Creamer AE, Gao B, Wang S. Carbon dioxide capture using various metal oxyhydroxide–biochar composites. Chem Eng J 2016;283:826–32.
[87] Liu W-J, Jiang H, Tian K, Ding Y-W, Yu H-Q. Mesoporous carbon stabilized MgO nanoparticles synthesized by pyrolysis of MgCl2 preloaded waste biomass for highly efficient CO_2 capture. Environ Sci Technol 2013;47:9397–403.
[88] Dalpisol M, Serrat BM, Motta ACV, Poggere GC, Bittencourt S, Barbosa JZ. Zinc, copper and manganese availability in soils treated with alkaline sewage sludge from Paraná state (Brazil). Ciência e Agrotecnol 2017;41:81–93.
[89] Lackner KS. The use of artificial trees p 80-104. In "Geoengineering of the climate system" Edited by RM Harrison and RE Hester. Vol. 38. 271 p. Royal Society of Chemistry. ISBN-13: 978-1849739535.2014.
[90] Holmes G, Nold K, Walsh T, Heidel K, Henderson MA, Ritchie J, et al. Outdoor prototype results for direct atmospheric capture of carbon dioxide. Energy Procedia 2013;37:6079–95.
[91] Rathi A. 2017. The world's first "negative emissions" plant has begun operation—turning carbon dioxide into stone. https://qz.com/1100221/the-worlds-first-negative-emissions-plant-has-opened-in-iceland-turning-carbon-dioxide-into-stone/ (Accessed 13 November 2017).
[92] Taylor LL, Quirk J, Thorley RM, Kharecha PA, Hansen J, Ridgwell A, et al. Enhanced weathering strategies for stabilizing climate and averting ocean acidification. Nat Clim Chang 2016;6:402–6.
[93] Rau GH, Carroll SA, Bourcier WL, Singleton MJ, Smith MM, Aines RD. Direct electrolytic dissolution of silicate minerals for air CO_2 mitigation and carbon-negative H2 production. Proc Natl Acad Sci 2013;110:10095–100.
[94] Li Z, Xu J, Teng HH, Liu L, Chen J, Chen Y, et al. Bioleaching of lizardite by magnesium- and nickel-resistant fungal isolate from serpentinite soils—implication for carbon capture and storage. Geomicrobiol J 2015;32:181–92.
[95] Thorley R, Taylor LL, Banwart SA, Leake JR, Beerling DJ. The role of forest trees and their mycorrhizal fungi in carbonate rock weathering and its significance for global carbon cycling. Plant Cell Environ 2015;38:1947–61.
[96] Schuiling RD. Olivine weathering against climate change. Nat Sci 2017;9:21.
[97] Rao AM, Malkin SY, Montserrat F, Meysman FJ. Alkalinity production in intertidal sands intensified by lugworm bioirrigation. Estuar Coast Shelf Sci 2014;148:36–47.
[98] De Beers Group. 2017. De Beers pioneers research programme to make carbon-neutral mining a reality. https://www.debeersgroup.com/en/news/company-news/company-news/de-beers-pioneers-research-programme-to-make-carbon-neutral-mini.html (Accessed 19 November 2017).
[99] Li J, Hitch M. Economic analysis on the application of mechanical activation in an integrated mineral carbonation process. Int Biodeterior Biodegrad 2016;https://doi.org/10.1016/jibiod201608026.

[100] Gerdemann SJ, O'Connor WK, Dahlin DC, Penner LR, Rush H. Ex situ aqueous mineral carbonation. Environ Sci Technol 2007;41:2587–93.

[101] Sanna A, Uibu M, Caramanna G, Kuusik R, Maroto-Valer M. A review of mineral carbonation technologies to sequester CO_2. Chem Soc Rev 2014;43:8049–80.

[102] Doney SC, Fabry VJ, Feely RA, Kleypas JA. Ocean acidification: the other CO_2 problem. Wash J Envtl L & Pol'y 2016;6:213.

[103] Meysman FJ, Montserrat F. Negative CO_2 emissions via enhanced silicate weathering in coastal environments. Biol Lett 2017;13:20160905.

[104] Schuiling R, De Boer P. Rolling stones; fast weathering of olivine in shallow seas for cost-effective CO_2 capture and mitigation of global warming and ocean acidification. Earth System Dynam Discuss 2011;2:551–68.

[105] Paquay FS, Zeebe RE. Assessing possible consequences of ocean liming on ocean pH, atmospheric CO_2 concentration and associated costs. Int J Greenhouse Gas Control 2013;17:183–8.

[106] Hauck J, Köhler P, Wolf-Gladrow D, Völker C. Iron fertilisation and century-scale effects of open ocean dissolution of olivine in a simulated CO_2 removal experiment. Environ Res Lett 2016;11:024007.

[107] Jones ISF. The cost of carbon management using ocean nourishment. Int J Clim Change Strateg Manage 2014;6:391–400.

[108] Jones ISF. 2016. Seven persistent misconceptions about Ocean Nourishment. Abstract #B34B-0367. AGU Fall Meeting Abstracts.

[109] Sinha E, Michalak A, Balaji V. Eutrophication will increase during the 21st century as a result of precipitation changes. Science 2017;357:405–8.

[110] Rabalais NN, Cai W-J, Carstensen J, Conley DJ, Fry B, Hu X, et al. Eutrophication-driven deoxygenation in the coastal ocean. Oceanography 2014;27:172–83.

[111] Smetacek V, Klaas C, Strass VH, Assmy P, Montresor M, Cisewski B, et al. Deep carbon export from a Southern Ocean iron-fertilized diatom bloom. Nature 2012;487:313–9.

[112] Martin P, Loeff MR, Cassar N, Vandromme P, d'Ovidio F, Stemmann L, et al. Iron fertilization enhanced net community production but not downward particle flux during the Southern Ocean iron fertilization experiment LOHAFEX. Glob Biogeochem Cycles 2013;27:871–81.

[113] Robinson J, Popova E, Yool A, Srokosz M, Lampitt R, Blundell J. How deep is deep enough? Ocean iron fertilization and carbon sequestration in the Southern Ocean. Geophys Res Lett 2014;41:2489–95.

[114] Güssow K, Proelss A, Oschlies A, Rehdanz K, Rickels W. Ocean iron fertilization: time to lift the research taboo. In: Climate change geoengineering: philosophical perspectives, legal issues. Cambridge, UK: Cambridge University Press; 2015. ISBN: 9781107502635. p. 242–62.

[115] Boyd PW, Bressac M. Developing a test-bed for robust research governance of geoengineering: the contribution of ocean iron biogeochemistry. Phil Trans R Soc A 2016;374:20150299.

[116] Lovelock JE, Rapley CG. Ocean pipes could help the Earth to cure itself. Nature 2007;449:403.

[117] Karl DM, Letelier RM. Nitrogen fixation-enhanced carbon sequestration in low nitrate, low chlorophyll seascapes. Mar Ecol Prog Ser 2008;364:257–68.

[118] Zhang D, Fan W, Yang J, Pan Y, Chen Y, Huang H, et al. Reviews of power supply and environmental energy conversions for artificial upwelling. Renew Sust Energ Rev 2016;56:659–68.

[119] Shepherd J, Iglesias-Rodriguez D, Yool A. Geo-engineering might cause, not cure, problems. Nature 2007;449:781.

[120] Oschlies A, Pahlow M, Yool A, Matear RJ. Climate engineering by artificial ocean upwelling: channelling the sorcerer's apprentice. Geophys Res Lett 2010;37.

[121] Kwiatkowski L, Ricke KL, Caldeira K. Atmospheric consequences of disruption of the ocean thermocline. Environ Res Lett 2015;10:034016.

[122] Calvin WH. 2013. Using the oceans to remove CO_2 from the atmosphere. http://geo-engineering.blogspot.fr/2013/03/using-the-oceans-to-remove-co2-from-the-atmosphere.html (Accessed 14 November 2017).

[123] Bauman SJ, Costa MT, Fong MB, House BM, Perez EM, Tan MH, et al. Augmenting the biological pump: the shortcomings of geoengineered upwelling. Oceanography 2014;27:17–23.

[124] Yang J, Zhang D, Chen Y, Fan W, Liang H, Tan M. Feasibility analysis and trial of air-lift artificial upwelling powered by hybrid energy system. Ocean Eng 2017;129:520–8.

[125] IPCC-AR5-WGIII. Climate change 2014: mitigation of climate change. contribution of working group III to the fifth assessment report of the Intergovernmental Panel on Climate Change [Edenhofer, O., R. Pichs-Madruga, Y. Sokona, E. Farahani, S. Kadner, K. Seyboth, A. Adler, I. Baum, S. Brunner, P. Eickemeier, B. Kriemann, J. Savolainen, S. Schlömer, C. von Stechow, T. Zwickel and J.C. Minx (eds.)]. Cambridge University Press, Cambridge, United Kingdom and New York, NY, USA. http://www.ipcc.ch/pdf/assessment-report/ar5/wg3/ipcc_wg3_ar5_full.pdf (Accessed 21 September 2017).

[126] Leung DY, Caramanna G, Maroto-Valer MM. An overview of current status of carbon dioxide capture and storage technologies. Renew Sust Energ Rev 2014;39:426–43.

[127] Gale J. CO_2 storage in deep geological formations: the concept. In: Geological storage of CO_2 in deep saline formations. Springer; 2017. p. 1–14.

[128] Lyngfelt A, Linderholm C. Chemical-looping combustion of solid fuels—status and recent progress. Energy Procedia 2017;114:371–86.

[129] Li J, Zhang H, Gao Z, Fu J, Ao W, Dai J. CO_2 capture with chemical looping combustion of gaseous fuels: an overview. Energy Fuel 2017;31:3475–524.

[130] Beget J. 2014. Antarctic pumpdown—a new geoengineering concept for capturing and storing atmospheric carbon dioxide. AGU Fall Meeting Abstracts.

[131] Agee E, Orton A, Rogers J. CO_2 snow deposition in Antarctica to curtail anthropogenic global warming. J Appl Meteorol Climatol 2013;52:281–8.

[132] Goldthorpe S. Potential for very deep ocean storage of CO_2 without ocean acidification: a discussion paper. Energy Procedia 2017;114:5417–29.

[133] Kim H-S, Cho G-C. Experimental simulation of self-trapping mechanism of CO_2 hydrates in marine sediments. ASME 2016 35th international conference on ocean, offshore and Arctic Engineering: American Society of Mechanical Engineers; 2016. pp. V008T11A3-VT11A3.

[134] World Energy Council. 2016. World energy resources—2016 Bioenergy. https://www.worldenergy.org/wp-content/uploads/2017/03/WEResources_Bioenergy_2016.pdf (Accessed 11 November 2017).

[135] Karlsson H, Byström L. 2011. Global status of BECCS projects 2010. Global CCS Institute & Biorecro AB. http://hub.globalccsinstitute.com/sites/default/files/publications/13516/gccsi-biorecro-global-status-beccs-110302-report.pdf (Accessed 11 November 2017).

[136] McCulloch S, International-Energy-Agency. 20 Years of carbon capture and storage: accelerating future deployment. Paris: IEA; 2016, https://www.iea.org/publications/

freepublications/publication/20YearsofCarbonCaptureandStorage_WEB.pdf. Accessed 29 October 2017.
[137] Herzog HJ. Scaling up carbon dioxide capture and storage: from megatons to gigatons. Energy Econ 2011;33:597–604.
[138] Global CCS Institute. The global status of CCS: summary report; 2016, ISBN: 978-0-9944115-6-3, https://hub.globalccsinstitute.com/sites/default/files/publications/201158/global-status-ccs-2016-summary-report.pdf. Accessed 11 November 2017.
[139] Sanchez DL. 2015. Deployment, design, and commercialization of carbon-negative energy systems. Ph.D. Dissertation, University of California, Berkeley. https://escholarship.org/content/qt0rs8n38z/qt0rs8n38z.pdf (Accessed 12 November 2017).
[140] McGlashan N, Shah N, Caldecott B, Workman M. High-level techno-economic assessment of negative emissions technologies. Process Saf Environ Prot 2012;90:501–10.
[141] McLaren D. A comparative global assessment of potential negative emissions technologies. Process Saf Environ Prot 2012;90:489–500.
[142] MacKerron G. 2014. Costs and economics of geoengineering. Clim. Geoeng. Govern. Working Paper Series. Working Paper 13. http://geoengineering-governance-research.org/perch/resources/workingpaper13mackerroncostsandeconomicsofgeoengineering.pdf (Accessed 13 November 2017).
[143] Kruger T. 2015. Dimensions of geoengineering: an analysis of the royal society's 'blob'diagram. Clim Geoeng Governance Working Paper Series No. 26.
[144] MacLeod M, Eory V, Gruère G, Lankoski J. 2015. Cost-effectiveness of greenhouse gas mitigation measures for agriculture. OECD Food, Agriculture and Fisheries Papers. 73 p.
[145] Canadell JG, Raupach MR. Managing forests for climate change mitigation. Science 2008;320:1456–7.
[146] Nielsen ASE, Plantinga AJ, Alig RJ. Mitigating climate change through afforestation: new cost estimates for the United States. Resour Energy Econ 2014;36:83–98.
[147] Kemena TP, Matthes K, Martin T, Wahl S, Oschlies A. Atmospheric feedbacks in North Africa from an irrigated, afforested Sahara. Clim Dyn 2017;1–21.
[148] Humpenöder F, Popp A, Dietrich JP, Klein D, Lotze-Campen H, Bonsch M, et al. Investigating afforestation and bioenergy CCS as climate change mitigation strategies. Environ Res Lett 2014;9:064029.
[149] Smith P. Soil carbon sequestration and biochar as negative emission technologies. Glob Chang Biol 2016;22:1315–24.
[150] Abalos D, Sanchez-Martin L, Garcia-Torres L, Van Groenigen JW, Vallejo A. Management of irrigation frequency and nitrogen fertilization to mitigate GHG and NO emissions from drip-fertigated crops. Sci Total Environ 2014;490:880–8.
[151] Wang G, Liang Y, Zhang Q, Jha SK, Gao Y, Shen X, et al. Mitigated CH4 and N2O emissions and improved irrigation water use efficiency in winter wheat field with surface drip irrigation in the north china plain. Agric Water Manag 2016;163:403–7.
[152] Roberts KG, Gloy BA, Joseph S, Scott NR, Lehmann J. Life cycle assessment of biochar systems: estimating the energetic, economic, and climate change potential. Environ Sci Technol 2009;44:827–33.
[153] Nanda S, Dalai AK, Berruti F, Kozinski JA. Biochar as an exceptional bioresource for energy, agronomy, carbon sequestration, activated carbon and specialty materials. Waste Biomass Valoriz 2016;7:201–35.
[154] Montserrat F, Renforth P, Hartmann J, Leermakers M, Knops P, Meysman FJ. Olivine dissolution in seawater: implications for CO_2 sequestration through Enhanced Weathering in coastal environments. Environ Sci Technol 2017;51:3960–72.

[155] Hitch M, Ballantyne SM, Hindle SR. Revaluing mine waste rock for carbon capture and storage. Int J Min Reclam Environ 2010;24:64–79.
[156] Renforth P, Jenkins B, Kruger T. Engineering challenges of ocean liming. Energy 2013;60:442–52.
[157] Davies P. Solar thermal decomposition of desalination reject brine for carbon dioxide removal and neutralisation of ocean acidity. Environ Sci: Water Res Technol 2015;1:131–7.
[158] Davies P, de Richter R, Yuan Q. 2018. Desalination as a negative emissions technology: a comparative study (Submitted).
[159] House KZ, Baclig AC, Ranjan M, van Nierop EA, Wilcox J, Herzog HJ. Economic and energetic analysis of capturing CO_2 from ambient air. Proc Natl Acad Sci 2011;108:20428–33.
[160] Socolow R, Desmond M, Aines R, Blackstock J, Bolland O, Kaarsberg T, et al. Direct air capture of CO_2 with chemicals: a technology assessment for the APS Panel on Public Affairs. American Physical Society; 2011.
[161] Realff MJ, Eisenberger P. Flawed analysis of the possibility of air capture. Proc Natl Acad Sci USA 2012;109:E1589.
[162] Broehm M, Strefler J, Bauer N. Techno-economic review of direct air capture systems for large scale mitigation of atmospheric CO_2; 2015. https://doi.org/10.2139/ssrn.2665702. (Accessed 27 November 2017).
[163] Pritchard C, Yang A, Holmes P, Wilkinson M. Thermodynamics, economics and systems thinking: what role for air capture of CO_2? Process Saf Environ Prot 2015;94:188–95.
[164] Wentworth J, Miller W. 2017 Greenhouse gas removal. Houses of Parliament, UK. Post notes-Research briefings no. 549. http://eprints.esc.cam.ac.uk/3819/1/POST-PN-0549.pdf (Accessed 22 October 2017).
[165] Wentworth J. 2013. Negative emissions technologies. Houses of Parliament. UK. Post notes-Research briefings no. 447. http://researchbriefings.files.parliament.uk/documents/POST-PN-447/POST-PN-447.pdf (Accessed 22 October 2017).
[166] Air Miners. 2017. The index of companies and projects mining carbon from the air. http://airminers.org/ (Accessed 27 November 2017).
[167] Jospe C. 2017. What is direct air capture? Part 3: technology comparison. http://carbonalist.com/2017/06/what-is-direct-air-capture-pt-3/ (Accessed 18 November 2017).
[168] Jospe C. 2017. What is direct air capture? (part 1). http://carbonalist.com/2017/05/what-is-direct-air-capture-pt1/ (Accessed 18 November 2017).
[169] Jospe C. 2017. Why the Carbfix2 project is a big deal. http://carbonalist.com/2017/10/why-carbfix-project-is-a-big-deal/ (Accessed 18 November 2017).
[170] Antoine de Ramon NY, Chynoweth DP, Capron ME, Stewart JR, Hasan MA. Negative carbon via ocean afforestation. Process Saf Environ Prot 2012;90:467–74.
[171] Brennan L, Owende P. Biofuels from microalgae—a review of technologies for production, processing, and extractions of biofuels and co-products. Renew Sust Energ Rev 2010;14:557–77.
[172] Moreira D, Pires JC. Atmospheric CO_2 capture by algae: negative carbon dioxide emission path. Bioresour Technol 2016;215:371–9.
[173] Greene CH, Huntley ME, Archibald I, Gerber LN, Sills DL, Granados J, et al. Geoengineering, marine microalgae, and climate stabilization in the 21st century. Earth's Future 2017;5:278–84.
[174] Doshi A, Pascoe S, Coglan L, Rainey TJ. Economic and policy issues in the production of algae-based biofuels: a review. Renew Sust Energ Rev 2016;64:329–37.

[175] Budzianowski WM. Negative carbon intensity of renewable energy technologies involving biomass or carbon dioxide as inputs. Renew Sust Energ Rev 2012;16:6507–21.
[176] Ed S, Da Silva T, Tokarski A, Ad N, De Carvalho J, Woiciecohwski A, et al. Screening of microalgae with potential for biodiesel production and nutrient removal from treated domestic sewage. Appl Energy 2011;88:3291–4.
[177] Arbib Z, de Godos I, Ruiz J, Perales JA. Optimization of pilot high rate algal ponds for simultaneous nutrient removal and lipids production. Sci Total Environ 2017;589:66–72.
[178] Harrison DP. Global negative emissions capacity of ocean macronutrient fertilization. Environ Res Lett 2017;12:035001.
[179] Strand SE, Benford G. Ocean sequestration of crop residue carbon: recycling fossil fuel carbon back to deep sediments. ACS Publications; 2009.
[180] Zeng N. Carbon sequestration via wood burial. Carbon Balance Manage 2008;3:1.
[181] Schuiling RD, de Boer PL. Six commercially viable ways to remove CO_2 from the atmosphere and/or reduce CO_2 emissions. Environ Sci Eur 2013;25:35.
[182] Flynn PC, Zhou JS. Can geoengineering sustain critical ocean currents? Macro-engineering seawater in unique environments. Springer; 2010. p. 583–93.
[183] Bates NR, Moran SB, Hansell DA, Mathis JT. An increasing CO_2 sink in the Arctic Ocean due to sea-ice loss. Geophys Res Lett 2006;33.
[184] Hanak DP, Jenkins BG, Kruger T, Manovic V. High-efficiency negative-carbon emission power generation from integrated solid-oxide fuel cell and calciner. Appl Energy 2017;205:1189–201.
[185] Kruger T. 2016. Energy generation process. U.S. Patent Application 2016/0181639 A1. Published 23 June 2016.
[186] House KZ, House CH, Schrag DP, Aziz MJ. Electrochemical acceleration of chemical weathering as an energetically feasible approach to mitigating anthropogenic climate change. Environ Sci Technol 2007;41:8464–70.
[187] Rau GH. Electrochemical splitting of calcium carbonate to increase solution alkalinity: implications for mitigation of carbon dioxide and ocean acidity. Environ Sci Technol 2008;42:8935–40.
[188] Boucher O, de Guillebon B, Abbadie L, Barré P, Bekki S, Bensaude-Vincent B, et al. 2014. Atelier de Réflexion Prospective REAGIR. Réflexion systémique sur les enjeux et méthodes de la géo-ingénierie de l'environnement, document de synthèse version courte.
[189] Kim D, Sakimoto KK, Hong D, Yang P. Artificial photosynthesis for sustainable fuel and chemical production. Angew Chem Int Ed 2015;54:3259–66.
[190] Acar C, Dincer I, Naterer GF. Review of photocatalytic water-splitting methods for sustainable hydrogen production. Int J Energy Res 2016;40:1449–73.
[191] de_Richter R, Ming T, Caillol S. Fighting global warming by photocatalytic reduction of CO_2 using giant photocatalytic reactors. Renew Sust Energ Rev 2013;19:82–106.
[192] Roy SC, Varghese OK, Paulose M, Grimes CA. Toward solar fuels: photocatalytic conversion of carbon dioxide to hydrocarbons. ACS Nano 2010;4:1259–78.
[193] Kondratenko EV, Mul G, Baltrusaitis J, Larrazábal GO, Pérez-Ramírez J. Status and perspectives of CO_2 conversion into fuels and chemicals by catalytic, photocatalytic and electrocatalytic processes. Energy Environ Sci 2013;6:3112–35.
[194] Chanmanee W, Islam MF, Dennis BH, MacDonnell FM. Solar photothermochemical alkane reverse combustion. Proc Natl Acad Sci 2016;113:2579–84.
[195] Gao Y, Deng L. Converting carbon dioxide into alkanes via alkane reverse combustion reaction. Sci Bull 2016;61:1160–2.

[196] Meinrenken CJ, Lackner KS. Options to dissociate CO_2 and H_2O for sustainable sunlight-to-fuel pathways: comparative assessment of current R&D hurdles and future potential. J Nat Sci 2014;2:19–54.

[197] de_Richter R, Caillol S. Fighting global warming: the potential of photocatalysis against CO_2, CH_4, N_2O, CFCs, tropospheric O3, BC and other major contributors to climate change. J Photochem Photobiol C: Photochem Rev 2011;12:1–19.

[198] Yazdanpour N, Sharifnia S. Photocatalytic conversion of greenhouse gases (CO_2 and CH_4) using copper phthalocyanine modified TiO_2. Sol Energy Mater Sol Cells 2013;118:1–8.

[199] Cuéllar-Franca RM, Azapagic A. Carbon capture, storage and utilisation technologies: a critical analysis and comparison of their life cycle environmental impacts. J CO_2 Util 2015;9:82–102.

[200] O'Brien KC, Lu Y, Patel V, Greenberg S, Locke R, Larson M, Krishnamurthy KR, Byron M, Naumovitz J, Guth DS, Bennett SJ. Creating markets for captured carbon: retrofit of Abbott Power Plant and future utilization of captured CO_2. Energy Procedia 2017;114:1263–72.

[201] Schuiling R. Serpentinite Slurries against Forest Fires. Open J Forestry 2015;5:255.

[202] Rau GH. 2014. Electrochemical production of metal hydroxide using metal silicates. U.S. Patent Application 13/008,238, filed January 18, 2011. U.S. Patent US2014/0234193A9 published August 21, 2014.

[203] Rau G, Baird J, Noland G. 2016. NEOTEC: negative-CO_2-emissions marine energy with direct mitigation of global warming, sea-level rise and ocean acidification. AGU Fall Meeting Abstracts.

[204] Rau G, Baird J. Negative-CO_2-emissions ocean thermal energy conversion. Renew Sust Energ Rev 2018; submitted.

[205] Rexed I, della Pietra M, McPhail S, Lindbergh G, Lagergren C. Molten carbonate fuel cells for CO_2 separation and segregation by retrofitting existing plants—an analysis of feasible operating windows and first experimental findings. Int J Greenhouse Gas Control 2015;35:120–30.

[206] Discepoli G, Milewski J, Desideri U. Off-design operation of coal power plant integrated with natural gas fueled molten carbonate fuel cell as CO_2 reducer. Int J Hydrog Energy 2016;41:4773–83.

[207] Barelli L, Bidini G, Campanari S, Discepoli G, Spinelli M. Performance assessment of natural gas and biogas fueled molten carbonate fuel cells in carbon capture configuration. J Power Sources 2016;320:332–42.

[208] Chacartegui R, Monje B, Sánchez D, Becerra JA, Campanari S. Molten carbonate fuel cell: towards negative emissions in wastewater treatment chp plants. Int J Greenhouse Gas Control 2013;19:453–61.

[209] Eisaman M, Schwartz D, Amic S, Larner D, Zesch J, Torres F, et al. in: Energy-efficient electrochemical CO_2 capture from the atmosphere. Technical proceedings of the 2009 clean technology conference and trade show; 2009. p. 175–8.

[210] Licht S, Ren J. 2015. Carbon nanofibers, precious commodities from sunlight & CO_2 to ameliorate global warming. https://arxiv.org/vc/arxiv/papers/1503/1503.06727v1.pdf (Accessed 21 November 2017).

[211] Ren J, Lau J, Lefler M, Licht S. The minimum electrolytic energy needed to convert carbon dioxide to carbon by electrolysis in carbonate melts. J Phys Chem C 2015;119:23342–9.

[212] Licht S, Wang B, Ghosh S, Ayub H, Jiang D, Ganley J. A new solar carbon capture process: solar thermal electrochemical photo (STEP) carbon capture. J Phys Chem Lett 2010;1:2363–8.

[213] Ren J, Li F-F, Lau J, González-Urbina L, Licht S. One-pot synthesis of carbon nanofibers from CO_2. Nano Lett 2015;15:6142–8.
[214] Licht S, Wang B, Wu H. STEP: a solar chemical process to end anthropogenic global warming. II: experimental results. J Phys Chem C 2011;115:11803–21.
[215] Eisaman MD, Parajuly K, Tuganov A, Eldershaw C, Chang N, Littau KA. CO_2 extraction from seawater using bipolar membrane electrodialysis. Energy Environ Sci 2012;5:7346–52.
[216] de Lannoy C-F, Eisaman MD, Jose A, Karnitz SD, DeVaul RW, Hannun K, et al. Indirect ocean capture of atmospheric CO_2: part I. Prototype of a negative emissions technology. Int J Greenhouse Gas Control 2018;.
[217] Willauer HD, DiMascio F, Hardy DR. Extraction of carbon dioxide and hydrogen from seawater by an electrolytic cation exchange module (E-CEM) part V: E-CEM effluent discharge composition as a function of electrode water composition. Washington DC, United States: Naval Research Lab; 2017.
[218] Fukuzumi S, Lee Y-M, Nam W. Fuel production from seawater and fuel cells using seawater. ChemSusChem 2017;.
[219] NERC. 2017. £8.6 million UK research programme on greenhouse gas removal. http://www.nerc.ac.uk/press/releases/2017/09-greenhousegas/ (Accessed 21 September 2017).
[220] Golub A, Hertel T, Lee H-L, Rose S, Sohngen B. The opportunity cost of land use and the global potential for greenhouse gas mitigation in agriculture and forestry. Resour Energy Econ 2009;31:299–319.
[221] Herrero M, Henderson B, Havlík P, Thornton PK, Conant RT, Smith P, et al. Greenhouse gas mitigation potentials in the livestock sector. Nat Clim Chang 2016;6:452–61.
[222] Hoornweg D, Sugar L, Trejos Gomez CL. Cities and greenhouse gas emissions: moving forward. Environ Urban 2011;23:207–27.
[223] Boucher O, Folberth GA. New directions: atmospheric methane removal as a way to mitigate climate change? Atmos Environ 2010;44:3343–5.
[224] Stolaroff JK, Bhattacharyya S, Smith CA, Bourcier WL, Cameron-Smith PJ, Aines RD. Review of methane mitigation technologies with application to rapid release of methane from the Arctic. Environ Sci Technol 2012;46:6455–69.
[225] Hossaini R, Chipperfield MP, Saiz-Lopez A, Fernandez R, Monks S, Feng W, et al. A global model of tropospheric chlorine chemistry: organic versus inorganic sources and impact on methane oxidation. J Geophys Res-Atmos 2016;121.
[226] Wittmer J, Zetzsch C. Photochemical activation of chlorine by iron-oxide aerosol. J Atmos Chem 2017;74:187–204.
[227] Oeste FD, de_Richter R, Ming T, Caillol S. Climate engineering by mimicking natural dust climate control: the iron salt aerosol method. Earth System Dynam 2017;8:1–54. https://doi.org/10.5194/esd-8-1-2017.
[228] Wang S, Schmidt JA, Baidar S, Coburn S, Dix B, Koenig TK, et al. Active and widespread halogen chemistry in the tropical and subtropical free troposphere. Proc Natl Acad Sci 2015;112:9281–6.
[229] Raso AR, Custard KD, May NW, Tanner D, Newburn MK, Walker L, et al. Active molecular iodine photochemistry in the Arctic. Proc Natl Acad Sci 2017;114:201702803.
[230] Myriokefalitakis S, Daskalakis N, Mihalopoulos N, Baker A, Nenes A, Kanakidou M. Changes in dissolved iron deposition to the oceans driven by human activity: a 3-D global modelling study. Biogeosciences 2015;12:3973.
[231] Li W, Xu L, Liu X, Zhang J, Lin Y, Yao X, et al. Air pollution–aerosol interactions produce more bioavailable iron for ocean ecosystems. Sci Adv 2017;3:e1601749.
[232] Gauci V, Dise NB, Howell G, Jenkins ME. Suppression of rice methane emission by sulfate deposition in simulated acid rain. J Geophys Res Biogeosci 2008;113.

[233] Jäckel U, Schnell S. Suppression of methane emission from rice paddies by ferric iron fertilization. Soil Biol Biochem 2000;32:1811–4.
[234] Chalmers N, Highwood EJ, Hawkins E, Sutton R, Wilcox LJ. Aerosol contribution to the rapid warming of near-term climate under RCP 2.6. Geophys Res Lett 2012;39:.
[235] Wang R, Balkanski Y, Bopp L, Aumont O, Boucher O, Ciais P, et al. Influence of anthropogenic aerosol deposition on the relationship between oceanic productivity and warming. Geophys Res Lett 2015;42:10745–54.
[236] Wang R, Balkanski Y, Boucher O, Bopp L, Chappell A, Ciais P, et al. Sources, transport and deposition of iron in the global atmosphere. Atmos Chem Phys 2015;15:6247–70.
[237] Wang R, Balkanski Y, Boucher O, Ciais P, Peñuelas J, Tao S. Significant contribution of combustion-related emissions to the atmospheric phosphorus budget. Nat Geosci 2015;8:48–54.
[238] Gauci V, Matthews E, Dise N, Walter B, Koch D, Granberg G, et al. Sulfur pollution suppression of the wetland methane source in the 20th and 21st centuries. Proc Natl Acad Sci U S A 2004;101:12583–7.
[239] Ming T, de_Richter R, Shen S, Caillol S. Fighting global warming by greenhouse gas removal: destroying atmospheric nitrous oxide thanks to synergies between two breakthrough technologies. Environ Sci Pollut Res 2016;23:6119–38.
[240] Obalová L, Šihor M, Praus P, Reli M, Kočí K. Photocatalytic and photochemical decomposition of N2O on ZnS-MMT catalyst. Catal Today 2014;230:61–6.
[241] Minami W, Kim H-J. Decomposition of halocarbons using TiO2 photocatalyst. Kagaku Kogaku Ronbunshu 2006;32(3):310.
[242] Sangchakr B, Hisanaga T, Tanaka K. Photocatalytic degradation of 1,1-difluoroethane (HFC-152a). Chemosphere 1998;36:1985–92.
[243] de_Richter R, Ming T, Davies P, Liu W, Caillol S. Removal of non-CO_2 greenhouse gases by large-scale atmospheric solar photocatalysis. Prog Energy Combust Sci 2017;60:68–96.
[244] Chen X, Li Y, Pan X, Cortie D, Huang X, Yi Z. Photocatalytic oxidation of methane over silver decorated zinc oxide nanocatalysts. Nat Commun 2016;7:.
[245] Ming T, Meng F, Liu W, Pan Y, de_Richter R. Analysis of output power smoothing method of the solar chimney power generating system. Int J Energy Res 2013;37:1657–68.
[246] Jacobson MZ. Strong radiative heating due to the mixing state of black carbon in atmospheric aerosols. Nature 2001;409:695–7.
[247] Shindell D, Kuylenstierna JC, Vignati E, van Dingenen R, Amann M, Klimont Z, et al. Simultaneously mitigating near-term climate change and improving human health and food security. Science 2012;335:183–9.
[248] Anenberg S, Talgo K, Arunachalam S, Dolwick P, Jang C, West J. Impacts of global, regional, and sectoral black carbon emission reductions on surface air quality and human mortality. Atmos Chem Phys 2011;11:7253–67.
[249] Storelvmo T, Kristjansson J, Muri H, Pfeffer M, Barahona D, Nenes A. Cirrus cloud seeding has potential to cool climate. Geophys Res Lett 2013;40:178–82.
[250] Bathgate S, Bosi S. A robust convection cover material for selective radiative cooling applications. Sol Energy Mater Sol Cells 2011;95:2778–85.
[251] Eriksson T, Lushiku E, Granqvist C. Materials for radiative cooling to low temperature. Solar Energy Mat 1984;11:149–61.
[252] Dyreson A, Miller F. Night sky cooling for concentrating solar power plants. Appl Energy 2016;180:276–86.

[253] Sun X, Sun Y, Zhou Z, Alam MA, Bermel P. Radiative sky cooling: fundamental physics, materials, structures, and applications. Nanophotonics 2017;6:997–1015.
[254] Rephaeli E, Raman A, Fan S. Ultrabroadband photonic structures to achieve high-performance daytime radiative cooling. Nano Lett 2013;13:1457–61.
[255] J-l K, Jurado Z, Chen Z, Fan S, Minnich AJ. Daytime radiative cooling using near-black infrared emitters. ACS Photon 2017;4:626–30.
[256] Zhai Y, Ma Y, David SN, Zhao D, Lou R, Tan G, et al. Scalable-manufactured randomized glass-polymer hybrid metamaterial for daytime radiative cooling. Science 2017;355:1062–6.
[257] Mochizuki M, Nguyen T, Mashiko K, Saito Y, Nguyen T, Wuttijumnong V. Challenges of heat pipe application for global warming. Proceedings of the 15th international heat pipe conference, Clemson, South Carolina, April 25–30, 2010; 2010.
[258] Lee J. 2017. List of some Weather modification and Geoengineering Patents. https://climateviewer.com/2014/03/24/geoengineering-weather-modification-patents/ (Accessed 23 October 2017).
[259] Diakun AT. 2015. Clearing the air on 'geoengineering'and intellectual property rights towards a framework approach. Master's Thesis. University of Waterloo, Ontario, Canada, 72 p.
[260] Parthasarathy S, Avery C, Hedberg N, Mannisto J, Maguire M. in: A public good?: geoengineering and intellectual property. Science, Technology, and Public Policy Program, Ann Arbor; 2010.
[261] Keith DW, Wagner G, Zabel CL. Solar geoengineering reduces atmospheric carbon burden. Nat Clim Chang 2017;7(9):617.
[262] Climate Analytics. 2016. Why negative CO_2 emission technologies should not be classified as Geoengineering. http://climateanalytics.org/files/why_net_is_not_geoengineering.pdf (Accessed 19 November 2017).
[263] Lomax G, Workman M, Lenton T, Shah N. Reframing the policy approach to greenhouse gas removal technologies. Energ Policy 2015;78:125–36.
[264] Robock A. 20 reasons why geoengineering may be a bad idea. Bull At Sci 2008;64:14–8.
[265] Gates WL, Boyle JS, Covey C, Dease CG, Doutriaux CM, Drach RS, et al. An overview of the results of the Atmospheric Model Intercomparison Project (AMIP I). Bull Am Meteorol Soc 1999;80:29–55.
[266] Kravitz B, Robock A, Boucher O, Schmidt H, Taylor KE, Stenchikov G, et al. The geoengineering model intercomparison project (GeoMIP). Atmos Sci Lett 2011;12:162–7.
[267] Keller DP, Lenton A, Scott V, Vaughan NE, Bauer N, Ji D, et al. The carbon dioxide removal model intercomparison project (CDR-MIP): rationale and experimental design. Geosci Model Dev Discuss 2017;2017:1–72.
[268] Tagliabue A, Dutkiewicz S. 2016. Iron Model Intercomparison Project (FeMIP). http://www.scor-int.org/Annual%20Meetings/2016GM/FeMIP.pdf (Accessed 19 November 2017).
[269] Caldeira K, Bala G. Reflecting on 50 years of geoengineering research. Earth's Future 2017;5:10–7.
[270] Long JC, Loy F, Morgan MG. Start research on climate engineering. Nature 2015;518:29.
[271] Dooley K, Kartha S. Land-based negative emissions: risks for climate mitigation and impacts on sustainable development. Int Environ Agreem Politics Law Econ 2018;18. 20 p, https://doi.org/10.1007/s10784-017-9382-9.

[272] Hansen J, Sato M, Kharecha P, Von Schuckmann K, Beerling DJ, Cao J, et al. 2016. Young people's burden: requirement of negative CO_2 emissions. arXiv preprint arXiv:160905878.

Further reading

[273] Pacella RM. 2007. Duct tape methods to save the earth: re-ice the Arctic. Popular Science. https://www.popsci.com/scitech/article/2007-06/duct-tape-methods-save-earth-re-ice-arctic (Accessed 16 November 2017).

Section F

Environmental and Human Issues

Normative issues of geoengineering technologies

21

Clare Heyward[1]
Institute for Advanced Sustainability Studies, Potsdam, Germany
[1]Corresponding author: clare.heyward@iass-potsdam.de

Chapter Outline

21.1 Introduction

The goal of international climate policy is to prevent "dangerous climate change," which for many years has been cashed out in terms of limiting the average global temperature rise to 2°C above preindustrial levels. The climate change that the planet is currently undergoing is caused by increased concentrations of greenhouse gases (GHGs), most notably carbon dioxide, in the atmosphere. For example, atmospheric concentration of carbon dioxide now stands at approximately 407 ppm, compared to an average of 280 ppm 2 centuries ago, whereas atmospheric concentration of methane has doubled [1]. GHGs are thus called because when in the atmosphere, they absorb a portion of solar energy that would otherwise be radiated back by the Earth into space. This results in warming. Reducing the concentration of GHGs in the atmosphere would thus reduce the amount of energy that can be absorbed, and correspondingly reduce the associated temperature increase.

There are two ways of reducing atmospheric GHG concentrations. The first is to reduce GHG emissions. The second is to increase the rate at which GHGs are drawn out of the atmosphere, i.e., to enhance sink capacity. These two broadly conceived strategies, emission reduction and sink capacity enhancement, have conventionally been grouped together under the term "mitigation." For example, in 2007, the

Managing Global Warming. https://doi.org/10.1016/B978-0-12-814104-5.00021-1

Intergovernmental Panel on Climate Change (IPCC) stated that mitigation means implementing policies to reduce GHG emissions and enhance sinks. However, in practice, discussions about mitigation of climate change have focused overwhelmingly on reducing global GHG emissions, particularly carbon dioxide emissions. This is understandable. To date, economic growth has been associated with increasing energy consumption, which typically involves GHG emissions. No state wishes to curtail its economic growth. Moreover, economic growth and development are necessary to overcome poverty, especially in developing countries. The overwhelming challenge is thus to "decarbonize" the world economy and to transfer to fuels that do not result in high GHG emissions so that growth and development can continue to occur, but without subjecting future generations to the risks of climate change.

Carbon sink capacity has featured in discussions about preserving carbon sinks by avoiding deforestation. This is the objective of the United Nations' reducing emissions from deforestation and forest degradation in developing countries, and the role of conservation, sustainable management of forests, and enhancement of forest carbon stocks in developing countries (REDD+) initiative [2]. REDD+ is, however, focused upon preserving the capacity of carbon sinks (mostly rainforests), not actively increasing it. Indeed, until recently, international or national climate policy discussions rarely featured in any detail the possibility of *enhancing* of carbon sinks. This is precisely the aim of negative emission technologies or NETs (sometimes called "greenhouse gas removal technologies" (GHGR technologies)). However, as carbon dioxide is the most important GHG, most NET proposals under consideration are aimed at enhancing carbon dioxide sinks, or removing more carbon dioxide from the atmosphere. Hence many discussions of NETs refer to "carbon dioxide removal" (CDR) technologies.

Even when confined to carbon dioxide removal, the term NETs covers a huge wide range of proposals. Here are some selected examples [3]. The interested reader can find several more comprehensive reviews and discussion, including in this volume:

- *Afforestation.* Arguably a favorite among members of the British public, afforestation is the large-scale planting of trees [4] (p. 2).
- *Bioenergy, carbon capture, and sequestration (BECCS).* Biomass is grown and used to generate energy. Carbon dioxide resulting from the process is not emitted, but captured and sequestered, in ways similar to conventional carbon capture and storage (CCS).
- *Direct air capture and storage (DACS).* This is the separation or "chemical scrubbing" of carbon dioxide from the air. It produces highly concentrated carbon dioxide gas which can then be sequestered.
- *Enhanced weathering* can be land based or ocean based. On land, processes aim to speed up or promote the reaction between atmospheric carbon dioxide and rocks and minerals, such as calcium carbonate and calcium silicate. Once the reaction has occurred, the resulting produce can be stored either in solid mineral form or in soluble form in the ocean.
- *Ocean fertilization* techniques add nutrients, most often iron, into parts of the ocean in order to stimulate phytoplankton growth, leading to an increase in ocean photosynthesis. It is the only CDR technique to have generated significant public controversy from both official experiments and similar activities [5,6].

Global GHG emissions—and with them, atmospheric GHG concentrations—are still increasing, which accounts for the increasing prominence of NETs in climate policy discussions. The more optimistic of the "Representative Concentration Pathway" scenarios used by the Intergovernmental Panel on Climate Change (IPCC) is characterized by widespread use of fossil fuels coupled with CCS, to reduce emissions, and the use of BECCS [7]. In RCP2.6, these measures are swiftly implemented so that carbon dioxide emissions peak in approximately 2020. By 2100, there is no overall increase in atmospheric carbon dioxide; instead approximately 1 gigaton of carbon dioxide is drawn down and sequestered from the atmosphere per year. A different way of putting that is that by 2100, global GHG emissions will be −1 GtC [8] (p. 103). In RCP2.6, the complete cessation of carbon dioxide emissions, plus concerted action on GHGs other than carbon dioxide, the average global temperature rise is limited to between 1.5°C and 2°C.

The *Fifth Assessment Report* (*AR5*) of the IPCC and the discussion it provoked made explicit that NETs are very likely to be needed if the 2°C target is to be achieved. Since then, in 2015, signatories to the Paris Agreement set an additional "aspirational" target of limiting global temperature rises to 1.5°C above preindustrial temperatures [9]. At the time of writing, the IPCC is compiling a Special Report on the impacts of and pathways to 1.5°C. We can expect NETs to feature, perhaps heavily.

Also increasing in prominence, and acknowledged in the IPCC's *AR5*, is another group of technological proposals, also aimed at counteracting climate change. They are sometimes called *sunlight reflection methods*, but more generally known as *solar radiation management* technologies (SRMs). SRM proposals do not aim to reduce atmospheric concentrations of GHGs, but rather seek to increase the amount of solar energy that is radiated back into space. Examples of SRMs can range from the seemingly mundane, for example, the brightening of human settlements by painting buildings white, to science fiction, for example, putting reflective materials into space. However, the most prominent SRM proposals are [10]:

- *Marine cloud-brightening*. This involves the introduction of aerosols, to increase the number of liquid drops in clouds, increasing the reflective surface area and thus overall reflectivity of cloud formations.
- *Stratospheric aerosol injection*. It has been known for many years that sulfur dioxide released during volcanic eruptions was largely the cause of decreased temperature and altered climate subsequently observed. Stratospheric aerosol injection (of sulfur dioxide, but perhaps other substances) thus aims to increase the amount of aerosols in the stratosphere in order that they scatter more sunlight, thus reducing the amount that reaches the Earth.

NETs and SRMs have often been discussed together under the term "geoengineering" or "climate engineering." However, the sheer diversity of specific proposals within the categories of NETs and SRMs should caution against presuming that NET technologies have much in common with SRMs. The main thing that unites them is that both are "technological imaginaries": i.e., no fully developed NET or SRM technology is yet implemented at scale. Another feature they have in common is that they, albeit in different ways, seek to avert dangerous climate change. Finally, as is the case with many other kinds of new technology, the development and use of both NETs and SRMs raise fundamental questions of values. Moreover, like any form of response to climate change, be it emission reductions, or adaptation measures, both NETs and SRMs can be justified and challenged on normative grounds.

Addressing climate change is a moral issue. We are concerned about climate change primarily because of its impacts on people's lives, both now and in the future. The sea-level rises, floods, droughts, and extreme weather events can be expected to cause much suffering, especially in places where poverty is endemic. This invokes questions of what people are entitled to, and the extent to which we have to alter our behavior if it affects them. Accordingly, while it might be necessary to devote time specifically to discussing climate change, at the end of the day, the challenge of addressing climate change should not be considered in isolation from other global issues such as poverty and development. Instead, what Simon Caney calls "integrationist" perspective is required [11]. More generally, this is true of *any* particular response to climate change. If any measure to combat climate change serves to worsen people's lives in general or is implemented in an unfair way, then this is potentially a cause for concern. Therefore, although in broad terms the following normative issues are raised by NETs and SRMs, they may be raised too by efforts in reducing emissions and adaptation measures [12].

21.2 Normative issues raised by geoengineering technologies: An overview

Before we begin, it is necessary to sound a note of caution: certain NETs and SRMs proposals *might* raise the following moral concerns, but it is, at this stage, by no means clear that they *will*, nor is it the case that any particular NET or SRM will be vulnerable to all the concerns listed below. We can expect some concerns (e.g., concern about side effects and issues of consent) to arise in the case of most proposals, for the simple reason that they are perennial issues in the development of any new technology. Finally, as already noted, the issues listed below are not uniformly applicable to all NET and SRM technologies, nor are they exclusive to them. With that caveat in mind, let us proceed. It might be helpful to divide the moral concerns raised by NETs and SRMs into the following four categories:

- Questions of distributive justice are about the distribution of social benefits and burdens within societies (domestic justice) across societies (global justice) and across generations (intergenerational justice). They cover the kinds of questions we might ask when we want to think about whether a society is fair or not. A full theory of distributive justice elucidates what kinds of social goods an individual is entitled to, how much of those goods an individual is entitled to, and which other individuals or institutions have duties to ensure those entitlements are met. For example, a *sufficientarian* theory of distributive justice might hold that all human beings have rights to a level of social and natural resources that enable them to be adequately nourished and to achieve an acceptable level of health. A rival *egalitarian* theory of distributive justice might hold that all individuals have a right to an equal share of social and natural resources.
- Questions of procedural justice are about the fairness of decision-making procedures, again, both domestically and globally. They include questions about who is entitled to have a say in a political decision, how they can best be represented, what kinds of decision-making rules

are most appropriate, and the place of norms such as transparency and fair representation in a decision-making process. For example, a theory of procedural justice might hold that all individuals who are affected by a political decision should have an equal say in it. A different theory of procedural justice might hold that the most affected should have more of a say, or even a veto, over the decision.

Distributive justice and procedural justice are analytically distinct categories, but in practice they can interact. Societies with greater procedural justice can often do better at the bar of distributive justice—for example, Amartya Sen has long argued that there has never been a famine in a modern democratic country [13]. However, to say that greater procedural justice will *always* lead to distributive justice is either to simply assume that political representatives will always make the morally correct decisions, or to ignore situations where democracies have voted for questionable policies.

Accounts of distributive and procedural justice set out the kinds of entitlements that individuals have. Should individuals be denied those entitlements, a theory of rectificatory justice is needed:

- Questions of rectificatory justice concern what to do when injustices occur: when entitlements are transgressed or violated. Discussions of rectificatory justice usually revolve around questions of retribution, restitution, compensation, and apology. Theories of rectificatory justice will typically take a position on what agents have duties to provide a remedy for an injustice, the nature of the appropriate remedy should take (retribution, restitution, compensation, and apology), and, especially when the remedy is retribution or compensation, the form and amount of it. When it comes to retribution, effectively all modern theories hold that *retributive* actions are morally justified only against the perpetrator of the injustice (they might vary on whether certain types of individual or groups can exhibit the kind of agency that allows us to talk of them committing an injustice). However, they differ between what kinds of retributive actions are appropriate and how onerous they may be to the perpetrator. Here, differing ideas of proportionality come into play. When it comes to compensation, some theories hold that compensation ought only be demanded from the perpetrators, whereas others hold that it can also be due from other agents, such as those who have benefitted from the causing of the injustice [14]. Theories of compensation will also differ in views about the form of compensation and amount that might be due.
- *Ethical* questions cover questions such as what kinds of things have value: what entities are morally considerable, what makes for a good human life, what characteristics make a good person, and what features of an action make it good. Religious doctrines are one source of views on ethical questions, but not the only one. In modern philosophical theory, *virtue theory* is a prominent approach. Rather than starting with the question "what should I do?," virtue theory starts with the question "what kind of person should I be" and offers an account of virtues (and vices) such as wisdom, courage, and temperance. Into this category also come questions about the value of human (and other life), including issues such as the permissibility of abortion, euthanasia, the appropriate treatment of animals, and the value of nature. On broadly liberal views of justice, a just society should aim to leave ethical questions to the conscience of the individual, providing that conscience does not lead him or her to commit an injustice. However, as individual's view on ethical questions may influence his or her view of justice (to a greater or lesser extent), this is not always possible. Indeed, questions about the legalization of abortion and euthanasia are some of the most controversial political issues in liberal societies.

With these broad categories in mind, let us consider some examples of common objections raised against the development of geoengineering technologies. Perhaps the most often discussed concern regarding geoengineering technologies is the so-called problem of "moral hazard." Also called the "mitigation obstruction argument," the "problem of trade-offs," and the "problem of risk-risk compensation," put simply, it is the concern that research into NETs or SRMs will detract from efforts to reduce GHG emissions [15–17]. The development of geoengineering technologies it is argued could be used as an excuse: some people might argue that there is less need to cut back on GHG emissions, because the use of NETs will allow atmospheric GHG concentrations to be maintained at a desirable level and the use of SRMs will allow global temperatures to be maintained at a desirable level.

Regardless of what it is called, behind this objection is an appeal to *distributive justice*. This might take some spelling out. The moral hazard objection is based on the idea that avoiding dangerous climate change is required by considerations of global and intergenerational distributive justice. However, the objection goes, if research into NETs and/or SRMs undermines support for emission reduction, then to avoid dangerous climate impacts, effective capacity to make up for the lack emission reduction of must be developed and implemented. Should the technologies turn out not to be as effective as hoped for, or cannot be implemented for some reason, then emission reduction strategies will have been reduced with nothing that can take their place. Accordingly, climate impacts will be worse than they would have been had emission reduction strategies been wholeheartedly implemented. In such a scenario, research and development of NETs and SRMs could result in distributive injustices, both globally and intergenerationally. It is important to emphasize that most formulations of this objection do not claim that research into NETs and SRMs will *inevitably* result in distributively unjust outcomes. Rather, research and development of NETs and SRMs *risks* disincentivizing emission reduction strategies and, accordingly, puts people at *risk* of greater climate impacts than would otherwise have been the case. Some people regard it as an "unjust and high-stakes gamble" with the well-being of others [18,19].

Another objection on the theme of a "high-stakes gamble" concerns the possibility of a "termination effect" should some SRM technologies come to be used without sufficient corresponding effort to reduce atmospheric concentrations of GHGs [10] (p. 63). If, for any reason, the use of the SRM technology is then suddenly halted, temperatures will rise sharply, subjecting people to climate impacts that are swifter (and hence harder to adapt to) than would have been the case without the use of the SRM. The harms of the resulting climate change would thus be even greater, which would be a matter of global and intergenerational distributive justice.

Objections that highlight undesirable *side effects* are also appeals to distributive justice. Physical side effects become problematic, from a normative point of view, if they interfere with, or put at risk, individuals' entitlements to social and natural goods. Some technologies will produce waste products which will adversely affect human health if not managed properly. For example, biochar—where biomass is burned and buried—might result in carcinogenic carbonized dust to be released [20] (p. 14). Ocean iron fertilization techniques change the distribution of ocean productivity, which could affect food sources of coastal communities [3] (p. 97). With respect to SRMs, the most serious of the possible side effects is that their use might change local and regional precipitation patterns [10] (pp. 7 and 40).

There are also normative issues with respect to the geographical location of the technologies. This is most obvious in the case of land-based techniques, particularly techniques such as BECCs which will require large amounts of land. Disadvantaged communities, especially indigenous peoples, have in the past been displaced from their traditional lands in order to create "conservation parks" and hydroelectric facilities [21,22]. Also, there are well-documented instances where heavy industries have been built in poor communities [23]. Another very important concern raised especially against land-based NETs is that, as happened in some cases with biofuel production, land use could change from food staple cultivation to crop-fuel cultivation, resulting in raised food prices and jeopardizing food security [24]. Both of these objections are based on competing ideas over claims to land, and what that land is used for—a perennial question of distributive justice.

Turning now to *ethics*, the fact that NETs and SRMs constitute intentional interventions in environmental systems in order to change the climate may make a moral difference. Common-sense morality holds that it is morally worse to cause harm intentionally than to do so unintentionally. Even if one disagrees with this, at the very least we can say that as a deliberate action, the decision to develop and use any NETs or SRM will signify something about mindset and attitudes of those who make it. Into this category, therefore, also comes the charge of *hubris*—that the development of technologies that are to be used to adjust the earth's climate it embodies humankind's arrogance—for example, a desire to "play God" or to "meddle with nature." Second, a supplementary ethical objection to the moral hazard argument is available. Those who believe that developing NETs or SRMs is a high-stakes gamble might go further to question the attitudes of societies who are prepared to advocate or implement it. For example, Stephen Gardiner argues that to gamble on an unproven SAI technology rather than to implement maximally strong emission reductions is an instance of moral corruption [25].

On certain views of *rectificatory justice*, intention can also play a role. As noted earlier, theories of rectificatory justice ask what victims of a distributive or procedural justice are owed: what ought to be done, by whom, to restore justice. It is commonplace for the mental states of the perpetrator to play a part in determining what the perpetrator has to do/or undergo for justice to be restored—for example, murder is punished more severely than manslaughter. It is an open question whether rectification for injustices attributed to the use of NETs—an intentional intervention into the climate system—should be treated any differently from injustices arising from the inadvertent climate caused by the burning of fossil fuels. In the context of NETs, questions of rectification might arise in at least the following two ways. First, should people suffer a loss because of the side effects of using NETs, then they might ask for some form of rectification, e.g., compensation. Second, it is possible that the use of NETs or SRMs might later be discovered to have adverse side effects that were unforeseeable at the time of implementation. Those who took the decision at the time cannot be accused of negligence. Rather, just like those responsible for developing and implementing technologies associated with coal, oil, and gas, they were "excusably ignorant" with respect to the eventual effects [26] (p. 761). In either scenario, it is necessary to assess how much rectification is due to those who suffer from the ill effects, and from whom such rectification should come.

There has always been disagreement over normative issues. In the case of NET and SRMs, we can expect disagreement in two areas. First, people will have different conceptions of what justice is, what is valuable, or being a good person entails. For example, some people might think that a just society is one which provides a "safety net" for its citizens: a level below which no one may fall. Other people might think that a just society must do more than that: it must strive for equality among citizens. Second, at least in this early stage of development we can expect disagreement about whether any particular technology will further or hinder the ideals set by views of justice and ethics. With diversity of opinion about various geoengineering technologies already evident, it will become necessary for societies to take decisions about whether to proceed or to cease with the development or implementation of any particular proposal.

Here we come to questions of procedural justice. Luke Tomlinson argues that these are less contested than questions of distributive justice or ethics [27]. Even accepting that is the case, there remain diverse ideas about what might count as a fair decision-making procedure, along the following dimensions. First, there is the classic "boundary problem": if "public participation" is called for, then who counts as "the public"? [28]. One standard answer is "the nation state," but this has obvious challenges. First, many states have national minorities or other groups who regard themselves as politically distinct. Second, in a globalized world, state sovereignty has changed considerably. Nation states can no longer (if they ever could) prevent global forces from having a significant impact on their citizen's lives. While these facts do not to consign the idea of sovereign statehood to history, they do suggest that other ways of defining the demos should be considered. Second, there is a question of the appropriate level of agreement, for example, must agreement be unanimous, or is a majority decision sufficient? Should the perspectives of different parties be treated equally, or might there be reasons for weighting them? Finally, consent from the relevant parties legitimates a decision only if it is appropriately informed. There is thus a question of what counts as "appropriately informed," and what, if any information be withheld.

Concerns that geoengineering technologies will be "imposed" on an unwilling population or that a population might be "co-opted" are appeals to procedural justice. So are the concerns expressed about *path dependency* and the *slippery slope* [29]. As Rose Cairns writes, development of technologies has frequently been driven by and suited the interests of "rather restricted social groups" [29] (p. 651). Some technologies are thought to be vulnerable to this possibility as research and development funding creates groups with reputational and financial interests in continuing development and eventual implementation. The end result of path dependency is "socio-technical lock-in," whereby it becomes extremely difficult for a society to stop using a given technology. Investments in research and especially infrastructure might cause such a lock-in. This may be assisted by the granting of patents which generate commercial interests in the development and use of technology [30]. Another worry concerning patents is that appeals to intellectual property and commercial confidentiality might also serve to justify the withholding of information from discussions about geoengineering technologies [30].

As the term geoengineering encompasses a very broad range of technological proposals, we can expect that any particular technology may be potentially vulnerable to only a subset of these objections. Moreover, out of those problems that do arise, some might be readily resolvable, and others prove more intractable. Rather than concluding that there should be an outright ban of research into either NETs or SRMs, most commentators hold that such objections raise important problems which ought to be taken into account in deciding what role any particular geoengineering technology should play in a society's response to climate change. These possible objections and problems should be the subject of discussion now, rather than as an afterthought after the technological development is done.

The following section offers some personal reflections on the issues of moral hazard, hubris, and participation.

21.2.1 Further discussion 1: The argument sometimes known as moral hazard

As noted earlier, the concern about moral hazard is perhaps the most discussed in the nascent literature on NETs. This is quite understandable: no scientist who is researching into methods to combat climate change wishes his or her work to ultimately result in increased risk of adverse climate impacts. However, there is relatively little empirical research available to tell us whether the prospect of NET research is disincentivizing emission reduction in the near term. Moreover, this is a difficult question to investigate. Most research on moral hazard has focused on individual behavior: e.g., whether wearing a seatbelt causes a driver to take more risks on the road. Whereas it is certainly possible to ask individuals whether the potential development of NETs or SRMs makes them more likely or less likely to mitigate—the form that research into the moral hazard in this context has largely taken—the conclusions we may draw from such studies are more limited. First, as Steve Gardiner points out, individuals' actions do not always track their stated intentions [31] (p. 167). Second, and more fundamentally, the extent to which individuals consume fossil fuels and emit GHGs is heavily constrained by their surrounding society: it is largely a structural issue. As Christian Baatz argues, to fully understand whether research into NETs or SRMs will prove to deter or delay emission reductions in any given society, it might be necessary to consider its possible effects on political structures and forces [16] (p. 36). However, as such investigations depend more heavily on essentially contested issues and concepts, it would be unwise to expect a definitive answer on this question in any policy-relevant timescale.

In a situation of such uncertainly, what, then, is the responsible thing to do? To answer this, let us see why the development of geoengineering technologies might be used to justify reduced efforts to make immediate emission reductions. If the development of a NET or SRM technology is to provide an adequate justification for a delay or a reduced effort in reducing emissions, then it must be the case that developing it will result in approximately the same outcome as reducing GHG emissions. It must in effect be presentable as a *plausible substitute* for reducing GHG emissions by some amount. And whether one object can be presented as a plausible substitute for another depends on the ultimate aims they are intended to serve. Therefore, the main way to reduce the possibility of any possible moral hazard is to emphasize that emission reduction strategies and NETs or SRM proposals are not equivalent: that, even if successfully developed, NETs or SRMs will not have some important benefits that would be gained from reducing emissions.

In the same way, if we move away from a narrow definition of the goals of climate policy to a broader one, the perceived "substitutability" of emission reductions and NETs may diminish somewhat. If emission reduction strategies and NETs are assessed purely in terms of "managing climate change," it appears that NETs and emission reduction strategies are fairly close substitutes. However, the criterion of "managing atmospheric concentrations of GHGs" is too narrow. The ultimate reason to be concerned about climate change, the reason why it is important to limit atmospheric concentrations of GHGs and global temperature rises, is human well-being and human development. The choice of responses to climate change should be guided by the impact of the various measures—emission reduction, NETs, SRM, and adaptation—on human development. And deciding whether any particular measure is a substitute for another should take into account all the ways in which the measures in question may contribute to, or detract from, human development goals. Therefore, I suggest that anyone concerned potential moral hazard highlights reasons to engage in emission reduction measures that are independent of their contribution to "atmospheric carbon dioxide management" and instead reflect broader human development goals. For example, switching to clean technologies will result in improvements in air quality, especially in cities in the developing world. Renewable technologies may also allow for greater energy security, both in the near and long term. Most countries have some capacity for renewable technologies, whereas coal and oil deposits are scattered very unevenly across the world. Moreover, despite the industries' abilities to find new economically viable sources, fossil fuels sources are finite at least on a human timescale. At some point, humans will have to learn to live without them. Conversely, should any technologies turn out to have advantages independent of their contribution to drawing down GHG or increasing albedo, then, those advantages should be highlighted. By emphasizing the differences, the idea that NETs or SRMs are substitutes for emission reduction measures is (further) challenged.

Of course, the above is not a guarantee that the potential of geoengineering technologies will never be used as a justification to delay or descale efforts in emission reduction. However, the more difficult it is to present geoengineering technologies and emission reduction measures as plausible substitutes for each other, the more difficult it will be to justify delayed or diminished efforts in emission reduction. Scientists and activists can both play a part in this. As David Morrow notes, some scientists might be uncomfortable with engaging in political discussions, but it can be done without overstepping the boundary between science and advocacy [32] (p. 11). The goal is to deliver accurate and easy to understand information: in this case,

I suggest to communicate that NET proposals and emission reduction proposals are not straightforward substitutes, especially when we consider the wider goals that climate response measures ought to serve.

21.2.2 Further discussion 2: Hubris

Ideas of hubris are frequently invoked in discussions of new technologies and geoengineering technologies are no exception. Hubris is the vice of having an exhalted sense of oneself—of excessive pride or arrogance. Like all vices, there are many different ways in which it can be manifested, but in the context of NETs and SRMs, concerns about hubris are most often voiced in conjunction with ideas of "messing with nature" [33]. The core of the hubris objection is that the development or use of a geoengineering technology is an embodiment of a problematic attitude (held by individuals or collectives) toward natural systems. However, there are two distinct problematic attitudes which the charge of hubris might identify, which are not always sufficiently distinguished. I suggest that we call them a *Promethean* and an *Icarian* sense of hubris.

In Greek mythology, both Prometheus and Icarus behaved arrogantly, but in two very different ways. Prometheus took it upon himself to steal fire from the gods, and to give it to mankind: to take something sacred and put it in the hands of humans for them to use for their ends. His actions are an attempt to assert himself above the commands of the gods and subvert the divine order, and it is for this that he is punished by Zeus. Icarus, by contrast, does not transgress any divine law or command when he steps out of the window with the wings that his father Daedalus has made for him. Neither he nor his father is punished for the act of flying. Icarus's downfall results because he becomes overconfident in his ability to fly and so, contrary to his father's advice, flies too close to the sun.

The reprehensible arrogance in the Promethean form of hubris is a refusal to stay in one's "appropriate place" in the world. People who invoke this objection believe that there is some kind of important relationship between human beings and the rest of the world, and it is wrong to change it or subvert it. This objection obviously fits in with many religious and spiritual worldviews: it might be argued that attempting to geoengineer the climate is a form of "playing God" [34]. However, the idea that human beings should "live in accordance with nature" can be invoked, as Dale Jamieson does, without presupposing avowedly religious view [35].

This is not to say that all religious perspectives or all "deep green" worldviews are automatically against any particular technological proposal. As Forrest Clingerman and Kevin O'Brien note, "hope that human beings can live harmoniously in nature without technological intervention resonates with stories of Eden shared by the three major monotheistic religions, while hope that human beings can innovate and engineer a way out of environmental problems is consistent with teachings in the same religions that God gave human beings authoritative dominion over other creatures and the Earth as a whole" [36] (p. 32). An individual judgment about whether any particular technology is hubristic thus depends on the precise details of their worldview, as well as how they understand the details of the technology itself. For example, it has been argued that the development of NETs might actually be a means of "atmospheric restoration"—an attempt to manage the Earth more responsibly than in the previous 2 centuries [37]. Such a view seems to agree that it is good to live in accordance with nature, and regards the utilization of NETS as a way of doing so. Having a religious or particularly "green" world view is thus a necessary, but not a sufficient condition for objecting to technologies on the grounds that they are "Promethian" or "playing God."

By contrast, the reprehensible arrogance in the Icarian form of hubris is excessive self-confidence in one's abilities, or, more widely, humankind's capacity to understand the world. Concerns that appeal to the complexity of natural processes, such as the possibility of unintended or unforeseen adverse consequences, are appeals to this Icarian sense. Researchers who advocate research into various geoengineering technologies, but who simultaneously caution against hubris and extoll the need for "humility" seem to have something close to the Icarian view in mind [38]. It is taken for granted that these technologies should not be dismissed on grounds that they encroach on the sacred, or subvert humankind's appropriate relationship with the world. However, caution is needed. The enormity and complexity of the task must not be underestimated, and in some cases, it might push human ingenuity to or beyond its limits. The Icarian form of the objection of hubris is, at essence, a skepticism about the limits of human intelligence and an attempt to check technological optimism.

As the charge of hubris is embedded in fundamental ideas about humankind's place in the world, and/or the limits of human understanding, we cannot expect there ever to be a final consensus on this matter. How then, may we proceed? In the case of Icarian hubris, there are some measures that could reduce its occurrence. For example, David Keith and colleagues suggested that a "red-team vs blue-team" approach, where one group of researchers tries to develop the technology and another searches for flaws [39] (p. 427). Additionally, Steve Rayner et al. suggest that lessons might be learned from "high reliability organizations" [40] (p. 507). Such suggestions for helping ensure that the technology works as planned should be complemented with robust investigations into the local effects, environmental and social, of its implementation. Investigations should be careful to include diverse perspectives, with an emphasis on local perspectives and where applicable, evidence from systems of traditional knowledge. As well as being mandated by considerations of procedural justice (see below) incorporating local and traditional systems of knowledge can highlight consequences which might otherwise be overlooked. Finally, the setting up compensation schemes may incentivize actors wishing to implement any technology to carry out rigorous risk assessments and forgo a "gung-ho" approach.

The Promethean form of the hubris objection is a more intractable challenge. The idea of the "right relationship with nature" one which has a great deal of intuitive appeal but elaborating on its content is extremely difficult. For this reason, it might be tempting to dismiss this concern as playing one of the most effective rhetorical tricks in the book [41]. Ultimately, the question of whether any particular NET or SRM technology is Promethean (and therefore unacceptable) will have to be settled by a fair procedure, informed by concrete information about how that technology operates. But I would stress that such concerns ought to be taken seriously, and not excluded from any discussion. The reason is that understandings of the human-nature relationship form a core part of individuals' sense of who they are and what is valuable to them.

Indeed they often comprise in part people's sense of identity. As such, they should certainly not be ignored or dismissed, and they should not be overridden without very good reason.

21.2.3 Further thoughts 3: Fair participation

Having considered one issue of distributive justice and one narrowly "ethical" issue, we come to procedural justice. As noted earlier, questions of procedural justice are those about the norms that should be adhered to ensure that a decision is legitimate. There are three reasons why procedural justice ought to be taken seriously. First, norms of procedural justice, such as fair participation, and transparency can often result in more informed decisions which turn out to be better on the whole. Second, people are more likely to accept a decision that goes against them if they believe that the procedure is fair. It is likely that this is linked to the third, most important reason: a person's life is his or her own, and it is right and proper for moral agents to decide for themselves on key issues that affect their lives. To make decisions that could change people's lives without their involvement is at best paternalistic. At worst, it is to subject them to the possibility of having their lives changed in a fundamental way because of the whim of a tyrant: effectively to deny that they and their lives matter.

If one accepts the view that procedural justice matters because moral agents should not stand to have their lives disrupted due to arbitrary decisions of others, it has possible implications for how a fair decision-making process is constituted. First, it supports a form of the "all-affected principle" [42]. As the name indicates, the all-affected principle states that all of those affected by a decision ought to have a say in it. Second, if what matters is that people are not subjected to decisions that change their lives without their having a say, a decision-making process must pay attention to how deeply and fundamentally people's lives are affected. This is for two reasons: first, if a process gave everyone who was affected in the slightest of ways a say in the decision, the possible constituency could be unworkably large. Second is the normative consideration that not all "effects" are equally disruptive of people's lives. Daniel Callies gives the example of a process deciding on the aesthetic standards of housing in a town. Those who live in the town are more greatly affected by the decision than someone who goes there for a yearly holiday [43] (p. 186). Accordingly, Callies endorses Brighouse and Fleurbaey's "proportionality principle," which states that "power or (influence) in any decision-making process should be proportional to individual stakes (or claims)" [44]. Another way of taking into account that the effects of a decision are not of equal normative weight is to modify the all-affected principle itself. Caney suggests that those who are affected in specific ways—i.e., those whose human rights will be affected—should have a say in a decision [45] (p. 158). Both Callies' and Caney's suggestions ensure that persons who stand to be deeply and profoundly affected by a decision will have priority over those who will be trivially affected: I shall not attempt to adjudicate between them here. I will instead make two points: first the definition of "affectedness" should include noneconomic as well as economic interests. If we are talking about profound disruption to people's lives, the loss of noneconomic goods, such as relationships, social standing, and understanding of the world, can cause at least as much disruption and distress, if not more than the loss of material goods.

Whereas it is deeply disruptive to lose *what you have*, it is more so to lose your understanding of *who you are*. Second, I will point out that there may be two communities who might said to be "affected," and therefore have a claim to participate by a decision concerning the *use* of a NET or SRM technology: those who might be exposed to side effects, or adverse consequences of any type should the technology be used, and those who are most vulnerable to climate change who would suffer if the technology was not used. Callies' suggestion of an index which combines vulnerability to climate change and population is a promising start, but needs to be supplemented with a similar index for exposure to side effects. If we are concerned with preventing suffering due to climate change, it would be ludicrous to include those who stand to be seriously affected by climate change, but exclude those who might suffer seriously if a certain technology was used.

21.3 From morality to governance: Developing social regulation for NETs and SRMs

The above were personal reflections on three of the questions that could be raised by research, development, and implementation of NETs and SRMs. They are presented as illustrations of some of the potential considerations and points of view that might be expressed when any given political community makes a decision on whether to start, proceed, or abandon the development of any geoengineering technology.

As no fully developed geoengineering technology yet exists, the illustrations are just that, and concrete conclusions about the permissibility of any NET or SRM technology cannot be determined. In general terms, all we can say at present is that each will at least raise questions about ethics, distributive, and procedural justice, but as to what the answers might be, we can only guess. Some readers may find this frustrating (some researchers certainly do!). Let us then consider how we might be able to come to more concrete conclusions over time. One promising suggestion is for an initially "bottom-up process," based on what have become known as the "Oxford Principles" [40]. The Oxford Principles are an initial attempt at capturing key societal values relevant to the development of geoengineering technologies. The short formulation of the Oxford Principles is:

Principle 1: Geoengineering to be regulated as a public good
Principle 2: Public participation in geoengineering decision making
Principle 3: Disclosure of geoengineering research and open publication of results
Principle 4: Independent assessment of impacts
Principle 5: Governance before deployment.

The first four principles ultimately appeal to ideas of distributive and procedural justice.[1] These leave ethical questions, as I understand them here, aside. Specifically,

[1] Principle 5 is different from the previous four. It highlights the need for governance structures to be in place before the implementation of any NET or SRM. Such governance structures, whatever form they eventually take, should ensure that the first four principles are adhered to.

the first and fourth principles make an appeal primarily to ideals of distributive justice, whereas Principles 2 and 3 appeal primarily to ideals of procedural justice. Specifically, the first principle expresses, among other things, the idea that "all of humanity has a common interest in a stable climate," and that the global climate should be "managed jointly, and for the benefit of all, with appropriate consideration given to future generations" [40] (p. 505). Specifying what counts for "the benefit of all" requires consideration of what are the most important interests common to humans, and in this case, including future generations. Similarly, Principle 4 states that people have adjudicate on the question of what counts as the most important type of impact. After all, the simple fact that a decision changes someone's life in *some* way is sufficient for it to be ruled impermissible. Therefore, Principle 4 requires that it is necessary to consider what kinds of human interests (and perhaps other interests) are important enough to be taken into consideration. The answering of the question of which are the most important (human) interests is one component of distributive justice.

Principles 2 and 3, by contrast, are concerned with procedural justice. Principle 2 states that there must be some recognizable form of public participation in decision making. As stated earlier, respecting persons as agents entails that they decide on key issues that affect their lives. The authors explain that there is a tacit appeal to the "all-affected principle" although as it stands, Principle 2 leaves open questions such as what kinds of interest must be affected to mandate a say, and what forms of consent are appropriate. Some scholars have assumed that in order to count as being "democratic" a decision-making process must rely on full deliberation, explicit consent, and allow those who withhold consent to "opt out" [46]. However, this is but one view of democratic participation and legitimacy and Principle 2 allows that other reasonable views exist [47] (p. 505).

The requirement of Principle 3 that parties have access to all relevant information during the decision-making process appeals to the procedural value of transparency [40] (p. 506). If key information is withheld from decision-making parties, then those parties have been hoodwinked, and their decision cannot be said to confer legitimacy.

While the principles are intended to be relevant to all kinds of NETs and SRMs, it does not follow that they are to be implemented by a single institution. Instead, the authors of the Oxford Principles envisioned a flexible "bottom-up" architecture in the first instance, with different forms of institutionalization depending on the scale and stage of research. They suggest that before embarking on a research project for any particular technology, "researchers should be required to prepare a research protocol explicitly articulating how the issues embodied in each of the Oxford Principles is to be addressed, to be interrogated by a competent third party as part of a stage-gate process" [40] (p. 509). In this way, over time, the values embodied in the principles will be concretized and transformed into more action-guiding recommendations and regulations, appropriate to the technology and stage of development. The third party would have the authority to withhold approval for the research until it is satisfied that the research will be conducted in accordance with a reasonable interpretation of the Oxford Principles. Depending on the stage of development, a "competent third party" might be anything from a university ethics committee, through independent review panels appointed by national scientific associations, to committees with international representation.

At present then, the primary task is with the scientific community to start research in a responsible manner and be prepared to be guided by the Oxford Principles, or some similar initiative, when proposing and undertaking any research [48]. However, new institutional developments may eventually be necessary, not only technology specific, but more broadly. In particular, as Albert Lin suggests, it will be highly desirable to implement a system for coordinating information about ongoing research projects and their results, which should feed into the various institutions that are responsible for a society's overall response to climate change [49] (p. 44). Technology-specific research protocols might well be sufficient to ensure that a piece of research is carried out in a justifiable and legitimate manner. Deciding whether to (further) pursue research into any particular technology is ultimately a decision about which kinds of responses (emission reductions, NETs, SRMs, adaptation, etc.) should be prioritized. Taking those decisions will certainly require reliable information and public trust in research projects, but also similar information about other options that a technology-specific research protocol might not be able to provide.

21.4 Conclusion

It is clear from the above, as it has been for many years, that the development of geoengineering technologies raises important questions of values: ethical values, plus questions of distributive, procedural, and rectificatory justice. We have looked at some of them here. Given that the technologies are in such an early stage of development, it is not yet possible to draw conclusions about their feasibility, much less their permissibility. Some researchers might be frustrated at all this talk about the ethics and governance of geoengineering technologies, when those technologies do not yet exist. Some scientists and engineers might fear that they are being overburdened, subjected to endless debate (and perhaps additional oversight) about ethics and politics that their colleagues in other fields are not. It would certainly be a shame if technologies with promise to address climate change were stifled through overregulation. However, we are nowhere near this stage yet. The kinds of discussions being had about geoengineering technologies are common when new technologies, especially those with environmental implications, come into the spotlight. These kinds of discussions cannot and should not be avoided, especially in the case of geoengineering technologies. After all, interest in NETs and SRMs exists because of the imperative to prevent the suffering and setbacks in human development that is expected should climate change continue unchecked. If scientists propose research into these kinds of technologies, the justification is that these technologies might be able to help address the problem of climate change. This immediately opens up the discussion of how societies ought to respond to climate change, and the place of NETs and SRMs in relation to emission reduction and adaptation strategies. This means in the short term that discussion about the normative issues is unavoidable: whether and how to respond to climate change simply is a normative matter. And in the long term, it means that any governance scheme for NETs and SRM technologies must be fully linked up with other kinds of efforts to avoid dangerous climate change.

Acknowledgments

This work was supported by the CEMICS 2 project (Contextualizing Climate Engineering and Mitigation: Illusion, Complement or Substitute) of the Priority Programme (SPP 1689) of the German Research Foundation.

References

[1] For up-to-date calculations of atmospheric GHG concentrations, see the Mauna Loa Observatory. Website: https://www.esrl.noaa.gov/gmd/ccgg/trends/ [Accessed 8 February 2018].

[2] For information more see http://www.unredd.net/index.php?option=com_docman&view=document&alias=15279-fact-sheet-about-redd&category_slug=fact-sheets&Itemid=134 [Accessed 8 February 2018].

[3] For general scientific facts and explanations about NETs and SRMs I have relied on two recent reports from the US National Academy of Sciences. For NETS, see National Academy of Sciences. Climate intervention: carbon dioxide removal and reliable sequestration. Washington, DC: The National Academies Press; 2015.

[4] An early public engagement exercise in the UK reported a strong preference for afforestation and more generally technologies perceived as "natural." See Ipsos-Mori. Experiment Earth? Report of a public dialogue on geoengineering, Swindon, National Environmental Research Council; 2010.

[5] For discussion of the first official scientific experiment, LOHAFEX, see Kintisch E. Hack the planet: science's best hope—or worst nightmare—for averting climate catastrophe. New Jersey: John Wiley and Sons; 2010.

[6] For discussion of the controversial dumping of iron nutrients by a controversial American businessman, Russ George, see R.S. Abate. Ocean iron fertilization and indigenous peoples' right to food: leveraging international and domestic law protections to enhance access to salmon in the Pacific Northwest. UCLA J Int Law Foreign Aff 2016;20:45–85.

[7] For more information about the development and use of the RCPS see van Vuuren DP, Edmonds J, Kainuma M, et al. The representative concentration pathways: an overview. Clim Change 2011;109:5–31.

[8] van Vuuren DP, Stehfest E, den Elzen MGJ, et al. RCP2.6: exploring the possibility to keep global mean temperature increase below 2°C. Clim Change 2011;109:95–116. p 103.

[9] See Article 2a of the Paris Agreement. The full text of the Agreement is available at https://unfccc.int/files/essential_background/convention/application/pdf/english_paris_agreement.pdf [Accessed on 8 February 2018].

[10] National Academy of Sciences. Climate intervention: reflecting sunlight to cool the earth? Washington, DC: The National Academies Press; 2015.

[11] Caney S. Just emissions? Philos Public Aff 2012;40:255–300.

[12] Heyward C. Situating and abandoning geoengineering: a typology of five responses to dangerous climate change. PS: Polit Sci Polit 2013;46:23–7.

[13] Sen A. Poverty and famines: an essay on entitlement and deprivation. Oxford: Oxford University Press; 1982.

[14] Different normative grounds for providing rectification for climate change were first discussed in Shue H. Global environment and international inequality. Int Aff 1999;75:531–45.
[15] Morrow DR. Ethical aspects of the mitigation obstruction argument against climate engineering research. Philos Trans R Soc A 2014;372.
[16] Baatz C. Can we have it both ways? On potential trade-offs between mitigation and solar radiation management. Environ Values 2016;25:29–49.
[17] Keith D. A case for climate engineering. Boston, MA: The MIT Press; 2013.
[18] Anderson K, Peters G. The trouble with negative emissions. Science 2016;354:182–3.
[19] Shue H. Climate dreaming: negative emissions, risk transfer and inevitability. J Human Rights Environ 2017;(2):203–16.
[20] See McClaren D. Negatonnes—an initial assessment of the potential for NETs to contribute safely and fairly to meeting carbon budgets in the 21st century. London: Friends of the Earth; 2011.
[21] See, for example, contributions to Chatty D, Colchester M, editors. Conservation and Mobile Indigenous Peoples: Displacement, Forced Settlement and Sustainable Development. New York; Oxford: Berghahn Books; 2002.
[22] For discussion see van Cleef A. Hydropower development and involuntary displacement: a global solution. Indiana J Glob Legal Stud 2016;3:349–76.
[23] Schrader-Frechette K. Environmental justice: creating equality, reclaiming democracy. Oxford University Press; 2002.
[24] See, for example, the Nuffield Council on Bioethics, Biofuels, London: Nuffield Council on Bioethics, 2011.
[25] Gardiner SM. Is "arming the future" with geoengineering really the lesser evil? Some doubts about intentionally manipulating the climate system. In: Gardiner SM, et al., editors. Climate ethics: essential readings. Oxford: Oxford University Press; 2010. p. 284–312.
[26] Caney S. Cosmopolitan justice, responsibility and global climate change. Leiden J Int Law 2005;18:747–75.
[27] Tomlinson L. Procedural fairness in the United Nations Framework Convention on Climate Change: negotiating fairness. Dordrecht: Springer; 2015.
[28] Whelan FG. Democratic theory and the boundary problem. In: Pennock JR, Chapman JW, editors. Liberal democracy. New York: New York University Press; 1983.
[29] Cairns R. Climate geoengineering: issues of path-dependence and socio-technical lock-in. Wiley Interdiscip Rev Clim Chang 2014;5:649–61.
[30] Kruger T, Rayner S, Redgewell C, Savulescu J, Pidgeon N. Memorandum submitted by T. Kruger et al, (GEO 07a) Available at: http://www.geoengineering.ox.ac.uk/oxford-principles/history/; 2010. [Accessed 8 February 2018].
[31] Gardiner SM. Some early ethics of geoengineering the climate: a commentary on the values of the Royal Society report. Environ Values 2010;20:163–88.
[32] See Morrow DR. Ethical aspects of the mitigation obstruction argument against climate engineering research. Philos Trans R Soc A 2014;372:20140062. https://doi.org/10.1098/rsta.2014.0062.
[33] Corner A, Parkhill K, Pidgeon N, Vaughan NE. Messing with nature? Exploring public perceptions of geoengineering in the UK. Glob Environ Chang 2013;23:938–47.
[34] For religious perspectives on geoengineering, see the contributions in Clingerman F. Theological and ethical perspectives on climate engineering: calming the storm. Rowman and Littlefield; 2017.
[35] Jamieson D. Ethics and intentional climate change. Clim Change 1996;33:323–36.

[36] Clingerman F, O'Brien K. Playing god: why religion belongs in the climate engineering debate. Bull At Sci 2014;70:27–37.
[37] Jackson R, Salzman J. Pursing geoengineering for atmospheric restoration. Issues Sci Technol 2010;67–76.
[38] For a possible example of this view, see Long JS, Scott D. Vested interests and geoengineering research. Issues Sci Technol 2013;29.
[39] Keith D, Parsons E, Morgan MG. Research on global sun block needed now. Nature 2010;436:426–7.
[40] See Rayner S, et al. The Oxford Principles. Clim Change 2013;121:499–512.
[41] Rayner S, Heyward C. The inevitability of nature as a rhetorical resource. In: Hastrup K, editor. Anthropology and nature. London: Routledge; 2013. p. 125–46.
[42] For a comprehensive discussion of the all-affected principle, see Rosenberg JH. The all-affected principle and its critics: a study on democratic inclusion. Uppsala: Uppsala University; 2016.
[43] Callies D. On the ethics and politics of climate engineering. Frankfurt: Goethe University of Frankfurt; 2018. Unpublished PhD thesis.
[44] Brighouse H, Fleurbaey M. Democracy and proportionality. J Polit Philos 2010; 18:137–55.
[45] Caney S. Justice beyond borders: a global political theory. Oxford: Oxford University Press; 2005.
[46] See Szerszynski B, Kearnes M, Macnaghten P, Owen R, Stilgoe J. Why solar radiation management geoengineering and democracy won't mix. Environ Plan A Environ Space 2013;45:2809–16.
[47] See also Heyward C, Rayner S. Uneasy expertise: social science and democracy in the anthropocene. In: Heazle M, Kane J, Patapan H, editors. Who makes policy? Knowledge and action in today's democracies. Routledge; 2015. p. 101–21.
[48] One notable effort to develop the Oxford Principles is "Code of Conduct for Responsible Geoengineering Research." See Hubert AM, Kruger T, Rayner S. Geoengineering: code of conduct for geoengineering, Nature 2016;537:488. The latest version of the Code is available at, http://www.ce-conference.org/system/files/documents/revised_code_of_conduct_for_geoengineering_research_2017.pdf. [Accessed 8 February 2018].
[49] Lin A. The missing pieces of geoengineering research governance. UC Davis Legal Studies Research Paper Series; 2015.

The social cost of carbon: capturing the costs of future climate impacts in US policy

22

Peter H. Howard[1]
Economics Director, Institute for Policy Integrity, New York University School of Law, New York, NY, United States
[1]Corresponding author: HowardP@mercury.law.nyu.edu

Chapter Outline

22.1 Introduction

Climate change is often recognized as one the most critical economic policy issues of our time. Well-known economists have referred to climate change as "the mother of all environmental externality problems" and the "greatest market failure [the world has] ever seen" [1,2]. Essentially, the absence of a market for greenhouse gases (GHGs) implies that damages from climate change—such as the spread of disease, decreased food security, and coastal destruction—are not reflected in the price of fossil fuels. This results in a tragedy of the commons whereby each individual in society acts to enhance his or her well-being through consuming fossil fuels without considering the full cost of his or her fossil-fuel burning behavior to other individuals. The end result is in an overconsumption of the commonly owned atmospheric good

Managing Global Warming. https://doi.org/10.1016/B978-0-12-814104-5.00022-3

[3]. If society continues down the Business as Usual (BAU) path of failing to account for these global GHG externalities, it is committing to costly climate impacts that will increase over time, potentially reaching catastrophic levels. Public policy is necessary to internalize the true cost of carbon or limit the emissions of carbon, thereby moving society away from our current inefficient market equilibrium.

The polar opposite of the BAU pathway is the socially optimal pathway whereby the full cost of each individual's decision to emit GHGs is internalized by that individual. Much of economics has focused on determining this first-best climate policy (e.g., [4]), usually in the form of the optimal carbon tax or cap-and-trade limit. This aim is understandable in that the optimal policy is both Pareto efficient (i.e., it is impossible to make an individual better off without making others worse off) and welfare maximizing (i.e., the Pareto efficient allocation of resources that maximizes social well-being) [4]. While it is helpful to know the final target, much of domestic and global policy is focused instead on iterative policy.

Instead of setting optimal GHG policies, many nations—and the global community in general—are agreeing to intermediate steps to begin the process of significantly reducing GHG emissions. Under the Obama Administration, US federal agencies promulgated policies that decrease US GHG emissions. However, in addition to its recent announcement of an intended future withdrawal from the Paris Climate Agreement, these policies are currently under review and targeted for repeal by the Trump Administration. Each of these repeals is evaluated using cost-benefit analysis (CBA), and courts will review and strike down any deemed irrational or insufficiently justified. In these analyses, the benefits of GHG-reducing regulations are valued using an estimate of the cost to society of climate change on the BAU pathway, known as the social cost of carbon (SCC).

This chapter focuses on the SCC, and its application in the United States' regulatory impact analysis framework. Dating back to the Reagan Administration, Executive Orders require that major regulations be justified by reference to CBA, with the implicit aim of determining whether they are welfare enhancing. Guidance for how to conduct these CBAs is laid out in the US Office of Management and Budget's Circular A-4 [5]. The OMB's (The Office of Management and Budget) Office of Information and Regulatory Affairs reviews these CBAs. In 2008, a federal court ruled that that all executive branch agencies must include the net benefits of GHG reductions in federal CBAs (Center for Biological Diversity vs National Highway Traffic Safety Administration) if other uncertain benefits and costs are included. The court also stated that climate costs are not too uncertain to monetize, and their value is certainly great than zero. In response, the Obama Administration convened an Interagency Working Group on the Social Cost of Carbon (IWG) to develop a scientifically rigorous estimate [6]. The IWG settled on using an estimate of the marginal cost to society of a CO_2 emission on the BAU pathway in a given year, known as the social cost of carbon. In 2010, the IWG [7] released a central SCC estimate of \$31/t CO_2 (\$31 per metric CO_2 tonne) (in 2017 USD) emitted in 2020. This number was updated in 2013 to \$51 /t CO_2 (\$51 per metric tonne) (in 2017 USD) [8], and reaffirmed in 2016 [9] (with a slightly lower value of \$50/t CO_2 due to model corrections between 2013 and 2016). Over this time frame, US regulators used these estimates in regulatory impact analyses to partially internalize the costs of climate change.

The IWG numbers have proven quite robust to economic and legal scrutiny. In economics, climate change raises unique challenges for the CBA framework. The intergenerational nature of climate impacts requires the models that estimate the SCC to make uncertain—and sometimes deeply uncertain—predictions in a complex scientific and socioeconomic setting characterized by nonlinearities and socially contingent responses [10]. This is further complicated by the need for modelers and regulators to make key ethical decisions over the temporal and spatial scope of the analysis. These challenges have led some economists to argue that current methodologies underestimate the costs of climate change by over relying on (relatively optimistic) modeler opinion and underemphasizing catastrophic consequences [11–13]. However, there is a general understanding that the IWG estimates are currently the best available [14], and, if anything, are biased downwards [13,15]. Despite this prevalent view within the academic community, the official US numbers have faced multiple challenges from industry groups and antiregulatory groups in courts and in the court of public opinion (e.g., [16–18]). However, the Government Accountability Office [19] found that "the working group's processes and methods…used consensus-based decision making, relied on existing academic literature and models, and took steps to disclose limitations and incorporate new information." The US Court of Appeals for the Seventh Circuit held the government estimates used to date by agencies are "reasonable" in 2016. The strong legal basis of the SCC is further evidenced by the numerous regulatory proceedings that apply the SCC [20].

Despite these findings, the Trump Administration in early 2017 disbanded the IWG and withdrew its estimates and documents. They called for new "scientifically rigorous" estimates and reliance on Circular A-4 in the choice of discount rates and geographic scope. Using assumptions counter to current economic consensus [21], the new interim domestic 2020 SCC estimates are \$8/t CO_2 and \$1/t CO_2 for 3% and 7% discount rates (2017 USD), respectively. These numbers will certainly face their own legal challenges. As of now, the Administration has not attempted to abandon the SCC altogether, further demonstrating the strong legal basis of these numbers [21].

This chapter is structured as follows. Section 22.2 discusses the SCC and explains its current use in CBA. Section 22.3 describes the challenges of including climate change in the CBA framework, and provides recommendations for future research. Section 22.4 lays out additional suggestions for improving the central measurement of the SCC, relative to the current reliance on the mean. Finally, it concludes with a survey of our key findings, and explains that dropping the SCC or making arbitrary assumptions is not an appropriate alternative.

22.2 Climate change in the CBA framework

The social cost of carbon (SCC) is the marginal cost (i.e., the additional cost) to society of producing one more tonne of carbon emissions in a given time period. Specifically, it is the present value of all future damages from adding one additional unit of emissions in a particular year to the BAU pathway. Since GHGs are a stock pollutant,

SCC estimates are dependent on future emissions paths and the climate's response to them. On the optimal emissions path, the SCC is equivalent to the Pigovian tax on carbon that would move the economy to a socially efficient economy, though the SCC and a carbon tax are distinct concepts [22]. The SCC trends upwards over time because global surface temperature (along with the stock of GHGs in the atmosphere) is expected to increase over time and damages are expected to intensify as natural systems worsen.

There are also social cost estimates for other GHGs (particularly methane and nitrogenous oxide) that differ in magnitude due to differing global warming potentials and lifespans in the atmosphere. They are calculated using the same models and have similar properties as the SCC. This chapter focuses on the SCC, though this discussion holds more generally.

22.2.1 How is the SCC estimated?

Economists estimate the SCC by linking together a global climate model and a global economic model to capture the feedback effects between these two spheres. The resulting models are called Economy-Climate Integrated Assessment Models (IAMs). This integration allows economists to capture each step in the climate-economic process that translates a metric tonne of CO_2 emissions into an estimate of the impact of that emission on well-being. To make this integration successful, modelers use simplified environmental and economic models. The simple climate model translates emissions into temperatures. The economic model consists of net damage functions to translate temperature changes into monetary damages and a social welfare function to translate monetary damages into a social welfare loss; a backstop (or abatement technology) is often modeled.

To calculate the SCC, the modeler must then specify a BAU scenario consisting of GDP, population, and GHG emission paths as an input. Some IAMs include growth and emission submodels making a population path the sole input. Generally, the modeler then:

(1) Runs each IAM to calculate climate damages under the BAU scenario,
(2) Reruns the model after adding an additional unit of CO_2 emissions in a specific base year to the current baseline to recalculate damages,
(3) Calculates net damages in each year, and
(4) Discounts these damages back to the initial base year, to calculate the net present value of damages

Nordhaus [22] instead calculates the SCC as the ratio of the shadow prices of emissions to consumption (see Fig. 22.1).

To calculate the official US SCC estimates, the 2010 IWG [7] modified three IAMs: Dynamic Integrated model of Climate and the Economy (DICE-2007) [23], climate Framework for Uncertainty, Negotiation and Distribution (FUND 3.5) [24], and Policy Analysis of the Greenhouse Effect (PAGE2002) [25]. The IWG recalibrated the IAMs to a common time frame of 2300 and replaced the climate sensitivity parameters, socioeconomic and emissions scenarios, and discount rates.

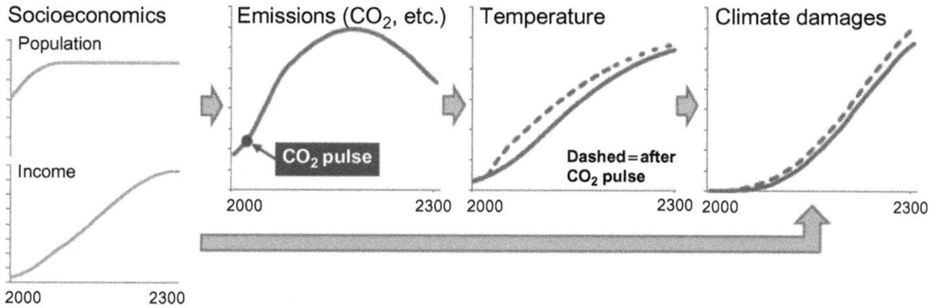

Fig. 22.1 Causal chain for computing the SCC in an IAM.
Reproduced from National Academies of Sciences, Engineering, and Medicine. Assessment of approaches to updating the social cost of carbon: Phase vol. 1. Report on a Near-Term Update. Washington, DC: National Academies Press; 2016; Rose SK, Turner D, Blanford G, Bistline J, de la Chesnaye F, Wilson T. Understanding the social cost of carbon: a technical assessment. Palo Alto, CA: Electric Power Research Institute; 2014.

The climate sensitivity parameter was calibrated to a Roe and Baker distribution to match the distribution described by the 4th Assessment Report of the Intergovernmental Panel on Climate Change (IPCC). Five socioeconomic and emissions scenarios were chosen from the Stanford Energy Modeling Forum exercise: four BAU trajectories and one optimistic (CO_2 concentrations stabilizing at 550 ppm) trajectory. Three constant discount rates were chosen to reflect consumption-based market rates: a 3% posttax "risk-free" rate of return (based on the average real return from Treasury notes); a 5% posttax risky rate of return (based on the pretax rates of household returns to risky investments); and a downward adjusted rate (as indicated by theory) of 2.5% to reflect discount rate uncertainty (discussed in Section 22.3). All other parameter values match those set by the model developers.

Running Monte Carlo simulations for each scenario-IAM-rate combination, the IWG generated 45 distributions of the global SCC for each year. Averaging over all IAM-scenario combinations produced a SCC distribution for each discount rate. Generally, the IWG focused on the mean SCC from each distribution with the exception of the 95th percentile from the central 3% discount rate to account for worse-than-expected impacts; see Table 22.1. Consistent with theory, estimates show the SCC to be decreasing in the discount rate and increasing over time.

The US government updated its SCC estimates on three occasions. In 2013, the IWG updated its estimates using the methodology mentioned earlier, except it updated the underlying IAMs: DICE-2010 [22], FUND 3.8 [27], and PAGE09 [28]. This led to a substantial increase in all four SCC estimates; see Table 22.1. In 2015, the IWG requested two studies by the National Academy of Sciences (NAS) to review its current methodology. The first report was to focus on possible short-term updates of the IWG's representation of uncertainty and modeling of the equilibrium climate sensitivity (ECS) parameter, while the latter report was to provide a more comprehensive review of potential modeling changes for long-run updates in 3 and 6 years time. In 2016, the IWG [9] reaffirmed the 2013 estimates, while providing more transparent reporting of uncertainty as

Table 22.1 **Federal SCC estimates (\$ per tonne of CO_2 in 2017 USD), by Year**

Version	IWG (2010) [7] (\$ /t CO_2)				IWG (2016) [9][a] (\$/t CO_2)				EPA (2017) [26][b] (\$/t CO_2)		
Discount rate	5.0%	3.0%	2.5%	3.0%	5.0%	3.0%	2.5%	3.0%	7.0%	3.0%	2.5%
Percentile	Avg	Avg	Avg	95th	Avg	Avg	Avg	95th	Avg	Avg	Avg
Geographic scale	Global	Global	Global	Global	Global	Global	Global	Global	Domestic	Domestic	Domestic
2015	7	28	45	86	13	42	66	124	1	7	10
2020	8	31	49	95	14	50	73	145	1	8	11
2025	10	35	54	107	17	54	80	163	1	9	12
2030	11	39	59	118	19	59	86	179	2	9	13
2035	13	42	64	129	21	65	92	198	2	10	14
2040	15	46	69	141	25	71	99	216	2	11	15
2045	17	50	73	151	27	76	105	232	3	12	16
2050	19	53	77	161	31	81	112	250	3	13	17

[a] The IWG [8] estimates are essentially identical to the IWG [9] estimates in that they only differ due to model corrections from 2013 to 2016.
[b] These are interim estimates [26].

requested in the first NAS report [29] released earlier in the year. A year later, the NAS [6] released its second report. However, soon after, Trump disbanded the IWG and called for updates to the SCC using methodologies consistent with "the best available science and economics" and OMB's Circular A-4 [5], despite the economic and legal support for these numbers discussed here. Later in the year, the Trump Administration released interim SCC estimates. While these estimates kept most of the IWG methodology [8,9] intact, the Administration ignored the NAS (2017) recommendations and instead made several questionable assumption changes: (1) replaced global SCC estimates with domestic (US damage only) SCC estimates, (2) replaced the 2.5% and 5% discount rates with a 7% discount rate based on the private return to capital, and (3) abandoned the 95th percentile estimate using the 3% discount rate [21]. These new estimates represent a significant decline in the cost of climate change; see Table 22.1.

Many other nations use the SCC or similar concepts in making regulatory decisions, including Canada, France, Germany, Mexico, Norway, and the United Kingdom. Some countries estimated a SCC independently from the US number. Others—like Mexico and Canada—formally aligned their estimates to the central US value, as do some US states (e.g., Minnesota and Washington). It is unclear how these other nations and US states will respond to the Trump Administration changes [20].

22.2.2 How is the SCC used?

The US government assesses policies and regulations using CBA. Simply put, CBA is a methodology whereby the analysts sum up the various costs and benefits of a regulation and subtract the former from the latter to assess whether the policy's net benefits exceed the net benefits of a predetermined baseline (i.e., cost benefit justified). In other words, regulators calculate net benefits of policy i (NB_i)

$$NB_i = \sum_{j=1}^{J} \text{Benefit}_{i,j} - \sum_{k=1}^{K} \text{Costs}_{i,k}$$

by calculating and summing all benefits in J measurable categories (improved health, safety, environment, climate damages, etc.) and all costs in K measurable categories (compliance costs, administrative costs, and adverse consequences). Usually benefits and costs are compared over time whereby the stream of benefits and costs is measured in present value terms

$$NB_i = \sum_{j=1}^{J} \sum_{t=1}^{T} \frac{\text{Benefit}_{i,j,t}}{(1+r)^t} - \sum_{k=1}^{K} \sum_{t=1}^{T} \frac{\text{Costs}_{i,k,t}}{(1+r)^t}$$

using discount rate r, which measures the value of a dollar today instead of tomorrow. Under uncertainty, regulators often conduct a sensitivity analysis

$$NB_{i,l} = \sum_{j=1}^{J} \sum_{t=1}^{T} \frac{\text{Benefit}_{i,j,l,t}}{(1+r_l)^t} - \sum_{k=1}^{K} \sum_{t=1}^{T} \frac{\text{Costs}_{i,k,l,t}}{(1+r_l)^t}$$

where they compare net benefits under L sets of assumptions. Ideally, they provide the probability distribution function of net benefits along with its probabilistic mean ($\overline{NB}_i = \frac{1}{L}\sum_{l=1}^{L} p_l \times NB_{i,l}$ where p_l is the probability of assumption l) and other moments.

Ideally, environmental regulators would identify the optimal environmental regulations by comparing the incremental benefits from the regulation with its corresponding incremental cost to set the policy equal to its socially optimal value (i.e., comparing all possible L policies). In the real world, this is not what occurs. Instead, to ensure the efficiency of a regulation relative to its baseline, the regulator compares the total net benefits of a policy to a baseline, sometimes considering several alternative policies. Under uncertainty, regulators make the comparison using a preferred set of assumptions (i.e., the median l) or they compare expected net benefits (i.e., a probabilistic mean). The CBA methodology requires that all benefits and costs are converted to the common metric of money to enable apples-to-apples comparison. A qualitative analysis is necessary for difficult to value benefits [30].

In benefit-cost analyses, agencies use the SCC to value the GHG emission reductions of a regulation. Just as the price of a good is multiplied by its quantity produced to calculate its value to society, the SCC is multiplied by the amount of GHG emissions prevented by a policy to value the policy's climate stability services to society. In other words,

$$NB_{i,l} = \sum_{j=1}^{J-1}\sum_{t=1}^{T}\frac{\text{Benefit}_{i,j,l,t}}{(1+r_l)^t} + \sum_{t=1}^{T}\frac{\text{SCC}_{l,t}\times \text{CO}_{2l,t}}{(1+r_l)^t} - \sum_{k=1}^{K}\sum_{t=1}^{T}\frac{\text{Costs}_{i,k,l,t}}{(1+r_l)^t}$$

where $SCC_{l,\,t}$ and $CO_{2l,\,t}$ are the SCC and the CO_2 emissions prevented by the policy, respectively, in period t under assumption set l. Using the SCC allows us to compare the costs of limiting our GHG pollution to the costs of climate change. The SCC is valid for valuing only marginal changes in emissions (relative to total global emissions) in a CBA, and not for nonmarginal changes; see Fig. 22.2.

22.3 Challenges of including climate change in the CBA framework

The uncertainty characterizing the climate-economic problem is not a reason to abandon the SCC and with it rational policymaking. The main challenge of *fully* integrating climate change into the CBA framework stems from the unprecedented scale of this uncertainty. While uncertainty is present in all CBAs, the scale of uncertainty underlying the climate-economic problem—which includes the potential collapse of human civilization—is unique. This high degree of uncertainty is fundamentally driven by the long-time horizon, complex, and nonlinear nature of the climate and socioeconomic systems, unobservability of large-scale nonmarket impacts, the uniqueness of human-driven climate change (particularly in the context of human civilization), and the ethical nature of some of the key parameters of the model. Instead of abandoning the SCC or underrepresenting (or even ignoring) this uncertainty, modelers should

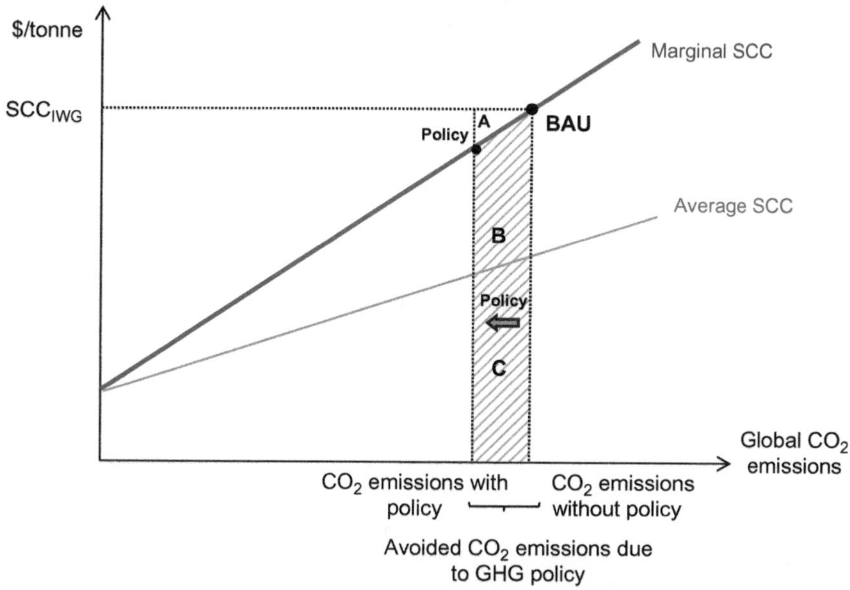

Fig. 22.2 Average and marginal SCC values in a static CO_2 emissions problem. This Figure shows the use of the SCC to value the benefits of a GHG reduction policy where the reduction in the stock of emissions starts from the right side of the figure. Calculating the area under the marginal SCC curve between the points BAU and Policy would give the damages avoided by the CES, and therefore the benefit of the GHG policy. The avoided emission benefits of the CES are represented by the sum of areas B and C. The cost-benefit analysis, which multiplies the avoided emissions due to the GHG policy by the SCC_{IWG}, calculates to the sum of A, B, and C. If the policy in question is small in its impact on emissions relative to the global stock of CO_2 emissions, such as in the case of most US sector-specific policies, the area A is very small. Thus, the upward bias from using the IWG's SCC estimate is negligible. Therefore, using the IWG's SCC estimate to value the benefits of a GHG reduction is clearly appropriate. The SCC is calculated using dynamic models, instead of the static representations of the SCC in Fig. 22.1. In a dynamic setting, these figures can be roughly interpreted as representing 1 year of the dynamic problem. For simplicity, we use linear marginal and average cost curves.

aim to more fully model uncertainty. The IWG [7] represents such an attempt using the methodologies of its time. However, the progression of the fields of economics and sciences since then enables the next generation of IAMs to more fully represent uncertainty. In doing so, regulators can more *fully* integrate climate change into the CBA framework. In fact, the SCC is likely to increase (as discussed in this section) leading to more efficient regulation.

Uncertainty can be classified as positive (parametric or stochastic) uncertainty over scientific and socioeconomic facts or normative uncertainty due to heterogeneous preferences. Parametric uncertainty covers uncertainty over model parameters, correct functional forms, appropriate probability distribution functions, and model structure. Key examples include equilibrium climate sensitivity (ECS)—the long-run temperature change from a doubling of CO_2 emissions relative to the preindustrial period—and climate damages. With learning, these uncertainties should decline over time as more information becomes available. Stochastic uncertainty is persistent randomness in the economic-climate system, including various environmental phenomena such as volcanic eruptions and sun spots [31]. Normative uncertainty arises due to heterogeneity in ethical or value judgments held by society [32,33]. Like stochastic uncertainty, normative uncertainty need not decline over time, as "reasonable people can reasonably disagree" over welfare parameters [32], such that future generations are likely to continue this debate [34]. While positive uncertainties are present in each component of the IAMs [32,35], normative uncertainty is present only in the social welfare function as policymakers must choose how to weight spatially and temporally heterogeneous individuals.

When modeling climate change uncertainty, scientists and economists have long emphasized the importance of accounting for the potential of catastrophic climate change [36,37]. Catastrophic outcomes combine several overlapping concepts including unlucky states of the world (i.e., bad draws), deep uncertainty (i.e., ambiguity), and climate-tipping elements [37]. Traditionally, IAM developers address uncertainty by specifying probability distributions over various climate and economic parameters. This allows for a bad draw to occur if multiple uncertain parameters turn out to be worse than expected, causing realized climate damages to greatly exceed expected damages. However, IAM modelers ignore "deep uncertainty" (i.e., ambiguity) by using thin-tailed distributions. Deep uncertainty is uncertainty over the true probability distributions for specific climate and economic parameters [37]. The mean and variance of many uncertain climate phenomena are unknown due to lack of data, resulting in fat-tailed distributions (relative to the normal distribution) as best guesses are derived under learning starting with subjectivity probabilities [38–40]. Fat tails increase the likelihood of a "very" bad draw with high economic costs. Unlike fat tails, some IAMs abstractly model a single representative tipping point (e.g., the versions of DICE used by the IWG and the PAGE model). Climate-tipping elements are irreversible and often ambiguous thresholds where a small change in climate forcing can lead to large, nonlinear shifts in the future state of the climate or society through positive feedback effects. Of these elements, tipping points refer to economically relevant thresholds after which change occurs rapidly, such that adaptation and intervention are limited [37]. There are many interacting tipping points, including environmental tipping points—like the reorganization of the Atlantic meridional overturning circulation—and social tipping points—like mass migration [37,41,42]. Together, these cataclysmic downside risks potentially imply a very high SCC, as indicated by Weitzman's Dismal Theory [43] (discussed later).

Due to the compounding nature of these uncertainties in the integrated framework of IAMs, the resulting SCC estimates are characterized by wide confidence intervals. As discussed later, these intervals may still underestimate the underlying uncertainty as IAMs modelers often fail to fully model uncertainties [11]. However, these substantial uncertainties are not a justification to drop the SCC entirely or to fill the gaps using modeler discretion. Instead, economists must use the latest scientific research to develop a next generation of IAMs, and fill in with formal expert elicitation where necessary. Later, the key future updates to IAMs are discussed—including (and in addition to) those recommended by the NAS [6,29]—many of which will likely lead to a higher SCC estimates (in line with expert opinion [13]).

22.3.1 Socioeconomic and emission scenarios

Socioeconomic and emission scenarios are critical inputs in IAMs. GDP and population projections determine the baseline level of social welfare. When GHG emissions are explicitly modeled as a function of GDP in IAMs using an emission factor—such as in DICE and FUND—they also determine the emissions paths. In other cases, such as the IWG [7], emissions are exogenously provided. Ideally, these paths are internally consistent as in EMF-22 selected by the IWG [7]. Inputting these baseline paths into IAMs, the models calculate baseline temperature and social welfare paths under BAU climate change.

The primary challenge with socioeconomic scenarios—along with many of the other components of IAMs—is making predictions over the long time horizon necessary to capture all relevant damage estimates. For example, the IWG [7] chose a time horizon of 300 years from the current time period to 2300 to capture all relevant damages. Since the EMF-22 models did not run past 2100 (as is common for socioeconomic scenarios [6]), the IWG estimates greatly relied on linear extrapolation, assuming that GDP per capita would stabilize in the long run. The use of such long time horizons that require long-run extrapolation of already century-long projects has raised the objections of some SCC critics. Industry groups, in a petition for correction to the IWG [7], incorrectly argued that the IPCC only makes century-long projections [16] such that longer time horizons are invalid. Using a similar logic about potential calculation errors over long time frames, a Minnesota administrative judge recommended that the Minnesota PUC apply a 200-year time horizon in calculating the SCC; the Minnesota PUC eventually decided on a range of 100–300 years [44].

Using a shorter time horizon is not the appropriate way to address uncertainty. By shortening the time horizon such that the present value of damages declines, Minnesota and others are implicitly assuming that climate damages are zero at the end of the horizon. While ignoring risks is never the appropriate way to address the uncertainty that society faces, it is a particularly poor strategy in the case of climate damages given that damages are expected to increase at an increasing rate with temperature [4,12] and temperatures are expected to continue to increase after 2100. Instead, the NAS [6] recommends that the time horizon should be defined to capture the "vast majority of the present value of damages." Thus, the relevant time horizon depends on the parameters of the model—including the discount rate—and no specific timeframe can be supported a priori. A good rule of thumb would be that the time horizon should capture 95% or more of benefits based on Circular A-4's requirements for the scope of analyses to be "long enough to encompass all important benefits."

An alternative to shortening the time horizon is to improve scenario projections beyond 2100. This is in fact what the NAS [6] proposes. In the near term, the NAS proposes that modelers use a combination of statistics and expert elicitation to develop internally consistent socioeconomic and emission scenarios out to 2300. In the long run, the NAS supports developing socioeconomic models explicitly for IAMs.

According to the NAS, these yet-to-be-developed models should go beyond the current state-of-the-art technique by modeling climate feedbacks of warming and damages. To address uncertainty, modelers should estimate probability distributions for scenario parameters.

Overall, the NAS [6] recommendations with respect to developing future scenarios are the appropriate way forward with the exception that scenarios should represent a BAU pathway instead of the most likely pathway (given expected policies). Consistent with the historical definition of the SCC as the marginal damages of CO_2 on the BAU path [22], federal agencies should only model BAU and then update the baseline of emissions as this BAU path shifts. This is also consistent with the US federal government, which often makes predictions accounting for only policies that have gone into effect (or at least been finalized); see the treatment of the Clean Power Plan by the US Energy Information Administration's Annual Energy Outlook. This recommendation differs from the NAS in that their methodology ensures that only policies expected (according to current knowledge) to be efficient would be adopted, while this paper's recommendation ensures that all policies that are efficient on the current emissions path would be adopted. Relative to the NAS recommendation, the alternative suggestion would ensure that modelers and the federal government take a more cautious position consistent with the possibility that expected climate policies may not come into force. These regulations can always be repealed if society learns that they are unnecessary.

22.3.2 Simple climate model

In IAMs, the climate model is necessary to translate CO_2 concentrations to temperature at the relevant spatial-temporal resolution for the model's corresponding damage function. To make IAMs tractable, a simple climate model must be chosen that replicates the results of a more sophistical global climate model. Using a simpler set of model equations, the simpler climate model ideally replicates our current understanding of the past and future climate, according to peer-reviewed journals, as well as their underlying uncertainties. This comes at the potential cost of inaccuracy from approximation [6]. None of the models meet all the criteria (scientific basis and uncertainty characterization, transparency and simplicity, and incorporation of non-CO_2 forcing) laid out by the NAS in order to reflect the best available science, as reflected by the IPCC [6].

Unlike with socioeconomic and emissions scenarios, the IWG [7] maintained the differing climate models of DICE, FUND, and PAGE with the exception of one parameter: the equilibrium climate sensitivity (ECS) parameter. Each IAM differs in its model structure of the climate, the carbon cycle, and sea-level rise—including representation of non-CO_2 forcing and inclusion of carbon cycle feedbacks—and their underlying parameterization. As a consequence, the models differ in the resulting temperature response to emission scenarios and pulses (Appendix E in [6]). To partially address these differences, the IWG [7] replaced each model's ECS parameter with a theoretically justified Roe and Baker distribution calibrated to match the distribution described by the 4th Assessment Report of the Intergovernmental Panel on Climate Change (IPCC). The resulting ECS distributions is a fat-tailed, right-skewed distribution with a median of 3°C of long-run warming from a doubling of CO_2, a two-thirds probability of between 2°C and 4.5°C, and a 0% probability of below 0°C and above 10°C. The resulting distribution implies roughly a 10% chance of 6°C or more of long-run warming from a doubling of CO_2. Even with this change, the IAMs differ in their temperature response to GHG emission paths and pulses (Appendix E in [6]).

Conservative critics have argued for a change in the modeling of the ECS based on new information. Specifically, the IPCC updated its assessment of the ECS distribution from 2007 [45] to 2013 [46]. Most notably, the IPCC eliminated the most likely value and slightly increased the probability of low levels of warming. Even with this update, the NAS [29] rejected suggestions that the federal government should update the ECS distribution to reflect the latest IPCC report and/or calibrate an alternative distribution to the Roe and Baker distribution. This was based on the conclusions that updating the ECS would not significantly change the SCC and that more attention should be focused on climate parameters that are relevant to the time scale reflected in IAMs (i.e., two to three centuries), such as the transient climate response, which measures the temperature change at the time of doubling (year 70) from doubling CO_2 concentration along a path that rises 1% per year. Like the ECS, the NAS [29] suggests that these relevant climate parameters should reflect the best available science in the IPCC.

Given that these estimates are based on well-understood physical models, the uncertainties underlying this portion of the model are less significant than those underlying the economic components [11]. Nonetheless, Pindyck [11] argues that the ECS parameter is on the verge of unknowable. Unlike most distributions, the Roe and Baker distribution is based on our current scientific understanding of the climate system, specifically that the ECS without climate feedbacks is 1.2°C. However, the distribution is further based on two common assumptions that: (1) "feedback factors are proportional to the change in surface temperature" and (2) "uncertainties in feedback factors are normally distributed" [7]. Pindyck [47] takes issue with the second of these assumptions. As he notes, we do not know the mean and variance of feedback factors, let alone its distribution. Additionally, the NAS [6,29] notes that current complex climate models used to estimate the ECS omit some feedbacks—dust, vegetation, and land ice feedbacks—systematically biasing the resulting estimates downwards, potentially by as much as 50%. As a consequence, the level of long-run warming from emissions is ambiguous, as is currently captured by the fat tails of the IWG's Roe and Baker distribution [7].

A partial solution to vast uncertainty is to represent the best available science in the simple climate model and update them frequently. The NAS [6] requests a more rigorous climate model that is transparent, matches up-to-date science (which is verified using a series of tests), and replicable. Specifically, the model should be calibrated to IPCC reports when available (due to their use of multiple lines of evidence), and experts (preferably authors of the relevant IPCC chapters) and studies when a key parameter was not reviewed by the IPCC. The NAS put forward FAIR as a simple climate model that meets its climate model criteria, and it indicated that none of the IAMs currently meet them despite being nested within the FAIR model. The NAS is also keen on specifying (joint) probability distributions for multiple climate parameters, including short-term warming and carbon pulse parameters, like the transient climate response, in addition to the ECS. The NAS specifically requires submodules of sea-level rise (already incorporated into some IAMs) and ocean acidification (currently an omitted damage). Temperature, sea-level rise, and ocean acidification should be downscaled to the appropriate temporal and spatial scales of IAMs using pattern scaling. The NAS recommended climate module addresses uncertainty by ensuring models reflect society's current scientific understanding and uncertainty, while allowing for frequent updating of the model as information improves to iterate toward the efficient policy. It does not capture ambiguity in the climate system.

The NAS [6] provides a partial solution to the Pindyck [47] critique in that it emphasized that the TCP and other short-run warming parameters are more relevant to calculating the SCC than the ECS. Furthermore, these short-run warming parameters are better constrained by existing data, and will be further constrained in the near future as more observations become available. Thus, focusing on improving modeling of short-term warming in IAMs rather than the ECS somewhat reduces the concerns about ambiguity over the ECS raised by Pindyck. To further address these concerns, modelers should include the omitted feedbacks in IAMs as permitted by their lower computational requirements relative to complex climate models. As in the warming parameters, attention should be given to tipping points rather than elements given their greater relevance over the next few centuries. Additionally, modelers should include risk and ambiguity aversion in IAMs (as discussed later), as well as fat-tailed distributions where deep uncertainties are present in the climate model.

22.3.3 Damage estimates

The damage function(s) translates global (or regional) average surface temperature into an annual loss of income. As with the simple climate model, the IWG [7] maintained the DICE, FUND, and PAGE damage functions. Each IAM implicitly assumes that climate damages directly affect annual income measured as a percentage of GDP. In the case of DICE, these damage functions capture damages net of adaptation, whereas PAGE captures total damages as the model explicitly models adaptation; FUND is a mix of these implicit and explicit approaches. Each model differs in the regional-sector breakdown of damages: DICE specifies a single global damage function and FUND and PAGE specify multiple regional-sector damage functions. Through the choice of damage sectors and the choice of calibration estimates, IAM developers determine what damages from climate change are included and excluded in the SCC.

The damage functions are generally calibrated at low temperatures (i.e., 1–3°C) using the enumerative strategy (i.e., a sector-regional analysis) whereby studies are found that provide sector-specific damage estimates by region. Extrapolation from observed regional damages to missing regions is often necessary. Currently, IAMs cover a number of direct effects of climate change, that is, a rise in global average surface temperature, on economic (that is, market) activity, and to a lesser extent the direct effects of climate change on the nonmarket services of the environment and human systems; see Table 22.2. The models also capture some climate benefits,

Table 22.2 Impacts included in IAMs

Damage type	Sector	Impacts	Model[a]	Description
Economic	Agriculture	Impacts on average crop yields due to average temperature increases and CO_2 fertilization effect	DICE, FUND, and PAGE	More optimistic than current observation, potentially due to optimistic assumptions about CO_2 fertilization effect
	Forestry	Shifting geographic range due to average temperature; CO_2 fertilization	FUND and partially PAGE	DICE does include forestry in other market systems, but assigns no impacts
	Fresh water resources	Changing precipitation	FUND and partially PAGE	Water supply system losses and disruptions
	Energy demand	Change in average temperature	DICE, FUND, and PAGE	In DICE, energy demand is included in other-market sector where Dr. Nordhaus assumes only impact in nontemperate climates. FUND models energy demand more explicitly
	Property and infrastructure loss	Storms, flooding, and sea-level rise	DICE, FUND, and PAGE	Coastal property losses due to storms, flooding, and sea-level rise
Nonmarket	Human health	Coastal mortality from flooding and storms; spread in geographic range of vector-borne diseases; heat- and cold-related deaths; water-borne diseases; morbidity: nonfatal illness and injury; air quality	DICE, FUND, and PAGE	Significant vector-borne diseases are included, though Lyme disease is excluded. Cardiovascular, respiratory disorders, diarrhea, and morbidity for some health impacts are included in FUND. Air quality is included in DICE, though does not account for changes due to pollen or wildfire

Continued

Table 22.2 Continued

Damage type	Sector	Impacts	Model	Description
	Terrestrial, freshwater, and marine ecosystems and wildlife	Shifting geographic ranges, seasonal activities, migration patterns, abundances, and species interactions; extinction and biodiversity loss; loss of habitat to sea-level rise	DICE, FUND, and PAGE	The value of ecosystems and biodiversity is included in general terms not specific to any one damage estimate. Wetland loss explicitly modeled in FUND, and thus partially in PAGE
	Climate amenities	Cultural services, such as aesthetics; outdoor leisure activities (i.e., nonmarket time use)	DICE	Only captures outdoor leisure activities, and omits all cultural values. However, estimates have come under considerable fire [48]
Catastrophic	Tipping points	Average temperature change on market-tipping point	DICE, PAGE	In early versions of DICE, Dr. Nordhaus estimates a certainty-equivalent catastrophic damage using a survey of experts [49]. In PAGE09, known tipping points are modeled as a single event, instead of multiple events
Socially contingent	Migration	Sea-level rise only	FUND	FUND models location-choice and the cost of climate-induced migration to the migrants and the receiving countries; the underlying parameters are highly subjective.

[a] We focus on DICE-1999 to DICE-2010, since DICE-2013 and DICE-2016 rely on meta-analysis of global climate damages; this makes the determination of included impacts near impossible. Dr. Hope calibrates the PAGE09 noncatastrophic damage functions using past estimates of DICE and FUND, and so PAGE includes each of these models' impacts.

like the impacts of CO_2 fertilization on agriculture and fewer cold-related deaths. More recently, Nordhaus [22,50] calibrated the DICE-2013R and DICE-2016R damage functions using meta-analyses of global damage estimates, making it less clear which impacts are included. The authors' choices of functional form determine how climate damages are projected to higher temperatures, and lack of damage estimates from climate change at high temperatures makes results unreliable at these high temperatures [51]. Each model disaggregates its damages to different sectoral-regional extents, allowing for extrapolation functions to differ by region and sector [48].

While some conservative politicians and industry groups question the magnitude of the SCC [17,52], if anything, the federal SCC estimates are biased downward, probably significantly so, due to outdated and omitted damage estimates. It is a relatively well-known fact that most of the impact estimates used in the calibration of the models date back to the 1990s [15,48]. Since then, the IPCC has released three Assessment Reports and a revolution in empirical damage estimates has occurred, resulting in an ever-growing wave of estimates, which surely overwhelm the teams of one to two modelers behind each IAM [53,54]. There is significant evidence that current models underestimate impacts in particular regions—such as the United States [55]—and sectors—like agriculture [56] and health [57].

IAMs also omit significant physical impacts of climate change. In general, the more difficult a climate impact is to estimate in the natural sciences (which measure the physical impact) and/or value in economics, the more likely that climate impact is to be excluded from IAMs [58]; see Fig. 22.3. With respect to the natural sciences, damages corresponding to more certain (that is, known) climate trends (for example, average temperature increases and sea-level rise) are included in IAMs; bounded trends, that is, climate change for which a range and/or distribution is specified, such as extreme weather events and weather variability (for example, droughts, floods, storms, and so on), are less likely to be included; and abrupt changes, in general, are the least likely to be included because they are the effects characterized by the greatest uncertainty. As a consequence, the models do a good job of measuring the costs of average temperature increases, but a poor job of capturing other climate-related drivers of impacts noted by the IPCC, such as ocean acidification, extreme weather events (e.g., droughts, heavy rains, and inland flooding), and weather variability; see Table 22.3. With respect to economics, damages that are easier to value are more likely to be included, such that many more market damages are included than nonmarket damages (i.e., damages affecting goods or services for which no established market exists, but which still provide value to humans). Those nonmarket damages that have been included in IAMS (for example, heat stress, loss of wetlands, biodiversity, and loss of life) require significant improvement. Socially contingent damages—damages that result from changes in social dynamics due to climate change (i.e., mass migration, interpersonal violence, and social and political conflict)—are outright omitted (with the exception of the relatively ad hoc and incomplete inclusion of migration in FUND); it is difficult to quantify the social effect, let alone value it. Finally, IAMs tend to only roughly capture catastrophic damages, which result from worse-than-expected outcomes or tipping points in general. IAMs have yet to model

		Uncertainty in valuation →		
		Market	**Nonmarket**	**(Social contingent)**
Uncertainty in predicting climate change ↓	**Projection** (e.g., sea level rise)	**I** Coastal protection Loss of dryland Energy (heating/cooling)	**IV** Heat stress Loss of wetland	**VII** Regional costs Investment
	Bounded risks (e.g. droughts, floods, storms)	**II** Agriculture Water Variability (drought, flood, storms)	**V** Ecosystem change Biodiversity Loss of life Secondary social effects	**VIII** Comparative advantage and market structures
	System change and surprises (e.g. major events)	**III** Above, plus Significant loss of land and resources Nonmarginal effects	**VI** Higher order Social effects Regional collapse	**IX** Regional collapse

Fig. 22.3 Map of types of damages in IAMS by level of scientific and economic uncertainty. Reproduced from Yohe G, Tirpak D. A research agenda to improve economic estimates of the benefits of climate change policies. Integr Assess 2008; 8:1–17.

fat-tailed events or nonmarket-tipping points [48,58]; see Tables 22.4 and 22.5. Thus, IAMs omit much of the climate impacts discussed by the IPCC [59].

An alternative critique leveled by Pindyck [10,11,47] is that damages are essentially unknowable. In Pindyck's view, current IAMs are flawed in that they focus on the most likely scenarios, instead of catastrophic damages. At high temperatures, even the most likely impacts are unknowable—partially due to climate nonlinearities—implying that the functional form of damages is also unknowable. In particular, Pindyck [11] calls out the arbitrariness of the quadratic damage function used in DICE, though it meets the sole requirement of convexity that he lays out [47]. Even how climate change impacts the economy has recently been called into question in that climate change may impact economic growth (e.g., a loss of capital stock, a decline in labor productivity, etc.) instead of or in addition to annual income levels, as damages are currently

Table 22.3 The status of IPCC's critical climatic and ecosystem changes—also known as climate-related drivers of impacts—within the US SCC estimates

Status	Climate-related drivers of impacts	Notes
Excluded	Extreme temperature	The health impacts of extreme temperatures are the only impact considered by IAMs
	Drying trend	–
	Extreme precipitation	–
	Snow cover	–
	Ocean acidification	–
Included	Warming trend	–
	Precipitation	–
	Damaging cyclones	–
	Carbon dioxide concentration	–
	Sea-level rise	–
Partially included	Flooding	Coastal flooding is included and inland flooding is excluded
	Storm surge	Partially included, fails to account for combine effect of sea-level rise and increased intensity of coastal storms

modeled in IAMs [60,61]. Essentially, climate damages are ambiguous. The combination of this deep uncertainty of damages with the possibility of unlimited damages (i.e., the possibility of the end of human life itself) can result in the Dismal Theory [43]: a very high (and potentially infinite) SCC when the probability of catastrophic climate damages is high enough to overwhelm the discount rate. In other words, the expected marginal utility of consumption can "explode under the possibility of fat-tailed damages." The Dismal Theory calls into question the use of the SCC in CBA, particularly if we accept the assumptions that lead to an infinite SCC: "very" fat tails or "very" risk adverse [38,39,43]. In this case, economists should instead focus on minimizing the cost of staying below a temperature limit of between 1.5°C and 2°C.

The NAS's [6] short-run solution to the shortcomings of damage functions is to essentially update net damages estimates to match the latest literature. Table 5.3 in NAS [6] lists the estimates that the NAS believes should be included in any comprehensive update. This is an endorsement of the enumerative (i.e., bottom up) approach over top-down and scientific-based approaches (discussed later). While, in the long-run, the NAS advocates for a more comprehensive approach (including better modeling of nonmarket impacts, interactions, and spillovers between sectors and regions, impacts to growth (not just consumption), adaptation (and its costs), and tipping points), the NAS essentially puts forth only the meta-analysis technique as the sole alternative to enumerative estimation.

Table 22.4 Taxonomy of omitted damages

Damage category	Missing damage
Market damages	Fisheries
	Pests, pathogens, and weeds
	Erosion
	Fire
	Energy supply
	Transportation
	Communication
	Ecological dynamics
	Decreasing growth rate
Nonmarket damages	Recreational goods and services
	Ecosystem services[a]
	Biodiversity and habitat[a]
	Health care costs[a]
	Relative prices
Socially contingent damages	Migration[a]
	Social and political conflict
	Violence and crime
Catastrophic damages	Tipping point[a]
	Fat tails
	Black swan events
Intersector damages	Intersector damages
Cross-sector damages	Interregional damages
	Destabilizers of existing nonclimate stressors
	Weather variability and climate extremes
	Ocean acidification

[a] Partially and/or insufficiently captured in current versions of DICE, FUND, and PAGE.

This solution is insufficient to address the Pindyck and Weitzman critique of the unknowability of future climate damages, particularly given that Howard and Sterner [62] find that damage estimates can greatly differ by estimation methodology. To address ambiguity, modelers should extend the NAS [6] damage module to consider multiple methodologies for calibrating damage function(s), instead of relying on just one: enumerative or meta-analysis. Specifically, the damage function(s) should consider evidence from the bottom-up methodologies (i.e., enumerative), top-down methodologies (expert elicitation, global damage meta-analysis, and statistics), and scientific-based methodologies (limits of human life and recommended temperature ceiling). This multiple-prong approach is necessary given that differing estimation strategies provide substantially different impact estimates [62], and models should include all valid estimation strategies that provide new information to the modeler. Improving damage estimates will likely take decades (as noted by Anthoff and Hope at the NAS's third committee meeting in Washington, DC on November 13, 2015), such that these differences are likely to continue. The use of multiple methodologies

Table 22.5 Alternative taxonomy of omitted damages

Damage category	Damage subcategory	Missing damage
Missing sector	Missing market and nonmarket sectors	Fisheries Energy supply Transportation Communication Recreational goods and services
	Missing interactions	Intersector damages Interregional damages
	Poorly/incompletely estimated sectors	Biodiversity and habitat[a] Ecosystem services[a] Health care costs[a]
Missing climate effects	Broad system changes	Tipping point[a] Fat tails Black swan events Weather variability and climate extremes Ocean acidification
	Specific impacts from broad system changes	Ecological dynamics Pests, pathogens, and weeds Erosion Fire
	Missing dynamic climate effects	Decreasing growth rate Relative prices Socially contingent damage

[a] Partially and/or insufficiently captured in current versions of DICE, FUND, and PAGE.

to estimate benefits of a regulation is also consistent with previous cost-benefit practices by federal agencies. For example, EPA developed a concentration response function to estimate mortality caused by particulate matter exposure by using published epidemiological studies and EPA's own expert elicitation [63]. Each of the employed methodologies should be transparent and systematic to make them easily updatable.

Several estimation strategies should be considered. The NAS [6] recommends an enumerative strategy as a bottom-up approach. Unlike past estimates, modelers should rely on state-of-the-art regional-sector meta-analyses and formal benefit transfers when regional data are unavailable. This is the approach taken by the Climate Impact Lab [55,57] and other recent research groups [56]. Of the top-down approaches, two are easily reproducible and less time intensive: expert elicitation [13,49,64] and global damage meta-analysis [1,22,62]. The NAS [6] cautions against using expert elicitation to estimate climate damages, despite this strategy's recent empirical acceptance [65]. Instead, the use of expert elicitation is a critical tool to account for omitted impacts—including tipping points and other catastrophic damages [41]—when combined with formal techniques [66] and frequent updates, though careful attention needs to be paid to the definition of "expertise." A meta-analysis of global damage estimates benefits from a rigorous treatment of the literature that is both transparent and reproducible, though the analysis needs to meet the standards set by the EPA [67] and the economic literature [62]. Panel estimates of the impact of climate change on macroeconomic-growth are another possible approach [60]. Finally, the scientific approach uses "physical thresholds—limits of human physical adaptability to heat stress [68] and the 2°C limit on global warming recommended by a large body of scientists and international organizations [22]—to derive global damage estimates from climate change" [62]. These methodologies appear better suited to providing damage estimates for high-temperature increases. Given the potential arbitrariness of the temperature limit [64], more weight should be given to the limits of the human respiratory system. For most of these methodologies (exceptions being the enumerative and explicit elicitation methods), modelers will need to modularize some IAMs to allow for a single, global damage function, unless estimates are used to inform a damage sensitivity analysis instead.

The extended damage module must provide a sensitivity analysis over non-catastrophic damages and functional form, as well as catastrophic damages (similar to the PAGE model). Ideally, an extended damage module would use a weighted combination of these estimation methodologies to derive a distribution of climate damages for a low-temperature level; one way to do this is a meta-analysis of damage estimates [62]. Due to the lack of damage estimates for high-temperature increases, sensitivity analysis over the functional form is also necessary to extrapolate to high temperatures. This functional form sensitivity analysis is essential given Kopp's [37] finding that functional form shape can greatly affect SCC estimates when the IAM's central planner is risk averse. An alternative is to conduct a sensitivity analysis to the various calibration points and functional forms, along the lines of CRED [69]. However, this type of modeling of uncertain damage parameters—which should be standard in IAMs [6]—is not an alternative to modeling catastrophic impacts explicitly [70]. Fat-tailed distributions are necessary given the level of ambiguity. Best economics and sciences also require modelers to explicitly model multiple, correlated tipping points: see [42,71] for examples in stochastic dynamic settings.

Some policymakers and analysts may argue that the IAMs need not worry about better estimating damages due to society's ability to adapt. In other words, adaptation implies that these higher costs will not be incurred. While adaptation must be accounted for when including these damage estimates, an altogether elimination of these damages (that is, such that they can be ignored) is unlikely [72]; current empirical evidence is mixed (and limited) and can be read as currently supporting moderate levels of effective adaptation. Adaptation is particularly limited for the case of nonmarket, socially contingent, and catastrophic damages, in general, where adaptation is likely to be less effective [28]. This is also increasingly the case for market damages as temperatures increase, though rising GDP over time may counteract some of these impacts. Adaptation will be particularly difficult for faster-than-expected temperature increases [27,28]. Furthermore, there is also the potential for maladaptation where climate risks actually increase as a result of adaptation decisions [72]. Even when adaptation is highly effective, adaptation is costly and these costs increase in temperature partially offseting some of these reduced damages [57,73,74]. Given that enumerative damages are likely to increase as omitted impacts are included [48] and that many top-down and scientific estimates imply higher damages than current enumerative estimates [62], the SCC will likely increase as a result of updating damage functions.

22.3.4 Social welfare function

The social welfare function translates climate damages into a social welfare loss. The IWG [7] maintains the structure of the social welfare function specified by the three IAMs, but not the underlying parameters. All three IAMs assume an isoelastic utility function with a discrete time approximation of exponential discounting. For example, DICE assumes that the present value of social utility $U(C)$ is defined as

$$U(C) = \sum_{t=0}^{T} \frac{1}{(1+\rho)^t} \mathrm{L_t} \frac{CPC_t^{\,1-\eta}}{1-\eta}$$

where ρ is the pure rate of time preference (i.e., the utility discount rate which measures social impatience), η is society's relative risk aversion, CPC_t is consumption per capita in period t, and L_t is population in period t. Contrary to the IAMs' assumption of a risk-averse social planner (i.e., $\eta > 0$), the IWG [7] implicitly assume a risk-neutral central planner ($\eta = 0$) by assuming a constant consumption discount rate (r) over time. Under this assumption, the IWG [7] social welfare function collapses to

$$U(C) = \sum_{t=0}^{T} \frac{1}{(1+\rho)^t} L_t \times CPC_t = \sum_{t=0}^{T} \frac{1}{(1+\rho)^t} C_t$$

where C_t is total consumption in period t and the pure rate of time preference equals the consumption discount rate (i.e., $\rho = r$); see the Ramsey equation in the next subsection. Essentially, the IWG [7] social welfare function equals the present value of total consumption over time.

The current welfare function is selected for empirical ease, despite known shortcomings of the isoelastic utility function. First, as currently applied in IAMs, the isoelastic utility function implies that the elasticity of intertemporal substitution (i.e., the determinant of intertemporal consumption smoothing) equals the inversion of relative risk aversion. As a consequence, society's preference for the intragenerational distribution of consumption, society's preference for the intergenerational distribution of consumption (see interpretation in the Ramsey equation), and risk aversion hold a fixed relationship. Modelers are asking too much from the η parameter in that they force these equalities despite evidence that they do not hold in the real world [75].

Second, the IAMs include only an aggregate consumption good, as measured by per capita consumption, in the social welfare function. On the consumer side of the economy, this assumption implies all goods and services, including market goods and nonmarket goods, are perfectly substitutable (even in the long run), and that they have constant relative prices [76]. Constant relative prices imply that the ratio of the prices of any two goods must remain constant, regardless of the amounts available of either good. This simplifying assumption biases the SCC estimates because climate change is predicted to affect nonmarket and market goods produced outdoors (such as agricultural, forestry, and environmental goods and services) more than market goods produced indoors. As a consequence, the current IAMs fail to capture the increase in value of outdoor produced goods and services relative to other traditional consumption goods [76].

Finally, IAM modelers currently assume that society is equally averse to known unknowns and unknown unknowns [71]. Empirical evidence indicates that individuals are ambiguity averse [77], such that they prefer known uncertainty to unknown uncertainty. There is an ongoing debate of whether ambiguity aversion is rational or a behavioral mistake. Given the strong possibility that this debate is unlikely to be resolved, Heal and Millner [77] recommend exploring both assumptions. Modeling ambiguity aversion is also a step toward addressing Pindyck's concerns that IAMs currently fail to address the impact of ambiguity over climate and damage parameters. Recent work in stochastic dynamic programming tends to better integrate fat tails—particularly with respect to tipping points—and address additional aversion to this type of uncertainty.

The solution is to adopt a more realistic utility function than the current isoelastic utility function. Adopting Epstein-Zin preferences that disentangle risk aversion and time preferences allows for these parameters to be calibrated separately using empirical evidence; doing so can significantly increase the SCC under uncertainty [42]. Similarly, to model the change in relative prices over time, a CES utility function can be nested within the isoelastic utility function [78]. While current research indicates that doing so should increases the SCC [76,79], the results are dependent on the level of consumption disaggregation, the empirical choice for the elasticity of substitution between types of consumption goods, and the choice of the total value of nonmarket goods in the base period. Like the previous two solutions, integrating ambiguity aversion into the isoelastic utility function requires modifying the utility function and increases the SCC [71]. It also requires IAM modelers to include fat-tailed distributions in their climate models, particularly where modelers currently use author discretion to determine parameters and functional forms. To calibrate fat-tailed distributions, modelers should use a wide survey of experts to calibrate the subjective priors from which they derive fat tails.

22.3.5 Discount rate

The discount rate is how economists measure the value of money over time—the tradeoff between what a dollar is worth to an individual today and what a dollar would be worth to him or her in the future. If offered \$1 now or \$1 in a year, almost everyone would choose to receive the \$1 now; most individuals would only wait until next year if they were offered more money in the future. The discount rate is how much more you would have to receive to wait until next year. Economists often measure the discount rate using various market interest rates (known the descriptive method), including the savings rate at your bank and the 30-year US Treasury bond. In other contexts—such as "when [a] new project is financed by an increase in savings from the current generation" [33]—an ethical approach is considered where economists weigh the intrinsic motivations for discounting: pure preference for current over future consumption and aversion to inequality across generations. Given that IAMs model intergenerational consumption and savings decisions, the latter methodology is appropriate for calculating the SCC.

All of the IAMs use the (standard) Ramsey discount rate equation. Under certainty, this isoelastic social welfare function implies a consumption discount rate (r) equal to

$$r = \rho + \eta \times g$$

where ρ is the pure rate of time preference (i.e., the utility discount rate, which measures social impatience), η now represents the society's aversion to intergenerational income inequality, and g is annual growth in per capita consumption. Economic consensus is that the Ramsey discount rate should be applied in the intergenerational context [6,80]. Instead of employing the theoretically correct Ramsey discount rate, the IWG [7] chose constant market discount rates of 2.5%, 3%, and 5%; essentially setting $\eta = 0$ (as discussed in the previous subsection). To avoid inconsistency, Marten et al. [80] note that we can interpret the constant discount rates as approximations of the Ramsey equation for a particular emission year (e.g., 2020 SCC). Given the IWG [7] growth rates, the 5% discount rate is roughly consistent with the DICE [4,50] parameter values of $\rho = 1.5$ and $\eta = 1.45$–2 and 2.5% and 3% discount rates are roughly consistent with the FUND [27] and (the mode values in) PAGE [28] parameter values of 1.0 for both parameters; Nordhaus calibrates ρ and η in DICE to replicate the global savings rate [22]. Unsurprisingly given its inappropriateness [21], the 7% private rate of return to capital recently used by the EPA (2017) to calculate the SCC requires outlier welfare parameters and implies a lower-than-observed savings rate, according to the Dasgupta [81] expression.

Both IAMs and the government should apply the theoretically appropriate discount rate expression. Assuming a risk-averse central planner and uncertain growth rates, an extended Ramsey rule applies rather than the standard Ramsey equation. If the future growth of consumption is uncertain with mean μ and variance σ^2, the extended Ramsey equation is

$$r = \delta + \eta \times \mu - 0.5\,\eta^2\sigma^2 = \delta + \eta \times g - 0.5\,\eta(\eta + 1)\,\sigma^2$$

where g is the growth rate of expected consumption and $\eta + 1$ is prudence. Economists have demonstrated that an extended Ramsey rule implies a declining discount rate when (1) the growth rate of per capita consumption is stochastic, and (2) consumption shocks are positively correlated over time (or their mean or variances are uncertain) [33,40,75]. While a constant adjustment downwards (known as the precautionary effect) can be theoretically correct when growth rates are independent and identically distributed [40], empirical evidence supports these two assumptions for the United States, thus implying a declining discount rate [40,75]. We should further expect this positive correlation to strengthen over time due to the negative impact of climate change on consumption, as climate change causes an uncertain permanent reduction in consumption [82]. This declining discount rate is time consistent [6,83]. If modelers adopt one of the more complex social welfare functions discussed in the previous subsection, IAM modelers will need to derive a consistent Ramsey-like discount rate expression under uncertainty [33,78].

In addition to updating the discount rate equation, Pindyck [11,64] challenges the IAM discount rates as representing only the opinion of the modelers. He notes that the DICE assumptions ($\rho = 1.5$ and $\eta = 1.45$–2) differ greatly from the Stern assumptions (i.e., $\rho \approx 0$ and $\eta = 1$) driving their differing opinions on optimal climate policy, specifically gradual versus immediate action. Consistent with this view that ρ and η represent modelers' views (i.e., the normative approach to discounting), the welfare parameters underlying discount rates are primitive and irreducible [32,34]. As a consequence, there is no, nor has there ever been, a consensus on the appropriate social discount rate and its underlying welfare parameters [84]. Nor may there ever be consensus as normative uncertainty may not decline over time, as "reasonable people can reasonably disagree" over welfare parameters [32] such that future generations are likely to continue this debate [34]. While there are those who may disagree and adhere to the position that social discount rate is a positive parameter that should reflect market rates, this position is also an "ethical view" over which reasonable people disagree given the lack of intergenerational safe market assets [33]. As demonstrated by a recent survey [85], experts on social discount rates disagree on the weight that the government should place on normative issues when determining the social discount rate, with a majority weighing normative issues more highly than positive issues in determining the social discount rate despite thinking that both types of issues are relevant. In a similar argument, some economists argue that discount rates should reflect the observed savings rate [22,81]. However, as Heal and Millner [86] note, people can reasonably differ on what is the optimal savings rate for society [86].

Given the subjectivity in selecting discount rates, an appropriate solution is to rely on the general opinion of experts gauged through expert elicitation, instead of relying exclusively on modeler opinion [12,13]. Respondents should be asked to provide an appropriate consumption-based discount rate, not a capital rate, to be consistent with economic theory [6,21]. Surveys of experts on discounts and climate change find strong support for consumption discount rates on the lower end of the IWG [7] range [12,13,84,85]. While Weitzman [84] found mean and median rates of 4% and 3%, respectively, more recent surveys find mean rates between 2% and 3% and median rates of ~2%. Interestingly, the 2% value supported by the latest surveys is supported by a recent US Council of Economic Advisers report [87], which finds that recent US government bond data supports a decline in the consumption discount rate from 3% (as currently employed by OMB-*Circular A-4* [5]) to 2%. This shift mirrors the decline in these surveys.

To allow for a declining rate based on the extended Ramsey equation (as recommended earlier), expert elicitation should instead be used to calibrate ρ and η. However, normative uncertainty (i.e., heterogeneity) over ρ leads to questions of how to find the equilibrium ρ to use for the representative agent when individuals differ in their respective values. While several efficient solutions for finding the collective pure rate of time preference exist, each solution is time-inconsistent because they imply a declining pure rate of time preference [83]. The second-best solution is to find a time-consistent policy. Millner and Heal [88] demonstrate that a voting procedure—whereby the median voter determines the collective preference—is: (1) time-consistent, (2) welfare enhancing relative to the noncommitment, time-inconsistent approach, and (3) preferred by a majority of agents relative to all other time-consistent plans. In a survey of several hundred experts on discounting, Drupp [85] finds median values of 0.5 and 1.0 for ρ and η, respecitively. Without considering uncertain growth rates, the IWG [7] scenarios imply average discount rates of 2.4% over the next 25 years, 2.2% over next 100 years, and 1.8% over next 200 years. Combining an uncertain growth rate and heterogeneous preferences together implies an even faster declining discount rate, particularly if a wider range of scenarios are considered as the NAS [6] suggests. Applying this technique implies a higher SCC.

22.3.6 Global versus domestic

All of the IAMs currently focus on the global SCC. Until recently, the federal SCC also focused on the global impacts of climate change, based on the unique nature of climate change as a global externality problem that the United States cannot solve alone. Even so, the IWG [7] developed a "high speculative" range of 7% (using FUND 3.5) to 23% (share of global GDP) for the ratio of US domestic to global SCC. Disregarding the concerns raised by the IWG with regards to the global nature of climate change, the need for global action, and the "highly speculative" nature of current domestic estimates (with little to no discussion), the EPA [26] recently moved its focus to a domestic SCC that focuses exclusively on impacts on US GDP. Using regional-based calculations within the IAMs, the EPA [26] developed a domestic SCC that is ~10% the global number. While the global SCC number has been held up in federal court as reasonable, this has yet to be seen for the domestic numbers.

Like the Trump Administration's decision to apply a 7% discount rate, there are reasons to believe that a court will reject the EPA's decision to focus on domestic SCC estimates. Expanding on the IWG's reasoning for applying the global SCC, Howard and Schwartz [20] and Revesz et al. [89] present a variety of economic and legal reasons for supporting a global SCC. From an economic perspective, a global SCC is efficient. Global adoption of the domestic SCC approach leads to an under-regulation of GHG emissions. Furthermore, the free-rider concern highlighted by many supporters of the domestic SCC fails to recognize that a globally efficient outcome is possible given the repeat nature of the problem and the small number of sophisticated actors involved [20,90]. As the sole country to withdraw from the Paris Climate Agreement, it is also difficult for the United States to apply the free rider argument. From a practical standpoint, IAMs were designed with a global SCC in mind. As a consequence, the domestic SCC cannot be currently calculated in any realistic or relevant way [6], as IAMs assume each model region is an island ignoring interregional spillovers like trade, international financial markets and asset ownership, global health, migration, and security. In fact, the United States is particularly vulnerable to these spillovers as a preeminent global power. Additional arguments for a global SCC include reciprocity of other countries to US emission reductions and altruism of US citizens for noncitizens.

There are academics who disagree with applying a global SCC [91], though even they support a strategic SCC that accounts for reciprocity and altruism. These individuals argue that a global SCC is inconsistent with the historical application of benefit-cost analysis. In applying a global SCC estimate, they controversially [20] argue that the federal government could set an undesirable legal precedent [91]. However, Gayer and Viscusi [91,92] agree that US altruism for noncitizens and reciprocity of other nations to US emission reductions as reasons for a "strategic" SCC exceeding a domestic-only SCC. The sole empirical work to address the strategic interactions of nations in setting a "strategic" SCC finds that the US "strategic" SCC should be at least 75% of the global value based solely on reciprocity [93]. This number represents a lower bound for the "strategic" SCC in that it does not address altruism or spillovers. Thus, there is at least some growing consensus that the domestic SCC estimated by the Trump Administration is too low.

Given the strong arguments for the appropriateness of a global SCC and the underestimation of the "strategic" SCC (due to the exclusion of altruism and spillovers), modelers should continue to focus on the global SCC as the central estimate. Additionally, they should conduct sensitivity analysis using the lower-bound estimate for the "strategic" SCC. Thus, a range between 75% and 100% of the global SCC is appropriate for use in CBA, though debate is likely to continue.

22.4 The appropriate central SCC value

IAMs and agencies should provide a "best" central SCC estimate along the BAU emissions path based upon the best scientific estimate. Though full transparent presentation of uncertainty is an admirable goal in CBA, only presenting a wide range of possibilities ignores the real-world applications of the SCC estimates by agencies. If agencies can select from a multitude of SCC estimates, agencies may use their discretion to determine which set of SCC estimates are valid for decision making when a proposed regulation is only cost-benefit justified under a subset of SCC estimates [94]. Similarly, a wide range of SCC estimates can lead agencies and the public to dismiss the SCC as meaningless [16,95]. Currently, for IAMs that run Monte Carlo simulations, the modelers generally present the mean SCC estimate as this central estimate. The IWG [7] presented the mean estimate from the 3% discount rate as the central estimate, while the EPA [26] does not advocate for any specific central estimate except to highlight those estimates it argues are consistent with OMB Circular A-4: domestic estimates with discount rates of 3% and 7%.

In a stochastic world, the certainty-equivalent SCC is the ideal central estimate of the SCC [96,97]. The certainty-equivalent SCC estimate is the theoretically correct measurement of social costs: under uncertainty, "the 'social cost of carbon' in a particular year is the decrease in aggregate consumption in that year that would change the current *expected value of social welfare* by the same amount as a one unit increase in carbon emissions in that year" [96] (emphasis added). The certainty-equivalent SCC is the weighted average of the deterministic SCC estimates where the marginal welfare of consumption is the appropriate weight [98]. Thus, only when the marginal utility of consumption is known with certainty [97] does the SCC under uncertainty equal the expected value of the deterministic SCC. Given that this assumption does not hold under the IWG's own assumptions, the IWG does not solve for the theoretically correct specification of the SCC by solving for the arithmetic mean [7]. Calculating the certainty-equivalent SCC is critical in that it addresses the IWG's omission of the risk premium corresponding to economic and climate change uncertainty that arises when the central planner is risk averse [7]. Accounting for this risk premium can significantly increase the IWG's current SCC estimates, providing it relaxes its theoretically incorrect assumption of risk neutrality [97].

The certainty-equivalent SCC can be calculated (see [97]). Using the theory developed for discounting the benefits of risky projects [33], the present value of benefits of a risky project is calculated by determining the certainty-equivalent benefits of the project over time, discounting these benefits using the risk-free discount rate (potentially a declining discount rate), and then summing across time periods. Thus, there is a simple series of steps for calculating the certainty-equivalent SCC: (1) calculate change in certainty-equivalent consumption over time due to climate change for the base scenario and the perturbation scenario, (2) take the difference, and (3) calculate the present value of the change in certainty-equivalent consumption (potentially using a declining discount rate). However, to make calculations tractable, it is necessary to make the common assumption that the utility function is exogenous such that the pure rate of time preference and the elasticity of marginal utility of consumption are known [98,99]; this is consistent with our earlier discussion of discount rates in the previous section. Alternatively, recent work by Christian Gollier demonstrates how to adjust the social discount rate directly by including a risk premium; see Further Reading.

Both the mean SCC from uncertainty propagation and the certainty-equivalent SCC exclude the (real) option value of preventing marginal CO_2 emissions [100,101]. Option value is the value of future flexibility due to uncertainty and irreversibility. In this case, the irreversibility corresponds to the emission of CO_2 due to its long life in the atmosphere [35], such that option value is more precisely defined as the information value of delaying an emission until more information is available on relevant scientific and economic uncertainties. If society exercises the option of emitting an additional unit of CO_2 emissions today, "we will lose future flexibility that the [mitigation] option gave" leading to possible "regret and…a desire to 'undo'" the additional emission because it "constraints future behavior." [35] Given that the SCC is calculated on the BAU emission pathway, option value will undoubtedly be positive for an incremental emission because society will regret this emission in most possible futures. Even under the IWG's current assumption of risk neutrality, option value will increase the SCC. Using a rough approximation of option value equal to 0.4 times the standard deviation of a distribution [101a,101b], the option value corresponding to the 2016 IWG's central 2020 SCC estimate is about $33 in 2017 USD, or roughly two-thirds of the IWG's central SCC estimate. Without adding option value to the SCC, modelers exclude the significant value that arises due to the irreversibility of emissions, a separate value from the risk premium [102].

22.5 Conclusions

Since 2009, the US federal government has used the SCC to measure the benefits of GHG emission reductions when assessing regulations. Large uncertainties characterize each step in the calculation of the SCC. Positive uncertainty underlies the socioeconomic and emission scenarios, simple climate model, damage and abatement cost functions, and the social welfare function. In the social welfare function, normative uncertainties also underlie the temporal and spatially weighting of individuals. This has led to criticism by some to abandon the official SCC altogether [17]. In a less extreme step, the EPA [26] changed two normative assumptions in ways inconsistent with the best science: allowing a discount rate equivalent to the private returns to capital and focusing exclusively on domestic damages. Abandoning the SCC or adopting relatively arbitrary modeling assumptions are not appropriate responses to uncertainty.

Climate change uncertainty generally warrants more stringent climate policy and raises the SCC. IAMs currently used to calculate the SCC show that the net effect of uncertainty about economic damage resulting from climate change, costs of mitigation, future economic development, and many other parameters raises the SCC compared to the case where models simply use our current best guesses of these parameters [1,9]. Even so, IAMs still underestimate the impact of uncertainty on the SCC by ignoring fundamental features of the climate problem: the irreversibility of climate change, society's aversion to risk and other social preferences, and many catastrophic impacts [35]. The next generation of numerical models designed to capture these features of the climate problem currently focus on the optimal tax (i.e., the SCC on the optimal emissions path) and require key simplifying assumptions, though existing results indicate that uncertainty leads to an increase in the optimal tax for realistic parameter values. Rather than providing a reason not to take action, if anything, uncertainty increases the SCC and should lead to more stringent policy to address climate change [35].

Instead, modelers should work relentlessly to ensure that the next generation of IAMs is consistent with the best economic evidence to date. The NAS [6,29] has offered several valuable recommendations to IAMs modelers and the federal government on how address the various positive and normative uncertainties:

- Employ a combination of statistics and expert elicitation to develop projections out to 2300;
- Keep the simple climate model up to date with respect to scientific knowledge, in particular the critical short-term warming parameters. Use expert elicitation when values of climate variables are unavailable from the IPCC;
- Keep the damage function up to date;
- Employ the theoretically correct discount rate, e.g., the extended Ramsey equation and consumption-based rates;
- Focus on the global SCC given the current state of knowledge.

As discussed here, to more fully correspond to the definition of the SCC and more fully represent uncertainty and ambiguity in this number—and society's aversion to them—modelers and the federal agencies should also:

- Calibrate IAMs to BAU projections;
- Use fat-tailed distributions where subjective priors are necessary to calibrate probability distribution functions (e.g., where author discretion was necessary in past IAMs);

- Rely on transparent expert elicitation results instead of opaque model assumptions where economic information on climate damages and intergenerational discount rates are unavailable, in addition to emission scenarios and climate parameters;
- Apply multiple damage estimates corresponding to multiple estimation approaches when calibrating the damage function;
- Adopt more realistic social welfare functions that allow for relative prices and risk and ambiguity aversion. Adjust the discount rate accordingly;
- Adopt the theoretically consistent certainty-equivalent SCC as the best central estimate modified to include the option value corresponding to the irreversibility of emissions.

Developers of the next generation of IAMs (including the federal government) should consider these recommendations, to address many of the key concerns raised by the critics of the SCC in legal and academic challenges in advocating for both a lower or higher number. Overall, these changes are expected to increase the SCC relative to its current value.

References

[1] Tol R. The economic effects of climate change. J Econ Perspect 2009;23:29–51.
[2] Stern N. The economics of climate change: the stern review. Cambridge: Cambridge University Press; 2007.
[3] Hardin G. The tragedy of the commons. J Nat Resour Policy Res 2009;1:243–53.
[4] Nordhaus W, Boyer J. Warming the world: economic models of global warming. Cambridge: MIT Press; 2000.
[5] U.S. Office Management Budget. Circular A-4. Washington, DC: U.S. OMB; 2003.
[6] National Academies of Sciences, Engineering, and Medicine. Valuing climate damages: updating estimation of the social cost of carbon dioxide. Washington, DC: National Academies Press; 2017.
[7] U.S. Interagency Working Group on Social Cost of Carbon. Technical support document: technical update of the social cost of carbon for regulatory impact analysis. Washington, DC: U.S. IWG; 2010.
[8] U.S. Interagency Working Group on Social Cost of Carbon. Technical support document: technical update of the social cost of carbon for regulatory impact analysis. Washington, DC: U.S. IWG; 2013.
[9] U.S. Interagency Working Group on Social Cost of Carbon. Technical support document: technical update of the social cost of carbon for regulatory impact analysis. Washington, DC: U.S. IWG; 2016.
[10] Pindyck R. The climate policy dilemma. Rev Environ Econ Policy 2013;7:219–37.
[11] Pindyck R. Climate change policy: what do the models tell us? J Econ Lit 2013;51:860–72.
[12] Pindyck R. The social cost of carbon revisited. National Bureau of Economic Research Working Paper No. 22807. 2016.
[13] Howard P, Sylvan D. Expert consensus on the economics of climate change. New York: Institute for Policy Integrity, New York University School of Law; 2015.
[14] Revesz R, Greenstone M, Hanemann M, Livermore M, Sterner T, Grab D, Howard P, Schwartz J. Best cost estimate of greenhouse gases. Science 2017;357: 655 pp.
[15] Revesz R, Howard P, Arrow K, Goulder L, Kopp R, Livermore M, Oppenheimer M, Sterner T. Global warming: improve economic models of climate change. Nature 2014;508:173–5.

[16] America's Natural Gas Alliance; American Chemistry Council; American Petroleum Institute; National Association of Home Builders; National Association of Manufacturers; Portland Cement Association; U.S. Chamber of commerce. Petition for correction: technical support document: social cost of carbon for regulatory impact analysis under executive order 12866 (February 2010) and technical support document: technical update of the social cost of carbon. Washington, DC: ANGA, ACC, API, NAHB, NAM, PCA, USCOC; 2013. September 3.
[17] U.S. Interagency Working Group on Social Cost of Carbon. Response to comments: social cost of carbon for regulatory impact analysis under executive order. vol. 12866. Washington, DC: U.S. IWG; 2015.
[18] Lesser J. New York's clean energy programs: the high cost of symbolic environmentalism. New York: Manhattan Institute; 2017.
[19] Governmental Accountability Office. Regulatory impact analysis: development of social cost of carbon estimates, GAO-14-663. Washington, DC: GAO; 2014.
[20] Howard P, Schwartz J. Think global: international reciprocity as justification for a global social cost of carbon. Columbia J Environ Law 2016;42:203–94.
[21] Newell R. Unpacking the administration's revised social cost of carbon. Resources for the Future [blog]. October 10. Available from: http://www.rff.org/blog/2017/unpacking-administration-s-revised-social-cost-carbon; 2017.
[22] Nordhaus W. Estimates of the social cost of carbon: concepts and results from the DICE-2013R model and alternative approaches. J Assoc Environ Resour Econ 2014;1:273–312.
[23] Nordhaus W. A question of balance: weighing the options on global warming policies. New Haven, CT: Yale University Press; 2008.
[24] Anthoff D, Tol R. The climate framework for uncertainty, negotiation and distribution (FUND). Technical Description, Version 3.5. FUND—Climate Framework for Uncertainty, Negotiation and Distribution [website]. May 17. Available from: http://www.fund-model.org/; 2010.
[25] Hope C. The marginal impact of CO2 from PAGE2002: an integrated assessment model incorporating the IPCC's five reasons for concern. Integr Assess J 2006;6:19–56.
[26] U.S. Environmental Protection Agency. Regulatory impact analysis for the review of the clean power plan: proposal. Office of air and radiation Office of air quality planning and standards. Washington, DC: US EPA; 2017.
[27] Anthoff D, Tol R. The climate framework for uncertainty, negotiation and distribution (FUND). Technical Description, Version 3.8. FUND—Climate Framework for Uncertainty, Negotiation and Distribution [website]. August 7. Available from: http://www.fund-model.org/; 2014.
[28] Hope C. The social cost of CO2 from the PAGE09 model. Economics 2011; Discussion Paper No 2011-39.
[29] National Academies of Sciences, Engineering, and Medicine. Assessment of approaches to updating the social cost of carbon: Phase. Vol. 1. Washington, DC: National Academies Press; 2016. Report on a Near-Term Update.
[30] Schwartz J. Approaches to cost–benefit analysis. In: Handbook of regulatory impact assessment; 2016. p. 33–51.
[31] Kann A, Weyant J. Approaches for performing uncertainty analysis in large-scale energy/economic policy models. Environ Model Assess 2000;5:29–46.
[32] Heal G, Millner A. Agreeing to disagree on climate policy. Proc Natl Acad Sci 2014;111:3695–8.
[33] Gollier C, Hammitt J. The long-run discount rate controversy. Annu Rev Resour Econ 2014;6:273–95.

[34] Millner A. Non-paternalistic social discounting. Social Science Research Network Working Paper No. 2832183. 2016.
[35] Pindyck R. Uncertainty in environmental economics. Rev Environ Econ Policy 2007;1:45–65.
[36] Nordhaus W. Economic policy in the face of declining productivity growth. Eur Econ Rev 1982;18:131–57.
[37] Kopp R, Shwom R, Wagner G, Yuan J. Tipping elements and climate–economic shocks: pathways toward integrated assessment. Earth's Future 2016;4:346–72.
[38] Nordhaus W. An analysis of the dismal theorem. Cowles Foundation Discussion Paper No 1686.
[39] Weitzman M. Fat-tailed uncertainty in the economics of catastrophic climate change. Rev Environ Econ Policy 2011;5:275–92.
[40] Cropper M, Freeman M, Groom B, Pizer W. Declining discount rates. Am Econ Rev 2014;104:538–43.
[41] Kriegler E, Hall J, Held H, Dawson R, Schellnhuber H. Imprecise probability assessment of tipping points in the climate system. Proc Natl Acad Sci 2009;106:5041–6.
[42] Cai Y, Lenton T, Lontzek T. Risk of multiple interacting tipping points should encourage rapid CO2 emission reduction. Nat Clim Change 2016;6:520–5.
[43] Weitzman M. On modeling and interpreting the economics of catastrophic climate change. Rev Econ Stat 2009;91:1–19.
[44] In the matter of the further investigation into environmental and socioeconomic costs under Minnesota Statutes Section 216B.2422, Subdivision 3: Hearings Before the State of Minnesota, Office of Administration of Hearings for the Public Utilities of Commission; 2016.
[45] Working Group I to the Fourth Assessment Report of the Intergovernmental Panel on Climate Change. Climate Change 2007: the physical science basis. Cambridge: Cambridge University Press; 2007.
[46] Working Group I contribution to the Fifth assessment report of the Intergovernmental Panel on Climate Change. Climate change 2013: the physical science basis. Cambridge: Cambridge University Press; 2013.
[47] Pindyck R. Coase lecture—Taxes, targets and the social cost of carbon. Economica 2017;84:345–64.
[48] Howard P. Omitted damages: what's missing from the social cost of carbon. Cost of Carbon Project: New York; 2014.
[49] Nordhaus W. Expert opinion on climatic change. Am Sci 1994;82:45–51.
[50] Nordhaus W. Revisiting the social cost of carbon. Proc Natl Acad Sci 2017;114:1518–23.
[51] Kopp R, Mignone B. The US government's social cost of carbon estimates after their first two years: pathways for improvement. Economics 2012;6:1–41.
[52] Nuccitelli D. Republican hearing calls for a lower carbon pollution price. It should be much higher. The Guardian 2017. March 1.
[53] Dell M, Jones B, Olken B. What do we learn from the weather? The new climate–economy literature. J Econ Lit 2014;52:740–98.
[54] Carleton T, Hsiang S. Social and economic impacts of climate. Science 2016;353: aad9837.
[55] Hsiang S, Kopp R, Jina A, Rising J, Delgado M, Mohan S, Rasmussen D, Muir-Wood R, Wilson P, Oppenheimer M, Houser T, Larsen K. Estimating economic damage from climate change in the United States. Science 2017;356:1362–9.
[56] Moore F, Baldos U, Hertel T. Economic impacts of climate change on agriculture: a comparison of process-based and statistical yield models. Environ Res Lett 2017;12:aa6eb2.

[57] Greenstone M. A New Path Forward for an Empirical Social Cost of Carbon [powerpoint]. Washington, DC: National Academies' Assessing Approaches to Updating the Social Cost of Carbon; 2016. May 5. Available from: https://sites.nationalacademies.org/cs/groups/dbassesite/documents/webpage/dbasse_172599.pdf.
[58] Yohe G, Tirpak D. A research agenda to improve economic estimates of the benefits of climate change policies. Integr Assess 2008;8:1–17.
[59] Environmental Defense Fund, the Institute for Policy Integrity, the Natural Resources Defense Council. Playing catch-up to the IPCC. The cost of carbon pollution. [blog] April 22. Available from: http://costofcarbon.org/blog/entry/playing-catch-up-to-the-ipcc; 2014.
[60] Burke M, Hsiang S, Miguel E. Global non-linear effect of temperature on economic production. Nature 2015;527:235–9.
[61] Dell M, Jones B, Olken B. Temperature shocks and economic growth: evidence from the last half century. Am Econ J Macroecon 2012;4:66–95.
[62] Howard P, Sterner T. Few and not so far between: a meta-analysis of climate damage estimates. Environ Resource Econ 2017;68:1–29.
[63] Walton T. Regulatory impact analysis: national emission standards for hazardous air pollutants from the Portland cement manufacturing industry. Research Triangle Park, NC: US Environmental Protection Agency; 2009.
[64] Pindyck R. The use and misuse of models for climate policy. Rev Environ Econ Policy 2017;11:100–14.
[65] Cooke R. Respectability looming for expert judgment. Resources for the future. [blog] May 22. Available from: http://www.rff.org/blog/2017/respectability-looming-expert-judgment; 2017.
[66] Oppenheimer M, Little C, Cooke R. Expert judgement and uncertainty quantification for climate change. Nat Clim Change 2016;6:445–51.
[67] U.S. Environmental Protection Agency, National Center for Environmental Economics. Report of the EPA Working Group on vsl meta-analysis. Report EE-0494. Washington, DC: EPA; 2006.
[68] Weitzman M. GHG targets as insurance against catastrophic climate damages. J Public Econ Theory 2012;14:221–44.
[69] Ackerman F, Stanton E, Bueno R. CRED v. 1.3 technical report. Stockholm Environment Institute: Somerville, MA; 2011.
[70] Lontzek T, Cai Y, Judd K, Lenton T. Stochastic integrated assessment of climate tipping points indicates the need for strict climate policy. Nat Clim Change 2015;5:441–4.
[71] Lemoine D, Traeger C. Economics of tipping the climate dominoes. Nat Clim Change 2016;6:514–9.
[72] Working Group II to the Fifth Assessment Report of the Intergovernmental Panel on Climate Change. Climate change 2014: Impacts, adaptation, and vulnerability. Cambridge: Cambridge University Press; 2014.
[73] Kelly D, Kolstad C, Mitchell G. Adjustment costs from environmental change. J Environ Econ Manag 2005;50:468–95.
[74] Guo C, Costello C. The value of adaption: climate change and timberland management. J Environ Econ Manag 2013;65:452–68.
[75] Arrow K, Cropper M, Gollier C, Groom B, Heal G, Newell R, Nordhaus W, Pindyck R, Pizer W, Portney P, Sterner T, Tol R, Weitzman M. Should governments use a declining discount rate in project analysis? Rev Environ Econ Policy 2014;8:145–63.
[76] Sterner T, Persson U. An even sterner review: introducing relative prices into the discounting debate. Rev Environ Econ Policy 2008;2:61–76.

[77] Heal G, Millner A. Reflections: uncertainty and decision making in climate change economics. Rev Environ Econ Policy 2014;8:120–37.
[78] Hoel M, Sterner T. Discounting and relative prices. Clim Change 2007;84:265–80.
[79] Cai Y, Judd K, Lenton T, Lontzek T, Narita D. Environmental tipping points significantly affect the cost–benefit assessment of climate policies. Proc Natl Acad Sci 2015;112:4606–11.
[80] Marten A, Kopits E, Griffiths C, Newbold S, Wolverton A. Incremental CH4 and N2O mitigation benefits consistent with the US Government's SC-CO2 estimates. Clim Policy 2015;15:272–98.
[81] Dasgupta P. Discounting climate change. J Risk Uncertain 2008;37:141–69.
[82] Gollier C. Should we discount the far-distant future at its lowest possible rate? Economics 2009;3:1–14.
[83] Gollier C. Pricing the planet's future: the economics of discounting in an uncertain world. Princeton, NJ: Princeton University Press; 2012.
[84] Weitzman M. Gamma discounting. Am Econ Rev 2001;91:260–71.
[85] Drupp M, Freeman M, Groom B, Nesje F. Discounting disentangled. University of Oslo Department of Economics Working Paper No. 20/2015. 2015.
[86] Heal G, Millner A. Discounting under disagreement. National Bureau of Economic Research Working Paper No. w18999. 2013.
[87] U.S. Council of Economic Advisers. Discounting for public policy: Theory and recent evidence. Issue Brief, Washington, DC: U.S. Council of Economic Advisers; 2017.
[88] Millner A, Gl H. Resolving intertemporal conflicts: economics vs politics. National Bureau of Economic Research Working Paper No. w20705. 2014.
[89] Revesz R, Schwartz J, Howard P, Arrow K, Livermore M, Oppenheimer M, Sterner T. The social cost of carbon: a global imperative. Rev Environ Econ Policy 2017;11:172–3.
[90] Nordhaus W. Climate clubs: overcoming free-riding in international climate policy. Am Econ Rev 2015;105:1339–70.
[91] Gayer T, Viscusi W. Determining the proper scope of climate change policy benefits in US regulatory analyses: domestic versus global approaches. Rev Environ Econ Policy 2016;10:245–63.
[92] Gayer T, Viscusi W. The social cost of carbon: maintaining the integrity of economic analysis—a response to Revesz et al. (2017). Rev Environ Econ Policy 2017;11:174–5.
[93] Kotchen M. Which social cost of carbon? A theoretical perspective. National Bureau of Economic Research Working Paper No 22246. 2016.
[94] U.S. Department of Agriculture, U.S. Forest Service. Supplemental Draft Environmental Impact Statement. Colorado roadless rule. Washington, DC: USDA; 2015.
[95] U.S. Department of the Interior, Office of Surface Mining, Reclamation, and Enforcement. Record of decision: four corners power plant and Navajo mine energy project. Washington, DC: US DOI; 2015.
[96] Newbold S, Griffiths C, Moore C, Wolverton A, Kopits E. A rapid assessment model for understanding the social cost of carbon. Clim Change Econ 2013;4:1350001.
[97] Kopp R, Golub A, Keohane N, Onda C. The influence of the specification of climate change damages on the social cost of carbon. Economics 2012;6:1–42.
[98] Gerlagh R. Calculating the social costs of carbon without knowing preferences comment on "a rapid assessment model for understanding the social cost of carbon" Clim Change Econ 2014;(5). 1450004.
[99] Newbold S, Griffiths C, Moore C, Wolverton A, Kopits E. Further comment on "a rapid assessment model for understanding the social cost of carbon" Clim Change Econ 2014;5:1450005.

[100] Arrow K, Fisher A. Environmental preservation, uncertainty, and irreversibility. Q J Econ 1974;88:312–9.
[101] Dixit A, Pindyck R. Investment under uncertainty. Princeton, NJ: Princeton University Press; 1994.
[101a] Anda J, Golub A, Strukova E. Economics of climate change under uncertainty: benefits of flexibility. Energy Policy 2009;37:1345–55.
[101b] Golub A, Brody M. Uncertainty, climate change, and irreversible environmental effects: application of real options to environmental benefit-cost analysis. J Environ Stud Sci 2017;7:519–26.
[102] Golub A, Brody M. Uncertainty, climate change, and irreversible environmental effects: application of real options to environmental benefit-cost analysis. J Environ Stud Sci 2017;7:519–26.

Further reading

[103] Rose SK, Turner D, Blanford G, Bistline J, de la Chesnaye F, Wilson T. Understanding the social cost of carbon: a technical assessment. Electric Power Research Institute: Palo Alto, CA; 2014.
[104] Dietz S, Gollier C, Kessler L. The climate beta. J Env Econ Manag 2018;87:258–74.
[105] Gollier C. Discounting and growth. Am Econ Rev 2014;104:534–37.
[106] Gollier C, Mahul O, Meddahia N, Nzesseua JT. Term structures of discount rates: An international perspective. Paper presented at the 2018 World Congress of Environmental and Resource Economists, Gothenburg, June 25–29; 2018.

Migration and climate change

23

Andrea C. Simonelli[1]
Department of Political Science, Virginia Commonwealth University, Richmond, VA, United States
[1]Corresponding author: Acsimonelli@gmail.com

Chapter Outline

23.1 Introduction

Human migration as a result of climate change has become an increasingly researched topic over the last decade, but not without some controversy. Descriptive labels such as "climate refugee" and "environmental refugee" dot the literature and monopolize news coverage along with apocalyptic imagery of sinking islands, villages in disrepair, and dark-skinned bodies struggling against rising waters; a Google news search performed on December 27, 2017 using the search term "climate refugee" yielded 349,000 results. This, paired with large-scale estimates of migrants using flimsy methodology [1], has often distracted from the real challenges that those who are losing their homes and homelands are facing. The dramatic portrayals of life on the edge of destruction and the millions of needy souls coming to a nearby border generate anxiety and excitement, but what is needed in the lives of those affected is not melodramatic prose, but effective policy solutions.

The settlements losing their physical integrity most rapidly are emitting the least amount greenhouse gasses into the atmosphere and yet have to depend on the nations who emit the most to fix their disintegration or allow their migration. There is a power imbalance among those nations being impacted and those most at fault for the damage. This has left many feeling helpless, seeking ill-fitting remedies wherever they can find an angle. Ioane Teitiota, a migrant from Kiribati, filed a legal claim with the High court of New Zealand, to "migrate with dignity" as a formal refugee [2] due to the deterioration of his homeland; his claim was eventually denied. Small island nations

Managing Global Warming. https://doi.org/10.1016/B978-0-12-814104-5.00023-5

have banded together to fight for an internationally binding resolution to their continued losses and ultimate survival.

Adapting to climate change for many will depend on social acceptability, institutional constraints, and a place in the wider landscape of economic development and social evolution [3]. No one person or society will be affected in the same way or will have the same options. For some, continuing their lives and livelihoods in place will become too difficult, but the perception of this will be subjective. The beliefs of self-efficacy, preferences, and perceptions of risk, knowledge, norms and values will determine an individual's limit to adaptation [4]. While adaptation processes involve an interdependence of agents through their relationships with each other, the institutions in which they reside, and the resource base on which they depend [3], some individuals are simply more resourceful and tolerant than others in the same situation. Migration will become a reasonable decision for those who are less willing to live with long-term uncertainty.

The focus of this chapter is to provide an overview of the main areas of contention within this nexus. The academic debate over using "migration" or "displacement," including the socioenvironmental drivers of such movement, has been a contentious debate of agency and choice. Whether people are "pushed" out of a locale due to inhabitability or are "pulled" out of their habitual place of residence due to better opportunities, impacts the need for protection and/or aid.

Simplifying causal mechanisms also obfuscates individual choice. Migration is also occurring or is being discussed in varied regional settings all over the globe—from the Arctic to the tropics. Climate change may impact the natural systems of the tundra and an atoll in different ways, but the processes that impact human habitability are similar and thus, when under stress, people will need to migrate due to similar conditions in diverse regions. Finally, there has been a coordinated international effort to govern human mobility and compensate for loss in this context for some time. Results have been mixed, moved slowly, and are bound by economic power differentials. To date, families and communities living on the front lines still have no funded remedy to their deteriorating living conditions.

23.2 Regional context and events

The majority focus of this research has been on small island states even though climate change has the potential to displace people in all areas of the globe. Small islands, specifically atoll nations, tend to be relatively more vulnerable in terms of economic structure and more food insecure than other small island developing states [5]. The Pacific island nations are internationally regarded as a barometer for the early impacts of climate change. Their geophysical characteristics, demographic patterns, and scattered positions throughout the Pacific make them particularly vulnerable to the effects of global warming [6]. Additionally, the acute awareness of the impact of climate change in the Pacific comes not only from books, but from first-hand knowledge and ongoing experiences with its effects [7]. Small, low lying, and bereft of nutrient-rich soils, residents rely heavily on the local reefs for provisions. However, supplying their populations with enough food is becoming increasingly difficult due to the overwash of seawater degrading fresh groundwater resources, rising sea surface temperatures, which are contributing to coral reef bleaching and degradation. This will only compound the coral reef ecosystem degradation, which is also impacting island communities' subsistence fisheries and tourism [7]. Continued degradation will negatively impact on livelihoods and long-term sustainability. It is likely that areas and some countries will become uninhabitable long before they are submerged [7].

Just as susceptible to these forces are the islands of the Indian Ocean. The Maldives, an island-nation archipelago of atolls, is also highly vulnerable to climate impacts and has endured much in recent years. Like the Pacific atolls, the country's islands are small and low lying, exposing them to increased rainfall and ocean-induced flooding [8]. About half of its human settlements are near the shoreline as well as almost three quarters of its critical infrastructure such as airports, power plants, landfills, and hospitals [8]. Adapting to rising seas, or at least coping with them, is not a new phenomenon. The Raa Atoll island of Kandholhudhoo fought them for years until the 2004 tsunami hastened their destruction and finally forced every single resident to a different island across the atoll. Residents sought relocation for many years prior, but with no government support, they were trapped for many years [9]. A new leader impressed the international community arguing fervently for the entire relocation of his entire nation; President Nasheed is still discussed for his efforts and a documentary [10] is still touring. Less impacted, in terms of migration, are the Seychelles and Mauritius, which have a higher elevations.

Similarly situated are the islands of the Caribbean; like the Seychelles they are larger in land mass, more populated, have higher elevations, and more diverse economies. This does not shield them from climate impacts, which might eventually serve to displace local populations and businesses. The beaches of Jamaica are already receding and when pollution, poor building practices, and prior hurricane damage is considered [11], local migration is going to be a concern. Beyond storm surge and erosion, climate change's intensification of hurricane strength can impact human injury, disease transmission, deterioration of water quality, and quantity [12]. All of these consequences have the potential to displace and force the migration of millions; hurricanes Irma and Maria did this in the fall of 2018 leaving 3.4 million in Puerto Rico without adequate water and electricity [13]. Due to these conditions, over 300,000 Puerto Ricans have moved to Florida, a number the Governor of Florida has publicized although it has been disputed. [14] As US citizens, these islanders can internally move to another state and start over again, but many cannot due to lack of funds; such as the Maldivians trapped on Kandholhudhoo, only those with money can migrate off an island.

Beyond the shore, glacial melt and its effect on nearby settlement is also a driver to migration. Across the Andean mountain ranges, glaciers have lost nearly a quarter of their surface area over the last 30 years [15]. Glacial melt feeds the high altitude wetlands, serves as the drinking water for major South American cities such as La Paz, and irrigates croplands for most of Peru [16]. A healthy ecosystem allows local peoples to use the water as it melts yearly for direct management and capture; higher regional temperatures either disallow the needed snow yields for adequate water supplies and/or can too quickly melt snow packs causing local flood hazards—both are a cause of migration and displacement. Bolivia defined many of its own resident as "climate migrants" as early as 2010 [17]. Across the globe, the Himalayas and surrounding Tibetan Plateau compromise the largest river runoff from any single location in the world filling the river basins of the Indus, Ganges, Brahmaputra, Irrawaddy, Salween, Mekong Yangtze, and Huang He to support a population of over 1.4 billion [18].

Large-scale ice melt is also an anxiety in the communities of the Alaskan Arctic where communities have been monitoring erosion due to permafrost loss since the 1980s [19] and have since been seeking their own relocation solutions. For the purposes of this overview, it is difficult to be thorough as to all of the locations and ways in which climate change is altering the living conditions of many—and so much so that either migration becomes a necessary option or that they are being forced to move. The intensification of local weather from winter storms to heat events to extended drought will scatter the impetus to move across all continents and for many reasons. How local and international governance could potentially manage these new flows of people will be explored here.

23.3 Academic background

The study of migration is broad and encompasses many forms of movement. Definitions of migration also tend to be expansive in nature, which can make the terms "environmental migration" or "climate migration" too general. Categorization is a way to differentiate between migration motives and to politically or legally define those moving within current systems of governance.

The field also considers within- and crossborder migration but suffers from a lack of consensus as to how to understand international migration outside of refugee frameworks. Theoretical studies of migration have focused on economic push-pull factors and larger spatial models versus individual journeys [20–26]. Spatial models are also employed to explain historical migrations such as the "Great Migration" of humans out of Africa, which is dated back to 130,000 BC [27]. These tend to assume that early human ancestors migrated great distances to follow big game and eventually occupied all continents because no one else existed there, suppositions are simplistic and tend to minimize agency, ingenuity, and connection to place. The history of humanity is almost entirely a story of migrations until the advent of the national border and the post WWI refugee flows. By the early 1920s, mass migration and the repatriation of refugees threatened to strain the newly developed social welfare systems of industrial societies [28]. This change in national redistribution systems transformed the way individual migrants were viewed through eyes of the state. There were now incentives to restrict refugee movement, use them as political pawns, and irrational and inaccurate opinions became influential [29]. Each nation tends to see its own unsolicited immigration in isolation. Thus, we have seen a modern global tightening of borders for all types of immigrants as well as asylum seekers. The right to migrate can be seen as a form of transnational racism, which posits the "naturalness" of violence in less developed regions and other perceived cultural incompatibilities with non-Western peoples [30]. This is a legitimizing ideology, which justifies unequal access to both migrants and asylum seekers alike and is visible on the nonwhite bodies of those rejected at border crossings.

A major conceptual divide exists in the literature, which subtly follows from aforementioned political historical change in attitude toward migrants. The term "migration" is often considered voluntary in that it is an active term used to describe those seeking to improve his/her lot in life and thus moves in order to do so. Without a social welfare state, it would be up to the individual to have to do so, on one's own; in the modern age it has been viewed as more opportunist. While exploitative economic and development policies by dominant states affect weaker ones [31], capitalist development causes both pull-and-push-based migration. It is often necessary to migrate as opportunities dry up at home and open up abroad and this can happen in tandem thus forcing people out rather than enticing them to leave.

"Forced migration studies" is a subfield of migration studies. It is concerned with the types of "push" factors, which drive migrants to leave their homes. This also includes studies on displacement types such as disaster-induced displacement, development-induced displacement, environmental displacement, and all those labeled refugees—whether deserving of the status or not. The main dispute with the subfield is between scholars who wish to separate refugee studies from forced migration as a whole—a rift, which is mirrored in policy as it relates to extending refugee protections to those being displaced by climate processes. It is argued that marrying refugee studies into a larger body of scholarship will take away the special circumstances of refugees. [32] A larger body of scholars contend that while refugees are a special category of forced migrants, this view underappreciates the larger field's contributions, poses no risk to the study of formal refugees, and since many refugees are not formal (they may flee generalized violence instead of direct persecution) [33–35], it excludes the experience of most forced migrants including those who are internally displaced. Refugee studies focus on many of the legal issues with refugee processing the determinants of becoming an asylum seeker, issues of resettlement, reconstruction, peace building, aid, and protracted crises.

Misplaced bureaucratic labels can blur the lines between refugee and other groups of forced migrants. As previously mentioned, many misuse such language implying that refugeehood, and the formal distribution apparatus, which accompanies it, could possibly be conferred onto "environmental refugees" or "climate refugees." Doing so confounds the important distinction between all those forced to migrate and the legal status designated for those who have been persecuted and which enables them the provision of aid.

"Forced migration studies" is a very broad field; beyond refugee studies, it also encompasses development-induced displacement and disaster-induced displacement, which are common and widely researched forms of forced migration—both of which are caused by environmental changes, attributed as "environmental migrants," and which can also be tied to climate impacts. There is a general agreement on the three main causes of environmental migration: natural disasters and environmental or industrial accidents, planned or unplanned relocation due to development, and health-related effects due to inadequate resources to maintain life [36]. As the impacts of climate change are layered onto these categories, the effect on those who will be environmentally displaced will only continue. Disasters are increasing in frequency and intensity and are becoming less "natural." More and more communities from Bangladesh to Alaska have been discussing the need for a managed retreat from coastlines; reclamation as well as reinforcement will take over current home sites. In other areas, continuing to simply exist will be too difficult. The environmental damage of human habitats can initiate a chain of events, which affect peoples' lives and livelihoods. The driver is not specifically the environment, but its effects; this is another reason that descriptive pose a challenge. Considering the label in isolation can oversimplify these situations and suggest a singular causation for migration/displacement. The majority of research in the field does acknowledge this; Dun and Gemenne [37] began arguing for a better definition of environmental migration a decade ago because it is often difficult to isolate environmental factors from other drivers of migration. People do no migrate simply because there is a drought, but because they cannot grow food anymore. They do not move because of a cyclone, but because the cyclone has destroyed their home. Environmental factors, especially climate impacts, are challenging to differentiate from other drivers as they are often underlying economic and labor choices. Additionally, they may not be clearly seen or understood by those affected by them in the same way that they are viewed by those studying an event after the fact.

The newest conception in forced migration is called "survival migration." Originated by Alexander Betts [38], it is defined as those persons outside their country or origin because of an existential threat to which they have no access to a domestic remedy. This threat has thus far been interpreted as anything from environmental/climate change to state fragility. This definition identifies the deprivation of socioeconomic rights, which may make many of those currently considered economic migrants, survival migrants. An existential threat can be economic if conditions are such that basic survival is not guaranteed and local governance structures can provide no remedy. This may very well be how climate migration/displacement plays out; slow-onset disasters such as rising sea levels and desertification interact with short-term weather events and changing economic pressures thus deteriorating some livelihoods faster than others—movement will then occur unevenly and sporadically over time, which may appear "as if" by economic pull, but in reality is due to survival needs and environmental push.

Climate migration/displacement will likely overlap the various categories of environmental migrants discussed earlier. With no academically agreed on definition or accepted bureaucratic label, any discussion of the general phenomenon of climate displacement in the academic literature cannot help but be a little fuzzy. Labels matter. With clear and widely acknowledged definitions, it is easy to discuss a topic, without them there is a constant need to fully define the subject before it can be debated. Much of the literature on climate migration or climate displacement defines one of these terms, uses others' definitions, or reviews them to clearly outline the concept in each article, book, or book chapter. Keeping the reader and researcher on the same page with each iteration of concept will consistently be a problem until a formal and legally accepted label is otherwise provided.

23.4 Migration governance

23.4.1 Climate "refugees"

In a practical sense, defining a phenomenon provides a basis for governance. Creating policy necessitates clear issue parameters, which will be used to formulate guidelines for action. Migration governance varies from state to state and with little international legal precedent. The plight of those being displaced is serious and those in such conditions are searching for the most similarly applicable international remedy. The global governance structure implicated most often is the United Nations High Commissioner for Refugees (UNHCR). The 1951 *Convention Relating to the Status of Refugees* outlined specific conditions, which would cause one to be a refugee including "race," "religion," "nationality," and "membership of a particular social group or political opinion," which are antecedents of the period between the two World Wars [39]. In 1967, the *Protocol Relating to the Status of Refugees* was confirmed, which formally applies the status of refugee to any person who fits the definition "as if" the date requirement relating to World War II had been omitted (Article 1, section 2). By the 1990s, UNHCR broadened their processing and began accepting *prima facie* group determination of refugee status largely taking the place of individual interviews. The circumstances, which offer protection in this situation, include persecution and insecurity as well as starvation and critical environmental conditions [40]. Specifically, it states, "The term 'refugee' shall also apply to every person who, owing to external aggression, occupation, foreign domination or events seriously disturbing the public order either in part or whole of the country" [41]. Drought, storm surge, or sea-level rise will certainly disrupt the public order; therefore if climate displacement is considered, one can extrapolate an increasing openness over time to assist more people even when it is not clearly mandated.

The 1951 *Convention* and the 1967 *Protocol* outline a specific legal definition, which is internationally recognized. It is also a confirmed obligation, which guarantees a certain set of rights and privileges. Refugees have the right to seek asylum, cannot be returned to their country of origin, and have a protected status; they are also in need care and maintenance, re-establishment, and legal and political protection. [42] However, not every desperate situation falls under the protection of the *Conventions;* this is why it is important to question the use of the label "refugee" on a certain group as they may not have a legally valid claim to it. A two-pronged analysis will demonstrate the difficulty with ascribing a legal refugee status to "climate change refugees."

23.4.1.1 *Convention analysis:* Strict definitional

The *Conventions*' definition of a refugee is structured around the concept of persecution with the only other clearly identified stipulation being that they must cross an international border. "...owing to well founded fear of *persecution* (emphasis mine) for reasons of race, religion, nationality, membership in a particular social group or political opinion is outside the country of his own nationality and is unable or, owing to such fear, is unwilling to avail himself of the protection of that country; or who, not having a nationality and being outside the country of his former habitual residence as a result of such events, is unable to or, owing to such fear, is unwilling to return to it" [43]. Thus, it needs to be established that "climate change refugees" are being persecuted. The general conception is persecution as an individual threat. It can be a threat to one's person because of who he/she is; a threat to one's safety because of an inclusion in a specific group. "Climate change refugees," then, need to be definable group; they are peoples living in the many places where the climate is shifting. This can include coastal communities, forest villages, and/or big cities. They will be people of varied cultures, societies, and economic conditions. Because climate change affects the entire globe, there will be few, if any, areas unaffected. However, the need to flee or become a "refugee" is only apparent in areas most severely affected. Therefore, there may be pockets of displacees; the only commonality among them will be a deterioration of living situations. Thus, the persecution could only be considered impersonal, as human environments are indiscriminately threatened in different ways.

If impersonal persecution is accepted, the question becomes, does an "Act of God" translate into persecution by a nonstate actor? In this case, for what reason would nature persecute? People already uprooted by famine and flood are not included in the UN definition [43]; therefore, the "Act of God" explanation is insufficient to offer any significant international protections. Can a case be made for impersonal persecution by the developed world or a conglomerate of states? While persecution here is equitable to negligence on the part of industrialized nations, the argument is still cloudy. Because many nations contribute to the problem, one could argue that all displacement in this fashion will be political. However, the *Conventions* do not conceive of refugees being created by the unfortunate by-products of industrialized nations. Carbon emissions span the globe and it is unlikely that any refugee can be sure *whose* carbon caused their predicament.

23.4.1.2 Convention analysis: Chain reaction

The *Conventions* are not constructed in such a way to offer any direct protections for situations such as a future with severe climate impacts. However, their interpretation is another thing. UNHCR does provide a *Handbook on the Procedures and Criteria for Determining Refugee Status under the 1951 Convention and 1967 Protocol relating to the Status of refugees*. This document outlines definitions of terms used in each of these agreements and how to interpret a person's situation into a status. As previously mentioned, the important thing to decipher is whether "climate change refugees" are being persecuted. The *Handbook* states, "There is no universally accepted definition of 'persecution', and various attempts…have met with little success" [44]. Because there is no universal definition, UNHCR workers on the frontlines have some room to interpret individual situations and can decide if the reasons for the persecution feared is met [45]. However, as in the case mentioned in the introduction, in-nation processors have not been willing to provide status on these grounds and appeals have been struck down.

However, the *Handbook* provides one last chance to include "climate change refugees," it is the concept of cumulative grounds. "In addition, an applicant may have been subjected to various measures not in themselves amounting to persecution...in such situations, the various elements involved may, if taken together, produce an effect on the mind of the applicant that can reasonably justify a well founded fear of persecution on "cumulative grounds.... Needless to say, it is not possible to lay down a general rule as to what cumulative reasons can give rise to a valid claim to refugee status. This will necessarily depend on all the circumstances, including geographical, historical, and ethnological context" [46]. Cumulative grounds can include the ways in which climate change will affect the lives of many; its complications, chain reactions, and refugee-causing catastrophes. Recent academic work is beginning to discuss these linkages. There are currently at least 40 case studies in which environmental resource scarcity has been cited as a contributing factor leading to violent conflict; environmental scarcity acts as an indirect cause of conflict by amplifying or triggering traditional causes of conflict [47]. Global climate change will impact a region's ability to produce agricultural goods, expose more people to floods and drought, and threaten the integrity of certain island chains. Through this chain reaction analysis, we can see how climate events can/will trigger many types of societal responses.

Climate Event → Destruction of Livelihood → Natural/Community Resource shortages → Conflict → Refugeehood

These event chains are long term in nature. Many of them will happen slowly and ultimately redistribute natural resources. Arable land and current sources of drinking water will have new geopolitical owners. This can easily create struggles for power and incite violence. Martin [47] argues there is a growing concern that scarcity-induced insecurities can contribute to the amplification of the perceived significance of ethnic differences. The model illustrates a natural progression of events, which can cause people affected by climate change to become legal refugees. However, they will have endured much hardship before that point. Two concrete examples are as follows:

Desertification → Loss of agricultural land → inadequate food supplies → Famine → War between those with and without food access.
Sea-Level rise → Increased storm surge → Salinization of drinking water → disintegration of assets due to increased prices of imported water → out migration → xenophobic back lash in new residence.

These are general events, which can take on complicated processes as they play out. The conceptions are an analytic frame that can be applied to any situation where a climate event is threatening to or has already created a deteriorating state of affairs.

Can "climate change refugees" benefit from the protections of the 1951 *Convention* and 1967 *Protocol*? In a strictly definitional sense, it is highly unlikely. There is no clear way that persecution can accurately describe what is happening to them. There may be a case if cumulative grounds are considered, but this is also very difficult. Because climate changes are slow, it makes it significantly more complicated for those seeking refugee status to explain the entirety of their predicament to a UNHCR processor. They would have to cite decade's worth of circumstances that led to their current need to flee, which is not only cumbersome, but possibly not fully understood by those being affected. Climate change migrants/displacees will appear in different forms depending on when, in the long-term process, they migrate. While many may claim to be "refugees," few will be able to prove they are legitimate victims of persecution.

23.4.2 Climate migration in the United Nations

Migration and displacement have long been concerns of the small island states at the United Nations Framework Convention on Climate Change (UNFCCC) meetings. For over a decade, a framework umbrella called "Loss and Damage" developed as a concept area through which to promote actionable items on this general topic. Loss and Damage is commonly regarded as the irreversible long-term destruction faced by people and states, which will occur due to historical fossil-fuel emissions, which may still not be apparent; it is based on the global estimate of warming beyond which any international mitigation or adaptation efforts can help. As such, many island nations are concerned about the loss of life, culture, and territory even if the UNFCCC negotiates an emissions treaty with strong targets and fully funds adaptation projects. Small island states are fighting against their ultimate loss—the permanent forced displacement of their homes and communities.

Loss and damage has been a priority of the Alliance of Small Island States (AOSIS) since the early 1990s. As a coalition of islands spanning the globe, AOSIS works together on issues of development and especially the shared vulnerabilities to climate change [48]. In 1991, AOSIS began devising an international instrument through which they could access funds immediately after future disasters; this sought to compensate beyond the short-term assistance of the current humanitarian disaster framework in hopes that in a world where losses would accumulate more frequently, building back, and building back better, could be funded with more immediacy. These efforts stalled until the 2007 talks in Bali. The Bali Action Plan (BAP) called for a means to address loss and damage in developing countries as well as an entire section on disaster risk management associated with climate change; however, any mention of compensation or liability for such loss was cause for discomfort for industrialized countries [49]. By 2008, the proposal had developed into a multipronged instrument with provisions for disaster risk management, compensation, and rehabilitation for unavoidable and irreversible damage [50]. Further institutionalization of a formal treaty instrument was again delayed at the COP 15 in Copenhagen due language [49]. "Compensation" and "liability," words that address fault and obligation continued to irk developed nations; their interpretation of the plight of island nations and displacement does matter because their money and floor votes are needed to remedy the situation. However, the island nations understood this and realized their needs could be used as bargaining chips during the negotiations and thus their leverage strategies contrasted. The Kiribati government pursued an international agreement that recognized that climate change contributed to their predicament but also that acknowledges relocation as a part of the international obligation [51]. Alternatively, the United States of Micronesia resisted the inclusion of relocation in an international agreement due to the fear that developed nations may see that as an out—the problem of sea-level rise could be solved by simply relocating populations instead of addressing greenhouse emissions [51].

Meaningful advancement commenced in 2010 at the COP16 meeting in Cancun. Bolivia argued on behalf of AOSIS for an expansion of the negotiating text, demanding stronger action, which would be legally binding. It already considered any of its residents "climate migrants," as the country defines them, and brought to the floor concerns for future [52]. The Cancun Adaptation Framework included approaches to loss and damage that considered impacts including sea-level rise, increasing temperatures, and ocean acidification. [50] Consensus on this niche issue transpired and the AOSIS nations along with their nonparty advocates, negotiated paragraph 14 (f) into the Cancun decision. The subsection invites parties to take nationally specific action nationally to enact: "Measures to enhance understanding, coordination and cooperation with regard to climate change-induced displacement, migration and planned relocation, where appropriate, at national, regional and international levels." [53] While the subsection is not legally binding, it asks individual nations to *acknowledge* the existence of climate change migration and displacement at several levels of governance. It also omits the terms "compensation" and "liability."

This decision launched the *Work Programme on Loss and Damage*, which initiated its work through preliminary workshops and expert meetings [50].

23.4.2.1 From "loss and damage" to the WIM

The decision emerging from the COP18 meeting represents a significant advance on the topic of loss and damage; it called for an advanced understanding of noneconomic loss and damage, patterns of migration and displacement, and recognizing the development of approaches to rehabilitation following climate-related loss and damage. The Doha Gateway also mandated the formation of an institutional mechanism at the next COP [50] and recognized sea-level rise as a threat [54].

Parties arrived in Warsaw to develop the institutional mechanism mandated in Doha. The Warsaw International Mechanism (WIM) was established at the COP19 to formally address loss and damage, including both slow-onset and extreme displacement events. The parties designated an Executive Committee, which would report annually to the COP [55]. The G-77 and China argued that the new international mechanism should be housed as a separate entity under the Convention itself [56], not under the Cancun Adaptation Framework where it was currently situated. This decision, however, initiated a point of controversy [57]. Loss and Damage was originally intended to address long-term irreparable losses and damages *beyond* adaptation; many nations thought loss and damage to be a practical measure of last resort, acknowledging that both mitigation and adaptation would not be enough to save certain vulnerable regions. If the issue could not stand on its own, many nations were concerned that it would be lost. The issue came to a head when the G-77 walked out of the negotiation in the middle of the night; bilateral talks did bring them back and a new compromise created the Loss and Damage Mechanism (or LDM) under the Cancun Adaptation Framework. The decision reinforced the negotiators' concerns beyond mitigation and adaptation, but the proposal did not promise compensation

[50]. It also did not address funding; AOSIS had argued that funding for Loss and Damage should be from a dedicated source and separate from that for adaptation [58].

The initial meeting of the Executive Committee (ExCom) of the WIM was held in Bonn in March 2014, where it adopted a 2-year work plan. Output from its continued work at the September Adaptation Committee consisted of a broader plan with eight specific work points and coordinating timelines, culminating in a 5-year rolling work plan for consideration by COP 22 in 2016; this did include an action item relating to migration and displacement. The COP 20 in Lima confirmed the work plan of the ExCom and outlined its reporting and operational procedures [59]. In February of 2015, the Adaptation Committee met in Geneva to produce the first version of the full negotiating text for that year's COP, which included three different options for the inclusion of Loss and Damage, all fully bracketed [60]. Tensions rose again as the options were conflicting, one even eliminated the WIM [61]. However, internal communications note that while the WIM was likely to survive, no agreement was possible for some nations if "compensation," which reappeared, remained in the text [62]. The process was complicated further in the output document from the June intercessional meeting; it now contained five options with the suggestion that "no new institutional arrangements on loss and damage are required for the purposes of this agreement"—also indicating that the WIM is to be strengthened separately [63].

The fight over institutional placement, "compensation," and the WIM's possible erasure was important in the lead up to COP21 because the participants in the Paris Accord (or second Kyoto Protocol: KP2) sought it to be legally binding; thus, any additional proposals being discussed prior to the meeting would also have a chance of becoming binding. Mitigation targets did monopolize the talks, but a compromise transpired and instead of omission from the text, Loss and Damage as well as the WIM was formalized in Article 8 as a third pillar of the climate regime. Consideration for migration and displacement due to climate change was formally integrated into the governance structure and linked to the institutional architecture of the Paris Agreement [64]. Specifically, the decision suggests that the WIM, *develop recommendations for integrated approaches to avert, minimize and address displacement related to the adverse impacts of climate change* [65].

In the next year, Loss and Damage, as well as its WIM, continued to develop. The ExCom advanced its mandated program by holding a high-level meeting relating to its Action Area 6 (AA6): *migration displacement and human mobility* in Casablanca. Aside from its July meeting, the ExCom submitted a full report of its activities from December 2015 to September of 2016 to the UNFCCC for the COP22. The report outlines the key deliverables of the ExCom and recommendations, organizational and procedural matters, progress toward implementation, and several annexes detailing its 5-year framework, membership, and actions taken under the initial 2-year work plan. [66] Additional updates to the WIM occurred at the COP22 in Marrakesh. The COP approved the outline in Annex I of the report of the 5-year rolling plan and encouraged the parties and relevant institutions to continue to consider the topics associated with loss and damage to strengthen cooperation. The COP also recommended that the next review of the WIM to be held in

2019 in the form of a technical paper to be available for review in June of that year [67].

Any additional progress was notably absent at the intercessional meeting in the spring of 2017 [68]. Similarly, no significant outcomes came from the sixth meeting of the ExCom held in October of the same year. [69] The new 5-year work plan does omit an important factor—money. The ministerial heads of the Least Developed Countries (LDC) group issued a communique regarding their priorities going into the COP23 in Bonn, which emphasized just that—that scaled up financial support for loss and damage is urgently required, including a permanent source of finance and a delivery mechanism [70].

23.5 Conclusion

Without an internationally recognized protectionary status and aid apparatus, it is likely that more individuals will seek refugee status as "climate refugees" even though they will not be entitled to it. The development of the WIM can be a viable alternative only if it functionally provides a mechanism through which help can be accessed for those who need to move due to adverse climate impacts. As it stands, the WIM is an embedded mechanism and has been developed to recommend management strategies on migration and displacement within a specific context nexus. The UNFCCC is a treaty-making body, and thus the WIM, as a nested instrument within such a body, has great potential to address climate-induced displacement with formal and binding consequences. It exists in a formal negotiating space with the potential to gain the consensus of all nations, but this is not an easy task. Any recommendations from the WIM will eventually face a full evaluation when reported back to the negotiating floor as a whole. From the account detailed here, it should be clear that when several larger nations dislike a concept, such as "liability," negotiations can stall for years; consensus is hard to build when fault and funding is on the line.

The UNFCCC, as a workspace for the WIM, also allows for civil society involvement. It is a sector, which seeks to persuade and which has had opportunities to push for strong language regarding safeguards, human rights, and equity. Human rights groups flock to the meetings every year to observe and bring along community members of those most affected to speak in hopes that their voices will swing votes on the main floor. However, even Fiji serving as the latest host of the COP23 did not lead to a greater progress in outcomes toward Loss and Damage and any protections for those being slowly displaced. Until global governance can catch up with the needs of those lives and livelihoods are being eroded, it is likely that individuals will weigh their options and seek to move when/if they still have the resources to do so; in those cases, they will appear to be economic migrants. Those who choose to wait may run the risk of becoming trapped, like the population of Kandholhudhoo hoping for the international community to make good on its promise to address the damages it has caused. They may have to wait a long while.

References

[1] Myers N. Environmental refugees: a growing phenomenon of the 21st century. Philos Trans R Soc B Biol Sci 2002;357(1420):609–13. https://doi.org/10.1098/rstb.2001.0953.

[2] Weiss KR. The making of a climate refugee. Foreign Policy 2015, http://foreignpolicy.com/2015/01/28/the-making-of-a-climate-refugee-kiribati-tarawa-teitiota/.

[3] Adger WN. Social capital, collective action, and adaptation to climate change. Econ Geogr 2003;79(4):397–404.

[4] Adger WN, Dessai S, Goulden M, et al. Are there social limits to adaptation to climate change? Clim Change 2009;93:335. https://doi.org/10.1007/s10584-008-9520-z.

[5] Barnett J, Adger WN. Climate dangers and atoll countries. Clim Change 2003;61:321–37.

[6] Kelman I. Hearing local voices from small island developing states for climate change. Local Environ 2010;15(7):605–19. https://doi.org/10.1080/13549839.2010.498812.

[7] Ferris E, Cernea MM, Petz D. On the front line of climate change and displacement: learning from and with Pacific Island countries. The Brookings-LSE Project on Internal Displacement.

[8] Sovacool BK. Perceptions of climate change risks and resilient island planning in the Maldives. Mitig Adapt Strateg Glob Chang 2012;1381-2386. 17(7):731–52.

[9] Simonelli AC. A good economy in rising seas: Kandholhudhoo, Dhuvafaaru, and the need for a development-based island migration policy. In: Migration, risk management and climate change: Evidence and policy responses. Global migration issues series. Springer; 2016. p. 131–48.

[10] Shenk J. The island president. Samuel Goldwyn Films; 2012.

[11] Simonelli Berringer AC. Climate change and emigration: comparing "sinking islands" and Jamaica. In: Miradas en Movimiento; 2012 Special volume: Naturally Immigrants.

[12] Nurse LA, McLean RF, Agard J, Briguglio LP, Duvat-Magnan V, Pelesikoti N, et al. Small islands. In: Barros VR, Field CB, Dokken DJ, Mastrandrea MD, Mach KJ, Bilir TE, Chatterjee M, Ebi KL, Estrada YO, Genova RC, Girma B, Kissel ES, Levy AN, MacCracken S, Mastrandrea PR, White LL, editors. Climate change 2014: impacts, adaptation, and vulnerability. Part B: regional aspects. Cambridge and New York, NY: Cambridge University Press; 2014. p. 1613–54 Contribution of Working Group II to the Fifth Assessment Report of the Intergovernmental Panel on Climate Change.

[13] Kelly C, Costa K, Edelman S. Safe, strong, and just rebuilding after hurricanes Harvey, Irma, and Maria: a policy road map for congress. Center for American Progress; 2017.

[14] Brinkmann P. How many Puerto Ricans have moved to Florida? State's numbers questioned. Orlando Sentinel 2018. January 8.

[15] Stern NH. The economics of climate change: the stern review part II; 2007. p. 55–143.

[16] Cushing L, Kopas J. A human crisis: climate change and human rights in Latin America. Interamerican Association for Environmental Defense; 2011. December.

[17] Solon P. Why Bolivia stood alone in opposing the Cancun climate agreement. The Guardian 2011. December 21.

[18] Warner K, Ehrhart C, de Sherbinin A, Adamo S, Chai-Onn T. In search of shelter: mapping the effects of climate change on human migration and displacement. CARE International (CI), Columbia University, United Nations High Commissioner for Refugees (UNHCR), United Nations University Institute for Environment and Human Security (UNU-EHS) World Bank, the (WB); 2009.

[19] Bronen R. Climate-induced displacement of Alaska native communities. Alaskan Immigration Justice Project; 2013.
[20] Anthony DW. Migration in archeology: the baby and the bathwater. Am Anthropol 1990;92(4):895–914.
[21] Clark WAV. Human migration. Beverly Hills, CA: Sage Publications; 1986.
[22] Lewis GC. Human migration: A geographical perspective. London: Croom Helm; 1982.
[23] Weildlich W, Haag G. Interregional migration. Dynamic theory and comparative analysis. Berlin: Springer-Verlag; 1988.
[24] Young DA. A. Space-time computer simulation method for human migration. Am Anthropol 2002;104(1):138–58 (SRC—Google Scholar).
[25] Petersen W. International Migration. Annu Rev Sociol 1978;4:533–75.
[26] Hyman G, Gleave D. A reasonable theory of migration. Transactions 1978;3(2):179–201.
[27] Balter M. Human evolutionary genetics. Genes confirm Europeans' blow to Native Americans. Science 2011;334(6061):1335.
[28] McNeill WH. Human migration. Bull Am Acad Arts Sci 1978;31(8):8–17.
[29] O'Brien P. Migration and its risks. Int Migr Rev 1996;30(4):1067–77.
[30] Castles S. Twenty-first-century migration as a challenge to sociology. J Ethn Migr Stud 2007;33(3):351–71.
[31] Belton KA, Morales WQ. In: The multi-faceted debate on human migration. Southeastern Council on Latin American Studies; 2009.
[32] Hathaway JC. Forced migration studies: could we just agree to date? J Refug Stud 2007;20 (3):349–69.
[33] DeWind J. Response to hathaway. J Refug Stud 2007;20(3):381–5.
[34] Adelman H, McGrath S. To date or to marry: that is the question. J Refug Stud 2007;20 (3):376–80.
[35] Cohen R. Response to hathaway. J Refug Stud 2007;20(3):370–6.
[36] Cardy WF. In: Environment and forced migration. Fourth International research and advisory panel conference; Sommerville College, University of Oxford; 1994.
[37] Dun O, Gemenne F. Defining 'Environmental Migration'. Forced Migr Rev 2008;31.
[38] Betts A. Survival migration: a new protection framework. Glob Gov 2010;16:361–82.
[39] Gallagher D. The evolution of the international refugee system. Int Migr Rev 1989;23 (3):579–98.
[40] Hakovirta H. The global refugee problem: and its applications. Int Political Sci Rev 1993;14(1):35–57.
[41] Protocol relating to the Status of Refugees. Geneva: UNHCR; (Article 1, number 2); 1951.
[42] Malin PM. The refugee: a problem for international organization. Int Organ 1947;1 (3):443–59.
[43] Robbins R. The refugee status: challenges and response. Law Contemp Probl 1956;21 (2):311–33.
[44] Handbook on procedures and criteria for determining refugee status under the 1951 convention and 1967 protocol relating to the status of refugees. Geneva: UNHCR; 1967 [Interpretation of terms, Section b, number 51].
[45] Handbook on procedures and criteria for determining refugee status under the 1951 convention and 1967 protocol relating to the status of refugees. Geneva: UNHCR; 1967 [Interpretation of terms, Section g, number 67].
[46] Handbook on procedures and criteria for determining refugee status under the 1951 convention and 1967 protocol relating to the status of refugees. Geneva: UNHCR; 1967 [Interpretation of terms, Section b, number 53].

[47] Martin S. Making the UN work: forced migration and institutional reform. J Refug Stud 2004;17(3):107–50.
[48] http://aosis.org/about/.
[49] Warner K, Zakieldeen SA. Loss and damage due to climate change: an overview of the UNFCCC negotiations. Oxford: European Capacity Building Initiative; 2011.
[50] Burkett M. Loss and damage. Clim Law 2014;4:119–30.
[51] McAdam J. Swimming against the tide: why a climate change displacement treaty is not *the* answer. Int J Refug Law 2011;23(1).
[52] Solon P. Why Bolivia stood alone in opposing the Cancun climate agreement. The Guardian 2010. December 21.
[53] UNFCCC. Decision -/CP.16.
[54] Bothe M. Doha and Warsaw: reflections on climate law and policy. Clim Law 2014;4:5–20.
[55] UNFCCC. Decision -/CP.19.
[56] McNamara KE. Exploring loss and damage at the international climate talks. Int J Disast Risk Sci 2014;5(3).
[57] Simonelli AC. Summary of rights-related developments at COP19. Loss and Damage: Human Rights and Climate Change Working Group; 2013.
[58] AOSIS, editor. Submission of Nauru on behalf of the alliance of small island states: views and information on elements of an international mechanism to address loss and damage from the adverse effects of climate change; 2013 November 12.
[59] UNFCCC. Decision -/CP.20.
[60] When the text it "bracketed" or in brackets, this indicates that the wording is under consideration and has not yet been agreed upon.
[61] FCCC/ADP/2015/1.
[62] Author's personal notes from internal communications at the meeting.
[63] Ad Hoc Working Group on the Durban Platform for Enhances Action. Second session. Part nine. Version 2015 11 June @16:30.
[64] Conway D. Loss and damage in the Paris agreement. Climate Focus, Briefing Note; 2015, December.
[65] UNFCCC. Decision/CP.21.
[66] UNFCCC. SB/2016/3.
[67] UNFCCC. Decision -/CP.22.
[68] Climate Action Network. Way forward on L&D. Eco Newsletter 2017. May 18.
[69] Rattani V. Climate-induced loss and damage continues to be neglected globally. Down to Earth 2017. October.
[70] Addis-Ababa LDC Ministerial Communiqué on Climate Change; 2017. October 9.

Social justice and climate change☆ 24

Alice Venn[1]
University of Bristol Law School, Bristol, United Kingdom
[1]Corresponding author: alice.venn@bristol.ac.uk

Chapter Outline

24.1 Introduction

> *"In the US as elsewhere, the injustice of climate change means the most vulnerable people in society—the poor, the young, the elderly, the unemployed and the marginalised—are most affected by climate impacts such as severe flooding and prolonged drought." [1] Mary Robinson.*

The remarks of the former UN High Commissioner for Human Rights and Special Envoy on Climate Change on the announcement of the withdrawal of the United States from the Paris Agreement in June 2017 poignantly illustrate the social injustice inherent in climate change impacts and, often, climate policy responses themselves. The adverse impacts of climate change, from sea-level rise, to extreme weather events and temperature rises [2] are felt most acutely by those groups in society who already experience marginalization and socioeconomic hardship [3], for whom adaptive capacity is limited. Climate change, in turn, represents a significant stress factor with the potential to compound poverty [4] and to jeopardize human security [5] and health [6]. Both direct and indirect climate impacts, including heightened food and water

☆Funded by the South West Doctoral Training Partnership of the UK Economic and Social Research Council.

Managing Global Warming. https://doi.org/10.1016/B978-0-12-814104-5.00024-7

insecurity [7], and even the increased risk of outbreaks of collective violence [8], are more likely to affect low-income countries and communities. Natural disasters intensified by climate change such as tropical cyclones and hurricanes [9] have been found to present graver risks to people with disabilities [10], women [11], racial minorities and the poorest members of society [12]. A self-perpetuating cycle of climate vulnerability and social injustice is thereby created in which those most affected by socioeconomic hardship and marginalization are most severely impacted by climate change and, in turn, see their adaptive capacity further reduced. Shue labels this "compound injustice" where the existence of an injustice lays the foundation for further injustice [13]. The inherent inequity in the manifestation of climate impacts is evident at both the global level, where Least Developed Countries and Small Island Developing States with limited economic resources face disproportionate climate loss and damage [14], and at the subnational level, where socioeconomic, cultural, and structural inequalities serve to exacerbate the climate vulnerability of particular groups and individuals [15].

The links between socioeconomic inequality, exclusion, and climate vulnerability pose important distributive and procedural justice questions, which the present chapter shall seek to elucidate to enable a critical examination of the international climate regime under the auspices of the United Nations Framework Convention on Climate Change (UNFCCC). The additional intergenerational justice implications of climate change shall be considered, before exploring the relationship between climate justice and the enjoyment of fundamental human rights. The reference to human rights in the preamble of the Paris Agreement, celebrated as the first formal consecration of rights in an international climate change treaty [16], contains a further acknowledgment of the social justice dimensions of climate impacts. The perambulatory clause notably refers expressly to "*the rights of indigenous peoples, local communities, migrants, children, persons with disabilities and people in vulnerable situations*" [17] in recognition of the particular susceptibility of these groups to climate impacts and holds that States Parties should promote these rights alongside gender equality in climate action [18]. The fact that this acknowledgment appears only in one clause of the preamble and not within the legally binding main body of the treaty however is reflective of the ancillary status of human rights as safeguarding guidelines in climate action, as opposed to integrated obligations [19]. Exposing the gaps in the human rights protection available to particularly climate-vulnerable groups represents an important first step in the development of a more inclusive approach to climate justice that is capable of providing redress for the inherent social injustice of climate impacts.

24.2 Conceptualizing climate justice

24.2.1 The distributive injustice of vulnerability, loss and damage

Justice in the field of climate change is primarily framed in terms of global distribution, with reference to the disproportionate environmental burdens being borne by developing countries which are particularly vulnerable to the impacts of climate change, yet are responsible for emitting comparatively low levels of greenhouse gases. Studies have demonstrated an inverse relationship between countries' level of greenhouse gas emissions and their vulnerability to climate change impacts, with many of the highest emitters being among the least vulnerable to adverse climate impacts, and vice versa [20]. This fundamental inequality is strikingly illustrated by Small Island Developing States, which it is estimated are responsible for less than 1% of global greenhouse gas emissions [21], yet face some of the severest threats from climate-induced sea-level rise and extreme weather events [22]. The relationship between exposure to climate change impacts such as temperature rises and socioeconomic vulnerability in developing countries [23] serves to further exacerbate this inequality. In order to explore the way in which distributive climate justice should be conceptualized in the present section, the origins and applicability of the concept of distributive justice will initially be explored, followed by the theoretical foundations of its application to climate change. The analysis will then turn to the way in which distributive justice has been concretely provided for within the international climate regime.

Distributive justice has its origins in Rawlsian concepts of distribution of social and economic inequalities [24], including the "difference principle" requiring that inequalities be distributed to the benefit of the least advantaged [25]. The term "inequalities" is defined broadly as encompassing both wealth and other "social goods" including civil liberties and institutional factors such as participation [26]. Although not originally conceived in either global or environmental terms, distributive justice has since formed one of the core pillars of environmental justice research, which has brought the relationship between environmental hazards and social justice into sharp focus. Originating in the United States as a social movement against the predominant placement of environmentally hazardous waste plants and polluting factories in minority ethnic communities [27], the movement has given rise to a substantial body of research exploring the way in which disproportionate environmental burdens are borne by marginalized, indigenous, and socioeconomically disadvantaged groups at both the subnational and global levels. Schlosberg describes distributive environmental justice as the exposition of how the distribution of environmental risk "mirrors the inequity in socio-economic and cultural status" [28]. This broad definition focusing on the interrelationship between risk and inequity has been further extrapolated to accommodate analyses of climate change as a complex global threat the environmental justice framework was not originally conceived to tackle.

In the context of global climate change, scholars including Shue and Caney have argued for a distributive approach to the conceptualization of climate justice based upon the redistribution of the burdens and costs of climate redress onto those who bear primary causal responsibility and are in the best economic position to bear them. Shue argues that climate justice should be founded upon fairness, with the distribution of burdens determined according to three core principles of equity, namely, the greatest contribution to the problem, the ability to pay, and the need to guarantee an adequate minimum of the resources necessary for a decent life to all [29]. Caney similarly examines climate change from the perspective of the global distribution of benefits and burdens and argues for a "hybrid" approach to the "polluter pays principle" based upon ability to pay with duties ascribed to those most able to pay to support the construction of institutions to discourage noncompliance [30]. While Shue argues for the clear attribution of collective responsibility to states, Caney argues that a more nuanced and individualistic evaluation of the contributions of actors at different levels, including inter alia, corporations, individuals, and international institutions [31]. For the purposes of the present Chapter, a state-level analysis of responsibility for climate change, benefits, and burdens is adopted in line with the international climate regime. This global distributive approach is based upon the primary responsibility of developed, high-emitting states to redress the burdens facing climate-vulnerable developing states, determined according to the emissions data made available through national reporting of the UNFCCC States Parties and in consideration of their historical emissions responsibility [32].

The attribution of responsibility for climate change mitigation within the UN Framework Convention regime mirrors the distributive injustice inherent in the manifestation of climate impacts in requiring developed "Annex I" states parties to take the lead in adopting measures to curb emissions [33]. The legal foundation of this distinction is embodied in the common but differentiated responsibility principle contained in Article 3(1), which makes clear the distinction between the responsibility and capabilities of developed and developing country parties in addressing climate change. This principle continues to find expression in the Paris Agreement [34] despite the development of a more inclusive regime in which developing country parties are participating alongside developed countries in progressively more ambitious action towards the stabilization of emissions in line with the temperature goals [35], albeit in the form of voluntary Nationally Determined Contributions (NDCs) to be reviewed periodically [36].

The principal attribution of responsibility for climate mitigation to developed country parties is accompanied by a series of obligations to provide increasing levels of financial and technological support to climate-vulnerable developing countries both in adapting to climate impacts and in addressing the loss and damage being suffered. Within the UNFCCC regime, there exists a strong focus on strengthening the adaptive capacity of developing and climate-vulnerable States Parties. Primarily, this is provided for in the form of financial assistance from the UNFCCC's financial mechanism, the Green Climate Fund, established by the Cancun Agreements of 2010 [37]. Alongside the Green Climate Fund, the pre-existing Global Environment Facility, together with the Least Developed Countries Fund and Special Climate Change Fund it manages, provide additional support for the development of adaptation programs, technology transfer, capacity building, and development assistance in key sectors such as energy and agriculture in developing and least developed countries [38].

The Paris Agreement reinforced the need for developed country parties to "take the lead in mobilizing climate finance" [39] more ambitiously and to take into account the needs of those developing countries most vulnerable to the impacts of climate change, including the need for public funding in the form of grants [40]. These provisions are accompanied by a number of obligations to provide assistance in addressing the loss and damage suffered by climate-vulnerable States Parties on a "cooperative and facilitative" basis in the form of the enhancement of understanding and support with aspects such as risk assessment and disaster preparedness [41]. The Warsaw International Mechanism for Loss and Damage associated with Climate Change Impacts, since COP21 in 2015, further incorporates a Task Force on Displacement with the aim of gathering data and developing recommendations on the climate-induced displacement of communities [42]. It is worth noting that although the inclusion of considerations such as those of "noneconomic losses" and the "resilience of communities" as requiring enhanced understanding and action [43] evidence an increasing focus upon the social impacts of climate change, the fact that compensation and liability were expressly excluded from the loss and damage provisions in the Paris Agreement [44] is reflective of States Parties' continuing resistance to more concrete distributive measures.

In order to identify which states and communities are most vulnerable to climate change impacts, enabling targeted assistance to be provided to alleviate the disproportionate environmental burdens, climate vulnerability must first be conceptualized. The Intergovernmental Panel on Climate Change (IPCC) has authoritatively defined climate vulnerability, in reliance upon expert evidence from multidisciplinary sources, in terms of the susceptibility of a system to climate impacts. Susceptibility is measured with respect to the sensitivity of the system to climate impacts and its adaptive capacity [45]. Systems themselves are defined broadly as including those which are geophysical, biological, and socioeconomic [46], while adaptive capacity has been defined as the extent to which a system is able to accommodate and absorb the relevant impacts [47].

The IPCC definition, which has informed international climate policy making, has been primarily based upon the exposure of systems to the physical impacts of climate change and on the magnitude of the risk posed. It is important to underline at this juncture that climate vulnerability, although conceptualized at the global level to inform international law- and policy-making, is not uniformly experienced at the subnational level as a result of the social and economic factors influencing the vulnerability of particular groups and individuals [48]. Indeed, the IPCC definition has been the subject of critique for its failure to adequately take into account the underlying social causes and human aspects of vulnerability, such as inequality, which play a significant role in determining the resilience and adaptive capacity of individuals and groups [49].

In the fifth assessment report, the IPCC took note of the ongoing development of vulnerability as a "multidimensional" concept, with particular reference to the relationship between vulnerability and "structural conditions of poverty and inequality" [50]. This is reflective of the increased focus in climate change scholarship upon social vulnerability [51], defined for example in terms of entitlement to resources broadly conceived as including not only material income, but the institutional and social factors impacting upon an individual's access to resources [52]. Climate vulnerability, broadly conceived in this manner, has strong links to procedural climate justice in that it carves out a space for the consideration of the underlying barriers which inhibit both adaptive capacity, and the ability of nations, groups, and individuals to participate effectively in policy making and institutional processes.

24.2.2 Procedural injustice and climate change

Procedural justice is most often defined in terms of equal participation in decision-making processes. It has its roots in theories of deliberative democracy and procedural legitimacy, notably including the work of Habermas exploring the value of "ideal speech" involving all affected parties in decision-making processes legitimized by consensus [53]. This has been complemented by an increasing focus upon gender, ethnicity, and other sociocultural factors, which serve to restrict the effective participation of particular groups in institutional processes, with calls for more nuanced articulations of procedural justice capable of taking account of the impact of engrained inequalities [54]. This involves more than prima facie access to institutions or decision-making processes, it requires the consideration of the consequences of the imbalances of power or social domination, which may lead to the "internal exclusion" [55] of groups, who, although they have access to decision-making processes, do not have an equal voice or share in the power to influence their outcomes.

The environmental justice literature has primarily focused upon participation in environmental decision making and on the barriers to participation faced by marginalized communities [56], alongside more established distributive justice considerations. This definition should be further broadened to encompass the ability to challenge procedural exclusion, the denial of rights, and environmental decision making through legal and administrative means. Access to justice, although only scantly acknowledged as a core component of procedural environmental justice in the literature [57], is of great importance in guaranteeing respect for both the procedural and substantive rights of affected communities. Procedural justice itself is largely overlooked in the prevalent climate justice discourses, which focus upon distributive injustice in the form of the burdens disproportionately borne by climate-vulnerable developing nations [58], and upon intergenerational equity, calling for greater consideration of the impacts of climate change for future generations [59]. Limited examples of the consideration of procedural justice in the context of the UNFCCC regime have focused upon the legitimacy and fairness of the process, measured in terms of the equal participation of all affected States Parties and regions of the world in the negotiations and drafting [60]. Questions surrounding global participation and the legitimacy of the climate regime are of great importance; however, the limited nature of the consideration of procedural issues, both in the climate justice literature and within the regime itself, is too narrowly conceived to effectively cover the multifaceted challenges facing climate-vulnerable states and communities.

There are strong links to be drawn between climate vulnerability and procedural justice, as Adger (2006) observes, "vulnerable people and places are often excluded from decision-making and from access to power and resources" [61]; thus, it is important that policy responses take into account the underlying causes of vulnerability and tackle "marginalization as a cause of social vulnerability" [62]. The concept of marginalization is defined here as entailing exclusion from institutional structures and decision-making processes. The "under-researched" nature of representation in decision making in the vulnerability literature [63] is reflective of the neglect of procedural justice considerations in the UNFCCC regime more broadly.

The failure of the UNFCCC to adequately provide for procedural justice has been effectively critiqued from a normative standpoint in light of the principles of equity, which are stated as forming the basis of the regime and the absence of participation by many of the States Parties in its initial drafting [64]. The failure of the regime to provide procedural safeguards through representatives for future generations who will be impacted by climate change has also been the subject of some interesting critique [65]; however, procedural justice in this context can be seen once again to be narrowly defined as prima facie participation in the process itself. There is evidence of progress in terms of the inclusivity of the process, with climate vulnerable developing nations exerting effective influence with respect to key aspects of the Paris Agreement negotiations, for example by successfully lobbying for the inclusion of the 1.5°C ambition threshold for climate action [66] and with the 23rd Conference of the Parties presided over for the first time by a Small Island Developing State [67]. Associated provisions contained within the Paris and UNFCCC frameworks on, for example, knowledge sharing, public awareness, and encouraging participation generally, including of NGOs [68], do have procedural justice ends in making up-to-date information available to States Parties and the public, thereby facilitating more effective engagement. Other core aspects of procedural environmental justice however remain absent from the international framework, most notably access to justice.

24.2.3 Intragenerational equity

Intergenerational equity represents a widely recognized principle of international environmental law providing for the preservation of natural resources and the environment for the benefit of future generations. It has roots in the 1972 Stockholm Declaration [69] and forms a core tenet of sustainable development frameworks [70]. The UNFCCC embeds intragenerational equity within the international climate change regime as a founding principle. Article 3 frames the concept in terms of the need to "*protect the climate system for the benefit of present and future generations of humankind*" [71], which is reinforced by the inclusion of sustainable development as a further core principle within the UNFCCC framework [72]. The continuing relevance of intergenerational equity as a guiding principle shaping climate action is reaffirmed by the Paris Agreement preamble [73], yet its precise conceptualization and implementation measures, beyond the implicit benefits of climate mitigation for future generations generally, remain unclear. Intergenerational equity, despite being widely referred to in the discourse and instruments of international institutions [74], is often provided for in the form of nonbinding "soft law" or remains undefined and open to interpretation, as in the UNFCCC. The status of the principle before the courts is contested and it is observed by Bell, McGillivray et al. that the inherent difficulty in defining intergenerational equity means that it is very seldom invoked in judicial decisions [75].

In climate justice literature, there is widespread agreement on the importance of taking the needs and rights of future generations into account. Caney for example argues that intergenerational duties represent a core facet of global climate justice due to the temporal dimension of climate change, which means that the adverse impacts of current greenhouse gas emissions will be experienced by future generations of people [76]. The way in which the principle should be defined and applied to climate change however lacks agreement. Brown Weiss has defined intergenerational equity in the form of three key principles of conservation, namely, of options, quality, and access [77]. These three principles, designed to protect natural resource diversity, the quality of the environment, and the ability of future generations to equitably access the benefits therefrom, are translated into the obligations to prevent and mitigate climate change, together with the obligation to provide adaptation assistance [78]. All three of these climate obligations can be found within the international climate change regime, which has established a warming threshold commensurate with the prevention of dangerous interference with the climate system, and ascribes mitigation and adaptation assistance obligations to developed nations.

It is however unclear which of the founding principles underpin each of these obligations or how the principles are being conceptualized. The application of liberal principles of fairness such as Rawls' "just savings" principle, described as "*an understanding between generations to carry their fair share of the burden of realizing and preserving a just society*" [79], whereby the present generation has a duty to save a "fair share" of resources for the next [80], has also been advocated for in the climate change context [81]. Attempts to fill the lacuna between principle and implementation have given rise to proposals for economic solutions in the form of "social discounting," based upon cost–benefit calculations of the value of acting to protect future generations. These contentious proposals and questions surrounding the values assigned to market, technological, and human factors have met with strong criticism on ethical grounds, including from fellow economists [82].

To what extent the interests of future generations should be provided for is itself a highly contested question and presents significant challenges, both ethically and legally. Some of the strongest critiques have argued that protecting future generations from climate impacts through stringent mitigation efforts is at the cost of improving the welfare of the present generation, particularly those living in poverty for whom resources are urgently required [83]. The false dichotomy between development and environmental protection has been particularly pervasive in the field of climate change, where, reinforced by a binary approach to common but differentiated responsibility, it has served to entrench the pre-Paris Agreement impasse on mitigation action between developed and high-emitting developing states [84]. A further common criticism of attempts to implement intergenerational equity and rights is the ability to identify the people to whom a duty is owed or by whom a right is held, known as the "non-identity problem" [85], discussed in greater detail in Section 24.3.3 below. The common factor in intergenerational approaches to climate change is their ambiguity in both the definition and application of the concept. Acknowledging these difficulties, it can pragmatically be argued that the best way to provide protection from adverse climate change impacts for future generations is to increase commitment to mitigation and overcoming social injustice in the present [86].

24.3 Linking climate justice and fundamental rights

24.3.1 Distributive climate justice and human rights

The impacts of anthropogenic climate change upon the enjoyment of a broad range of fundamental human rights are widely acknowledged by UN bodies including the Human Rights Council [87], the Office of the High Commissioner for Human Rights (OHCHR) [88], and the Special Rapporteur on Human Rights and the Environment [89]. The Human Rights Council has poignantly acknowledged that "*climate change poses an immediate and far-reaching threat to people and communities around the world and has adverse implications for the full enjoyment of human rights*" [90]. Climate change, both through direct impacts, including extreme weather events and temperature increases, and indirect impacts, such as food and water insecurity, has adverse consequences for the enjoyment of a broad range of civil, political, economic, social, and cultural rights. Climate change increases rates of mortality, disease and malnutrition, threatening the right to life and the right to health [91], while other impacts have been shown to jeopardize the rights to self-determination and an adequate standard of living, including inter alia, the rights to adequate food and housing [92]. The extent to which these human rights are threatened depends upon both the vulnerability of the individuals to climate change, defined in terms of their exposure and adaptive capacity, and upon the socioeconomic and cultural factors which underpin capacity and susceptibility to harm, including the enjoyment of rights.

In addition to the two core covenants on civil and political rights (ICCPR) and socioeconomic and cultural rights (ICESCR), international human rights law provides specialized instruments, treaty bodies, and complaints mechanisms for the rights of women under the Convention on the Elimination of All Forms of Discrimination Against Women (CEDAW), for people with disabilities under the Convention on the Rights of Persons with Disabilities (CRPD), and for children under the Convention on the Rights of the Child (CRC). All of these groups have independently been found to be particularly vulnerable to the impacts of climate change and disasters [93], while socioeconomic factors and the marginalization they encounter already serve to render them more susceptible to harm and abuse [94]. The enjoyment of fundamental human rights of these particularly susceptible groups is therefore subject to compound threats. Climate-related loss and damage can augment other threats; studies in the Pacific for example have shown that women are far more likely to suffer from violence and abuse in the aftermath of a disaster [95]. Indigenous peoples also experience greater vulnerability to climate change impacts and human rights threats, both as a result of the exposure of the regions they inhabit, and their increased dependency upon natural resources for livelihoods and cultural practices [96].

Established distributive climate justice debates center principally upon the need to address the disproportionate burdens borne by particularly climate-vulnerable states. Distributive injustice is nevertheless evident at multiple scales, subnationally it is found in the inequitable distribution of harms to vulnerable groups. The strong links established between social injustice, marginalization, poverty, and vulnerability to climate change impacts underpin the need to guarantee a minimum level of protection for fundamental rights. This has been framed in the climate justice literature as the need to guarantee an "adequate minimum" of the resources required for a decent life [97] and as securing minimum moral thresholds [98]. These moral arguments pair naturally with existing human rights as legally consecrated mechanisms for the enforcement of such standards, leading some scholars to invoke the rights to life, health, subsistence, and physical security [99] as the basis of duties to alleviate disproportionate climate burdens. The increasing focus upon the human rights impacts of climate change in research and policy responses however has not led to the integration of rights into the international mechanisms designed to provide redress for climate burdens, including notably, the Warsaw Mechanism on Loss and Damage [100].

24.3.2 Developing links between procedural justice, rights, and climate change

The relationship between climate vulnerability, social marginalization, and structural exclusion, discussed in Section 24.2.2, has significant implications for the enjoyment of fundamental rights. Factors such as discrimination on the grounds of gender and disability have been found to reinforce climate vulnerability [101] while, in turn, climate change is likely to exacerbate the socioeconomic factors already restricting the access of climate-vulnerable groups to decision-making processes and legal redress. Indigenous communities who are particularly vulnerable to climate change impacts for example often face marginalization and exclusion from decision-making processes [102]. The mutually reinforcing factors of procedural exclusion and climate vulnerability call for the firm integration of procedural rights into the climate framework with provision made for corresponding implementation programs at the national level. International human rights law contains a number of provisions safeguarding democratic participation [103], the pursuit of information [104], and rights to an effective remedy [105] capable of forming the foundations for procedural climate justice. The International Covenant on Civil and Political Rights (ICCPR) further provides in Article 2 that the rights within it shall be guaranteed without distinction on any grounds including, inter alia, "*race, colour, sex... national or social origin, property, birth or other status*" [106] comprising a prohibition of the type of discrimination found to increase vulnerability to climate change. Similar nondiscrimination provisions are established in respect of socioeconomic rights in ICESCR [107], the rights of women in CEDAW [108], and the rights of persons with disabilities in CRPD [109].

The importance of procedural justice guarantees in developing an effective response to climate change capable of providing concrete protections to those communities and individuals most vulnerable to its impacts should not be underestimated. The procedural justice safeguards in the UNFCCC regime at present provide for the rights of States Parties to vote [110] and for bodies and agencies to participate as observers in the negotiations [111]. Article 6 of the UNFCCC provides that States Parties should "promote and facilitate" public access to information and participation in a general provision on education and awareness. The climate regime additionally contains knowledge sharing and enhanced understanding provisions [112], which serve to provide for the availability of essential climate information for adaptation and mitigation action. The Paris Agreement with respect to adaptation measures provides that the actions taken by States Parties should be "gender-responsive, participatory, and fully transparent" taking into account vulnerable groups and indigenous knowledge [113]. These provisions provide for prima facie participation and access to information at the international level; however, concrete measures to uncover the underlying barriers to participation and access to justice encountered by the most climate-vulnerable groups and the requisite strategic support for this are absent. Increasingly, human rights bodies and experts at the international level are recognizing the importance of the integration of procedural guarantees for the benefit of securing increased human rights protection [114], yet this has not led to the development of an effective rights-based response by the UNFCCC.

24.3.3 Rights of future generations?

The status of intergenerational equity as a longstanding principle within the UNFCCC [115] and a much-debated facet of climate justice has not yet led to the establishment of any legal rights for future generations, either within or outside of the international climate regime. Developing precise rights or duties for the benefit of future generations involves first overcoming the challenges to their establishment, the most commonly cited of which is the so-called nonidentity problem. The nonidentity problem, articulated by Parfit [116], is premised upon the birth of future people being determined by present actions, meaning that changes in behavior now will necessarily affect the conception and birth of particular individuals. The core assumption is that if it is not possible to determine which specific individuals will be born, it is not possible to assign rights or owe duties to them. It is correspondingly considered to be impossible to harm specific future individuals by way of current actions, unless the argument can be made that they would be better off not existing at all [117]. These conceptual challenges to the establishment of rights for future generations are reinforced by the legal challenges associated with the attribution of responsibility for harm and the enforceability of rights.

The consecration of fundamental human rights in law has important ramifications in terms of enforceability for the individuals affected. The individualistic nature of current human rights laws however inherently restricts their intergenerational and transboundary scope as individuals must be able to prove the existence of an immediate threat or demonstrable harm in respect of the enjoyment of their rights, and to establish a causal link to the acts or omissions of the state, primarily with territorial jurisdiction [118]. In the absence of either the establishment of new rights, such as a right to an equitable share of ecological space [119] or a right to subsistence emissions [120], or the development of more liberal legal doctrines of "extrajurisdictional duties" [121] and legal standing for future generations [122] for the purposes of exercising legal rights, concrete enforceability will remain limited.

A number of scholars have nevertheless presented convincing ethical arguments for the allocation of legal rights to protect future generations from climate change

[123]. The most compelling of these arguments rests upon the foreseeability of harm arising in the future as a result of the failure of the present generation to mitigate the effects of anthropogenic climate change. This shifts the focus away from the identification of specific individuals, to the universality of the fundamental rights owed to future people, whomever they may be, as human beings [124]. Dietzel has convincingly argued, based upon a cosmopolitan assumption of the equal moral worth of individuals, that as the time at which a person is born is determined by chance, it should have no moral bearing upon the entitlements of the individual, so the rights of future people should be considered of equal worth to those of the present [125]. This underlying moral assumption ties in closely both with the overarching aims of the international human rights framework to provide for the "*inherent dignity and [...] the equal and inalienable rights of all members of the human family*" [126], and, indeed, of the UNFCCC to protect the climate system for the benefit of present and future generations [127].

24.4 Conclusion

This chapter has outlined the need to move beyond purely distributive conceptions of climate justice in order to effectively accommodate the complex underlying social injustices, which serve to render particular groups more vulnerable to adverse climate change impacts. Renewed efforts to integrate procedural justice into climate law- and policy-making are called for, with a strong focus upon barriers to participation, access to information, and access to justice at the subnational level, where climate-vulnerable groups will suffer most acutely from the impacts of compound socioeconomic and climate injustice. Fundamental human rights, as minimum moral standards for the protection of human life, dignity, and equality, should form the bedrock of the climate justice framework informing future action. In light of the theoretical and legal challenges associated with the granting of rights to future generations, and given the need for substantial legal reform, it is argued that the interests of future generations are best served by establishing a strong climate justice framework to guide action in the present.

Substantive rights to life, health, and an adequate standard of living can form the basis of measures to address the disproportionate loss and damage being suffered by climate-vulnerable groups, providing a minimum level of protection beneath which the international community's response, and states' national measures as the principal actors within it, should not be allowed to fall. Procedural justice safeguards similarly find reinforcement in human rights provisions for legal remedies, democratic participation, and nondiscrimination in the exercise of fundamental rights, which can offer guidance in the implementation of measures designed to tackle the underlying social justice barriers to essential procedural engagement, particularly access to justice as a neglected component of international climate responses. The reference made to the rights of indigenous communities, children, persons with disabilities, and other climate-vulnerable groups in the Paris Agreement preamble represents a tentative first step toward the integration of human rights and social injustice into the climate framework. More focused efforts are however required to meaningfully address the underlying socioeconomic injustices, and the procedural barriers which serve to reinforce them at the subnational level, if climate justice is truly to be attained.

References

[1] Statement by Mary Robinson on the Announcement by President Trump to Exit the Paris Agreement (1 June 2017) The Mary Robinson Foundation—Climate Justice. Available at: https://www.mrfcj.org/resources/us-withdrawal-from-paris-climate-agreement-is-an-injustice/.

[2] Intergovernmental Panel on Climate Change (IPCC), Contribution of Working Group I to the Fifth Assessment Report Climate Change 2013: The Physical Science Basis (2013), Summary for Policymakers.

[3] IPCC Working Group II Contribution to the Fifth Assessment Report, Climate Change 2014: Impacts, Adaptation and Vulnerability (2014), Summary for Policymakers, at 6.

[4] Hallegatte S, Bangalore M, Bonzanigo L, Fay M, Kane T, Narloch U, et al. Shock waves: Managing the impacts of climate change on poverty. The World Bank: Washington; 2016.

[5] IPCC AR5 WGII Summary, supra note 3, at 20.

[6] World Health Organization. Protecting Health from Climate Change: Connecting Science, Policy and People. Geneva: WHO Press; 2009.

[7] IPCC AR5 WGII Summary, supra note 3, at 6.

[8] Levy BS, Sidel VW. Collective violence caused by climate change and how it threatens health and human rights. Health Hum Rights 2014;16(1):32–40.

[9] IPCC Fourth Assessment Report: Climate Change 2007, Working Group I: The Physical Science Basis (2007), 10.3.6.3 Tropical Cyclones (Hurricanes).

[10] The Global Partnership for Disability & Development and The World Bank, The Impact of Climate Change on People with Disabilities (8 July 2008). Available at: https://www.ucl.ac.uk/lc-ccr/centrepublications/staffpublications/Impact_of_Climate_Change_on_Disability-Report-2010.pdf.

[11] Alston M, Whittenbury K, editors. Research, action and policy: addressing the gendered impacts of climate change. Dordrecht: Springer; 2013 at 9–10; and UN Women Fiji, *Why is Climate Change a Gender Issue?* (2014), Available at: http://asiapacific.unwomen.org/en/digital-library/publications/2015/1/why-is-climate-change-a-gender-issue.

[12] See for example research conducted on Hurricane Katrina: Byrnes WM. Climate Justice, Hurricane Katrina, and African American Environmentalism, J Afr Am Stud 2014;18 (3):305–14; and Katrina, Climate Change and the Poor, Editorial, CMAJ, 2005; 173: 8, available at: http://www.cmaj.ca/content/173/8/837.

[13] Shue H. Climate Justice: Vulnerability and Protection. Oxford: Oxford University Press; 2014 at 4.

[14] Tol RSJ, Downing TE, Kruik OJ, Smith JB. Distributional aspects of climate change impacts, Glob Environ Chang 2004;14:259–72. and United Nations Office of the High Representative for the Least Developed Countries, Landlocked Developing Countries and Small Island Developing States (UN-OHRLLS), Small Island Developing States in Numbers: Climate Change Edition 2015 (2015), Available at: https://sustainabledevelopment.un.org/content/documents/2189SIDS-IN-NUMBERS-CLIMATE-CHANGE-EDITION_2015.pdf.

[15] IPCC Working Group II Contribution to the Fifth Assessment Report, Summary for Policymakers, supra note 3, at 6; andAdger WN, Kelly M. Social vulnerability to climate change and the architecture of entitlements. Mitig Adapt Strateg Glob Chang 1999;4:253–66.
[16] Mayer B. Human rights in the Paris agreement. Clim Law 2016;6:109–17.
[17] Paris Agreement (signed 12 December 2015) entered into force 4 November 2016, C.N.92.2016.TREATIES-XXVII.7.d of 17 March 2016, Preamble.
[18] Ibid, Paris Agreement Preamble.
[19] Venn A. Universal human rights? Breaking the institutional barriers facing climate-vulnerable small-island developing states. Clim Law 2017;7(4):322–46 at 323.
[20] Althor G, Watson JEM, Fuller RA. Global mismatch between greenhouse gas emissions and the burden of climate change. Sci Rep 2016;6(1)20281.
[21] UN-OHRLLS 2015 Report, supra note 14, at 18.
[22] Ibid, UN-OHRLLS 2015 Report.
[23] Miranda ML, Hastings DA, Aldy JE, Schlesinger WH. The environmental justice dimensions of climate change. Environ Just 2011;4(1):17–25 and Tol et al., supra note 14.
[24] Rawls J. A theory of justice. Cambridge Massachusetts: The Belknap Press of Harvard University Press; 1971.
[25] Daniels N. Reading Rawls: Critical studies on Rawls' a theory of justice. Stanford: Stanford University Press; 1989.
[26] Ibid, Daniels (1989), at 192–193.
[27] Bell K. Achieving environmental justice: a cross-national analysis. Bristol: Policy Press; 2014, at 15.
[28] Schlosberg D. Reconceiving environmental justice: global movements and political theories. Environ Politics 2004;13(4):517–40 at 522.
[29] Shue H. (2014), supra note 13, at 180–194.
[30] Caney S. Cosmopolitan justice, responsibility, and global climate change. In: Gardiner S, Caney S, Jamieson D, Shue H, editors. Climate ethics: essential readings. Oxford: Oxford University Press; 2010. p. 122–45 at 136–137.
[31] Ibid. Caney (2010), at 126–127.
[32] The United Nations Framework Convention on Climate Change [1992] 1771 UNTS 107, Preamble, contains the following clause: '*Noting that the largest share of historical and current global emissions of greenhouse gases has originated in developed countries*'.
[33] Ibid. UNFCCC [1992], Article 4(2).
[34] Paris Agreement (2015), supra note 17, Article 2(2).
[35] Ibid. Paris Agreement, Article 3.
[36] Ibid. Paris Agreement, Articles 4 and 14.
[37] UN Framework Convention on Climate Change, Decision 1/CP.16, The Cancun Agreements: Outcome of the work of the Ad Hoc Working Group on Long-term Cooperative Action under the Convention (15 March 2011), FCCC/CP/2010/7/Add.1.
[38] UN Framework Convention on Climate Change, Decision 7/CP.7, Funding under the Convention (10 November 2001) 8th Plenary Meeting, FCCC/CP/2001/13/Add.1.
[39] Paris Agreement, supra note 17, Article 9.
[40] Ibid. Paris Agreement Article 9.
[41] Ibid. Paris Agreement, Article 8.
[42] UNFCCC, Taskforce on Displacement at a Glance (2017) Available at: http://unfccc.int/files/adaptation/groups_committees/loss_and_damage_executive_committee/application/pdf/tfd_brochure_nov_2017.pdf.
[43] Paris Agreement, supra note 17, Article 8.

[44] UNFCCC Conference of the Parties, Twenty-First Session, Adoption of the Paris Agreement, Decision 1/CP.21, FCCC/CP/2015/10/Add.1 (29 January 2016), Paragraph 51.
[45] IPCC Contribution of Working Group II to the Fifth Assessment Report (AR5) Chapter 1 Point of Departure, Available at: http://www.ipcc.ch/pdf/assessment-report/ar5/wg2/WGIIAR5-Chap1_FINAL.pdf, at 179.
[46] IPCC Fourth Assessment Report, Climate Change 2007, Working Group II: Impacts, Adaptation and Vulnerability, 19.1.2 Conceptual framework for the identification and assessment of key vulnerabilities, Available at: https://www.ipcc.ch/publications_and_data/ar4/wg2/en/ch19s19-1-2.html.
[47] Adger WN. Vulnerability. Glob Environ Chang 2006;16:268–81 at 270.
[48] Tol et al. supra note 14.
[49] Adger (2006), supra note 47, at 274.
[50] IPCC AR5 WGII, supra note 3, Chapter 1, at 179.
[51] See for example Adger WN, Kelly M. Social vulnerability to climate change and the architecture of entitlements. Mitig Adapt Strateg Glob Chang 1999;4:253–66; Otto IM, Reckien D, Reyer CPO, Marcus R, Masson V, Jones L, Norton A, Serdeczny O. Social vulnerability to climate change: a review of concepts and evidence. Reg Environ Change 2017;17:1651–62.
[52] Ibid. Adger and Kelly (1999), at 256–257.
[53] Habermas J. The Theory of Communicative Action, Volume 1, Reason and the Rationalization of Society, translated by Thomas McCarthy. Cambridge: Polity Press; 1984.
[54] Young IM. Justice and the politics of difference. Princeton: Princeton University Press; 1990.
[55] Young IM. Inclusion and democracy. Oxford: Oxford University Press; 2000 at 55.
[56] See for example Schlosberg D. Reconceiving environmental justice: global movements and political theories. Environmental Politics 2004;13(4):517–40.
[57] See for example Bell K. (2014) supra note 27.
[58] See for example Shue H, Caney S, discussed above.
[59] See for example Lawrence P. Justice for future generations: climate change and international law. Cheltenham: Edward Elgar; 2014 and Page EA, Climate change, justice and future generations. Cheltenham: Edward Elgar; 2006 and Gardiner SM, A perfect moral storm: climate change, intergenerational ethics and the problem of corruption. In Gardiner S, Caney S, Jamieson D, Shue H, editors. Climate ethics: essential readings, Oxford: Oxford University Press; 2010.
[60] Ibid. Lawrence (2014), at 16-17; and Vanderheiden S, Atmospheric justice: a political theory of climate change. Oxford: Oxford University Press; 2008, at 57–60.
[61] Adger WN. (2006), supra note 47, at 276.
[62] Ibid. Adger (2006), at 276.
[63] Ibid. Adger (2006), at 277.
[64] Vanderheiden (2008), supra note 60, at 58–59.
[65] Lawrence (2014), supra note 59, at 17 and 188–190.
[66] See for example Ourbak T, Magnan AK. The Paris Agreement and climate change negotiations: Small Islands, big players, Reg Environ Chang 2017; available at: https://doi.org/10.1007/s10113-017-1247-9.
[67] COP 23 Fiji UN Climate Change Conference, Bonn, 2017, webpage, available at: https://cop23.com.fj/fiji-and-the-pacific/.
[68] See for example Article 4 of the UN Framework Convention [1992], Articles 7, 8 and 10 of the Paris Agreement, supra notes 17 and 32.

[69] Declaration of the United Nations Conference on the Human Environment (Stockholm Declaration) [1972], Principle 1.
[70] See notably, Report of the United Nations Conference on Environment and Development, Rio Declaration on Environment and Development [1992], Principle 3.
[71] UNFCCC [1992], supra note 32, Article 3.
[72] Ibid. UNFCCC, Article 3(4).
[73] Paris Agreement [2015], supra note 17, Preamble.
[74] See also United Nations Educational, Scientific and Cultural Organization (UNESCO), Declaration on the Responsibilities of the Present Generations Towards Future Generations (1997). Available at: http://portal.unesco.org/en/ev.php-URL_ID=13178&URL_DO=DO_TOPIC&URL_SECTION=201.html.
[75] Bell S, McGillivray D, Pedersen OW, Lees E, Stokes E. Environmental law, Ninth Edition. Oxford: Oxford University Press; 2017, at 64–65.
[76] Caney (2010) supra note 30, at 123-124.
[77] Brown Weiss E. Climate change, intergenerational equity and international law: an introductory note. Clim Chang 1989;15:327–35 at 330.
[78] Ibid. Brown Weiss, at 330.
[79] Rawls (1971) supra note 24, at 289.
[80] Vanderheiden (2008), supra note 60, at 114–121.
[81] McKinnon C, Climate change and future justice: Precaution, compensation, and triage. Abingdon: Routledge; 2012, at 31–46.
[82] Stern N. The economics of climate change: the stern review. Cambridge: Cambridge University Press; 2007 Chapter 2 and Hayward T, Climate change and ethics, Nature Climate Change, 2012; 2(12): 843–848, at 845–846.
[83] Page EA. Climate change, justice and future generations. Cheltenham: Edward Elgar Publishing; 2007, at 43–44.
[84] See for example the US position described inUnderdal A, Hovi J, Kallbekken S, Skodvin T. Can conditional commitments break the climate change negotiations deadlock? Int Polit Sci Rev 2012;33(4):475–93 at 485 and Okereke C, Schroeder H. How can justice, development and climate change mitigation be reconciled for developing countries in a post-Kyoto settlement?, Clim Develop 2009;1:10–15.
[85] Parfit D. Reasons and Persons, 1986, Oxford Scholarship Online Monographs, Part four, 16; and Vanderheiden (2008), supra note 60, at 121–125.
[86] Hayward (2012), supra note 82, at 846.
[87] Resolution 18/22, Human Rights and Climate Change (17 October 2011), UN Human Rights Council Eighteenth Session, A/HRC/RES/18/22.
[88] Office of the UN High Commissioner for Human Rights (OHCHR), Report of the Office of the United Nations High Commissioner for Human Rights on the relationship between climate change and human rights (15 January 2009) Human Rights Council Tenth Session, A/HRC/10/61.
[89] Report of the Special Rapporteur on the issue of human rights obligations relating to the enjoyment of a safe, clean, healthy and sustainable environment, Human Rights Council Thirty-first session (1 February 2016) A/HRC/31/52.
[90] HRC Resolution 18/22, supra note 87, at 1.
[91] OHCHR (2009), supra note 88, at 8–13.
[92] OHCHR (2009), supra note 88, at 8–15; and Wewerinke M, Doebbler C. Exploring the legal basis of a human rights approach to climate change, Chin J Int Law, 2011; 10(1):141–160, at 147–156.

[93] See for example GPDD and World Bank, The Impact of Climate Change on People with Disabilities, supra note 10, UNICEF 'Children's Vulnerability to Climate Change and Disaster Impacts in East Asia and the Pacific' (2011) UNICEF East Asia and Pacific regional office; and Levy BS, Patz JA. Climate change, human rights, and social justice. Ann Glob Health 2015;81(3):310–22.
[94] Nifosi-Sutton I. The protection of vulnerable groups under international human rights law. Abingdon: Routledge; 2017 at 1.1.
[95] UN Women Fiji. Climate change, disasters and gender-based violence in the Pacific. UN Women Fiji: Suva; 2014.
[96] International Labour Office (ILO), Indigenous peoples and climate change: from victims to change agents through decent work (2017) Geneva: ILO, at 7.
[97] Shue (2014) supra note 13, at 189–194.
[98] Caney S. Climate change, human rights and moral thresholds. In: Gardiner S, Caney S, Jamieson D, Shue H, editors. Climate ethics: Essential readings. Oxford: Oxford University Press; 2010. p. 163–77.
[99] Bell D. Does anthropogenic climate change violate human rights? Crit Rev Int Soc Pol Phil 2011;14(2):99–124 and Caney, ibid. at 166.
[100] UN Framework Convention on Climate Change, Warsaw international mechanism for loss and damage associated with climate change impacts, Decision 2/CP.19 (31 January 2014) FCCC/CP/2013/10/Add.1.
[101] IPCC WGII AR5 Summary, supra note 3, at 6.
[102] ILO (2017), supra note 96, at 7.
[103] See for example International Covenant on Civil and Political Rights (ICCPR) [1966] UN treaty series, Vol. 999, 171, Article 25.
[104] See for example ICCPR [1966], ibid. Article 19(2).
[105] See for example ICCPR [1966], ibid. Article 2(3).
[106] ICCPR [1966], ibid. Article 2(1).
[107] International Covenant on Economic, Social and Cultural Rights (ICESCR) [1966] UN treaty series, Vol. 993, 3, Article 2(2).
[108] Convention on the Elimination of All Forms of Discrimination Against Women. UN Treaty Ser 1979;1249:13.
[109] Convention on the Rights of Persons with Disabilities (CRPD). UN Treaty Ser 2006;2515:3.
[110] UNFCCC [1992], supra note 32, Article 18.
[111] Paris Agreement [2015], supra note 17, Article 16(8).
[112] See for example UNFCCC [1992], supra note 32, Article 4(g), Paris Agreement [2015], supra note 17, Articles 7, 8.
[113] Ibid. Paris Agreement [2015], Article 7(5).
[114] See Report of the Special Rapporteur on the issue of human rights obligations relating to the enjoyment of a safe, clean, healthy and sustainable environment, Human Rights Council Thirty-first session (1 February 2016) A/HRC/31/52 and Office of the High Commissioner for human rights (OHCHR), *Understanding Human Rights and Climate Change*, Submission of the Office of the High Commissioner for Human Rights to the 21st Conference of the Parties to the United Nations Framework Convention on Climate Change (27 November 2015), Available at: http://www.ohchr.org/Documents/Issues/ClimateChange/COP21.pdf, at 3.
[115] UNFCCC [1992] supra note 32, Article 3.
[116] Parfit (1986), supra note 85.

[117] Hayward (2012), supra note 82, at 845.
[118] See for example ICCPR Article 2(1), supra note 103; European Convention for the Protection of Human Rights and Fundamental Freedoms [1950] ETS 5, article 1; and Milanovic M. Extraterritorial application of human rights Treaties: Law, principles, and policy (2011). Oxford: Oxford University Press.
[119] Hayward T. Human rights versus emission rights: climate justice and the equitable distribution of ecological space. Ethics Int Affairs 2007;21(4):431–50.
[120] Shue H. Subsistence emissions and luxury emissions. Law Policy 1993;15:39–59 and Shue H. Global environment and international inequality. In Gardiner S, Caney S, Jamieson D, Shue H, editors. Climate ethics: essential readings, Oxford: Oxford University Press; 2010. p. 101–111.
[121] Knox JH. Climate change and human rights law. Virginia J Int Law 2009;50:1 at 40.
[122] Johnston R. Lacking rights and justice in a burning world: the case for granting standing to future generations in climate litigation. Tilburg Law Rev 2016;21:31–51.
[123] Bell (2011), supra note 99, 99–124.
[124] Nelson G. Future generations and climate change. Int Soc Sci J 2013;64:211–2 89–97, at 91.
[125] Dietzel A. The Paris agreement—protecting the human right to health? Global Pol 2017;8(3):313–21 at 315–316.
[126] Universal Declaration of Human Rights. General assembly resolution. Preamble: A (III); 1948217.
[127] UNFCCC [1992], supra note 32, Article 3.

The economics of geoengineering 25

Anthony Harding[1], *Juan B. Moreno-Cruz*
School of Economics, Georgia Institute of Technology, Atlanta, GA, United States
[1]Corresponding author: tony.harding@gatech.edu

Chapter Outline

25.1 Introduction

In the climate change debate, mitigation of emissions is historically the sole method researchers and policymakers have seriously considered as a strategy to constrain anthropogenic climate change. This is primarily because the contribution of emissions to stocks of greenhouse gases (GHGs) in the atmosphere is the best-known driver of climate change, and because it was the only method considered feasible. Mitigation efforts, however, have not worked to the extent needed to control anthropogenic climate change.

Since mitigation is a global public good, it embodies the classic free-rider problem. There is little incentive for countries to cut their own emissions because the costs are high and are borne internally while the majority of the benefits are external and captured globally. Thus, countries have repeatedly failed to find ways to cooperate; the Kyoto Protocol or even the more recent Paris Climate Accords have either been unsuccessful or not ambitious enough to seriously control growth in emissions. Recent climate trajectories suggest the likelihood of limiting climate change to below the concerted goal of 1.5°C, or even 2°C, global temperature increase above preindustrial levels is quickly diminishing [1]. Even if countries were to agree on stringent mitigation efforts, the amount of warming committed is already too large and likely to cause large disruption in the economy and the environment, even if GHG

Managing Global Warming. https://doi.org/10.1016/B978-0-12-814104-5.00025-9

emissions were to drop to zero. The lack of cooperation and the committed risks have motivated researchers to turn to alternative approaches for restraining anthropogenic climate change or at least the anticipated damages from rising temperatures.

One alternative to mitigation that has gathered momentum and attention in the research and policymaking realms over the past decade is geoengineering. Geoengineering is defined as the large-scale man-made manipulation of the global climate. Strategies for geoengineering can be placed into two broad categories: carbon dioxide removal (CDR) and solar radiation management (SRM). Carbon dioxide removal techniques involve the removal and storage of CO_2 either during the emissions process or after they have entered the atmosphere. Solar radiation management techniques reflect incoming solar radiation to reduce global temperatures. From a technical standpoint, these two categories of geoengineering widely differ in their methods for controlling the sources and impacts of climate change as well as in their economic properties and timescales for broad impact. Thus, the two categories play different roles in their potential to disrupt the current climate change debate and are analyzed separately.

Over the past decade, economists have primarily focused their research on SRM. This is because of research exhibiting the positively contrasting properties of SRM over mitigation such as lower implementation costs and impacts on the time scale of months rather than decades [2,3]. However, SRM has important drawbacks and limitations. It is unable to counteract all aspects of climate change and has the potential to introduce additional problems that could require further climate and Earth system manipulations to correct. SRM has thus been identified as a quick, cheap, but imperfect alternative climate change strategy [4,5]. Intrigued by these characteristics, economists are keen to determine if SRM's favorable quick and cheap properties can outweigh its imperfections [6]. Economists have focused on the implications of SRM for optimal climate policy as well as in international climate negotiations.

Contrasting with quick, cheap, but imperfect SRM, CDR strategies have the opposite characteristics. They are slow, expensive, but perfect for counteracting impacts of climate change. By directly reducing the stocks of GHG and slowing the accumulation in the atmosphere, CDR strategies perfectly counteract emissions. However, removal and storage of GHGs can be costly and works on the same timescale as the global climate, taking up to decades to negate or reverse climate damages. Since the properties of CDR are more akin to mitigation in both costs and effects, which has been heavily studied, less research has focused on CDR. However, CDR does have the unique property of being able to generate net negative emissions, which differentiates it from mitigation. As negative emissions technologies (NETs), CDR strategies have become an integral part of most climate projections that maintain the goal of limiting warming to below 2°C.

While geoengineering has existed for decades as a theoretical concept, a call for further research by Nobel Laureate Paul Crutzen has spurred analyses of geoengineering strategies over the past decade to test the feasibility of geoengineering as a serious alternative in the near future [2,7]. These initial studies propose different implementation strategies, examining the technological requirements, costs of implementation, and impacts and possible risks of large-scale deployment.

25.2 Feasibility

High costs are a primary driver of mitigation's free-riding problem and an obvious barrier to large-scale deployment of an alternative climate change strategy; so, the first test of an alternative is typically its financial feasibility. For geoengineering, researchers proposed a variety of implementation strategies to evaluate and compare costs.

25.2.1 *Solar radiation management*

The cheapest SRM technique is stratospheric aerosol albedo modification. For this technique, sulfate particles are released into the stratosphere to reduce radiative forcing. Studies concur that this can be extremely effective for a fraction of the cost of mitigation [8]. Proposed implementation methods include the repurposing of existing technologies such as naval rifles, airplanes, balloons, or the construction of large towers. Costs of implementation for these different technologies are in the range of $\$0.003–1\,t(CO_2)^{-1}$ [7,9,10]. Recent studies updating these cost analyses estimate yearly costs on the order of $\$10 \times 10^9\ a^{-1}$ where a refers to annum ($10 billion per year) with costs varying depending on the technologies used and preferred particle levels [8,11].

Another SRM technique is marine cloud brightening, where small particles are added to clouds over oceans to increase their albedo. Studies of marine cloud brightening find that the technique can counteract temperature changes induced by up to 103 $Gt(CO_2)\ a^{-1}$, at an implementation cost of around $\$0.03–1\,t(CO_2)^{-1}$ [7]. Further research has reinforced that marine cloud brightening can have a moderate effectiveness against climate change, again for a fraction of the cost of mitigation [3,8]. This has placed marine cloud brightening closely behind stratospheric aerosol albedo modification as the most probable SRM strategy to be implemented.

Less feasible SRM strategies include the repurposing of land use for modification of albedo and the construction of space mirrors to reflect incoming sunlight. While the repurposing of land use by changing cropland or desert albedo can be effective for low implementation costs, smaller surface areas and limited potential for albedo modification limit the scalability of these strategies compared to stratospheric aerosol albedo modification and marine cloud brightening [8]. Similarly, research in the use of space mirrors finds that the strategy can be effective in reducing global temperatures; however, costs for the technology required are currently prohibitively high [7,8,12].

25.2.2 *Carbon dioxide removal*

As with SRM, researchers have compared estimates of a variety of CDR strategies and technologies in terms of their respective implementation costs and feasibility. The most commonly discussed CDR strategy is Direct Air Capture (DAC) where CO_2 is directly removed from the atmosphere and then stored. Initial estimates for direct air capture at an industrial level find implementation costs of around

$30–100 $t(CO_2)^{-1}$ [13]. Follow-up research incorporating the innovation of technologies, learning, and commercial expansion find higher costs ranging from $200 $t(CO_2)^{-1}$ to as high as $1000 $t(CO_2)^{-1}$ initially, but also finds they can fall to between $30 $t(CO_2)^{-1}$ and $300 $t(CO_2)^{-1}$ later in the century [14–16]. Researchers agree that direct air capture has the potential for expansion and significant impact on climate change, but the costs may be prohibitively high at least until further innovation occurs [8,17].

Other CDR strategies have similar or higher implementation costs. Bioenergy with carbon capture and sequestration (BECCS) involves using biomass as fuel and capturing and sequestering the emitted CO_2. It can be implemented using existing technology. Cost estimates of BECCS have a wide range of $0–1000 $t(CO_2)^{-1}$ to remove 14,000 Gt (CO_2) a^{-1} [8,18]. Ocean fertilization increases limiting nutrients such as iron or nitrogen to oceans to increase algal growth on the surface to sequester CO_2. It is estimated to have a more practical cost of $1–15 $t(CO_2)^{-1}$ up to around $22–119 $t(CO_2)^{-1}$ to remove 7–11 $Gt(CO_2)$ a^{-1} [7,8,19]. Reforestation to increase CO_2 sequestration is estimated to cost from $3–10 $t(CO_2)^{-1}$ to sequester 240 Mt. (CO_2) a^{-1}, but costs rise to $20–40 $t(CO_2)$ to sequester 400–800 Mt.(CO_2) a^{-1} [7,8]. Enhanced weathering and different techniques for increasing the ocean's alkalinity have comparably high cost [8,20].

25.2.3 Other factors

While SRM implementation costs are low, researchers have identified several potential drawbacks. The most common critique of SRM is its inability to account for all the impacts of increased GHG stocks in the atmosphere. It only reduces radiative forcing to counteract rising global temperatures and has little or no effect on other climate change impacts such as increased ocean acidification [21]. There are also drawbacks that can be specific to SRM strategies. Stratospheric aerosol albedo modification may increase ozone decay and terrestrial sulfate deposition [22–25]. Both stratospheric aerosol albedo modification and marine cloud brightening can generate changes in the hydrological cycle that can have heterogeneous impacts such as altering precipitation and vegetation based on where particles are released [26–30]. These issues are not accounted for in simplistic financial analyses of implementation costs.

Similarly, researchers have identified several risks and drawbacks to proposed CDR strategies not incorporated in estimates of implementation costs. Limits to scalability and additional risks are two of the most important. For example, BECCS and ocean fertilization have been identified to have limited abilities to capture and store high quantities of CO_2. Additionally, some strategies such as ocean fertilization can have detrimental impacts on local ecosystems [31]. While there are risks and drawbacks to both SRM and CDR, there is a consensus that CDR is the safer alternative.

For SRM, low costs and the potential to repurpose existing technologies for large-scale deployment indicate that it is feasible in the near future. Further, costs may be so low that many countries will want to implement it, even when this could potentially harm other countries. This has raised additional concerns that will be discussed later. For CDR, if lower-cost estimates are accurate, it may be part of an efficient strategy to control climate change, disrupting current climate policy. However, if higher estimates are more accurate, researchers may be better off looking elsewhere for alternatives. While low implementation costs are a critical feature for a serious alternative to mitigation, there are other important factors to consider. For geoengineering to disrupt the climate change debate and become an integral factor in a larger climate change portfolio, researchers must also consider the social costs and benefits that come with these technologies as well as potential risks [32].

25.3 Efficiency vs. equity in geoengineering

There is still a large uncertainty in the implementation costs of geoengineering strategies as none have been deployed at large enough scales to provide concrete evidence of costs. However, estimates have provided sufficient evidence to necessitate further analysis. As potentially feasible alternatives to mitigation, it is critical to understand the socioeconomic and political impacts of geoengineering or, in certain cases, simply the existence of the requisite technology.

Specifically, researchers need to determine geoengineering's potential role in climate policy. This requires careful consideration on two frontiers: efficiency and equity. It is important first to identify the most efficient geoengineering deployment pathways to achieve optimal equilibria. It is then necessary to examine what policies, if any, can induce that best-case scenario. While efficient outcomes are meaningful for obvious reasons, there often exists a tradeoff between the efficiency and equity of policies. Efficient policies are optimal, but policymaking is inherently political, so equity often gains higher weight in debates [33]. Thus, research must describe any heterogeneity embedded in policies, including the most efficient policy, to determine their equity.

25.3.1 Efficiency and geoengineering

To evaluate the efficient role of various geoengineering strategies in a climate portfolio designed to control climate change, researchers use both theoretical and quantitative methods.

25.3.1.1 Theory

The depth of theoretical work, while small, has been important in informing tradeoffs between mitigation and geoengineering. Models intertwining climate dynamics with economic growth show that mitigation and geoengineering may act as strategic complements with the former being present independent of geoengineering if damages from atmospheric CO_2 are high [34,35]. This is because SRM can only counteract rises in temperature. When incorporating uncertainty in the impacts of emissions and SRM, SRM is deployed at a higher level if emissions have a large impact, even if economic damages from SRM are high [5,35]. Reductions in the uncertainty of the side-effects of SRM can significantly reduce the total costs of climate change [5].

In addition to identifying these tradeoffs, in an optimal control model incorporating SRM, CDR, abatement, and adaptation optimal carbon taxes, theory shows that an optimal carbon tax will be equal to the marginal cost of geoengineering rather than the social cost of carbon as predicted when mitigation is the sole alternative [36].

25.3.1.2 Simulations

The primary quantitative method for evaluating efficient climate pathways and policies is the simulation of integrated assessment models (IAMs). These IAMs are numerical simulation models that integrate Earth climate system models and economic growth models to evaluate the economic and climate consequences of different climate trajectories. These models have been used to evaluate mitigation pathways, but over the past decade, researchers have augmented them with geoengineering strategies. For example, the schematic in Fig. 25.1 displays the modules and flows in the Dynamic Integrated Climate-Economy (DICE) model, an IAM, augmented with SRM. In the model, an agent chooses flows along the blue arrows in the economy with the goal of maximizing their utility. The agent's choices influence the environment through emissions and SRM, the red arrows reducing and the green arrows supplementing. Reciprocally, environmental changes damage the economy through the purple arrows.

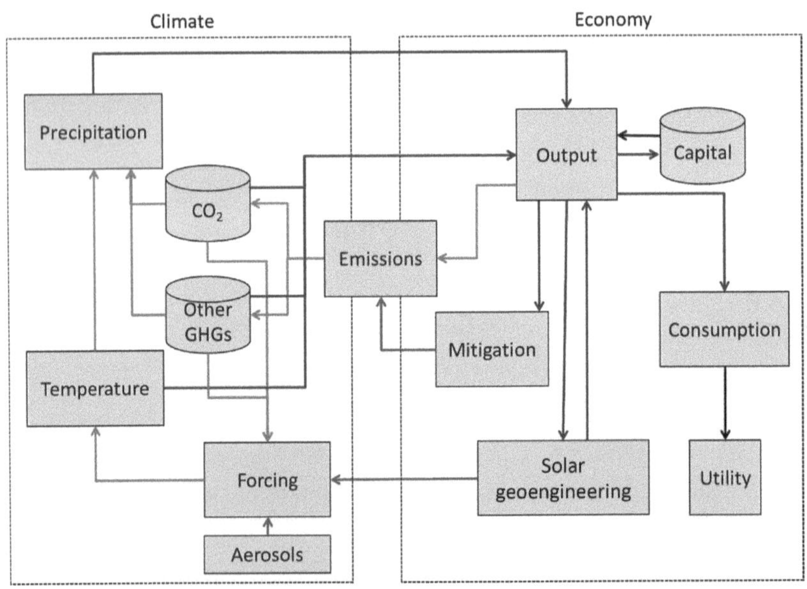

Fig. 25.1 Schematic of the modules and flows in a DICE model augmented with SRM. The model integrates the climate and the economy through emissions. Blue arrows represent the agent's choices, red arrows are contributing flows, green arrows are reducing flows, and purple arrows reflect damages. (For interpretation of the references to color in this figure legend, the reader is referred to the web version of this article.)

IAMs are used in two ways. Some research simulate specific pathways to assess whether they are welfare improving relative to a business-as-usual pathway. This type of evaluation is akin to a cost-benefit analysis. Other research uses dynamic optimization to identify the efficient levels of mitigation and geoengineering. This type of evaluation identifies optimal policy.

Simulating different climate pathways in a cost-benefit analysis that incorporates SRM strategies with abatement, IAMs estimate that SRM deployment increases welfare and performs better in conjunction with abatement [37,38]. In evaluations of afforestation and BECCS with abatement, afforestation is effective but is limited in scalability while BECCS significantly reduces abatement costs with higher value and deployment levels later in the century [18,39,40]. When carbon taxes are included in these model, significant levels of BECCS can lead to negative emissions later in the century. While this is beneficial for climate change, it can generate economic issues if subsidies exceed revenues from taxes [41]. When simulating SRM, CDR, and abatement together in a single model, SRM is the only geoengineering strategy that passes the cost-benefit test, but it still has large uncertainty [42]. These simulations can provide evidence for whether implementation of a strategy is beneficial, but more complex techniques are required to identify the optimal levels of deployment.

Using dynamic optimization, IAMs can simulate multiple pathways through backward induction to identify efficient deployment at each time period to construct optimal climate and economic pathways. With a climate change portfolio consisting of abatement and SRM, research suggests that SRM is a component of the optimal climate portfolio; but because it is an imperfect substitute, relative levels of SRM and mitigation depend on the size of climate damages unaccounted for by SRM [43,44]. Additionally, including adaptation as an alternative improves outcomes with an optimal portfolio that mixes all three strategies [45]. Moving to CDR strategies, analysis of ocean sequestration identifies it as an optimal strategy only in the short run unless costs are below a critical level [46]. Both direct air capture and BECCS strategies are found to be implemented in the long run with direct air capture augmenting BECCS once BECCS has been fully exploited [47,48]. In general, both SRM and direct air capture are consistently identified as part of a larger climate change policy portfolio in an efficient outcome, but exact levels of each are sensitive to costs and damages. Other strategies depend more heavily on optimistic cost estimates.

IAMs are currently the best strategy for quantitatively estimating the climate and economic impacts of climate change portfolios, but because of the complexities in climate systems and economic systems, they require simplifications for computational tractability and interpretability. For example, Nordhaus's famous DICE model consolidates the climate and economic system to less than 20 equations [49,50]. These simplifications combined with gaps in knowledge about the climate system and certain geoengineering strategies have been the focus of criticism by opponents of IAMs. Critics argue that IAMs can be highly sensitive to arbitrary parameters, that they overstated understanding of climate sensitivity and impacts from climate change, and that they fail to incorporate the possibility of a catastrophic climate outcome [51–53]. While IAMs are consistently the best approach for quantitative assessment, more research is needed to improve the accuracy of parameter inputs and validate functional forms to solidify confidence in their results.

25.3.2 *Equity, heterogeneity, and geoengineering*

Climate change is known to affect regions of the world heterogeneously. This raises the question of equity in climate change policy design. Left unchecked, poorer, equatorial countries will carry the heaviest burden of economic consequences from persisting climate change [54]. These inequalities are compounded by inequalities in the costs of mitigation, creating issues for equitable policy design. With the introduction of geoengineering, these regional disparities have the potential to grow wider or to shift focus.

Though SRM strategies can be implemented locally at small scales, impacts from the long-term, large-scale SRM deployment required would be nearly impossible to contain locally [55]. At a global scale, SRM can introduce new regional inequalities [56]. For example, SRM can cool different regions of the world more than others and differentially alter hydrological cycles, as is displayed in Fig. 25.2. While SRM reduces mean and variance of temperature changes and precipitation changes, changes are heterogeneous across the different regions. Additionally, residual consequences of climate change can still have significant and unequal effects. For example, ocean

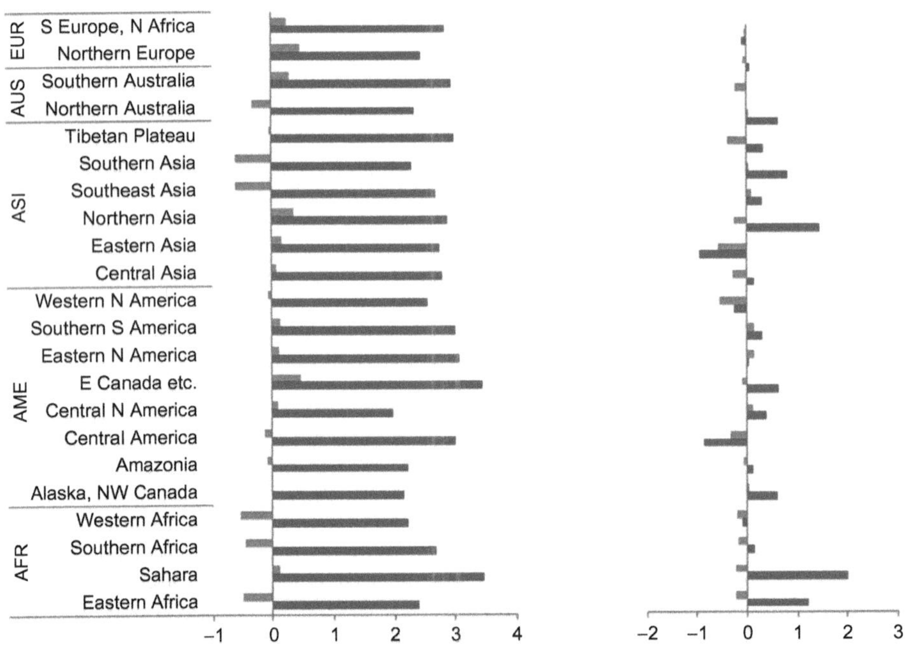

Fig. 25.2 Regional estimation of changes in temperature and precipitation relative to 1990s level due to a doubling of CO_2 with and without SRM available. The left panel displays temperature changes and the right panel displays precipitation changes. With SRM, mean and variance are reduced for both temperature and precipitation changes. This is modified from data and running the analysis in [57].

acidification can cause coral bleaching. This has been attributed to the collapse of reefs, which disproportionately impacts those who more prominently depend on them [58]. Thus, large-scale SRM deployment will have unevenly distributed costs and benefits, causing different regions of the world to prefer different levels of SRM [59,60].

Quantitative research of disparities from SRM has been small. Research using residual climate response models that compute damages not accounted for by SRM has found that disparities, while existent, may not be as great as previously thought. Also, optimized SRM deployment is predicted to be able to restore up to 97%–99% of population-weighted temperature or precipitation changes, but not both simultaneously [57]. SRM deployment of up to 85% of levels that reverse global temperatures to preindustrial levels have been found to increase welfare in all regions, but not necessarily equally [61]. While optimistic, these results are sensitive to inputs and damage specifications and do not account for other known and unknown side effects of SRM besides changes in temperature and precipitation [53].

Since impacts of CDR closely mirror mitigation, it exhibits many of the equity issues present in abatement efforts with additional regional variation in costs and side effects of certain CDR strategies. Countries with different levels of technological development will have different costs of developing and implementing CDR technology. CDR strategies can also have adverse effects. For example, BECCS, reforestation, and afforestation are predicted to cause increases in food prices [41,62]. This can have a differential impact based on a country's food production and consumption. Alternatively, ocean fertilization can have a heterogeneous impact on local ecosystems. This will have a larger effect on regions that are more dependent on those local ecosystems [8].

Aside from regional inequities, the deployment or even the existence of some geoengineering strategies can influence intergenerational equity. In the case of SRM, some argue that the existence of the technology gives current generations excessive power over future generations in their ability to deploy SRM and increase emissions to their benefit. This can be detrimental to future generations and raises ethical concerns. However, it can be avoided if SRM costs are sufficiently high or if there are worries about a pro-SRM bias in future generations [63]. In the latter case, current generations may even increase abatement efforts to disincentivize future generations from relying on SRM [64]. For these potential intergenerational inequalities, preferences can be important in determining when a generation should research SRM technologies [65].

As with cost analyses, the use of IAMs has produced initial efforts for identifying efficient policy, but it requires further work to minimize large uncertainties and to incorporate currently unknown factors. For practical policy, since climate change and geoengineering have global impacts, sources of inequality will be important considerations in coordination with efficiency analyses when it comes to the politics of policy design [33]. To emphasize fairness, some argue that those affected most by geoengineering should have a higher weight in global decision-making [66]. Ultimately, many agreements may compromise suboptimal outcomes to better maintain equity.

25.4 Strategy

In current climate change policy, the prohibiting factor for developing an effective, cooperative agreement to quell harmful emissions is a lack of proper incentives. Even though countries would experience welfare gains if they cooperate, they have not had sufficient incentives in historically proposed agreements to reduce their own emissions when others receive most of the rewards. This lack of proper incentive structure creates a collective action problem where countries free ride on others' abatement efforts, implementing suboptimal levels of abatement. This demonstrates that, in addition to an understanding of efficient and equitable deployment strategies and policies, it is important to understand incentives and rational decisions in a strategic decision-making framework. It is commonly understood that abatement as a climate change policy embodies a classic free-rider problem, but the introduction of geoengineering strategies can alter existing incentives.

SRM's cheap and fast but imperfect properties have ideally placed the geoengineering strategy as a backstop or insurance policy against a failure to cooperate and improve global abatement levels sufficiently to avoid disastrous climate change. For example, it is often proposed that SRM is used to avoid anticipated climate-tipping points if they become unavoidable by other means [67,68]. While this would be the ideal deployment strategy, this is not necessarily the outcome in a rational decision-making framework. A primary apprehension to the development of SRM technologies is the potential for the free-rider problem of abatement to quickly turn into a "free-driver" problem for SRM [69,70]. The cheap and fast properties have caused concern that certain countries or a small coalition of countries could deploy SRM to the detriment of others because it is individually rational [69]. Additionally, this free-driver problem may be the same size as the existing free-driving problem [71]. This quickly changes the cooperation problem of abatement into a coordination problem and has generated repeated demands for the immediate governance of SRM technologies [72–76].

To properly analyze the disparities between ideal decision-making and rational decision-making with regards to climate change and geoengineering strategies, researchers have utilized well-known game-theoretical techniques. This research again typically augments models originally designed around mitigation with geoengineering, allowing researchers to identify the noncooperative Nash and subgame perfect Nash equilibria for comparison with the cooperative analyses for the efficient outcomes. Using this framework, research has analyzed geoengineering technologies throughout their lifecycle, from research and development through commercial deployment.

SRM strategies have been the focus of most strategic analyses because of their unique properties. Beginning with the development of SRM technologies, strategic analyses have identified the possible presence of a free-driver effect. If countries are sufficiently worried about high damages from the widespread use of SRM following the development of the necessary technologies, they may begin to invest in counteracting the free-driver's innovation efforts [77]. Regardless of the intergenerational equity arguments about the existence of SRM technology discussed earlier, a research and development conflict may arise simply based on the expectation of free-driving following development.

Following the development of SRM technologies, simplified two country analyses find that rational abatement efforts and incentives to deploy the technology are contingent on asymmetries in preferences and sensitivities between countries. These asymmetries are based on the expected damages from climate change and from side-effects of SRM. In a static scenario, if countries are adequately symmetric or if expected damages are sufficiently low, mitigation efforts will be reduced in expectation of SRM deployment in the future [70,78]. Contrasting with worries about the free-driver effect, if countries' preferences are highly asymmetric or damages from SRM are expected to be large, availability of SRM technologies may lead to higher mitigation efforts by one country in the short run to avoid future use of SRM by the other [70]. Expanding to a dynamic analysis leads to similar results [79]. Incorporating uncertainty in SRM damages can reduce the scale of deployment. With asymmetric knowledge about the effects of SRM, only the more confident country will use it [80].

Expanding these game-theoretic analyses to incorporate more countries has little impact on conclusions about tradeoffs between SRM and mitigation efforts. In a multiple country scenario with SRM and abatement, geoengineering will be overprovided; the free-driving effect will scale nearly linear in the number of countries interacting. This can be seen in panel A of Fig. 25.3. Panel B shows that quantitative estimates of the free-driving effect can be of similar magnitude to the current free-riding effect for mitigation [71]. Panel B also shows that free riding has a much larger impact on mitigation efforts than the introduction and partial substitution of SRM. By incorporating more countries, analyses of coalitions, such as international environmental agreements (IEAs), become possible. In a traditional IEA augmented with the option of SRM, there exists the possibility that the free-driver threat can induce higher mitigation levels and broader participation in an IEA [81]. When SRM is the only alternative for climate change in an exclusive coalition that requires broad participation to overcome political frictions, asymmetries incentivize small but powerful coalitions, but the benefits of exclusivity are small [82]. This is the result is shown in Fig. 25.4, which displays an example breakdown for an exclusive coalition with SRM. Panels A and B show that all coalition members benefit from being members of the coalition, but not all regions with positive benefits are able to join the coalition because it is exclusive. However, given the SRM implemented, coalition members are not much better off than nonmembers as shown by panel D.

There are fewer analyses investigating the strategic implications of CDR technologies because the properties are so similar to mitigation, which has been studied heavily, but several postulations can be made with confidence. Because CDR exemplifies comparably high costs to abatement as well as a perfect ability to counteract climate change, if costs are too high, the existing prisoners' dilemma game for abatement will persist [83]. However, if CDR costs fall adequately below the costs of abatement, a harmony game might occur in which countries have proper incentive to deploy CDR. Therefore, it becomes clear that strategic use of CDR is heavily dependent on the relative costs of mitigation and CDR strategies. In a strategic framework, the expected development of CDR technologies can have a downside. As seen in the

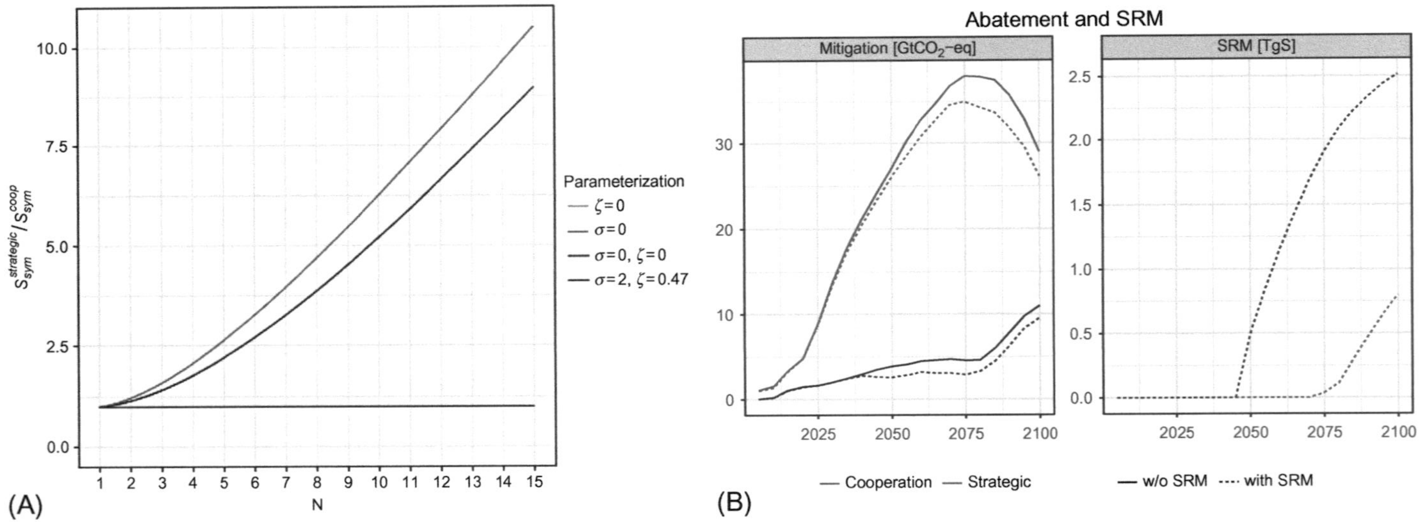

Fig. 25.3 Analysis of cooperative versus noncooperative decision-making with mitigation and SRM. Panel A shows the free-driver effect measured as the ratio of SRM implemented in the noncooperative setting relative to cooperative levels. ζ represents the marginal impacts from SRM. σ represents the marginal costs of SRM. Panel B shows mitigation and SRM levels across noncooperative and cooperative settings with and without SRM. Reproduced from Emmerling J, Tavoni M. Quantifying non-cooperative climate engineering, 2017. with permission of Authors.

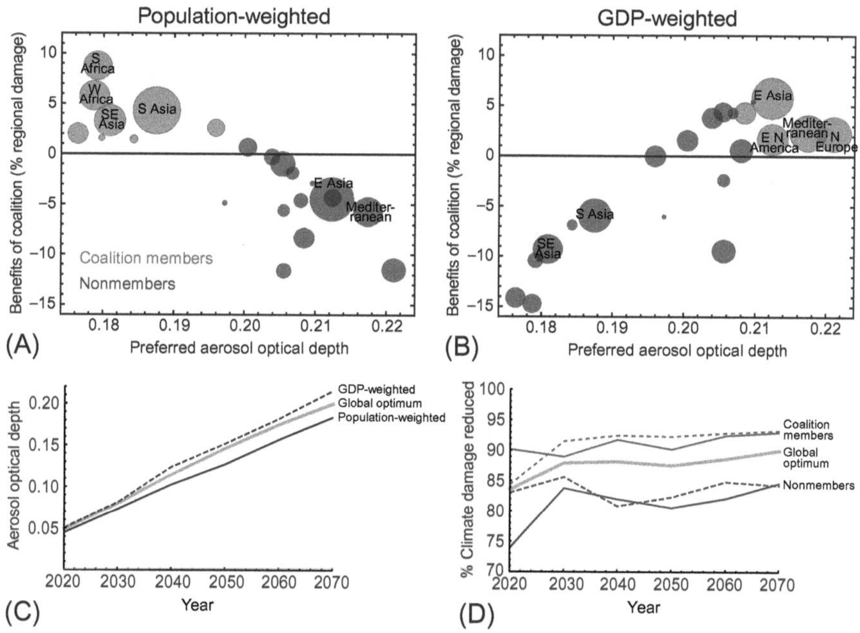

Fig. 25.4 Analysis of an exclusive coalition around SRM. Panel A shows population-weighted benefits of joining the coalition. Panel B shows GDP-weighted benefits of joining the coalition. Panel C shows resulting aerosol optical depth from SRM. Panel D shows estimated reductions in damages for members and nonmembers. This estimation uses data and methodology from [82].

quantitative analyses of efficient outcomes discussed here, countries may increase emissions in the short run with the expectation of CDR technologies compensating for foregone abatement efforts in the future.

Further research is crucial to inform how to align incentives so that countries will adopt the preferred efficient or equitable policies in a strategic framework. Specifically, more quantitative studies of strategic decision-making by countries as well as improved estimates of key model parameters can be instrumental in supporting policy design and the probability of geoengineering success in a public choice model when it is desired [38].

25.5 Risk and uncertainty

Prominent obstacles in the economics of climate change are the numerous sources of risk and uncertainty. Uncertainties, both known and unknown, can create compound issues for climate and economic modeling, as the Earth's climate system and the global economy are extremely complex and nonlinear. Some known sources of uncertainties such as uncertainty in model parameters are measurable. These uncertainties can be directly included in analyses. Other known uncertainties such as the proper damage function for climate change may not be measurable. These Knightian uncertainties require flexible estimation of different functional forms or the use of probability functions to encompass a variety of possible realizations. Unknown uncertainties are more complicated because they are inherently unidentified and unmeasurable. These necessitate more drastic action to resolve.

In quantitative analyses, measurable uncertainty can prove difficult. When forecasting economic growth, macroeconomic models are typically accurate for short time spans of a few years up to a decade, but they begin to lose predictive power when forecasting in the ranges necessary to analyze climate change models. This is largely due to high levels of uncertainty. There are similar issues for modeling of the Earth's climate system and the impacts of geoengineering strategies. When climate and economy models are integrated for long run, climate change analyses uncertainties can compound, further limiting the accuracy of existing climate-economy models.

Risks and uncertainties that can be incorporated in models are often assigned probability distributions to describe possible outcomes and their probability of occurrence. Analyses of these models will typically estimate expected outcomes and construct a confidence interval around the expected outcome using either the distributions in the model or various sampling techniques. While uncertainty is being more frequently included in analyses to acknowledge the importance and prominence of uncertainty in climate change, some critics argue that we still do not know what the proper distributions are and that the distributions currently used are not accurate. Maintaining that extreme outcomes have a higher probability of occurrence than currently accounted for, these critics believe the use of fat-tailed distributions would be more accurate [85,86]. Fig. 25.5 shows climate forecasts based on a variety of climate sensitivity distributions. This figure clearly shows that assumptions about distributions can have a large impact on quantitative results. As the tails of distributions in panel A get fatter, optimal policy recommends a conservative policy of more SRM to maintain lower temperatures and mitigate the risk of high damages.

Minimizing different risks and uncertainty in climate change and geoengineering caused by gaps in knowledge or imprecise estimations can have a measurable effect on the future impacts of climate change. Improving knowledge of SRM side effects through learning could substantially reduce the expected costs of climate change [5]. A better understanding of the potential effects of SRM on temperature as well as its side effects and costs can also inform the construction of governance structures and address concerns of a potential free-driver effect [4]. Increased precision of estimates for key model parameters, validation of underlying equations, and better incorporation of risks and uncertainty can boost confidence in the results of quantitative analyses [51–53]. This permits more accurate evaluations of different climate change alternatives as well as sharpened identification of efficient, equitable, and strategic outcomes.

Some risks and uncertainties can be mitigated or better understood through theoretical research, historical case studies, or lab tests. However, some measure of uncertainty, especially unknown uncertainty, can only be inferred by using geoengineering in the real world. Though it may be politically difficult, the use of field testing could be the best way to learn about geoengineering short of full-scale deployment [75].

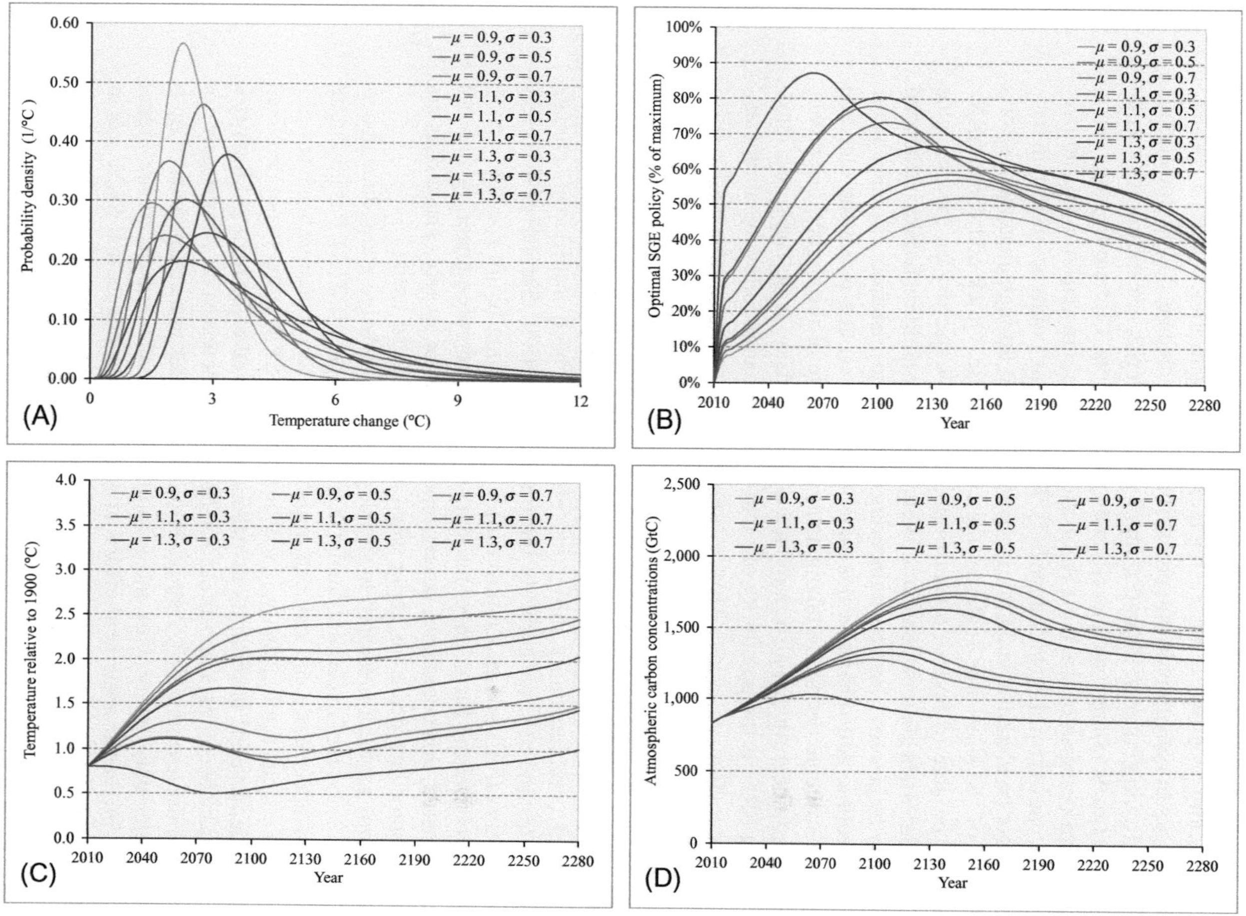

Fig. 25.5 Panel A shows alternative climate sensitivity distributions. The distribution is lognormal with mean and variance μ and σ, respectively. Panel B shows the corresponding optimal SRM policy. Panel C shows the temperature responses following SRM policies. Panel D shows the resulting carbon concentrations. Fatter tail distributions correspond with higher SRM policies to maintain temperatures below 3°C. Figures are estimated using the Geo-DICE model developed by [84].

Small-scale field testing of geoengineering can minimize political and social concerns while providing concrete evidence for the costs of implementation and improving understanding of the precise impacts and potential side effects.

The potential for risk can also drive the use of geoengineering. As mentioned earlier, SRM is ideally proposed as a backstop technology or insurance policy to protect society from risks of catastrophic outcomes from climate change. This proposition is derived from growing concern for uncertain or unforeseen risk of catastrophic consequences from climate change such as climate-tipping points. Climate-tipping points, like the possible collapse of the West Antarctic Ice Sheet, are catastrophic and irreversible events caused by climate change [87]. These tipping points are uncertain risks and they may not be identifiable until it is too late to avoid through abatement efforts alone because of the slow time scales of the climate cycle. Quick and cheap SRM, however, may be able to reverse a catastrophic climate trajectory on the precipice of a tipping point when abatement cannot. Weighing in the additional risks introduced by SRM, it may be best to delay geoengineering until a major risk from climate change appears catastrophic and otherwise unavoidable. However, this is not the outcome often found in strategic decision-making analyses because of misaligned incentives.

Use of geoengineering can also introduce additional risks. A potential risk introduced by using SRM is the possibility of rapid climate change following abrupt suspension or termination in its use. Most serious implementation strategies for SRM require continuous maintenance. For strategies like stratospheric aerosol albedo modification and marine cloud brightening, particles injected into the atmosphere or clouds have a limited lifespan before descending. Thus, regular injections are required. If maintenance is ignored or SRM operations are suddenly stopped for any reason and complementary mitigation efforts have been insufficient, climate change could resume at a dangerously more rapid pace [88]. Similarly, if society delays abatement efforts and geoengineering, either expecting future geoengineering to compensate or waiting for a dangerous climate event, and geoengineering does not work as anticipated, this could have severe repercussions for society. These risks are important to factor into decisions about when SRM should be used.

It is important for future research to shrink gaps in knowledge and reduce uncertainties, but some level of uncertainty and risk will always persist. Thus, it is important to identify how uncertainty and risk can impact each of the economic analyses discussed here. In efficient policy analysis with uncertainty, SRM is likely not deployed or minimally deployed if uncertainty or risks are perceived to be high. At high-risk levels, mitigation becomes a substitute for SRM and increases to compensate for lower expected deployment [89]. As Fig. 25.6 demonstrates in a comparison of optimal policy under uncertainty with and without SRM, abatement efforts are higher without SRM. Increased abatement efforts at a lower rate suggest that SRM acts as an insurance against the risk that does not exist when abatement is the only alternative. However, if there is initially uncertainty in climate change and realized climate outcomes are drastic, SRM will be deployed even if risks are high to avoid extreme damages from climate change [5]. Alternatively, as the expectation of positive outcomes from geoengineering increases, mitigation levels decrease until geoengineering is a viable alternative [90]. In a strategic scenario with uncertain risks from

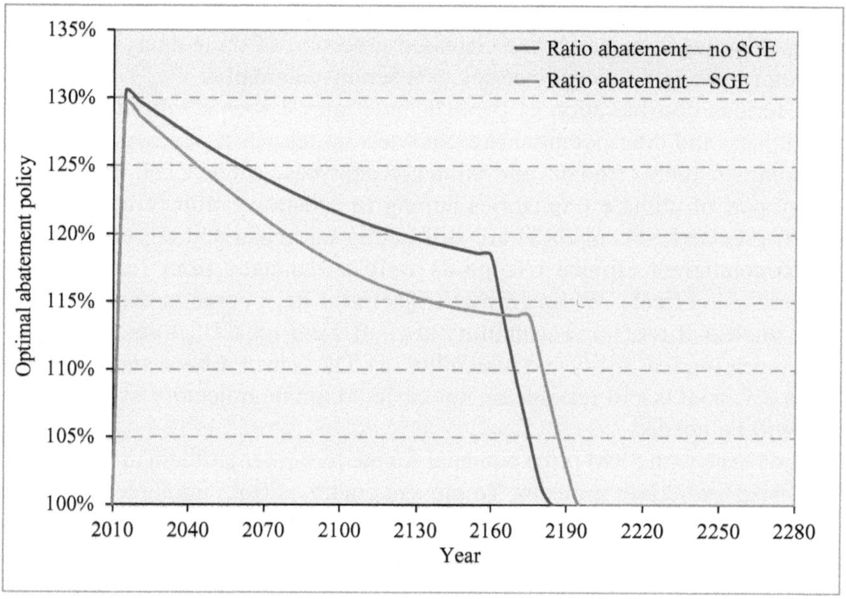

Fig. 25.6 Estimation of optimal abatement policy with and without SRM under uncertainty relative to abatement in the deterministic case. The blue line represents abatement without SRM and the gray line represents abatement with SRM. Without SRM, optimal abatement efforts are higher to protect from a climate risk when abatement is the only option. With SRM, abatement is lower, suggesting SRM acts as an insurance against risk. (For interpretation of the references to color in this figure legend, the reader is referred to the web version of this article.)

geoengineering, if there is asymmetric ambiguity aversion, only the country confident in their understanding of geoengineering will deploy it [80].

25.6 Conclusion

Over the past decade, there has been growing momentum in the research of geoengineering as an alternative to mitigation. While countries cooperation and abatement efforts continue to struggle under the free-rider problem, climate change continues. Researchers and policymakers have been intrigued by the unique properties of geoengineering alternatives, hopeful that they can overcome the issues of abatement. SRM is quick, cheap, but imperfect, reflecting the opposite of mitigation characteristics. CDR is slow, expensive, but perfect, mirroring the properties of mitigation, but with the additional benefit of the potential for negative emissions.

The unique properties exhibited by geoengineering have the potential to radically disrupt the economics of climate change as it currently stands. And these alternatives may not be far off with several proposed strategies exhibiting potential for deployment by repurposing existing technologies. Initial cost analyses of these strategies validate the notion that geoengineering strategies may be feasible, especially for cheaper SRM strategies. Continued research has increased precision of these analyses and future innovation can solidify geoengineering as a serious alternative.

With serious consideration of these technologies comes attention to their economic impacts and their potential to create new issues. SRM deployment is consistently found in both efficient and strategic analyses, while CDR has become a necessary part of climate trajectories hoping to constraint climate change to 2°C. However, precise levels of SRM are still highly uncertain and dependent on their ability to counteract climate change as well as damages from residual climate change and side effects. While climate trajectories have come to depend on CDR, it is still unclear if costs and scalability are sufficient for CDR to become a viable option. Accessing feasibility and scalability of CDR is incredibly important to determine if a 2°C goal is still feasible on our current climate trajectory or if other alternatives will be needed.

A major issue with SRM is the potential for the free-rider problem to transition into an equal-size free-driver problem. To prevent countries from implementing SRM to the detriment of others, a governance structure needs to be developed. This will require global coordination among countries to create governance policies that provide proper incentives. Determining what the proper incentives are and how to create them will require more accurate quantitative and strategic analyses. With the availability of these technologies in the near future, coordination to design a governance structure will be needed quickly.

Over the past decade, significant gains have been made in our understanding of geoengineering and the economics of geoengineering, but there are still risks and uncertainties. Some of these stand as a barrier to the economics of geoengineering and require further research to understand. Others can and need to be incorporated into analyses to accurately inform policy. Future research to validate key parameters and model functional forms can have a significant impact on the precision of understanding as well as impacts of future climate change. Some of this research can continue to come from modeling and lab research, but moving forward may necessitate small-scale field tests. This will require significant political and societal support.

References

[1] IPCC, Fifth Assessment Report: Climate Change. 2013.

[2] Crutzen PJ. Albedo enhancement by stratospheric sulfur injections: a contribution to resolve a policy dilemma? Clim Change 2006;77(3–4):211.

[3] National Research Council. Climate Intervention: Reflecting sunlight to cool earth. Washington, DC: National Academies Press; 2015.

[4] Keith DW, Parson E, Morgan MG. Research on global sun block needed now. Nature 2010;463(7280):426–7.

[5] Moreno-Cruz JB, Keith DW. Climate policy under uncertainty: a case for solar geoengineering. Clim Change 2013;121(3):431–44.

[6] Harding A, Moreno-Cruz JB. Solar geoengineering economics: from incredible to inevitable and half-way back. Earths Future 2016;4(12):2016EF000462.

[7] National Academy of Sciences. Policy implications of greenhouse warming: Mitigation, Adaptation, and the Science Base; 1992.
[8] Shepherd JG. Geoengineering the climate: science, governance and uncertainty. Royal Society; 2009.
[9] Robock A, Marquardt A, Kravitz B, Stenchikov G. Benefits, risks, and costs of stratospheric geoengineering. Geophys Res Lett 2009;36(19).
[10] Katz JI. Stratospheric albedo modification. Energ Environ Sci 2010;3(11):1634–44.
[11] McClellan J, Keith DW, Apt J. Cost analysis of stratospheric albedo modification delivery systems. Environ Res Lett 2012;7(3):034019.
[12] Angel R. Feasibility of cooling the earth with a cloud of small spacecraft near the inner Lagrange point (L1). Proc Natl Acad Sci 2006;103(46):17184–9.
[13] Sachs J, Lackner KA. Robust strategy for sustainable energy. Brook Pap Econ Act 2005;2.
[14] Lackner KS. Capture of carbon dioxide from ambient air. Eur Phys J-Spec Top 2009;176 (1):93–106.
[15] House KZ, Baclig AC, Ranjan M, van Nierop EA, Wilcox J, Herzog HJ. Economic and energetic analysis of capturing CO2 from ambient air. Proc Natl Acad Sci 2011;108 (51):20428–33.
[16] Socolow R, Desmond M, Aines R, Blackstock J, Bolland O, Kaarsber T, et al. Direct air capture of CO2 with chemicals: a technology assessment for the APS panel on public affairs. American Physical Society; 2011.
[17] Lackner KS, Brennan S, Matter JM, Park AHA, Wright A, Van Der Zwaan B. The urgency of the development of CO2 capture from ambient air. Proc Natl Acad Sci 2012;109 (33):13156–62.
[18] Kriegler E, Edenhofer O, Reuster L, Luderer G, Klein D. Is atmospheric carbon dioxide removal a game changer for climate change mitigation? Clim Change 2013;118(1): 45–57.
[19] Rickels W, Rehdanz K, Oschlies A. Economic prospects of ocean iron fertilization in an international carbon market. Resour Energy Econ 2012;34(1):129–50.
[20] Renforth P, Henderson G. Assessing Ocean alkalinity for carbon sequestration. Rev Geophys 2017.
[21] Matthews HD, Cao L, Caldeira K. Sensitivity of ocean acidification to geoengineered climate stabilization. Geophys Res Lett 2009;36(10).
[22] Tilmes S, Garcia RR, Kinnison DE, Gettelman A, Rasch PJ. Impact of geoengineered aerosols on the troposphere and stratosphere. J Geophys Res Atmos 2009;114(D12).
[23] Tilmes S, Kinnison DE, Garcia RR, Salawitch R, Canty T, Lee-Taylor J, et al. Impact of very short-lived halogens on stratospheric ozone abundance and UV radiation in a geo-engineered atmosphere. Atmos Chem Phys 2012;12(22):10945–55.
[24] Pitari G, Aquila V, Kravitz B, Robock A, Watanabe S, Cionni I, et al. Stratospheric ozone response to sulfate geoengineering: Results from the geoengineering model Intercomparison project (GeoMIP). J Geophys Res Atmos 2014;119(5):2629–53.
[25] Kravitz B, Robock A, Oman L, Stenchikov G, Marquardt AB. Sulfuric acid deposition from stratospheric geoengineering with sulfate aerosols. J Geophys Res Atmos 2009;114(D14).
[26] Trenberth KE, Dai A. Effects of mount Pinatubo volcanic eruption on the hydrological cycle as an analog of geoengineering. Geophys Res Lett 2007;34(15).
[27] Robock A, Oman L, Stenchikov GL. Regional climate responses to geoengineering with tropical and Arctic SO2 injections. J Geophys Res Atmos 2008;113(D16).
[28] Kravitz B, Rasch PJ, Forster PM, Andrews T, Cole JNS, Irvine PJ, et al. An energetic perspective on hydrological cycle changes in the geoengineering model Intercomparison project. J Geophys Res Atmos 2013;118(23).

[29] Tilmes S, Faullo J, Lamarque J-F, Marsh DR, Mills M, Alterskjær MH, et al. The hydrological impact of geoengineering in the geoengineering model Intercomparison project (GeoMIP). J Geophys Res Atmos 2013;118(19).

[30] Glienke S, Irvine PJ, Lawrence MG. The impact of geoengineering on vegetation in experiment G1 of the GeoMIP. J Geophys Res Atmos 2015;120(19).

[31] Barrett S. The coming global climate—technology revolution. J Econ Perspect 2009;23 (2):53–75.

[32] MacKerron G. Costs and economics of geoengineering. Clim Geoeng Govern 2014;.

[33] Dietz S, Atkinson G. The equity-efficiency trade-off in environmental policy: evidence from stated preferences. Land Econ 2010;86(3):423–43.

[34] Moreno-Cruz JB, Smulders S. Geoengineering and economic growth: making climate change irrelevant or buying time?; 2007.

[35] Moreno-Cruz JB, Smulders S. Revisiting the economics of climate change: the role of geoengineering. Res Econ 2017;71(2):212–24.

[36] Moreno-Cruz JB, Wagner G, Keith D. An economic anatomy of optimal climate policy; 2017.

[37] Wigley TM. A combined mitigation/geoengineering approach to climate stabilization. Science 2006;314(5798):452–4.

[38] Gramstad K, Tjøtta S. Climate engineering: cost benefit and beyond; 2010.

[39] Edmonds J, Luckow P, Calvin K, Wise M, Dooley J, Kyle P, et al. Can radiative forcing be limited to 2.6 Wm-2 without negative emissions from bioenergy AND CO2 capture and storage? Clim Change 2013;118(1):29–43.

[40] Van Vuuren DP, Deetman S, van Vliet J, van den Berg M, van Ruijven BJ, Koelbl B. The role of negative CO2 emissions for reaching 2 C—insights from integrated assessment modelling. Clim Change 2013;118(1):15–27.

[41] Muratori M, Calvin K, Wise M, Kyle P, Edmonds J. Global economic consequences of deploying bioenergy with carbon capture and storage (BECCS). Environ Res Lett 2016;11(9):095004.

[42] Bickel JE, Lane L. An analysis of climate engineering as a response to climate change. Smart Clim Solut 2009;40.

[43] Goes M, Tuana N, Keller K. The economics (or lack thereof) of aerosol geoengineering. Clim Change 2011;109(3–4):719–44.

[44] Heutel G, Moreno-Cruz JB, Shayegh S. Solar geoengineering, uncertainty, and the price of carbon, National Bureau of economic research; 2015.

[45] Bahn O, Chesney M, Gheyssens J, Knutti R, Pana AC. Is there room for geoengineering in the optimal climate policy mix? Environ Sci Policy 2015;48:67–76.

[46] Rickels W, Lontzek TS. Optimal global carbon management with ocean sequestration. Oxf Econ Pap 2011;64(2):323–49.

[47] Chen C, Tavoni M. Direct air capture of CO2 and climate stabilization: a model based assessment. Clim Change 2013;118(1):59–72.

[48] Fuss S, Reuter WH, Szolgayová J, Obersteiner M. Optimal mitigation strategies with negative emission technologies and carbon sinks under uncertainty. Clim Change 2013;118 (1):73–87.

[49] Nordhaus WD. An optimal transition path for controlling greenhouse gases. Science 1992;258(5086):1315–9.

[50] Nordhaus WD. Revisiting the social cost of carbon. Proc Natl Acad Sci 2017; 201609244.

[51] Pindyck RS. Climate change policy: What do the models tell us? J Econ Lit 2013;51 (3):860–72.

[52] Pindyck RS. The use and misuse of models for climate policy. Rev Environ Econ Policy 2017;11(1):100–14.
[53] Heyen D, Wiertz T, Irvine PJ. Regional disparities in SRM impacts: the challenge of diverging preferences. Clim Change 2015;133(4):557–63.
[54] McCarthy JJ, Canziani OF, Leary NA, Dokken DJ, White KS. Climate Change 2001: Impacts, Adaptation, and Vulnerability. In: Contribution of working group II to the third assessment report of the intergovernmental panel on climate change. Cambridge: Cambridge University Press; 2001. p. 1032.
[55] Quaas J, Quaas MF, Boucher O, Rickels W. Regional climate engineering by radiation management: prerequisites and prospects. Earths Future 2016;4(12):618–25.
[56] Klepper G, Rickels W. Climate engineering: economic considerations and research challenges. Oxford University Press; 2014.
[57] Moreno-Cruz JB, Ricke KL, Keith DW. A simple model to account for regional inequalities in the effectiveness of solar radiation management. Clim Change 2012;110(3):649–68.
[58] Hoegh-Guldberg O, Mumby PJ, Hooten AJ, Steneck RS, Greenfield P, Gomez E, et al. Coral reefs under rapid climate change and ocean acidification. Science 2007;318(5857): 1737–42.
[59] Ban-Weiss GA, Caldeira K. Geoengineering as an optimization problem. Environ Res Lett 2010;5(3):034009.
[60] Ricke KL, Morgan MG, Allen MR. Regional climate response to solar-radiation management. Nat Geosci 2010;3(8):537.
[61] Kravitz B, MacMartin DG, Robock A, Rasch PJ, Ricke KL, Cole JNS, et al. A multi-model assessment of regional climate disparities caused by solar geoengineering. Environ Res Lett 2014;9(7):074013.
[62] Kreidenweis U, Humpenöder S, Bodirsky BL, Kriegler E, Lotze-Campen H, Popp A. Afforestation to mitigate climate change: impacts on food prices under consideration of albedo effects. Environ Res Lett 2016;11(8):085001.
[63] Burns WC. Climate geoengineering: solar radiation management and its implications for intergenerational equity; 2011
[64] Goeschl T, Heyen D, Moreno-Cruz JB. The intergenerational transfer of solar radiation management capabilities and atmospheric carbon stocks. Environ Resource Econ 2013;56(1):85–104.
[65] Quaas MF, Quaas J, Rickels W, Boucher O. Are there reasons against open-ended research into solar radiation management? A model of intergenerational decision-making under uncertainty. J Environ Econ Manage 2017;84:1–17.
[66] Corner A, Pidgeon N. Geoengineering the climate: the social and ethical implications. Environ Sci Policy Sustain Dev 2010;52(1):24–37.
[67] Lenton TM. Early warning of climate tipping points. Nat Clim Change 2011;1(4):201.
[68] Barrett S, Dannenberg A. Sensitivity of collective action to uncertainty about climate tipping points. Nat Clim Change 2014;4(1):36–9.
[69] Weitzman ML. A voting architecture for the governance of free-driver externalities, with application to geoengineering. Scand J Econ 2015;117(4):1049–68.
[70] Moreno-Cruz JB. Mitigation and the geoengineering threat. Resour Energy Econ 2015;41:248–63.
[71] Emmerling J, Tavoni M. Quantifying non-cooperative climate engineering, 2017.
[72] Victor DG. On the regulation of geoengineering. Oxf Rev Econ Policy 2008;24(2):322–36.
[73] Barrett S. The incredible economics of geoengineering. Environ Resource Econ 2008;39 (1):45–54.

[74] Victor DG, Morgan MG, Apt J, Steinbruner J, Ricke K. The geoengineering option: A last resort against global warming? Foreign Aff 2009;64–76.
[75] Blackstock JJ, Long JCS. The politics of geoengineering. Science 2010;327(5965):527.
[76] Betz G. The case for climate engineering research: An analysis of the 'arm the future' argument. Clim Change 2012;111(2):473–85.
[77] Heyen D. Strategic conflicts on the horizon: R&d incentives for environmental technologies. Clim Change Econ 2016;7(4):1650013.
[78] Urpelainen J. Geoengineering and global warming: a strategic perspective. Int Environ Agreem Polit Law Econ 2012;12(4):375–89.
[79] Manoussi V, Xepapadeas A. Cooperation and competition in climate change policies: mitigation and climate engineering when countries are asymmetric. Environ Resource Econ 2017;66(4):605–27.
[80] Emmerling J, Manoussi V, Xepapadeas A. Climate engineering under deep uncertainty and heterogeneity, 2016.
[81] Millard-Ball A. The Tuvalu syndrome. Clim Change 2012;110(3):1047–66.
[82] Ricke KL, Moreno-Cruz JB, Caldeira K. Strategic incentives for climate geoengineering coalitions to exclude broad participation. Environ Res Lett 2013;8(1):014021.
[83] Sandler T. Collective action and geoengineering. Rev Int Organ 2017;1–21.
[84] Heutel G, Moreno-Cruz J, Shayegh S. Climate tipping points and solar geoengineering. J Econ Behav Organ 2016;132:19–45.
[85] Pindyck RS. Fat tails, thin tails, and climate change policy. Rev Environ Econ Policy 2011;5(2):258–74.
[86] Weitzman ML. Fat-tailed uncertainty in the economics of catastrophic climate change. Rev Environ Econ Policy 2011;5(2):275–92.
[87] Lenton TM, Held H, Kriegler E, Hall JW, Lucht W, Rahmstorf S, et al. Tipping elements in the Earth's climate system. Proc Natl Acad Sci 2008;105(6):1786–93.
[88] Matthews HD, Caldeira K. Transient climate–carbon simulations of planetary geoengineering. Proc Natl Acad Sci 2007;104(24):9949–54.
[89] Manoussi V, Xepapadeas A. Mitigation and Climate Engineering under Deep Uncertainty, 2015.
[90] Emmerling J, Tavoni M. Climate engineering and abatement: a 'flat'relationship under uncertainty. Environ Resource Econ 2017;1–21.

Justice in managing global climate change

26

Ivo Wallimann-Helmer[1]
Advanced Studies in Applied Ethics, Center for Ethics, University of Zurich, Zurich, Switzerland
[1]Corresponding author: wallimann@philos.uzh.ch

Chapter Outline

26.1 Ethical evaluation in managing climate change

Global warming or global climate change is one of the biggest threats humankind is facing today. Changing climatic conditions is expected to lead to rising sea levels, higher frequency of natural hazards, extended phases of drought, and a greater risk of many other sudden and slow-onset events. These threats will impact not only economic development but also the livelihood and cultures of many communities and regions. Managing climate change to minimize these increased risks concerns not only understanding the science of climate change but also analyzing policy options with methods from social sciences and economics. Moreover, it means making evaluative judgments about what policy measures can be considered just and fair. These kinds of judgments are normative; they concern ethics. The aim of this chapter is to lay out some of the most important aspects of ethical evaluation, justice, and responsibility in managing global climate change.

Ethics in managing climate change most often involves two issues that are tightly connected. The first involves considerations about the just distribution of entitlements and burdens, and the second concerns the fair differentiation of responsibilities. The distribution of entitlements and burdens can be assessed by relying on one or combinations of principles of climate justice. This is why ethics in climate change is often identified with justice. Although the fairness of any differentiation of responsibilities must rely on these principles of justice, the applicability of these principles and the demands they make depend strongly on the agents bearing the responsibility and what

Managing Global Warming. https://doi.org/10.1016/B978-0-12-814104-5.00026-0

policy domains are at issue. Not all agents can be ascribed the same responsibilities, and not all measures for climate protection can or should be realized by the same differentiation of responsibilities.

Section 26.2 explains the most important ethical implications of international climate politics. It shows why most ethical discussion revolves around the principle of common but differentiated responsibilities and is closely linked to considerations of justice. Thereafter, I introduce the main domains of climate justice and their most salient principles (Section 26.3). These are the domains of historical, global, and intergenerational justice. The various climate policy domains of mitigation, adaptation, and climate loss and damage (L&D) may require different combinations of principles of climate justice for the just management of climate change. As these combinations require varying differentiations of responsibilities, Section 26.4 discusses how responsibilities can be differentiated. These differentiations not only depend on which principles of justice are employed but also on the particular climate policy area and level. For instance, international agreements on mitigation demand another differentiation of responsibilities than local adaptation or L&D policy. The final section concludes the chapter (Section 26.5).

26.2 The ethical challenges in international climate politics

International climate politics, under the umbrella of the United Nations Framework Convention on Climate Change (UNFCCC), has produced a myriad of agreements, decisions, and institutional bodies dedicated to managing and financing global climate action. One of the most recent developments is the Paris Agreement. As in many other agreements and decisions taken under the UNFCCC, the parties to the Paris Agreement agree that the consequences of climate change are dangerous and should be kept to a minimum if not prevented entirely [1]. What is a historical breakthrough in the Paris Agreement is that the parties agreed to keep the rise in global mean temperature "well below 2°C above preindustrial levels." This is an important political breakthrough not because a target has been set for the maximum increase in global mean temperature but because this target sets a very low acceptable temperature rise before climate change is considered to become dangerous. While the target of a maximal 2°C rise in global mean temperature demands emission reductions, in Article 2(b) the Paris Agreement also declares that, beyond such mitigation efforts, it is essential to strengthen the adaptive capacities of communities, regions, and countries threatened by the adverse effects of climate change. This is an important achievement, because according to the Paris Agreement strengthening adaptive capacities not only concerns measures to adapt to changing climatic conditions but also addressing those climate impacts that cannot or will not be avoided by adaptation: climate loss and damage.

In my view, both these political developments and all other political achievements before the Paris Agreement concerning the three climate policy domains of mitigation, adaptation, and L&D face different ethical challenges, but all of these are closely connected to questions of justice and responsibility. While since the establishment of the

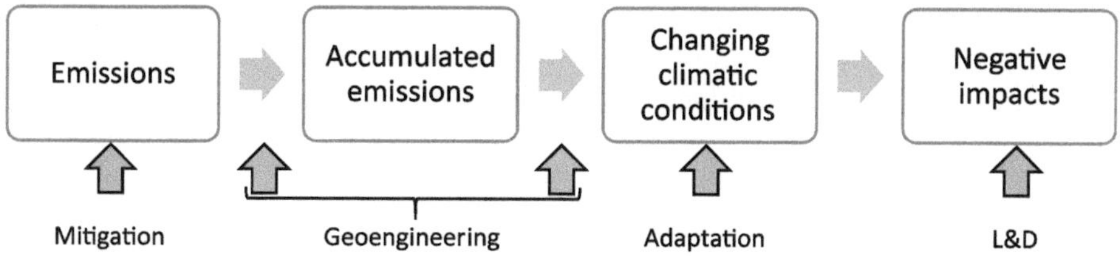

Fig. 26.1 Illustration of the process of climate change and how different areas of climate action can interact to avoid or minimize negative climate impacts (for a more detailed discussion of this figure, see Chapter 4 in Ref. [7]).

Warsaw International Mechanism in 2013 the challenges of adaptation are kept politically separate from those of L&D, they are not readily kept apart conceptually [2,3]. By contrast, mitigation concerns two ethical issues that are conceptually clearly distinct but are not negotiated explicitly in international climate politics [4]. To keep the global temperature rise well below 2°C, reducing current emissions is not enough. Some kind of technical intervention in the climate system is needed too [5,6]. Accordingly, the area of climate action to keep the global mean temperature well below the 2° C also concerns geoengineering. These are the four areas of international climate politics with distinct ethical implications that I explore in the reminder of this section: mitigation, adaptation, L&D, and geoengineering (Fig. 26.1).

Considerations of justice very often investigate the distribution of entitlements and burdens of the members of communities toward a shared goal, and this is also true of climate ethics. In this case, the most relevant community is all parties to the UNFCCC. The shared goal is avoiding or minimizing dangerous climate change and its negative effects. The first principle in Article 3 of the Framework Convention holds that all parties to it have common but differentiated responsibilities to protect the climate system [8].

> *The Parties should protect the climate system for the benefit of present and future generations of humankind, on the basis of equity and in accordance with their common but differentiated responsibilities and respective capabilities. Accordingly, the developed country Parties should take the lead in combating climate change and the adverse effects thereof.*

According to this principle, the parties have accepted to share the burdens of mitigating climate change. In later decisions, the parties adopted the same principle to share the burdens for adaptation and L&D action [2,9]. How these burdens must be shared among the different parties is a typical question of justice [10–12]. However, justice is not only about distributing burdens. It also concerns the just distribution of entitlements. In the case of climate change, these entitlements not only include financing measures for adaptation and L&D but also rights to emitting. Similarly to the burdens of climate action, the appropriate distribution of these entitlements also concerns justice. But principles of justice used to secure a just distribution of entitlements need not be the same as those used to distribute the burdens of climate action.

An example can illustrate this point. Together with two friends, you are organizing a birthday party for a fourth friend, who will soon be 40. When it comes to the question of who bakes the birthday cake, then it is very probable that you will let whoever of the three who is most experienced do the baking. However, the ingredients can be bought by whoever has a shop nearby and owns a car. Perhaps one friend both bakes and shops, or two friends do one of these tasks each. To distribute preparations for the party fairly, after these two tasks are allocated, the third friend should volunteer to do other things. When discussing this distribution of tasks, all three friends will bring forward distinct criteria or principles of justice. But these principles will not be necessarily the same as those employed when it comes to the distribution of the cake and other tasks at the party. There, the friend celebrating his birthday will most probably receive the first slice, while the three who have organized the party will wait until all the other guests have been served. Perhaps the hungriest will get two slices, and those who had dinner just before the party will only take a small slice. Although principles of justice will be employed in preparing the party, in baking the cake, and in distributing slices at the party, these principles are not necessarily the same.

This example of the cake most clearly illustrates the case of mitigation. If the target is not to be overshot, setting a target for the maximum increase in global average temperature means limiting the accumulation of emissions. The volume of emissions that can be produced is restricted to what will leave the global mean temperature well below 2°C above preindustrial levels. The total sum of cumulative emissions that leaves the average global temperature below this target establishes a budget to be distributed among all the parties. This is the cake that becomes relevant to the just distribution of burdens and entitlements with mitigation. The slices to be distributed are the entitlements to produce a certain amount of emissions. Whether these emission entitlements must vary, and if so according to what criteria, is defined by the principles of justice adopted.

Although these principles of justice can be the same when sharing the burdens of mitigation, such as emission reductions that may well restrict economic development, they need not be identical. As Article 3 of the Convention states, one might consider it fair that developed countries take the lead in reducing their emissions. But one might also argue that countries should only be expected to reduce emissions if they can secure subsistence conditions for their citizens. All countries are entitled to the emissions necessary to secure subsistence for their citizens. Thus, the principles of justice to distribute the burdens of mitigation might differ from those adopted to distribute the entitlements to a certain amount of emissions. In Section 26.3, I introduce some principles of justice that could become relevant here and regarding the other two areas of climate action.

The cakes to be produced and distributed in the cases of adaptation and L&D are not the same as in mitigation. Adaptation and L&D concern measures to minimize the risks of climate impacts that cannot be prevented by mitigation efforts. Measures to minimize these risks do not concern a cake of emissions to be distributed but a set of measures to be taken [13,14]. First of all, this concerns the fair distribution of burdens for financing, implementing, and maintaining these measures. However, distributing the burdens in adaptation and L&D not only means contributions from those responsible for the fact that these measures become necessary, it can also mean obliging those best able to contribute irrespective of their responsibility for climate change. The distribution of financing, implementing, and maintaining burdens once again uses principles of justice to assign these responsibilities.

The same holds when it comes to the distribution of the two cakes: how to most effectively and efficiently provide adaptation and L&D action for the communities, regions, and countries that need such measures. But in securing adaptation and L&D measures, it is important to clarify who is entitled to a slice of the cake. According to agreements under the UNFCCC, this is most probably the developing countries, because they contributed less to climate change and are usually less well prepared and able to take action [9,15]. Furthermore, countries that apply for adaptation finance probably have to prove that the measures they want to implement will be effective and efficient [9]. Again, the principles of justice that are employed to distribute assistance in adaptation and L&D action need not be identical to those adopted when distributing the burdens of financing, implementing, and maintaining adaptation and L&D measures.

However, distributing the burdens and entitlements of adaptation and L&D measures is complicated by another challenge. Although the parties to the Paris Agreement confirmed that adaptation and L&D are two distinct areas of climate action, it is not at all clear how to distinguish the two. This is because all measures taken in reaction to changing climatic conditions can be understood as coping with the risks of negative climate impacts. Although L&D seems only to concern the climate impacts that cannot or will not be avoided [16], it is far from clear which specific measures are relevant only to L&D and which ones to adaptation [3]. Early warning systems offer an illustrative example. An early warning system that alerts people living below a glacial lake about an outburst flood might be understood as adapting to the new risk because it allows them to flee from a flood when occurring. However, since such a measure cannot prevent potential loss and damage but only minimize these, it can also be deemed a measure relevant only if a flood could not have been prevented—in other words, an L&D measure.

Certainly, adaptation measures are usually understood as being capable of preventing negative climate impacts. But since these measures are always a reaction to cope with potential L&D and because some risk always remains that they will not be successful, they can always also be viewed as being or as being linked to L&D measures. This is because the cake to be distributed when dealing with climate impacts not preventable by mitigation always concerns the finance, implementation, and maintenance of whatever measures to cope with potential loss and damage. Depending on whether the risk of loss and damage can be reduced to a tolerable minimum or not, the measures can be deemed relevant either for adaptation or L&D [17].

However, there is one crucial difference between the issues of justice involved in adaptation and those in L&D. In adaptation, compensatory claims are only relevant to the production of the cake, but in L&D, compensation plays an important role in choosing and implementing which measures are taken. If adaptation is possible, only a small risk of loss and damage remains. By contrast, L&D measures are implemented because loss and damage are expected. They become relevant when loss and damage have materialized. Consequently, the optimal L&D measures after a hazardous event are those that most fairly rectify the loss and damage the victims now face. This concerns considerations of compensatory justice that are distinct from those of distributive justice, to some extent at least [18].

While the distinction between adaptation and L&D is difficult to draw because the cakes to be produced and distributed are quite similar in both areas of climate action, the cake in geoengineering is rather different from that in mitigation. Although the goals of mitigation measures and geoengineering are identical, the measures to be taken are clearly distinct. To avoid overshooting the global mean temperature rise target, it is essential that emission reductions are accompanied by technical interventions into the climate system [6]. This makes it necessary to acquire scientific knowledge about geoengineering techniques and to finance implementation and maintenance of these interventions. Thus, the cake to be baked here consists of knowledge production about geoengineering techniques and their governance. This is quite different from measures to reduce emissions by producing less [19].

The same holds for the distribution of the cake once geoengineering solutions are in place. While with emissions the burdens to be distributed consist of reducing emissions, in geoengineering, it is the distribution of risks stemming from geoengineering interventions that become relevant. Whatever geoengineering measures are taken, they will have negative side effects that, depending on the techniques employed, will be more or less severe and more or less controllable. Thus, in case of geoengineering, another two sets of justice principles might become relevant to international climate politics. The first of these are principles relevant to the fair distribution of burdens in developing knowledge about geoengineering. The second will be needed to decide the fair distribution of risks arising from technical interventions in the climate system. However, while the various justice issues of adaptation and L&D are already on the political agenda, those of geoengineering do not yet play such a great role [4]. As we have seen in this section, implications of justice can be found in all areas of international climate politics. Section 26.3 discusses in more detail the most prominent domains and principles of climate justice that may become relevant.

Before introducing these domains and principles of climate justice, one final remark on international climate politics is in order. The parties to the Convention and to the Paris Agreement and other decisions under the UNFCCC are sovereign countries. To ensure their compliance, it might be advisable to respect their sovereignty and let them take whatever measures they think appropriate to meet their responsibilities [20]. This means that although in the different areas of climate action there might be agreement on an international level about which combination of principles of justice is most appropriate, this does not mean that all parties to these decisions will implement the same combination of principles domestically. And as I have argued elsewhere, it might even be the case that some political systems and their cultures undermine the implementation of certain principles of justice more than others [21].

26.3 Domains of climate justice

Section 26.2 explained why international climate politics and especially the UNFCCC is fueled by considerations and claims of justice. It also indicated why differing principles of justice might become relevant in the various areas of climate policy [22–24]. They might differ between mitigation, adaptation, and L&D. L&D demands comprehensive compensatory measures, while adaptation probably does not. And finally, although geoengineering shares with mitigation the target of keeping global mean temperature rise well below 2°C, the burdens and entitlements relevant for principles of justice are not the same in both policy areas. Here, I introduce the three main domains of climate justice and explain their relevance for climate policy in general and the different areas of climate action more specifically. Before doing so, however, I want to explain how the three domains of historical, global, and intergenerational justice relate to each other (see Fig. 26.2).

According to the latest reports of the Intergovernmental Panel on Climate Change (IPCC), many regions and communities are facing the adverse effects of climate change already today, but the frequency and severity of such events will heavily increase in the future [15]. Similarly, greenhouse gases remain in the atmosphere for thousands of years and affect the climate long into the future. This means that whenever climate agreements and policies are negotiated, what we owe to the future is also more or less explicitly under scrutiny [7]. Depending on what measures we take today to reduce our emissions, the climatic conditions in the future will differ. For instance, the steps we take today to adapt to or to prepare for climate loss and damage lead to more or less severe conditions for those living in the future and probably threatened by climate impacts. This is why our duties toward the future define the scope of tolerable policy decisions and obligatory climate action.

Principles of intergenerational justice aim to define what we owe to future generations and so to determine what action we have to take today [25,26]. However, the duties of the currently living are also determined by their ancestors or their countries' past emissions, since these emissions are the cause of changing climatic conditions today and in the future. Principles of historical justice define what duties those living today have acquired from their ancestors. Since climate change is a challenge that can only be settled by a global effort, considerations of global justice are also relevant to fairly distribute the entitlements and burdens between the parties today.

At least three principles can become relevant when defining our duties toward the future [27–29]: equity, sufficiency, and the precautionary principle. According to the principle of equity, we have a duty to ensure that future generations will find the same conditions of well-being as we do today [30–32]. The principle of sufficiency

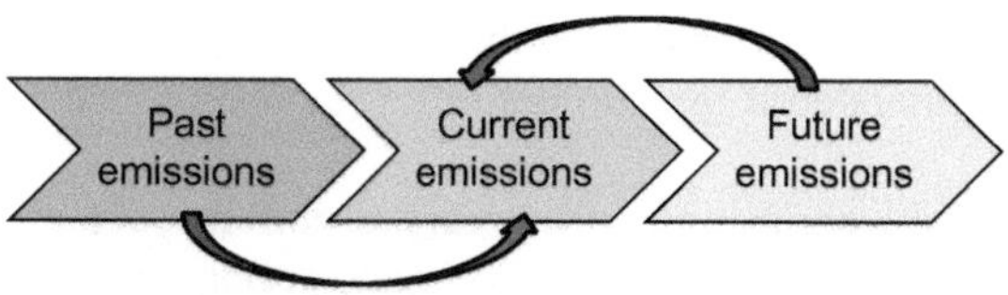

Fig. 26.2 Illustration of the relation between the three domains of climate justice.

demands we ensure that future generations will have enough to live decently [33,34]. The precautionary principle demands that we consider the worst effects of climate change and requires us to do the best we can to prevent them from happening [27,35]. The precautionary principle is the most general one of the three, because it can also be used to decide which measures are most appropriate to meeting our responsibilities by the other two principles. It asks us to conceive the risk of failing to fulfill their demands and therefore to take precautionary measures to prevent this from happening.

I believe that whether or not the currently living are obliged to ensure equal conditions of well-being for the future or only a level of sufficiency is a matter of degree and changes according to the time horizon within which we frame the demands of intergenerational justice. The further in the future that we consider the generations entitled to what we owe them, the more our duties will tend to taper toward the sufficiency level and so the less demanding they will be. I believe this to be so for two reasons. First, we seem to know very well what will count as equivalent conditions of well-being in the near future. However, the further into the future we look, the more difficult this becomes. So many factors could change the preferences and ideals of the good life far in the future that it is very difficult, if not impossible, to know what we have to ensure for these future generations. Second, it seems to be clear that for our closer descendants we should ensure that they find at least equal if not better living conditions than we enjoy today. But the further into the future we look, the more demanding this duty becomes, because ensuring equal conditions of well-being far in the future involves considering so many uncertainties that our living conditions today would have to be significantly restricted. Especially, if we consider the precautionary measures needed to avoid failing our duties toward the future, we would have to restrict our own lives even more severely.

Both these challenges point toward the intergenerational principle of sufficiency. If anything is certain with regard to future generations, it is the fact that human beings in the future will need some basic minimum to live a decent life [24,33,34]. How to define this sufficiency level is a matter of extensive debate. Some argue that basic needs or fundamental interests must be fulfilled [36–38]. Others claim that what must be protected are basic human rights [24]. However, whatever sufficiency level is defined as necessary for future generations, it would contradict this principle to demand that the currently living should sacrifice the same sufficiency conditions to do their duty towards future generations. This is an important consideration of global justice when distributing the burdens of climate protection and measures among the currently living. It can be formulated in two slightly different principles. The formulation, closer to considerations of sufficiency, calls for the distinction between sufficiency emissions and luxury emissions [39]. While sufficiency emissions are the emissions necessary for a decent life, luxury emissions are those that are produced by a lifestyle beyond the sufficiency level. According to this distinction, emission reductions can only legitimately be demanded if they do not infringe upon the emissions needed to lead a decent life.

The distinction between sufficiency emissions and luxury emissions establi shes a threshold defining among whom the burdens of climate action may be legitimately distributed. The same holds for a second principle of global justice relevant to climate change: the ability-to-pay principle [22,40,41]. According to this principle, it is only legitimate to demand people carrying burdens for climate action if they are able to do so. In terms of the distinction between sufficiency emissions and luxury emissions, this would be those polluters producing more emissions than they need for a decent life. However, the distinction between subsistence emissions and luxury emissions only discriminates between legitimate and illegitimate bearers of responsibility for climate action, but the ability-to-pay principle can also be understood more gradually. Ability-to-pay can also mean that climate action should be taken where its goals can be reached most efficiently and effectively and by those who are best able to bring them about. For instance, this can justify a distribution of burdens to support adaptation depending on geographic proximity or the possession of specialized knowledge [14]. Similarly, it has been argued that emissions should be reduced where it is most efficient for the global economy, for example, by transferring clean technology to developing countries [42].

Arguments from ability-to-pay are dedicated to ensuring that only those that can or are best able to meet them actually acquire responsibilities for climate action. This demand is part of the first principle in Article 3 of the Framework Convention, which states that the parties should not only protect the climate system according their common but differentiated responsibilities but also according to their various capabilities. However, what has become more prominent in international climate politics is another part of this principle of the Convention. This part of the principle holds that "the developed country Parties should take the lead in combating climate change and the adverse effects thereof" [8]. From an ethical point of view, this duty can be justified in two ways, one relating back to the ability-to-pay principle, and hence to global justice, and the other to the domain of historical justice and its principles. According to the first way, this part of the principle means that developed countries have a duty to take the lead because they have more resources and know-how than developing countries. In the second, developed countries should take the lead because they are most responsible for climate change and its negative impacts in the first place. This leads to the domain of historical justice.

Historical justice is important to ethical considerations of climate change because, since the beginning of industrialization in the 18th century, it has been the developed countries that have produced most of the emissions that have triggered climate change. For this reason, they can be said to have a duty to take the lead in climate action and to carry heavier burdens than developing countries. Arguing this way, amounts to the polluter-pays principle [22,40,43]. This principle claims that emitters should shoulder burdens of climate action in proportion to their contributions to climate change. Consequently, developed countries have a duty to foster climate action in proportion to the emissions they have produced since the beginning of industrialization [44,45].

The polluter-pays principle has been criticized for at least two reasons. First, scientific knowledge about climate change only developed during the 20th century and became a widely acknowledged scientific fact only with the first assessment report of the IPCC in 1990 [28]. Usually, we assign responsibility only for consequences that could have been known by the actor. This is why many argue that only emissions since

1990 should count in distributing the burdens of climate action. Second, citizens of industrialized countries neither contributed to the emissions before their existence nor had they any opportunity to influence policy decisions allowing uncontrolled pollution [46]. The beneficiary-pays principle is a possible response to these and other criticisms of the polluter-pays principle. According to this principle, developed countries and their citizens do not have a duty for climate action because they contributed to climate change but because they profit from the emissions of their ancestors [47]. Industrialization allowed the economic and technological development that led to the favorable conditions of most citizens in developed countries enjoy today. This is why these countries should take the lead and shoulder proportionally heavier burdens. Interestingly, both the polluter-pays principle and the beneficiary-pays principle need not only be understood as principles of historical justice in such a strict sense. They can also be applied to those living today [48]. Following the polluter-pays principle, those emitting more should pay more for their contribution to climate change. The beneficiary-pays principle could be understood as making those who have a more emission-intensive lifestyle shoulder heavier burdens.

The principles of global justice and historical justice introduced so far all concern the fair distribution of burdens necessary to meet our intergenerational duties toward the future but not any entitlements. The only principle at least partially concerned with entitlements is the distinction between sufficiency emissions and luxury emissions. It entitles all living human beings to the emissions they need to lead a decent life. Another principle aiming at a fair distribution of entitlements is the equal-per-capita principle [49]. According to this principle, all human beings are entitled to an equal amount of emissions. That is, all are entitled to an equal slice of the cake of emissions permitted without overshooting the target of an increase in average global mean temperature of well below 2°C. However, as plausible as this principle may appear at first sight, it becomes problematic if the right to equal emissions is extended in time and all over the world. The more emitters in the past and the future are considered in addition to the currently living, the smaller the slice of emission allowances becomes for those living today [30]. In addition, change of climatic conditions with latitude means individuals need different and not equal amounts of emissions to lead a decent life. For instance, in northern countries more heating is need during winter than in southern countries with no winter at all.

Emission trading schemes allow these negative implications to be cushioned when applying the equal-per-capita principle [50,51]. Emission trading schemes start from the assumption that each human being is entitled to an equal amount of emissions. Accordingly, these schemes assign countries emission permits in proportion to their population. Those countries consuming more than their equal share of emissions are allowed to buy emission permits from those not using their own allowances. From the point of view of global justice, this can be seen as a positive effect, because most often those not using their fair share of emission allowances are the today's poor, often living in developing countries. This would allow a global redistribution of resources that ought to lead to less global poverty. Moreover, having to buy emission permits when overshooting one's fair share can also be understood as a sort of a compensatory fine making those who unfairly polluted too much pay.

Certainly, such an argument for compensation also holds with the principles of historical justice. Making those who polluted or benefited from past emissions pay is some sort of compensatory measure, because those who pay are in fortunate conditions due to past or current emissions beyond their fair share. However, in contrast to mitigation and adaptation, compensation for L&D and the distribution of risks stemming from geoengineering plays a more general role than in these historical principles of justice [3,18]. Compensation here means making those whole again who undeservingly face the negative impacts of climate change or climate engineering. Whether compensation is owed by those who contributed to this harm or by those who are best able to assist seems to be only of secondary importance. If we accept that certain regions and communities are already facing more or less acute situations of emergency due to these impacts, it is more important to assist as effectively and efficiently as possible rather than to clarify who caused them.

This broader understanding of compensation is less relevant in adaptation because in international climate politics adaptation is usually assumed to minimize the risk of climate loss and damage to a level that they can be said to have been prevented [52]. With adaptation, it is more important to ensure that finance, implementation, and maintenance of adaptation measures is fairly distributed. But depending on the weight that efficiency and effectivity are given, ability-to-pay considerations become more or less important than considerations of historical justice [14,53]. While with both L&D and adaptation it seems plausible to carefully weigh considerations of ability-to-pay and historical justice, this might be considered less fair with mitigation. With mitigation, it might be considered less fair to allow current or past high emitters to avoid shouldering heavier burdens because in fact they contributed significantly more to climate change. What combination of principles of justice is fair in these cases remains a matter of ethical investigation and political negotiations, the question cannot be decided here. Instead, we move on to another normative issue in international climate politics: the differentiation of responsibilities.

26.4 Differentiating responsibilities

Section 26.3 introduced the three domains of climate justice: historical justice, global justice, and intergenerational justice. Intergenerational justice always determines the scope of legitimate policy decisions of those living today in light of their duties toward the future. Section 26.2 illustrated how the common responsibilities of all parties to the UNFCCC can be differentiated according to different combinations of principles of climate justice. It also indicated why these combinations might vary by climate policy area. However, whatever distribution of entitlements and burdens is deemed most appropriate for the fair distribution of burdens and entitlements, in my view, it is of key importance to ensure that the main actors in international climate politics are and remain compliant and motivated to act for climate protection [21]. As a consequence, I believe that it is essential to understanding international climate politics as a system with at least two different levels of policy-making [20,54,55]. On the first, international level, compliance can only be reached if any combination of principles adopted by international agreements can be accepted as fair by all parties involved. For the same reason, on the second, domestic level, national self-determination in implementing internationally agreed responsibilities should be guaranteed. This model of two different levels of climate policy leads to complex differentiations of responsibilities. It is to this topic that this section now turns.

To show why the differentiation of responsibilities for climate change adaptation is complex, it is helpful to consider the four aspects of responsibility [14,56]. Someone, the subject of responsibility (i) is always responsible for something, the object of responsibility (ii), answerable to some institution (iii), and held accountable to some norm (iv). This section first explains why the subject of responsibility varies depending on the level of climate policy for which responsibilities are differentiated. Second, I also clarify why the object of responsibility in climate policy must change according to the area of climate action at issue. Third, elaborating on the first two helps to understand why, depending on the policy level and the object of responsibility, the institutions that the responsibility bearers are accountable to may also differ. Fourth, these considerations do not help to clarify what combination of principles of climate justice is most appropriate in specific areas of climate policy. But they help explain why the same combinations of principles cannot plausibly be applied to all levels, objects, and institutions of responsibility (see Fig. 26.3).

In order to understand why the subjects of responsibility change depending on the policy level at issue, we must be clear about the conditions an agent requires to be responsible in the first place. Following common sense, agents can only be held responsible if they are able to form intentions, generate knowledge, and reach and implement decisions. Usually, most human beings fulfill these conditions. When having to choose whether or not to eat a second slice of the birthday cake, we are most often able to form an intention that we want to eat that extra slice or not. In most circumstances, we know what the consequences of our decisions are, and we are most

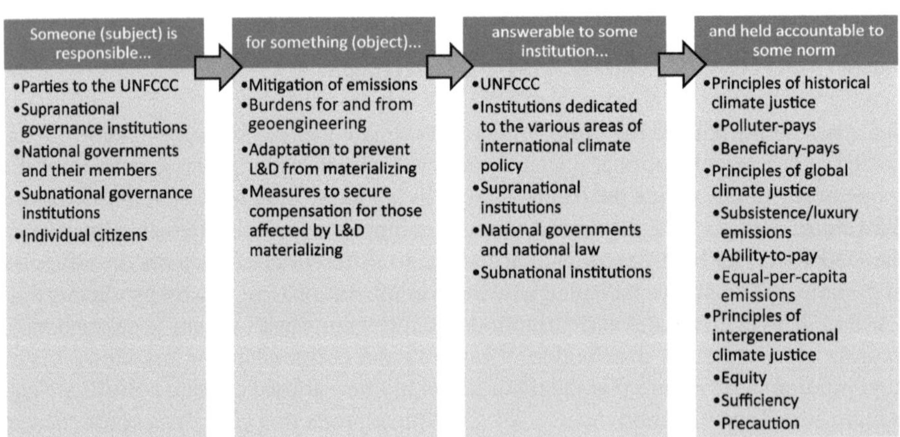

Fig. 26.3 Overview of the most important subjects, objects, institutions, and norms relevant in climate politics and ethics.

often also able to decide and ask for an extra slice or not. Transferred to the two levels of international climate policy, this means that the responsibility bearers must also be able to fulfill these conditions. In international climate politics, these are the parties that negotiated the agreements, the signatory countries. In domestic politics, it is the governments that decide which policies should be implemented to fulfill the agreements negotiated at the international level.

Sticking to these two levels of climate politics, it becomes clear that the subject of responsibility varies [54,55]. In international climate politics, the countries are first and foremost the responsible subjects. They are conceived as collective agents able to meet responsibilities irrespective of the human individuals actually filling the positions of government. It is a complex task to understand in what sense collectives such as countries can be deemed responsible agents. But if they are perceived as capable of shouldering responsibility as collectives, they must, in some sense at least, be capable of fulfilling the four conditions just mentioned. Assigning responsibility seems to be less complex at the domestic level, where individual members of a government can be conceived as the responsibility bearers. However, this is more complex than it seems, because members of a government are usually assigned responsibilities for all impacts their ministries as collectives cause irrespective of whether their decisions and actions as individuals directly caused these effects.

However, the subject of responsibility also changes depending on the policy level at issue. Since there are not only two levels of policy-making that can become relevant to climate change and the areas of climate politics, many more relevant agents can be deemed responsible. Different subjects might be assigned specific responsibilities for climate action. Subnational and supranational communities and regions can be assigned varying responsibilities too. Different countries with differing constitutions and political cultures might not be able to carry the same burdens of climate action. Furthermore, since all these potentially relevant subjects of responsibility do not exist in isolation of each other, their individual responsibilities are part of a complex net of responsibility dependences, which justifies further differentiations of responsibilities.

This already complicated picture becomes even more complex when considering the varying objects of responsibility in climate politics. Consider the different goals of mitigation and adaptation. If we follow the polluter-pays principle, developed countries can be assigned responsibility for their past and current emissions, from 1990 at least. On this account, the quantity of past and current emissions for which countries are responsible decides how responsibilities for mitigation and adaptation have to be differentiated. However, applying this same principle to both policy areas misunderstands the goal of adaptation measures in contrast to those of mitigation [14]. It emphasizes the idea that polluters should pay for their emissions but ignores the fact that adaptation is meant to enable vulnerable countries or communities to cope with the adverse effects of climate change that cannot be prevented by mitigation. The goal of adaptation is different from that of mitigation. Accordingly, while with mitigation it makes perfect sense to differentiate responsibilities according to emissions, this is less convincing with adaptation.

It makes sense to differentiate responsibilities for mitigation regarding differences in emissions because the goal of mitigation policy is to prevent dangerous climate change from happening, chiefly by ensuring that an emission budget is not exceeded. With adaptation, the goal is different. Adaptation measures are not meant to avoid dangerous climate change by reducing emissions. Adaptation to climate change aims to cope with the threat of climate impacts. For instance, building seawalls prevents or at least minimizes the risk of negative climate impacts. With adaptation, then, the object of responsibility is different; what matters here are not emissions but financing, implementing, and maintaining appropriate adaptation measures where human beings and natural systems are under threat of climate impacts [14]. To put it in terms of the birthday cake, the cakes to be produced and distributed are simply not the same in these two areas of climate policy. As indicated when discussing the birthday cake earlier, the same seems to hold for L&D and geoengineering. Thus, depending on the policy area at issue, the object of responsibility seems to greatly vary.

Turning now to the third aspect of responsibility, it seems clear that the institutions the responsibility bearers are answerable to varies with the subject of responsibility as well as with the object of responsibility. The institutions the subjects of responsibility are answerable to change depending on the specific subject, because every subject is always embedded in differing relations of answerability. On the level of international climate politics, countries are answerable to the institutions established under the umbrella of the UNFCCC. On the domestic level, governments are answerable to their people. And individual citizens are accountable to national law. How these relations of answerability are shaped depends on the nature of the institutions considered. This is where the object of responsibility becomes relevant, because the constitution of political institutions to secure appropriate climate action most often depends on the aim for which they are established. Funds for adaptation or L&D are established to finance corresponding climate action. Those receiving these funds are answerable to these institutions and must prove that the money has been invested in appropriate provisions.

Evaluating whether or not the money received for adaptation or L&D measures has been invested properly or whether the emission reductions undertaken have been as expected cannot be decided without relying on any norm. This brings us to the fourth aspect of responsibility; the norm according to which the responsibility bearers are held accountable. In climate politics, these norms most often correspond with one or a combination of principles of climate justice. In assigning responsibilities, these norms or principles can take two different conceptual roles. Either they assign responsibility for an outcome or for remedy of harm [57]. Outcome responsibility denotes responsibility for bringing about a certain state of affairs, and remedial responsibility denotes responsibility for remedying harm. While the first kind of responsibility is backward-looking, the second looks forward [58]. That is, the first concerns the domain of historical justice and the second the domain of intergenerational justice. In principle, both these conceptions of responsibility and domains of justice are independent, but in case of historical justice they are linked. In order to legitimately claim that the bearer of outcome responsibility is also remedially responsible, some kind of normatively relevant tie between the two must be established.

Miller suggests moral failure, significant contribution to the outcome, and mere causal contribution as legitimate reasons for assigning remedial responsibilities based on outcome responsibility [57]. If this kind of connection is or can be established, then we are dealing in the domain of historical justice. In this case, it is the assignment of responsibility for a certain state of affairs that justifies remedial responsibility. This is the most obvious way to distribute responsibilities in the case of climate action. Responsibilities would be assigned in proportion to the contribution to climate change and its effects, like levels of greenhouse gas emissions. Across all areas of climate action, the reasons linking outcome responsibility to remedial responsibility would amount to wrongful emitting, nonwrongful but significant contribution to climate impacts, or causal contribution [58]. Whichever of these three reasons justifies a connection between outcome responsibility and remedial responsibility, it operates within the framing of historical justice. Those bringing about harm are assigned the responsibility to carry heavier burdens than those less responsible for the harm. But as argued earlier with regard to the varying objects of responsibility, such a backward-looking assignment of responsibilities might not be equally appropriate in all areas of climate policy.

When harm has materialized, considerations of historical justice aim to identify those responsible for the harm in order to assign remedial responsibilities. However, such a purely backward-looking assignment of remedial responsibilities may either not be fully possible or fully appropriate in an ethical sense [3,42,59,60]. The first is probably the case because attribution of a hazardous event to climate change might not be feasible. The second might be the case because only a portion of experienced climate impacts would be covered were remedial responsibility to be based on outcome responsibility [58]. Natural climate variability and socioeconomic factors significantly contribute to much of the climate impacts as well. These impacts would not be covered if responsibilities for climate action would be assigned by a purely backward-looking approach. Such an approach only captures climate impacts that are anthropogenically caused but not those occurring or being intensified by naturally changing climatic conditions, economic misfortune or social disadvantage.

26.5 Conclusion

The discussion in this chapter started with an examination of international climate politics. It focused on the principle of common but differentiated responsibilities and its ethical implications. International climate politics is fueled by issues of justice because differentiating the various responsibilities and entitlements related to climate action demands principles of climate justice. These principles can stem from three domains: historical, global, and intergenerational justice. While the third defines what we owe toward the future, the other two domains can help to specify the various responsibilities borne by differing responsibility bearers in the diverse areas and levels of climate action. Considering these areas and levels of climate action shows that it is not only the subjects of responsibility that may change, depending on whether mitigation, adaptation, L&D, or geoengineering policy is at issue. They may well also have different objects of responsibility. And since responsibility bearers are embedded in complex nets of responsibilities, the area and level of climate action in which they are involved will also render them answerable to different institutions and accountable to varying norms.

However, decisions about the appropriate differentiation of responsibilities cannot be made from the drawing board. These decisions demand political negotiations that must be secured by appropriate political institutions. Thus, examining ethical implications in international climate politics should also extend to considerations about the nature of the institutions to be established [21,61,62]. The involvement of those affected may be more or less important, depending on the policy area. Similarly, the level of climate policy might define who has to bear what duties to ensure that the responsibility bearers meet their responsibilities and to decide what those responsibilities should be. But these are issues that could not be addressed at length here and must be addressed in another paper.

References

[1] United Nations. COP 21: adoption of the Paris agreement; 2015.

[2] United Nations. Report of the conference of the parties on its nineteenth session, held in Warsaw from 11 to 23 November 2013: part two: action taken by the conference of the parties at its nineteenth session; 2014.

[3] Wallimann-Helmer I. Justice for climate loss and damage. Clim Change 2015;133 (3):469–80.

[4] Honegger M, Münch S, Hirsch A, Beuttler C, Peter T, Burns W, et al. Climate change, negative emissions and solar radiation management: it is time for an open societal conversation, St. Gallen: Stiftung Risiko-Dialog; 2017. Available from: https://doi.org/10.5167/uzh-137059.

[5] Geden O, Löschel A. Define limits for temperature overshoot targets. Nat Geosci 2017;10 (12):881–2.

[6] Geden O. The Paris agreement and the inherent inconsistency of climate policymaking. Wiley Interdiscip Rev Clim Change 2016;7(6):790–7.

[7] Roser D, Seidel C. Climate justice: an introduction. London, New York: Routledge Taylor & Francis Group; 2017.

[8] United Nations. Framework convention on climate change; 1992.

[9] United Nations. Report of the conference of the parties on its sixteenth session, held in Cancun from 29 November to 10 December 2010: part two: action taken by the conference of the parties at its sixteenth session; 2011.

[10] Moellendorf D. Climate change justice. Philos Compass 2015;10(3):173–86.

[11] McKinnon C. Climate justice in a carbon budget. Clim Change 2015;133(3):375–84.

[12] Pottier A, Méjean A, Godard O, Hourcade C. A survey of global climate justice: from negotiation stances to moral stakes and back. Environ Resour Econ 2017;11:1–53.

[13] Jagers S, Duus-Otterstrom G. Dual climate change responsibility: on moral divergences between mitigation and adaptation. Env Polit 2008;17(4):576–91.

[14] Wallimann-Helmer I. Differentiating responsibilities for climate change adaptation. Archiv Rechts Sozialphilos Beih 2016;149:119–32.

[15] IPCC. Climate change 2014: impacts, adaptation, and vulnerability: WG II AR5: technical summary; 2014.

[16] Mace MJ, Verheyen R. Loss, damage and responsibility after COP21: all options open for the Paris agreement. Rev Eur Comp Int Environ Law 2016;25(2):197–214.

[17] Dow K, Berkhout F, Preston BL, Klein RJ. Limits to adaptation. Nat Clim Change 2013;3(4):305–7.
[18] Page EA, Heyward C. Compensating for climate change loss and damage. Polit Stud 2017;65(2):356–72.
[19] Honegger M, Reiner D. The political economy of negative emissions technologies: consequences for international policy design. Clim Policy 2017;18(3):306–21.
[20] Miller D. Global justice and climate change: how should responsibilities be distributed? The Tanner lectures on human values, vol. 28; 2008. p. 117–56.
[21] Wallimann-Helmer I. Kollektive Verantwortung für den Klimaschutz. Z Prakt Philos 2017;4(1):211–38.
[22] Page E. Distributing the burdens of climate change. Environ Polit 2008;17(4):556–75.
[23] Gardiner SM, Caney S, Jamieson D, Jamieson D, Shue H, editors. Climate ethics: essential readings. Oxford, New York: Oxford University Press; 2010.
[24] Birnbacher D. Klimagerechtigkeit—Verursacher oder Leistungsfähigkeitsprinzip? In: Kallhoff A, editor. Klimagerechtigkeit und Klimaethik. Berlin/Boston: de Gruyter; 2015. p. 67–80.
[25] Gosseries A, Meyer LH, editors. Intergenerational justice. Oxford, New York: Oxford University Press; 2009.
[26] Page E. Intergenerational justice and climate change. Polit Stud 1999;47(1):53–66.
[27] Meyer L, Roser D. Intergenerationelle Gerechtigkeit: Die Bedeutung von zukünftigen Klimaschäden für die heutige Klimapolitik. [July 18, 2013].
[28] Tremmel J. Klimawandel und Gerechtigkeit. Jahr Wiss Ethik 2012;16(1):115–39.
[29] Moellendorf D. Justice and the assignment of the intergenerational costs of climate change. J Soc Philos 2009;40(2):204–24.
[30] Singer P. One world: the ethics of globalization. 2nd ed. New Haven, CT: Yale University Press; 2004.
[31] Gosseries A. Constitutions and future generations. Good Soc 2008;17(2):32–7.
[32] Moellendorf D. Common atmospheric ownership and equal emissions entitlements. In: - Arnold DG, editor. The ethics of global climate change. Cambridge: Cambridge University Press; 2011. p. 104–23.
[33] Page EA. Justice between generations: investigating a sufficientarian approach. J Glob Ethics 2007;3(1):3–20.
[34] Meyer LH, Roser D. Enough for the future. In: Gosseries A, Meyer LH, editors. Intergenerational justice. Oxford, New York: Oxford University Press; 2009. p. 219–48.
[35] Sunstein CR. The catastrophic harm precautionary principle. Issues Legal Scholarsh 2007;6(3):1–29.
[36] Wolf C. Intergenerational justice, human needs, and climate policy. In: Gosseries A, Meyer LH, editors. Intergenerational justice. Oxford, New York: Oxford University Press; 2009. p. 349–79.
[37] Lacey J. Climate change and Norman Daniels' theory of just health: an essay on basic needs. Med Health Care Philos 2012;15(1):3–14.
[38] Schuppert F. Climate change mitigation and intergenerational justice. Environ Polit 2011;20(3):303–21.
[39] Shue H. Subsistence emissions and luxury emissions. In: Gardiner SM, Caney S, Jamieson D, Jamieson D, Shue H, editors. Climate ethics: essential readings. Oxford, New York: Oxford University Press; 2010. p. 200–14.
[40] Shue H. Global environment and international inequality. Int Aff 1999;75(3):531–45.
[41] Caney S. Cosmopolitan justice, responsibility, and global climate change. Leiden J Int Law 2005;18(04):747–75.

[42] Huggel C, Wallimann-Helmer I, Stone D, Cramer W. Reconciling justice and attribution research to advance climate policy. Nat Clim Change 2016;6(10):901–8.
[43] Gardiner SM. Ethics and global climate change. Ethics 2004;114:555–600.
[44] Meyer LH, Roser D. Distributive justice and climate change: the allocation of emission rights. Anal Krit 2006;28:223–49.
[45] Caney S. Climate change and the duties of the advantaged. Crit Rev Int Soc Polit Philos 2010;13(1):203–28.
[46] Gosseries A. Historical emissions and free-riding. Ethical Perspect 2004;11(1):36–60.
[47] Baatz C. Responsibility for the past? Some thoughts on compensating those vulnerable to climate change in developing countries. Ethics Policy Environ 2013;16(1):94–110.
[48] Meyer LH, Roser D. Climate justice and historical emissions. Crit Rev Int Soc Polit Philos 2010;13(1):229–53.
[49] Caney S. Justice and the distribution of greenhouse gas emissions1. J Glob Ethics 2009;5 (2):125–46.
[50] Ohndorf M, Blasch J, Schubert R. Emission budget approaches for burden sharing: some thoughts from an environmental economics point of view. Clim Change 2015;133 (3):385–95.
[51] Schuppert F, Seidel C. Feasibility, normative heuristics and the proper place of historical responsibility—a reply to Ohndorf et al. Clim Change 2017;140(2):101–7.
[52] Roderick P, Verheyen R. Beyond adaptation—the legal duty to pay compensation for climate change damage. WWF-UK, Climate Change Programme discussion paper 2008, 2008.
[53] Kaswan A. Climate change adaptation and theories of justice. Archiv Rechts Sozialphilos Beih 2016;149:97–118.
[54] Wallimann-Helmer I. The republican tragedy of the commons: the inefficiency of democracy in the light of climate change. Ancilla Iuris 2013;1–14.
[55] Wallimann-Helmer I. The liberal tragedy of the commons: the deficiency of democracy in the light of climate change. In: Birnbacher D, Thorseth M, editors. The politics of sustainability. philosophical perspectives. New York: Routledge; 2015. p. 20–35.
[56] Bayertz K. Eine kurze Geschichte der Herkunft der Verantwortung. In: Bayertz K, editor. Verantwortung: Prinzip oder Problem?. Darmstadt: Wissenschaftliche Buchgesellschaft; 1995. p. 3–71.
[57] Miller D. National responsibility and global justice. Oxford: Oxford University Press; 2007.
[58] Wallimann-Helmer I, Meyer L, Mintz-Woo K, Schinko T, Serdeczny O. Ethical challenges in the context of climate loss and damage. In: Mechler R, Bouwer L, Linnerooth-Bayer J, Schinko T, Surminski S, editors. Loss and damage from climate change: concepts, principles and policy options. Heidelberg, New York: Springer; 2018 [in print].
[59] Huggel C, Stone D, Eicken H, Hansen G. Potential and limitations of the attribution of climate change impacts for informing loss and damage discussions and policies. Clim Change 2015;133(3):453–67.
[60] Stone D, Auffhammer M, Carey M, Hansen G, Huggel C, Cramer W, et al. The challenge to detect and attribute effects of climate change on human and natural systems. Clim Change 2013;121(2):381–95.
[61] Wallimann-Helmer I. The ethics of waste policies. In: Lever A, Poama A, editors. The Routledge handbook of ethics and public policy. New York: Routledge; 2018 [in print].
[62] Wallimann-Helmer I, Meyer L, Burger P. Democracy for the future: a conceptual framework to assess institutional reform. In: Sturma D, Honnefelder L, Fuchs M, editors. Jahrbuch für Wissenschaft und Ethik. Berlin: Walter de Gruyter; 2016. p. 197–220.

Local action that changes the world: Fresh perspectives on climate change mitigation and adaptation from Australia

27

Phil Ireland[*,1], *Declan Clausen*[†]
*Online Progressive Engagement Network, Newcastle, NSW, Australia, †Deputy Lord Mayor, City of Newcastle, Newcastle, NSW, Australia
[1]Corresponding author: phil@the-open.net

Chapter Outline

27.1 Introduction

The Earth's climate is in crisis. In the past 12 months alone humanity has witnessed its hottest year in recorded history, a record Atlantic hurricane season, and atmospheric concentrations of carbon dioxide passing 400 molecules CO_2 per 10^6 molecules (400 ppm) for the first time in four million years. Global efforts to address climate change are now many decades old with the World Climate Conference in 1979, the formation of the Intergovernmental Panel on Climate Change (IPCC) in 1988, and the United Nations Convention on Climate Change (UNFCCC) in 1992. For the past 25 years, the world's national governments have come together every year to negotiate a solution to this problem. Despite these efforts, global emissions have

Managing Global Warming. https://doi.org/10.1016/B978-0-12-814104-5.00027-2

continued to increase, the world has continued to warm, and extreme weather events are becoming more frequent and intense [1]. Over the past 25 years, scholars and policymakers have tended to focus on the national and supranational levels of governance far more than substate climate governance. In this chapter, we explore some reasons why this might be the case and subsequently outline a range of current local policy options and actions.

The scale of local government and household action has been neglected in many climate change policy discussions, yet it is one of the most significant opportunities for mitigation and adaptation. Global institutions such as the UNFCCC and IPCC provide very structured and formalized forums to coordinate action across national boundaries. The lack of attention to opportunities at local and household scales for climate change action has resulted in missed opportunities and misinformation. In this chapter, we argue that the local government and household levels are one of the most significant sites of climate action, which also provide opportunities to build resilience.

It is not new or novel to argue that local and household actions on climate change are important and essential. It should be noted that in recent years, mainly since the widely perceived failure of the Copenhagen Conference of Parties, there has been an increasing focus on understanding and implementing substate climate governance [2–5].The attention of some policymakers has also been shifting. For example, C40 is arguably the largest and most influential grouping of large cities that are supporting each other take action on climate change. Following the Copenhagen climate summit, the chair of C40, Mayor Michael Bloomberg stated: "While international negotiations continue to make incremental progress, C40 Cities are forging ahead. Collectively they have taken more than 4700 actions to tackle climate change, and the will to do more is stronger than ever. As innovators and practitioners, our cities are at the forefront of this issue—arguably the greatest challenge of our time." [6]. Efforts such as C40, alongside the continued advocacy of American cities following the Trump Administration's exit from the Paris Climate Agreement, demonstrate substate actors are taking climate action into their own hands [7]; see Fig. 27.1.

However, it should be noted that around 15 years of supranational climate negotiations unfolded before strong networked substate-level advocacy emerged. There are also numerous examples of literature cutting across climate change issues that have argued for a greater focus on local contexts. For example, alternative development scholars have argued that the local scale is the most significant and postdevelopment scholars reject scalar framing that places any scale as implicitly more significant than another [3,8–14].

Despite recent efforts and attention at the local scale, it is evident that these scales have not received the same policy and political attention as the national and supranational scales over the past 25 years. As a result, many opportunities have been missed to reduce greenhouse gas emissions and develop resilience for an uncertain future. Molony et al. state that "to achieve the ambitious goal of retaining global warming to below two degrees, it is increasingly clear that cascading, top-down action from the international scale is going to be insufficient" (p. 2) [4]. We argue that the neglect of the substate over most of the past 25 years of climate change policy and research has resulted in significant missed opportunities to reduce emissions and prepare for the

Fig. 27.1 C40 Infographic of action and impact.

impact of climate change. Understanding why these scales have been neglected is an essential part of ensuring humanity is taking appropriate steps to prepare for the future.

In this chapter, we identify some possible reasons why substate and household actions have been obscured and misguided, to the detriment of climate mitigation and adaptation. Firstly, we explore how dominant framings of scale have obscured and misguided climate change mitigation and adaptation at the local and household scale. Secondly, we identify a range of simple evidence-based policy options for local government that demonstrate the potential volume of emissions reductions for local government. Thirdly, we identify several simple household actions that contribute to climate change mitigation and adaptation. Finally, we make an argument that one of the most leveraged opportunities for individuals to affect significant emissions reductions is through advocacy and campaigning. It should be noted that many climate change mitigation and adaptation policy options and actions are context specific. As such, for this chapter, we draw on examples from Australia.

27.2 Climate change and the local scale

For the past 25 years, climate change policy and politics have been dominated by the national and the supranational scales. We understand the supranational scale to be that which is beyond the Nation State and involves all or many nations. The most prominent examples of at the supranational scale are the UNFCCC and IPCC. Climate change has also been a key topic of discussion at a range of other supranational bodies and forums including the United Nations General Assembly (UNGA), Group of Twenty (G20), and Asia-Pacific Economic Cooperation (APEC). Recently, Molony et al. conducted a literature review and concluded that "research, policy practice and politics of climate change (climate governance) have for a long time had a strong focus on the global level and formation of international regimes" (p. 2) [4]. In this section, we explore several possible explanations for the emphasis on the supranational scale. We also present several arguments that demonstrate why the local scale requires greater focus.

There are several possible explanations for the focus on the supranational scale in climate change discourses. Firstly, a focus on the supranational scale reflects long-held theoretical analysis such as Taylor's world systems theory [15,16]. This theoretical perspective pervades society and places global and national actors as more powerful and influential than local actors. For example, national decision-making bodies are more powerful that substate decision-making bodies (such as state or regional government), which are then more powerful than local government decision-making bodies. This theory holds true for some aspects of governance such as those that involve the military and national taxation. Often national decision-making bodies can overrule local decision-making bodies. However, to extend these examples to all aspects of society is an oversimplification that obscures the significance of local action. There are many critical aspects of society where state and local decision makers have greater power. These differ from country to country but include critical areas such as childcare, waste management, and water.

Secondly, the atmosphere is a global commons (i.e., what goes into the atmosphere in one part of the globe is theoretically spread across the whole globe), and as such it is a common assumption that addressing emissions must primarily occur at a global level. For example, Aalst et al. argue that the top-down thinking pervades climate policy and research is "derived from the original characterization of the issue as a global environmental pollution problem" (p. 166) [17]. Sellers et al. break this down and state "this dominant paradigm begins with a global atmospheric climate model, which is being rapidly elaborated to include more hydrological, oceanographic, and vegetation features" [18]. Thirdly, with such great diversity in local sites around the world, analyzing local climate change responses is a complex task. An action that results in greenhouse gas mitigation in one place may have a null or negative effect in another (for example, the electrification of transport networks may be beneficial in an area with renewable energy sources but less beneficial in an area with a carbon-intensive energy system). Similarly, an action that results in adaptation in one place may be maladaptive in another (for example the development of flood mitigation measures is heavily dependent on local rainfall, local fluvial geomorphology, and surface runoff features). Wilbanks and Kates describe this problem by stating "only some local actions have universal relevance, and the majority of mitigation, and even more so, adaptation measures are necessarily highly customized to local context" (p. 6) [19]. Similarly, Ireland and McKinnon argue that climate change will play out differently in every locality [3,14].

While the recent Paris Accord is an example of important and significant decisions having been made at the supranational scale [20], to date, the combination of emission reduction decisions from substate actors is also significant. [1] argues that for the ambitious goals of the Paris agreement to be achieved "it is increasingly clear that cascading top-down action from the international to national scale is going to be insufficient" (p. 1). Supranational debates have obscured the fact that most of the world's greenhouse gas emissions result from actions at the local level from the use of combustion engines for transport to lighting, to garbage to water heating. For example, while cities only occupy 2% of the world's landmass, they account for more than 70% of global CO_2 emissions [21]. Local policy can address many of these sources. The embedded assumption that the supranational level is the most significant scale has delayed action at other levels of governance.

27.3 Local government policy options

Today, more than half the world's population live in cities, an increase from just 13% in 1900. This is a trend that is not expected to stop, with continued mass urbanization predicted to see more than 60% of the world's population living in cities by 2030 [22].

This transformation poses some significant challenges. For much of humanity's development, rural lifestyles have ensured that populations remain connected with agriculture and other key resources that provide sources of food, clothing, and energy. Today, surveys of children highlight the growing disconnection from food production such as the misidentification of where milk comes from, that wheat is a grain, and the origins of cheese [23]. While such comments can be easily dismissed as insignificant, they highlight the disconnection of entire populations from the natural environment. As discussed elsewhere in this book, global climatic change is of serious concern to the longevity of humanity and the natural environment that we understand.

While the exact form and composition vary from country to country, most national governments, through constitutional recognition, devolved power or charter, grant policy making and regulatory powers to local authorities such as local city councils. Many of these forums are democratic bodies, with a remit covering local planning, waste and energy policy, parks and reserves, and local transportation including the provision of roads, cycleways, and public transport. Globally, the OECD estimates that the average population of each local government area is 56,000 people [24].

While small in scale, cumulatively local authorities have a large economic and social impact. The OECD estimates that subnational government entities are responsible worldwide for 25% of all government spending, equating to 9% of global gross domestic product (GDP) [24]. While the OCED notes that the vast majority of this spending is tied to specific policy outcomes (such as the delivery of transport and waste collection), "discretionary" spending categories such as local spending on housing, recreation, culture and religion, and environmental protection by subnational governments worldwide equates to around 0.5% of Gross World Product (GWP) approximately US$375 $\times$ 10^9 (US$375 billion) [24].

The growing role and influence of local authorities is recognized at the highest level of international governance. The ratification of the New Urban Agenda at the United Nation's Habitat III Conference in 2016, solidified the role local authorities will play worldwide in addressing the challenges of the 21st century to make cities and human settlements inclusive, safe, resilient, and sustainable [25].

One of the greatest benefits of allowing city governments to lead this change is the opportunities for cities themselves to act as laboratories for new approaches, providing examples that can be scaled up to national levels, and with greater flexibility in the design and implementation of policy measures than that available to nation states.

There are a wide variety of measures that can be implemented at a local authority level. These include:

27.3.1 Climate resilience policies

Climate change will have varying effects on different communities. However, there is little debate that climate-based extremes are likely to test the design standards of infrastructure in cities worldwide. Increased periods of drought, more intense storms, and sea-level rise are predicted to have many negative effects [1]. As far back as 2006, it was estimated that the economic cost of unmitigated climate change would be equivalent to losing 5% of GWP each year [26]. A significant portion of this expense is expected to be borne by local communities seeking to upgrade stormwater systems to prevent flooding of urban areas, in the delivery of new dams to collect water to survive prolonged periods of drought, and in deploying assets that prevent erosion. City leaders play a key role in deploying technology and making investment decisions, which actively deliver increased resilience. Harnessing this process, US philanthropists created the Rockefeller Foundation's 100 Resilient Cities program, which aims to connect a variety of cities worldwide, which are seeking to create resilient communities [27]. The leadership of the 100 Resilient Cities network provides a range of resilience-based responses to common urban problems.

27.3.2 Transport

With increased urbanization, there has been a rapid increase in the need for transportation; the International Energy Agency estimates that global transport energy use has doubled in the 30-year period to 2012, and is expected to double again between now and 2050 [28]. Since the industrial revolution transportation has almost exclusively been powered by fossil fuels, which have a significant impact on global greenhouse gas emission. Globally, transportation equates to approximately 15% of all anthropogenic greenhouse gas emissions, making transportation one of the largest single contributors to anthropogenic climate change [29]. The disruption caused by new technologies is quickly changing regulatory frameworks and our understanding of the future of transport, with governments and private enterprise worldwide preparing for the likely introduction of autonomous electric vehicles, and for systematic changes brought about by changes in ownership structures due to the sharing economy.

While there is still uncertainty as to the exact implications of these changes, they are likely to significantly change the way we interact with transportation and derive benefit. This includes dramatic reductions in congestion. For example, some studies predict that ride-sharing apps such as Uber Pool and Lyft Line could reduce the number of taxi vehicles on the road in major cities such as New York by up to 75% [30]. Also, there is a reduction in greenhouse gas emissions through increased electrification of transport networks through the widespread uptake of electric cars.

There are a range of policy opportunities for local authorities. Local authorities will need to be creative and adaptive in responding to these new technologies. For example, increased automation of cars will necessitate reconsideration of long-understood principles such as the function of basic road intersections. Pool ridesharing services such as Uber Pool and Lyft Line will require a shift in thinking in the role of big public transport infrastructure systems, including how to adapt public transport networks to suit increased custom journey expectations. These technological changes will also require local authorities to make shift changes in the deployment of new infrastructure, such as working with others to deliver networks of electric-vehicle-charging stations.

Local authorities will also need to adjust planning mindsets to prepare transportation opportunities to suit their future communities better. Better integration of land-use planning and transport planning is required, which considers changing usage patterns and community expectations. Current best practice is dominated by approaches that focus on efficient, high-frequency public transport systems (metro style train, express bus or light rail) linked with well-integrated "last-mile" services including facilitated active transport such as cycling or walking, or use of taxi/ride sharing. Greenhouse gas reductions are possible through the integration of land use and transport planning, which can eliminate the need for transport and reduce journeys through greater utilization of mixed-use development.

27.3.3 *Water and wastewater*

Most local authorities have a role to play in the provision of water and wastewater services to their communities. This could be either through direct provision or ownership of water and wastewater utilities or through influence over-regulation. These utilities provide communities with potable water from dams, aquifers, or desalination, and collect and treat wastewater (sewage) before discharging back into the environment or recycling for industrial or potable reuse. These processes are energy and greenhouse gas intensive. For example in 2005 the California Energy Commission found that: "water-related energy use [in California, USA] consumes 19 percent of the state's electricity, 30 percent of its natural gas, and 400×10^9 L (88 billion gallons) of diesel fuel every year—and this demand is growing" [31].

The relationship between water usage and energy demand has been described as the energy-water nexus, and describes a simple relationship whereby actions to save water also save energy. Additionally, saving water also reduces carbon emissions by saving energy otherwise generated to move and treat water.

Within these water/wastewater systems, significant portions of greenhouse gas emissions are generated through the treatment of wastewater. At a primitive level wastewater treatment has existed since the early 1840s, with technological advancement driven in succeeding decades by significant health events such as Britain's great stink of 1858 [32]. While incremental improvements have occurred across the last century, the process has remained relatively unchanged, with primary screening, followed by nutrient removal (secondary treatment) and in recent decades disinfection (tertiary treatment in sensitive receiving environments). Secondary treatment processes tend to be the most greenhouse gas intensive, with microbes decaying organic matter within the wastewater to carbon dioxide gas [33].

Modern wastewater treatment processes, however, can be optimized through anaerobic digestion to become energy powerhouses, utilizing new designs that harness microbes that produce methane gas rather than carbon dioxide to decompose nutrients in wastewater, and processes that collect this methane as a source of energy for cogeneration. Using this technology, wastewater treatment plants can now be designed to be near energy self-sufficient [34].

Although the availability of such significant technological advancement uptake tends to be slow with path dependency remaining for the use of technologies that are equally effective regarding nutrient reduction being utilized however fail to generate the other positive byproducts (namely, methane for generation of heat and energy). Local authorities play a significant role in investment in these technologies and must be cognizant of the decisions made to ensure that long-term options are considered.

27.4 Household participation in emission reductions

Direct household actions to reduce emissions are subject to significant attention in climate change policy research. National emission reduction targets are often realized as national schemes like mandatory renewable energy targets, cap-and-trade systems, or direct action plans with large corporate emitters or environmental programs. Australia is unique in that national action has had a "start-stop-rewind" experience making national schemes largely ineffective in meeting national targets. However, over the same period, the residential sector has been very active, particularly in the area of distributed generation and energy efficiency.

Rooftop solar photovoltaic (PV) has dramatically increased across Australia, leveraging private capital from home loans to build a ~6 GW distributed power station (see Fig. 27.2). Although these systems have been partly funded through the national Small-scale Renewable Energy Scheme (SRES), and some have received feed-in tariff subsidies, the take-up rates, and quantity of household capital leveraged has dramatically exceeded expectations of policymakers [35]. These investments have been made by households who would have otherwise been unlikely to make significant investments into emission-reducing ventures.

It is interesting to explore the motivations of the households behind this PV boom. It is clear that the high level of solar radiation in most of Australia, the large proportion of owner/occupiers, and detached houses with plenty of roof space have been underlying advantages in this market. An ethic of self-sufficiency, changing consumer norms, and frustration with energy retailers have also played significant roles [36].

While these energy technologies remain a small capital outlay as a portion of a home, uptake in households can remain high, even if economically irrational. PV early adopters, and, more recently, early battery system owners, have been willing to make

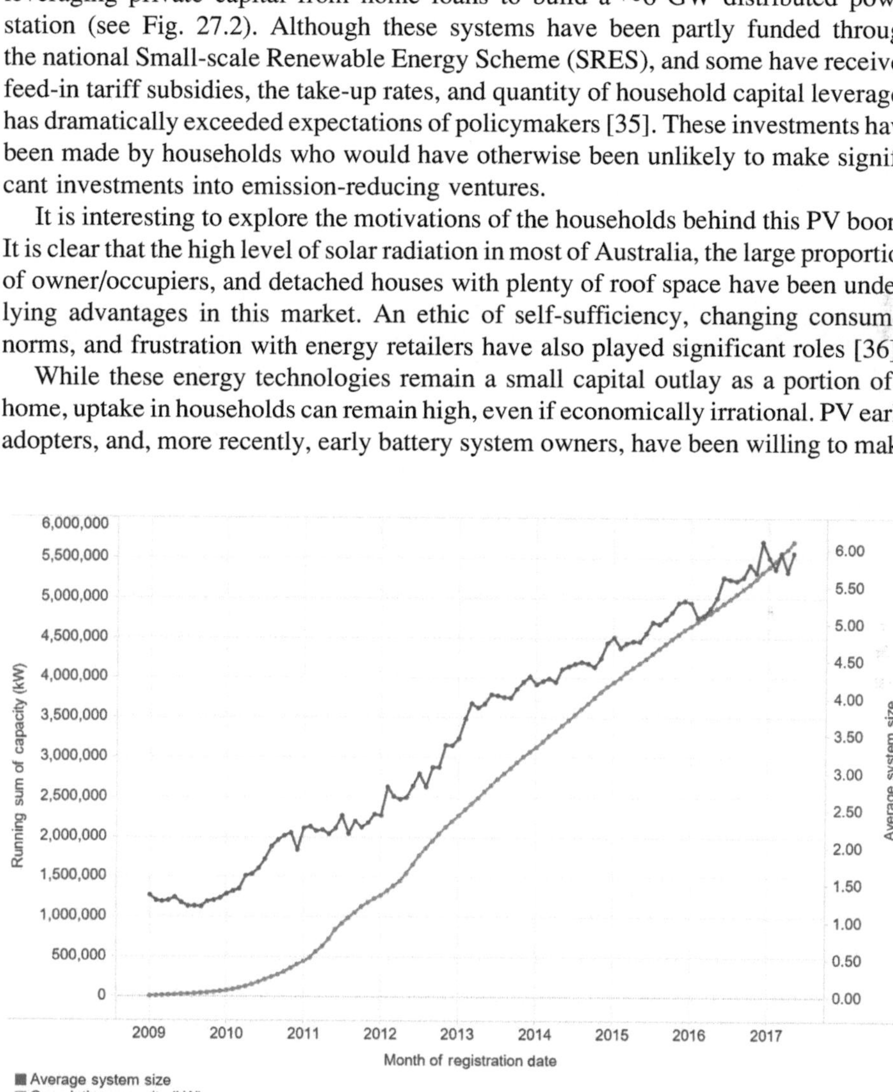

Fig. 27.2 Cumulative capacity and average system size for SRES systems 2009–16. *Source*: PV in Australia Report 2016, APVI, July 2017.

small or negative returns on new technologies. Eco-friendly home builders are often willing to go above and beyond when challenged to minimize energy use. Some households have aspirations to "go off grid" to satisfy their frustration with power prices, even if the resulting levelized cost of energy is higher than grid prices [37]. While some households are motivated by an approach of frugality, others react negatively to this and feel liberated by the idea that their solar panels provide "free" energy, despite their significant capital outlay. Industry and commercial sectors are less likely to accept small returns, or losses, on emissions-reducing ventures unless accompanied by publicity or marketing opportunities.

Many households can be motivated by the promise of lower energy usage or costs. However, there is often little understanding of the difference between reducing energy use and reducing emissions. Australian households more interested in emission reductions can become Greenpower (Australian renewable energy) customers, where they opt into a scheme whereby their retailer buys enough renewable energy certificates (RECS) from the national market and extinguishes them, resulting in the equivalent of 100% renewable energy for that customer. These actions are not allowed to be accounted for in the national targets, resulting in additional national emission reductions due to local commitment to climate action. With 300,000 households joining the program, resulting in 18 Mtonnes of emission reductions [38], this clear "additionality" is attractive to some Australians, driven by personal values, and disappointment with national targets.

We argue that households are willing and able investors in a safe climate. Many households can continue to be engaged in climate action not just by economic opportunities, but by a sentiment of reducing waste, or a vision of abundant, "home-made," clean energy.

27.5 Individual advocacy and campaigning

Individual advocacy and campaigning at the household level is one of the most underutilized and highly leveraged opportunities for individuals to impact upon greenhouse gas emissions in democratic countries. In its most basic form, this involves individuals spending their time pressuring governments, corporations, or other powerful actors to reduce their emissions. When successful, this action can enable an individual, who emits a relatively small volume of greenhouse gases, to leverage a much higher volume of greenhouse gas emissions from an institution, organization, or government. The following explores several case studies that demonstrate the impact of individual advocacy and campaigning.

27.5.1 Saving Australia's renewable energy agency

In 2016 the Australian campaigning organization, GetUp, facilitated 3040 individuals to save AUD800 million (800,000,000 AUD) of government funding for renewable energy from being cut [39]. GetUp estimated that for every member that took action,

AUD263 thousand (263,000 AUD) of funding was saved for renewable energy. Over a period of several weeks, 3040 GetUp members signed petitions, crowd-funded advertisements, and visited their members of parliament [39].

27.5.2 Stopping the development of one of the world's largest new coal mines

For many years, Indian company Adani has been planning to develop one of the world's largest coal mines in the Australian state of Queensland. Over a projected 60-year lifetime, the mine could produce around 2.3 billion tonnes (2,300,000,000 tonnes) of coal [40]. This mine development has been criticized by mainstream actors across the political spectrum in Australia and internationally. The campaign involved around 1000 volunteers and 160 local #stopadani groups. These groups demonstrated out the front of banks, made over 130,000 phone calls, knocked on doors, and disrupted political events. In November 2017, the Queensland Premier reversed her position and committed to vetoing a federal loan that would enable the mine to be built [41]. If this mine doesn't go ahead, the potential emissions avoided per volunteer would be around 46 million tonnes (CO_2) (46,000,000 tonnes of CO_2) {this assumes 1000 volunteers and a lifetime output of coal of 2.3 billion tonnes (2,300,000,000 tonnes), which would result in 4.6 billion tonnes (CO_2) (4,600,000,000 tonnes of CO_2)}. To put this impact in perspective, the average per capita emissions of the average Australian are around 20 t(CO_2) a^{-1} (20 t CO_2 per annum). This example demonstrates the leveraged potential of individual advocacy.

27.5.3 Local government advocacy

Given the scale of local government authorities, and their relative connection to local areas, local residents have immense power to challenge decision makers and deliver outcomes that are more sustainable. In New South Wales, there are numerous examples of the role residents can play in creating this change. This includes divestment campaigns [42], direct advocacy to reduce city energy usage [43], solar bulk buys [44], and campaigns to stop inappropriate development. By lobbying elected representatives, and holding local government authorities to account on delivering promised change, individual local residents have the opportunity to play a significant role in changing the future of their cities.

These examples challenge hierarchical perspectives on the agency of individuals in climate change adaptation and mitigation. A hierarchical view of scale assumes that the higher the level of governance, the greater the agency. These examples challenge this normative frame to demonstrate that agency is spread unevenly and unpredictably across scales. For example, it is conceivable that the head of state in a middle-income country or senior UN bureaucrat may have less impact on emissions than a "stop Adani" volunteer in Australia or shareholder advocate.

27.6 Conclusion

Decades of international negotiations have confirmed the immense challenges faced by nation states in affecting the necessary change in behavior and actions required to address anthropogenic climate change. With such an enormous proportion of the impacts of climate change being experienced by individuals and subnational governments, action at these levels must be seriously considered as one of the most encouraging opportunities to reduce greenhouse gas emissions and build resilience for an uncertain future.

Leaders at a substate level are utilizing a variety of policy levers and experiences, which can make a measurable impact in reducing climate change impacts.The individual decisions made by citizens or by levels of government close to citizens have never before been so powerful.

References

[1] IPCC. In: Team CW, Pachauri RK, Myer LA, editors. Climate Change 2014: Synthesis Report. Geneva, Switzerland: IPCC; 2014. p. 151.
[2] Chan S, et al. Reinvigorating international climate policy: a comprehensive framework for effective nonstate action. Global Pol 2015;6(4):466–73.
[3] Ireland P, Keegan P. Climate change adaptation: challenging the mainstream. In: Sygna L, O'Brien K, Wolf J, editors. A changing environment for human security. Oxon, USA: Routledge; 2013.
[4] Moloney S, Fünfgeld H, Granberg M. Climate change responses from the global to the local scale: an overview, In: Moloney S, Fünfgeld H, Granberg M, editors. Local action on climate change opportunities and constraints. Routledge; 2018. p. 1–16 https://www.routledge.com/Local-Action-on-Climate-Change-Opportunities-and-Constraints/Moloney-Fuenfgeld-Granberg/p/book/9781138681521.
[5] IPCC. Summary for policymakers. In: Field CB, et al., editors. Intergovernmental panel on climate change special report on managing risks of extreme events and disasters to advance climate change adaptation. Cambridge: Cambridge University Press; 2012. p. 1–19.
[6] C40. History of the C40. 2017; Available from: http://www.c40.org/history.
[7] Jordans, F. and D. Thiesing, US cities, states defy Trump, still back Paris climate deal, in AP. 2017, AP: Online. https://www.apnews.com/365b032c1e4140c8a07021245f87d66c/US-cities,-states-defy-Trump,-still-back-Paris-climate-deal.
[8] Escobar A. Imagining a post-development era? Critical thought, development and social movements. Social Text 1992;(31/32):20–56.
[9] Escobar A. Imagining a post-development era. In: Crush JS, editor. Power of development. London: Routledge; 1995. p. 211–27.
[10] Gibson-Graham JK. Surplus possibilities: postdevelopment and community economies. Singapore Journal of Tropical Geography 2005;26(1):4–26.
[11] McKinnon K. Postdevelopment, professionalism, and the politics of participation. Annals of the Association of American Geographers 2007;97(4):772–85.
[12] Ireland P. Climate change adaptation: Business-as-usual aid and development or an emerging discourse for change? Development and Change 2012;11(2):92–110.

[13] Ireland P. Nepalganj, the Centre of the world: Local perceptions of environmental change and the roles of adaptation actors. Local Environment 2012;17(2):187–201.
[14] Ireland P, McKinnon K. Strategic localism for an uncertain world: a postdevelopment approach to climate change adaptation. Geoforum 2013;47:158–66.
[15] Taylor PJ. World-systems analysis and regional geography. Professional Geographer 1988;40(3):259–65.
[16] Taylor PJ. Political geography: world economy, nation-state and locality. 3rd ed. Harlow: Longmans; 1993.
[17] Aalst M, Cannon T, Burton I. Community level adaptation to climate change: the potential role of participatory community risk assessment. Glob Environ Chang 2008;18:165–79.
[18] Sellers P, et al. Modeling the exchanges of energy, water, and carbon between continents and the atmosphere. Science 1997;275(5299):502–9.
[19] Wilbanks TJ, Kates RW. Global change in local places: how scale matters. Clim Change 1999;43(3):601–28.
[20] Bulkeley H, Newell P. Governing climate change. Routledge; 2015.
[21] C40. Why cities?. 2017; Available from: http://www.c40.org/why_cities.
[22] United Nations Department of Economic and Social Affairs, P.D., The World's cities in 2016v-data booklet. 2016.
[23] Hillman, K. and S. Buckley, Food, fibre and the future: report on surveys of students' and teachers' knowledge and understanding of primary industries. The Australian Council of Educational Research (http://research.acer.edu.au/national_surveys/1, Accessed 9 April 2012), 2011.
[24] OECD/UCLG, Subnational governments around the world: structure and finance, OECD, Editor. 2016.
[25] Nations, U. The New Urban Agenda. 2016 [cited 17 December 2017]; Available from: http://habitat3.org/the-new-urban-agenda/.
[26] Stern N. Stern review report on the economics of climate change; 2006.
[27] Spaans M, Waterhout B. Building up resilience in cities worldwide—Rotterdam as participant in the 100 resilient cities programme. Cities 2017;61:109–16.
[28] Dulac, J. Global transport outlook to 2050: Targets and scenarios for a low-carbon transport sector. 2012 [cited 2017 10 December]; Available from: https://www.iea.org/media/workshops/2013/egrdmobility/DULAC_23052013.pdf.
[29] EPA, U. Global greenhouse gas emission data. 2017 [cited 12 December 2017]; Available from: https://www.epa.gov/ghgemissions/global-greenhouse-gas-emissions-data.
[30] Alonso-Mora J, et al. On-demand high-capacity ride-sharing via dynamic trip-vehicle assignment. Proc Natl Acad Sci 2017;201611675.
[31] Schwarzenegger A. California's water–energy relationship; 2005, http://www.energy.ca.gov/2005publications/CEC-700-2005-011/CEC-700-2005-011-SF.PDF.
[32] Ashton J, Ubido J. The healthy city and the ecological idea. Soc Hist Med 1991;4(1):173–80.
[33] Kiparsky M, et al. The innovation deficit in urban water: the need for an integrated perspective on institutions, organizations, and technology. Environ Eng Sci 2013;30(8):395–408.
[34] Jenicek P, et al. Energy self-sufficient sewage wastewater treatment plants: is optimized anaerobic sludge digestion the key? Water Sci Technol 2013;68(8):1739–44.
[35] Mountain B, Szuster P. Solar, solar everywhere: opportunities and challenges for Australia's rooftop PV systems. IEEE Power Energy Mag 2015;13(4):53–60.
[36] Haines F, McConnell D. Environmental norms and electricity supply: an analysis of normative change and household solar PV in Australia. Environ Sociol 2016;2(2):155–65.

[37] Anderson S. Power prices: AusNet Services warns customers going off the grid will push up costs for others. Sydney: Australian Broadcasting Corporation; 2017.

[38] Government, A. GreenPower accredited renewable energy: program impacts. 2017 [cited 12 December, 2017]; Available from: http://www.greenpower.gov.au/About-Us/Program-impacts/#.http://www.couriermail.com.au/news/queensland/queensland-government/adani-veto-is-premier-annastacia-palaszczuks-first-official-act/news-story/2e2f8c99714763675e196629bd3f7681

[39] GetUp! We Saved ARENA. [cited 17 December 2017]; Available from: https://www.getup.org.au/campaigns/renewable-energy/we-saved-arena/we-saved-arena; 2016, http://www.smh.com.au/business/mining-and-resources/newcastle-home-of-coal-joins-divestment-push-20150826-gj7xz1.html.

[40] Aston H. Adani announces 'green light' for Carmichael mega coal mine, In: Sydney morning herald. Fairfax: Sydney; 2017, https://www.governmentnews.com.au/2017/11/councils-sign-climate-action/.

[41] McCarthy JL. Stuart, Adani veto is Premier Annastacia Palaszczuk's first official act, In: Courier Mail; 2017, http://reneweconomy.com.au/first-bulk-buy-program-launched-to-slash-costs-of-battery-storage-61156/.

[42] Saunders A. Newcastle, home of coal, joins fossil fuel divestment push. In: Sydney Morning Herald. Fairfax: Sydney; 2015.

[43] Phillipson G. More councils sign up for climate action. In: GovernmentNews. Sydney; 2017.

[44] Parkinson G. First bulk-buy program launched to slash costs of battery storage. In: Renew Economy; 2016.

Index

Note: Page numbers followed by *f* indicate figures, *t* indicate tables, and *b* indicate boxes.

C

D

F

H

I

J

K

L

N

O

P

R

T

U

V

Kolibri 360 GmbH
Ferdinand-Jühlke-Straße 7
99095 Erfurt, Deutschland
produktsicherheit@kolibri360.de